Advances in Intelligent Data Analysis and Its Applications

Advances in Intelligent Data Analysis and Its Applications

Editors

Chao Zhang
Wentao Li
Huiyan Zhang
Tao Zhan

MDPI

Basel • Beijing • Wuhan • Barcelona • Belgrade • Novi Sad • Cluj • Manchester

Editors

Chao Zhang
Shanxi University,
Taiyuan, China

Wentao Li
Southwest University,
Chongqing, China

Huiyan Zhang
Chongqing Technology and
Business University,
Chongqing, China

Tao Zhan
Southwest University,
Chongqing, China

Editorial Office
MDPI
St. Alban-Anlage 66
4052 Basel, Switzerland

This is a reprint of articles from the Special Issue published online in the open access journal *Electronics* (ISSN 2079-9292) (available at: https://www.mdpi.com/journal/electronics/special_issues/771L15O65G).

For citation purposes, cite each article independently as indicated on the article page online and as indicated below:

Lastname, A.A.; Lastname, B.B. Article Title. *Journal Name* **Year**, *Volume Number*, Page Range.

ISBN 978-3-03928-615-7 (Hbk)
ISBN 978-3-03928-616-4 (PDF)
doi.org/10.3390/books978-3-03928-616-4

Contents

electronics

MDPI

Editorial

Recent Advances in Intelligent Data Analysis and Its Applications

Chao Zhang [1,*], Wentao Li [2], Huiyan Zhang [3] and Tao Zhan [4]

[1] Key Laboratory of Computational Intelligence and Chinese Information Processing of Ministry of Education, School of Computer and Information Technology, Shanxi University, Taiyuan 030006, China
[2] College of Artificial Intelligence, Southwest University, Chongqing 400715, China; drliwentao@gmail.com
[3] National Research Base of Intelligent Manufacturing Service, Chongqing Technology and Business University, Chongqing 400067, China; huiyanzhang@ctbu.edu.cn
[4] School of Mathematics and Statistics, Southwest University, Chongqing 400715, China; zhantao@swu.edu.cn
* Correspondence: czhang@sxu.edu.cn

1. Introduction

In the current rapidly evolving technological landscape, marked by transformative advancements such as cloud computing, the Internet of Things (IoT), and industrial internet, the complexity of data analysis tasks is escalating across the socio-economic spectrum. Within this dynamic environment, the challenges faced by current problem-solving programs when handling big data primarily revolve around the effective management, modeling, and processing of extensive datasets.

This surge in data intricacy necessitates a proactive approach towards researching and developing intelligent models and methods for efficient data analysis and its application. It is crucial to explore innovative solutions that can navigate the intricacies of large datasets while ensuring not only the accuracy of analyses but also the timely extraction of valuable insights. Such research endeavors have become indispensable in addressing the growing demand for robust data processing capabilities in diverse sectors.

Moreover, as the technological landscape continues to evolve, the importance of staying at the forefront of data analysis methodologies becomes evident. This involves not only adapting to existing challenges but also anticipating future complexities. By delving into research on intelligent data models and methods, we pave the way for advancements that are not only responsive to current demands but also resilient in the face of emerging technologies and data-related challenges in our ever-changing socio-economic landscape.

Presently, the domain of intelligent data analysis [1] has experienced a rise in the number of scholars and professionals working within it. Innovative methods have been proposed from diverse perspectives, including data mining, machine learning (ML), natural language processing, granularity computation, social networks, machine vision, cognitive computing, and more. These approaches are intricately woven into the fabric of intelligent data analysis, presenting expansive and profound application scenarios for the field of data mining.

Data mining technology [2] plays a crucial role in dealing with large-scale data by extracting valuable information from massive datasets. It provides essential training data for ML algorithms, enabling the construction of more accurate models. Simultaneously, the development of natural language processing [3] allows machines to better understand and parse human language, imparting more practical meaning to the results of data analysis. Advancements in granularity computing [4] have improved the effectiveness of data analysis by simplifying information into fundamental concepts, facilitating swift and in-depth analysis. Social network analysis [5] uncovers patterns in interpersonal relationships and group behavior, offering substantial groundwork for the development of marketing strategies and policy formulation. The progression of machine vision [6]

Citation: Zhang, C.; Li, W.; Zhang, H.; Zhan, T. Recent Advances in Intelligent Data Analysis and Its Applications. *Electronics* **2024**, *13*, 226. https://doi.org/10.3390/electronics13010226

Received: 2 January 2024
Accepted: 3 January 2024
Published: 4 January 2024

broadens the horizons of data analysis to encompass image and video processing, providing strong support for applications such as intelligent surveillance and autonomous driving. Concurrently, the integration of cognitive computing [7] emulates the functions of the human brain, enhancing the innovation and intelligence of data analysis.

These intelligent data analysis methods have broadened the comprehension of intricate data processing at the theoretical research level, concurrently yielding positive effects on socio-economic development. Especially within the era of big data [8], these methods have shown considerable importance in tackling practical challenges across diverse domains, presenting fresh perspectives and innovative solutions for the complexities posed by intricate data. They not only make data analysis more intelligent and efficient but also drive the development of socio-economics, providing more comprehensive and viable strategies for solving practical issues. The research on these intelligent data analysis methods is becoming a crucial engine for advancing the integration of technology and society.

By conducting in-depth research and widely applying these methods, one can better address the challenges posed by the increasingly vast and diverse data streams, further propelling technological innovation. Not only is this innovation exhilarating, it is also playing an increasingly crucial role in solving practical problems. Further in-depth research and widespread application of newly emerging models and methods in the field of intelligent data analysis are anticipated to drive continuous progress in societal digital transformation and innovation.

To advance research in the field of computer science and engineering, new methods for intelligent data analysis and their applications must be persistently explored. Throughout this explorative process, the focus will be on the practicality, reliability, and effectiveness of innovative technologies and methods, ensuring their maximum impact in real-world applications. By closely integrating theoretical research with practical applications, there is the potential to advance the forefront of the intelligent data analysis field, contributing more beneficial insights to the development of a data-driven society in the future.

Overall, research on intelligent data analysis [9] and its applications holds significant value in the era of big data [10]. Through interdisciplinary approaches and technological innovations, it is possible to better address the challenges posed by complex data in the real world, further advancing the field of computer science and engineering. In the ongoing exploration in this field, attention is directed towards enhancing the practical applicability of intelligent data analysis methods to address real-world challenges. This endeavor aims to provide more reliable and innovative solutions for technological progress and societal development by resolving issues in practical scenarios. Through these efforts, there will be a continual contribution of greater depth and breadth of knowledge to propel the development of the field of data science, continuously pushing the boundaries of technological innovation.

One of the core tasks of intelligent data analysis is to effectively handle vast amounts of data and extract insightful information that informs decision making [11]. The essence of this article is to delve into the latest developments in the field of intelligent data analysis and explore how these technological innovations can be applied to address real-world challenges in the realms of society, economy, and science. By comprehensively understanding the latest developments in this field, one can better grasp the trends in technological advancement. This knowledge enables a more flexible application of these innovative technologies in practical scenarios.

Proactively exploring and implementing forward-looking approaches is pivotal for advancing intelligent and efficient data processing methods across diverse fields. This adaptability is indispensable for navigating the ever-evolving landscape of emerging complex challenges. Immersing oneself in the dynamic realm of intelligent data analysis facilitates not only better adaptation but also leadership in the unfolding trends of data science.

This proactive stance plays a crucial role in fostering innovation and formulating practical solutions that make significant contributions to the sustainable development of society, the economy, and the scientific domain. Delving deeper into the intricacies of intelligent data analysis not only enhances our capacity to address current issues but also positions us at the forefront of anticipating and responding to future challenges.

In this context, keeping abreast of emerging technologies and methodologies is paramount, allowing us to harness the full potential of data-driven insights. Embracing a forward-thinking mindset empowers us to not only meet present demands but also to shape and propel the future of data science. This proactive engagement acts as a catalyst for developing and implementing innovative solutions with far-reaching implications for the betterment of our global community.

Within this Special Issue, twenty-eight papers are published, encompassing diverse aspects of decision making, recommendation systems, intrusion detection, question answering, as well as topics in ML and deep learning (DL).

2. Overview of Contributions

For diverse domain requirements, numerous intelligent granular computing models have been established. The utilization of knowledge distance serves to quantify distinctions between granular spaces, representing an uncertainty metric with robust discriminative capabilities in rough set theory. However, the existing knowledge distance metric falls short when considering the relative disparities between granular spaces within the context of uncertain concepts. To address this gap, Yang et al. (Contribution 1) explored the concept of relative knowledge distance for intuitionistic fuzzy concepts.

Air pollution poses a significant environmental threat that could have potential consequences for human health. The emergence of IoT devices enables instantaneous and ongoing surveillance of atmospheric contaminants in metropolitan regions. However, the presence of uncertainty and inaccuracy in IoT sensor data present challenges in the effective utilization and fusion of these data. Additionally, divergent opinions among decision-makers regarding air quality evaluation (AQE) can impact final decisions. Addressing these issues, Li et al. (Contribution 2) systematically investigated a method utilizing hesitant trapezoidal fuzzy information, examining its application in AQE.

The multigranulation rough set (MGRS) model, extending the Pawlak rough set, describes uncertain concepts using optimistic and pessimistic upper/lower approximate boundaries. However, existing information granules in MGRS lacked sufficient approximate descriptions of uncertain concepts. In response, Yang et al. (Contribution 3) introduced the cost-sensitive multigranulation approximation of rough sets, encompassing optimistic and pessimistic approximations, grounded in approximation set theory. The associated properties of these approximations are scrutinized. Additionally, a cost-sensitive selection algorithm is proposed for optimizing the multigranulation approximation.

A myriad of research endeavors have extensively explored diverse facets within the field. In this context, Liu and his colleagues (Contribution 4) investigated the utilization of contextual information and users' interest preferences within location-based social networks to propose the subsequent point-of-interest for users in the IoT environment. Their study demonstrated that their model, named CGTS-HAN, could more accurately capture the contextual features of users' POI compared to alternative models.

Addressing the tendency of recommender systems to overlook diverse neighbor views in collaborative filtering, Zheng et al. (Contribution 5) proposed a multiscale neighbor-aware attention network. This approach integrates overarching semantics from various neighbor types with significant local embeddings of multiscale neighbors. The collaborative signals for predicting user ratings of items are derived from a range of neighbors, encompassing both attribute views and interaction views.

Modeling users' dynamic preferences is a challenging yet crucial task in recommendation systems. Hu et al. (Contribution 6) systematically addressed this challenge by considering both local fluctuations in user interests and the need for global stability.

Coping with vast amounts of data requires sophisticated methodologies. Variations in procedures and protocols across healthcare services and facilities has resulted in the incomplete and erroneous documentation of medical backgrounds. Rectifying these discrepancies is imperative to establish a singular, accurate record moving forward. A widespread strategy for tackling this concern includes utilizing imputation methods to anticipate absent data values by relying on established information within the dataset. A widespread strategy for tackling this concern is utilizing imputation techniques to forecast missing data values by leveraging known values within the dataset. In this regard, Wilcox et al. (Contribution 7) introduced a neighborhood similarity measure-based imputation technique.

Network clustering held significance in the fields of data mining and bioinformatics. Wang et al. (Contribution 8) introduced a specialized algorithm in this domain, targeting the detection of protein complexes by integrating features of power-law distributions.

A frequency synthesizer serves the fundamental function of generating a desired frequency through the manipulation of a reference frequency signal. Across diverse sectors, including communication, control, surveillance, medical, and commercial applications, the demand for stable and precise frequency generation is critical to ensuring the dependable performance of mechanical equipment. In response to this imperative, Park et al. (Contribution 9) undertook the design and implementation of a highly reliable frequency synthesizer specifically tailored for use in railway track circuit systems. This synthesizer exclusively utilized audio frequency (AF) and was integrated into the logic circuit of a field-programmable gate array, eliminating the need for a microprocessor. The deployment of this exceptionally precise AF-class frequency synthesizer significantly elevated the safety and efficiency of braking and signaling systems in transportation equipment, such as railways and subways.

Data analysis has greatly benefited from the pivotal role played by ML and DL models. For example, mining machinery heavily relies on picks for the automated extraction of coal, and the condition of these picks significantly influences the effectiveness of mining equipment. Facing the task of accurately assessing the overall wear status of cutting tools during coal mining operations, Song and colleagues (Contribution 10) introduced a data-centric model for recognizing the wear condition of these tools. The model employed sophisticated optimization techniques for long short-term memory networks, integrating Bayesian algorithms.

Various devices within the smart home environment experience different levels of susceptibility to attacks. Devices characterized by lower attack frequencies encounter challenges in amassing sufficient attack data, thereby limiting the capacity to train effective intrusion detection models. In response, Chen et al. (Contribution 11) presented an innovative approach termed the improvement technique, which leverages feature enhancement and data augmentation to generate ample training data, particularly for broadening few-shot datasets. The utilization of an expanded dataset in training intrusion detection models significantly enhanced detection performance.

Yet, determining whether the augmented dataset truly enhanced model performance poses a challenge. Relying on the training model for each assessment instance to confirm the data augmentation and dataset quality incurs considerable time and resource costs. To tackle this issue, Cui et al. (Contribution 12) proposed a straightforward and pragmatic method to assess the effectiveness of data augmentation in image classification tasks, making a valuable contribution to the theoretical research on assessing data augmentation quality. Bieliński et al. (Contribution 13) delved into the exploration of how specific ML algorithms tackle the challenge of establishing a virtual mental health index.

Exploring the issue of flight delays, traditional DL algorithms encounter difficulties marked by low accuracy and heightened computational complexity. This poses a practical challenge in directly deploying deep flight delay prediction algorithms to mobile terminals. In response, Qu et al. (Contribution 14) studied the lightweight neural network MobileNetV3 algorithm and the improved ECA-MobileNetV3 algorithm. Their methodology included preprocessing real flight information and weather data.

Zhang et al. (Contribution 15) tackled the underutilization of relationships among words in a question through the introduction of a question-answering methodology for a knowledge base. This approach employed graph convolutional networks, facilitating the effective pooling of information across diverse dependency structures. The result was a heightened efficacy in the representation of sequence vectors.

Amidst efforts to control healthcare expenses and adapt to changing regulations, pharmaceutical laboratories aim to prolong the longevity of crucial equipment, particularly fluid bed dryers crucial for drug manufacturing. Barriga et al. (Contribution 16) proposed a pioneering solution that incorporates exploration data analysis and a Catboost ML model to tackle challenges associated with older dryers lacking real-time temperature optimization sensors. The integration of the Catboost algorithm resulted in a noteworthy decrease in initial heating time, leading to substantial energy conservation. The ongoing surveillance of essential parameters signified a departure from traditional fixed-time models, indicating a paradigm shift in the industry.

Recognizing orphan genes (OGs) can be a labor-intensive process. To address this challenge, Gao et al. (Contribution 17) introduced XGBoost-A2OGs, an automated predictor specifically designed for the identification of OGs in seven angiosperm species. The methodology involves the utilization of hybrid features and XGBoost.

Accurately classifying imbalanced data classes poses a formidable challenge due to the inherent uneven distribution in datasets. To tackle this obstacle, the incorporation of sampling procedures into ML and DL algorithms has underscored its indispensability. In this context, Sadaiyandi et al. (Contribution18) conducted a study that employed sampling-based ML and DL approaches to automate the identification of deteriorating trees within a forest dataset.

In the process of feature learning, conventional models for abnormal state detection frequently neglect the variation in position and orientation system data within the frequency domain. This neglect results in the forfeiture of vital feature details, hindering the possibility for additional improvements in detection capability. To overcome this limitation and with the goal of improving UAV flight safety, Yang et al. (Contribution 19) introduced a technique for detecting abnormal UAV states.

Autonomous underwater vehicles (AUVs) encounter challenges in underwater navigation due to the considerable costs associated with inertial navigation devices and Doppler velocity logs, which impede the acquisition of essential navigation data. In addressing this issue, methodologies such as underwater simultaneous localization and mapping are employed. These approaches, coupled with navigation methods reliant on perceptual sensors like vision and sonar, aim to enhance self-positioning precision. In the field of machine learning (ML), extensive datasets play a crucial role in improving algorithmic performance. Wang et al. (Contribution20) introduced an underwater navigation dataset derived from controllable AUVs.

A network intrusion detection (NID) tool grapples with network data characterized by high feature dimensionality and an imbalanced distribution across categories. Presently, certain detection models exhibit suboptimal accuracy in practical detection scenarios. In response to these challenges, Zhang et al. (Contribution 21) introduced an NID model leveraging multi-head attention and bidirectional long short-term memory.

To address the accuracy limitations of the traditional interacting multiple-model (IMM) algorithm in target tracking, Wang (Contribution 22) proposed an innovative algorithm named VSIMM-CS. This algorithm adopts a variable structure interacting multiple-model approach. The real-time construction of its model ensemble is based on the initial set, considering both the error characteristics of a linear system and the inherent symmetry in the structure of the model set.

Semi-supervised classification stands as a fundamental approach for addressing incomplete tag information without manual intervention. Nevertheless, prevailing algorithms necessitate the storage of all unlabeled instances, leading to iterative processes with potential drawbacks, such as slow execution speed and substantial memory requirements, particularly for large datasets. While previous solutions have primarily concentrated on supervised classification, Song et al. (Contribution 23) presented a novel approach aimed at reducing the size of the unlabeled instance set in the context of semi-supervised classification algorithms.

To enhance scatter quality without a notable reduction in the lateral resolution of the delay multiply and sum (DMAS) beamforming coherence factor, Guo (Contribution 24) introduced an adaptive, spatio-temporally smoothed coherence factor combined with DMAS. In this research, the generalized coherence factor was applied to identify local coherence and dynamically ascertain the subarray length for spatial smoothing. Incorporating this parameter to assess the results improved scatter quality without a substantial compromise in lateral precision, making it particularly advantageous in intricate clinical environments.

In the field of intelligent manufacturing, the proficient use of industrial robots faces a hurdle due to the issue of low absolute positioning accuracy. Tao et al. (Contribution 25) presented an algorithm for precise compensation in the absolute positioning of industrial robots, leveraging deep belief networks through an offline compensation approach. They employed deep belief networks through an offline compensation approach, optimizing these networks using a differential evolution algorithm. Additionally, they introduced a position error mapping model incorporating evidence theory. The aim is to streamline the process of precision compensation, specifically targeting the enhancement of absolute positioning accuracy in industrial robots.

The detection of blind spot obstacles in intelligent wheelchairs holds significance, particularly within semi-enclosed environments of elderly communities. Current solutions relying on LiDAR and 3D point clouds are costly, difficult to implement, and demand substantial computing resources and time. Du et al. (Contribution 26) introduced an optimized lightweight obstacle detection model called GC-YOLO, based on YOLOv5 architecture.

While sentiment analysis has been extensively researched, the majority of studies have concentrated on analyzing individual corpora. Yang et al. (Contribution 27) introduced a pioneering framework, CNEC, tailored for conducting sentiment analysis on bilingual text that includes emojis, commonly found on social media platforms.

Knowledge graph question answering supported users without mandating data structure comprehension, addressing challenges such as semantic understanding, retrieval errors, word abbreviation, object complement, and entity ambiguity. To tackle these issues, Zuo (Contribution 28) presented the innovative Chunked Learning Network method. The model incorporated vector representations of entities and predicates into the question, fully leveraging embeddings derived from the knowledge graph. Adapted for diverse scenarios, the model utilizes a variety of approaches to acquire vector representations for the subject entities and relationships within the question.

Author Contributions: C.Z., W.L., H.Z. and T.Z. worked together in the whole editorial process of the Special Issue, "Advances in Intelligent Data Analysis and Its Applications", published by the journal *Electronics*. H.Z. and T.Z. drafted this editorial introduction. C.Z. and W.L. reviewed, edited, and finalized the manuscript. All authors have read and agreed to the published version of the manuscript.

Funding: This editorial was supported in part by the Natural Science Foundation of Chongqing (No. CSTB2023NSCQ-MSX0152), the Special Fund for Science and Technology Innovation Teams of Shanxi (202204051001015), the Science and Technology Research Program of Chongqing Education Commission (Nos. KJZD-K202300807, KJQN202300202, KJQN202100206), the Training Program for Young Scientific Researchers of Higher Education Institutions in Shanxi, the Cultivate Scientific Research Excellence Programs of Higher Education Institutions in Shanxi (CSREP) (2019SK036), and the China Postdoctoral Science Foundation (No. 2023T160401).

Conflicts of Interest: The authors declare no conflicts of interest.

List of Contributions

1. Yang, J.; Qin, X.; Wang, G.; Zhang, X.; Wang, B. Relative Knowledge Distance Measure of Intuitionistic Fuzzy Concept. *Electronics* **2022**, *11*, 3373.

2. Li, W.; Zhang, C.; Cui, Y.; Shi, J. A Collaborative Multi-Granularity Architecture for Multi-Source IoT Sensor Data in Air Quality Evaluations. *Electronics* **2023**, *12*, 2380.
3. Yang, J.; Kuang, J.; Liu, Q.; Liu, Y. Cost-Sensitive Multigranulation Approximation in Decision-Making Applications. *Electronics* **2022**, *11*, 3801.
4. Liu, X.; Guo, J.; Qiao, P. A Context Awareness Hierarchical Attention Network for Next POI Recommendation in IoT Environment. *Electronics* **2022**, *11*, 3977.
5. Zheng, J.; Jing, T.; Cao, F.; Kang, Y.; Chen, Q.; Li, Y. A Multiscale Neighbor-Aware Attention Network for Collaborative Filtering. *Electronics* **2023**, *12*, 4372.
6. Hu, J.; Liu, Q.; Zhao, F. Local-Aware Hierarchical Attention for Sequential Recommendation. *Electronics* **2023**, *12*, 3742.
7. Wilcox, C.; Giagos, V.; Djahel, S. A Neighborhood-Similarity-Based Imputation Algorithm for Healthcare Data Sets: A Comparative Study. *Electronics* **2023**, *12*, 4809.
8. Wang, J.; Jia, Y.; Sangaiah, A.K.; Song, Y. A Network Clustering Algorithm for Protein Complex Detection Fused with Power-Law Distribution Characteristic. *Electronics* **2023**, *12*, 3007.
9. Park, C.; Han, E.; Kim, I.; Shin, D. A Study on the High Reliability Audio Target Frequency Generator for Electronics Industry. *Electronics* **2023**, *12*, 4918.
10. Song, D.; Zhao, Y. A Data-Driven Approach Using Enhanced Bayesian-LSTM Deep Neural Networks for Picks Wear State Recognition. *Electronics* **2023**, *12*, 3593.
11. Chen, Y.; Wang, J.; Yang, T.; Li, Q.; Nijhum, N.A. An Enhancement Method in Few-Shot Scenarios for Intrusion Detection in Smart Home Environments. *Electronics* **2023**, *12*, 3304
12. Cui, X.; Li, Y.; Xie, Z.; Liu, H.; Yang, S.; Mou, C. ADQE: Obtain Better Deep Learning Models by Evaluating the Augmented Data Quality Using Information Entropy. *Electronics* **2023**, *12*, 4077.
13. Bieliński, A.; Rojek, I.; Mikołajewski, D. Comparison of Selected Machine Learning Algorithms in the Analysis of Mental Health Indicators. *Electronics* **2023**, *12*, 4407.
14. Qu, J.; Chen, B.; Liu, C.; Wang, J. Flight Delay Prediction Model Based on Lightweight Network ECA-MobileNetV3. *Electronics* **2023**, *12*, 1434.
15. Zhang, C.; Zha, D.; Wang, L.; Mu, N.; Yang, C.; Wang, B.; Xu, F. Graph Convolution Network over Dependency Structure Improve Knowledge Base Question Answering. *Electronics* **2023**, *12*, 2675.
16. Barriga, R.; Romero, M.; Hassan, H. Machine Learning for Energy-Efficient Fluid Bed Dryer Pharmaceutical Machines. *Electronics* **2023**, *12*, 4325.
17. Gao, Q.; Zhang, X.; Yan, H.; Jin, X. Machine Learning-Based Prediction of Orphan Genes and Analysis of Different Hybrid Features of Monocot and Eudicot Plants. *Electronics* **2023**, *12*, 1433.
18. Sadaiyandi, J.; Arumugam, P.; Sangaiah, A.K.; Zhang, C. Stratified Sampling-Based Deep Learning Approach to Increase Prediction Accuracy of Unbalanced Dataset3. *Electronics* **2023**, *12*, 4423.
19. Yang, T.; Chen, J.; Deng, H.; Lu, Y. UAV Abnormal State Detection Model Based on Timestamp Slice and Multi-Separable CNN. *Electronics* **2023**, *12*, 1299.
20. Wang, C.; Cheng, C.; Yang, D.; Pan, G.; Zhang, F. Underwater AUV Navigation Dataset in Natural Scenarios. *Electronics* **2023**, *12*, 3788.
21. Zhang, J.; Zhang, X.; Liu, Z.; Fu, F.; Jiao, Y.; Xu, F. A Network Intrusion Detection Model Based on BiLSTM with Multi-Head Attention Mechanism. *Electronics* **2023**, *12*, 4170.
22. Wang, Q.; Li, G.; Jin, W.; Zhang, S.; Sheng, W. A Variable Structure Multiple-Model Estimation Algorithm Aided by Center Scaling. *Electronics* **2023**, *12*, 2257.
23. Song, Y.; Zhang, J.; Zhao, X.; Wang, J. An Accelerator for Semi-Supervised Classification with Granulation Selection. *Electronics* **2023**, *12*, 2239.
24. Guo, Z.; Geng, X.; Yao, F.; Liu, L.; Zhang, C.; Zhang, Y.; Wang, Y. An Improved Spatio-Temporally Smoothed Coherence Factor Combined with Delay Multiply and Sum Beamformer. *Electronics* **2023**, *12*, 3902.

25. Tao, Y.; Liu, H.; Chen, S.; Lan, J.; Qi, Q.; Xiao, W. An Off-Line Error Compensation Method for Absolute Positioning Accuracy of Industrial Robots Based on Differential Evolution and Deep Belief Networks. *Electronics* **2023**, *12*, 3718.
26. Du, J.; Zhao, S.; Shang, C.; Chen, Y. Applying Image Analysis to Build a Lightweight System for Blind Obstacles Detecting of Intelligent Wheelchairs. *Electronics* **2023**, *12*, 4472.
27. Yang, T.; Liu, Z.; Lu, Y.; Zhang, J. Centrifugal Navigation-Based Emotion Computation Framework of Bilingual Short Texts with Emoji Symbols. *Electronics* **2023**, *12*, 3332.
28. Zuo, Z.; Zhu, Z.; Wu, W.; Wang, W.; Qi, J.; Zhong, L. Improving Question Answering over Knowledge Graphs with a Chunked Learning Network. *Electronics* **2023**, *12*, 3363.

References

1. Chen, Y.H.; Yao, Y.Y. A multiview approach for intelligent data analysis based on data operators. *Inf. Sci.* **2008**, *178*, 1–20. [CrossRef]
2. Yang, J.; Li, Y.; Liu, Q.; Li, L.; Feng, A.; Wang, T.; Zheng, S.; Xu, A.; Lyu, J. Brief introduction of medical database and data mining technology in big data era. *J. Evid.-Based Med.* **2020**, *13*, 57–69. [CrossRef] [PubMed]
3. Young, T.; Hazarika, D.; Poria, S.; Cambria, E. Recent trends in deep learning based natural language processing. *IEEE Comput. Intell. Mag.* **2017**, *13*, 55–75. [CrossRef]
4. Lin, T.Y. Granular computing: From rough sets and neighborhood systems to information granulation and computing with words. In Proceedings of the European Congress on Intelligent Techniques and Soft Computing, Aachen, Germany, 8–11 September 1997; pp. 1602–1606.
5. Abkenar, S.B.; Kashani, M.H.; Mahdipour, E.; Jameii, S.M. Big data analytics meets social media: A systematic review of techniques, open issues, and future directions. *Telemat. Inform.* **2020**, *57*, 101517. [CrossRef] [PubMed]
6. Kaur, H.; Pannu, H.S.; Malhi, A.K. A systematic review on imbalanced data challenges in machine learning. *ACM Comput. Surv.* **2019**, *52*, 1–36. [CrossRef]
7. Gupta, S.; Kar, A.K.; Baabdullah, A.M.; Al-Khowaiter, W. Big data with cognitive computing: A review for the future. *Int. J. Inf. Manag.* **2018**, *42*, 78–89. [CrossRef]
8. Buxton, B.; Goldston, D.; Doctorow, C.; Waldrop, M. Big data: Science in the petabyte era. *Nature* **2008**, *455*, 8–9. [PubMed]
9. Zhang, C.; Li, D.Y.; Liang, J.Y. Multi-granularity three-way decisions with adjustable hesitant fuzzy linguistic multigranulation decision-theoretic rough sets over two universes. *Inf. Sci.* **2020**, *507*, 665–683. [CrossRef]
10. Chen, G.Q.; Li, Y.L.; Wei, Q. Big data driven management and decision sciences: A NSFC grand research plan. *Fundam. Res.* **2021**, *1*, 504–507. [CrossRef]
11. Lei, Y.; Jia, F.; Lin, J.; Xing, S.; Ding, S.X. An intelligent fault diagnosis method using unsupervised feature learning towards mechanical big data. *IEEE Trans. Ind. Electron.* **2016**, *63*, 3137–3147. [CrossRef]

electronics

MDPI

Article

A Study on the High Reliability Audio Target Frequency Generator for Electronics Industry

Changsik Park [1,2], Euntack Han [1,3], Ikjae Kim [1,4] and Dongkyoo Shin [1,5,*]

1 Department of Computer Engineering, Sejong University, Seoul 05006, Republic of Korea; pcs5130@hanmail.net (C.P.); shrd9600@daum.net (E.H.); kij397@mnd.go.kr (I.K.)
2 Juyoung Electronic Co., Ltd., Gunpo-si 15844, Republic of Korea
3 Hyukshin Engineering Co., Ltd., Seoul 04623, Republic of Korea
4 R.O.K Cyber Operations CMD, Seoul 13834, Republic of Korea
5 Department of Convergence Engineering for Intelligent Drones, Sejong University, Seoul 05006, Republic of Korea
* Correspondence: shindk@sejong.ac.kr

Abstract: The frequency synthesizer performs a simple function of generating a desired frequency by manipulating a reference frequency signal, but stable and precise frequency generation is essential for reliable operation in mechanical equipment such as communication, control, surveillance, medical, and commercial fields. Frequency synthesis, which is commonly used in various contexts, has been used in analog and digital methods or hybrid methods. Especially in the field of communication, a precise frequency synthesizer is required for each frequency band, from very low-frequency AF (audio frequency) to high-frequency microwaves. The purpose of this paper is to design and implement a highly reliable frequency synthesizer for application to a railway track circuit systems using AF frequency only with the logic circuit of an FPGA (field programmable gate array) without using a microprocessor. Therefore, the development trend of analog, digital, and hybrid frequency synthesizers is examined, and a method for precise frequency synthesizer generation on the basis of the digital method is suggested. In this paper, the generated frequency generated by applying the digital frequency synthesizer using the ultra-precision algorithm completed by many trials and errors shows the performance of generating the target frequency with an accuracy of more than 99.999% and a resolution of mHz, which is much higher than the resolution of 5 Hz in the previous study. This highly precise AF-class frequency synthesizer contributes greatly to the safe operation and operation of braking and signaling systems when used in transportation equipment such as railways and subways.

Keywords: frequency synthesizer; direct frequency synthesizer; indirect frequency synthesizer; railway track circuit

Citation: Park, C.; Han, E.; Kim, I.; Shin, D. A Study on the High Reliability Audio Target Frequency Generator for Electronics Industry. *Electronics* **2023**, *12*, 4918. https://doi.org/10.3390/electronics12244918

Academic Editors: Chao Zhang, Wentao Li, Huiyan Zhang and Tao Zhan

Received: 7 September 2023
Revised: 20 November 2023
Accepted: 28 November 2023
Published: 6 December 2023

1. Introduction

Today, most devices, such as electronics, telecommunications, medical, transportation, and industrial devices, require an RF (reference frequency) inside for their original operation. This is essential for modulation and demodulation if it is a communication device, and for transmission, reception, or monitoring control processing of signals if it is a motoring control device [1]. This RF is usually generated by oscillation using the vibration of the device itself or by using the LC circuit, but the concept of a FS (frequency synthesizer) has been introduced to generate a specific frequency quickly and efficiently. There is a traditional analog method called PLL (phase-locked loop), a digital method, and a hybrid method that combines the two [2–8]. In this way, various technologies have been developed, from low-frequency to microwave, depending on the frequency range of frequency synthesis that produces the specific frequency desired. The output of the frequency synthesizer is very important in terms of the performance index and the accuracy of the generated frequency, which maintains stable output without shaking the generated frequency, like phase noise. For the frequency synthesizer, DDFS (direct digital frequency synthesizer), which is faster than PLL, is used to synthesize the desired frequency quickly [2]. However, the digital frequency synthesizer can synthesize the desired frequency quickly, but due to the nature of the microprocessor operated by the program mainly used here, it has a fatal flaw, such as malfunction or inoperability due to some external factors and environmental variables. For that reason, it is intentionally stuck to analog methods in industrial or highly stable special applications, not for general commercial or personal use. However, due to the various convenient characteristics of frequency synthesis, digital frequency synthesizers that can operate stably in disturbance environments such as Surge are sometimes implemented and used only by pure logic circuits without a microprocessor. Therefore, in this paper, to generate the target frequency used in the railway track circuit, the target frequency is generated using the pure logic of the FPGA to ensure the convenience, excellent performance, and safety of the digital frequency synthesizer. FPGA-based frequency synthesizers have been studied as shown in [8], but most of them deal with relatively high frequencies, and it is rare to use very low frequencies such as audio frequency bands as used in railway track circuits. In this paper, we investigate the technical development stage of frequency synthesis and its theoretical structure and design, fabricate, and simulate a frequency synthesizer with mHz deviation without a processor using only FPGA logic on the traditional structure of the digital frequency synthesizer DDFS.

2. Research and Technology Trends

DFS (direct frequency synthesis) and indirect frequency synthesis can be used to make frequency synthesizers that are essential for electronic communication devices. Direct frequency synthesis methods include direct analog synthesizers such as DAS (direct analog synthesizer) and DDFS (direct digital frequency synthesizer). Indirect frequency synthesis methods include PLL (phase-locked loop) and DLL (delay-locked loop) [6]. In addition, a hybrid FS in which two methods are mixed is used. Among the frequency generation methods, the analog method has high accuracy but difficult control, while the digital method has a simple control but low accuracy. In this section, the major frequency synthesis methods that have been studied so far, their advantages and disadvantages are examined, and the theoretical basis for implementing the improved AF frequency synthesizer is provided through the conceptual basis of DDFS using FPGA to be studied in this paper.

2.1. Analog Frequency Synthesizers

Analog frequency synthesizers can be divided into direct and indirect methods according to frequency generation methods.

The direct method is a method of generating a frequency directly from a frequency source, and the indirect method is a method of generating a necessary frequency by modulating a separately generated frequency. A direct analog frequency synthesizer is a method of generating a desired frequency by combining a reference frequency generator, a mixer, a frequency up/down converter, and a frequency doubler/multiplier, and it is a method of applying a frequency circle itself. The indirect analog frequency synthesizer generates a desired frequency using a simple technique such as inversion and frequency division and does not go through a process such as increasing another frequency source.

Andrzej Rokita [9] presented a PLL design that reduces the phase noise generated when new frequencies are generated by operations such as multiplication, mixing, filtering, and segmentation performed in direct analog synthesis. It states that very fast switching is an advantage of direct analog.

Figure 1 shows a conceptual diagram of the analog direct frequency synthesizer of [9]. In Figure 1, (a) is a method of selecting one of four oscillation frequencies as a SP4T (Single Pole 4 Transfer) switch and then multiplying it by four to obtain the desired frequency. Meanwhile, (b) is an improved method in which phase noise is reduced by (b) compared to (a) by first passing four oscillation frequencies through each of the four generators and then selecting them as switches.

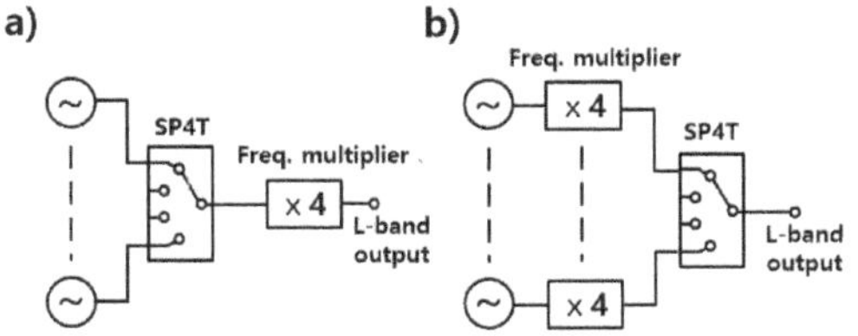

Figure 1. Block diagram of analog direct frequency synthesizer.

In contrast, indirect frequency synthesizers are widely known and widely used PLL. PLL is a technique that compares the output signal of a VCO (voltage-controlled oscillator) with respect to an input signal and adjusts the frequency of the VCO to maintain a constant phase difference of the output signal of the VCO with respect to the input signal. In an indirect frequency synthesizer, a PLL is used to generate the desired frequency. At this time, the components of the PLL determine the performance of the frequency synthesizer.

The PLL system proposed in [1] by Yoon Kwang-sup and others includes the components of the integer-N PLL. The reference clock generated in the reference divider is compared with the VCO output signal in a PFD (phase-frequency detector) to generate an up/down signal, and a CP (charge pump) converts the up/down signal into a current and transmits it to a LF (loop filter). The LF is used to convert current to voltage and control the frequency of the VCO; the 1/N Divider in Figure 2 divides the output signal of the VCO by N to finally produce the desired frequency. Fractional-N (N) dividers are also used to combine integers and fractions to enable finer frequency control.

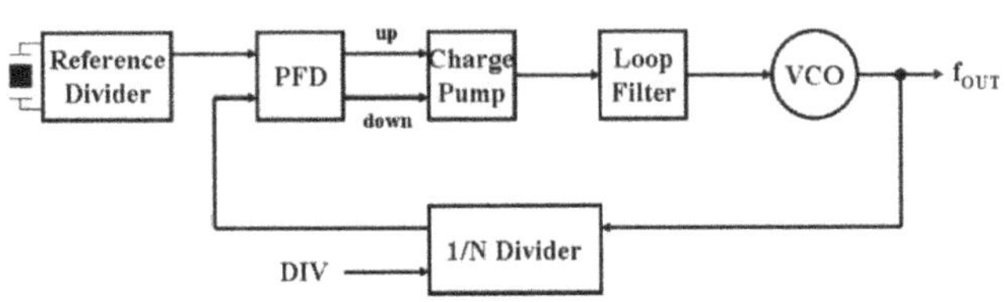

Figure 2. Basic structure of PLL frequency synthesizer.

The advantage of this PLL system is that the spurious signal level is reduced due to the LF operation, and it is simpler than the direct analog frequency synthesizer. However, it is a disadvantage that the frequency switching time is increased and the phase noise is higher than in the direct analog method.

However, it is a disadvantage that the frequency switching time is increased and the phase noise is higher than in the direct analog method. The phase noise performance of the frequency synthesizer in the LF bandwidth can be represented by $\lambda = \lambda PFD + 10 logN$ and λPFD is the accumulated phase noise of the reference frequency, phase detector, LF, and feedback 1/N divider inputted to the phase detector.

Yuchen Wang, Xuguang Bao, and Wei Hua have applied PLL to accurately determine the rotor position of a PMSM using a permanent magnet using the excellent phase lock capability of [10] PLL. Phase analysis of a three-phase signal is generally based on a synchronous reference system. PLL (SRF-PLL), a synchronous reference system, is the most widely used technique for extracting phase, frequency, and amplitude in a three-phase system. In this thesis, a phase shift PLL is used to map an asymmetric phase shift signal to a two-phase fixed coordinate system. In the study [11] of Kim Sang-woo and others applied to the design of a low-power frequency synthesizer for a GPS receiver using PLL, a frequency synthesizer was studied by applying a traditional fractional-N divider. Figure 3 shows the block diagram of the frequency synthesizer studied in [11]: PFD as a phase detector, CP as a charge pump, active low-pass filter, VCO, fractional-N divider, and sigma-delta modulator.

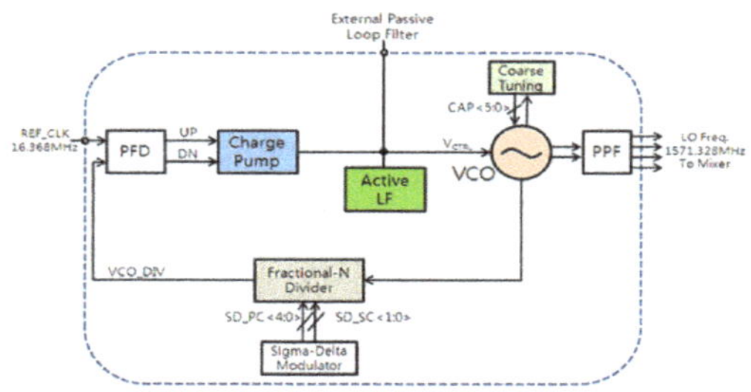

Figure 3. Application structure of PLL frequency synthesizer.

Figure 4 is a DLL-based FS block diagram, which is very similar to PLL, except that it has a VCDL (Voltage-Controlled Delay Line) instead of the VCO of PLL, and some researchers define it as a class of PLL. The idea of DLL is basically designed to solve errors related to delays that inevitably occur as the clock signal of the system goes through several stages. Despite the advantages of low noise and no phase accumulation, DLL systems are generally not recommended for FS applications due to unprogrammable, limited multiplicative factors, and high power consumption during operation [6].

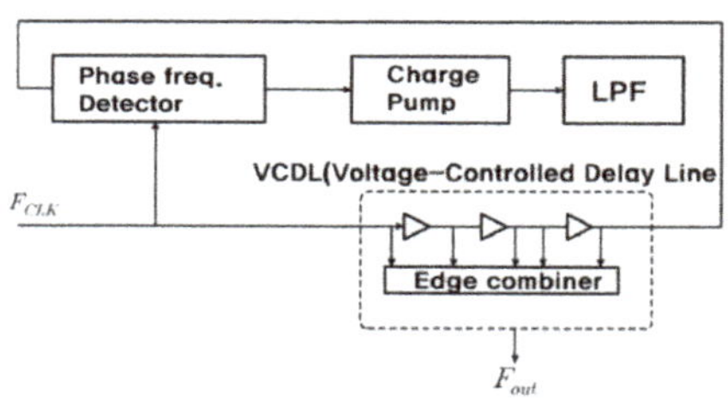

Figure 4. Frequency synthesizer block diagram based on DLL.

2.2. *Digital Frequency Synthesizers*

DDFS, which takes advantage of the development of digital technology, is commonly referred to as a DDS (direct digital frequency synthesizer) and shows a simple basic configuration diagram in Figure 5 [8].

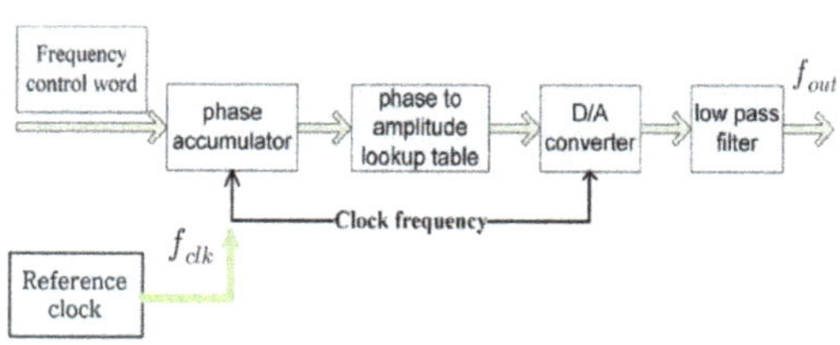

Figure 5. Direct digital frequency synthesizer basic diagram.

The DDS consists of a reference clock, a phase accumulator, a phase-to-amplitude LUT (look-up table), a DAC (digital-to-analog converter), a LPF (low-pass filter), and a FCW (frequency control word) that controls the output frequency.

The DDS finally produces an output signal (f_{out}) at the reference clock frequency (f_{clk}). This process, which mainly consists of digital control, is very fast and provides a high switching speed compared to the direct analog frequency synthesis method. DDS exhibits low-phase noise characteristics, even though the phase noise of the clock source itself is included.

In Figure 5, the frequency control word is added to the sum of the accumulator input from the phase accumulator and calculated, and the value is implemented in the bit adder and the result is supplied to the accumulator register.

On the one hand, the waveform data, in which the value sent as the sample address corresponds to the phase-amplitude conversion circuit, is outputted according to the address value. It is transformed through the D/A converter and LPF to the analog waveform and waveform data is outputted

The biggest advantage of this DDS is that the output frequency of the Heltz (Hz) level is generated by the fine frequency resolution due to the phase accumulator, but the limitation of available bandwidth and spurious performance are disadvantages. At this time, the highest possible frequency is limited to less than half of the clock frequency by the Nyquist theorem, and spurious noise is higher than the analog frequency synthesis method due to quantization and DAC conversion errors.

A.A. Alsharef et al. implemented the typical DDS of Figure 6 in FPGA (field programmable gate array) in [12]. FPGA is a device that is composed of unit blocks called CLB (Configurable Logic Block) rather than individual logic devices. It can be used as a desired input and output by the user, thereby reducing the complexity of hardware circuits and increasing reliability. The DDS using FPGA is designed as a LUT consisting of a Verilog code, which is also composed of PA, LUT, and D/A and simulated by the RTL (register transfer language) model.

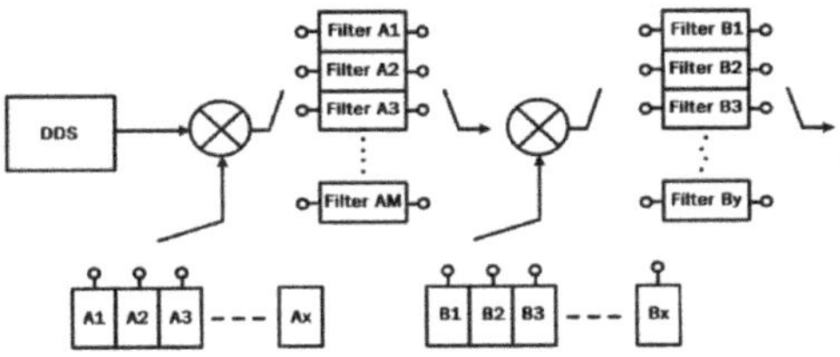

Figure 6. Conceptual diagram of DAFS with DDS added.

Matt Bergeron and Alan N. Willson, Jr. studied 1 GHz DDS on FPGA in [13]. The fast orthogonal DDS implemented with the FPGA is based on a new multiplier-based angle rotation algorithm that does not distort the magnitude of the sine and cosine outputs. The algorithm is designed to be well mapped to the DSP slices present in the FPGA. The device is implemented in the Xilinx Virtex-7 device and consumes 54.9 mW at 1 GHz, a performance previously only achieved in the ASIC design.

Another study using FPGA proposed a frequency synthesizer with a frequency resolution of 1.5 kHz, with a power consumption of 3.96 mW and a spurious performance of 59 dBc in the quadrature DDS of [14] studied by M. Saber Saber, M. Elmasry, and M. Eldin Abo-Elsoud.

In this study, ROM is not used for low-power implementation during operation on the FPGA. A simple approach that compensates for the shortcomings of the converter from phase to amplitude in the structure of a typical DDS, as shown in Figure 6, is to use a ROM with a function called LUT; however, as shown in the following formula:

$$f_{out} = \frac{W}{2L} Fclk \tag{1}$$

L: bit number of accumulator
W: bit value of input frequency word

In general, to achieve fine frequency tuning, several techniques have been devised to limit the ROM size while maintaining adequate performance to require large values, one of which is to reduce the number of angles required for the sine amplitude to four using a quarter-wave symmetry of the sine function. Cutting out the output of the phase accumulator leads to spurious noise, but this approach is commonly used because it achieves a fine frequency resolution that requires a very large value for L.

To reduce the memory size in LUT-based FS, various angular decomposition methods have been proposed, which typically consist of dividing the ROM into several small units, each of which is processed as part of the truncated phase accumulator output. Data searched in each low-rank ROM is added, and the sine curve approximation value is produced. Therefore, the proposed structure in [14] states that the sine function is divided into linear segments, each segment has a linear equation, and the value of this equation is obtained through additional hardware.

Wenjun Chen et al. [15] studied how to implement DDS performance improvement with the CORDIC (coordinated rotation digital computer) algorithm. They used XILINX's FPGA to reduce the output delay by repeatedly merging into a small amount of ROM, which can be seen to realize a sinusoidal wave with an SFDR of 86.76 dB at a high frequency of 350 MHz. Yixiong Yang et al. propose the LUT-ROT (rotation) architecture of traditional DDS in [16]. In order to optimize the speed and area of 2 GHz DDS, a performance of 11.7 mW/GHz is implemented in an area of 0.016 mm^2 by pipelined LUT.

2.3. *Hybrid Frequency Synthesizers*

A mixture of analog and digital frequency synthesizer structures has been studied in both direct and indirect methods, first using DDS in the direct analog frequency synthesis mentioned above, as shown in Figure 6.

In the analog method, DDS can be added to the input unit to reduce the complexity of the design and the overall components, and then DDS can be inserted instead of the fractional-N divider in the PLL system. It is a kind of mixed system of PLL and DDS. In the study [3] analyzing the phase noise of the digital hybrid PLL frequency synthesizer, the input noise to obtain the minimum phase noise, the D/A conversion noise due to the quantization error, and the mathematical model of the VCO noise source were derived and analyzed.

In order to improve the performance of the high-resolution beam-forming receiver based on DDS and PLL, Ref. [4] implements high-phase resolution by applying 14 bits of DDS.

In [4], the implementation of beamforming using only the existing DDS was implemented using DDS-PLL, as shown in Figure 7.

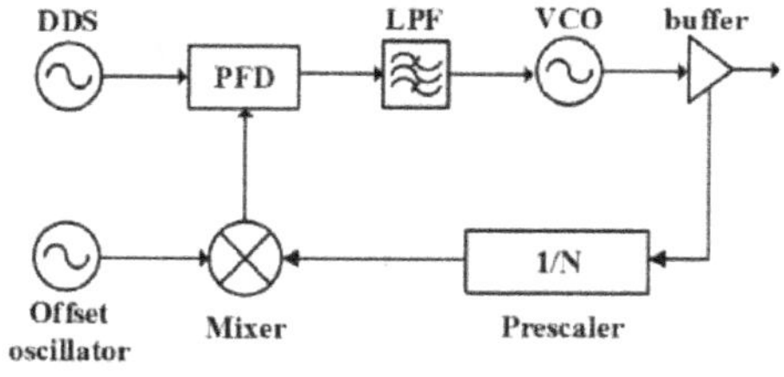

Figure 7. DDS-PLL block diagram.

In the high-performance PLL FS (Frequency Synthesizer) [17] applied to the radar system studied by Kim Song-sik and others, the D/A of the DDS was applied for the PLL modeling design to realize excellent phase noise and high-speed frequency synthesis time in the broadband characteristics. In order to compensate for the disadvantages of general PLL FS, we provide a coarse tune through D/A of DDS. Akila Gothandaraman and Syed K. [18] proposed an all-digital frequency locked loop (ADFLL) that eliminates analog shortcomings by implementing all PLL digitally.

In this study, an algorithm to frequency synthesize ADFLL capable of high-speed frequency acquisition is proposed to reduce hardware cost and architecture, enable full digitization, and become a pulse-output DDFS that is easy to design and implement. In addition, an adaptive phase estimator is proposed to show that the DDFS has a 16-bit binary weighting control, and the simulation results show that the ADFL can operate in the frequency range between 50 MHz and 500 MHz.

3. Design of DDFS for Low Frequency Using FPGA

3.1. Necessity

As we have seen so far in related studies, various methods of frequency synthesizers are used for their advantages and disadvantages. As discussed in Section 2, the application range of FPGA-based DDS to be studied in this paper is gradually increasing due to reliability and convenience of development. Railways and subways use AF (audio frequency) track circuit devices that accurately detect the track driving of a train and perform train control and monitoring in a specific section.

Currently, the AF track circuit device used in railways is a modulation-demodulation transmission system that modulates and transmits a specific audible frequency of 30 Hz or less. It has been 20 to 30 years since it was used as an analog method using LC oscillation. It is time to change to a more precise digital method. In this paper, we study the frequency synthesizer for the design and fabrication of audio frequency generation equipment that generates the desired audible frequency in order to produce a frequency synthesizer that generates the audible frequency accurately and stably. In this paper, we propose and implement DDS using an FPGA and a pure logic circuit without a microprocessor in order to implement a stable device free from external noise. The accuracy of frequency synthesis is confirmed by simulation.

As shown in Figure 8, the reference clock A and the frequency control word (FCW) input to the PA (phase accumulator) are determined by the input of the frequency control word. The output of the PA is converted into a sinusoidal amplitude value in a look-up table (LUT) stored in the ROM, and a pure sinusoidal frequency is generated using a digital–analog converter (DAC) and an LPF. The frequency at this time is defined by the following equation [13].

$$F_{out} = \frac{FCW}{2L} F_{CLK} \tag{2}$$

FCW: Frequency control word
L: The number of bits of PA

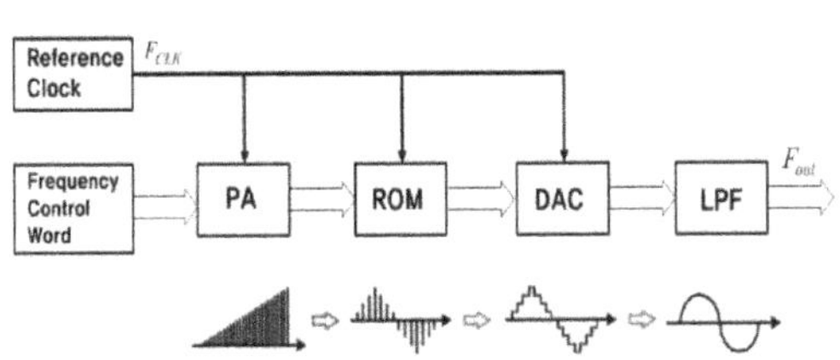

Figure 8. Basic structure of the proposed system.

In this case, the resolution of the frequency tends to increase as the reference clock frequency is small and the number of PA bits is large, as shown in the following equation. The frequency resolution of DDS can be defined as the reference clock frequency divided by the bit of the accumulator, which can be expressed by the following equation:

$$\Delta F = \frac{F_{CLK}}{2^L} \tag{3}$$

Therefore, it is necessary to properly limit the size of the ROM with a large *L* value for fine tuning and precise frequency generation.

3.2. Target Frequency

In the proposed system in Figure 9, the stability of the reference clock is very important, so a high-frequency X-tal must be used to generate the reference clock. This crystal oscillator, which is supplied for industrial use, is very stable due to its low temperature coefficient in the range of −40–120 °C. The reference frequency composed of the crystal oscillator generates the frequency of 1 kHz∼6 kHz required for the track circuit and changes it to a TWS (thumb-wheel switch) attached to the side of the simulation device. The oscillation frequency is generated within 0.05% of the target accuracy, and the desired frequency can be obtained by changing the TWS value.

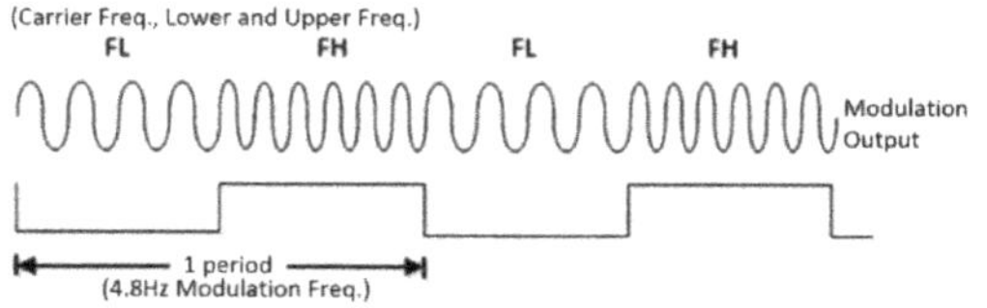

Figure 9. Track circuit frequency composition used in railways.

The frequency generated in this way can be applied to various fields, but in this study, the frequency used for the railway AF track circuit is generated. The AF track circuit frequency modulates and transmits FSK (frequency shift keying), and the reception demodulates it to detect and analyze the transmitted frequency to determine whether there is a train in the corresponding track circuit section. That is, when the frequency is detected, it is judged that there is no train in the corresponding track circuit section, and if the detection is not performed, it is judged that there is a train. FSK is a frequency shift modulation method in which data selects different frequencies between 0 and 1 according to a frequency having a constant amplitude.

If the frequency shift is A, the following equation becomes the FSK modulated signal.

$$S(t) = Acos2\pi(f_C - \Delta f)t{:}0 \leq t \leq T{:}1 \tag{4}$$

$$S(t) = Acos2\pi(f_C - \Delta f)t{:}0 \leq t \leq T{:}0 \tag{5}$$

where A is the amplitude of the FSK, f_C is the center frequency of the carrier frequency, and Δf is the deviation frequency.

The receiver recognizes two sinusoidal frequencies, which are carriers, in advance, extracts the corresponding frequencies, and restores the modulation frequency, which is called demodulation. This method is less defective than ASK (amplitude shift keying), and the circuit is relatively simple, so it is widely used in transmission equipment such as the AF track circuit device used in railways. In the track circuit device of the railway, the modulation frequency is fixed at 4.8 Hz, and the carrier frequency has eight center frequencies in consideration of the subway line and the upper and lower lines and modulates two frequencies of 17 Hz for the upper shift frequency and 17 Hz for the lower shift frequency of the center frequency. The track circuit frequency combination of the railway is shown in Figure 9.

These track circuit frequencies are used in Europe and Asia, and the Bombardier specification [19] specifies the transmission and receiver frequency arrangement, as shown in Table 1 below.

Table 1. Bombardier track circuit frequency array.

Frequency Name	Center Frequency (Hz)	Frequency Separation	Frequency (Hz) ±5%	Note
A	1699	Lower Freq.(FL)	1682	1699 − 17
		Upper Freq.(FH)	1716	1699 + 17
B	2296	Lower Freq.(FL)	2279	2296 − 17
		Upper Freq.(FH)	2313	2296 + 17
C	1996	Lower Freq.(FL)	1979	1996 − 17
		Upper Freq.(FH)	2013	1996 + 17
D	2593	Lower Freq.(FL)	2576	2593 − 17
		Upper Freq.(FH)	2610	2593 + 17
E	1549	Lower Freq.(FL)	1532	1549 − 17
		Upper Freq.(FH)	1566	1549 + 17
F	2146	Lower Freq.(FL)	2129	2146 − 17
		Upper Freq.(FH)	2163	2146 + 17
G	1848	Lower Freq.(FL)	1831	1848 − 17
		Upper Freq.(FH)	1865	1848 + 17
H	2445	Lower Freq.(FL)	2428	2445 − 17
		Upper Freq.(FH)	2462	2445 + 17

(1) Algorithm design

In this paper, we find and implement the optimal frequency generation algorithm after several trials and errors. This algorithm first generates the target frequency of Hz unit resolution by passing through D-FF 13 times, which has the function of delaying the desired generation frequency 8192 times by the time interval of the clock pulse. Specific methods and configurations, such as circuits and FPGA blocks, for simulating this method are described below. The circuit required for the algorithm generating the AF track circuit frequency is shown in Figure 10.

The M2S010 of the upper left part consists of an FPGA, and the right equivalent part uses an X-tal of 67.108864 Mhz as a reference clock generator. The lower left part represents a digital analog converter and an LPF. In the middle of the right side is a TWS that selects the frequency as the rotary switch.

The 8-bit FPGA output is converted to an analog signal using AD7541 from analog devices, a D/A converter.

The FPGA output is converted to 8 bits by a D/A converter called AD7541 from an analog device company.

According to the selection of the TWS, one frequency corresponding to the orbital frequency (upper and lower sides of A–H) according to the Bombardier standard is generated, and the upper frequency and the lower frequency can be composed of two frequency outputs generating the same method.

A logic block called ADC_A_OUT with a built-in look-up table (LUT) value is implemented as an FPGA, and a digital output of 256 steps is generated by this logic block. This result is applied to the AD7541 input, which is a digital-to-analog conversion IC, and is output as an analog signal.

In order to obtain the analog sine wave output signal, the D-FF (Flip Flop) is performed 13 times in the logic block called DCOUNT13 in the FRGEN block inside the FPGA to have a duty ratio of 50:50. This output is the SW2 of Figure 11, and 16 frequencies can be selected from 0 to F, and the frequency output according to the switch position is designed as shown in Table 1.

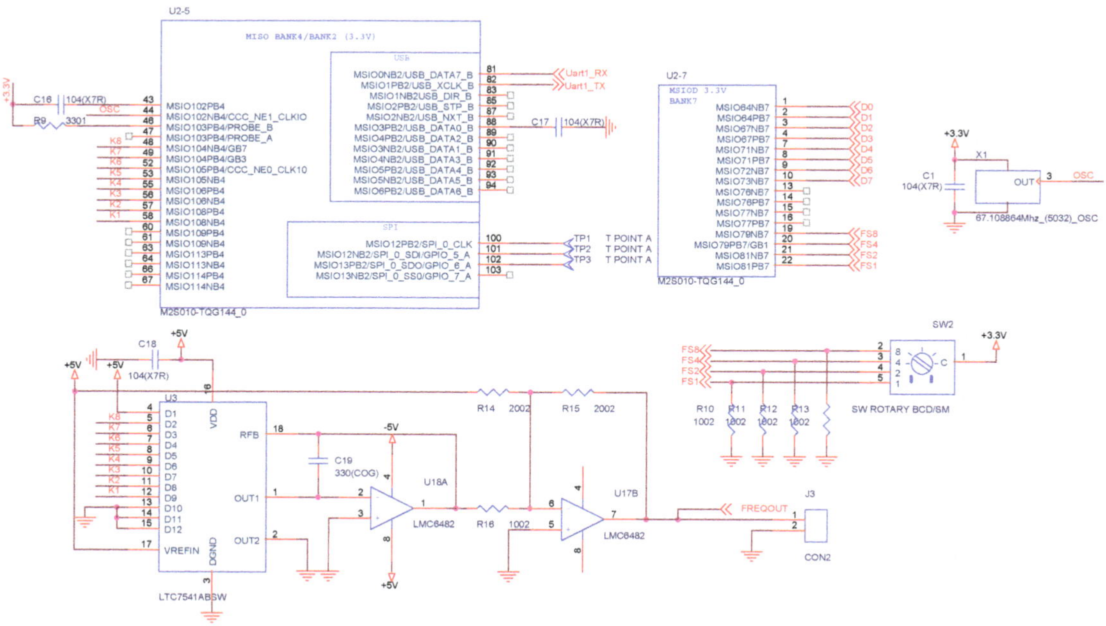

Figure 10. Circuit diagram required to implement the algorithm.

(2) FPGA logic blocks

A. Configuring I/O ports

The input/output port has an oscillator clock, a reset, and a frequency selection switch input as input units, and an output unit has an 8-bit digital frequency and three test terminals.

```
entity PCSFRGEN is
  port(Clk          : in std_logic; -- 67108864MHz Clock
       ACLR_IN      : in std_logic; -- Reset
       SEL_IN       : in std_logic_vector(3 downto 0);
       ROM_OUTA     : out std_logic_vector(7 downto 0);
       TEST_P1      : out std_logic;
       TEST_P2      : out std_logic;
       TEST_P3      : out std_logic
       );
end PCSFRGEN;
```

B. Component configuration

The components of the internal logic consist of PCSFR, Value_filter, ADC_A and clock buffer, and the components except for the clock buffer are configured as follows:

```
component PCSFR is
  port( Aclr     : in std_logic;
        CLK      : in std_logic;
        WREN     : in std_logic;
        SEL      : in std_logic_vector(3 downto 0);
        Q        : out std_logic_vector(7 downto 0)
        );
end component;
```

```
component Value_filter is
    port(
        Aclr        : in std_logic;
        Clk         : in std_logic; -- 4Mhz
        SEL_IN      : in std_logic_vector(3 downto 0);
        SEL_OUT     : out std_logic_vector(3 downto 0)
    );
end component;
```

```
component ADC_A_BLK is
    port ( ADC_CT : in std_logic_vector(7 downto 0);
           ADC_A  : out std_logic_vector(7 downto 0);
           PCOUNT : out std_logic
           );
end component;
```

The connection process of the component is as follows:

```
U1 : CLKINT              port map(Clk4M_f, Clk4M);
U2 : CLKINT              port map(Clk, Clk64M);
U3 : Value_filter        port map(
                                  Aclr     => Aclr,
                                  Clk      => Clk4M,
                                  SEL_IN   => not SEL_IN,
                                  SEL_OUT  => SEL
                                  );
U4 : PCSFR          port map(Afr_Reset,Clk64M,WREN, not SEL,QA);
U5 : ADC_A_BLK   port map(QA,ROM_DAT,FCOUNT);
end behav;
```

C. Configuration logic compilation

Compiled blocks of the entire FPGA internal use are captured and shown in the figure below.

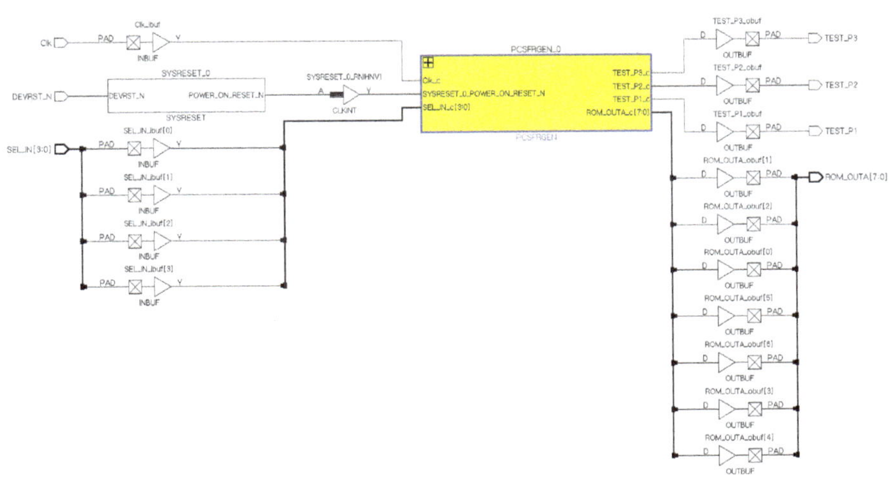

Figure 11. Block diagram of compilation.

In the entire compile block of Figure 11, terminals for input and output are connected to the PCSFRGEN block. The inside of the PCSFRGEN block is configured as shown in Figure 12 below.

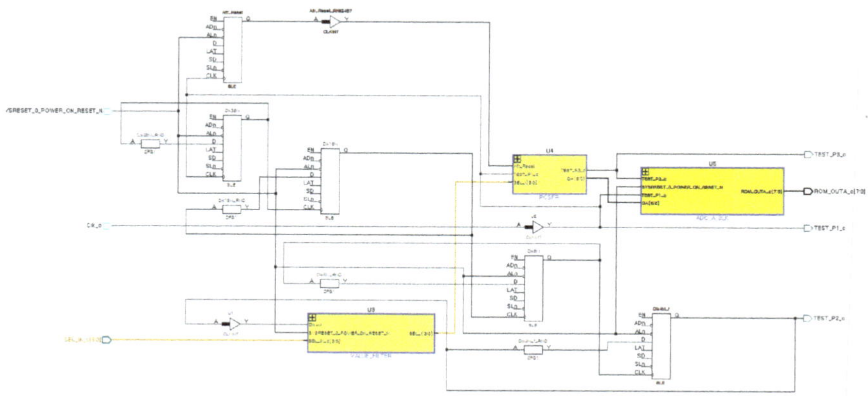

Figure 12. PCSFRGEN compilation block diagram.

Figure 13 is composed of the value filter logic block created for chattering, protecting the switch input of the input unit, and the frequency generator, Figures 14 and 15 show the FRGEN logic block and the ADC A OUT logic block having an 8-bit look-up table for the sine wave output. Each of the configured compile blocks is shown below.

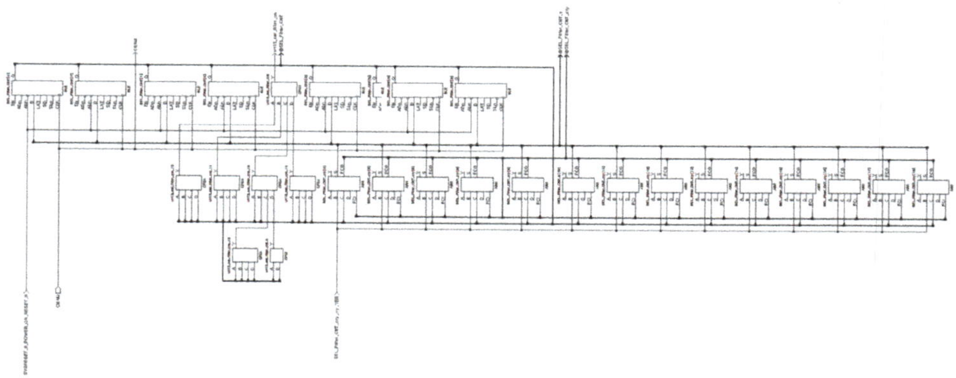

Figure 13. Value_filter compilation block diagram.

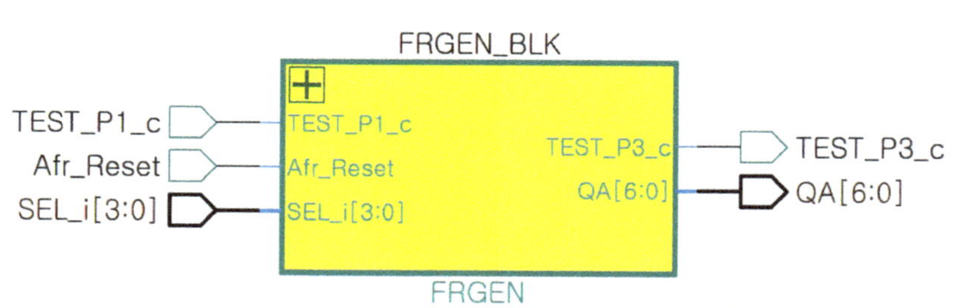

Figure 14. FRGEN compilation block diagram.

Figure 15 is a detailed compiled block of this block, and the details are divided into Figures 16 and 17.

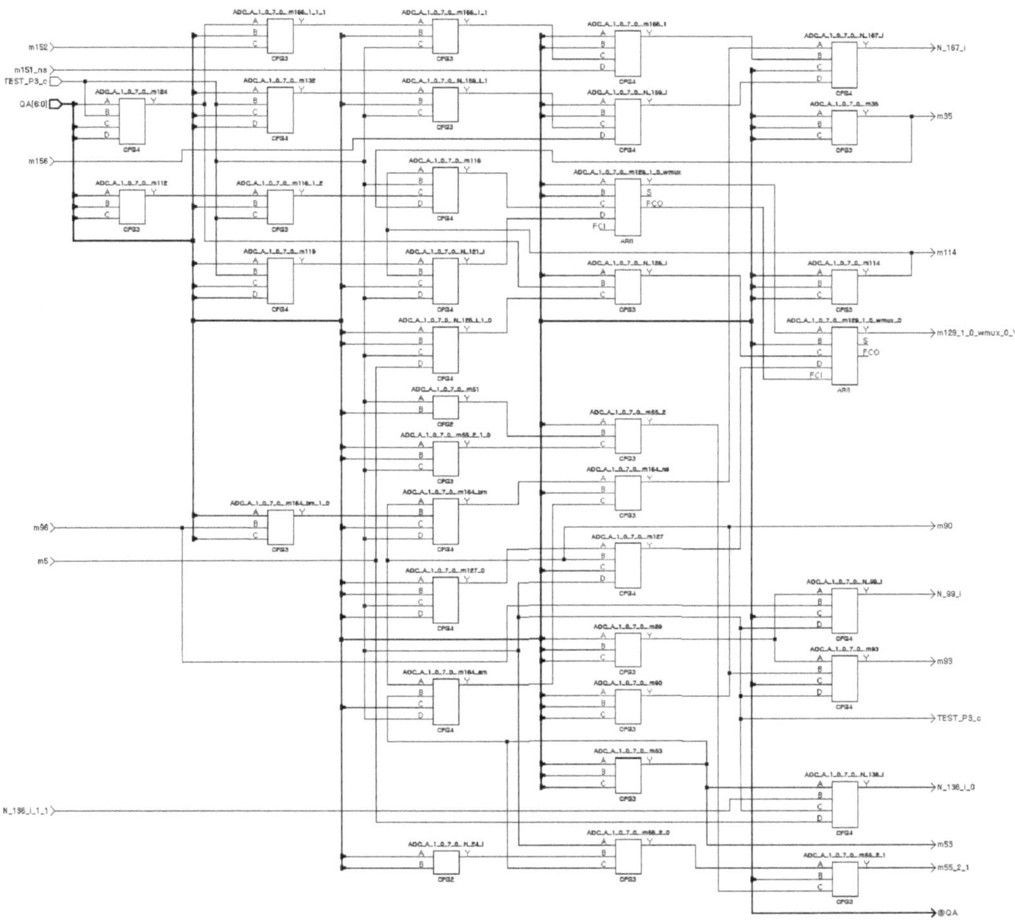

Figure 15. ADC_A_OUT (look-up table) compilation block diagram.

The FRGEN block may be referred to as a set of initial logic for frequency generation (Figure 14). The logic configuration and the compiled capture circuit of each part thereof are as follows:

```
entity FRGEN is
  port( RESETn : in std_logic;
        CLK    : in std_logic;
        SEN    : in std_logic;
        SI     : in std_logic_vector(3 downto 0);
        Q      : out std_logic_vector(12 downto 0)
        );
end FRGEN;
```

The split block diagram of Figure 16 is shown in Figures 17 and 18.

The detailed composition of Figure 16 FRGEN consists DCOUNT26 in Figure 19, DEC_Y in Figure 20, MuX26x8 in Figure 21, COUNT13 in Figure 22, OR2 in Figure 23 and R3 in Figure 24, logic blocks and each picture and compilation source is shown below.

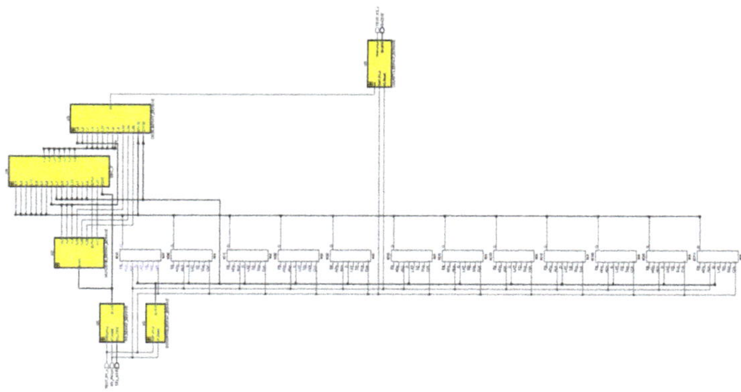

Figure 16. FRGEN compilation detail block diagram.

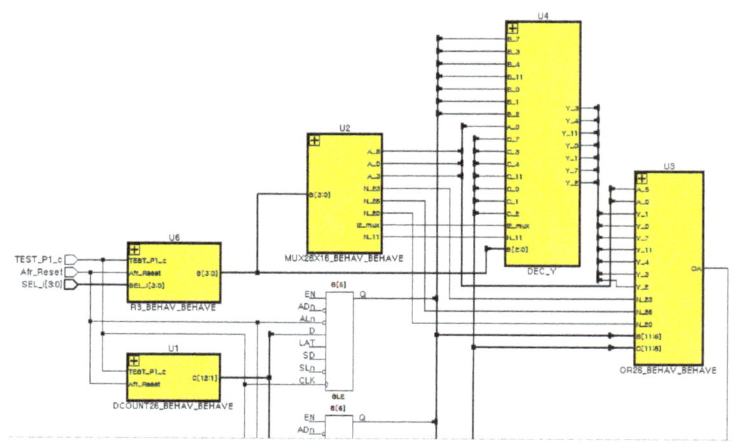

Figure 17. FRGEN compilation partial block Diagram 1.

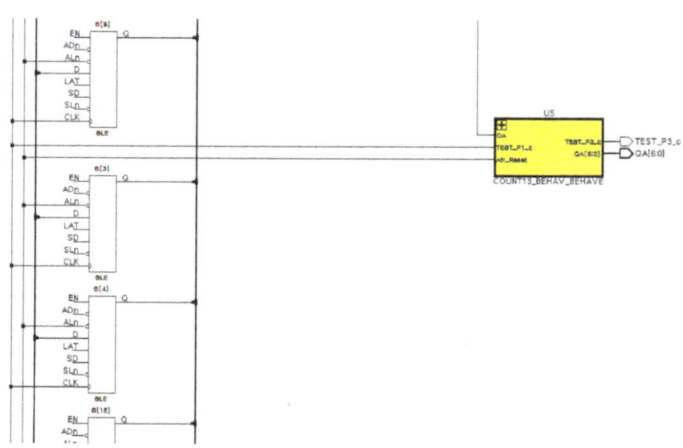

Figure 18. FRGEN compilation partial block Diagram 2.

The composition of the internal components is as follows:

```
entity DCOUNT26_behave is
  port(RESETn : in std_logic
        Clock : in std_logic;
        Q : out std_logic_vector(25 downto 0);
        );
end DCOUNT26_behave;
architecture behavioral of DCOUNT26_behave is
  signal Qaux : UNSIGNED(25 downto 0);
begin
  process(Clock, RESETn)
  begin
    if (RESETn = '0') then
      Qaux <= (others => '0');
    elsif (Clock'event and Clock = '1') then
      Qaux <= Qaux - 1;
    end if;
  end process;
  Q <= std_logic_vector(Qaux);
end behavioral;
```

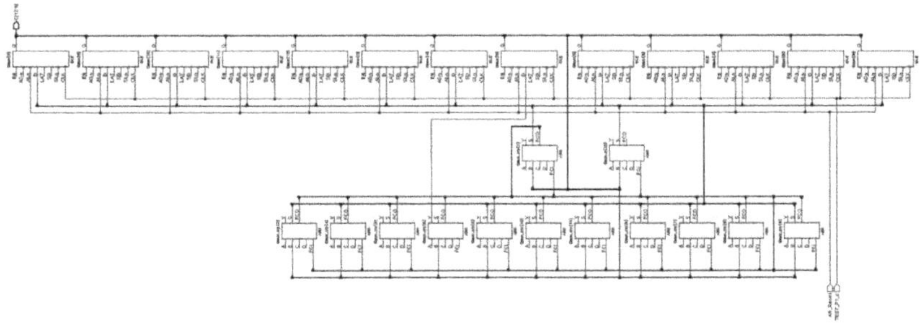

Figure 19. DCOUNT26 compilation block.

```
entity DEC_Y_behave is
  port(A : in std_logic_vector(25 downto 0);
        B : in std_logic_vector(25 downto 0);
        C : in std_logic_vector(25 downto 0);
        Y : out std_logic_vector(25 downto 0));
end DEC_Y_behave;
architecture behavioral of DEC_Y_behave is
begin
  process (A,B,C)
  begin
    Y(0) <= A(0) and B(0) and not C(0);
    Be described repeatedly from Y(1) to Y(24).
    Y(25) <= A(25) and B(25) and not C(25);
  end process;
end behavioral;
```

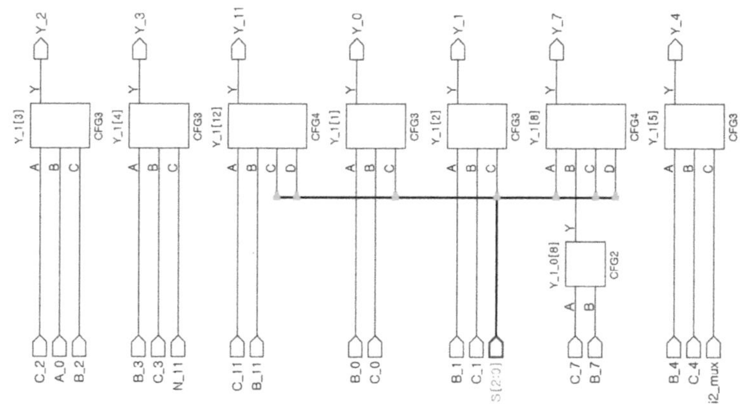

Figure 20. DEC_Y compilation block diagram.

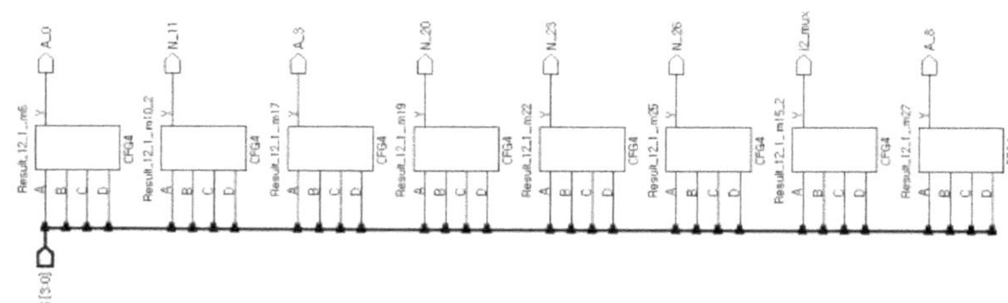

Figure 21. MUX26x8 compilation block diagram.

```
entity MUX26X16_behave is
  port(Data0_port : in std_logic_vector(25 downto 0);
       Data1_port to Data15_port are designed with
Data0_port..
       Sel0 : in std_logic;
       Sel1 : in std_logic;
       Sel2 : in std_logic;
       Sel3 : in std_logic;
       Result : out std_logic_vector(25 downto 0));
end MUX26X16_behave;
architecture behavioral of MUX26X16_behave is
begin
  process (from Data0_port, Data15_port, Sel3, Sel2, Sel1, Sel0)
    variable sel : std_logic_vector(3 downto 0);
  begin
    sel := Sel3 & Sel2 & Sel1 & Sel0;
    case (CONV_INTEGER(UNSIGNED(sel))) is
      when 0 => Result <= Data0_port;
      when Design from 1 to 15 as 0.
      when others => Result <= "--------------------------";
    end case;
  end process;
end behavioral;
```

```
entity count13_behave is
  port(Clock : in std_logic;
       Q      : out std_logic_vector(12 downto 0);
       RESETn : in std_logic;
       Enable : in std_logic);
end count13_behave;
architecture behavioral of count13_behave is
  signal Qaux : std_logic_vector(12 downto 0);
begin
  process(Clock, RESETn)
  begin
    if (RESETn = '0') then
      Qaux <= (others => '0');
    elsif (Clock'event and Clock = '1') then
      if (Enable = '0') then
        Qaux <= Qaux + 1;
      end if;
    end if;
  end process;
  Q <= Qaux;
end behavioral;
```

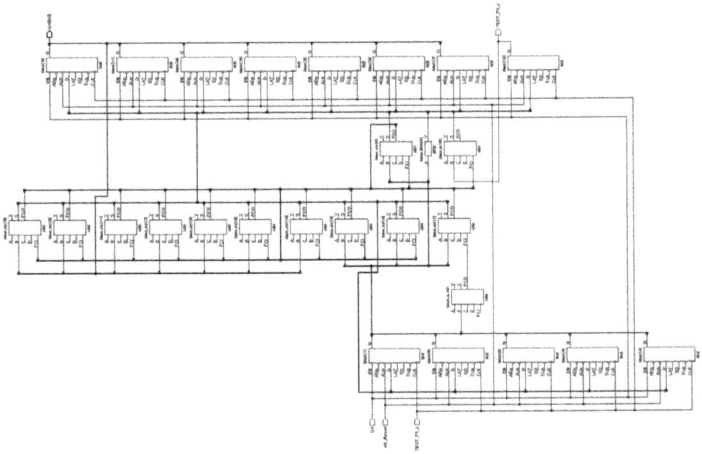

Figure 22. COUNT13 compilation block diagram.

(3) Track circuit use frequency generation

The 13 bits used to generate the target frequency are generated more accurately toward the MSB (Most Significant Bit) of the up counter treated with the FPGA of Figure 23, so the upper 8 bits are used and the duty ratio is designed to be 50:50 using the LSB (Least Significant Bit) 5 bits. Thus, the generated frequency follows the following equation:

$$LC\ Frequency(F_{LCF}) = k/2^5 \quad (6)$$

LC: Track circuit as a line circuit
k: Bits extracted with the target frequency

```
entity OR26_behave is
  port(Data : in std_logic_vector(25 downto 0);
        Result : out std_logic);
end OR26_behave;
architecture behavioral of OR26_behave is
begin
  process (Data)
    variable aux : std_logic;
  begin
    aux := '0';
    for I in 0 to 25 loop
      aux := aux or Data(I);
    end loop;
  Result <= aux;
  end process;
end behavioral;
```

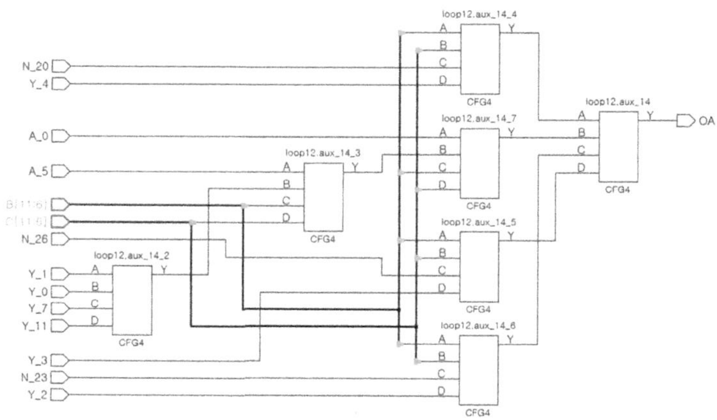

Figure 23. OR26 compilation block diagram.

```
entity R3_behave is
  port(Data : in std_logic_vector(3 downto 0);
        Enable : in std_logic;
        RESETn : in std_logic;
        Clock : in std_logic;
        Q : out std_logic_vector(3 downto 0));
end R3_behave;
architecture behavioral of R3_behave is
begin
  process(Clock, RESETn)
  begin
    if (RESETn = '0') then
      Q <= (others => '0');
    elsif (Clock'event and Clock = '1') then
      if (Enable = '0') then
        Q <= Data;
      end if;
    end if;
  end process;
end behavioral;
```

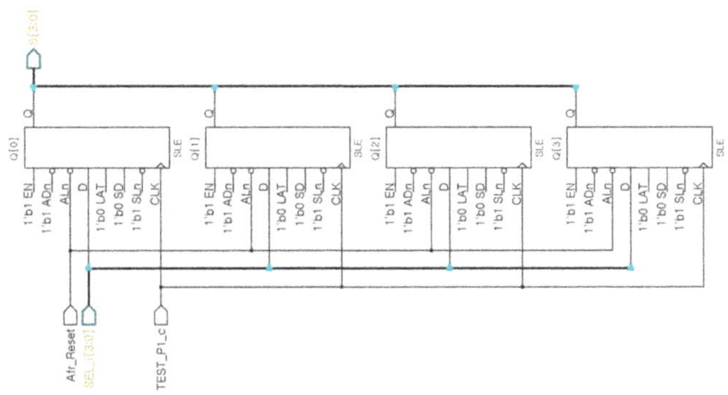

Figure 24. R3 compiles internal block diagram.

The output upper 8 bits are input to a DAC circuit (AD7541) and converted into an analog sine wave output, and the converted signal is generated as an audio frequency output to output a complete sinusoidal frequency.

The FPGA chip of the proposed system uses the Smart -Fusion2 SoC M2S010 [19] of Microchip Co., Ltd. The maximum usable logic device of this chip is composed of 12,084 highly integrated system chip ICs.

The M2S010 is designed for low power consumption and is a chip device that provides excellent reliability and security for multipurpose applications such as video image processing, I/O expansion and conversion, and Gigabit Ethernet. The ARM series MPU is also built-in, but this proposed system is not used because of its high reliability. Figure 25. Shows the internal block diagram of the M2S010.

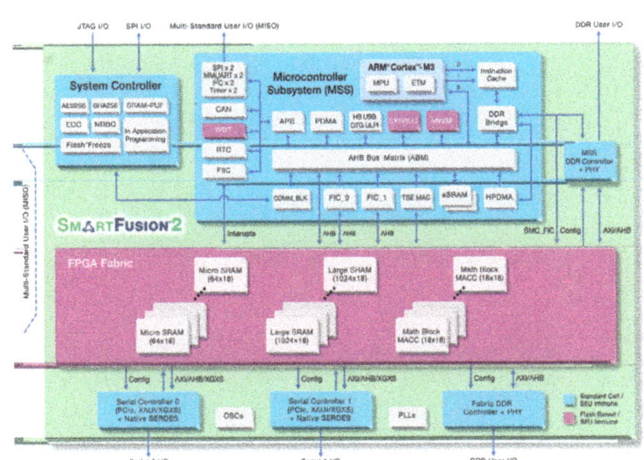

Figure 25. Internal block diagram of M2S010.

It shows a product composed of circuits designed in Figure 26.

Refer to the description of the main part in the right clockwise direction of the used substrate, FPGA chip, test terminal, JTAG interface terminal, and X-tal oscillator. From the upper left, the description is the power input terminal, the frequency change SW, the reset SW, the output terminal, and the analog waveform output IC. The 10-pin JTAG Interface Terminal on the lower right is used to input the program to the FPGA.

In implementing the frequency accuracy, we show that the circuit used in this simulation works very well, as shown in the results of the following Section 4. The circuit configuration is very simple, but the result shows that the method of verifying the envisioned algorithm is suitable. Therefore, it can be seen from the related picture that the desired precise frequency is output by this configuration. The results of the experiment were measured and recorded by selecting the frequency with the octal frequency change switch, and the output frequency was described in detail in the experimental results.

Figure 26. Circuit board used for testing.

4. Experimental Results

After repeated trials and errors in the method of optimally generating the target frequency processed by the FPGA logic block, the results of the following equations were verified.

$$Y = \frac{1}{CLOCK} \times 2^{13} \times 2^{26} \times Frequency \tag{7}$$

By implementing the practical structure and algorithm that precisely creates the AF band DDFS used in railway track circuits with FPGA, we show that 16 frequencies currently used in Europe and Korea are implemented with a precision of 99.9980%~99.9996%, and this is shown as a simulation result.

As a result of the simulation, it was confirmed that a stable and accurate frequency output with a frequency deviation superior to the target error range was made. Compared to the results of previous studies using FPGAs with a deviation of 5 Hz in the range of [8] 0–160 kHz, it shows that it is much better than the deviation of 1~2 Hz, which is [20] of the Bombardia product specification. The results can be seen in Tables 2 and 3.

Table 2. Simulation result.

Item	Target Design Frequency (Hz) (±0.05%)	Test Result Frequency (Hz)	Accuracy (%)	Note
0	1682	1682.0007	99.99995838	A
1	1716	1716.0009	99.99994755	
2	2279	2279.0043	99.99981132	B
3	2313	2313.0069	99.99970169	
4	1979	1979.0036	99.99981809	C
5	2013	2013.0009	99.99995529	
6	2576	2576.0030	99.99986025	D
7	2610	2610.0052	99.99980077	
8	1532	1532.0026	99.99983029	E
9	1566	1566.0047	99.99969987	
A	2129	2129.0021	99.99990136	F
B	2163	2163.0037	99.99982894	
C	1831	1831.0037	99.99979792	G
D	1865	1865.0038	99.99979625	
E	2328	2428.0057	99.99976904	H
F	2462	2462.0056	99.99977254	

Table 3. Simulation result comparison.

Comparison With Existing Research and Products	Frequency Deviation	Accuracy	Note
A Study on the High Reliability Audio Target Frequency Generator	0.001~0.006 Hz	0.001%	This Research Paper
Design and Implementation of a FPGA-based Direct Digital Synthesizer	+5 Hz~−5 Hz	0.30%	Reference [11]
Bombardier TI21 Track Circuit Test and Investigation Guideline	1 Hz~2 Hz	0.06~0.12%	Reference [21]

Simulation

As shown in Figure 27, the board was connected and operated.

The output results observed while turning the TWS for frequency change on the right side of the test board are shown in Figures 28 and 29:

(1) Track frequency A test results and waveforms

The signal waveform of the ADC7541 A/D converter output is 100 − (100 × (1682.0007 − 1682)/1682) = 99.9999%, and the upper frequency accuracy is shown in Table 2.

The simulation results and waveforms from the track circuit generation frequency B to H can be confirmed in the attached Appendix A.

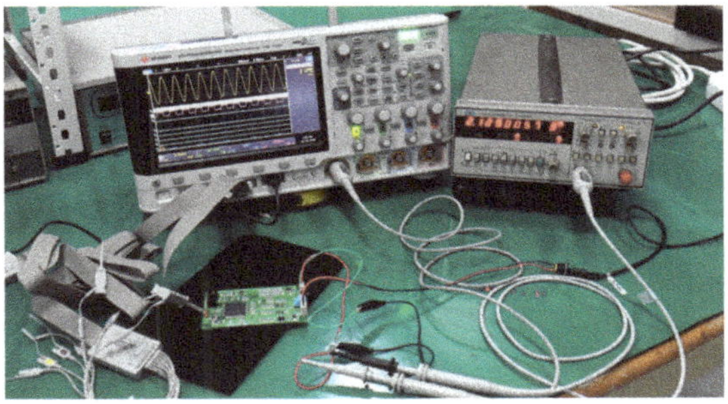

Figure 27. Simulation.

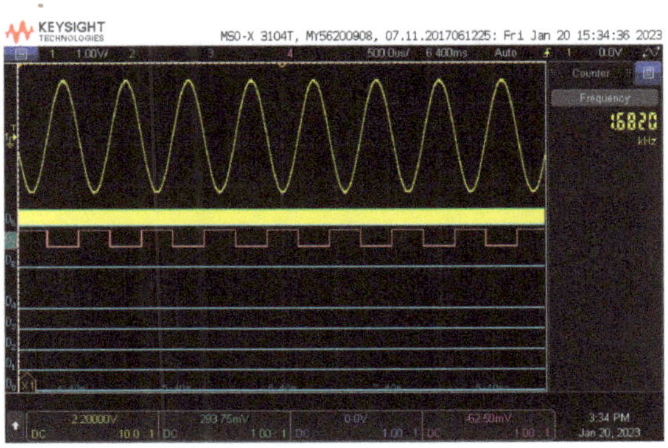

Figure 28. Lower frequency 1682 Hz generated waveform.

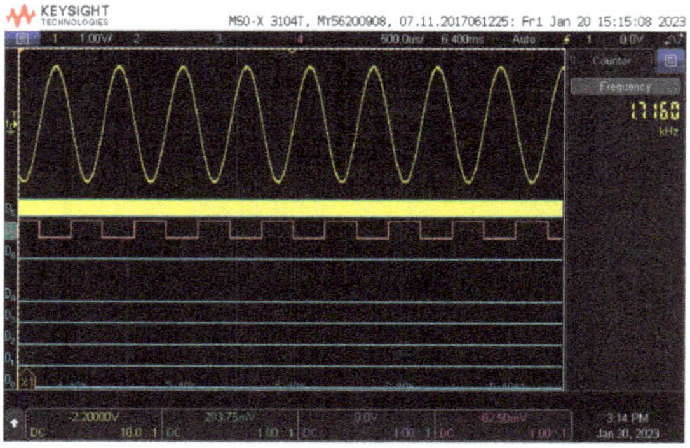

Figure 29. Upper frequency 1716 Hz generated waveform.

5. Conclusions

In this paper, we propose a method to implement AF for railway track circuits in DDS using a microchip FPGA. This frequency generator is composed of pure logic circuits without using a general CPU, minimizing the factors of malfunction and suggesting the possibility of increasing safety in key industries. By proposing a practical structure and algorithm that precisely creates DDFS in the AF band, 16 frequencies currently used in railway track circuits are implemented with a precision of 99.9980–99.9996%, and these are shown as simulation results. This proves that the performance is superior to the 5 Hz deviation of the previous study [8].

It was able to generate very stable and accurate frequency output, and it is judged that it will be possible to make a precise frequency generator with high reliability in the field of key industries such as railways. These results are expected to enhance the safety and user convenience of the control system in key industries such as railways. In the future, it is expected that the research on multiple AF DDFS that generate multiple frequencies at the same time by developing this study will be practically useful in various industries.

Author Contributions: Conceptualization, C.P. and E.H.; methodology, E.H.; software, E.H.; validation, C.P. and E.H.; formal analysis, C.P.; investigation, C.P.; resources, C.P.; data curation, C.P.; writing—original draft preparation, C.P.; writing—review and editing, I.K.; visualization, D.S.; supervision, D.S.; project administration, D.S. All authors have read and agreed to the published version of the manuscript.

Funding: This work was supported by the National Research Foundation of Korea (NRF) grant funded by the Korea government (MSIT) (No. 2022R1F1A1074773).

Data Availability Statement: Data are contained within the article.

Conflicts of Interest: The authors declare no conflict of interest. The companies had no role in the design of the study; in the collection, analyses, or interpretation of data; in the writing of the manuscript; or in the decision to publish the results.

Appendix A

In this section, measurement results from Group B to Group H among the simulation results in Table 2 are described.

(1) Track frequency B test results and waveforms

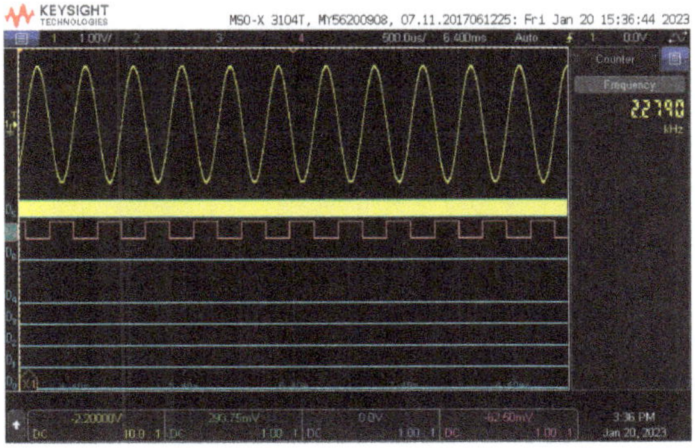

Figure A1. Lower frequency 2279 Hz generated waveform.

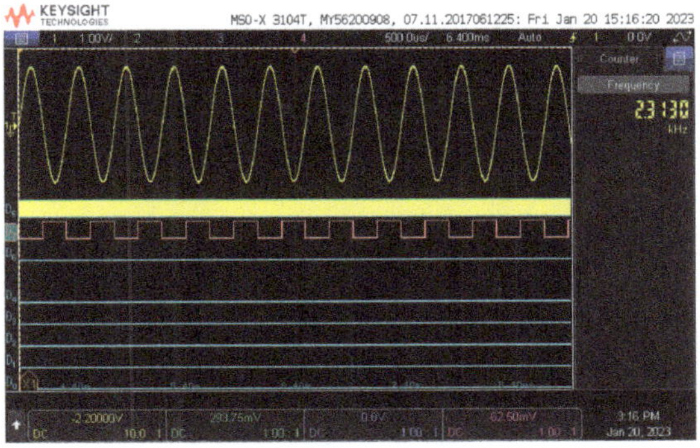

Figure A2. Upper frequency 2313 Hz generated waveform.

The signal waveform shown in the ADC7541 A/D conversion circuit output is 100 − (100 × (2279.0043 − 2279)/2279) = 99.9998% compared with the lower frequency value designed with the frequency of 2279 Hz when the TWS (thumb-wheel switch) is located at 1, and the upper frequency accuracy is shown in Table 2.

(2) Track frequency C test results and waveforms

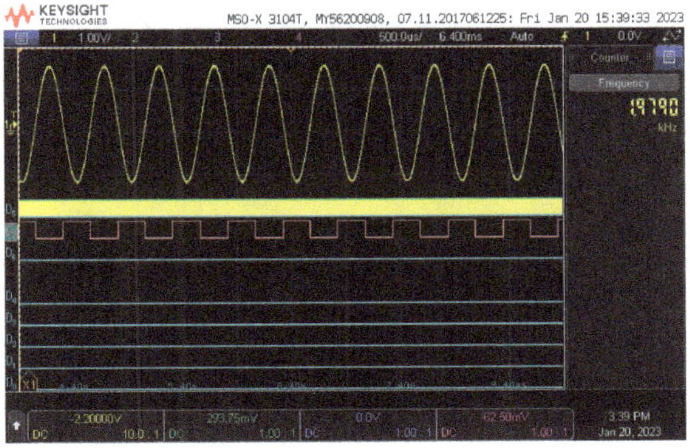

Figure A3. Lower frequency 1979 Hz generated waveform.

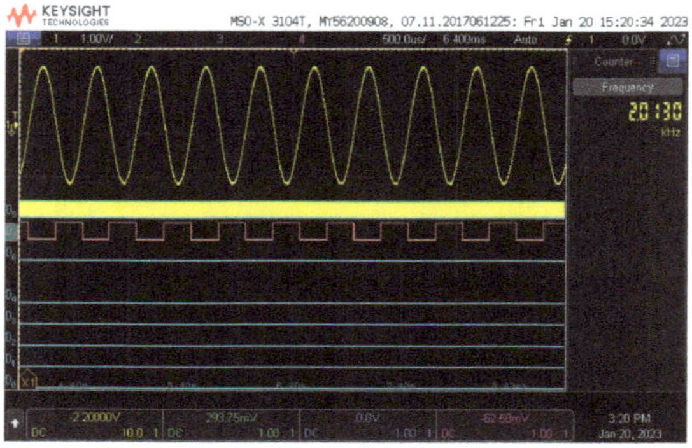

Figure A4. Upper frequency 2013 Hz generated waveform.

The signal waveform shown in the ADC7541 A/D conversion circuit output is $100 - \{100 \times (1979.0036 - 1979)/1979\} = 99.9998\%$ compared with the lower frequency value designed with the frequency of 2279 Hz when the TWS (thumb-wheel switch) is located at 1, and the upper frequency accuracy is shown in Table 2.

(3) Track frequency D test results and waveforms

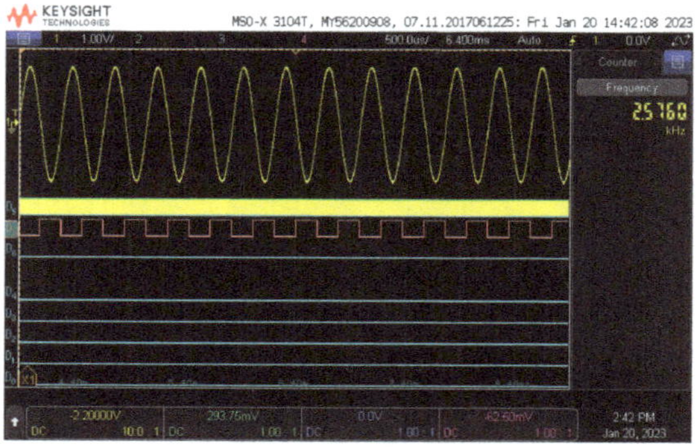

Figure A5. Lower frequency 2576 Hz generated waveform.

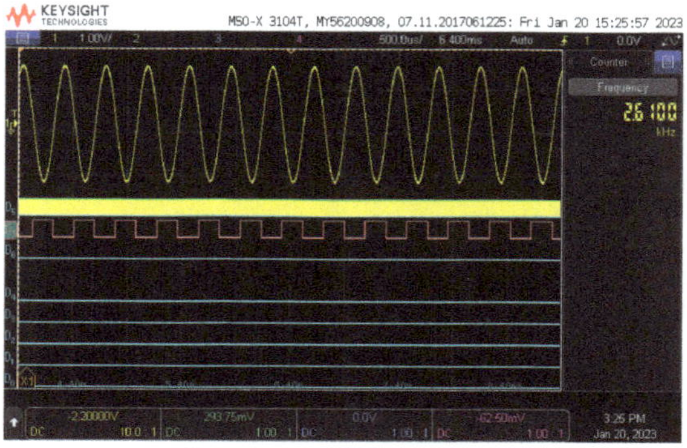

Figure A6. Upper frequency 2610 Hz generated waveform.

The signal waveform shown in the ADC7541 A/D conversion circuit output is $100 - \{100 \times (2576.0030 - 2576)/2576\} = 99.9998\%$ compared with the lower frequency value designed with the frequency of 2279 Hz when the TWS (thumb-wheel switch) is located at 1, and the upper frequency accuracy is shown in Table 2.

(4) Track frequency E test results and waveforms

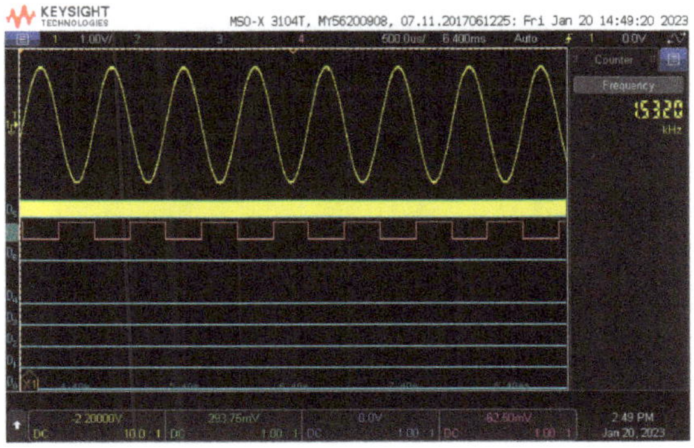

Figure A7. Lower frequency 1532 Hz generated waveform.

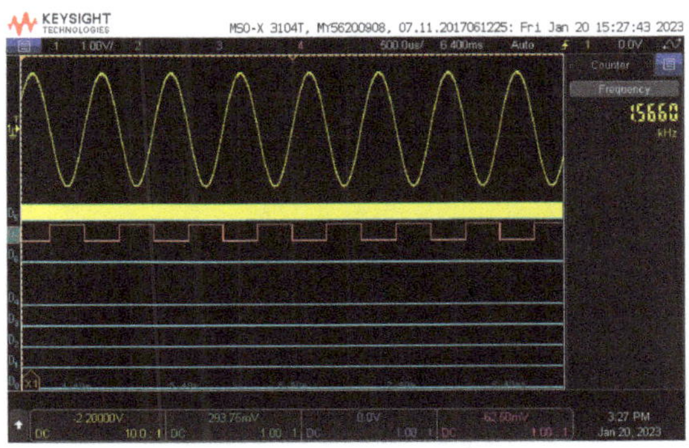

Figure A8. Upper frequency 1566 Hz generated waveform.

The signal waveform shown in the ADC7541 A/D conversion circuit output is $100 - \{100 \times (1532.0026 - 1532)/1532\} = 99.9998\%$ compared with the lower frequency value designed with the frequency of 2279 Hz when the TWS (thumb-wheel switch) is located at 1, and the upper frequency accuracy is shown in Table 2.

(5) Track frequency F test results and waveform

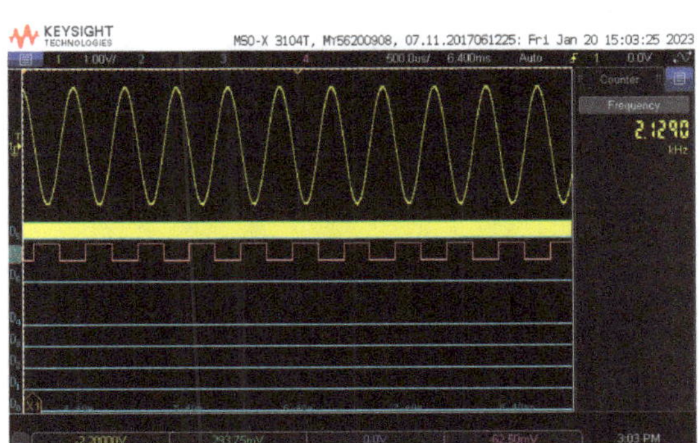

Figure A9. Lower frequency 2129 Hz generated waveform.

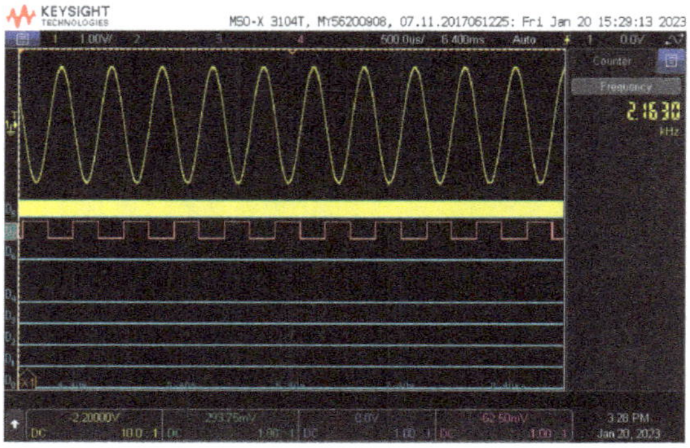

Figure A10. Upper frequency 2163 Hz generated waveform.

The signal waveform shown in the ADC7541 A/D conversion circuit output is 100 − {100 × (2129.0021 − 2129)/2129} = 99.9999% compared with the lower frequency value designed with the frequency of 2279 Hz when the TWS (thumb-wheel switch) is located at 1, and the upper frequency accuracy is shown in Table 2.

(6) Track frequency G test results and waveforms

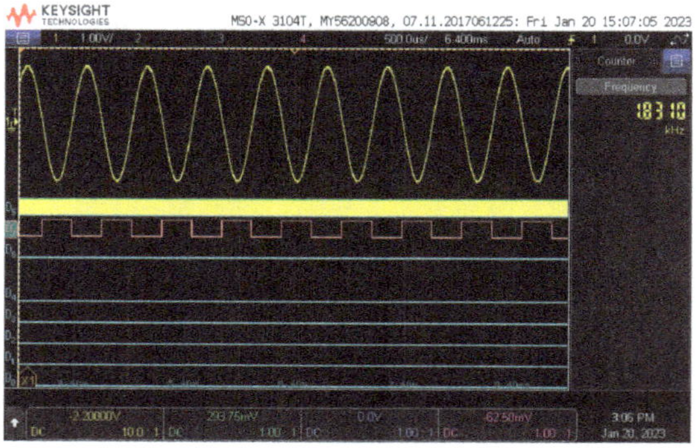

Figure A11. Lower frequency 1831 Hz generated waveform.

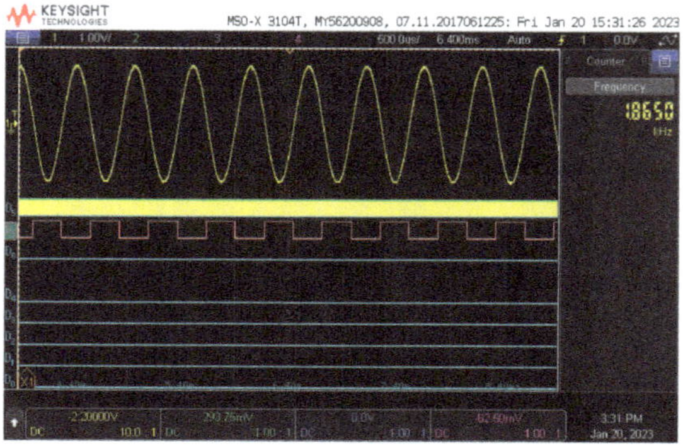

Figure A12. Upper frequency 1865 Hz generated waveform.

The signal waveform shown in the ADC7541 A/D conversion circuit output is 100 − {100 × (1831.0037 − 1831)/1831} = 99.9998% compared with the lower frequency value designed with the frequency of 2279 Hz when the TWS (thumb-wheel switch) is located at 1, and the upper frequency accuracy is shown in Table 2.

(7) Track frequency H test results and waveforms

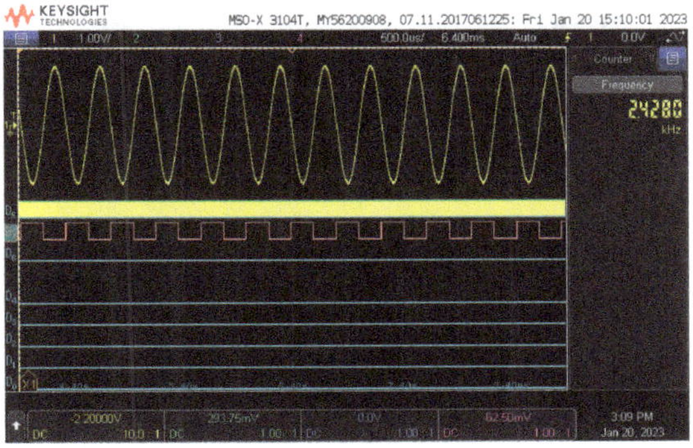

Figure A13. Lower frequency 2428 Hz generated waveform.

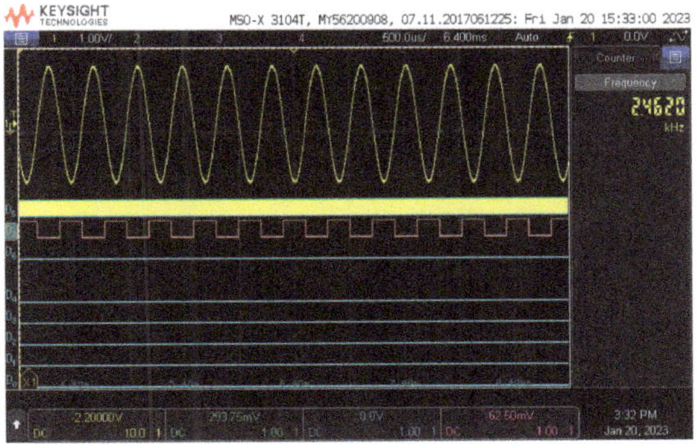

Figure A14. Upper frequency 2462 Hz generated waveform.

The signal waveform shown in the ADC7541 A/D conversion circuit output is $100 - \{100 \times (2428.0057 - 2428)/2428\} = 99.9997\%$ compared with the lower frequency value designed with the frequency of 2279 Hz when the TWS (thumb-wheel switch) is located at 1, and the upper frequency accuracy is shown in Table 2.

References

1. Yoon, K.; Song, M.; Noh, J.; Lee, K. *Design of Data Converters and PLL*; Hongneung Science Publishing House: Daejeon, Republic of Korea, 2013; pp. 299–324. [CrossRef]
2. Tierney, J.; Radre, C.M.; Gold, B. A Digital Frequency Synthesizer. *IEEE Trans. Audio Electro Acoust.* **1971**, *AU-19*, 49–50. Available online: https://ieeexplore.ieee.org/abstract/document/1162151 (accessed on 27 November 2023).
3. Ryu, H.G.; Lee, H.S. Analysis and Minimization of Phase Noise of The Digital Hybrid PLL Frequency Synthesizer. *IEEE Trans. Consum. Electron.* **2002**, *48*, 305–306. Available online: https://ieeexplore.ieee.org/abstract/document/1010136 (accessed on 27 November 2023).
4. Kim, D.C.; Chi, Y.E.; Park, J. High-Resolution Digital Beamforming Receiver Using DDS–PLL Signal Generator for 5G Mobile Communication. *IEEE Trans. Antennas Propag.* **2022**, *70*, 1429–1430. Available online: https://ieeexplore.ieee.org/abstract/document/9539071 (accessed on 27 November 2023). [CrossRef]
5. Queiroz, E.D.; Ota, J.I.Y.; Pomilio, J.A. State-Space Representation Model of Phase-Lock Loop Systems for Stability Analysis of Grid-connected Converters. In Proceedings of the 14th IEEE International Conference on Industry Applications 2021, São Paulo, Brazil, 15–18 August 2021; pp. 388–389. Available online: https://ieeexplore.ieee.org/abstract/document/9529609 (accessed on 27 November 2023).
6. Akurwy, S.H. *A Novel ROM Design for High Speed Direct Digital Frequency Synthesizer*; Lap Lambert Academic Publishing: Saarbrücken, Germany, 2014; pp. 6–15.

7. Gao, S.; Barnes, M. Phase-locked loops for grid-tied inverters: Comparison and testing. In Proceedings of the 8th IET International Conference on Power Electronics, Machines and Drives (PEMD 2016), Glasgow, UK, 19–21 April 2016. Available online: https://digital-library.theiet.org/content/conferences/10.1049/cp.2016.0304 (accessed on 27 November 2023).
8. Shan, C.; Chen, Z.; Yuab, H.; Hu, W. Design and Implementation of a FPGA-based Direct Digital Synthesizer. In Proceedings of the 2011 International Conference on Electrical and Control Engineering, Yichang, China, 16–18 September 2011; pp. 614–615. Available online: https://ieeexplore.ieee.org/abstract/document/6057152 (accessed on 27 November 2023).
9. Rokita, A. *Direct Analog Synthesis Modules for an X-Band Frequency Source*; Telecommunications Research Institute: Daejeon, Republic of Korea, 1997; pp. 63–64. Available online: https://ieeexplore.ieee.org/abstract/document/737920 (accessed on 27 November 2023).
10. Wang, Y.; Bao, X.; Hua, W. Implementation of Embedded Magnetic Encoder for Rotor Position Detection Based on Arbitrary Phase Shift Phase Lock Loop. *IEEE Trans. Ind. Electron.* **2002**, *69*, 2035–2037. Available online: https://ieeexplore.ieee.org/abstract/document/9369043 (accessed on 27 November 2023). [CrossRef]
11. Kim, S.; Kim, J.; Oh, H.; Cheon, J.; Park, G.; Go, S.; Lee, K. Design of Low Power Frequency Synthesizer for GPS Receiver. *Korea Inst. Intell. Transp. Syst.* **2008**, *11a*, 165–168.
12. Alssharef, A.A.; Ali, M.A.M.; Sanusi, H. Direct Digital Frequency Synthesizer Design and Implementation on FPGA. *Res. J. Appl. Sci.* **2012**, *7*, 387–390. [CrossRef]
13. Bergeron, M.; Willson, A.N. A 1-GHz Direct Digital Frequency Synthesizer in an FPGA. In Proceedings of the 2014 IEEE International Symposium on Circuits and Systems (ISCAS), Melbourne, VIC, Australia, 1–5 June 2014; pp. 329–332. Available online: https://ieeexplore.ieee.org/abstract/document/6865132 (accessed on 27 November 2023).
14. Saber, M.S.; Elmasry, M.; Abo-Elsoud, M.E. Quadrature Direct Digital Frequency Synthesizer Using FPGA. In Proceedings of the 2006 International Conference on Computer Engineering and Systems, Cairo, Egypt, 5–7 November 2006; pp. 14–15. Available online: https://ieeexplore.ieee.org/abstract/document/4115478 (accessed on 27 November 2023).
15. Chen, W.; Wu, T.; Tang, W.; Jin, K.; Huang, G. Implementation Method of CORDIC Algorithm to Improve DDFS Performance. In Proceedings of the IEEE 3rd International Conference on Electronics Technology 2020, Chengdu, China, 8–12 May 2020; pp. 58–61. Available online: https://ieeexplore.ieee.org/abstract/document/9119621 (accessed on 27 November 2023).
16. Yang, Y.; Wang, Z.; Yang, P.; Chang, M.F.; Ho, M.S.; Yang, H.; Liu, Y. A 2-GHz Direct Digital Frequency Synthesizer Based on LUT and Rotation. In Proceedings of the 2018 IEEE International Symposium on Circuits and Systems (ISCAS), Florence, Italy, 27–30 May 2018; pp. 1–3. Available online: https://ieeexplore.ieee.org/abstract/document/8351207 (accessed on 27 November 2023).
17. Kim, D.; Lee, H.; Kim, J.; Kim, S. Design and Modeling of a DDS Driven Offset PLL with DAC. *Korea Internet Broadcast. Commun. Soc.* **2012**, *12*, 1–9. [CrossRef]
18. Gothandaraman, A.; Islam, K.S. An All-Digital Frequency Locked Loop (ADFLL) with a Pulse Output Direct Digital Frequency Synthesizer (DDFS) and an Adaptive Phase Estimator. In Proceedings of the IEEE Radio Frequency Integrated Circuits Symposium 2003, Philadelphia, PA, USA, 9–10 June 2023; pp. 303–305. Available online: https://ieeexplore.ieee.org/abstract/document/1213949 (accessed on 27 November 2023).
19. Bombardier. *EBI Track 200 TI21 Audio Frequency Track Circuit Technical Manual*; Bombardier: Montreal, QC, Canada, 2019. Available online: https://docplayer.net/28867426-Ebi-track-200-ti21-audio-frequency-track-circuit.html (accessed on 27 November 2023).
20. Microchip. *FPGA and SoC Product Families*; Microchip Technology Inc.: Chandler, AZ, USA, 2019; pp. 3–5. Available online: http://ww1.microchip.com/downloads/en/DeviceDoc/00002871B.pdf (accessed on 27 November 2023).
21. Transport RailCorp. *TI21 Track Circuit Test and Investigation Guideline*; Transport RailCorp: Sydney, NSW, Australia, 2016; pp. 13–14. Available online: https://www.transport.nsw.gov.au/industry/asset-standards-authority/find-a-standard/ti21-track-circuit-test-and-investigation (accessed on 27 November 2023).

electronics

Article

A Neighborhood-Similarity-Based Imputation Algorithm for Healthcare Data Sets: A Comparative Study

Colin Wilcox [1], Vasileios Giagos [2] and Soufiene Djahel [3,*]

1 Department of Computing and Mathematics, Manchester Metropolitan University, Manchester M15 6BH, UK; colin.r.wilcox@stu.mmu.ac.uk
2 Department of Mathematical Sciences, University of Essex, Colchester CO4 3SQ, UK; v.giagos@essex.ac.uk
3 Centre for Future Transport and Cities, Coventry University, Priory Street, Coventry CV1 5FB, UK
* Correspondence: ae3095@coventry.ac.uk

Abstract: The increasing computerisation of medical services has highlighted inconsistencies in the way in which patients' historic medical data were recorded. Differences in process and practice between medical services and facilities have led to many incomplete and inaccurate medical histories being recorded. To create a single point of truth going forward, it is necessary to correct these inconsistencies. A common way to do this has been to use imputation techniques to predict missing data values based on the known values in the data set. In this paper, we propose a neighborhood similarity measure-based imputation technique and analyze its achieved prediction accuracy in comparison with a number of traditional imputation methods using both an incomplete anonymized diabetes medical data set and a number of simulated data sets as the sources of our data. The aim is to determine whether any improvement could be made in the accuracy of predicting a diabetes diagnosis using the known outcomes of the diabetes patients' data set. The obtained results have proven the effectiveness of our proposed approach compared to other state-of-the-art single-pass imputation techniques.

Keywords: healthcare; imputation algorithms; incomplete data; neighborhood similarity

Citation: Wilcox, C.; Giagos, V.; Djahel, S. A Neighborhood-Similarity-Based Imputation Algorithm for Healthcare Data Sets: A Comparative Study. *Electronics* **2023**, *12*, 4809. https://doi.org/10.3390/electronics12234809

Academic Editors: Wentao Li, Huiyan Zhang, Tao Zhan and Chao Zhang

Received: 6 October 2023
Revised: 22 November 2023
Accepted: 23 November 2023
Published: 28 November 2023

1. Introduction

Due to widespread computerization, medical services have embarked on moving their historic paper-based medical data onto computer systems [1]. This has raised a number of technical and societal issues. Generations of paper-based medical records need to be digitally encoded in a way that is not only capable of handling the large information backlog, but must also be accurate, sensitive, and, most importantly for many financially stretched services, cost effective [2,3]. Historic medical data has highlighted the inconsistencies of the previous recording and transcription practices and processes used by both medical practitioners and regional authorities such that, in many cases, data may be incomplete, incorrectly encoded, or just erroneous. This is not just a legacy issue, as modern recording techniques also suffer from similar issues of data incompleteness that emphasize the need to find a robust solution to this wider problem [4,5].

In the future, legacy data will form the basis of a much wider medical profile describing an individual and will include more granular and real-time information. This data may include a person's movements, access to medical facilities, data from personal fitness trackers, and other biometric devices. Data from all such sources need to be recorded in a consistent manner. By ensuring high quality and the accuracy of such data, these medical data sources become points of truth when identifying the individual to which they relate and can thereby be used as a means of individual identification.

Imputation is the overarching term used for describing the range of techniques used to replace missing data in a data set. The techniques can range from very simple numerical replacement to more complex statistical approaches. They can be broadly split into several types of approaches [6]:

- **Normal imputation**: When the data is numerical, we can use simple techniques, such as mean or modal values for a feature, to fill in the missing data. For data that is more categorical (i.e., they have a defined and limited range of possible values), then the most frequently occurring modal value for this feature can be used.
- **Class-based imputation**: Instead of replacing missing data with a calculated value based on existing feature values above, the replacement is done based on some internal classification. This approach determines the replacement value based on the values of a restricted subclass of known feature values.
- **Model-based imputation**: A hybrid approach where the missing value is considered as the class, and all the remaining features are used to train the model for predicting values.

The problem we aim to address concerns the rapidly growing amount of incomplete personal medical data that exists. The rapid increase in volume and complexity of this data has highlighted potential problems and issues caused by our current reliance on this incomplete or inaccurate information. Such unqualified use may lead to a loss or misinterpretation of critical medical information. This problem is not limited to a medical domain and equally applies to any problem domain that uses incomplete personal information in a technology-driven environment. The focus of this paper is on a medical context, but the solution should be readily generalizable to other problem domains. The existence and use of incomplete medical data may lead to a loss or misrepresentation of critical medical information [6]. The increasing amount and variety of stored data about individuals in the smart healthcare era only emphasizes the urgency in finding solutions to this problem [7]. Our approach will select imputed data values in a more localized manner, thus applying a more intelligent selection of candidate values rather than one of the more simplistic, and widely used, imputation methods.

In this paper, we propose a neighborhood-based imputation algorithm that uses the idea of feature value similarity in similar data records to predict missing feature values in incomplete records. This subset of candidate records is specific to a single incomplete record and so is recreated for each incomplete record found in a data set. This differs from other imputation techniques, which may consider all records in the data set and give a more general and less localized result, or other approaches, which determine neighborhood values based on other criteria such as using weighted average or variance estimation techniques [7].

Our algorithm aims to improve on some of the limitations of existing imputation algorithms, especially kNNs, by providing a fast, yet accurate imputation process suitable for use on, initially, medical data, but also on more generic incomplete data sets from other similar problem domains. The main contributions of this work can be summarized as follows:

- Reducing the speed degradation of the algorithm as the size of the data set increases.
- The way imputed values are selected is more localized rather than potentially using all similar values in the data set.
- Reducing the negative impact of outlying values by making imputed values selection more localized.
- Providing a solution that can be extended for use with textual and categorical data, as well as numeric data.

The remainder of this paper is organized as follows. In Section 2, we present the background to understanding the problem being studied in this paper. Section 3 presents our proposed algorithm to improve prediction accuracy, and Section 4 evaluates the performance of our proposed technique in comparison with other imputation methods. Section 5 discusses our conclusions and findings during this work, and, finally, Section 6 indicates some directions for future work.

2. Background and Related Work

In this section, we present the background of incomplete medical data and the reasons why data integrity and completeness are important.

Imputation is the name given to the range of techniques that attempt to restore missing information in a data set with values based on the feature values of complete data records. The complexity of this process can range from merely replacing missing values with fixed absolute values, thus applying some mathematical function to known feature values for a given missing feature, or, in the simplest scenario, incomplete data records may be completely removed from consideration [6,8]. The choice of the technique used depends on a number of factors, including the nature of the source data, the amount of missing or erroneous data in the data set, and the time needed to create a suitably complete set of data. More complex approaches, such as time-series-based methods, attempt to rebuild potential structures within the data set by considering wider factors such as patterns in the data and relationships between the values of related features rather than just individual value replacement. Examples of such approaches include linear interpolation techniques, which take two known feature values and use a weighted distance between these endpoints to calculate intermediate values [9], and the use of adjacent known feature values as candidates for replacing missing feature values [10]. Such techniques tend to be more time consuming, and their effectiveness is reliant upon the intended use and ability to identify suitable structures to recreate within the source data [11]. Many of these restoration techniques have analogies in the non-digital world, which may be considered as possible approaches for imputing sets of data. In the following section, we briefly discuss three common approaches to single-pass imputation [12].

2.1. Imputation by Mean/Mode/Median and Others

If the missing values in a data set's feature column are numeric, they can be imputed by using the mean value of the existing values for that feature variable. The mean imputed value could be replaced by the median feature value if the feature is suspected to have outlying values. For a categorical feature, the missing values could be replaced by the mode of the existing values for that feature. The major drawback of this method is that it reduces the variance of the imputed variables. This method also reduces the correlation between the imputed variables and other variables, because the imputed values are just estimates and will not be related to other values inherently [13].

Another algorithm worthy of note is the k-nearest neighbors (kNNs) algorithm [14]. In a similar manner to our proposed algorithm, kNNs attempts to impute missing feature values by using the mean value of the corresponding known feature values for the k-closest records. The kNNs algorithm has a number of limitations, which our algorithm attempts to resolve. The kNNs is a robust algorithm belonging to a family of *nearest neighbor* algorithms used to predict unknown classifications based on a data set of known classifications. It is commonly used because it is intuitive and easy to implement and is nonparametric, meaning that it makes no prior assumptions about the nature of the data set. It may be used for both classification and regression problems, thus making it a widely used and popular choice of algorithm.

The kNNs algorithm has a number of disadvantages, which our solution attempts to improve upon and include the following:

- The kNNs is a relatively slow algorithm, with its performance decreasing as the size of the data set increases.
- The kNNs suffers from the curse of dimensionality [15]. As the number of feature values (dimensions) per record increases, the amount of data required to predict a new data point increases exponentially.
- The manner in which kNNs measures the closeness of a pair of records is quite simple, by using Euclidean or Manhattan distances for example.
- The kNNs algorithm needs homogeneity such that all the features must be measured using the same scale, since the distance is taken as an absolute measure.

- The kNNs does not work well with imbalanced data. Given two potential choices of classification, the algorithm will naturally tend to be biased towards a result taken from the largest data subsets, thus leading to potentially more misclassifications.
- The kNNs is sensitive to outlying values, as the choice of closest neighbors is based on an absolute measure of the distance.

Our algorithm aims to improve on these drawbacks, especially in the areas of outlier sensitivity, thereby reducing the likelihood of misclassification and the choice of imputed feature values. Since the kNNs uses the mean of the k-nearest feature values, this could lead to a value being calculated that does not appear in any of the actual complete records; our algorithm removes this scenario by only choosing imputed feature values from a pool of candidate values taken from the actual feature values of the most similar complete records.

The class of nearest neighbour predictive algorithms can make accurate predictions, which do not require a human-readable model [16]. The quality of these predictions depends on the measure of the distance between the data values [17]. There are several advantages to this class of algorithms, including a robustness for noisy data and the ability to be tuned quite easily. However, the kNNs has some drawbacks, such as the need for all the feature values for any missing value to be considered. This was a motivation and opportunity to use a more localized approach for determining missing data values [16].

2.2. Simple Statistical Imputation Techniques

Statistical techniques are usually applied because they tend to be fast, have low memory overhead, and are applicable in isolation to any surrounding data. These simpler approaches involve determining the value of a missing feature by applying a simple functional calculation on the set of known feature values [15]. In our comparison, these are represented by the mean (MAV) and modal (MDAV) value algorithms. Calculations tend to be linear in nature and applied independently from other data fields in the same data set. Calculations may range from setting missing data values to a known fixed value to finding an average of those values that exist in other records in the data sets, or some trivial manipulation of existing data values from other records [18]. More involved algorithms have been developed, which try to use wider information about the nature of the data values and any relationships that may exist between features as a way of more accurately determining missing feature values. In our discussion, we highlight two such algorithms, kNNs imputation and empirical Bayes inference; however, there are many more that could be considered. This approach can be extended to use multiple imputation techniques, which involves repeatedly applying simple mathematical techniques to improve the missing feature value prediction, as defined by the pseudo flow below:

1. Identify missing values in the source data set.
2. Iterate through the data set. For each record with missing values, replace each missing value with a statistical measure based on values for the same field found in other records where this field is not missing.
3. Once all the records have been completed, if the nature of the data set meets the criteria for its intended use, then stop; otherwise, repeat Step 2.

2.3. Multistage Techniques

Multiple imputation is a general approach to the problem of missing data that is available in several commonly used statistical packages such as R [19,20]. Single-pass imputation is the process of "filling in" gaps representing missing values in data sets. An imputation method is a function that takes a number of known feature values as inputs and uses them to calculate a potential value for a missing feature value. Single-pass imputations apply such a mapping only once to the original set of known feature values. Multiple imputation, however, is a technique for reducing the uncertainty of missing values in a data set by creating several different viable imputed data sets and appropriately combining the results obtained from each of them to determine a suitable replacement value. We will compare the performance of our N-Similarity (NSIM) algorithm against that of three simple single-pass imputation algorithms, which either replace the missing feature value with the mean (MAV) and modal (MDAV) values of the known feature values or just remove all incomplete records from the processed data set.

Using single values carries with it a level of uncertainty about which values to impute. Multiple imputation reduces this uncertainty by calculating several different possible values ("imputations"). Several versions of the incomplete data sets are created, which are then combined to make the "best" value selections. Such an approach has several advantages such as reducing bias and minimizing the likelihood of errors being introduced to the rebuilt data sets, thus improving the validity of the data and increasing the precision or closeness between two or more imputed values, which makes the data set more resistant to outlying values [21,22].

The second stage is to use common statistical methods to fit the model of interest to each of the imputed data sets. Estimated associations in each of the imputed data set will differ because of the variation introduced in the imputation of the missing values, and they are only useful when averaged together to give overall estimated associations. Valid inferences are obtained because we are averaging over the distribution of the missing data given the observed data [23,24].

Other data-focused approaches using machine learning and deep data analysis techniques are being used as a means of predicting medical events from incomplete medical data sets. The use of such automated tools in the identification and prediction of medical conditions is becoming increasingly important due to the shortage of skilled medical professionals, as well as their ability to increase the prediction accuracy, thus reducing the burden on medical staff [25,26].

3. Proposed Algorithm

In this section, we outline our approach to improving the effectiveness of predicting binary outcomes based on a series of numerical feature values. We used a suitably anonymized diabetes diagnosis data set, which identified whether a patient with diabetes has been positively diagnosed (*true positive*) or whether one who does not have diabetes has been negatively diagnosed (*false positive*).

3.1. Proposal Main Steps

Our algorithm aims to improve on a number of traditional single-pass imputation techniques to achieve a higher percentage of correct predictions when applied to an incomplete diabetes data set, D. The approach will consist of the following steps.

- Apply our imputation technique to fill in each missing attribute f_i in turn, where i corresponds to the ith feature in each patient record, for the current record r to create a complete record in D. This will become the basis of the later comparisons. Incomplete records r are given by

$$\forall r \in D, r = (f_0, f_1, f_2, \ldots f_i - 1, f_i + 1, \ldots) \quad (1)$$

- Use the k-fold (with k = 10) [27,28] technique to partition D into non-intersecting subsets. In turn, each subset (fold) will be considered to be the *test fold*, and the remaining folds will be used as *training folds*. For each record in the test fold, we apply a comparison function $F()$, which is in our case the cosine similarity, to obtain a numerical measure of how similar the test record is to the current record in the training folds. An ordered *similarity table*, S, is maintained and stores details of each training record and how similar it is to the current test record. This is repeated until the test record has been compared against all the records in all the training folds. After each change to the contents of S, it will be sorted in such a way that the most similar training record will appear as the first item in the list. This could be more complicated depending on the comparison function used, but in our case, the sort order is merely used to maintain the n-closest items (defining the neighborhood) in S in an increasing cosine similarity order. The contents of S must be cleared once all the training set records have been compared and are ready for subsequent cycles.

Folds containing a large number of records can increase the time needed to compare all the combinations of these records against a given test record. This could result in a relatively large similarity table. To address this issue of similarity table size, our proposed algorithm introduces the concept of a *neighborhood* containing the most similar n records in the training set. The size of this neighborhood limits the maximum size of the similarity table and is used as a means of calculating the new replacement value for a missing attribute.

Considering S_t to be the set of test records and S_{tr} to be the set of training records for a given cycle, such that $t \in S_t$ and $tr \in S_{tr}$, we can say that

$$\forall t \in S_t, \forall tr \in S_{tr}; S_t \cap S_{tr} = 0, S_t \cup S_{tr} = D \tag{2}$$

If there are less than n records in the similarity table, then add the current training record, tr, into the next freely available position p. If the similarity table already contains n records and the current test record, t is more similar than the last record in the similarity table (at position $n-1$ for zero-based arrays); then, we replace the last entry in the similarity table with the current training record tr. This can be shown with the pseudocode below.

```
clear SimilarityTable, S
FOR EACH t IN testFold DO
    p <- 0
    FOR EACH tr IN trainingFolds DO
        size = count(S)
        IF size < n THEN
            S[p] <- F (t, tr)
        ELSE
            IF F(t, tr) > S[n-1] THEN
                S[n-1] <- F(t,tr)
```

Each time the contents of the similarity table are changed, they should be immediately sorted based on decreasing similarity value to maintain a list of the most similar training records for the current test record. In order to build a complete data set D, we need to calculate each of the missing data values across all the records in D. This is achieved by comparing each row that contains missing values against all the complete rows that exist in D. By doing this, we build up a similarity table containing the most similar complete records from which the candidate values for the missing data values may be selected. Once all the complete records in the data set have been compared against the current incomplete record, we are in a position to impute the missing values for the current record in order to make it complete. This record can then be used as a candidate record for matching the other incomplete records in later cycles of the process. The end result will be a completely imputed data set, which can then be used for comparison purposes with the different imputation techniques.

3.2. Similarity Model Behavior

Our proposed model is built around the idea that patients with the same sets of symptoms (features) will result in the same diagnosis. A patient with an unknown diagnosis will have a number of recorded symptoms, which may or may not be complete. Our algorithm takes those features that are known and uses them to find those diagnosed patients (neighborhood) that are the closest match in terms of the most similar features. This neighborhood is then used to determine what the likely diagnosis of the target patient may be. This has the advantage over other techniques in the fact that only similar patient records are used to build the picture of the diagnosis rather than a much wider spread of patients who may have less correlation with the patient in question.

This similarity model is based on the splitting of the source data set as previously described. The idea is to take the source data set and split it into two disjoint subsets—the training data set and the test data set. The splitting of the source data ensures that the number of records in the test data set is a fixed proportion of the total number of records according to the supplied parameters.

Each record in the test data subset is compared in turn with each record in the training data subset. A comparison of each pair of records is made using the concept of cosine similarity to obtain a measure of how similar the corresponding pairs of field attributes are with each other, thus yielding a numeric measure of their similarity. During this process, a similarity table is built giving a similarity measure of each training record in the training set against a single test record. This table is maintained such that the record with the most similar value (i.e., the most similar) is the first record in the table. The rationalization is that the training set records that are considered to be a similar match to the test record, and therefore the initial best-matching training record, will have very similar values for their input arguments, and, as such, they are the best candidates to determine whether the outcome given by the closest-matching test record was in fact valid.

Finally, a replacement value for the missing attribute, f_i, is determined by applying a prioritized set of rules to choose the most appropriate value from the candidate value set C. This approach may be extended to include '*categorical variables*', which describe those features that take a value from a limited set of possible values. Since the feature value set C, used as the pool of possible replacement values, is constructed from known feature values of the most similar records, then the selection rules are equally applicable and will select a suitable replacement value from C.

Considering the process diagram shown in Figure 1, the similarity modeling process is split into two main subflows. The colors used are unimportant and just used for highlighting purposes. The blue flow describes the processing steps of loading external data and standardizing it into a form that can be used by the second (green) flow through the application of the k-fold technique to split the source data set into folds. The green flow indicates the application of the N-Similarity algorithm. The key points of the algorithm flow are to take each fold as a test record in turn and apply cross correlation against each of the remaining training folds to generate the similarity table of the most similar training records for each record in the test fold. This is repeated for each training record until all comparison combinations have been performed. For each incomplete record, the missing feature value is determined by considering the properties of the closest records in the similarity table, and a candidate is selected based on a number of rules and criteria. The results of these comparisons are shown in Table 1.

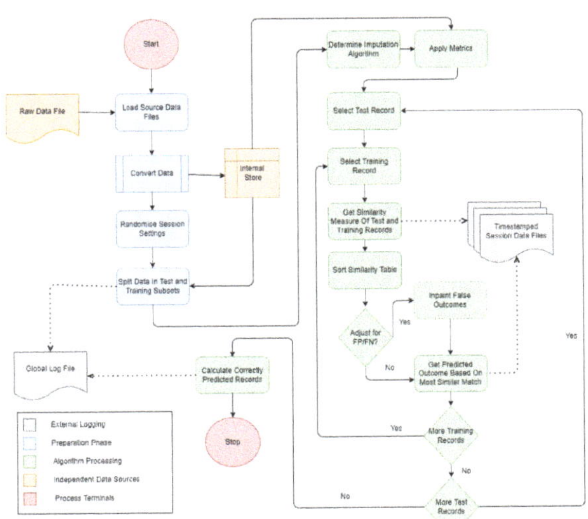

Figure 1. Main steps of the similarity modeling process.

Table 1. Relative prediction accuracy of our N-Similarity algorithm compared to the average prediction accuracy across all selected single imputation techniques for different neighbourhood sizes N.

	N = 1	N = 2	N = 3	N = 4	N = 5	N = 6	N = 7	N = 8	N = 9	N = 10
Accuracy	55.64%	73.37%	58.01%	69.84%	58.84%	70.82%	60.13%	66.81%	60.74%	67.41%
Correlation	76.97%	88.91%	89.65%	89.55%	89.57%	89.06%	89.30%	89.63%	89.33%	88.99%
Precision	31.12%	58.41%	30.79%	55.59%	32.28%	59.07%	32.30%	46.98%	33.53%	48.58%
Recall	33.01%	55.13%	28.71%	39.62%	25.50%	29.73%	23.53%	32.57%	24.23%	28.60%
Specificity	66.18%	81.47%	71.34%	83.76%	74.22%	89.57%	77.04%	82.43%	77.49%	85.12%
TPR	23.03%	38.12%	20.34%	28.12%	18.19%	20.44%	16.48%	23.07%	17.10%	19.80%
FPR	33.82%	18.53%	28.66%	16.24%	25.78%	10.43%	22.96%	17.57%	22.51%	14.88%
Average MCC	**−0.0495**	**0.4582**	**0.0792**	**0.3419**	**0.0842**	**0.3069**	**0.0383**	**0.2326**	**0.0123**	**0.1771**

The colour coding scheme used in Table 1 reflects how, for different neighbourhood sizes, the prediction accuracy of our N-Similarity algorithm compares to the average prediction accuracy of the other imputation algorithms under consideration. The green values indicate those measures where our algorithm performs better than the average of the other imputation algorithms, red values indicate those measures where our algorithm performs worse, and the blue values indicate those measures where there is marginal difference between the algorithms.

3.3. Empirical Bayes Correction

Dealing with missing data and its mechanism is of paramount importance in statistics [29], and in this section, we propose a correction for imputing numerical variables motivated by a normal-normal hierarchical model (see [30], Section 3.3.1). Let $D = \{Y, X\}$ be our observations, where Y is the part that contains missing values, and X (a $N_{\text{Obs}} \times N_X$ matrix) is fully observed. We consider the following correction term for the imputed candidate value $\hat{\theta}_m$ given the (observable) sample mean $\bar{Y}$ and the most similar value Y^*_m:

$$\hat{\theta}_m = \alpha\bar{Y} + (1-\alpha)Y^*_m, \quad \alpha = \frac{s^2_y}{s^2_y + (\hat{\tau}^2_{Y|X})^+}, \tag{3}$$

where s^2_y is the sample variance of $y = (y_1, \ldots, y_l)$ for the l-most-similar observations (comparing X_m to X_{obs}), and $(\hat{\tau}^2_{Y|X})^+$ is an approximation of the Empirical Bayes estimate of [30]:

$$(\hat{\tau}^2_{Y|X})^+ = \max\left[0, \lambda \times s^2_Y - s^2_y\right], \tag{4}$$

where λ is a fixed hyperparameter, and s^2_Y the sample variance of the observable Y.

Since $0 \leq \alpha \leq 1$, and (3) is a weighted average between Y^*_m and $\bar{Y}$, which essentially shrinks the proposal towards the mean $\bar{Y}$, the amount of shrinking is determined by α. When $\alpha = 0$, (3) suggests a direct imputation with Y^*_m, whereas $\alpha = 1$ suggests an imputation using $\bar{Y}$. Generally, our candidate imputed value shrinks towards $\bar{Y}$ when the variance associated with Y^*_m exceeds the sample variance of Y.

Motivation

We motivate (3) by considering an empirical Bayes approach to our hierarchical model. We introduce two types of random variables: one expressing the missing values Y_m and one $\theta_{m|X}$ expressing the neighboorhood-similarity-based guesses (can also be thought of as model-based guesses) that rely on a relation between Y and X. For each missing value Y_m, we assume that it is a normal random variable with mean $\theta_{m|X}$ and a variance $\sigma^2_{m|X}$. This allows us to express the "true" missing value in relation to our similarity-based guesses: for ms with small variances ($\sigma^2_{m|X}$), the similarity-based guesses are informative, and for large variances, they are not.

For each $\theta_{m|x}$, we again assume a normal distribution with a common mean and variance $(\mu_{Y|X}, \tau^2_{Y|X})$:

$$Y_m \Big| X, \theta_{m|X}, \sigma^2_{m|X} \sim N\left(\theta_{m|X}, \sigma^2_{m|X}\right) \tag{5}$$

$$\theta_{m|X} \Big| X, \mu_{Y|X}, \tau^2_{Y|X} \sim N\left(\mu_{Y|X}, \tau^2_{Y|X}\right), \tag{6}$$

which expresses the overall relation of Y given X as a normal distribution with its mean and variance varying according to X. In other words, instead of considering the similarity-based guess of the missing value as a single point, we introduce a normal-distributed kernel centered around it, which depends on the fully observed X. Our two-level hierarchical model uses (5) locally to express the distribution of Y_m and (6) to express the associated mean $\theta_{m|X}$ using a global model between X and Y. Given a candidate value Y^*_m, we can impute Y_m with the posterior empirical Bayes mean $\hat{\theta}_{m|X}$ [30], which is a point estimate of $\theta_{m|X}$:

$$\hat{\theta}_m = \alpha\mu_{Y|X} + (1-\alpha)Y^*_m,$$

where $\alpha = \sigma^2_{m|X}/(\sigma^2_{m|X} + \tau^2_{Y|X})$. Linear and nonlinear regression models have been used for the conditional mean $\mu_{Y|X}$ in a Bayesian setting [31], whereas [32] used a nonparametric kernel regression, but in our performance evaluations, we also considered the weighted sample mean and sample variance, e.g., $s^2_y = \sum_i w_i(y_i - \bar{y})^2$, with weights approximated by a Gaussian kernel with a minimal RMSE improvement. The empirical Bayes estimate of [30] for $\tau^2_{Y|X}$ is based on sample estimates for $\sigma^2_{m|X}$ and $\tau^2_{Y|X}$:

$$(\tau^2_{Y|X})^+ = \max(0, \lambda\hat{\tau}^2_{Y|X} - \hat{\sigma}^2_{Y|X}).$$

If we consider the case that Y and X are independent, any similarity between X_{obs} and X_m provides no information about the missing Y_m. This also implies that $\mu_{Y|X}$ and $\sigma^2_{Y|X}$ become the marginal μ_Y and σ^2_Y, respectively. Furthermore, the y sample becomes a random sample of Y, with $\bar{y}$ and s^2_y being unbiased estimates of μ^2_Y and σ^2_Y, respectively. Therefore, we can use $\bar{Y}, s^2_Y$, and s^2_y as approximations for $\mu_{Y|X}$, $\hat{\tau}^2_{Y|X}$, and $\hat{\sigma}^2_{Y|X}$, respectively, which, under independence, set α towards one and can serve as a warning for noninformative imputation. Finally, if Y and X are not independent, y will be a conditional sample from $Y|X_m$, and we expect $\text{var}(Y) \geq \mathbb{E}[\text{var}(y)]$ to lead to to smaller shrinkage ($\alpha < 1$) towards $\bar{Y}$.

4. Performance Evaluation

In this section, we evaluate the performance of our similarity-based approach, using the sample diabetes data set, in comparison with a number of other imputation techniques.

4.1. Implementation Overview

The algorithm is made up of three steps: the *first step* is to partition the raw data set, D, into two disjoint subsets: one containing all the complete records, S_c, and the other containing records that are missing one or more feature values, S_i. The incomplete records are then checked in order. Whenever an incomplete record $S_i(k)$ contains a missing feature value $f_{k,i}$, the nearest N-Similarity algorithm (*second step*) is applied to create a similarity table of the closest n records from S_c. The missing feature value, $f_{k,i}$, is then determined by applying a series of rules below during the *third* and final step.

$$S_c \cup S_i = D, S_c \cap S_i = \emptyset$$

Considering the corresponding feature values of the n-most-similar complete records in the similarity table created by the stage above, the algorithm creates a set of candidate values, C, that will be used to replace the current missing feature value. The algorithm uses a number of simple rules, applied in strict order, to determine which of these candidate values is the *most likely* to be used as the replacement value for the missing feature in the current incomplete record.

$$\forall k \in S_i, f_{k,i} = S_c(j), 0 \leq jn$$

where j is the index of the best candidate value in C.

The set of rules applied to C in determining a predicted value are derived from both an evaluation of the corresponding feature values in the most similar diabetes records together with the nature of the values in the candidate set C. The rules are applied in order, with the most specific selection criteria applied first and moving down to the most general selection criteria applied last. For the candidate value set, C, apply the following rules in order of decreasing priority:

1. If there is a unique modal value in C, then use this value as the imputed feature value.
2. For those modal values which occur in C with equal highest frequency, if one of these modal values has the same feature value as the actual feature value of the most similar complete record in S_c, then select this modal value as the new imputed feature value for the current incomplete record.
3. Determine whether one of the values in C lies closer to the median value of the candidate set than the others. If such a value is found, select this as the imputed feature value.
4. If none of the previous rules have been satisfied, then select the mean value of C.

By comparing the prediction accuracy of the algorithm on the training data set (training folds), we can determine that the results are not noticeably different than the results obtained by applying the algorithm on the test data set (test fold), and therefore, we can ascertain that the algorithm does not overfit the diabetes data set.

This is repeated for each missing feature in the current partial record $S_i(k)$, after which the now complete record is moved from S_i to S_c to become a potential candidate for the completion of the next incomplete record in S_i.

4.2. Evaluation of RMSE

The evaluation was performed using a simulation-based approach that consists of repeatedly using a random selection of M records from the complete data records subset S_c. Since they were complete, each of these records had a known *actual* value for each feature, which could be used later for comparison purposes. The selected M values of each feature, f_i, were ignored and imputed using our N-Similarity algorithm in order to provide a more reliable estimate for the RMSE. These *predicted* values were then compared against the *actual* values to provide an estimate for the root mean squared error measure (RMSE) to determine the predictive performance of our algorithm [33]. In Sections 4.3 and 4.4.3, we used the three methods, i.e., similarity (NSIM), similarity with empirical Bayes correction (NSIM-EB), and k-nearest neighbors (kNNs) to repeatedly impute each feature and report the corresponding RMSEs.

4.3. Simulated Dataset

We proceeded to simulate 1000 datasets (of 1000 observations each) based on $x_1, \ldots, x_6$ (7) random variables. The x_1, x_2, and x_3 are independent Poisson, uniform, and exponential-distributed random variables, respectively, whereas the z_1, z_2, and z_3 are independent standard normal variables. The remaining (x_4, x_5, and x_6) are functions of the previous ones, with their relations outlined in (7). Overall, the simulated data sets contain an independent random variable (x_1), as well as noisy nonlinear relationships (e.g., x_6 with x_2).

$$\begin{aligned} z_1, z_2, z_3 &\sim \text{Normal}(0,1) \\ x_1 &\sim \text{Poisson}(1) \\ x_2 &\sim \text{Uniform}(18,83) \\ x_3 &\sim \text{Exponential}(1/30) \\ x_4 &= z_1 \times x_3 + 3 \\ x_5 &= x_4 \times 3 + z_2 * \sqrt{10} \\ x_6 &= \exp(-x_2 \times 0.2 + z_3) \end{aligned} \tag{7}$$

Table 2 shows the imputation RMSE of the three methods assuming 1.50 and 100 missing observations (M) per each simulated data set. Overall, the RMSE for the NSIM-EB was consistently lower than the rest. For x_1, as the number of M increased, the RMSE increased too for all the methods, which is expected, as x_1 is independent from the rest. Generally, the RMSE of the NSIM was similar, if not slightly reduced, compared to the RMSE of the kNNs. Both similarity-based methods were faster (NSIM performed in 126 s and NSIM-EB performed in 167 s; both were implemented in R) compared to the kNNs (215 s) using the implementation (with Mahalanobis distance) of the yaImpute package [34].

Table 2. Imputation of RMSE for simulated data using similarity (NSIM), similarity with empirical Bayes correction (NSIM-EB), and k-nearest neighbors (kNNs) methods.

M	Method	x_1	x_2	x_3	x_4	x_5	x_6
1	NSIM	1.382	1.402	1.414	1.378	1.345	1.333
	NSIM-EB	0.996	1.054	1.068	1.052	1.052	1.035
	kNNs	1.399	1.422	1.508	1.453	1.456	1.386
50	NSIM	1.421	1.401	1.398	1.402	1.401	1.380
	NSIM-EB	1.047	1.035	1.041	1.042	1.042	1.027
	kNNs	1.417	1.413	1.407	1.409	1.405	1.399
100	NSIM	1.420	1.396	1.385	1.386	1.385	1.373
	NSIM-EB	1.046	1.031	1.034	1.038	1.038	1.014
	kNNs	1.413	1.418	1.411	1.415	1.410	1.417

4.4. *Pima Indians Diabetes Data Set*

Another data set that was used extensively in this paper is the *Pima Indians Diabetes* data set [35], which is originally from the National Institute of Diabetes and Digestive and Kidney Diseases. The complete data set contains information of 768 women from a population located around Phoenix, Arizona, USA. The outcome tested was for diabetes, with 258 testing positive and 510 testing negative. The data was structured as follows: there was one target (dependent) variable and eight (feature) attributes: the number of pregnancies, oral glucose tolerance test, blood pressure, skin thickness, insulin, body mass index, age, and pedigree diabetes function. More technical details of the file used can be seen in Table 3. The Pima population has been under study by the National Institute of Diabetes and Digestive and Kidney Diseases at intervals of 2 years since 1965. As epidemiological evidence indicates that type 2 diabetes results from the interaction of genetic and environmental factors, the Pima Indians Diabetes data set includes information about attributes that could be related to the onset of diabetes and associated future complications.

The original data used zero as the marker for a missing feature value, because it was deemed that this could never be a valid value based on the nature of the features being represented. The obvious exception to this is the final binary outcome field, which may have a value of zero (for a negative diagnosis). The diagnosis outcome was a binary integer value indicated by a one for a positive diagnosis and a zero for a negative diagnosis, although in actuality, any nonzero integer would equally be interpreted as a positive diagnosis. Where it was possible, we converted this encoding convention to the programming language's standard missing value mechanism (e.g., Section 4.4.3), or we adapted our implementation (e.g., in Section 4.4.2, the similarity calculations are based only on valid feature values).

Out of the total number of records, 336 were complete (no missing feature values) (43.75%), and there were 763 missing feature values spread across the data set out of the total number of 6144 feature values (12.42%).

Table 3. Structure of PIMA diabetes data file.

Feature	Data Type	Value Range (Zero Indicates Missing Value)
Number of Times Pregnant	Positive Integer	0…17
Plasma Glucose Concentration	Real	0…199
Diastolic Blood Pressure	Real	0…122
Triceps Skinfold Thickness	Real	0…99
Serum Insulin Levels	Real	0…846
Body Mass Index	Real	0…67.1
Diabetes Pedigree Function	Real	0.078…2.42
Age	Positive Integer	21…81
Classification	Binary	1 = positive diagnosis, 0 = negative diagnosis

4.4.1. Comparison with Popular Imputation Methods

Three popular imputation techniques were used to provide a comparative baseline for the results obtained from applying our N-Similarity algorithm [36]. *Listwise deletion* is the process of removing all incomplete records from a data set prior to imputation [37]. If the original data is incomplete, then its application will naturally result in a smaller data set being produced for analysis. Depending on the sparsity of the original data, this may impact any ongoing analysis, thereby making it an unviable option for comparison against other imputation techniques that attempt to restore missing feature values without removing data. The statistical power [38] relies in part on a high sample size, and this is helped by having a relatively complete data set with few incomplete records. The other possible drawback to using listwise deletion is when the missing feature values may not be randomly distributed. For example, this occurs if a certain feature has missing values based on the nature in which the values for that feature were collected (questions aiming to extract sensitive information that the individual just skipped). As a result, and again depending on the level of sparsity of such missing data, the results may introduce bias into later analysis. One possible way to address these limitations and reduce the bias is to use multiple imputation techniques [39,40]. An extension of this approach, which was considered as a technique for comparison, was *pairwise deletion* [41]. This approach allows for the use of incomplete data but only allows for analysis on those features that have complete data. This introduces bias and makes like-for-like analysis more difficult, so it was was rejected as an option.

Some analysis has been undertaken [42] to determine the most popular imputation methods since 2000 (Figure 2). *Popularity* has been measured based on the number of times each imputation algorithm is mentioned in Google Scholar articles and papers. The results are somewhat surprising, since simpler, older techniques seem to be more popular than more recent approaches:

- **Remove Incomplete Records (Listwise Deletion)**: Any records in D that have one or more missing feature values are removed from the data set prior to processing. The removal of any incomplete records will lead to a smaller but complete data set D. It is not recommended that this technique is used arbitrarily as a means of direct comparison with other techniques used in the paper, since factors, such as the initial completeness of D, need to be assessed. It has been included due to its general popularity only (Figure 2).

- **Replace Missing Data With Mean Attribute Value**: Any missing feature values are replaced with the *average* value calculated from the corresponding feature values in all the complete records in the data set.
- **Replace Missing Data With Modal Attribute Value**: Any missing feature values are replaced with the *most common* value gathered from the corresponding feature values from all the complete records in the data set.
- **Replace Missing Data Using Empirical Bayes Algorithm**: This method is for statistically inferring missing feature values using a prior distribution of known values in a data set.
- **Replace Missing Data With N-Similarity Algorithm**: Any missing feature values are replaced with the *best candidate* value calculated from the corresponding feature values in the *N*-most-similar complete records in the data set.

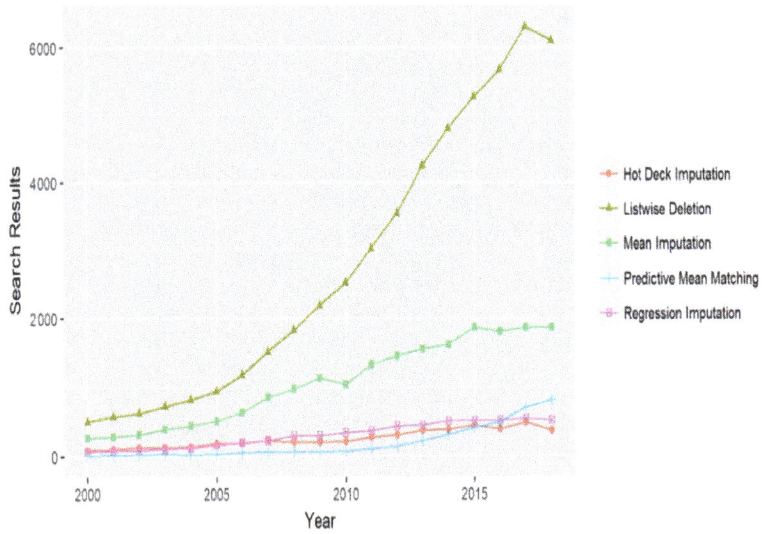

Figure 2. Google Scholar search results (Statistics Globe 2019).

Cross validation is a sampling procedure used to evaluate models that use a limited data sample. The procedure has a single parameter called *k* that refers to the number of equal sized groups (or folds) over which the data sample will be equally divided. The procedure is often called *k-fold cross validation* and is used to estimate the ability of a machine learning model to make predictions based on unseen data; it uses a limited sample in order to estimate how the model is expected to perform in general when used to make predictions on data not used during the training of the model.

The average cross validation over *n* folds is given by

$$\frac{1}{n}\sum_{k=1}^{k=n} Similarity_k$$

where $Similarity_k$ is the measure of similarity between the current test and the training folds for the session run *k*.

4.4.2. Results, Limitations, and Discussion

Table 1 shows the improvement in predicting true positive cases when using our proposed N-Similarity algorithm compared against several other single-pass imputation algorithms. Entries highlighted in *green* show the improvement achieved by our algorithm over other popular techniques. Those highlighted in *red* show a worsened prediction accuracy. Those entries highlighted in *blue* indicate no or negligible change in the prediction accuracy.

The testing process splits the data sets into **ten** approximately equally sized folds. The arbitrary partitioning of records in each fold, in any given run, meant that each fold could contain a combination of complete and incomplete records. The proportion of incomplete records was allowed to vary so as to not impose any potentially restrictive classification on the fold contents. Should we have wanted to impose a limiting proportion of incomplete records in a fold, for some reason, then a stratified k-fold approach or similar would have been used. When applying cross correlation techniques, some of the ratio calculations shown in Table 1 had no correctly predicted positive outcomes ($TP = 0$), thus leading to incomplete runs being produced. Similarly, in some folds, it may also be possible that the number of true positive (TP) and true negative (TN) training records are not predictable for a given fold, thus meaning that the *precision* metric was indeterminate for specific pairings of test and training data folds, since $TP + TN = 0$. The likelihood of these eventualities could be reduced somewhat by reducing the number of folds for the given data sets, thereby increasing the number of records in each fold. However, the missingness of the data sets (proportion of incomplete to complete records) will be the ultimate determinate of how likely such scenarios were to occur. By introducing an error tolerance, indicated in *blue* for those results that varied by less than $\pm 5.0\%$, we can see that the only metric where the other techniques produced better results than our algorithm was *Correlation*; the results for *TPR* and *Recall* changed marginally, and the other metrics showed good improvements achieved by our algorithm. Applying the MAV and MDAV imputation techniques shows very similar results, which may have been caused by the relatively sparse data sets, the size of the data sets, or the nature of the data itself.

As shown in Table 4, the results differed depending on which imputation method was used. When incomplete records were removed as part of the imputation process prior to the application of our N-Similarity algorithm, all of the metrics, apart from accuracy, were worsened, albeit on a restricted data set. The results of using either the mean or modal replacement approaches were very similar and could be due to the relatively small data sets used in our tests. What can be taken from this is the importance of fine tuning expectations based on which metrics are the most important to the end user. Considering our neighborhood-similarity-based approach (Table 4), we obtained better results for accuracy (+9.33%), precision (+9.67%), specificity (+13.84%), and FPR (13.86%), but this has to be tempered against worse results for correlation (−6.07%). The recall and TPR were roughly unchanged and remained within a 5% tolerance. What has become apparent is that the metrics used are very susceptible to the neighborhood size (N) and nature of the data to which they are being applied. The best results may be achieved by balancing the size of the neighborhood considered against the imputation algorithm that will be used to identify the most suitable compromise between true positive and true negative outcomes. In our testing, we ran our algorithm using different-sized neighborhoods and found that a neighborhood of size four ($N = 4$) gave the most balanced results. For comparison, the results obtained using other neighborhood sizes can be seen in Table 1.

Table 4. Performance of our proposed N-Similarity algorithm compared against other single imputation techniques.

	Remove Incomplete Records (Listwise Deletion)	**Replace Missing Data with MAV**	**Replace Missing Data with MDAV**	**Average N-Similarity Algorithm (N = 1...10)**
Number Of Perfect Tests	10	10	10	10
Accuracy	54.76%	54.85%	54.88%	64.16% (+9.33%)
Correlation	92.48%	94.92%	95.00%	88.06% (−6.07%)
Precision	36.94%	31.31%	31.32%	42.86% (+9.67%)
Recall	31.26%	36.65%	37.35%	32.06% (−3.03%)
Specificity	68.96%	63.28%	62.82%	78.86% (+13.84%)
True Positive Rate (TPR)	22.23%	25.17%	25.21%	22.47% (−1.73%)
False Positive Rate (FPR)	31.04%	36.72%	37.18%	21.12% (−13.86%)
Average MCC	**0.0891**	**0.0160**	**−0.0413**	

4.4.3. Benchmarking with kNN

Using the PIMA dataset (Table 3), we also compared our similarity-based imputation (NSIM), its enhanced version NSIM-EB (with the empirical Bayes (EB) correction, $\lambda = 1$), and the Mahalanobis distance based k-nearest neighbors (kNNs) [34] (See Table 5 for the obtained RMSEs). Both the kNNs imputation and NSIM are nonparametric and rely on the k-nearest and N-similar observations, respectively. Apart from the use of neighborhood observations, the Mahalanobis distance uses the covariance matrix, while our EB correction (3) uses two estimates of sample variance as the weight of the imputed proposal. We used the three schemes (i.e., NSIM, NSIM-EB, and kNNs) for 1000 imputations per variable—for a total of M combinations (Table 5). As seen in Table 5, the RMSE performance of the NSIM was comparable to the kNN imputation, whereas the NSIM-EB outperformed both in all scenarios with minimal computational time overhead (the NSIM, NSIM-EB, and kNNs took approximately 68, 103, and 477 s, respectively, in our R implementation).

Table 5. Imputation RMSEs for simulated data using our similarity method (NSIM), our similarity method with empirical Bayes correction (NSIM-EB), and the k-nearest neighbors (kNNs) method.

M	Method	Pregnancy	Glucose	BP	Triceps	Insulin	BMI	DPf	Age
	NSIM	0.875	0.963	1.051	0.937	0.942	0.892	1.103	0.848
1	NSIM-EB	0.737	0.770	0.777	0.816	0.726	0.782	0.791	0.752
	kNNs	0.872	0.948	1.101	1.043	0.896	0.968	1.013	0.884
	NSIM	1.134	1.114	1.275	1.128	1.148	1.065	1.288	1.089
5	NSIM-EB	0.900	0.899	0.962	0.948	0.882	0.899	0.937	0.894
	kNNs	1.343	1.328	1.265	1.315	1.230	1.322	1.272	1.289
	NSIM	1.172	1.149	1.335	1.167	1.235	1.096	1.372	1.125
10	NSIM-EB	0.942	0.928	0.992	0.958	0.956	0.927	0.984	0.891
	kNNs	1.382	1.356	1.331	1.406	1.293	1.360	1.349	1.263
	NSIM	1.177	11.54	1.350	1.181	1.249	1.109	1.388	1.140
15	NSIM-EB	0.944	0.940	1.004	0.971	0.961	0.936	1.030	0.917
	kNNs	1.419	1.370	1.376	1.418	1.333	1.359	1.404	1.379
	NSIM	1.187	1.167	1.356	1.185	1.269	1.121	1.393	1.160
20	NSIM-EB	0.959	0.942	1.006	0.969	0.995	0.950	1.014	0.928
	kNNs	1.399	1.359	1.378	1.397	1.336	1.367	1.345	1.372

Our algorithm performed better when the source data set had a small percentage of missing data values, due to our blind random selection of data values across all the folds. The larger the number of missing data values, the higher the likelihood would be that some of the folds would be more sparsely populated. The choice of the number of data partitions in the k-fold step needs to be carefully selected; otherwise, we risk the possibility of introducing bias into the selection of data values put in any given fold. We settled on k = 10, as much of the academic literature indicated that this was a commonly used value. One way of limiting the impact of this problem is to use a stratified approach as mentioned above. We left this direction as a line of potential future work. The choice of the size of the neighborhood, N, and, as a direct result, the number of candidates in the set of values for selecting imputed values, was also sensitive. We spent considerable trial-and-error effort looking for the best selection for this parameter against the PIMA data set; we tried a much wider range of potential values for N than are shown. The results for these higher values were negligibly different in our case.

Table 1 shows that N = 4 was the best choice in our case, although this could vary for different data sets. Further research is required to determine whether the choice of value for N could be automated by looking at all the possible potential values for N and whether this approach would even be practical for large data sets in terms of processing time and improvements in the results.

5. Conclusions

Our neighborhood-based algorithm was able to provide noticeably improved results when compared against other techniques, but the degree of this improvement was sensitive to the size of the neighborhood, with some features being more readily improved than others for smaller neighborhood sizes and other metrics being noticeably less well predicted as the size of the neighborhood increased. This paper proposes a technique to provide a more accurate prognosis of possible patient diabetes based on a number of key patient characteristics. Our approach creates a similarity neighborhood using the most similar diagnosed patient records and uses the feature set values of these patients to help with the diagnosis of undiagnosed patients. By comparing our N-Similarity algorithm against several widely used single-pass imputation techniques using the same collection of data sets, both real-world and simulated, we found that it produces better results against several of our performance metrics (Table 4). However, we observed that the size of the neighborhood had an impact on the performance of our algorithm. We also noticed that the limited data set sizes and degrees of missingness of the initial source data could impact the results, and more extensive work would be necessary using a wider range of different data sets in order to see how these measures are related. The empirical Bayes correction of the neighborhood-based algorithm offered consistently smaller RMSEs over the simple algorithm and the k-nearest neighbors imputation, with minimal computational overhead. In addition to the performance advantages, we recommended it as a general method, since the shrinking parameter α indicated a degree of certainty between our inputted value and the sample mean (with zero indicating certainty of the similarity of the inputted value and one indicating most uncertain).

6. Future Work

The main limitation of our current work is that the PIMA data set contains only numeric feature values. Future work could include support for both categorical and textual data. Both types of information are widely found in medical data sets and would help to support the usefulness of our algorithm in this domain, as well as in other similar domains. The implementation of our algorithm has been deliberately developed to be loosely coupled to the source data to allow for different file formats and structures in the source data to be supported with minimal effort, thus allowing for generalization of the code for different future uses.

To aid with future development of this algorithm, we have provided the full source code to the software we used to generate the presented results. The source code, written in the Go programming language, can be freely used and modified, and it has been designed to be modular and loosely coupled to any data set, thereby making it easier to extend as required.

Author Contributions: Conceptualization, C.W., V.G. and S.D.; Methodology, C.W., V.G. and S.D.; Software, C.W.; Formal analysis, V.G.; Investigation, C.W.; Data curation, C.W.; Writing—original draft, C.W.; Writing—review & editing, V.G. and S.D.; Supervision, V.G. and S.D.; Project administration, S.D. All authors have read and agreed to the published version of the manuscript.

Funding: This research received no external funding.

Data Availability Statement: The main Github repository can be found here: https://github.com/ColinWilcox1967/PhD-DataSetAnalysis-Diabetes (accessed on 22 November 2023); An example of how this code may be used with other data sets is given here: https://github.com/ColinWilcox1967/PHD-DataSetAnalysis-Traffic (accessed on 22 November 2023).

Conflicts of Interest: The authors declare no conflict of interest.

Abbreviations

The following abbreviations are used in this manuscript:

RMSE	**R**oot **M**ean **S**quared **E**rror
NSIM	Our **N**eighbourhood **SIM**ilarity algorithm
NSIM-EB	Our **N**eighbourhood **SIM**ilarity algorithm with **E**mpirical **B**ayes Correction
kNN	Classification of **N**earest **N**eighbour algorithms
TP	**T**rue **P**ositive
FP	**F**alse **P**ositive
TN	**T**rue **N**egative
FN	**F**alse **N**egative
TPR	**T**rue **P**ositive **R**ate
FPR	**F**alse **P**ositive **R**ate
MAV	**M**ean **A**verage **V**alue
MDAV	**M**odal **A**verage **V**alue
MCC	**M**atthews **C**orrelation
BMI	**B**ody **M**ass **I**ndex
BP	**B**lood **P**ressure
DPf	**D**iabetes **P**edigree **F**unction

References

1. Tang, J.; Zhang, X.; Yin, W.; Zou, Y.; Wang, Y. Missing data imputation for traffic flow based on combination of fuzzy neural network and rough set theory. *J. Intell. Transp. Syst. Technol. Plan. Oper.* **2019**, *5*, 439–454. [CrossRef]
2. Agrawal, R.; Prabakaran, S. Big data in digital healthcare: Lessons learnt and recommendations for general practice. *Heredity* **2020**, *124*, 525–534. [CrossRef] [PubMed]
3. Adam, K. Big Data Analysis And Storage. In Proceedings of the 2015 International Conference on Operations Excellence and Service Engineering, Orlando, FL, USA, 10–11 September 2015; pp. 648–658.
4. Ford, E.; Rooney, P.; Hurley, P.; Oliver, S.; Bremner, S.; Cassell, J. Can the Use of Bayesian Analysis Methods Correct for Incompleteness in Electronic Health Records Diagnosis Data? Development of a Novel Method Using Simulated and Real-Life Clinical Data. *Public Health* **2020**, 8, 54. [CrossRef] [PubMed]
5. Xiaochen, L.; Xia, W.; Liyong, Z.; Wei, L. Imputations of missing values using a tracking-removed autoencoder trained with incomplete data. *Neurocomputing* **2019**, *266*, 54–65. [CrossRef]
6. Singhal, S. Defining, Analysing, and Implementing Imputation Techniques. 2021. Available online: https://www.analyticsvidhya.com/blog/2021/06/defining-analysing-and-implementing-imputation-techniques/ (accessed on 22 November 2023).
7. Beretta, L.; Santaniello, A. Nearest neighbor imputation algorithms: A critical evaluation. *BMC Med. Inform. Decis. Mak.* **2016**, *16*, 197–208. [CrossRef] [PubMed]
8. Khaled, F.; Mahmoud, I.; Ahmad, A.; Arafa, M. Advanced methods for missing values imputation based on similarity learning. *Clim. Res.* **2022**, *7*, e619. [CrossRef]
9. Huang, G. Missing data filling method based on linear interpolation and lightgbm. *J. Phys. Conf. Ser.* **2021**. [CrossRef]
10. Peppanen, J.; Zhang, X.; Grijalva, S.; Reno, M.J. Handling bad or missing smart meter data through advanced data imputation. In Proceedings of the 2016 IEEE Power & Energy Society Innovative Smart Grid Technologies Conference (ISGT), Ljubljana, Slovenia, 9–12 October 2016; pp. 1–5. [CrossRef]
11. Jackobsen, J.; Gluud, C.; Wetterslev, J.; Winkel, P. When and how should multiple imputation be used for handling missing data in randomised clinical trials—A practical guide with flowcharts. *BMC Med. Res. Methodol.* **2017**, *17*, 162. [CrossRef]
12. Hayati Rezvan, P., Lee, K.J.; Simpson, J.A. The rise of multiple imputation: A review of the reporting and implementation of the method in medical research. *BMC Med. Res. Methodol.* **2015**, *15*, 30. [CrossRef]

13. Nguyen, C.; Carlin, J.; Lee, K. Practical strategies for handling breakdown of multiple imputation procedures. *Emergent Themes Epidemiol.* **2021**, *18*, 5. [CrossRef]
14. Guo, G.; Wang, H.; Bell, D.; Bi, Y.; Greer, K. KNN Model-Based Approach in Classification. In *Confederated International Conferences "On The Move To Meaningful Internet Systems 2003"*; Lecture Notes in Computer Science; Springer: Berlin/Heidelberg, Germany, 2003; Volume 2888, pp. 986–996. [CrossRef]
15. Pohl, S.; Becker, B. Performance of Missing Data Approaches Under Nonignorable Missing Data Conditions. *Methodology* **2018**, *16*, 147–165. [CrossRef]
16. Ali, N.; Neagu, D.; Trundle, P. Evaluation of k-nearest neighbour classifier performance for heterogeneous data sets. *SN Appl. Sci.* **2019**, *1*, 1559. [CrossRef]
17. Abu Alfeilat, H.A.; Hassanat, A.B.A.; Lasassmeh, O.; Tarawneh, A.S.; Alhasanat, M.B.; Eyal Salman, H.S.; Prasath, V.S. Effects of Distance Measure Choice on K-Nearest Neighbor Classifier Performance: A Review. *Big Data* **2019**, *7*, 221–248. [CrossRef]
18. Khan, S.; Hoque, A. SICE: An improved missing data imputation technique. *J. Big Data* **2020**, *7*, 37. Available online: https://journalofbigdata.springeropen.com/articles/10.1186/s40537-020-00313-w (accessed on 22 November 2023). [CrossRef] [PubMed]
19. Misztal, M. Imputation of Missing Data Using R. *Acta Univ. Lodz. Folia Oeconomica* **2012**, *269*, 131–144.
20. Kowarik, A.; Templ, M. Imputation with the R Package VIM. *J. Stat. Softw.* **2016**, *74*, 1–16. [CrossRef]
21. Choi, J.; Dekkers, O.; Le Cessie, S. A comparison of different methods to handle missing data in the context of propensity score analysis. *Eur. J. Epidemiol.* **2019**, *34*, 23–36. [CrossRef]
22. Cetin-Berber, D.; Sari, H. Imputation Methods to Deal With Missing Responses in Computerized Adaptive Multistage Testing. *Educ. Psychol. Meas.* **2018**, *79*, 495–511. [CrossRef]
23. Alwohaibi, M.; Alzaqebah, M. A hybrid multi-stage learning technique based on brain storming optimization algorithm for breast cancer recurrence prediction. *J. King Saud Univ. Comput. Inf. Sci.* **2021**, *34*, 5192–5203. [CrossRef]
24. Kabir, G.; Tesfamariam, S.; Hemsing, J.; Rehan, S. Handling incomplete and missing data in water network database using imputation methods. *Sustain. Resilient Infrastruct.* **2020**, *5*, 365–377. [CrossRef]
25. Mujahid, M.; Rustam, F.; Shafique, R.; Chunduri, V.; Villar, M.G.; Ballester, J.B.; Diez, I.D.L.T.; Ashraf, I. Analyzing Sentiments Regarding ChatGPT Using Novel BERT: A Machine Learning Approach. *Information* **2023**, *14*, 474.
26. Mujahid, M.; Rehman, A.; Alam, T.; Alamri, F.S.; Fati, S.M.; Saba, T. An Efficient Ensemble Approach for Alzheimer's Disease Detection Using an Adaptive Synthetic Technique and Deep Learning. *Diagnostics* **2023**, *13*, 2489. [CrossRef] [PubMed]
27. Nti, I.; Nyarko-Boateng, O.; Aning, J. *Performance of Machine Learning Algorithms with Different K Values in K-Fold Cross Validation*; MECS Press: Hong Kong, China, 2021. [CrossRef]
28. Brownlee, J. *How to Configure k-Fold Cross-Validation*; Machine Learning Mastery: San Juan, PR, USA, 2020.
29. Little, R.J.; Rubin, D.B. *Statistical Analysis with Missing Data*; John Wiley & Sons: Hoboken, NJ, USA, 2019; Volume 793.
30. Carlin, B.; Louis, T. *Bayes and Empirical Bayes Methods for Data Analysis*, 2nd ed.; Chapman and Hall CRC: Boca Raton, FL, USA, 2000. [CrossRef]
31. Zhou, X.; Wang, X.; Dougherty, E.R. Missing-value estimation using linear and non-linear regression with Bayesian gene selection. *Bioinformatics* **2003**, *19*, 2302–2307. [CrossRef] [PubMed]
32. Cheng, P.E. Nonparametric Estimation of Mean Functionals with Data Missing at Random. *J. Am. Stat. Assoc.* **1994**, *89*, 81–87. [CrossRef]
33. Root Mean Squared Error Definition. 2022. Available online: https://www.sciencedirect.com/topics/engineering/root-mean-squared-error (accessed on 22 November 2023).
34. Crookston, N.L.; Finley, A.O. yaImpute: An R package for kNN imputation. *J. Stat. Softw.* **2008**, *23*, 1–16. [CrossRef]
35. PIMA Indian Diabetes Database. 2016. Available online: https://www.kaggle.com/datasets/uciml/pima-indians-diabetes-database (accessed on 22 November 2023).
36. Lin, W.C.; Chih-Fong, T. Missing value imputation: A review and analysis of the literature (2006–2017). *Artif. Intell. Rev.* **2020**, *53*, 1487–1509. [CrossRef]
37. Dong Y, P.C. Principled missing data methods for researchers. *Springerplus* **2013**, *2*, 222. [CrossRef]
38. Huang, L.; Wang, C.; Rosenberg, N.A. The Relationship between Imputation Error and Statistical Power in Genetic Association Studies in Diverse Populations. *Am. J. Hum. Genet.* **2009**, *85*, 692–698. [CrossRef]
39. Pepinsky, T.B. A Note on Listwise Deletion versus Multiple Imputation. *Political Anal.* **2018**, *26*, 480–488. [CrossRef]
40. Lall, R. How multiple imputation makes a difference. *Political Anal.* **2006**, *24*, 414–433. [CrossRef]
41. Allison, P. Listwise Deletion: It's NOT Evil. 2014. Available online: https://statisticalhorizons.com/listwise-deletion-its-not-evil/ (accessed on 22 November 2023).
42. Joachim Schork, S.G. Imputation Methods (Top 5 Popularity Ranking). 2019. Available online: https://statisticsglobe.com/imputation-methods-for-handling-missing-data/ (accessed on 22 November 2023).

electronics

Article

Applying Image Analysis to Build a Lightweight System for Blind Obstacles Detecting of Intelligent Wheelchairs

Jiachen Du [1], Shenghui Zhao [2,*], Cuijuan Shang [2] and Yinong Chen [3]

[1] School of Computer Science and Engineering, Anhui University of Science and Technology, Huainan 232000, China
[2] School of Computer and Information Engineering, Chuzhou University, Chuzhou 233100, China
[3] Computer Science and Engineering, Arizona State University, Tempe, AZ 85287, USA
* Correspondence: zsh@chzu.edu.cn

Abstract: Intelligent wheelchair blind spot obstacle detection is an important issue for semi-enclosed special environments in elderly communities. However, the LiDAR- and 3D-point-cloud-based solutions are expensive, complex to deploy, and require significant computing resources and time. This paper proposed an improved YOLOV5 lightweight obstacle detection model, named GC-YOLO, and built an obstacle dataset that consists of incomplete target images captured in the blind spot view of the smart wheelchair. The feature extraction operations are simplified in the backbone and neck sections of GC-YOLO. The backbone network uses GhostConv in the GhostNet network to replace the ordinary convolution in the original feature extraction network, reducing the model size. Meanwhile, the CoordAttention is applied, aiming to reduce the loss of location information caused by GhostConv. Further, the neck stem section uses a combination module of the lighter SE Attention module and the GhostConv module to enhance the feature extraction capability. The experimental results show that the proposed GC-YOLO outperforms the YOLO5 in terms of model parameters, GFLOPS and F1. Compared with the YOLO5, the number of model parameters and GFLOPS are reduced by 38% and 49.7%, respectively. Additionally, the F1 of the proposed GC-YOLO is improved by 10% on the PASCAL VOC dataset. Moreover, the proposed GC-YOLO achieved mAP of 90% on the custom dataset.

Keywords: target detection; YOLOV5s; attention mechanism; lightweighting

Citation: Du, J.; Zhao, S.; Shang, C.; Chen, Y. Applying Image Analysis to Build a Lightweight System for Blind Obstacles Detecting of Intelligent Wheelchairs. *Electronics* **2023**, *12*, 4472. https://doi.org/10.3390/electronics12214472

Academic Editor: Arkaitz Zubiaga

Received: 24 August 2023
Revised: 25 October 2023
Accepted: 26 October 2023
Published: 31 October 2023

1. Introduction

With the aging of the population, senior care communities have become an important place to meet the needs of the elderly in their daily lives. The use of wheelchairs among the elderly and individuals with physical disabilities is increasing. The cited paper [1] provides a statistical test of wheelchair usage. However, due to the deterioration of the elderly's physical functions and diminished perceptual abilities, the barriers on both sides of the wheelchair can pose a number of safety risks when using a wheelchair. Smart wheelchairs offer significant assistance to individuals in such situations. To improve performance, they distinguish the gaze of electric wheelchair passengers by introducing the distance between objects in the visual field as a new feature vector. This addition helps to reduce errors caused by unintentional gaze [2]. When image recognition is combined with the wheelchair control system, facial visual recognition is primarily employed to govern the movement direction of intelligent wheelchairs [3]. In terms of smart care, remote intelligent systems are designed to provide care [4,5]. They have achieved significant progress, and this paper focuses primarily on the impact of the blind spots on both sides of the wheelchair, emphasizing the areas not easily perceived by vision. To ensure the safety and comfortable use of smart wheelchairs for the elderly in senior care communities, the detection of dangerous obstacles in the blind zones on both sides of the smart wheelchair has become an important task. In target detection algorithms across various fields, diverse sensor-based methods, including lidar, millimeter-wave radar, and ultrasonic radar sensors, are employed to address different scenarios. Employing the 3D point cloud encoding of lidar for the purpose of detection [6], which is characterized by high computational complexity and data sparsity, and has limitations on small mobile devices; 3D target detection from LiDAR data for autonomous driving has shown good performance in 3D vision detection such as autonomous driving [7], but has the limiting issues of high cost and complexity of deployment on lightweight mobile devices, which can be a challenge for real-time applications or resource-limited devices. In regard to model lightweight design [8], the guide dog robot realizes traffic light and motion target detection based on the actual scene requirements using the MobileNet algorithm. The algorithm's lightweight advantage is effectively utilized to address the problem, highlighting the importance of lightness on mobile devices. The deep learning target detection algorithm used in this paper has made significant contributions to the field of computer vision, providing an effective method for solving the problem of detecting safety hazards in a wheelchair's blind field of view. This algorithm, which is based on deep learning and designed to detect targets, analyzes real-time visual information around wheelchair users, helping the elderly avoid accidentally hitting dangerous obstacles in the blind zones on either side of the wheelchair, such as dogs, cats, potholes, and human bodies that are incompletely represented at low angles, such as feet, legs, and wheels. These targets were gathered to construct a custom dataset. Collaborative annotation and video data management tools can be utilized for curation purposes [9]. In this paper, the approach to handling video data involves processing it frame by frame, resulting in an image dataset. When used in these areas, the following issues must be addressed.

- Target Specificity. The targets displayed on both sides of the wheelchair are incomplete. In the case of oversized targets, only part of the target's feature map is captured, such as feet, legs, and wheels.
- Model lightweighting issues. Adapting to resource-constrained environments and meeting the needs of resource-constrained devices.
- The issue of heavy performance loss caused by lightweighting. Losing model performance while lightweighting is difficult to balance.

Aiming at the above problems, this paper mainly starts from three parts, the first is to collect the target information under the low view angle to form the unique dataset, the second in the model is to go through the Ghost [10] module to obtain the sparse feature maps, and at the same time, utilize the CoordAtt [11] attention to obtain the channel information and the position information, and finally, the two parts of the features will be integrated, then, through the residual block adjustment in the neck part, the channel information of the feature map is enhanced by the SE [12] attention, and more features are captured to compensate for the feature loss problem caused by the convolution in the GhostNet idea, and a richer feature output is obtained through the residual connection, finally, trained on the PASCAL VOC dataset, the number of model parameters is nearly 3/5 of the original, and the GFLOPS are equivalent to 1/2 of the original, with almost the same detection time, but the overall accuracy and F1 value are significantly improved.

2. Related Work

Target detection models generally have a complex network structure and a large number of parameters, resulting in slow operation, large memory occupancy, and power consumption on low-end devices when the model is deployed on the mobile side. To solve these problems, in recent years, researchers have proposed that the study of lightweight target detection models focuses on two aspects: one is a lightweight model based on the network structure, and the other is based on some special tricks to reduce the computational and parametric quantities of the model, so that it can be efficiently operated on low-end devices.

Network-based lightweight modeling is a common technique used to compress lightweight models. Among these, the most seminal lightweight model designs involve a deep learning approach for MMR utilizing the SqueezeNet V1.1 architecture [13], while SqueezeNet mainly uses the Fire module to reduce the computation and parameter size of the network, MobileNets (V1 [14], V2 [15], V3 [16]) proposed by the Google team, which uses DSConv [17] (Depthwise Separable Convolution depth-separable convolution) and lightweight bottleneck structure to reduce the computation and number of parameters, ShuffleNet [18], proposed by the Megvii Inc (Face++) team, introduces a channel rearrangement mechanism to accelerate the convolutional computation, and EfficientNet [19], published by Google, uses a composite scaling factor to balance the depth, width, and resolution of the network, and GhostNet [10], proposed by Noah is Ark Lab, Huawei Technologies, introduces a Ghost module to enhance the feature representation capability, etc. These networks are usually designed as lightweight networks that focus on reducing the number of parameters and computational complexity to achieve efficient inference, but they all suffer from common limitations in terms of reduced detection performance, inability to adequately capture complex features in the image, especially in complex scenes or with small targets, lower performance in terms of accuracy, and possible limitations in terms of multiscale detection that does not fully exploit multiscale information.

Based on some special techniques to reduce the computational and parametric sizes of the model, which include model pruning, quantization, distillation, and so on. Among them, model pruning can reduce the computation and parameter count of the model by deleting unimportant connections, for example, the indoor target detection task, Zhang et al. used a specific channel pruning strategy in the YOLOv3 model to achieve up to 40% computational compression [20], but it also relies on the training model, which is not very suitable for the scenarios that require retraining. The authors of the [21] used a block perforation pruning method to achieve a 14 $\times$ compression rate with 49.0 mAP for YOLOv 4. However, the implementation needs to be adapted to different hardware architectures and device characteristics, and may face limitations in terms of computational resources, memory, and power consumption on older or low-end mobile devices. Quantization can reduce the storage and computational overhead of the model by reducing the bit width of the weights and activation values, for example, for the target detection task, Liu et al. [22] used 8-bit quantization in Faster R-CNN, which reduced both the storage and computation to one-fourth of the original, but it requires more complex training processes and optimization techniques, and may be more complicated to implement and debug. Moreover, knowledge distillation can effectively enhance the performance of compact models, the Adaptive Reinforcement Supervised Distillation (ARSD) framework to improve the recognition of lightweight models [23], but it requires a large model to be used as a baseline, which may require more computational resources and training time.

Different lightweight models have their own advantages and disadvantages. Network-based models usually have better speed, but may require more storage space. Model reduction based on some special tricks can greatly reduce the storage space and computation of the model, but it may have an impact on the accuracy of the model, the model of distillation technique can obtain a smaller model without decreasing the accuracy, but it requires larger computational resources to train the large model. Hence, it is crucial to achieve a balance between model lightweighting and model performance, and flight delays can be predicted using the ECA-MobileNetV3 algorithm [24], which balances model performance and weight by improving the feature extraction capability of the lightweight algorithm through the attention module. This module has been reported as performing well in the paper. This paper addresses the aforementioned issues by focusing on wheelchair blind obstacle detection in the elderly community environment. It combines both the advantages and disadvantages of the model to prioritize performance, reduce model complexity, and enhance the feature extraction through a better attention mechanism. The aim is to strike a balance between model compression, detection accuracy, and speed to comprehensively improve the model's performance.

3. Questions and Methods

3.1. Problem Description

In the semi-enclosed environment of an elderly community, real-time obstacle detection for the blind spots of mobile intelligent wheelchairs is achieved. Traditional algorithms often require substantial computational resources, which makes them unsuitable for mobile devices with limited resources, secondly, in much of the current lightweight research, the reduction in model size often comes at the cost of decreased model performance. This trade-off makes it challenging to achieve a balance between detection accuracy and model lightweightness. Therefore, there is a need to design a lightweight target detection algorithm that can perform the target detection task quickly and accurately on mobile devices.

The goal of this thesis is to design a lightweight target detection algorithm with high detection performance for smart wheelchair devices. In this paper, we will explore deep-learning-based target detection algorithms and reduce the complexity and computational overhead of the algorithms by optimizing the network structure, reducing model parameters, and using quantization techniques. Meanwhile, in this paper, the detection accuracy and speed of the algorithm will be evaluated in experiments and compared with existing lightweight target detection algorithms. Specifically, our research will cover the following issues:

- Reduce model parameters and computational complexity by controlling network depth and width. Design a lightweight target detection network structure suitable for mobile devices.
- While lightweighting the model, the performance of the model feature extraction is improved to compensate for the feature loss problem caused by lightweighting.
- Experimental evaluations were performed on the publicly available PASCAL VOC dataset. Targets at low viewing angles were collected to construct a custom dataset and to test the experimental effectiveness of the custom dataset.

Through the above research, an efficient, accurate and lightweight target detection algorithm for mobile devices can be proposed.

Model Quantification

To reduce the model parameters and computation for network depth and width, the model is mathematically modeled using two metrics, GFLOPS (the model's floating point operations, which denotes the amount of computation in billions of floating point operations required by the model to perform inference) and Parameters (the model's parameter count, which denotes the total number of parameters to be trained in the model). Backbone partially improves the efficiency of residual feature extraction in the C3 module, reducing computational complexity and the number of parameters. Assuming that the GFLOPS of the original Backbone with C3 module is F_{backbone} and the number of parameters is P_{backbone}, $\alpha(0 < \alpha < 1)$ is a scaling factor for reducing the computational complexity and the number of parameters, the GFLOPS and Parameters of the improved Backbone module are $\alpha \times F_{\text{backbone}}$ and $\alpha \times P_{\text{backbone}}$, respectively. The original GFLOPS of the Neck part is Fneck and the number of parameters is Pneck, and the computational overhead is reduced by a scaling factor $\beta(0 < \beta < 1)$, so that the quantized GFLOPS and Parameters are $\beta \times F_{\text{neck}}$ and $\beta \times P_{\text{neck}}$, respectively. In summary, the parameters and computational quantities of the model before and after definition are

$$F = \alpha \times F_{\text{backbone}} + \beta \times F_{\text{neck}} \quad (1)$$

$$P = \alpha \times P_{\text{backbone}} + \beta \times P_{\text{neck}} \quad (2)$$

3.2. Model Method Description

This paper is inspired by Ghost convolutions in Ghostnet, one of the lightweight state-of-the-art models designed for efficient inference on mobile devices. Its main component is the Ghost module, which uses low-cost operations to generate more feature maps instead of the original convolution. Given an input feature $X \in R^{H \times W \times C}$ with height H, width W, and number of channels C, a typical Ghost module can replace the standard convolution in two steps. First, a 1×1 convolution is used to generate the original features, i.e.,

$$Y' = X \times F_{1 \times 1} \tag{3}$$

$F_{1 \times 1}$ is a point-by-point convolution, and $Y' \in R^{H \times W \times C'_{out}}$ are intrinsic features whose size is usually smaller than the original output features. Then, the cheap operation (F_{dp} for depth-separated convolution) is used to generate more features based on the intrinsic features. The two parts of the feature are linked along the channel dimension, so that

$$Y = \text{Concat}\left(\left[Y', Y' \times F_{dp}\right]\right) \tag{4}$$

In the Ghost module, only half of the features, the essential features, are smaller than the original output features, which will lose the captured spatial and position information, and to consider this loss, this paper will use the attention module to enhance its spatial and position features.

4. Model Structure

4.1. Yolov5 Algorithm Principle

The YOLOv5 network structure consists of four main parts: Input, Backbone, Neck and Head. The four parts, respectively, perform data input processing, feature learning, feature enhancement processing, and target detection and classification.

Input performs Mosic operations on the input data, mainly cutting, splicing, resizing, and optimizing the input image data to compute the anchor frames. Mosic data augmentation is used to increase the diversity of the dataset, thus increasing the robustness and generalizability of the model.

Backbone is mainly used for feature learning, and the main constituent modules are C3 and SPPF (Spatial Pyramid Pooling—Fast).

The C3 module is similar to the original CSP (Cross-Stage Partial Network) structure, which is mainly used to simplify the network structure, reduce the number of convolutional layers and channels, and maintain the performance, and the SPPF module is the fusion of deep and shallow information to improve the feature extraction ability of the network.

The Neck structure uses a PANet structure to achieve feature enhancement through multi-layer feature fusion of top-down and bottom-up deep and shallow features, thereby increasing the robustness of the model and improving the accuracy of the target detection.

The Head structure obtains the position of the prediction frame target in the input image as well as the category information by designing three detection heads for detecting targets of different scales, each of which acquires feature information of different scale sizes from different layers of the Neck.

4.2. Improve the Structure of the Model

The most time-consuming part of the model is the C3 module, which is used to extract features and enhance the receptive field by reducing the number of channels of the input feature map with a 1×1 convolutional layer, then using a set of 3×3 sequential convolutions to extract the features, and finally using 1×1 convolutions with residual linking to sum the output of the previous step with the output of that layer. In this paper, the C3 modules of the Backbone and Neck sections are quantified. The Backbone part reduces the computational complexity and the number of parameters to the model through the Ghost Module idea and uses Coordinate Attention (CoordAtt) to focus on the global information. Coordinate attention has some unique advantages over other attention mechanisms such as SE (Squeeze-and-Excitation), CBAM [25], SAM [26], ECANet [27], and others. Spatially adaptive: it is able to focus on different locations of the input feature map to capture important contextual information in the image.

Parameter-efficient: it is more advantageous in terms of parameter efficiency compared to SE Attention, which is realized by simple linear transformations and softmax operations, making it more feasible in the case of limited computational resources. The design of Coordinate Attention makes it more flexible and can be used in combination with other attention mechanisms. Combining the various features mentioned above, Coordinate Attention has good assistance in improving the capture of spatial and positional information of features, improving the ability of module feature extraction, and reducing the number of parameters of modules. In the Neck part of the input feature map, $X \in R^{H \times W \times C}$ already contains a large amount of feature information, in the original Neck, the SE attention and Ghost module are used to improve the C3 module, reduce the number of parameters of the module and the extraction of channel features, and the overall structure of the model is shown in Figure 1.

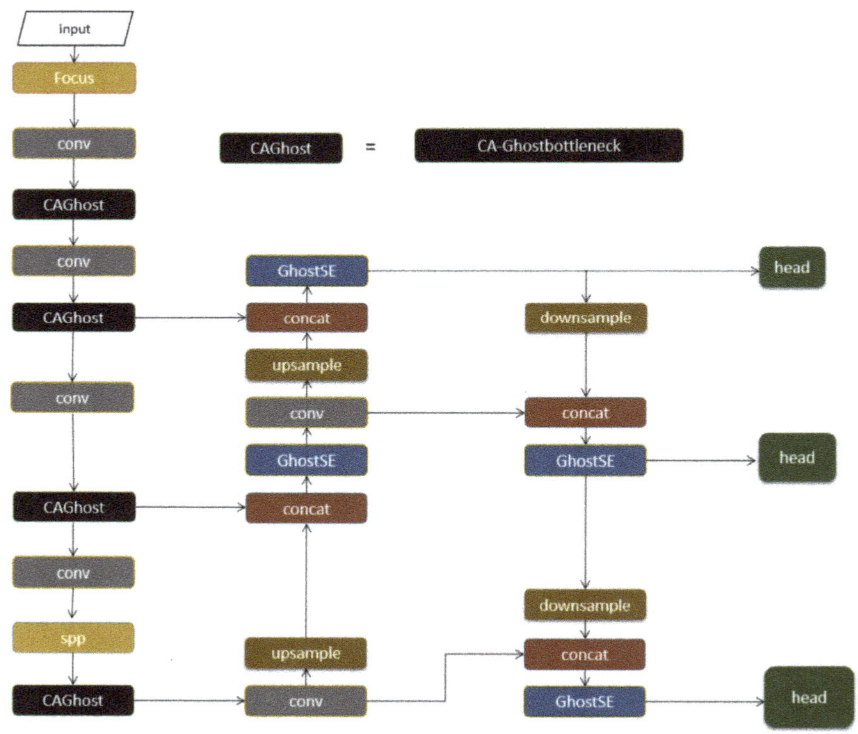

Figure 1. Improved GC-YOLO model diagram. Compared to the native YOLOv5 model, the enhanced GC-YOLO model replaces the original C3 module in the backbone section with the CAGhost and replaces the original C3 module in the Neck section with the GhostSE module.

4.3. CA-Ghostbotelneck

CA-GhostBotelneck (shown in Figure 2), as a key network module in the backbone network, adopts ideas from GhostnetV2 [28]. The CA-GhostBotelneck in this paper takes into account the fact that the Ghost module is only half functional and the nature features are smaller than the original output features, and when extracting the features for the input feature map $X \in R^{H \times W \times C}$, the output $Y \in R^{H \times W \times C_{out}}$ is obtained, and the features of Y are lost in both the channel information and the position information. In this paper, the input X is processed in two stages. First, the sparse feature map Y is obtained by the Ghost module, and second, the channel information and position information are obtained by the CoordAttention module, and finally, the two parts of the features are integrated to obtain a new output. The benefits of using CA-GhostBotelneck are as follows:

- Reduce the number of parameters: the Ghost module can use sparse convolution to obtain the nature features, improving the lightness of the effect.
- Improve model expressiveness: CoordAttention captures channel and position information, allowing more flexible access to global feature information and improving model expressiveness.

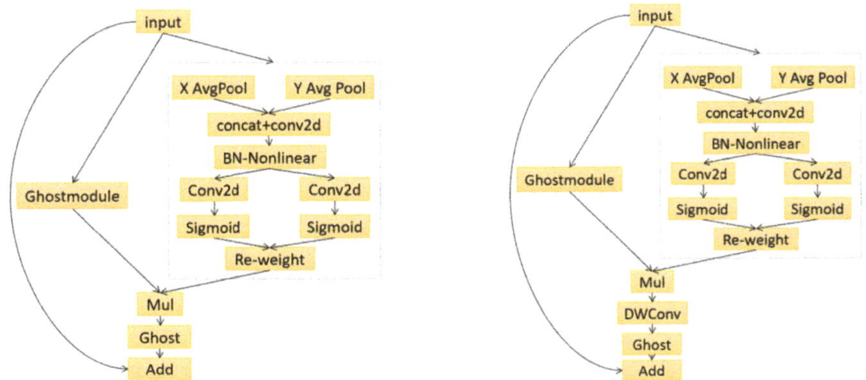

(a) Step size of 1 **(b) Step size of 2**

Figure 2. CA-GhostBotelneck with step size 1 on the left, CA-GhostBotelneck with step size 2 on the right.

Given an input feature $X \in R^{H \times W \times C}$ with height H, width W, and number of channels C, the CA-GhostBotelneck module can replace the normal convolution in two steps. First, a 1×1 convolution is used to generate the intrinsic features, i.e.,

$$Y' = X \times F_{1\times1} \tag{5}$$

$Y' \in R^{H \times W \times C'_{out}}$ are intrinsic features whose sizes are usually smaller than the original output features, which compensate for the lack of original channel and position information by having stronger feature information from CoordAttention than Deep-WiseConv, i.e.,

$$F_{\text{coordAtt}} > F_{dp} \tag{6}$$

$$Y = \text{Concat}\left(\left[Y', Y' \times F_{\text{coordAtt}}\right]\right) \tag{7}$$

4.4. GhostSE

In this paper, the GhostSE structure is used in the Neck part (as shown in Figure 3), and the intrinsic features obtained by 1×1 convolution have fewer output features than those obtained by ordinary convolution. Improved access to channel information of feature maps using SE attention to capture more features to compensate for the feature loss problem caused by convolution in the Ghost idea, second, residual joining is used to obtain richer feature output, and finally residual joining is performed using the GhostConvSE module and GhostBottelneck, reducing the number of parameters and float calculations while keeping as much feature-rich information as possible.

Given an input feature $X \in R^{H \times W \times C}$ with height H, width W, and number of channels C, output Y' via 1×1 Conv,Y' via SE Attention to output Y'', input Y'' in GhostSE, and output Z' after Add, and finally output the feature map Z, i.e.,

$$Y' = X \times F_{1\times1} \tag{8}$$

$$Y'' = \text{Concat}\left(\left[Y', Y' \times F_{SE}\right]\right) \tag{9}$$

$$Z' = \text{Concat}\left(\left[Y'', Y'' \times F_{\text{GhostBottelneck}}\right]\right) \tag{10}$$

$$Z = Z' \times F_{\text{GhostConv}} \tag{11}$$

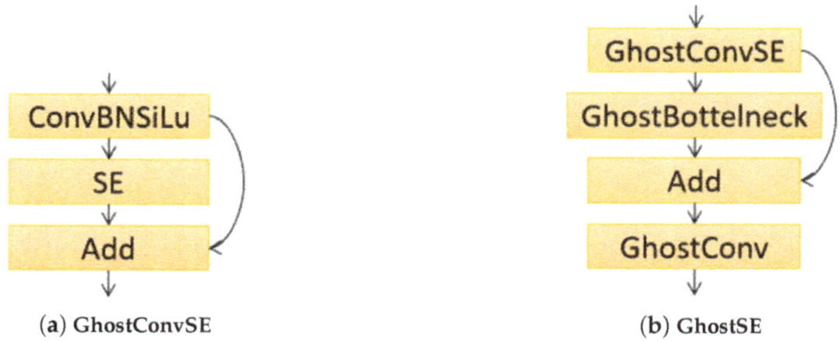

Figure 3. The left image shows GhostConvSE, which uses SE attention to obtain more feature information; the right image shows GhostSE.

5. Experiment

5.1. Experimental Environment

The experiment selects the PASCAL VOC dataset commonly used for target detection for training, which is mainly used for detecting the four major classes of vehicle, household, animal and person in the environment, and the detection target samples are relatively abundant. The computer configuration for the experiment is GPU: RTX 3060, CPU: I5-10400, 16G RAM; the training network environment is Python: 3.9, CUDA12.1.

5.2. Model Evaluation

In target detection tasks, it is often necessary to compare the predicted results with the true labels, and three metrics are used to evaluate model performance in this process.

1. *TP*: Means labeled as a positive sample and predicted as a positive sample.
2. *FP*: Means that the label is a negative sample and the prediction is a positive sample.
3. *FN*: Refers to samples labeled as positive samples but predicted to be negative.
4. *TN*: Means that the label is a negative sample and the prediction is a negative sample.
5. *Precision*: Indicates the percentage of samples that were correctly predicted out of those predicted as positive examples.

$$P = TP/(TP + FP) \tag{12}$$

6. *Recall*: indicates the proportion of positive samples that are true positive samples.

$$R = TP/(TP + FN) \tag{13}$$

7. *mAP*: Used to evaluate overall model detection performance in multiple categories. Where n is the number of categories, AP_i is the average precision of the i-th category, and r is the *recall*.

$$AP = \int_0^1 P(r)dr, r \in (0,1) \tag{14}$$

$$mAP = \frac{1}{n}\sum_{i=1}^{n} AP_i \tag{15}$$

8. *F1 score*: Combines Precision and Recall to evaluate the performance of the model and is defined as the harmonic mean of Precision and Recall.

$$F1 = 2 \times P \times R/(P + R) \tag{16}$$

By balancing the lightness, detection accuracy and detection speed of the model, this paper improves the model. By calculating the Efficient value, the model M that balances the detection efficiency and speed is finally obtained.

$$\text{Efficient} = E(F, P, mAP, R, F1) \tag{17}$$

$$M = \max_{\mathcal{M} \in \mathbb{M}} \text{Efficient}_i^{\mathcal{M}} \tag{18}$$

5.3. Experiment

In order to verify the overall improvement of GC-YOLO of the design model, this paper designs several comparative experiments with typical lightweight networks. The PASCAL VOC dataset is selected, and the experimental dataset is divided into the training set and the validation set with a ratio of 9:1, the image size is 640 × 640, the training batch is set to 32, and all reference models are trained for 300 epochs according to this parameter. The experiments compared the number of model parameters, GFLOPS, mean accuracy mAP: 0.5, and harmonic mean $F1$.

As shown in Table 1, the original Yolov5s had 7.28 M parameters and 17.16 G GFLOPS. With CA-GhostBotelneck and GhostSE's GC-YOLO, the number of parameters is 2.8 M less and GFLOPS are 8.53 G less with a slight increase in mAP and $F1$. The results show that the model's feature extraction capability is significantly improved and at the same time the model's parameter number is reduced.

Table 1. Comparative testing of models.

Model	Parameter (M)	GFLOPS (G)	mAP (%)	$F1$	FPS
YOLOV5s	7.28	17.16	84.06	0.62	25
Yolov4-MobileNetv3	11.73	18.22	69.13	0.68	28
Yolov4-tiny	6.1	6.96	64	0.54	30
Yoloxs	8.95	26.73	83.8	0.74	18
Yolov7-tiny	6.23	13.86	80.83	0.76	26
GC-Yolo(our)	4.48	8.63	84.19	0.72	24

The partial detection results of the GC-YOLO model are shown in figures. Figure 4 shows the harmonic mean F1 after training the model on the VOC dataset with the threshold set to 0.5. The F1 value combines the accuracy and completeness of the model, and is particularly useful for dealing with category imbalance or focusing on improving both accuracy and recall. Higher F1 values are an indication of better model performance in positive sample detection and negative sample exclusion, from the twenty categories in the figure, it is evident that the F1 values of 12 categories are higher than the average value of 0.72, and there are only a few major fluctuations, which proves that the model achieves a relatively balanced performance in different categories, has a better generalization ability in each category, can adapt to different categories, and has an advantage in dealing with multicategory problems; Figure 5 shows the average accuracy of the model mAP on each category, combining the prediction accuracy and recall of the model on different target categories, which is used to measure the overall performance of the model on multiple target categories, as shown in the figure, the data are more tightly clustered, with 13 categories exceeding 84.19%, five categories surpassing 90%, and two categories falling below 70%. This suggests that the model is successful in accurately localizing and identifying the target object across multiple categories, without excessively focusing on certain categories and disregarding others, and with exceptional overall performance. Figure 6 shows the "loss rate" of the model, especially in the security and surveillance area. It reflects the proportion of targets missed by the model during target detection, and a lower leakage rate indicates that the model performs better in target detection and is able to capture targets more comprehensively, from the figure, it is evident that the leakage rate is mainly below 0.3, with five categories exceeding this threshold. However, the highest leakage rate is only 0.56, indicating that the model has a high recall rate and can detect most of the target objects. It is also robust to the size and location of different targets, ensuring a consistently low leakage rate.

In addition to testing GC-YOLO detection on images outside the VOC dataset, in this paper, an image downloaded from the Internet was used for detection, and the comparative detection experiment is shown in Figure 7, and it is evident that (a) enhances target recognition accuracy by approximately 0.05 in unobstructed and approximately 0.1 in obstructed targets compared to (b). The model's overall accuracy for recognizing categories is enhanced, including the recognition of yellow cars.

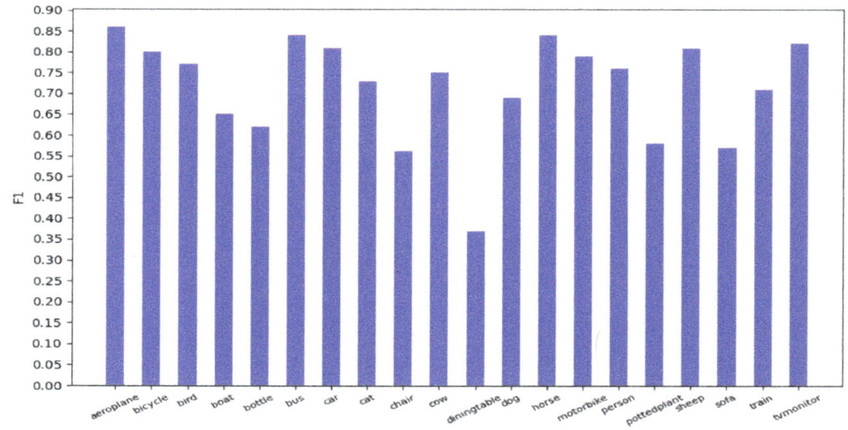

Figure 4. Harmonic mean F1 values of the GC-YOLO model for the VOC dataset.

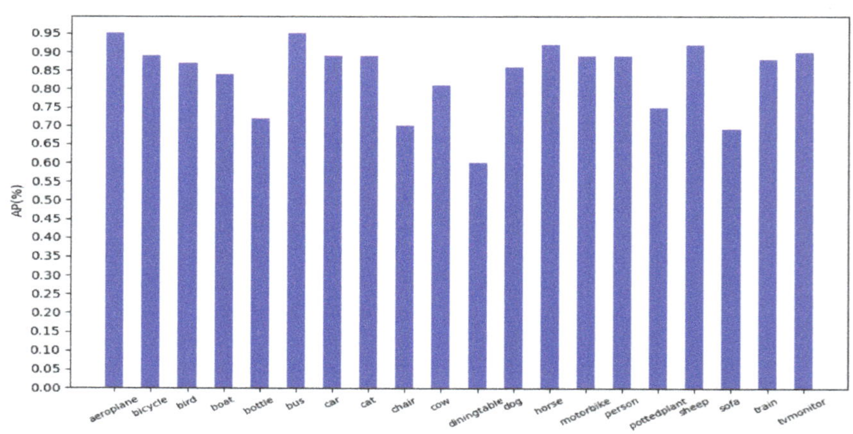

Figure 5. Average accuracy of the GC-YOLO model on the VOC dataset (mAP = 84.19%).

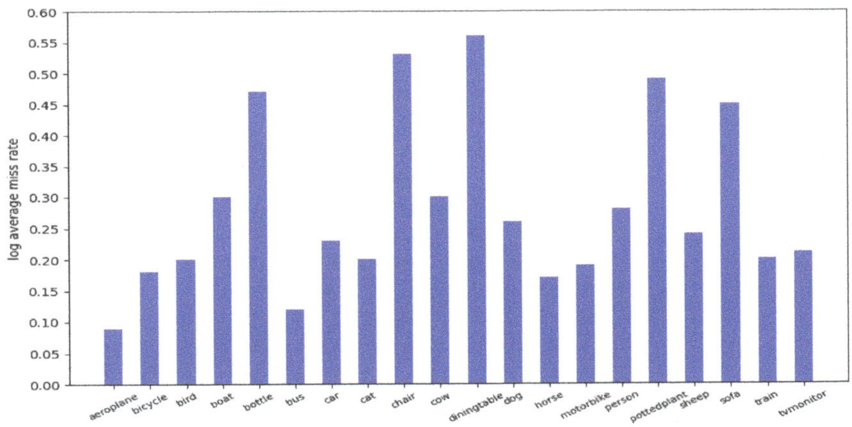

Figure 6. GC-YOLO model miss rate in VOC dataset.

(**a**) (**b**)

Figure 7. (**a**) GC-YOLO model detection results; (**b**) original model detection results. In this context, blue boxes indicate people, while green boxes represent cars. Compared to Figure (**b**), Figure (**a**) displays superior accuracy in identifying individuals, successfully detecting the concealed yellow car, and demonstrating an increased confidence in identifying the black car. Furthermore, there is a decreased likelihood of misidentifying a person as a vehicle.

Comparison experiments were also performed between the GC-YOLO model and the FPS of Yolov5's real-time detection, as shown in Figure 8. The model introduces attention to improve the extraction of features, while ensuring the real-time performance of the lightweight model, and the model's FPS is relatively smooth.

5.4. Scenario Experiments with Custom Datasets

This GC-YOLO lightweight model is used in the realization of intelligent wheelchair devices on the blind spot obstacle detection to help the user to avoid the blind spot on both sides of the obstacles caused by safety issues.

It was found that the visual angle of wheelchair blindness is different from the unusual visual angle, and is more a differentiator caused by the low viewing angle, which is reflected in the overhead angle and the incompleteness of the display target. In this paper, dangerous obstacles for wheelchair blindness in the community are defined as cats, dogs, potholes, water bottles, and feet, legs, and wheels at the top view, and the seven categories are designed to avoid physical injury from animal attacks, falls, and wheel accidents. For this purpose, field targets were collected for dataset production, mainly through mobile phone simulation of wheelchair heights and perspectives in different scenarios, based on different age groups, scenarios, and time periods, to build diverse sample data. In this paper, the captured video data are processed, and the video file is sliced by frames, and one image is selected every ten frames to obtain the foot and leg photos under different angles and different poses. To ensure the diversity of the dataset and the amount of data, some of the categories of the dataset were obtained from publicly available datasets and web crawling, respectively, totaling 7916 image data. To solve the problem of unbalanced data volume in the target dataset, this paper sets the data enhancement rate of mosaic to fifty percent, which expands the data diversity and improves the generalization ability of the training model. The experiment is shown in Figure 9, and the precision for various categories exceeds 0.85, some even surpassing 0.95. The model demonstrates excellent robustness in its capacity to generalize across targets of differing sizes such as feet, legs, and wheels. which shows that the performance of the overall model on the custom dataset is relatively stable, with a mAP of 90.34%, and Figure 10 shows that the average accuracy of the F1 values is 0.84 at score threshold = 0.5, on the custom dataset, the F1 score is no worse than the average of the model trained on the VOC dataset (0.72), and the model can adapt to different categories, demonstrating its ability to generalize.

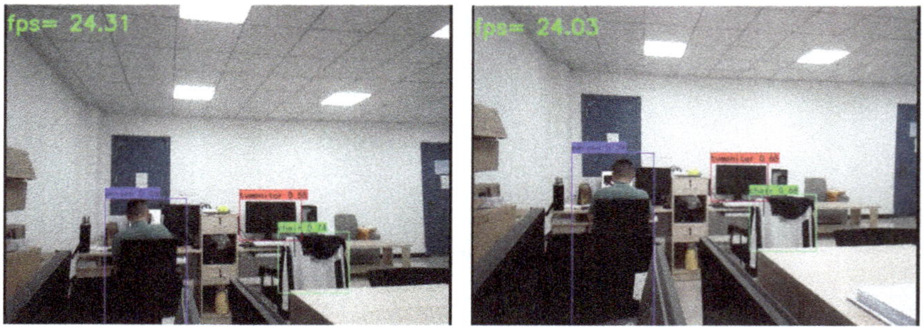

(a) (b)

Figure 8. (**a**) Shows the FPS detection effect of the GC-YOLO model; (**b**) shows the FPS detection effect of the original model. The detection rates of the two graph algorithms have been compared, and the modified algorithm consistently maintains high performance without any reduction in detection rates.

The trained model is tested in real scenarios, and the results of the test scenarios are shown in Figure 11. For intelligent wheelchair obstacle detection in the blind zones on both sides of the wheelchair in a senior living community environment, side safety is judged mainly based on the display of incomplete targets. The four images reflect wheels, legs, feet, and potholes in low vision, and the first three images judge obstacle targets based on human targets with incomplete displays in low vision.

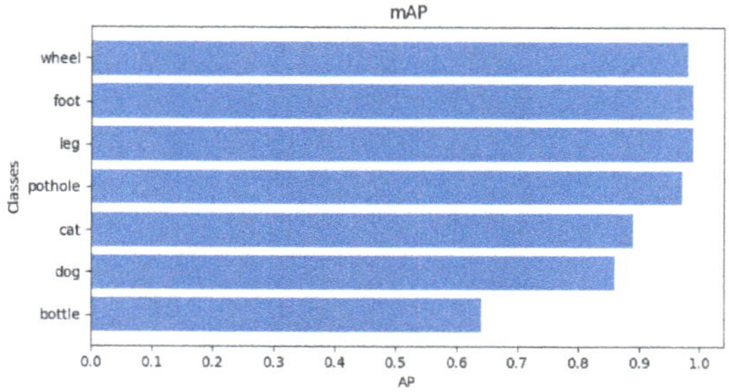

Figure 9. Average accuracy of the GC-YOLO model on the custom set (mAP = 90.34%).

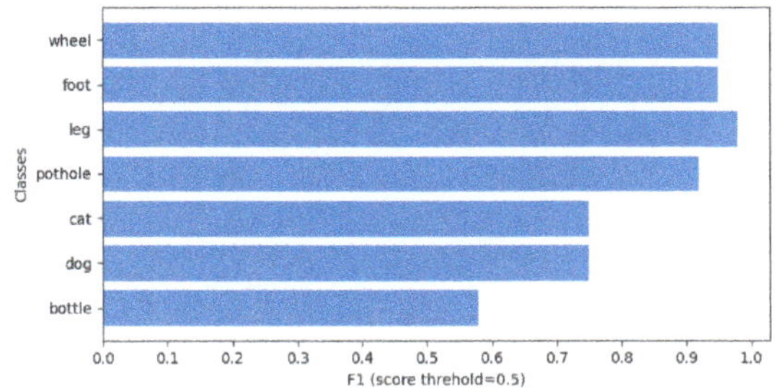

Figure 10. F1 of the GC-YOLO model on the homebrew set (score threshold = 0.5).

(**a**) Identify the target through specific regions, including the wheels.

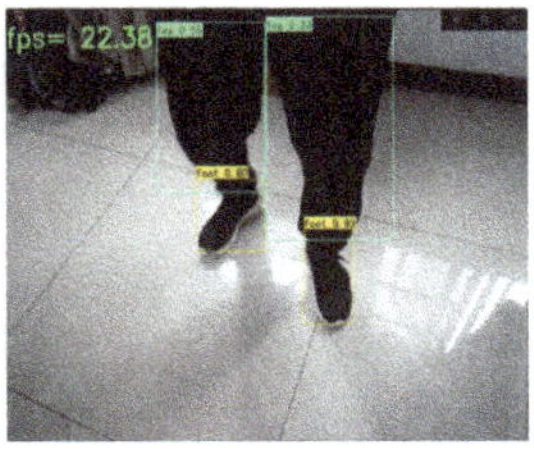

(**b**) Indoor experimental trials.

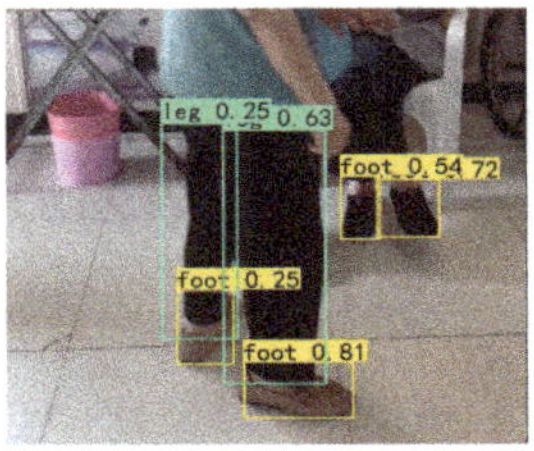

(**c**) Data within the elderly community, illustrating the algorithm's detection efficacy.

(**d**) From a low angle, the algorithm's efficacy in detecting potholes.

Figure 11. The four panels show the real scene model detection effect. Distinctive colors serve to discern and display diverse categories: green boxes denote legs, purple boxes symbolize wheels, yellow boxes indicate feet, and red boxes represent potholes.

The above experimental results show that compared with YOLOv5s, YOLOv4-mobilenetv3 and other lightweight model algorithms, the real-time performance of the GC-YOLO model is as stable as that of the native YOLOv5s, but it has been improved in the number of parameters, GFLOPS, mAP, and the evaluation of the F1 value, and it also has a very good performance in the custom dataset to perform the safety supervision of the blind zones of intelligent wheelchair detection.

6. Conclusions

In this paper, we propose a lightweight target detection algorithm GC-YOLO based on YOLOv5. By improving the network, the model is able to achieve good detection performance while being lightweight, balancing the relationship between lightweight and detection performance. In intelligent wheelchairs for the elderly community that have a good lightweight performance, the research found that the model has good detection performance in blind spot obstacle detection to avoid potential safety threats. In future work, the algorithm will deploy on *Nvidia Jetson Nano*, and cameras will install on both sides of the wheelchair to detect each side independently. Subsequent experiments will aim to further improve and optimize the system. However, limitations may arise during the experimental process, as well as during the maintenance and retraining of the model after deployment on the mobile terminal. When major environmental changes occur, the model's performance may diminish. In our future studies, we will explore the integration of multi-modal or unsupervised learning approaches to improve the model's responsiveness to environmental fluctuations and continue our research in this area.

Author Contributions: Conceptualization, S.Z. and J.D.; methodology, Y.C. and J.D.; validation, S.Z. and C.S.; formal analysis, J.D. and S.Z.; investigation, S.Z.; data curation, S.Z., C.S. and J.D.; writing—original draft preparation, J.D.; writing—review and editing, S.Z., Y.C., C.S. and J.D. All authors have read and agreed to the published version of the manuscript.

Funding: This research was funded by Chuzhou University. This work was supported by Anhui Higher Education Research Program Project under Grant 2022AH010067 (Title: Smart Elderly Care and Health Engineering Scientific Research Innovation Team).

Data Availability Statement: Data sharing not applicable. Further research is needed.

Conflicts of Interest: The authors declare no conflict of interest.

References

1. Ahmadi, A.; Argany, M.; Neysani Samany, N.; Rasooli, M. Urban Vision Development in Order To Monitor Wheelchair Users Based on The Yolo Algorithm. *Int. Arch. Photogramm. Remote. Sens. Spat. Inf. Sci.* **2019**, *XLII-4/W18*, 25–27. [CrossRef]
2. Okuhama, M.; Higa, S.; Yamada, K.; Kamisato, S. *Improved Visual Intention Estimation Model with Object Detection Using YOLO*; IEICE Technical Report; IEICE Tech: Tokyo, Japan, 2023; Volume 122, pp. 1–2.
3. Chatzidimitriadis, S.; Bafti, S.M.; Sirlantzis, K. Non-Intrusive Head Movement Control for Powered Wheelchairs: A Vision-Based Approach. *IEEE Access* **2023**, *11*, 65663–65674. [CrossRef]
4. Hashizume, S.; Suzuki, I.; Takazawa, K. Telewheelchair: A demonstration of the intelligent electric wheelchair system towards human-machine. In Proceedings of the SIGGRAPH Asia 2017 Emerging Technologies, Bankok, Thailand, 27–30 November 2017; p. 1.
5. Suzuki, I.; Hashizume, S.; Takazawa, K.; Sasaki, R.; Hashimoto, Y.; Ochiai, Y. Telewheelchair: The intelligent electric wheelchair system towards human-machine combined environmental supports. In Proceedings of the ACM SIGGRAPH 2017 Posters, Los Angeles, CA, USA, 30 July–3 August 2017; p. 1.
6. Lang, A.H.; Vora, S.; Caesar, H.; Zhou, L.; Yang, J.; Beijbom, O. Pointpillars: Fast encoders for object detection from point clouds. In Proceedings of the IEEE/CVF Conference on Computer Vision and Pattern Recognition, Long Beach, CA, USA, 15–20 June 2019; pp. 12697–12705.
7. Meyer, G.P.; Laddha, A.; Kee, E.; Vallespi-Gonzalez, C.; Wellington, C.K. Lasernet: An efficient probabilistic 3D object detector for autonomous driving. In Proceedings of the IEEE/CVF Conference on Computer Vision and Pattern Recognition, Long Beach, CA, USA, 15–20 June 2019; pp. 12677–12686.
8. Chen, Q.; Chen, Y.; Zhu, J.; De Luca, G.; Zhang, M.; Guo, Y. Traffic light and moving object detection for a guide-dog robot. *J. Eng.* **2020**, *13*, 675–678. [CrossRef]
9. Ferretti, S.; Mirri, S.; Roccetti, M.; Salomoni, P. Notes for a collaboration: On the design of a wiki-type educational video lecture annotation system. In Proceedings of the IEEE International Conference on Semantic Computing (ICSC 2007), Irvine, CA, USA, 17–19 September 2007; pp. 651–656.
10. Han, K.; Wang, Y.; Tian, Q.; Guo, J.; Xu, C.; Xu, C. Ghostnet: More features from cheap operations. In Proceedings of the IEEE/CVF Conference on Computer Vision and Pattern Recognition, Seattle, WA, USA, 13–19 June 2020; pp. 1580–1589.
11. Hou, Q.; Zhou, D.; Feng, J. Coordinate attention for efficient mobile network design. In Proceedings of the IEEE/CVF Conference on Computer Vision and Pattern Recognition, Nashville, TN, USA, 20–21 June 2021; pp. 13713–13722.
12. Hu, J.; Shen, L.; Sun, G. Squeeze-and-excitation networks. In Proceedings of the IEEE Conference on Computer Vision and Pattern Recognition, Salt Lake City, UT, USA, 18–23 June 2018; pp. 7132–7141.
13. Lee, H. J.; Ullah, I.; Wan, W.; Gao, Y.; Fang, Z. Real-time vehicle make and model recognition with the residual SqueezeNet architecture. *Sensors* **2019**, *19*, 982. [CrossRef] [PubMed]
14. Sheng, T.; Feng, C.; Zhuo, S.; Zhang, X.; Shen, L. A quantization-friendly separable convolution for mobilenets. In Proceedings of the IEEE 2018 1st Workshop on Energy Efficient Machine Learning and Cognitive Computing for Embedded Applications (EMC2), Williamsburg, VA, USA, 25 March 2018.
15. Sandler, M.; Howard, A.; Zhu, M.; Zhmoginov, A.; Chen, L.-C. Mobilenetv2: Inverted residuals and linear bottlenecks. In Proceedings of the IEEE Conference on Computer Vision and Pattern Recognition, Salt Lake City, UT, USA, 18–23 June 2018; pp. 4510–4520.
16. Howard, A.; Sandler, M.; Chu, G.; Chen, L.C.; Chen, B.; Tan, M.; Wang, W.; Zhu, Y.; Pang, R.; Vasudevan, V.; et al. Searching for mobilenetv3. In Proceedings of the IEEE/CVF International Conference on Computer Vision, Seoul, Republic of Korea, 27 October–2 November 2019; pp. 1314–1324.

17. Nascimento, M.G.; Fawcett, R.; Prisacariu, V.A. Dsconv: Efficient convolution operator. In Proceedings of the IEEE/CVF International Conference on Computer Vision, Seoul, Republic of Korea, 27 October–2 November 2019; pp. 5148–5157.
18. Zhang, X.; Zhou, X.; Lin, M.; Sun, J. Shufflenet: An extremely efficient convolutional neural network for mobile devices. In Proceedings of the IEEE Conference on Computer Vision and Pattern Recognition, Salt Lake City, UT, USA, 18–23 June 2018; pp. 6848–6856.
19. Tan, M.; Le, Q. Efficientnet: Rethinking model scaling for convolutional neural networks. In Proceedings of the 2019 International Conference on Machine Learning, Long Beach, CA, USA, 10–15 June 2019; pp. 6105–6114.
20. He, Y.; Zhang, X.; Sun, J. Channel pruning for accelerating very deep neural networks. In Proceedings of the IEEE 2017 International Conference on Computer Vision, Venice, Italy, 22–29 October 2017; pp. 1389–1397.
21. Cai, Y.; Li, H.; Yuan, G.; Niu, W.; Li, Y.; Tang, X.; Ren, B.; Wang, Y. Yolobile: Real-time object detection on mobile devices via compression-compilation co-design. In Proceedings of the AAAI Conference on Artificial Intelligence, Virtually, 2–9 February 2021; pp. 955–963.
22. Jacob, B.; Kligys, S.; Chen, B.; Zhu, M.; Tang, M.; Howard, A.; Adam, H.; Kalenichenko, D. Quantization and training of neural networks for efficient integer-arithmetic-only inference. In Proceedings of the IEEE Conference on Computer Vision and Pattern Recognition, Salt Lake City, UT, USA, 18–23 June 2018; pp. 2704–2713.
23. Yang, Y.; Sun, X.; Diao, W.; Li, H.; Wu, Y.; Li, X.; Fu, K. Adaptive knowledge distillation for lightweight remote sensing object detectors optimizing. *IEEE Trans. Geosci. Remote Sens.* **2022**, *60*, 1–15. [CrossRef]
24. Qu, J.; Chen, B.; Liu, C.; Wang, J. Flight Delay Prediction Model Based on Lightweight Network ECA-MobileNetV3. *Electronics* **2023**, *12*, 1434. [CrossRef]
25. 25 Woo, S.; Park, J.; Lee, J.-Y.; Kweon, I.S. Cbam: Convolutional block attention module. In Proceedings of the European Conference on Computer Vision (ECCV), Munich, Germany, 8–14 September 2018; pp. 3–19.
26. Zhu, X.; Cheng, D.; Zhang, Z.; Lin, S.; Dai, J. An empirical study of spatial attention mechanisms in deep networks. In Proceedings of the IEEE/CVF International Conference on Computer Vision, Seoul, Republic of Korea, 27 October–2 November 2019; pp. 6688–6697.
27. Wang, Q.; Wu, B.; Zhu, P.; Li, P.; Zuo, W.; Hu, Q. ECA-Net: Efficient channel attention for deep convolutional neural networks. In Proceedings of the IEEE/CVF Conference on Computer Vision and Pattern Recognition, Seattle, WA, USA, 13–19 June 2020; pp. 11534–11542.
28. Tang, Y.; Han, K.; Guo, J.; Xu, C.; Xu, C.; Wang, Y. GhostNetv2: Enhance cheap operation with long-range attention. *Adv. Neural Inf. Process. Syst.* **2022**, *35*, 9969–9982.

electronics

Article

Stratified Sampling-Based Deep Learning Approach to Increase Prediction Accuracy of Unbalanced Dataset

Jeyabharathy Sadaiyandi [1], Padmapriya Arumugam [1,*], Arun Kumar Sangaiah [2,3] and Chao Zhang [4,*]

[1] Department of Computer Science, Alagappa University, Karaikudi 630003, India; jeyabharathys_phdscholar@alagappauniversity.ac.in
[2] International Graduate School of AI, National Yunlin University of Science and Technology, Douliu 64002, Taiwan; arunks@yuntech.edu.tw
[3] Department of Electrical and Computer Engineering, Lebanese American University, Byblos 13-5053, Lebanon
[4] School of Computer and Information Technology, Shanxi University, Taiyuan 030006, China
* Correspondence: padmapriyaa@alagappauniversity.ac.in (P.A.); czhang@sxu.edu.cn (C.Z.)

Abstract: Due to the imbalanced nature of datasets, classifying unbalanced data classes and drawing accurate predictions is still a challenging task. Sampling procedures, along with machine learning and deep learning algorithms, are a boon for solving this kind of challenging task. This study's objective is to use sampling-based machine learning and deep learning approaches to automate the recognition of rotting trees from a forest dataset. Method/Approach: The proposed approach successfully predicted the dead tree in the forest. Seven of the twenty-one features are computed using the wrapper approach. This research work presents a novel method for determining the state of decay of the tree. The process of classifying the tree's state of decay is connected to the issue of unequal class distribution. When classes to be predicted are uneven, this frequently hides poor performance in minority classes. Using stratified sampling procedures, the required samples for precise categorization are prepared. Stratified sampling approaches are employed to generate the necessary samples for accurate prediction, and the precise samples with computed features are input into a deep learning neural network. Finding: The multi-layer feed-forward classifier produces the greatest results in terms of classification accuracy (91%). Novelty/Improvement: Correct samples are necessary for correct classification in machine learning approaches. In the present study, stratified samples were considered while deciding which samples to use as deep neural network input. It suggests that the proposed algorithm could accurately determine whether the tree has decayed or not.

Keywords: machine learning; deep learning; imbalanced datasets; stratified sampling; prediction; classification; accuracy; wrapper classes

Citation: Sadaiyandi, J.; Arumugam, P.; Sangaiah, A.K.; Zhang, C. Stratified Sampling-Based Deep Learning Approach to Increase Prediction Accuracy of Unbalanced Dataset. *Electronics* **2023**, *12*, 4423. https://doi.org/10.3390/electronics12214423

Academic Editor: Heung-Il Suk

Received: 12 September 2023
Revised: 21 October 2023
Accepted: 25 October 2023
Published: 27 October 2023

1. Introduction

In data mining and machine learning, classification analysis is a well-researched method. Because of its ability to forecast future outcomes, it is used in a wide range of real-world scenarios. However, classification accuracy is directly proportional to the training data quality utilized. Real-world data frequently has an imbalanced class distribution, with the dominating majority class and ignoring the least ones.

When dealing with an imbalanced class distribution problem, selecting appropriate training data becomes crucial for improving classification accuracy. When all the available data are used for training, the resulting classifier tends to predict most of the incoming data as belonging to the majority class. This leads to the misclassification of minority class instances. Hence, careful selection of training data is essential to address the challenges posed by imbalanced class distributions in classification problems. In the context of forest ecosystems, the need for accurate classification algorithms cannot be overstated. Forests are a critical component of the planet's ecological balance, sequestering and storing massive amounts of carbon from the atmosphere. The carbon stored in forest biomass is a crucial element of healthy forest ecosystems and the global carbon cycle.

Forests store carbon in various forms that can be challenging to accurately quantify. The estimation of carbon storage in forests depends on several factors, including the density of tree wood, decay class, and density reduction factors. Accurate estimations of carbon storage in forests are essential for effective carbon flux monitoring. Moreover, the classification of forest data is critical in determining the health and productivity of forest ecosystems. Forest classification algorithms can help identify various features of forests, such as tree species, forest density, and biomass, which are essential in monitoring changes in forest structure and function.

Forest-based accurate classification can also help to predict the occurrence and spread of forest disturbances like wildfires, insect infestations, and diseases. Such disturbances can cause significant losses of carbon from forests, negatively impacting the planet's ecological balance. Therefore, the development of accurate and robust classification algorithms for forest datasets is critical for maintaining healthy forest ecosystems and mitigating the impact of natural disasters on the environment. In the realm of predicting tree decay rates in forests, past research has mainly focused on using regression techniques. However, these methods may not be suitable for distinguishing individual dead trees within a forest.

Deep neural network (DNN) architecture is aimed at detecting individual dead trees within the forest more accurately in this study. For that, this research work proposed a novel approach to deal with imbalanced datasets using sampling techniques. The imbalanced nature of forest datasets can make predictions less accurate, particularly when most data points belong to a single class (e.g., living trees). Therefore, by employing sampling techniques, we balanced the dataset, which improved the accuracy of predictions for both dead and living trees. This ultimately improves the accuracy of predictions made with unbalanced forest datasets. The organization of this research work is as follows. The dataset used for this research work is described first. Then, we employ a DNN with sampling techniques to forecast both dead and living trees. This method was then compared to other techniques for its efficacy. Finally, we present our findings and future directions.

Overall, the development of DNN architecture for predicting individual dead trees in forests, coupled with sampling techniques to handle imbalanced datasets, can raise prediction accuracy and contribute to better forest management. It enables forest managers to conserve and protect the forest ecosystem by making informed decisions.

2. Literature Review

In general, the process of classifying unbalanced datasets consists of three steps: selecting features, fitting the data distribution, and training a model. The review of the literature is presented below in Table 1.

Table 1. Background study.

Reference	Methodology Used	Observations
[1]	CNN	Outlines the open research problems like enhancing the accuracy of tree species classification, applying the approach to various forest types, exploring its potential for estimating forest characteristics, and creating an easy-to-use tool for forest managers and conservationists.
[2]	SMOTE	The approach neglects to consider the computational cost and resource requirements of various algorithms. These resource requirements could be critical in real-world deployment scenarios.
[3]	Stratified with SVM	Limit its scalability to large datasets.
[4]	Classification using SVM and DNN	DNN shows low accuracy.

Table 1. *Cont.*

Reference	Methodology Used	Observations
[5]	Undersampling	Undersampling may lead to the loss of some useful information by removing significant patterns.
[6]	Oversampling	The performance of this approach may be influenced by the hyperparameters selected for the DCGAN and CNN models. The hyperparameters used in this model were not extensively optimized in this study.
[7]	Synthesizing data using Variational Auto Encoders (VAE) on raw training samples.	Detailed analysis of the computational cost of the proposed method was provided, which may be a concern for large datasets.
[8]	SMOTE	This approach did not consider the impact of SMOTE on real-world data.
[9]	Snag persistence Forest inventory model	This research work did not address the impact of tree species or decay stage on volume estimation accuracy.

The goal of feature selection is to identify subsets of features that are most suited for classifying the unbalanced data while considering the feature class imbalance. This contributes to the development of a more efficient classifier [10–13]. To limit the impact of class imbalances on the classifier, most data preparation procedures, such as various resampling techniques, are used to adjust the data distribution [14–17]. These techniques significantly balance the datasets.

Model training to accommodate unequal data distribution requires primarily adding an enforcement algorithm to an existing classification approach or applying ensemble learning. Standard cost-sensitive learning is an example of the latter [18–20]; it improves minority class classification accuracy by increasing the weights of the class samples. Classification accuracy can be achieved via ensemble learning techniques like boosting and bagging [21–23].

Distribution-level data resampling will resolve the class imbalance. The most significant advantage of this methodology is that the sampling method and the classifier training procedure are independent of one another. Typically, the sample distribution of the training set is changed at the data preprocessing stage to decrease or eliminate class imbalance. The representative methods consist of a few resampling strategies, with the two main categories being oversampling and undersampling.

Oversampling entails adding appropriately created new points to increase the sample points in a minority class to attain sample balance. The synthetic minority oversampling method (SMOTE) and several of its variants, as well as ROS, are examples of prevalent algorithms [24]. SMOTE generates synthetic samples and inserts them between a given sample and its neighbors, whereas datasets are balanced by ROS by adding minority sample points at random.

$$X_{new} = X_j + rand(0,1)\cdot(X_i - X_j)$$

In Equation (1), X_jj (j = 1, 2, 3, 4, 5) represents a minority class point, X_{new} represents the generated virtual samples based on the nearest neighbors X_i, and rand (0, 1) is a random number between 0 and 1 [4].

The earlier study relied heavily on local data to increase sample sizes. Although the number of samples is equalized, since the information on the overall distribution of the data is not taken into consideration, the data distribution of the new dataset following oversampling cannot be guaranteed. Furthermore, utilizing an oversampling approach may result in a big amount of redundant information, increasing the classifier's calculation and training time.

Undersampling decreases the sample size in a majority class by eliminating some of them, and therefore has the apparent benefit of shortening training time. The most basic undersampling approach is RUS [24], which discards majority class samples at random. To balance the magnitude of primary class samples with the least class samples, another undersampling strategy uses appropriate majority class samples. The training set will be more evenly distributed because of this method, which will also improve the classification accuracy of minority class samples. The disadvantage is that a sizable portion of the majority class sample characteristics could be lost, and the model might not fully acquire the majority class sample properties. As a result, it is crucial to make sure that the learning process is set up so that the bulk of the information covered in class is retained.

3. Materials and Methods

This research work is aimed at predicting the decay information of forest trees. Healthy trees absorb the harmful carbon dioxide and emit oxygen. Trees are the carbon sink of our planet. At the same time, decayed and fallen trees emit carbon dioxide. So, the identification of the decay level of a tree is essential to preserve the ecological condition of our planet. In this research work, details about trees in a forest are examined. Several attributes are associated with forest trees. The age of a tree is usually determined by its wood density. During the initial years, the wood density is increasing, and after attaining normal growth, the wood density starts decreasing. Based on the wood density, the trees are classified into five different decay classes ranging from "freshly killed" to "extremely decayed". The dead trees fall, may cause forest fires, and it may take several years to decompose. Here, the dataset is preprocessed first to compute wood density and identification of decay class (either Not yet decayed or Decayed) using the wrapper method. Due to the imbalance in the dataset after decay class identification, stratified sampling is used to overcome this issue without losing any inputs [25]. The stratified sampled input is fed to the DNN network for drawing predictions about the decay class of a tree.

This section contains a description of the proposed methodology, a description of the forest tree dataset, and the preparation process. The architecture of the proposed stratified sampling-based deep neural networks approach for increasing the prediction accuracy of the unbalanced dataset is shown in Figure 1.

The proposed methodology can be categorized into three phases.

- Data preprocessing phase;
- Training phase;
- Test phase.

The neural network is chosen for classification in this research work over the SVM, Random Forest, and Naïve Bayes because of its ability to handle imbalanced data, feature learning capabilities, model nonlinear relationships, and the ability to fine-tune hyperparameters for optimal performance.

3.1. Description of the Dataset

The dataset was obtained from the USDA repository [26]. Data collection began in 1985 and is expected to last until 2050. The Douglas fir, red cedar, Pacific silver fir, and Western hemlock tree are the four species used for investigation. The data gathered for this study compare the breakdown of tiny logs (20–30 cm in diameter and 2 m in length) in a stream channel at the H.J. Andrews Experimental Forest to that of logs on an adjacent upland site. Above the intersection of Mack Creek and Lookout Creek, the stream is of the third order. A portion of the logs are periodically resampled to assess changes in volume, bark cover, density, and nutrient reserves. Dry mass and volume, as determined by dimensional measures, are used to calculate wood density. Table 2 shows the attributes in the dataset.

For training and testing, different proportions of the dataset were employed. The decay class and wood density of the relevant species are in the training dataset. Also, the wood density threshold value is present in the training dataset. The test data includes information on four species, including circumference, tree's age, volume, dry weight, and moisture. A total of 54,000 instances with 21 attributes are available in the test dataset.

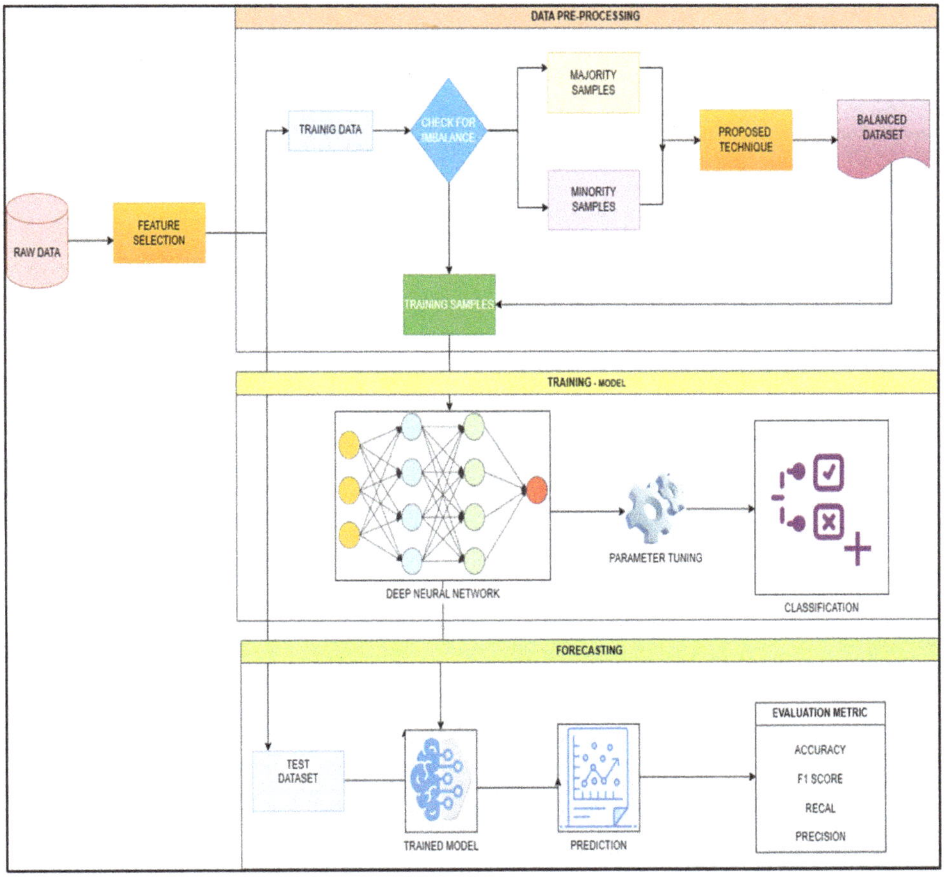

Figure 1. Stratified sampling-based deep neural network (SSDNN) approach for predicting decay class of forest trees.

3.2. *Preprocessing the Dataset*

The dataset is preprocessed before the technique is applied. In this forest dataset, the data distribution is uneven among the live and decayed trees. A tree may belong to a not-yet-rotted or a decayed tree group. Out of the 11,387 trees in the dataset, 9132 belong to the not-yet-rotted group, whereas only 2255 trees belong to the decayed trees group. The data can be either overfit or underfit. This kind of uneven data distribution will have a critical impact on the problem of prediction and categorization, so the data need to be preprocessed.

The preprocessing stage consists of feature selection and checking the skewness of the data. This process will help to reduce the time consumption in handling the unbalanced forest dataset.

3.2.1. Feature Selection Method

The dataset is preprocessed with the feature selection approach back elimination for identifying the optimal subset attributes for forecasting the tree's wood density (Kusy and Zajdel, 2021). Six of the twenty-one features that are essential for prediction were chosen via the wrapper method–back elimination.

Table 2. Dataset specifications.

Attributes	Description
Log num	Log number
Species	Four categories of trees in this region
Time	The tree's age in years
Year	Year of the tree
Subtype	Hard, soft, and other tree types
Rad pos	The location of the measurement
D1	Tree circumference
D2	Tree's circumference in various positions
D3	Tree's circumference in various positions
D4	Tree's circumference in various positions
VOL1	Tree's volume
VOL2	Tree's volume
Wet Wt	Weight of the water content in the tree
DRYWT	The dried weight of the tree
MOIST	Wood's moisture content
Decay	The tree's level of decay
WDENSITY	The tree's wood density with respect to vol1
Den2	The tree's wood density with respect to Vol2
Knot Vol	The wood's volume at a knot
Sample Date	Sample collected date
Comments	Other features of the tree

The model is iteratively trained on several subsets of features using the wrapping technique, and the best subset of features is chosen. The choice of the feature subset selection is based on the inferences from the model. A feature selection strategy called backward elimination starts with a model that incorporates all the available features and gradually eliminates the least significant ones until a stopping requirement is met. This strategy, also known as a wrapper, is typically combined with statistical models to choose a subset of important features. By repeatedly removing elements that are the least significant based on the selected significance level, backward elimination assists in identifying the most pertinent characteristics. Table 3 shows the extracted features using feature selection methods for further processing. Before assessing the feature subsets, these strategies train and test the model using a variety of feature combinations. The strategy reduces overfitting and eliminates pointless or unnecessary features to enhance the model's performance and interpretability.

Table 3. Reduced attributes after preprocessing the dataset.

Attributes	Description
Species	Four categories of trees in this region
Year	Tree's age
D1	Tree's circumference
VOL1	Tree's volume
DRYWT	The dried weight of the tree
WDENSITY	Tree's wood density based on vol1

In the experimental dataset, the explanatory variables Species, Diameter, Volume, Wet Weight, Dry Weight, and Decay are considered for multiple linear regression, and the target variable is Wood density W_i of the tree.

The prediction equation is given below.

$$W_i = \beta_0 + \beta_1 Species + \beta_2 Diameter + \beta_3 Volume + \beta_4 Wetwt + \beta_5 Drywt + \beta_6 Decay + \in$$

where, for n observations

W_i is the dependent variable, and Species, Diameter, Volume, Wetwt, DryWt, and Decay are the explanatory variables,

β_0 is the y-intercept (constant term)

β_j are the slope coefficients for each explanatory variable (j indicates attribute index)

$\in$ is the model's error term (also known as the residuals)

3.2.2. Checking the Skewness of the Data

Classifiers are built up in machine learning to eliminate misclassification errors and, as a result, optimize predictive accuracy. The class imbalance problem, which refers to an uneven distribution of response variable values, is one of the most prevalent issues that influence raw data.

An unbalanced dataset is one in which the number of samples in different classes is highly uneven, making classification difficult. With uneven data, modern machine learning techniques struggle because they focus on reducing error rates serving the dominant class while disregarding the underrepresented group. Classification becomes extremely difficult because the results may be skewed by dominant class values.

As per the experimental dataset, a tree may belong to any one of the five different decay levels ranging from 1 to 5. If a tree belongs to class 1, it means it is not yet decayed; otherwise, it has a decaying component. Since our aim is to classify trees, we considered only two classes, namely "Not yet Decayed" trees and "Decayed" trees. The dataset is considered for the experimental study of the class imbalance problem. As mentioned earlier, there are possibilities of overfit or underfit.

The class details are given below.

Class 0: 9132 (Not yet Decayed)

Class 1: 2255 (Decayed Trees)

The class imbalance problem in the experimental dataset is depicted in Figure 2.

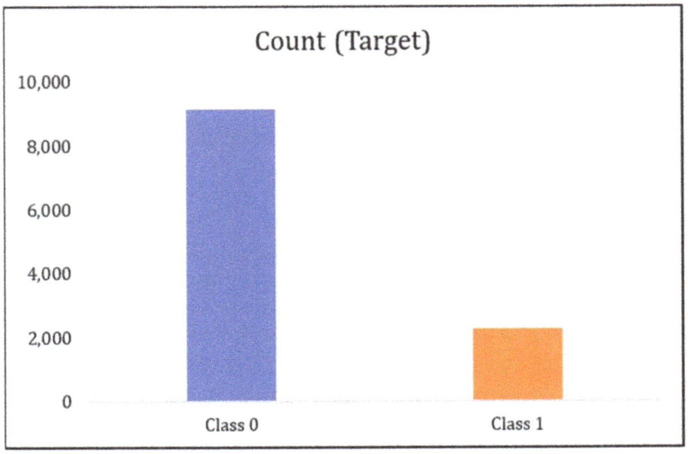

Figure 2. Depiction of class imbalance problem in the experimental dataset.

3.3. Stratified Sampling-Based Deep Neural Network (SSDNN) Approach

The process of classifying unbalanced datasets involves three main steps: feature selection, fitting the data distribution, and model training. Feature selection helps to identify the most suitable subsets of features for classifying unbalanced data while considering the class imbalance among the features. Various resampling approaches that minimize the impact of class inequality on the classifier can be used to fit the data distribution.

The most common resampling strategies are oversampling and undersampling. These strategies aim to balance the datasets by increasing or decreasing the sample points in the minority and majority classes, respectively. However, oversampling algorithms may generate duplicate information and increase the training time of the classifier, while undersampling may result in the loss of the majority of class information.

Both random oversampling (ROS) and random undersampling (RUS) violate the law governing data distribution. The generated samples might not be helpful in illustrating the distribution. SMOTE has drawbacks like supersampling the noisy samples and uninformative data. It is highly challenging to determine the closest neighbors of anonymous synthesized samples. Also, the SMOTE samples are always contained within the samples, and pruning them will lead to an increase in misclassification rate.

We proposed stratified random sampling method to resolve said issue, which will perform the task of test input selection for DNNs. According to the sampling theory, stratified random sampling involves dividing a population into smaller groups without any duplication and avoiding records. The proposed method increases the computation efficiency in the reliability evaluation of the model.

The stratified sampling approach divides the data into blocks based on specified values to extract the structural facts of the data and then draws samples at random from these distinct data blocks. Stratification makes it simple to find representative samples. In the case of forest datasets, stratified sampling can be applied to guarantee that the number of samples for each class is balanced and that the variance of the data within each class is considered when choosing the optimum number of samples. This helps to preserve the original data structure feature information while also ensuring a balance in the number of samples for the majority and minority classes. The specific procedure is to randomly select some examples from both positive and negative occurrences and then combine the training samples for classification. Stratified sampling is best suited for the uneven distribution of data, and it is applied to different domains [25,27–29].

The diversified dataset N is split into similar groups, $S0$, $S1$, and so forth. For data selection, Sn, also known as strata, utilizes uniform random or systematic sampling in each stratum. The reduction in estimation error is the primary advantage of stratified sampling over other sampling techniques. Within strata, a sample for data analysis is taken via random sampling after relative homogenous data objects are grouped together based on the necessary parameters.

The Stratified sampling-based deep neural networks approach is shown below in Figure 3.

Deep learning is a feed-forward neural network with one or more hidden layers. Deep learning is a subfield of machine learning that emphasizes the use of numerous linked layers to transform input into features and predict associated outputs. Artificial neural networks are the core of it. Input, output, and numerous hidden layers are all present in deep neural networks (DNNs). The hidden layer is in the middle, after the input layer and preceding the output layer. In the training of a deep neural network, the following steps are taken: first, initialization is performed according to requirements, and the structure of the DNN is set; second, the layer is then communicated between layers to obtain an error using forward; and finally, the layer is transferred between layers to obtain an error using forward.

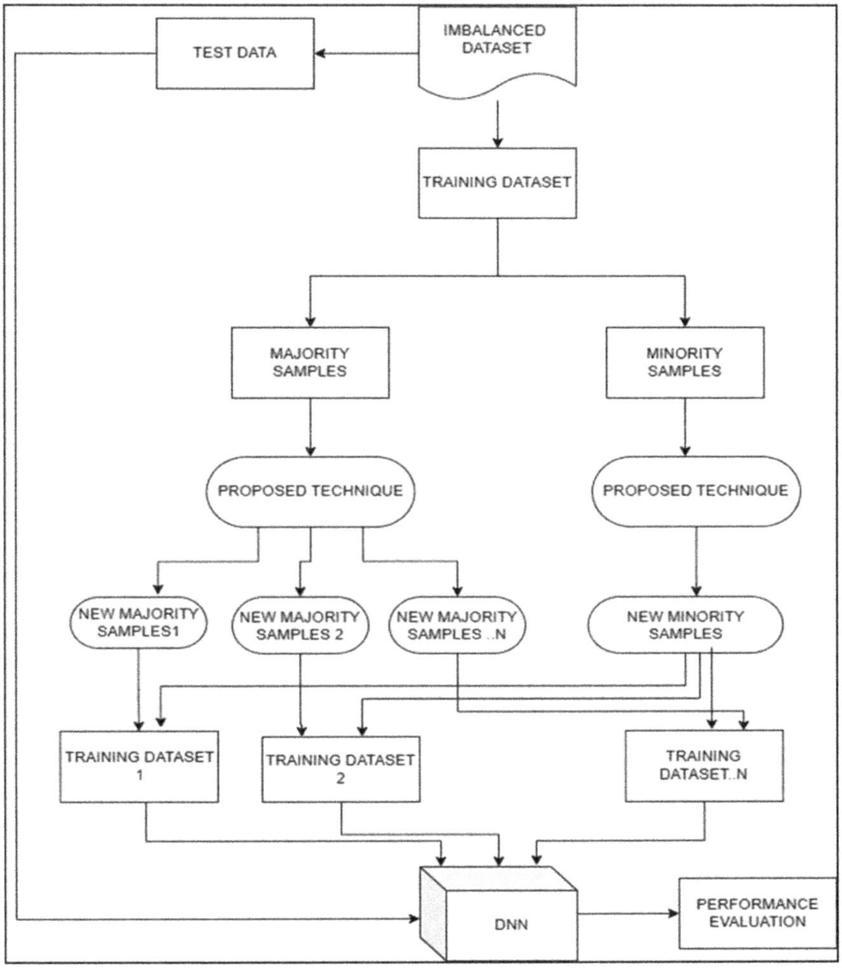

Figure 3. SSDNN model.

DNNs can handle both linear and nonlinear issues by monitoring the probability of each output layer by layer with an appropriate activation function. In essence, DNNs are fully linked neural networks. A deep neural network is sometimes known as a multi-layer perceptron (MLP). The hidden layer alters the input feature vectors, which eventually arrive at the output layer, where the binary classification result is obtained.

Environments have been interested in determining functional links between carbon storage and plant uncertainty of wood density; an appropriate technique is required. Developing empirical models to forecast the DECAY CLASS of the tree is the focus of this research. A deep neural network, a subset of expert systems, predicts the DECAY CLASS of the tree more accurately than standard models. Because there was no constraint for constructing models in DNN, the outcomes are more accurate predictions than the ensemble model. The data loss was achieved using the training data, as shown in the topology of the model, implying that there was no overfitting.

The suggested work's learning model has four layers: one input layer, two hidden layers, and one output layer is shown in Figure 4. At the last three layers, the ReLu activation function was utilized, and the sigmoid function was used at the output layer. The binary cross-entropy loss between the input was used to establish the objective function, which should be minimized in the NN. Adam's optimization was chosen above other existing optimization techniques because it was more efficient. To create a model, each dataset was first randomly divided into two parts: a 75% training set and a 25% test set.

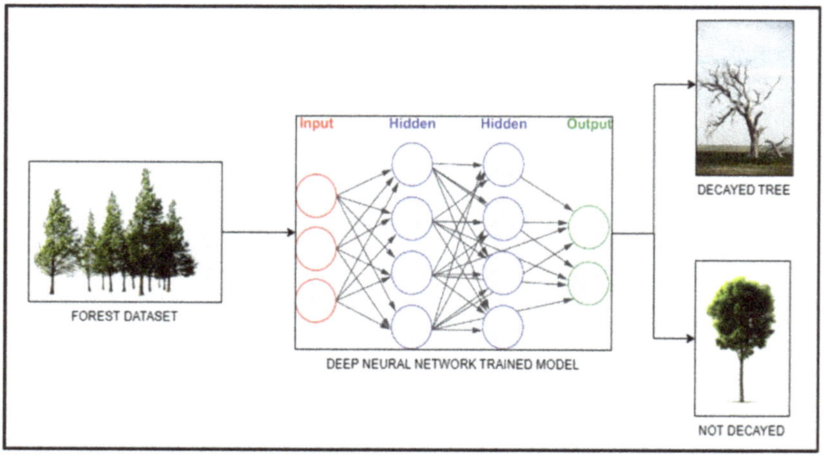

Figure 4. Prediction of tree decay class using DNN model.

The training set is examined for skewness and, if necessary, balanced using a stratified sampling procedure. The balanced training set is then used to develop DNN models and train them, while the test sets are utilized to evaluate the performance of the predictive models. We used the following easy method to choose the best threshold. The curve of balanced accuracy as a function of prediction is first plotted. The best threshold was finally determined to be where the DNN achieved the most balanced accuracy. The unbalanced learn library from Python was then used to apply each data-balancing technique to each training batch. The model has been tried with different numbers of mini-batches as 10, 50, 25, and 100 and determined 100 as the best choice with epoch sizes as 10, 25, and 50.

3.3.1. Algorithm for SSDNN Model

The algorithm for the proposed SSDNN model is given in Algorithm 1. This proposed SSDNN model will first extract the features required for the job and verify whether the ratio of the dataset is unbalanced or not. Next, it will choose the right samples for prediction. Below is a representation of the suggested model algorithm.

Algorithm 1. Proposed Algorithm for SSDNN Model

1. Import the dataset
2. Perform the Wrapper method (Back Elimination Method)
3. Check the Skewness of the dataset
4. Apply Stratified
5. Update the imbalanced dataset
6. Load the training dataset
7. Train the DNN
8. Shuffle and Split as 75% and 25%
9. Use SVM-Kernel for classification
10. Tune the Parameters
11. Apply to the test dataset
12. End

3.3.2. DNN with Hyperparameter Tuning

Deep neural network hyperparameter tuning employs a random search to identify the ideal hyperparameter combination from a set of hyper parameter values. Random search resulted in a set of 20 hyperparameter combinations. The following are the best hyperparameters found via random search.

Finally, the model is hyper-tuned using a random search approach, where the optimum parameters are 2 hidden layers, 400 neurons, ReLu, 50 epochs, and a batch size of 100 as listed in Table 4.

Table 4. DNN hyperparameters.

Hyperparameter	**Value/Type**
Hidden Layers	2
Neurons	400
Optimizer	Adam
Hidden Layer	ReLu
Output Layer	Softmax
Epochs	10, 25, 50
Batch size	100

When the number of epochs increases, the accuracy of the proposed method also increases, and we obtain maximum accuracy when the epoch is closer to 100. The built model is compared with the existing models, and the performance is analyzed in the results and discussion section.

4. Experiment Results and Discussion

To recognize dead and live trees, we used the forest tree dataset to perform our classification. The dataset was preprocessed to determine the relevance of the variables for categorization. The dataset was split into two parts: training and testing. We used a training dataset to train the DNN and a test set to evaluate classifier performance. We conducted a huge number of trials to discover the ideal DNN design and parameters, using various combinations of batch sizes, number of hidden units, and learning rate.

Because of the imbalanced dataset, DNN accuracy is good, but other performance metrics like F1Score, Precision, and Recall value are low. As a result, the dataset is balanced via stratified sampling, and the resulting strata are supplied to DNN as a training set. The result of the proposed model is compared with the previous model SVM, Naïve Bayes, and Random Forest. Earlier, we tried to perform the classification using these three models with different datasets. Each model has its own credits and pitfalls. For the smaller datasets, SVM produces better results but is not promising for larger ones. At the same time, Random Forest is one of the best choices for larger datasets but is time consuming. Naïve Bayes is simple, and it is not preferred for large datasets. It assumes that the variables are independent.

It is evident from the results that the proposed model gives high accuracy in addition to performing well in the case of large datasets. The proposed approach is written in Python using Jupyter Notebook and uses the Keras package on a 64-bit OS with an X64 CPU, and the model worked well on the Google Lab platform. Thus, by combining DNN with a stratified sampling-based deep learning model, the prediction and classification of dead trees in the forest are successfully completed. Forest managers will be able to predict the early stages of decaying trees with this information. The proposed method can also be applied to similar datasets belonging to different domains.

4.1. Performance Metrics

The efficiency of the proposed method is analyzed using classification accuracy, Precision, Recall, and F1 Score. The performance of the three approaches, namely SVM, Random Forest, and Naïve Bayes, with different sampling techniques, are depicted in the following figure.

4.2. Results and Discussion

Table 5 shows the comparison of test accuracy among the proposed DNN models with sampling methods. The performance in terms of accuracy of the existing and proposed algorithms along with different sampling techniques are shown in Figure 5.

Table 5. Comparison between proposed and existing methods.

Methods	Accuracy	Precision	Recall	F1 Score
DNN	80	0.7	0.75	0.72
DNN+OVERSAMPLING	76	0.75	0.75	0.75
DNN+UNDERSAMPLING	69	0.74	0.69	0.71
DNN+SMOTE	78	0.76	0.78	0.77
DNN+STRATIFIED	91	0.88	0.87	0.87

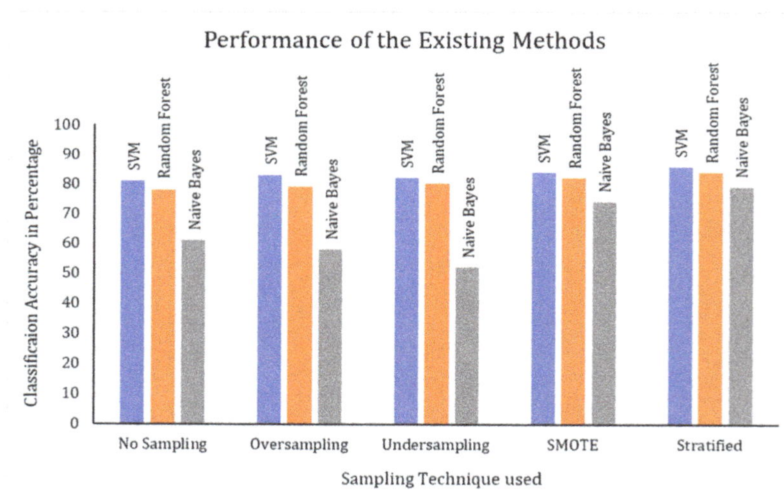

Figure 5. Performance of the existing approaches with different sampling methods.

The performance of the proposed SSDNN method with different existing sampling techniques is shown in Figure 6.

The DNN, DNN+ oversampling, DNN+ undersampling, DNN+ SMOTE, and DNN+ stratified sampling yields test accuracy of 80%, 76%, 69%, 78%, and 91%, respectively. First, the DNN model was created and tested on the prepared dataset, yielding low accuracy. The DNN model was analyzed for the reason of yielding low accuracy, and it was found that the dataset was unbalanced. The imbalanced dataset was subsequently handled using a stratified sampling technique, which divided the training dataset into groups of distinct strata for each class. The data from each stratum was distributed uniformly to the deep neural network, resulting in good accuracy, precision, recall, and F1 score. Several tests using the tree dataset were carried out to determine the optimal deep neural network.

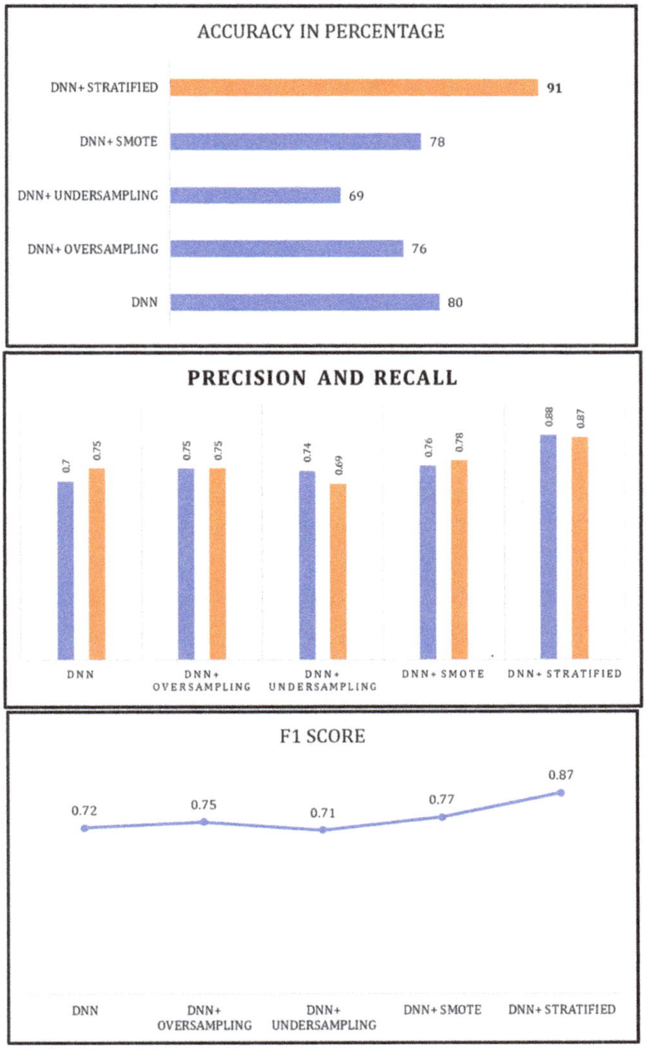

Figure 6. Comparison between existing methods with proposed SSDNN.

The training, as well as testing accuracy and loss of the proposed SSDNN, is visualized in Figure 6. From the figure, during the initial epochs, accuracy is not appreciable, and at the same time, the loss is highly noticeable. But in the subsequent epochs, the results are more promising. Similarly, the same parameters are analyzed for the testing phase. The testing phase also has the same impact on model accuracy and model loss. To observe the variations more clearly, the chart is prepared up to 25 epochs.

Also, the training/testing accuracy and loss of the proposed method is shown in Figure 7. The proposed DNN + stratified sampling results in an accuracy of 91% with higher efficiency. The proposed model was compared to the ensemble SVM kernel algorithm used in prior work, and the results show that the proposed DNN + stratified model is more efficient. The proposed method is robust compared to the traditional methods due to hyper-tuning, low false positive, and high recall.

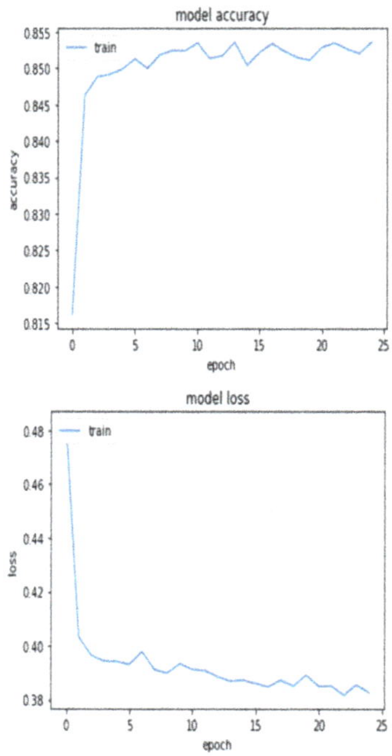

(a) Training accuracy and training loss of SSDNN

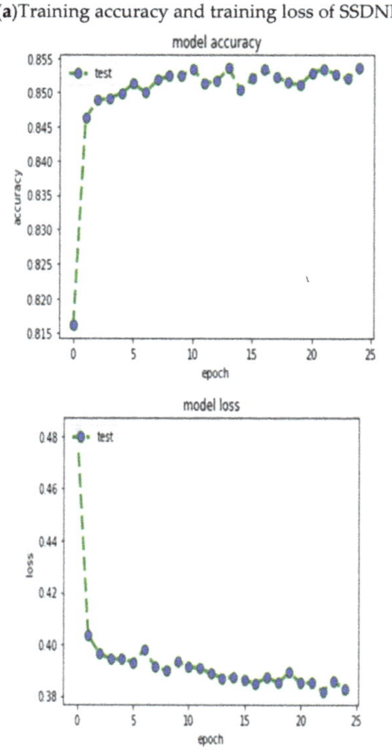

(b) Testing accuracy and testing loss of SSDNN

Figure 7. Performance in terms of training/testing accuracy, as well as loss of the proposed SSDNN.

5. Conclusions

In this research, we experimented to find the best model to classify the forest tree as a dead or live tree. For predicting the decay class of a tree, the classification models DNN, DNN+ oversampling, DNN+ undersampling, DNN+ SMOTE, and DNN+ stratified sampling were applied to the dataset. The results show that DNN+ stratified sampling offers better performance with high accuracy.

The proposed method correctly classifies a tree as either dead or alive compared to other models. The proposed model will be suitable to handle any imbalanced dataset for classification. In deep learning, classification accuracy often increases when the amount of data used for training increases; thus, using a larger dataset for training can be a good research direction to continue improving our classification accuracy of forest tree classification. This paper suggests that identifying decaying trees earlier will help forest managers in removing them before they begin to emit carbon back into the atmosphere.

This research promotes reforestation by planting a new tree after removing a dead tree to reduce pollution and forest fires. In the case of stratified sampling, the research gap discovered is that the number of records in both classes is not equal; hence, deficit records occur when training the model. To address this issue, the deficit class is oversampled, strata are shuffled, and the model is trained to increase model efficiency. In future work, the proposed method can be applied to smart forest management. Since there may be uneven data or irrelevant data during data collection, we can use IOT-based RFID for each tree to automate data collection for the tree and also to indicate its level of decay and carbon absorption.

Author Contributions: Conceptualization, P.A.; methodology, J.S. and P.A.; validation, A.K.S. and C.Z.; writing—original draft preparation, P.A. and J.S.; writing—review and editing, A.K.S. and C.Z. All authors have read and agreed to the published version of the manuscript.

Funding: This work was supported by the Rashtriya Uchchatar Shiksha Abhiyan (RUSA) Phase 2.0 [grant sanctioned vide Letter No.F.24-51/2014-U, Policy (TNMulti-Gen), Department of Education, Government of India, Date 9 October 2018].

Data Availability Statement: https://andrewsforest.oregonstate.edu/data (accessed on 11 September 2023).

Conflicts of Interest: The authors declare that they have no conflict of interest.

References

1. Briechle, S.; Krzystek, P.; Vosselman, G. Silvi-Net—A dual-CNN approach for combined classification of tree species and standing dead trees from remote sensing data. *Int. J. Appl. Earth Obs. Geoinf.* **2021**, *98*, 102292. [CrossRef]
2. Karatas, G.; Demir, O.; Sahingoz, O.K. Increasing the performance of machine learning-based IDSs on an imbalanced and up-to-date dataset. *IEEE Access* **2020**, *8*, 32150–32162. [CrossRef]
3. Cao, L.; Shen, H. CSS: Handling imbalanced data by improved clustering with stratified sampling. *Concurr. Comput. Pr. Exp.* **2020**, *34*, e6071. [CrossRef]
4. Li, K.; Chen, X.; Zhang, R.; Pickwell-MacPherson, E. Classification for Glucose and Lactose Terahertz Spectrums Based on SVM and DNN Methods. *IEEE Trans. Terahertz Sci. Technol.* **2020**, *10*, 617–623. [CrossRef]
5. Mînăstireanu, E.-A.; Meșniță, G. Methods of Handling Unbalanced Datasets in Credit Card Fraud Detection. *BRAIN. Broad Res. Artif. Intell. Neurosci.* **2020**, *11*, 131–143. [CrossRef]
6. Shoohi, L.M.; Saud, J.H. DCGAN for Handling Imbalanced Malaria Dataset based on Over-Sampling Technique and using CNN. *Medico-Legal Update* **2020**, *20*, 1079–1085.
7. Sheikh, T.S.; Khan, A.; Fahim, M.; Ahmad, M. Synthesizing data using variational autoencoders for handling class imbalanced deep learning. In Proceedings of the International Conference on Analysis of Images, Social Networks and Texts, Kazan, Russia, 17–19 July 2019; pp. 270–281.
8. Elreedy, D.; Atiya, A.F. A Comprehensive Analysis of Synthetic Minority Oversampling Technique (SMOTE) for handling class imbalance. *Inf. Sci.* **2019**, *505*, 32–64. [CrossRef]
9. Oberle, B.; Ogle, K.; Zanne, A.E.; Woodall, C.W. When a tree falls: Controls on wood decay predict standing dead tree fall and new risks in changing forests. *PLoS ONE* **2018**, *13*, e0196712. [CrossRef]

10. Tallo, T.E.; Musdholifah, A. The Implementation of Genetic Algorithm in Smote (Synthetic Minority Oversampling Technique) for Handling Imbalanced Dataset Problem. In Proceedings of the 2018 4th International Conference on Science and Technology (ICST), Yogyakarta, Indonesia, 7–8 August 2018; pp. 1–4. [CrossRef]
11. Moayedikia, A.; Ong, K.-L.; Boo, Y.L.; Yeoh, W.G.; Jensen, R. Feature selection for high dimensional imbalanced class data using harmony search. *Eng. Appl. Artif. Intell.* **2017**, *57*, 38–49. [CrossRef]
12. Maldonado, S.; López, J. Dealing with high-dimensional class-imbalanced datasets: Embedded feature selection for SVM classification. *Appl. Soft Comput.* **2018**, *67*, 94–105. [CrossRef]
13. Maldonado, S.; Weber, R.; Famili, F. Feature selection for high-dimensional class-imbalanced data sets using Support Vector Machines. *Inf. Sci.* **2014**, *286*, 228–246. [CrossRef]
14. Ng, W.W.; Hu, J.; Yeung, D.S.; Yin, S.; Roli, F. Diversified sensitivity-based under-sampling for imbalance classification problems. *IEEE Trans. Cybern.* **2014**, *45*, 2402–2412. [CrossRef] [PubMed]
15. Sáez, J.A.; Krawczyk, B.; Woźniak, M. Analyzing the oversampling of different classes and types of examples in multi-class imbalanced datasets. *Pattern Recogn.* **2016**, *57*, 164–178. [CrossRef]
16. Gónzalez, S.; García, S.; Lázaro, M.; Figueiras-Vidal, A.R.; Herrera, F. Class Switching according to Nearest Enemy Distance for learning from highly imbalanced data-sets. *Pattern Recognit.* **2017**, *70*, 12–24. [CrossRef]
17. Cao, L.; Shen, H. Imbalanced data classification using improved clustering algorithm and under-sampling method. In Proceedings of the 20th International Conference on Parallel and Distributed Computing, Applications and Technologies, Gold Coast, Australia, 5–7 December 2019.
18. Cheng, F.; Zhang, J.; Wen, C.; Liu, Z.; Li, Z. Large cost-sensitive margin distribution machine for imbalanced data classification. *Neurocomputing* **2016**, *224*, 45–57. [CrossRef]
19. Cao, C.; Wang, Z. IMCStacking: Cost-sensitive stacking learning with feature inverse mapping for imbalanced problems. *Knowl.-Based Syst.* **2018**, *150*, 27–37. [CrossRef]
20. Ohsaki, M.; Wang, P.; Matsuda, K.; Katagiri, S.; Watanabe, H.; Ralescu, A. Confusion-Matrix-Based Kernel Logistic Regression for Imbalanced Data Classification. *IEEE Trans. Knowl. Data Eng.* **2017**, *29*, 1806–1819. [CrossRef]
21. Sun, Z.; Song, Q.; Zhu, X.; Sun, H.; Xu, B.; Zhou, Y. A novel ensemble method for classifying imbalanced data. *Pattern Recognit.* **2015**, *48*, 1623–1637. [CrossRef]
22. Feng, W.; Huang, W.; Ren, J. Class Imbalance Ensemble Learning Based on the Margin Theory. *Appl. Sci.* **2018**, *8*, 815. [CrossRef]
23. Chen, Z.; Lin, T.; Xia, X.; Xu, H.; Ding, S. A synthetic neighborhood generation based ensemble learning for the imbalanced data classification. *Appl. Intell.* **2018**, *48*, 2441–2457. [CrossRef]
24. Japkowicz, N. The class imbalance problem: Significance and strategies. In Proceedings of the 2000 International Conference on Artificial Intelligence (IC-AI'2000), Las Vegas, NV, USA, 26–29 June 2000.
25. Zhao, X.; Liang, J.; Dang, C. A stratified sampling based clustering algorithm for large-scale data. *Knowl.-Based Syst.* **2019**, *163*, 416–428. [CrossRef]
26. Available online: https://www.nal.usda.gov/data/find-data-repository (accessed on 10 October 2023).
27. Wang, W.; Zhao, Y.; Zhang, T.; Wang, R.; Wei, Z.; Sun, Q.; Wu, J. Regional soil thickness mapping based on stratified sampling of optimally selected covariates. *Geoderma* **2021**, *400*, 115092. [CrossRef]
28. Alogogianni, E.; Virvou, M. Handling Class Imbalance and Class Overlap in Machine Learning Applications for Undeclared Work Prediction. *Electronics* **2023**, *12*, 913. [CrossRef]
29. Wu, Z.; Wang, Z.; Chen, J.; You, H.; Yan, M.; Wang, L. Stratified random sampling for neural network test input selection. *Inf. Softw. Technol.* **2023**, *165*, 107331. [CrossRef]

electronics

MDPI

Article

Comparison of Selected Machine Learning Algorithms in the Analysis of Mental Health Indicators

Adrian Bieliński, Izabela Rojek * and Dariusz Mikołajewski

Faculty of Computer Science, Kazimierz Wielki University, 85-064 Bydgoszcz, Poland; adrian.bielinski.pl@gmail.com (A.B.); dariusz.mikolajewski@ukw.edu.pl (D.M.)
* Correspondence: izabela.rojek@ukw.edu.pl

Abstract: Machine learning is increasingly being used to solve clinical problems in diagnosis, therapy and care. Aim: the main aim of the study was to investigate how the selected machine learning algorithms deal with the problem of determining a virtual mental health index. Material and Methods: a number of machine learning models based on Stochastic Dual Coordinate Ascent, limited-memory Broyden–Fletcher–Goldfarb–Shanno, Online Gradient Descent, etc., were built based on a clinical dataset and compared based on criteria in the form of learning time, running time during use and regression accuracy. Results: the algorithm with the highest accuracy was Stochastic Dual Coordinate Ascent, but although its performance was high, it had significantly longer training and prediction times. The fastest algorithm looking at learning and prediction time, but slightly less accurate, was the limited-memory Broyden–Fletcher–Goldfarb–Shanno. The same data set was also analyzed automatically using ML.NET. Findings from the study can be used to build larger systems that automate early mental health diagnosis and help differentiate the use of individual algorithms depending on the purpose of the system.

Keywords: computer science; artificial intelligence; machine learning; burnout; clinical reasoning

1. Introduction

The development of machine learning (ML) is driven by the vast amount of data available (so-called big data), which are used to train algorithms to adapt them to solve scientific, clinical and industrial problems quickly and efficiently [1,2]. ML is a data-driven approach in which rules are extracted automatically based on associations between input and output data sets, and their relevance is tested against validation data. Models learned in this way (mainly traditional and deep artificial neural networks) can then be trained to better fit new data. Machine learning is increasingly being used to solve clinical problems in diagnosis, therapy and care [3–5]. The number of publications on clinical applications of machine learning increased rapidly after 2010, with the main areas of research being in diagnostics and prediction, and less often in classical clinical problem solving (Figure 1a–d).

In recent years, there has been a growing interest in the application of ML in the diagnosis (less frequently: therapy) of mental health (Figure 1e) [6,7]. This is due to a number of factors, but above all to the fact that this group of conditions is becoming common as a new group of diseases of civilization in adults, children and adolescents, while at the same time representing very complex and stigmatizing disease entities that are difficult to combat with limited resources and numbers of specialists. Automation of certain procedures is therefore possible and desirable for both patients and medical staff.

The main aim of the study was to see how the selected ML algorithms deal with the problem of determining a virtual mental health index.

Citation: Bieliński, A.; Rojek, I.; Mikołajewski, D. Comparison of Selected Machine Learning Algorithms in the Analysis of Mental Health Indicators. *Electronics* **2023**, *12*, 4407. https://doi.org/10.3390/electronics12214407

Academic Editors: Wentao Li, Huiyan Zhang, Tao Zhan and Chao Zhang

Received: 12 September 2023
Revised: 21 October 2023
Accepted: 23 October 2023
Published: 25 October 2023

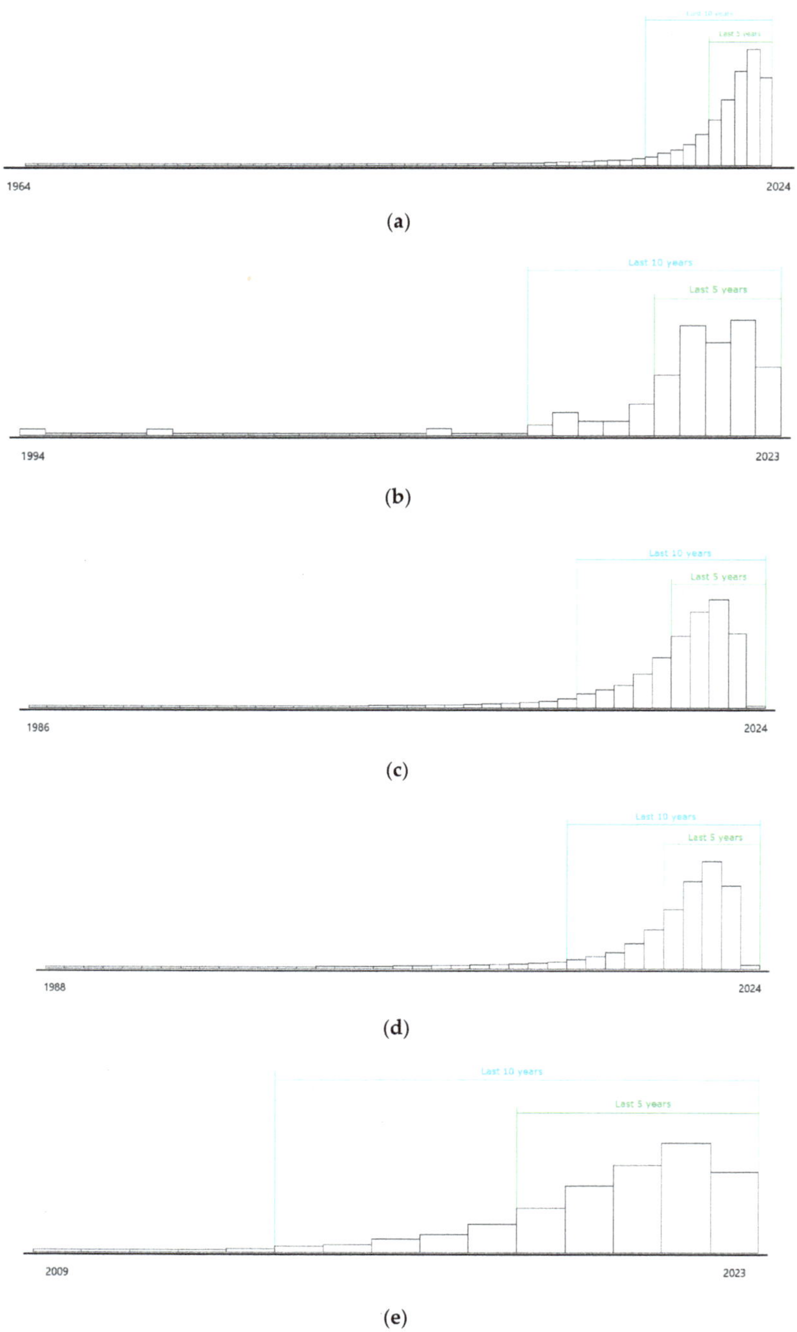

Figure 1. Number of scientific publications: (**a**) concerning clinical applications of machine learning (total number of publications: 103,017), (**b**)with keywords "machine learning" and "clinical problem solving" (total number of publications: 113), (**c**) with keywords "machine learning" and "diagnosis" (total number of publications: 37,242), (**d**) with keywords "machine learning" and "prediction" (total number of publications: 50,619), and (**e**) with keywords "machine learning" and "mental health" (2332).

Related Publications

There are many articles in the literature on the virtual mental health index. Each of them stands out from the others, approaching the topic from a different point of view. One article addresses the topic of e-health and modern technologies used in mental health care [8,9]. It is indicated that the aim of the article is to present issues related to e-health, and its elements used in the diagnosis and treatment of patients with mental disorders. The article points out that there is a lot of enthusiasm for e-health issues around the world, which may be related to the transformation potential of the healthcare system [8,9]. The article points out that e-health solutions have been shown to be effective in preventing, diagnosing and treating patients with a variety of illnesses, both physical and mental [9], including substance abuse, depression, bipolar disorder, anxiety, stress and/or suicidal thoughts. This article adopts the World Health Organisation's (WHO) definition of e-health. In addition, differences between the original and the newer definition are pointed out, as the newer definition describes it as the use of electronic means of communicating health-related information, resources and services, whereas the original definition presented the concept as the use of information technology, locally and remotely, in support of health and related fields. The newer definition according to the WHO also includes electronic health records, mobile health and health analytics. An important change was also indicated in the context of the patient–professional relationship, i.e., the patient participates as a partner in the diagnosis and treatment process, rather than being merely a passive figure. An increase in patients' responsibility for their own treatment, an increase in their involvement in treatment decisions or a tendency to use strengthening and improvement exercises were also noted. It was also mentioned that inviting the patient into the e-health system does not imply patient involvement. The studies mentioned in this article identified three different types of involvement: active, partner and submissive [8,9]. Mobile apps used in practice were also identified, including for practicing stress management skills, in the diagnosis and treatment of depression, and as an aid to screening. The cited authors indicated that apps could be used to monitor mental status and mood, as well as bipolar affective disorder [8,9]. This article presents modern technology as an opportunity for the development of medicine, including in the context of mental health. The article draws on a number of sources, indicating that these are not isolated, exceptional situations. It is noteworthy that it was written before the onset of the problems associated with the COVID-19 pandemic. This article provides an interesting insight into the applications of technology not only in treatment but also in prevention. In contrast, another article [10] deals with the use of ML techniques to predict stress in active workers. As an introduction, the prevalence of mental disorders among the working class was highlighted, with a clear upward trend when looking at the percentage of employees who experience depressive and anxious states. It was concluded that the greatest emphasis must be placed on maintaining a stress-free atmosphere in order to achieve better productivity and well-being of employees. The authors [10] used the results of a survey of technology employees in 2017, with which they trained various models for their analyses. The original data consisted of 750 responses from people from different technical departments in the form of 68 attributes related to private life and work. A data cleaning exercise was carried out, which left 14 parameters, in addition to which a one-hot encoding (1 of n) was used to represent some fields as numeric. In addition, the text responses 'Yes' were given a value of 1, 'No' a value of 0, and 'Maybe' a value of 0.5. NaN values were replaced by 0, and nominal data were converted to numeric using a label encoder. The authors chose models for training that had already been tested in classification problems, implementing them in Python using the Scikit-learn library:

- Logistic regression;
- K-nearest-neighbormethod;
- Decision trees;
- Random forest;
- Boosting (increasing the effectiveness of existing models);
- Bagging.

The following were used as metrics of model performance:

- Classification accuracy;
- False Positive Rate, which indicates how many negative cases were classified as positive;
- Precision, i.e., the fraction of cases predicted to be positive that were actually positive;
- Area Under the Curve (AUC) score;
- Cross-validation AUC score [10].

Each model assessed whether a person required treatment. These tests resulted in model accuracies ranging from 69.43% to 75.13%, with the bagging algorithm achieving the lowest level and the boosting algorithm the highest. The greatest influence on stress and mental health was the gender of the individual, as well as family history and the services provided by the employing entity for mental health care. As further research opportunities, the authors [10] suggested using deep learning (DL) techniques, seeking a broader and more detailed dataset. They also consider the possibility of modifying the questionnaire to make the responses in a suitable format, to increase the number of attributes used, and they suggest the inclusion of questionnaires from organizations such as the WHO (World Health Organization) related to stress and mental health. They also suggest formulating a homogeneous scale to assess stress levels. The article [11] mentions that people with common mental illnesses usually do not seek medical help, which makes attempts to monitor them to create opportunities for early intervention extremely difficult. The documented use of continuous digital monitoring to reach people with common mental illnesses among communities was noted as a strategy with some potential. At the same time, the limitations of monitoring systems based on assessments of mental health at specific points in time, on the basis of self-assessment and control by an expert, have been highlighted. These concern [11]:

- Impact of memory problems;
- Possibility to perform only in limited time windows;
- Possibility to perform only under controlled conditions;
- Frequent requirement for the patient to move to a medical setting in order to receive a diagnosis.

This raises further issues:

- Inability to assess the impact of interaction with the environment in the context of the mental state in real time;
- This undermines progress towards understanding and classifying mental illness and its treatment.

The authors also compare the use of mobile phones, in the context of dedicated solutions and solutions based on already available applications and devices. They point out that the second method has greater potential, as it significantly reduces costs and the risk of behavioral deformities associated with traditional forms of behavioral research [11]. In particular, they highlighted activity-tracking apps and wearable devices, which have received little attention in the context of research [11]. The study involved 53 of 120 recruited Australian volunteers aged between 18 and 25 years. They provided data in the form of a detailed health and lifestyle questionnaire and access to recorded information on activity-tracking apps. The Depression and Anxiety Scale-21 (DASS-21), which examines depression, anxiety and stress, was used to assess mental health. In addition, data on the duration of daily activities were included as a key point of interest. These were determined using data from miniature motion sensors, including location-based accelerometers, which were collected by various connected applications and fed into a cloud-based API, from where they were then stored in a database [11]. Based on the DASS-21, it was found that those monitored had symptoms of depression, anxiety and stress at intermediate levels. In contrast, the apps or devices that were linked to the API for the study were several:

- Fitbit;
- Garmin;
- Healthkit;

- Misfit;
- Moves;
- Myfitnesspal;
- Strava [11].

Based on the data collected, it was discovered that:

- Examined daily activity time received from wearable devices was greater than that derived from the mobile phone app;
- Of the 43 participants from whom at least three daily activity observations were obtained, 11 of them had at least 20% missing data between the first and last observation, but this did not show a relationship with DASS-21 scores;
- For the remaining 32 participants, entropy techniques were used, which initially showed no significant relationship between data and DASS-21 scale scores. It was not until splitting into two equal groups in relation to the amount of data that a significant, positive correlation was detected between the DASS-21 anxiety subscale and entropy in those with more data [11].

The authors [11] point to the lack of standardized systems for continuous mental health monitoring, which, together with continued monitoring in specific time windows, has contributed to the escalation of the problem. They note that people with mental health conditions are generally willing to share information from their mobile phones to help with research into these conditions, including serious illnesses. The authors present their work as a proof of concept for continuous mental health monitoring of mental health, but note the challenges of privacy, assessment and clinical integration and inclusion that would need to be addressed before it is more widely accepted. Another article [12], which deals with the determination of a voice-based mental health indicator using a mind-state observation system, explores the validity of such an approach. It draws attention to the huge cost of mental illness in developed countries and the need for early detection technology for depression and stress. Light is also shed on the current state of screening methods in the context of mental illness, including general health questionnaires (questionnaires including the General Health Questionnaire (GHQ) or the Beck Depression Index (BDI)). The effectiveness of such approaches in assessing disease conditions in the early stages was highlighted, and the problems of reporting bias, i.e., the effect of consciously or unconsciously under- or overestimating a patient's self-report, as well as the problem of reduced detection rates of mental illness in organizations with established hierarchies, were also noted. The authors of [12] report on their active research and work on voice-based mental health estimation. They list additional advantages of this approach:

- Ease of application;
- Possibility to monitor day by day, which conventional methods do not allow.

They have developed a software development toolkit (SDK) called MIMOSYS (https://medical-pst.com/en/products/mimosys/ accessed on 11 September 2023), whose features include:

- Recording a voice from a microphone;
- Analyzing this voice;
- Determining a health indicator based on this.

To enable daily monitoring, the authors developed a mobile app using MI-MOSYS. The aim of the study was to compare the indicator defined in the app with the BDI indicator. The study was carried out with the support of the local authority, which provided mobile phones with the mobile app installed for 50 company employees. The test participants had to record their voices by reading out ready-made phrases and talking using the device they were given. In addition, a BDI test was conducted at the beginning of the experiment. The voice analysis was based on the fact that people with mental illness show changes in the expression of emotions and changes in the proportions of the components of the voice. The four components hidden in the voice—anger, sadness, joy and calmness—were calculated from the characteristics of the recorded voice. In addition, the degree of excitement of the respondent was determined. Taking these values into account, a short-term and a medium-term index of psychological well-being was determined, the latter based on short-term indices collected over a two-week period. As a result of the experiment, the correlation value was determined to be negative, with a value of 0.208 for the short-term value and 0.285 for the medium-term value. A lower correlation coefficient value was obtained for telephone calls, below 0.2 [12]. For the optimal cut-off, the following values of sensitivity, specificity and accuracy were obtained when analyzing the ROC curve:

- 0.795; 0.643; 0.660 for the short-term indicator;
- 1.000; 0.605; 0.646 for the medium-term indicator [12].

In the context of this research, the weak negative correlation between the indices from the app and the BDI was understandable, as a lower mental health index was associated with a higher rate of depression. Finally, the performance of the method in distinguishing between individuals with a high BDI was shown to confirm the appropriateness of the method. The efficiency of data accumulation was also noted, and furthermore, the results indicated that such a system could complement routine screening. However, the authors have set their sights on the commercialization of the product, as they do not disclose details in the form of the algorithms used or the scheme of operation of the system. Furthermore, it is not possible to download this toolkit without first contacting them via a form, which presumably means that it is made available for a fee. In addition, the library (Sensibility Technology) underpinning this software is also unavailable.

In [13], mental health before and during the COVID-19 pandemic was compared using a large probability sample from the UK population. The coronavirus and the methods used to slow its spread had a serious impact on people's livelihoods, incomes and debts, and was associated with serious concerns about an uncertain future. The authors of this publication [13] drew attention to the limited research on mental health during the pandemic, due to problems such as:

- Use of incomplete samples;
- Use of unverified or modified assessment tools;
- Lack of comparable pre-pandemic data to measure change.

Their study [13] was based on a large-scale survey conducted since 2009, including people aged 16 years and older. In addition, invitations to participate in the COVID-19 online survey were sent to participants in the last two series of surveys via emails, text messages and even letters. The pre-pandemic health assessment was based on data collected since 2014, and the data included results from the GHQ-12 questionnaire (a valid tool for assessing general mental health problems in the past two weeks, particularly effective in large-scale surveys). This scale was scored in two ways, the first based on a mean value and the second based on a binary threshold above which individuals were judged to have a significant level of mental health problems. The rating scale of this questionnaire for each question ranged from 0 to 3 (from no deviation to significant deviation). The authors [13] also carried out analyses by gender, age ranges, geographical location, or looking at the data from an ethnic perspective. Estimates of total annual income, employment status, living with a partner, age of the youngest child in the family were also analyzed, and a group of people at risk and those involved in COVID-19 was identified. Years with a small number of observations were excluded from the study, which may have led to less accurate estimates. Changes in mental health were also assessed using regression [13]. These models only included people for whom data from both the COVID-19 survey and at least one pre-pandemic data set were available, therefore 16- and 17-year-olds were excluded from this section. The value of the GHQ-12 index was constructed during the pandemic and placed in a time-variable model where average scores were used as the baseline, instead of using a binary index, as this would affect the statistical power of the results and their generalization. The final model included the following factors:

- Age;
- Sex;

- Family income;
- Employment status;
- Living with a partner;
- Presence of risk factors [13].

Various patterns related to variables have been detected, including [13]:

- Higher GHQ-12 scores in women;
- Higher scores in younger age groups;
- Slight differences in ethnicity (apart from the difference between Asians and white British—Asians scored higher);
- Slightly lower results were recorded outside cities;
- Higher scores in low-income families;
- Unemployed and professionally inactive people scored higher than employed and retired people;
- People without a partner and with young children had higher scores, as did the risk groups;
- Significant increase in average scores was noticed comparing the state before and during the pandemic [13].

The authors present their publication as one of the first in their country to measure the impact of the pandemic on the mental health of the population. The increase in mental health problems was not even among the designated groups. However, towards the end, they conclude that the increase was not significant, but point out the need for further studies spread over time, even postponed by half a year. They note that although GHQ-12 is a screening tool, it is not a clinical diagnosis. In the publication [14], it was mentioned that in the coming years, a radical change will be needed, consisting of attaching the patient's mental health profile to provide him with better treatment and help him recover faster. It was also noted that there has already been discussion about how medical predictive analytics could revolutionize healthcare globally. Factors affecting mental health include:

- Globalization;
- Pressures in the workplace;
- Competition [14].

The authors of [14] claim that the K-nearest neighbor's method, the naive Bayes classifier, or regression can be used to build the model. In their approach to identifying mental health, they used classification and clustering algorithms. They note the need for early diagnosis of deviations in mental health. The WHO report urged the nations of the world to harness the power of knowledge and technology to tackle mental health. They list some of the mental health assessment tools:

- Questionnaires;
- Sensors of wearable devices;
- Biological signals [14].

They also mention work on statistical relationships between mental health and other parameters, including:

- Educational achievements;
- Socioeconomic achievements;
- Satisfaction with life;
- Quality of interpersonal relations;

They also list various assessment methods [14] appearing in other works:

- Regression analysis;
- K-nearest neighbors method;
- Decision trees;
- Support vector method;
- Fuzzy logic;
- K-means method [14].

In their work [14], they started the analysis with clustering in order to better understand the data—obtaining certain groups, however, without any interpretation. They list and describe commonly used clustering methods:

- K-means;
- Hierarchical;
- Based on density;
- And their variants [14].

In addition, they presented frequently used indicators for validating clustering and applied the concept of the Mean Opinion Score (MOS) scale, used for subjective quality assessment. Their questionnaire consisted of 20 questions, posed to two populations: the first included 300 people aged 18 to 21, and the second 356 people aged 22 to 26. The rating scale for each question was five-point, from 1 (almost never) to 5 (almost always). The division into a set of training and test data was in the ratio of 80:20. In terms of validity, the best of all models were: bagging and random forest (0.90), slightly worse support vectors and K-nearest neighbors (0.89), and even worse logistic regression (0.84) and decision tree (0.81). The worst result was achieved by the naive Bayes classifier (0.73). It should be noted that the bagging algorithm uses multiple decision trees, trained on the basis of subsets of data selected by sampling with return. The remaining, undrawn data becomes the testing set. For already-built tree models, voting is used to get the final answer. The authors [14] pointed out that the quality of the features affects the reliability of the produced models, and they also propose the use of a feature subset selection strategy to shorten the learning time, or fuzzy logic when the number of classes is increased. In addition, they propose recursive neural networks as a possible option for larger data sets, also ensuring high accuracy. The authors of the publication [15], on the other hand, note the lack of a global definition for positive mental health, presenting various approaches to this issue. They mention the observation that definitions of good mental health are, and should be, to some extent context-dependent. The Public Health Agency of Canada, mentioned by the authors of [15], refers to positive mental health as the ability to feel, think and act in a way that strengthens the ability to enjoy life and cope with the problems encountered. Keyes describes it in a slightly different way, suggesting a definition of the syndrome of signs of positive feelings and positive functioning in life. The authors [15] note that a positive state of mental health is not synonymous with the absence of mental illness. This is the short version of the Mental Health Continuum (MHC), based on the concept of two related but distinguishable dimensions. The authors cite successful tests of this scale in countries such as Poland, Italy, Brazil and the United States. Many indicators of positive mental health have been identified in populations, including aspects such as general health, physical activity, sleep, substance use, violence or discrimination. For young people, factors such as relationships with peers or support from teachers are particularly important. Similarly, income, employment and place of residence were positively associated with good mental health. In their study, the authors [15] examined 5399 students from grades 8 and 10. All of them were willing to answer questions, and 92% of students answered all of them. The questionnaire used in the study was based on the Swedish version of the Survey of Adolescent Life in Vestmanland, which also included a short version of the MHC and other questions related to general health, substance abuse, exposure to technology, school life and socioeconomic background. Changed the wording of several questions to better fit the Chinese context. The data obtained were analyzed using SPSS 22 software, using multivariate logistic regression, likelihood ratios and 95% confidence intervals for the analysis of variables related to positive mental health as a dependent variable. In the beginning, the collinearity of the variables was checked by Spearman's correlation analysis. Further, insignificant indicators were dropped until the model was statistically significant. Nagelkerke's Pseudo-R2 statistic and model fit were also calculated. Their research [15] extends knowledge about the prevalence of positive mental health among Chinese minors, as well as about the indicators of positive mental health. As a result, information was obtained that the surveyed group of Chinese people was significantly healthier in terms of mental health than in similar studies in other countries. The authors acknowledge that their study covered only one city in China, so further research in different regions will be needed. On the other hand, the authors of the publication [16] on economic difficulties and reported mental health problems during the COVID-19 epidemic point to the problem of isolation increasing the risk of loneliness, or the need to assess the links between the labor market and mental health, also in order to understand the impact of the pandemic on existing the socioeconomic inequalities. Their considerations [16] include factors related to changes in workload, income decline and job loss, as well as three mental health issues:

- Depression;
- Loneliness;
- Fear for your health [16].

The data came from employee surveys in Italy, Spain, the Czech Republic, Slovakia, the Netherlands and Germany from March and April 2020. The research also took into account the International Socio-Economic Index (ISEI). It expresses the relative position of the profession in the labor market, on a scale of 10 to 89 points. During the analyses [16], it was noted that occupations with an ISEI index below 30 points were characterized by a much higher risk of economic difficulties—about twice as high as medium and high-rated occupations (ISEI up to about 80 points). In addition, freelance and self-employment increased the likelihood of a reduction in workload by more than 32 percentage points, a decrease in income by 42 percentage points, and a loss of a job by just under 20 percentage points, compared to typical workers. Similarly, in the comparison between employees and employers, reductions in workload and income were more pronounced in the first group. In the final part of the work [16], they point out that the indicators used by them are not clinically confirmed, which makes it impossible to compare them on an equal basis, but they are an assessment of feelings about mental health. In addition, they consist of single questions, which makes them a non-detailed assessment of mental health. The authors explain that this is due to the data in the questionnaires not being designed to capture mental health, so researchers have had to rely on crude indicators. On the other hand, in the paper [17] attention was drawn to incomplete or partial evidence of the connection between mental illnesses and work. Therefore, the authors assumed that the mental health of an individual depends on characteristics such as:

- Personality;
- Sex;
- Own results at work;
- Loss of a job by a family member [17].

They developed [17] two models, one for the issue of the impact of job loss by a partner on the spouse, and the other describing the effects of parental job loss on underage children. They also sought to limit biasing effects in their study, based on data from around 7700 Australian households. The data consisted of responses to the Household, Income and Labor Dynamics in Australia (HILDA) survey. In order to develop two models, two separate data samples were created [17]—one for married couples, the other for parent–child pairs. Part of the data included answers to the Self-Completion Questionnaire (SCQ), which the researchers used in both the first data sample and the second. The MHI-5 (MHI—Mental Health Inventory) was used as the output variable [17], consisting of five questions on a 6-point scale. These questions were as follows:

- Were you a nervous person?
- Have you felt so down that nothing could cheer you up?
- Did you feel calm and composed?
- Have you felt depressed?
- Were you a happy person? [17].

The scores on this scale ranged from 0 to 100, where the lower the value, the worse the mental health. As a result of these studies [17], it turned out that the impact of losing a wife's job had no greater effect on husbands, while wives whose spouses lost their jobs had between 2 and 2.7 lower scores than women whose husbands still had jobs. However, the authors, taking into account other factors, indicate that this is not a statistically significant result. It was only when differentiating between groups with persistent unemployment, financial stress and dissatisfaction with relationships that a significant effect of losing a job by husbands was found. They found that continued unemployment caused a significant decline in mental health between studies and that the financial stress situation did not significantly contribute to worse mental health, while both women and men experienced worse mental health as dissatisfaction with their partner increased compared to previous answers. However, looking at the results [17] regarding the mental health of children after the loss of a job by one of the parents, it did not have a significant impact on its deterioration. A drop of 6.6 points was recorded when the mother was unemployed between examinations, which has a much higher impact than was observed for other variables. Comparing the mental state of boys and girls, it was shown that the deterioration of mental health was greater in girls, especially when the mother was unemployed. However, in the work [18], the impact of natural disasters on the mental health of minors is compared with their peers who have not experienced such events. Their study uses data on students from two Canadian cities located in the same province (Fort McMurray and Red Deer). In the surveys conducted in these cities, six questionnaires common to both studies were used, including:

- Patient Health Questionnaire, Adolescent version (PHQ-A);
- Hospital Anxiety and Depression Scale (HADS);
- CRAFFT questionnaire;
- Tobacco Use Questionnaire;
- Rosenberg's self-esteem scale;
- Kidscreen questionnaire [18].

The authors [18] performed a statistical analysis based on these questionnaires, and also compared the percentage odds of:

- Depression;
- Thoughts of suicide;
- Medicines;
- Using alcohol/stimulants;
- Tobacco use;
- Any of the options: about depression, about fears or use of alcohol/stimulants.

An additional limitation was the use of only complete answers for each measure, i.e., without omitted questions. A comparison [18] of indicators between the two regions found significant differences in 8 out of 12 measures of mental health status. The rates of possible depression were significantly higher in the city that experienced a natural disaster, as were those for suicidal thoughts and tobacco use. On the other hand, the self-esteem and quality of life scales (Rosenberg and Kidscreen, respectively) were much lower, but this is related to the nature of their questions. The conclusions [18] include the observation that this research reinforces the need for policies and programs to care for mental health among minors, especially after natural disasters, in order to reduce their vulnerability and build a positive state of mental health. They also note that it would be useful to compare these studies with data for post-traumatic stress symptoms from both cities, as the authors did not have such data from the city of Red Deer. They also indicate that minors are very vulnerable to the adverse impact of natural disasters. Summing up the studied literature, it can be noted that these are extremely diverse studies, they address many aspects related to mental health indicators, both from the side of positive and negative mental health. In addition, a variety of approaches were used, including voice data analysis, conducting surveys using many different questionnaires, random forests, bagging algorithm, support vector method, K-nearest neighbor's method, and statistical analysis. However, it should be borne in mind the need to expand research in the search for more effective algorithms that can be used in this area.

The proposed solution can be used in a prototype preventive mental health medicine system (Figure 2) for healthy people to monitor and detect the first symptoms of chronic stress and burnout as early as possiblebased on a combination of a generic standard and a dynamic standard generated directly from the data set. Given the second opinion offered by the ML system, it will support the activities of primary care physicians and psychology and psychiatry specialists in their daily efforts to provide early diagnosis and treatment of this group of conditions and will allow the selection and application of prevention and, if necessary, minimize the duration of potential therapy and reduce its cost [19].

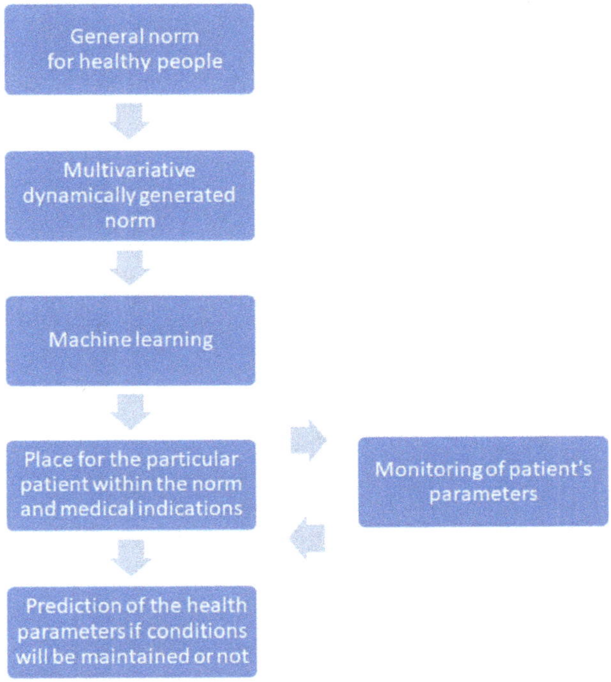

Figure 2. Prototype of preventive mental health medicine system [19].

Novelty and contribution lie in the application and matching of ML methods to the form and characteristics of test data describing chronic stress and job burnout. Pre-selection of methods and their initial facilitated matching to presumed criteria is key, which will support the development of preventive mental health medicine systems.

The research aims to determine a virtual indicator of mental health using selected ML algorithms, as well as to determine their effectiveness in this task by checking the learning time, operation and accuracy. In addition, research hypotheses will be verified, i.e.,:

- Choice of the ML method affects the regression accuracy, learning time and running time;
- Differences in accuracy are relatively small—up to about 10 percentage points difference between methods.

2. Materials and Methods

2.1. Material

The results of 99 patients (36 women and 63 men, mean age 27.93, SD = 4.64, mean seniority 3.78, SD = 2.94) with suspected chronic stress and burnout were analyzed using ML (Table 1).

Table 1. Data set distribution.

Parameter	Mean	SD	Min	Q1	Median	Q3	Max
PSS item 1	2.96	0.79	1	2	3	4	4
PSS item 2	3.14	0.74	2	3	3	4	4
PSS item 3	2.87	0.92	1	2	3	4	4
PSS item 4	2.66	1.05	0	2	3	3	4
PSS item 5	3.06	0.65	1	3	3	3	4
PSS item 6	2.90	0.85	1	2	3	3	4
PSS item 7	3.08	0.97	1	3	3	4	4
PSS item 8	2.67	0.90	0	2	3	3	4
PSS item 9	2.94	0.71	1	3	3	3	4
PSS item 10	2.49	0.93	1	2	2	3	4
MBI item 1	3.27	1.96	0	2	3	5	6
MBI item 2	2.73	1.73	0	2	3	4	6
MBI item 3	2.49	1.70	0	1	3	3	5
MBI item 4	2.24	2.32	0	0	1	5	6
MBI item 5	1.50	1.68	0	0	1	3	6
MBI item 6	1.53	1.48	0	0	1	3	6
MBI item 7	3.37	1.78	0	2	3	5	6
MBI item 8	1.69	1.68	0	0	1	3	6
MBI item 9	2.86	2.57	0	0	3	6	6
MBI item 10	1.56	1.35	0	1	1	3	6
MBI item 11	2.09	1.55	0	0	3	3	6
MBI item 12	2.55	1.66	0	1	3	3	6
MBI item 13	2.09	1.52	0	1	2	3	6
MBI item 14	2.17	1.86	0	1	1	3	6
MBI item 15	2.36	2.03	0	0	2	4	6
MBI item 16	1.68	1.77	0	0	1	2.5	6
MBI item 17	2.76	2.04	0	1	3	3	6
MBI item 18	1.64	1.45	0	0	1	3	5
MBI item 19	2.56	1.95	0	1	3	3	6
MBI item 20	1.87	2.23	0	0	1	3	6
MBI item 21	1.21	1.48	0	0	0	3	4
MBI item 22	2.43	1.45	0	2	3	4	6
SWLS item 1	4.08	0.93	2	4	4	5	5
SWLS item 2	3.24	1.53	1	2	3	4	6
SWLS item 3	3.30	1.66	1	1	4	5	6
SWLS item 4	3.20	1.66	1	2	2	4	6
SWLS item 5	2.51	1.59	1	1	2	4	5

Mental well-being data were used, including people's gender, age, length of service and their responses to the three questionnaires: Perceived Stress Scale (PSS), Maslach Burnout Inventory (MBI) and Satisfaction with Life Scale (SWLS).

The subject of the study was data from a set of 99 people, information about which was divided into 4 subgroups, each in a separate MS Excel sheet: "Patient data", "PSS10", "MBI", and "SWLS". The first of the above sheets includes the patient's gender, age and work experience. The second sheet contains answers to 10 questions from the PSS set, on a scale of 0 to 4, where 0 corresponds to "never", 1—"almost never", 2—"sometimes", 3—"quite often", and 4—"very often". The third sheet contains answers to 22 questions from the MBI set, on a scale of 0 to 6, where 0 corresponds to "never", 1—"several times a year", 2—"once a month", 3—"severaltimes a month", 4—"once a week", 5—"several times a week", and 6—"every day". The fourth sheet contains answers to 5 questions from the SWLS set, on a scale of 1 to 7, where 1 corresponds to "strongly disagree", 2—"disagree", 3—"slightly disagree", 4—"neither agree nor disagree", 5—"agree slightly", 6—"agree", and 7—"strongly agree". Based on these four sheets, a CSV (Comma Separated Values) file was created, which was used in the application due to the inability to directly load an Excel file with the .xls extension, also taking into account the available NuGet packages—they are satisfactorily documented for use in the project. This CSV file uses a semicolon (;) as the delimiter, which has been included in the app as the default delimiter value. The total is based on all answers from PSS, MBI and SWLS sets. All but the first column of the CSV file contain numeric values, while the first column can only contain two options: M (Male) or F (Female).

The study was approved by the Bioethics Committee No. KB 391/2018 at the Ludwik Rydygier Collegium Medicum in Bydgoszcz, Nicolaus Copernicus University in Toruń. Each participant in the study gave informed consent.

2.2. Methods

Two languages were used to develop the application: C# in .NET and Extensible Application Markup Language (XAML), whereby:

- C# language was used to describe the actions performed by the program;
- XAML was used to develop the layout of the user interface in a Universal Windows Platform (UWP) application, along with the naming of elements (which allows them to be used in C# as variables), or the binding of events to specific functions in the code behind the interface (code behind).

A number of ML models based on Stochastic Dual Coordinate Ascent (SDCA), limited-memory Broyden–Fletcher–Goldfarb–Shanno, Online Gradient Descent, etc., were built based on a clinical dataset (PSS, MBI and SWLS) and compared based on criteria in the form of learning time, running time during use and regression accuracy. The rationale for choosing these particular algorithms lies in their popularity and the authors' previous experience and previous research on measuring long-term stress and burnout using the aforementioned group of tests and AI [19–22]. Knowledge in the area of matching AI/ML tools for the analysis, inference and prediction of stress and burnout measurements is still nascent and no computational or theoretical basis can be cited as yet.

The predicted value was a virtual mental health index.

The data set has been divided into a training set (70% of samples) and a test set (30% of samples).

SDCA algorithm is a linear algorithm, meaning that it generates a model that calculates results based on a linear combination of the input data and a set of weights. The model weights are those parameters that are determined during training. In the general case, linear algorithms are scalable, fast and have a low cost during training and during prediction. This class of algorithms goes through the training dataset many times [23]. It is devoid of parameters for manual tuning and has a clearly defined stopping criterion. This algorithm has good implicit performance. It combines some of the best features, such as:

- Possibility of streaming learning, i.e., operating on data without having to put it all in memory at once;
- Achieving satisfactory results with a small number of circuits through the entire data set;
- Not wasting computing power on zeros in sparse datasets [24].

It should be borne in mind that the results obtained with this algorithm are dependent on the order of the training data, but the solutions obtained can be treated as equally good between different executions of the algorithm [25]. This algorithm is a stochastic version of DCA. The basic version of the algorithm (DCA) performs optimization on a single variable in each iteration without affecting the others. The SDCA version of the algorithm performs a pseudo-random selection of a double coordinate for optimization based on a uniform probability distribution [26].

LBFGS is an abbreviation for limited-memory Broyden–Fletcher–Goldfarb–Shanno, an optimization algorithm based on BFGS, but using limited Random Access Memory (RAM) [27,28], as it does not store a matrix approximating the inverse of the Hessian $\nabla 2$ f(x), instead using an intermediate approximation [28,29]. The calculation is based on an initial approximation and an update rule that models local curvature information [27,28]. The original Broyden–Fletcher–Goldfarb–Shanno method, called full BFGS (BFGS), proposed by these four authors in 1970, keeps the aforementioned matrix in memory, whose computational cost of updating is high, of the order of O(n2) [28,29]. As for the convergence of the BFGS method, if the function has a continuous second derivative and the function is strongly convex, the sequence of successive values of x_k tends towards the global minimizer, and furthermore, when it is assumed that the Hessian satisfies the Lipschitz condition, the rate of convergence is super linear [28,29]—i.e., faster than linear. The convergence of the LBFGS algorithm depends on the quality of the Hessian approximation, which is difficult to achieve, and it has been observed in numerical observations that an appropriate guess of the initial Hessian has a significant impact on the search direction and convergence [27,28].

The Online Gradient Descent (OGD) algorithm is a variation of the Stochastic Gradient Descent (SGD) method used for online training—i.e., training by learning concepts incrementally by processing examples from the training set one at a time, one after the other, with the algorithm not storing the last occurrence after each update, but based on the next sample [29,30]. SGD uses an iterative technique based on error gradients, in addition to providing the ability to update the weight vector using the average of the observed data vectors as the algorithm progresses [31]. SGD is popular for its simplicity, computational efficiency, or convergence independent of the training dataset, and the performance of DL methods depends heavily on this algorithm. However, it is susceptible to the effects of noisy data, especially noticeable in robotics, where robots do not have the capacity to collect enough data to negate these effects [32].

Three questionnaires were used to determine a virtual mental health index: PSS, MBI and SWLS. Data from these questionnaires were used in the application to train the models and determine metrics and statistics.

PSS is a scale developed by Cohen, Kamarck and Mermelstein in 1983, which aimed at respondents' self-assessment of the unpredictability of their life, their lack of control over it and the overload they feel. The original version has fourteen general questions on a four-point scale, and the final score is obtained by reversing the scale for positively valenced questions and then adding up the scores for all questions. In addition, two shorter versions of the scale have been developed, the ten-question scale used in this work, as well as a four-question scale [33]. Research on this instrument is massively carried out all over the world, including China, Ethiopia, Iran and Greece, and the results indicate that this scale can be relied upon to be used in these countries. To validate the scale, Cohen studied the responses of people of different ages, both genders and a variety of racial backgrounds [34]. Similar information is presented by the authors of a Czech study, where they briefly describe that all versions of the scale had previously been compared in a variety of cultural and linguistic contexts and that these researchers agreed that the ten-question scale was at least comparable to or better than the original version in terms of internal consistency while noting a significant decrease in reliability of the four-question version, which was attributed to it simply being too short [33].

The MBI (Maslach Burnout Inventory) was developed by Christina Maslach and her team. In her article, she explains the concept of professional burnout (burnout)—a syndrome of emotional exhaustion and cynicism often found in people who work with people, with a key component being the increased sense of emotional exhaustion mentioned earlier. It indicates that with the depletion of their emotional resources, employees begin to feel that they are not able to give their best; furthermore, they develop negative, even cynical attitudes and feelings about their clients. The two aspects seem to be linked, and a tendency to evaluate oneself negatively, especially in relation to one's work, not feeling satisfied with one's achievements, is mentioned as a third effect related to professional burnout [34]. Occupational burnout is characterized by high levels of emotional exhaustion, dehumanization and low feelings of personal fulfillment. In addition, they point out that occupational burnout and depressive states are related, but they are not the same concepts, i.e., their characteristics do not overlap and thus cannot be used interchangeably [35]. The version used in this thesis consists of three groups of questions regarding these issues:

emotional exhaustion (nine questions), sense of personal accomplishment (eight questions) and dehumanization (five questions).

In terms of SWLS acceptability, reliability, validity as well as gender independence have been demonstrated, as indicated by the authors of the article [36]. The scale was first presented in 1985 and was summarized as narrowly focused on the issue of overall satisfaction with life, without addressing issues such as loneliness or positive affect [37]—which is described as the feeling experienced when a certain goal is achieved, or a source of danger is averted, or the person is satisfied with the current state of affairs [30]. It was developed as a response to a number of scales that contained only one question, and to scales that went beyond life satisfaction. The process of shaping the questions in this set began with a list of 48 questions, and, after eliminating questions about affect and questions with a Factor Loading of less than 0.60 and omitting those with a high degree of similarity, yielded five questions [36], scored on a scale of 1 to 7, which in effect generated a score range of 5 to 35.

Test results and calculations were recorded in an MS Excel spreadsheet.

Statistical analysis was performed using Statistica 13 (StatSoft, Tulsa, OK, USA). The Shapiro–Wilk test was used to check the normality of the distribution of the studied data. The *p*-value was set at 0.05. Where possible, analyzed values with distributions close to normal were presented as mean values and standard deviation (SD). The analyzed values with distributions different from the normal distribution were presented using the minimum value, the lower quartile (Q1), the median, the upper quartile (Q3) and the maximum value.

Selected ML algorithms were compared on the basis of:

- Metrics: mean absolute error, mean squared error, mean squared error, coefficient of determination;
- Learning time, expressed in milliseconds: minimum, average, maximum;
- Prediction time, expressed in milliseconds: minimum, average, maximum.

For each of the compared values, the best algorithm was determined, choosing the one with:

- Minimum value for learning times, prediction, mean absolute error, mean squared error;
- Maximum value for the coefficient of determination.

In a similar way, the worst algorithm in terms of a given criterion was determined, this time by:

- Maximum value for learning times, prediction, mean absolute error, mean squared error, mean squared error,
- Minimum value for the coefficient of determination.

In order to make more accurate use of the software's capabilities, each learning was performed four times in order to make the resulting parameters more meaningful and, in addition, each time the steps were carried out on a new occurrence of the application. To compare the algorithms, we selected the best hyperparameters for each optimizer and for each data set using a validation procedure with a learning set and a validation set. On each data set, for each hyperparameter, we calculated the accuracy after a certain number of epochs for a range of values and a certain validation set. The study considered the following hyperparameters of each of the algorithms tested:

- SDCA: c (regularization strength) and stopping time;
- LBFGS: solver, penalty, max_iter, c, tol, fit_intercept, intercept_scaling, class_weight, random_state, multi_class, verbose, warm_start, and l1_ratio;
- OGD: learning rate and diameter of the decision set.

The same data set was also analyzed automatically using ML.NET (Visual Studio 2022, Microsoft, Redmond, WA, USA).

3. Results

The algorithm with the highest accuracy was Stochastic Dual Coordinate Ascent, but although its performance was high, it had significantly longer training and prediction times (Figure 3a).

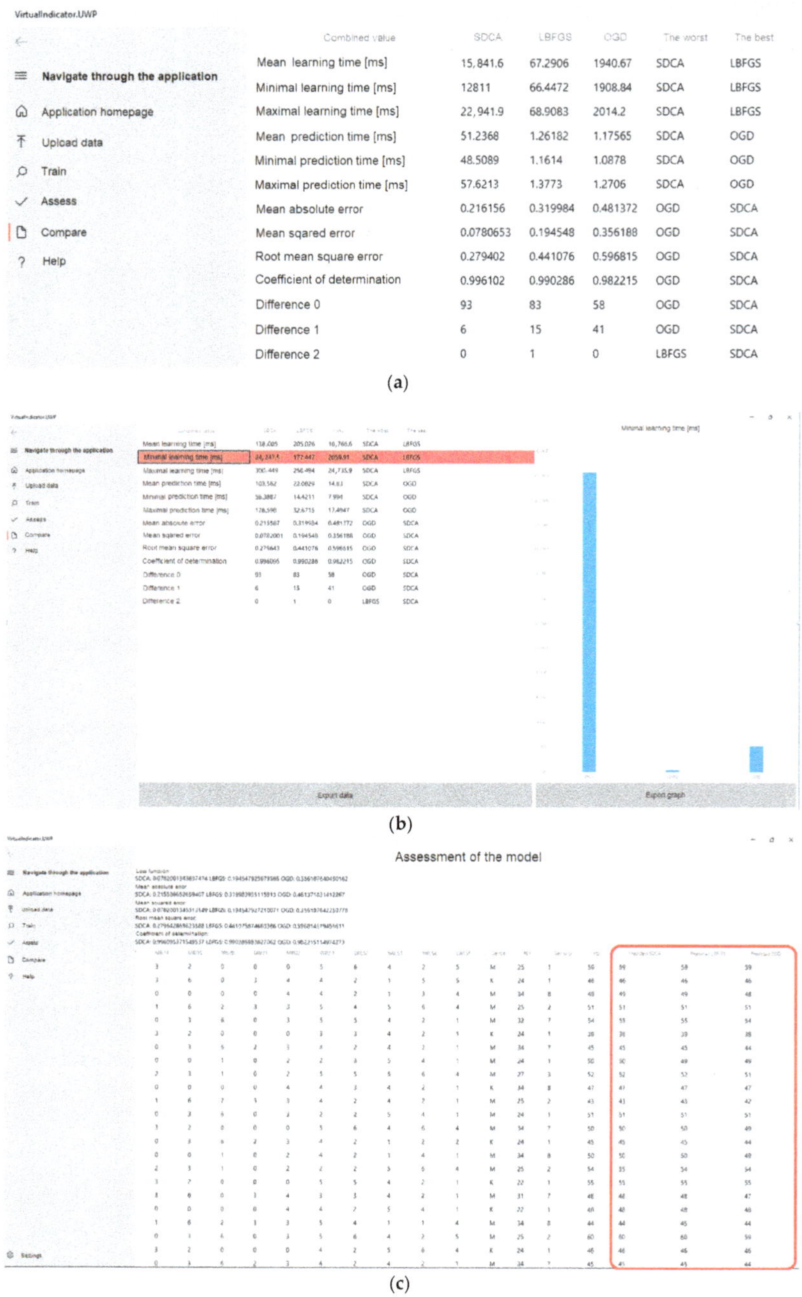

Figure 3. (**a**) General comparison of metrics, training times and predictions, (**b**) comparison of selected metrics and (**c**) assesment of models (three columns on right side).

The fastest algorithm looking at learning and prediction time, but slightly less accurate, was the limited-memory Broyden–Fletcher–Goldfarb–Shanno (Figure 3b).

The first criterion considered was the model learning time, expressed in milliseconds. The average, minimum and maximum values were taken into account. Both the average, minimum and maximum times were the longest for the SDCA model and the shortest for the LBFGS model. This means that the SDCA model performed the worst in this ranking, and the LBFGS model performed best. It should be noted that while for the LBFGS and OGD models, the difference between their maximum and minimum values was relatively small (about 4% of the average, both for LBFGS and OGD), for the SDCA model it was about 64% of the average. Another criterion was the prediction time for the entire data set, expressed in milliseconds. The average, minimum and maximum values were taken into account. This time was the lowest for the OGD model, but it differed only slightly from the LBFGS model, both models reached a time slightly above 1 ms. On the other hand, the average time for the SDCA model was about 44 times longer than for the OGD model, and again there were larger differences between the maximum and minimum values for the SDCA model (approximately 18% of the mean value). The average absolute error was the lowest for the SDCA model and amounted to about 0.216, while it was the highest for the OGD model, amounting to about 0.481 (more than twice as much). On the other hand, for the LBGFS model, it was around 0.320, which corresponds to an increase of 48%. For this criterion, as well as for the mean squared error and the mean squared error, the best results were achieved by the SDCA model, and the worst by the OGD model. The ranking for the coefficient of determination looks similar, the value of which is closer to 1, the better for the model. The last lines of the comparison show the number of occurrences for which the absolute value of the difference between the rounded prediction and the value from the dataset was 0, 1 or 2, respectively. Looking at the difference equal to 0, the best result was obtained by the SDCA model, and the worst by the OGD model. For a difference of 1, the best result was obtained by the SDCA model (6 occurrences), and the worst by the OGD model (41 occurrences). However, for the difference equal to 2, there was one such occurrence for the LBFGS model (Figure 3c).

In this particular problem of determining a virtual mental health index, all three models considered achieved comparable final results. Based on the criterion of model learning time, and considering other factors (e.g., prediction time), the LBFGS model would be the best choice. On the other hand, looking at metrics in the form of, among other things, mean absolute error or coefficient of determination, the SDCA model, whose biggest drawbacks are learning time and prediction time, would prove to be the best choice. Although the OGD model achieved the best prediction time, it achieved the worst of the results when looking at the metrics.

Looking at the results obtained, the differences between the ML methods used are clearly visible, especially for learning time and metrics. Furthermore, bearing in mind that the accuracy of the model increases as the value of the coefficient of determination approaches one, the differences between the methods amounted to a maximum of around 1.4 percentage points, looking at the difference between the maximum and minimum value in relation to the maximum possible value, i.e., 1 (which can be understood as 100%).

We have compared the aforementioned results with automated analysis using ML.NET results (249 models checked, Tables 2 and 3).

Table 2. Results of ML-based classification.

Parameter	Micro Accuracy (%)	Macro Accuracy (%)	Best Trainer
Gender	75.16	69.32	FastTreeOva
Age	71.24	62.82	FastTreeOva
Seniority	78.73	72.23	FastForestOva
Total pts.	17.29	14.79	LightGbmMulti

Table 3. Results of ML-based prediction.

Parameter	Accuracy (%)	Best Trainer
Gender		Not possible
Age	93.32	LbfgsPoissonRegressionRegression
Seniority	97.57	FastTreeRegression
Total pts.	97.42	LbfgsPoissonRegressionRegression

Despite the fact that the data lends itself to both prediction and classification, it has not been possible to find one algorithm that is good at everything—a thoughtful combination of different algorithms must be used in automated analysis.

4. Discussion

A comparison of the three ML algorithms showed small differences in regression accuracy (about 1.4 percentage points, or, according to the thesis, less than 10 percentage points), which, in relation to the work [10], which nevertheless dealt with the classification problem, but revealed differences in accuracy between six different methods of about 5.5 percentage points, probably means a small impact of the method used on regression accuracy or classification accuracy.

The results of the paper [14] are similar, where all the algorithms used, except for the naive Bayes classifier—which is the simplest one used and probably for this problem did not have a strong connection to reality—obtained accuracy differences of at most 9 percentage points. Looking at the proposal in that paper, continuing research would need to use a feature subset selection strategy so that the solution is based on the highest quality features. Applying such an approach successfully would mean a reduction in learning time and potentially an increase in model reliability. In addition, the inclusion of the patient's mental profile mentioned can be considered to have been done, as the data contains answers to a set of questions to assess the patient's mental health status.

When comparing with studies [13,16–18], it is important to note the lack of analysis of the impact of individual factors on the virtual mental health index, considering particular attributes such as age, gender, or length of service. This implies an opportunity for further research to be able to establish some trends, for example among different age groups, as in the article [6]. In addition, further data would have to be collected, not only more numerous but possibly also including the ISEI index, which expresses the relative position of the occupation in the labor market, as in the study [16]. Regarding the study [17], the dataset could be extended to include information on the dynamics of employment, or also the household of the person surveyed. Looking at the study [18], it would be valuable to assess the risk of problems such as depression, anxiety or the use of stimulants, which could be baseline variables for the trained models.

Referring to the work [11], which addresses the problems of assessing health status in discrete moments in time, mainly in terms of not being able to assess the impact of the environment on the patient in real time, one could use data from apps and activity monitoring devices of potential volunteers to derive models based on measured data. On the one hand, this would make it possible to assess mental health on a continuous basis, and on the other hand, it would make it independent of the patient's self-assessment.

As described in the paper [12], voice-based mental health determination, while it appears to be a promising solution, the authors did not present the ML methods used, which, combined with the commercialization of the developed library and system, does not allow for the extension of these analyses. On the other hand, the idea is intriguing, but in order to be realized, it would require an appropriate selection of libraries and ML methods, as well as the disposal of voice data, together with the determination of the patient's mental health status for these data.

On the other hand, the observation in the article [15] that a positive mental health status does not imply the absence of mental illness, which was taken into account in the Mental Health Continuum scale, whose tests in various countries have been successful.

This is something to bear in mind, as it happens that mental illnesses are able to be hidden, both consciously and unconsciously. It is also important to consider factors that are often indicative of a patient's mental state, such as their physical activity, sleep, use of stimulants, and relationships with peers in the case of adolescents or relationships with co-workers among adults.

In the study, learning time or prediction time is an evaluation criterion. The performers' tasks do not require real time, but with large databases and a large number of simultaneous system users, the value of this parameter can be very important.

It is noteworthy that a variety of tools have been used in these papers, whether in the form of questionnaires, such as the Depression Anxiety Stress Scale-21, Beck Depression Inventory or algorithms (logistic regression, K-nearest-neighbor method, decision trees, bagging, support vector method) and technology, including the Python language, Scikit-learn library, physical activity tracking mobile apps and wearable devices. In addition, many of these papers did not present the programming language used, making it impossible to make a technology choice based on them.

Key findings in the area of ML-supported human mental health analysis have shown that, despite the variety of tools that have been used in these papers, one leading approach is lacking, both in the selection of tests and in the selection of ML-based aggregation and analysis methods. This makes it difficult both to compare different approaches and to extract the best ones (based on common criteria) for further development and use in both simple predictive systems within preventive medicine and complex diagnostic and monitoring systems within more complex specialized studies. This results in the unique contribution of the current study compared to the existing literature, which includes how to aggregate test results into a virtual mental health index and how to select optimal ML methods for its further use providing a basis for further research, including for other groups of clinicians and researchers. Our experience to date shows that this element of technological support is lacking in clinical practice, hence interdisciplinary teams are needed for further research.

4.1. Limitations of Studies

Research on determining a virtual index of mental health using ML algorithms may encounter a number of limitations and challenges that should be considered:

- Lack of unequivocal measures of mental health (patients and healthy people)—mental health is a subjective concept and difficult to define unambiguously, which complicates the process of creating ML models;
- Population diversity—individual healthy individuals differ from each other in terms of mental health as well as in different life contexts, which makes general modeling difficult and it will be necessary to adapt models to different population groups;
- Lack of qualitative data—most of the available data is quantitative, which can hinder a fuller understanding of mental health;
- Lack of historical data—it is often important to consider the historical context of the patient's illness;
- Data privacy—mental health data are very sensitive, so it is necessary to maintain appropriate standards of data privacy and security;
- Cultural differences—Mental health can be understood and experienced differently in different cultures.
- Interpretability of models—understanding why a model made certain decisions can be a problem for mental health diagnosis and treatment;
- Importance of experts—ML models will not replace human expertise, but will only support it [38–42].

4.2. Directions for Further Research

Research on the determination of a virtual index of mental health using ML algorithms is an area that can bring many benefits in the field of health care and mental well-being, as well as their objective, partially automated assessment and monitoring of changes. A summary of research directions that can be explored in this context is presented in Table 4.

Table 4. Directions of research on the virtual index of mental health with the use of ML algorithms [43–45].

Area	Description and Detailed Tasks
Data collection and analysis	The use of many different data sources, including multi-modal ones, such as behavioral data (e.g., online activity, phone calls), biometric data (e.g., heart rate, sleep monitoring), survey data, photos and videos, as well as test results collected automatically, etc.
Collaboration with field experts	Collaboration with physicians and mental health professionals can help understand the mechanisms and create and evaluate the effectiveness of models.
Ethics and privacy	The manner in which data are collected, stored, used and destroyed should comply with relevant regulations and ethical standards.
Data preparation	Data preparation may include data normalization, removal of erroneous, uncertain, incomplete and outlier data, coding of categorical variables, etc.
Selection of ML algorithms and hyperparameters	Selection and adaptation of algorithms and hyperparameters of models to a specific problem from among possible solutions, such as decision trees, neural networks, support vector machines (SVM) or clustering algorithms.
Evaluation/cross-validation of models	Define model performance metrics (accuracy, sensitivity, specificity, F1-score, Receiver Operating Characteristic (ROC) curves, etc.) and analyze model performance using them.
Interpretability of models	Understanding how the model makes its predictions (why the model made certain decisions).
Checking the learning time	Model training time can be a critical factor in clinical practice—it needs to be investigated how long it takes to train different models and whether this can be optimized.
	The model should be adapted to real-time operation (including learning on new patients) in order to be used in clinical practice.
Validation on a large sample of patients	The effectiveness of the models should be tested on a large sample of patients to ensure that the model generalizes well to different cases.

This research can be a long and complicated process, but it can have significant benefits in diagnosing, monitoring and managing patients' mental health [46,47].

5. Conclusions

The ability of ML to identify burnout using passively collected electronic health record (EHR) data and predict future health status with an accuracy of more than 70% (for some traits: more than 90%) accounts for the usefulness of this group of methods in daily clinical practice, which is worth developing.

The algorithms did not differ significantly from each other in terms of accuracy (about 1.4 percentage points) but differed more strongly in other parameters. The algorithm with the highest accuracy was Stochastic Dual Coordinate Ascent, but although its performance was high, it had a significantly longer training and prediction time. In contrast, the fastest algorithm looking at learning and prediction time, but slightly less accurate, was the limited-memory Broyden–Fletcher–Goldfarb–Shanno.

Findings from the study can be used to build larger systems that automate early mental health diagnosis and help differentiate the use of individual algorithms depending on the purpose of the system.

Author Contributions: Conceptualization, A.B., I.R. and D.M.; methodology, A.B., I.R. and D.M.; software, A.B., I.R. and D.M.; validation, A.B., I.R. and D.M.; formal analysis, A.B., I.R. and D.M.; investigation, A.B., I.R. and D.M.; resources, A.B., I.R. and D.M.; data curation, A.B., I.R. and D.M.; writing—original draft preparation, A.B., I.R. and D.M.; writing—review and editing, A.B., I.R. and D.M.; visualization, A.B., I.R. and D.M.; supervision, I.R.; project administration, I.R.; funding acquisition, I.R. and D.M. All authors have read and agreed to the published version of the manuscript.

Funding: The work presented in the paper has been financed under a grant to maintain the research potential of Kazimierz Wielki University.

Data Availability Statement: Data are unavailable due to privacy and cyber security.

Conflicts of Interest: The authors declare no conflict of interest.

References

1. Asatryan, B.; Bleijendaal, H.; Wilde, A.A.M. Toward advanced diagnosis and management of inherited arrhythmia syndromes: Harnessing the capabilities of artificial intelligence and machine learning. *Heart Rhythm.* **2023**, *20*, 1399–1407. [CrossRef]
2. Kannampallil, T.; Dai, R.; Lv, N.; Xiao, L.; Lu, C.; Ajilore, O.A.; Snowden, M.B.; Venditti, E.M.; Williams, L.M.; Kringle, E.A.; et al. Cross-trial prediction of depression remission using problem-solving therapy: A machine learning approach. *J. Affect. Disord.* **2022**, *308*, 89–97. [CrossRef] [PubMed]
3. Hong, N.; Liu, C.; Gao, J.; Han, L.; Chang, F.; Gong, M.; Su, L. State of the Art of Machine Learning-Enabled Clinical Decision Support in Intensive Care Units: Literature Review. *JMIR Med. Inform.* **2022**, *10*, e28781. [CrossRef]
4. Lopez-Jimenez, F.; Attia, Z.; Arruda-Olson, A.M.; Carter, R.; Chareonthaitawee, P.; Jouni, H.; Kapa, S.; Lerman, A.; Luong, C.; Medina-Inojosa, J.R.; et al. Artificial Intelligence in Cardiology: Present and Future. *Mayo Clin. Proc.* **2020**, *95*, 1015–1039. [CrossRef]
5. Reid, J.E.; Eaton, E. Artificial intelligence for pediatric ophthalmology. *Curr. Opin. Ophthalmol.* **2019**, *30*, 337–346. [CrossRef] [PubMed]
6. Mentis, A.A.; Lee, D.; Roussos, P. Applications of artificial intelligence–machine learning for detection ofstress: A critical overview. *Mol. Psychiatry* **2023**, 1–13. [CrossRef]
7. Galatzer-Levy, I.R.; Onnela, J.P. Machine Learning and the Digital Measurement of Psychological Health. *Annu. Rev. Clin. Psychol.* **2023**, *19*, 133–154. [CrossRef] [PubMed]
8. Sutrisno, S.; Khairina, N.; Syah, R.B.Y.; Eftekhari-Zadeh, E.; Amiri, S. Improved Artificial Neural Network with High Precision for Predicting Burnout among Managers and Employees of Start-Upsduring COVID-19 Pandemic. *Electronics* **2023**, *12*, 1109. [CrossRef]
9. Adapa, K.; Pillai, M.; Foster, M.; Charguia, N.; Mazur, L. Using Explainable Supervised Machine Learning to Predict Burnout in Healthcare Professionals. *Stud. Health Technol. Inform.* **2022**, *294*, 58–62. [CrossRef]
10. Srinivasulu Reddy, U.; Thota, A.; Dharun, A. Machine Learning Techniques for Stress Prediction in Working Employees. In Proceedings of the 2018 IEEE International Conference on Computational Intelligence and Computing Research (ICCIC), Madurai, India, 13–15 December 2018; pp. 1–4.
11. Knight, A.; Bidargaddi, N. Commonly available activity tracker apps and wearables as a mental health outcome indicator: A prospective observational cohort study among young adults with psychological distress. *J. Affect. Disord.* **2018**, *236*, 31–36. [CrossRef]
12. Hagiwara, N. Validity of Mind Monitoring System as a Mental Health Indicator using Voice. *Adv. Sci. Technol. Eng. Syst. J.* **2017**, *2*, 338–344. [CrossRef]
13. Pierce, M. Mental health before and during the COVID-19 pandemic: A longitudinal probability sample survey of the UK population. *Lancet Psychiatry* **2020**, *7*, 883–892. [CrossRef] [PubMed]
14. Srividya, M.; Mohanavalli, S.; Bhalaji, N. Behavioral modeling for mental health using machine learning algorithms. *J. Med. Syst.* **2018**, *42*, 88. [CrossRef] [PubMed]
15. Guo, C.; Tomson, G.; Keller, C.; Söderqvist, F. Prevalence and correlates of positive mental health in Chinese adolescents. *BMC Public Health* **2018**, *18*, 263. [CrossRef] [PubMed]
16. Witteveen, D.; Velthorst, E. Economic hardship and mental health complaints during COVID-19. *Proc. Natl. Acad. Sci. USA* **2020**, *117*, 27277–27284. [CrossRef]
17. Bubonya, M.; Cobb-Clark, D.A.; Wooden, M. Jobloss and the mental health of spouses and adolescent children. *IZAJ. LaborEcon.* **2017**, *6*, 6.
18. Brown, M.R.G. After the Fort McMurray wild fire there are significant increases in mental health symptoms ingrade 7–12 students compared to controls. *BMC Pyschiatry* **2019**, *19*, 18.
19. Pal, S.; Xu, T.; Yang, T.; Rajasekaran, S.; Bi, J. Hybrid-DCA: A double asynchronous approach for stochastic dual coordinate ascent. *J. Parallel Distrib. Comput.* **2020**, *143*, 47–66. [CrossRef]
20. Spiridonoff, A.; Olshevsky, A.; Paschalidis, I.C. Robust Asynchronous Stochastic Gradient-Push: Asymptotically Optimaland Network-Independent Performance for Strongly Convex Functions. *J. Mach. Learn. Res.* **2020**, *21*, 58.

21. Pu, S.; Olshevsky, A.; Paschalidis, I.C. A Sharp Estimate on the Transient Timeoff Distributed Stochastic Gradient Descent. *IEEE Trans. Automat. Contr.* **2022**, *67*, 5900–5915. [CrossRef]
22. Pu, S.; Olshevsky, A.; Paschalidis, I.C. Asymptotic Network Independence in Distributed Stochastic Optimization for Machine Learning. *IEEE Signal Process. Mag.* **2020**, *37*, 114–122. [CrossRef]
23. Mohsen, F.; Al-Saadi, B.; Abdi, N.; Khan, S.; Shah, Z. Artificial Intelligence-Based Methods for Precision Cardiovascular Medicine. *J. Pers. Med.* **2023**, *13*, 1268. [CrossRef]
24. Price, M.J. *Hello, C#! Welcome,. NET! In C# 8.0 and.NET Core 3.0—Modern Cross-Platform Development*, 4th ed.; Packt Publishing Ltd.: Birmingham, UK, 2019; pp. 1–69.
25. Perkins, B.; Hammer, J.V.; Reid, J.D. Introducing C#. In *Beginning C# 7 Programming with Visual Studio 2017*; Wiley: Hoboken, NJ, USA, 2018; pp. 3–13.
26. Shalev-Shwartz, S.; Tong, Z. Stochastic Dual Coordinate Ascent Methods for Regularized Loss Minimization. *arXiv* **2013**, arXiv:1209.1873.
27. Lu, X.; Yang, C.; Wu, Q.; Wang, J.; Wei, Y.; Zhang, L.; Li, D.; Zhao, L. Improved Reconstruction Algorithm of Wireless Sensor Network Based on BFGS Quasi-Newton Method. *Electronics* **2023**, *12*, 1267. [CrossRef]
28. Aggrawal, H.O.; Modersitzki, J. Hessian Initialization Strategies for L-BFGS Solving Non-linear Inverse Problems. *arXiv* **2021**, arXiv:2103.10010.
29. Asl, A.; Overton, M.L. Behavior of limited memory BFGS when applied to nonsmooth functions and their nesterov smoothings. *arXiv* **2020**, arXiv:2006.11336.
30. Bousbaa, Z.; Sanchez-Medina, J.; Bencharef, O. Financial Time Series Forecasting: A Data Stream Mining-Based System. *Electronics* **2023**, *12*, 2039. [CrossRef]
31. Benczúr, A.A.; Kocsis, L.; Pálovics, R. Online Machine Learning in Big Data Streams. *arXiv* **2018**, arXiv:1802.05872.
32. Ilboudo, W.E.L.; Kobayashi, T.; Sugimoto, K. Robust stochastic gradient descent with student-t distribution basedfirst-order momentum. *IEEE Trans. Neural Netw. Learn. Syst.* **2020**, *33*, 1324–1337. [CrossRef]
33. Figalová, N.; Charvat, M. The Perceived Stress Scale: Reliability and validity study in the Czech Republic. *Ceskoslovenská Psychol.* **2021**, *65*, 46–59. [CrossRef]
34. Prasetya, A.; Purnama, D.; Prasetyo, F. Validity and Reliability of The Perceived Stress Scale with RASCH Model. *PSIKOPEDAGOGIA J. Bimbing. Konseling* **2020**, *8*, 48–51. [CrossRef]
35. Maslach, C.; Jackson, S.E. The measurement of experienced burnout. *J. Occup. Behav.* **1981**, *2*, 99–113. [CrossRef]
36. Schaufeli, W.B.; Bakker, A.B.; Hoogduin, K.; Kladler, A.; Schaap, C. On the clinical validity of the Maslach Burnout Inventory and the Burnout Measure. *Psychol. Health* **2001**, *16*, 565–582. [CrossRef] [PubMed]
37. Checa, I.; Perales, J.; Espejo, B. Measurement in variance of the Satisfaction with Life Scale by gender, age, marital status and educational level. *Qual. Life Res. Int. J. Qual. Life Asp. Treat. Care Rehabil.* **2019**, *28*, 963–968. [CrossRef]
38. Diener, E.; Emmons, R.A.; Larsen, R.J.; Griffin, S. The Satisfaction with Life Scale. *J. Personal. Assess.* **1985**, *49*, 71–75. [CrossRef]
39. Prokopowicz, P.; Mikołajewski, D.; Mikołajewska, E. Intelligent System for Detecting Deterioration of Life Satisfaction as Tool for Remote Mental-Health Monitoring. *Sensors* **2022**, *22*, 9214. [CrossRef]
40. Rojek, I. Neural networks as prediction models for water intake in water supply system. In *Artificial Intelligence and Soft Computing—ICAISC 2008. Lecture Notes in Computer Science, 5097*; Rutkowski, L., Tadeusiewicz, R., Zadeh, L.A., Zurada, J.M., Eds.; Springer: Berlin/Heidelberg, Gemany, 2008; pp. 1109–1119. Available online: https://link.springer.com/chapter/10.1007/978-3-540-69731-2_104 (accessed on 31 August 2023).
41. Spoor, J.M.; Weber, J. Evaluation of process planning in manufacturing by a neural network based on an energy definition of Hopfield nets. *J. Intell. Manuf.* **2023**, 1–19. [CrossRef]
42. Teixeira, I.; Morais, R.; Sousa, J.J.; Cunha, A. Deep Learning Models for the Classification of Cropsin Aerial Imagery: A Review. *Agriculture* **2023**, *13*, 965. [CrossRef]
43. Rojek, I.; Mikołajewski, D.; Macko, M.; Szczepański, Z.; Dostatni, E. Optimization of Extrusion-Based 3D Printing Process Using Neural Networks for Sustainable Development. *Materials* **2021**, *14*, 2737. [CrossRef]
44. Rojek, I.; Mikołajewski, D.; Kotlarz, P.; Macko, M.; Kopowski, J. Intelligent system supporting technological process planning for machining and 3D printing. *Bull. Pol. Acad. Sci. Tech. Sci.* **2021**, *69*, e136722.
45. Mohammadi, E.K.; Talaie, H.R.; Azizi, M. A healthcare service quality assessment model usinga fuzzy best–worst method with application to hospitals within-patient services. *Healthc. Anal.* **2023**, *4*, 100241. [CrossRef]
46. Gajos, A.; Wójcik, G.M. Independent component analysis of EEG data for EGI system. *Bio-Algorithms Med-Syst.* **2016**, *12*, 67–72. [CrossRef]
47. Kawala-Janik, A.; Podpora, M.; Pelc, M.; Piatek, P.; Baranowski, J. Implementation of an inexpensive EEG headset for the pattern recognition purpose. In Proceedings of the 2013 IEEE 7th International Conference on Intelligent Data Acquisition and Advanced Computing Systems (IDAACS), Berlin, Germany, 12–14 September 2013; Volume 1, pp. 399–403.

electronics

MDPI

Article

A Multiscale Neighbor-Aware Attention Network for Collaborative Filtering

Jianxing Zheng [1], Tengyue Jing [2], Feng Cao [3,*], Yonghong Kang [3], Qian Chen [3] and Yanhong Li [3]

1 Institute of Intelligent Information Processing, Shanxi University, Taiyuan 030006, China; jxzheng@sxu.edu.cn
2 North Automatic Control Technology Institute, Taiyuan 030006, China; 202022408027@email.sxu.edu.cn
3 School of Computer and Information Technology, Shanxi University, Taiyuan 030006, China; 202322405009@email.sxu.edu.cn (Y.K.); chenqian@sxu.edu.cn (Q.C.); liyh@sxu.edu.cn (Y.L.)
* Correspondence: caof@sxu.edu.cn

Abstract: Most recommender systems rely on user and item attributes or their interaction records to find similar neighbors for collaborative filtering. Existing methods focus on developing collaborative signals from only one type of neighbors and ignore the unique contributions of different types of neighbor views. This paper proposes a multiscale neighbor-aware attention network for collaborative filtering (MSNAN). First, attribute-view neighbor embedding is modeled to extract the features of different types of neighbors with co-occurrence attributes, and interaction-view neighbor embedding is leveraged to describe the fine-grained neighborhood behaviors of ratings. Then, a matched attention network is used to identify different contributions of multiscale neighbors and capture multiple types of collaborative signals for overcoming sparse recommendations. Finally, we make the rating prediction by a joint learning of multi-task loss and verify the positive effect of the proposed MSNAN on three datasets. Compared with traditional methods, the experimental results of the proposed MSNAN not only improve the accuracy in MAE and RMSE indexes, but also solve the problem of poor performance for recommendation in sparse data scenarios.

Keywords: multiscale neighbors; attentional mechanism; collaborative embedding; recommendation

Citation: Zheng, J.; Jing, T.; Cao, F.; Kang, Y.; Chen, Q.; Li, Y. A Multiscale Neighbor-Aware Attention Network for Collaborative Filtering. *Electronics* **2023**, *12*, 4372. https://doi.org/10.3390/electronics12204372

Academic Editor: Dimitris Apostolou

Received: 25 September 2023
Revised: 11 October 2023
Accepted: 17 October 2023
Published: 22 October 2023

1. Introduction

Collaborative filtering has become a fundamental technology in e-commerce platforms, which has made remarkable achievements. Collaborative filtering assumes that similar users have similar interests, and makes recommendation services for the target user based on similar users' interests. In most e-commerce scenarios, the interaction between users and items reflects users' interest preferences, which is often used to find similar neighbors to model collaborative representations for users and items [1,2].

However, the e-commerce platform generates a large number of new users and product items every day. Some users have not rated or purchased new products, which leads to fewer interactive records. Thus, it is difficult to learn high-quality embedding representations for users and items. As a result, collaborative filtering still faces the problem of interaction data sparsity in recommender systems [3]. Aiming to solve the sparse recommendation problem, most recommender systems leverage auxiliary attribute information of users and items to establish users' preferences and mine similar neighbors for collaborative filtering [4]. However, similar neighbors mainly come from an attribute preference of uniform scale. In fact, different users have various social attributes, and different items have multiple product descriptions. Different combinations of attributes are conducive to generating various types of neighbors to describe semantic characteristics of different granularity for nodes. That is, when some users have multiple attributes and the other users have a few attributes, we can make collaborative recommendations according to their multiscale attribute neighbors. In addition, the interaction rating of users reflects their differentiated preference for items. We can leverage differentiated rating preference via different rating neighbors to explore the fine-grained preference motivation. Thus, how to deal with collaborative filtering by combining multiscale attribute neighbors with rating neighbors is an important task. Multiscale node embedding describes the fine-grained semantic representations from multiple perspectives, and can effectively solve the problem of poor performance of sparse recommendation, which is of great significance for industrial applications.

In the e-commerce recommender systems, multiscale attribute combinations can produce single-attribute-view neighbors and multi-attribute-view neighbors. For example, in movie recommender systems, users with the same gender and age have more similar interest behaviors than users of the same gender. As a result, different types of attribute-view neighbors can be constructed in various attribute combination spaces. In addition, different interaction-view neighbors can be obtained according to the types of interaction ratings, such as 1–5. We model the interaction-view neighbor embedding of nodes on various interactive views. The attention mechanism [5] is used to focus on specific input features, analyze the importance of all aspects of input features, and improve the expression ability of the model, which has been widely applied in the fields of natural language processing and image processing. Inspired by [6], this paper captures a different type of attribute neighbor embedding and interactive neighbor embedding and mine their collaborative signals with the attention mechanism to model multiscale node embedding. The multiscale node embedding can effectively capture diverse semantics of nodes from different types of neighbors to enhance sparse recommendation.

To summarize, the main contributions of this paper are as follows.

- Various neighbor graphs of attribute tag and rating tag are designed to learn attribute neighbor embedding and interaction neighbor embedding, which capture embedding signals of various neighbors at different levels.
- An attention network is developed to refine the collaborative semantics of multiscale neighbors, which is utilized to filter the irrelevance signals of various types of neighbors.
- A joint learning of multiscale neighbor embedding is proposed for rating prediction, which solves the problem of poor accuracy in the context of sparse recommendation.

The rest of this paper is structured as follows. Section 2 outlines related work, including the state-of-the-art of neighbor-based recommendation and attention mechanism. In Section 3, we present the framework of the proposed multiscale neighbor-aware attention network. Section 4 provides the methodology of the MSNAN recommendation. Section 5 describes the experimental setup and evaluation. Section 6 gives the experimental results and analysis. Finally, Section 7 concludes this work.

2. Related Work

In this section, the related work on neighbor-based recommendations of collaborative filtering and attention mechanisms are briefly reviewed.

2.1. Traditional Collaborative Filtering

Traditional collaborative filtering utilizes the nearest neighbors of a user or an item to generate recommendation results. Most popular methods leverage matrix factorization to learn the latent factors of users and items in terms of their historical information, such as BiasedMF [7], PMF [8], SVD++ [9], and LLORMA [10]. The decomposed low-dimensional factor vectors can be used to predict the user's rating preference for items. These methods mainly depend on the rating interactions between the user and item and have low performance in the case of sparse data. Thus, some side information is merged into matrix factorization to alleviate the sparsity of interaction records [11,12]. SSLIM [13] develops a sparse aggregation coefficient matrix by considering the user–item profiles and side information of items. The auto-encoder and -decoder techniques are used to learn the latent factors of nodes for collaborative filtering [14]. Park et al. [15] developed a group recommender system to select the suited recommendation items for store product placement. In recent years, some matrix decomposition models of deep learning have been studied [16,17]. DeepFM learns low-order and high-order interaction features of compressed interaction neighbors through a neural network [18]. Cai et al. [19] leveraged various multi-grained sentiment features and latent factors of matrix factorization to obtain sufficient representations of users and items to make rating predictions. Although these models have the ability to handle the sparsity recommendation problem, they leverage raw neighbors to learn their high-order interaction features and have limitations on processing different contributions of fine-grained neighbors.

2.2. Deep Learning-Based Collaborative Filtering

Deep learning-based methods can utilize the interaction neighbors between users and items to extract their latent vectors, which achieves excellent performance. Deep neural networks take advantage of both side information and feedback information to model linear and nonlinear features. For example, NFM [20] encodes interaction neighbor IDs of users, items, and their features into different vectors. Wide and Deep learning [21] combines the linear model and the deep neural network to implement efficient recommendations for scenarios with sparse data. Some works transfer diverse interactions of neighbors to learn rich features of nodes [1]. MCCF decomposes and recombines the latent components of user–item interaction graph to capture fine-grained user preference [22]. Aiming to solve the cold-start problem, Magron et al. [23] considered content information of acoustic features to learn the interaction between users and songs and proposed a neural content-aware collaborative filtering framework for music recommendation. Graph neural networks can learn the interaction characteristics of multi-order neighbors. MBGCN leverages multiple types of user-to-item interactions and the similarity item-to-item to propagate neighbor semantics [24]. Multiple neighbors on the path are used to capture the diversity of user interests and improve the accuracy of personalized recommendations [25]. User–item neighbor interaction and item–item neighbor relevance were leveraged to model a two-hop paths-based deep network to improve user engagement [26]. Tai et al. [27] designed a user-centric two-level path network in terms of entities of knowledge graph to generate user portfolio information. Duan et al. [28] learned the features of nodes from different dimensions of time and position and investigated the consistency of two representations for sequential recommendation. Most of neighbor embedding algorithms exploit interactive neighbors and do not distinguish the representations of different types of neighbors.

2.3. Attention Network for Collaborative Filtering

An attention mechanism is applied in collaborative filtering recommendations by identifying different importances of neighbors [5,29]. DGCF identifies the importance of diverse user–item interaction neighbors and models fine-grained intent-aware graph collaborative filtering [30]. By integrating content-based with collaborative filtering, ACCM considers the importance of end-to-end features and traditional attributes adaptively via an attention mechanism to handle the cold-start problem [3]. Chen et al. designed both item-level and component-level attention networks for multimedia recommendation [31]. Different auxiliary information can be learned to enhance important signals for neural networks. By leveraging intra-entity interaction and inter-entity interaction, AKUPM explores the relationships between users and other entity neighbors for alleviating the sparsity recommendation problem [32]. The attentional mechanism can provide different functions of multi-type entities on the knowledge graph and help obtain important neighbor information on different paths. KGAT learns the importance of higher-order neighbors of nodes in knowledge graphs by considering various features of entities [33]. A knowledge-aware attention mechanism is adopted to discriminate the contributions of different collaborative neighbors for recommender systems [34]. There are considerable advantages to applying the attention model in graph-based neural networks. A neural co-attention model utilizes auxiliary information of meta-based neighbors for top-n recommendation of heterogeneous information networks [35]. By leveraging the higher-order friends in the social network,

Xiao et al. [36] designed a social explorative attention network to make personal interest recommendations. Ye et al. [37] utilized both influence graph and the preference graph to fuse different user and item embeddings to make rating predictions. However, most attentional models deal with the role of the same type of neighbors, which limits the discriminative contributions of various neighbors to the user's overall decision making.

3. Framework of Multiscale Neighbor-Aware Attention Network

Figure 1 shows the framework of the proposed multiscale neighbor-aware attention network. The framework comprises four components: (1) attribute-view neighbor node embedding, (2) interaction-view neighbor node embedding, (3) attentional multiscale neighbor node embedding, and (4) rating prediction. In the framework, attribute-view neighbor node embedding is used to learn the different roles that similar neighbors with different attribute sets play in collaborative filtering. The interaction-view neighbor node embedding learns the role of similar neighbors with the same rating behavior in collaborative filtering. Neighbors with different attribute sets can form multiscale similar neighbors. Considering that users with different attributes have different rating behavior habits, we model the preferences of attribute sets for rating behavior through a multiscale neighbor-aware attention network. The attention network measures the collaborative contribution of coarse-scale attribute neighbors and fine-scale attribute neighbors to the rating decisions of target users.

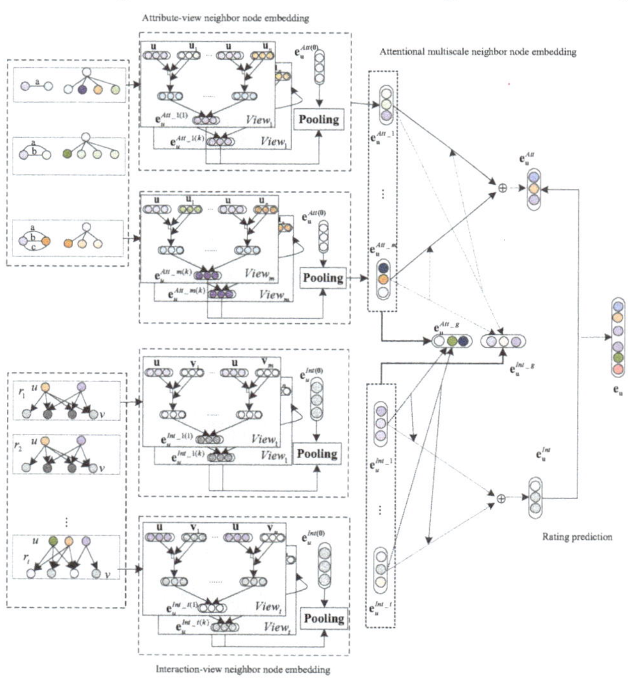

Figure 1. The framework of the multiscale neighbor-aware attention network.

In a nutshell, the framework works as follows. For the attribute-view neighbor node embedding, we first construct an attribute-view neighbor graph according to the association of a node on an attribute set such as {a} or {a,b}. Attribute sets of different scales induce multiple types of attribute neighbors. Based on various attribute-view neighbors, graph neural networks are used to obtain attribute-view neighbor node embedding.

For the interaction-view neighbor node embedding, we divide different rating tag spaces according to different rating grades. Under different rating tag spaces, we form similar neighbors with different rating behaviors. Then, graph neural networks are leveraged on different user–item interaction neighbor graphs to obtain various interaction-view neighbor node embedding.

Then, an attention mechanism is utilized to estimate the interactive contributions between various attribute neighbors and interaction neighbors. For the attribute-view and interaction-view neighbors, we calculate the global attribute neighbor collaborative signal and interactive neighbor collaborative signal, respectively. Meanwhile, we match the local collaborative signals between different-scale attribute neighbors and interactive neighbors. Considering the global and local semantic information provided by different neighbors, multiscale neighbor node embedding is computed to capture rich collaborative signals.

Finally, based on multiscale user embedding and item embedding, the inner product can be used to predict the rating score of the user to the item.

4. Methodology

4.1. Attribute-View Neighbor Embedding

In e-commerce networks, some attribute descriptions describe characteristics of users or products, which are helpful to discover various types of similar users or similar items. In this subsection, we utilize various types of attribute sets to calculate similar neighbors of different scales. Then, we learn the nodes' attribute-view neighbor embedding in terms of different-scale neighbor graphs.

Usually, users have various kinds of attributes, such as gender, age, occupation, and so on, which reflects users' interest preference to a certain extent. For example, users with the same gender can form a neighbor graph with a coarse-grained perspective, while users with the same gender and age can build a neighbor graph with a fine-grained attribute space. A coarse-grained neighbor can provide robust interest preferences for cold-start recommendation. A fine-grained neighbor helps discover refined similar preferences and model accurate collaborative recommendation. Based on this assumption, we can construct different views of attribute neighbor graphs to incorporate signals of various neighbors for modeling the embedding of nodes.

Given an attribute a, we can define a neighbor set N_u^a of user u on an attribute a as follows:

$$N_u^a = \{u^{'} | f_a(u) = f_a(u^{'})\} \tag{1}$$

where $f_a(u)$ is the attribute value of the user u on attribute a. N_u^a describes the collaborative neighbors with the same attributes as user u. Considering all attribute value types in the set A, we can construct the user-attribute relation matrix as $\mathbf{M}_{U\times A}$. Then, based on the user distribution of the set A, we can establish the neighbor relationship matrix of users by $\mathbf{MM}^T$, labeled as $\mathbf{M}_{U\times U}$. Here, various attribute-view neighbors reflect multiscale collaborative preference, which can affect the decision-making tendency of the target user.

Based on the idea in [33,38], we can define the collaborative signal of a first-order neighbor $u^{'}$ for user u on the attribute a as follows.

$$\mathbf{m}^a_{u\leftarrow u'} = \frac{\mathbf{u} + (\mathbf{u} \odot \mathbf{u}')}{|N_u^a||N_{u'}^a|} \tag{2}$$

Here, $\mathbf{m}^a_{u\leftarrow u'}$ represents the influence of similar neighbors in terms of attribute a on the target user u. $\mathbf{u}$ is the initialized embedding vector. Thus, considering all the first-order neighbors, attribute a-view collaborative signal for user u can be defined as $\mathbf{e}_u^{a(1)} = \sum_{u' \in N_u^a} \mathbf{m}^{a(1)}_{u\leftarrow u'}$. As is known, neighbors with the same attributes tend to spread their preferences in the social network. Considering the spread contributions of $k-1$ hop neighbors, the recursive collaborative signal of neighbor $u^{'}$ for user u on the attribute a can be formulated as follows [33,38].

$$\mathbf{m}^{a(k)}_{u\leftarrow u'} = \frac{\mathbf{u}^{(k-1)} + (\mathbf{u}^{(k-1)} \odot \mathbf{u}^{'(k-1)})}{|N_u^a||N_{u'}^a|} \tag{3}$$

Further, attribute a-view recursive collaborative signal for user u can be defined as $\mathbf{e}_u^{a(k)} = \sum_{u' \in N_u^a} \mathbf{m}_{u \leftarrow u'}^{a(k)}$. We adopt the average pooling to obtain the attribute a-view neighbor-aware node embedding for user u as $\mathbf{e}_u^{Att\_a} = agg(\mathbf{e}_u^{a(1)}, \cdots, \mathbf{e}_u^{a(k)})$. Here, some aggregation strategies can be used to fuse different orders of neighbor embedding.

Different types of attributes can induce different similar neighbors. Considering various types of neighbors in other attribute views, we can obtain attribute A-view neighbor-aware user embedding as $\mathbf{e}_u^{Att\_A} = avg(\mathbf{e}_u^{A(1)}, \cdots, \mathbf{e}_u^{A(k)})$. Similarly, given an item v, we adopt different types of item neighbors to obtain the attribute-view neighbor-aware item embedding.

Different users have special behavioral perceptions of rating labels, which can be used to explore users' behavioral preferences in fine granularity. Thus, according to the types of rating labels, we also divide different interaction view spaces with rating labels and construct various rating interaction graphs for learning user and item embeddings. For example, we can think of item groups with the same rating as neighbors of a user with the same scale preference. Based on different rating labels, we model interaction-view neighbor-aware embedding with different rating neighbors, which can be defined as $\{\mathbf{e}_u^{Int_1}, \cdots, \mathbf{e}_u^{Int_r}\}$. Then, interaction-view neighbor-aware item embedding of an item v can be defined as $\{\mathbf{e}_v^{Int_1}, \cdots, \mathbf{e}_v^{Int_r}\}$.

4.2. Cross Attention-Based Multiscale Neighbor Embedding

In the homogeneous attribute view, the multiscale neighbor-aware embedding signals with different granularity can provide diversified collaborative signals for node representation. In addition, various neighbor-aware embeddings of attribute view and interactive view can be used to model the heterogeneous collaborative signals, which can enrich and enhance the representation ability of node embedding.

According to neighbor-aware node embedding on different attribute-view spaces, we can model global attribute neighbor-aware node embedding for user u as follows.

$$q_i = \mathbf{h}_i \mathbf{e}_u^{Att\_i} \tag{4}$$

Equation (4) describes the influence of user embedding in the attribute i space on the global attribute neighbor-aware user embedding. $\mathbf{h}_i$ is the parameter vector. Considering different user embeddings of m spatial types, the normalized weights are defined in Equation (5).

$$\alpha_i = \frac{e^{q_i}}{\sum_{s \in \{1,\ldots,m\}} e^{q_s}} \tag{5}$$

Then, the global attribute neighbor-aware node embedding for user u is defined as follows.

$$\mathbf{e}_u^{Att\_g} = \sum_{i \in \{1,\ldots,m\}} \alpha_i \mathbf{e}_u^{Att\_i} \tag{6}$$

In Equation (5), different neighbor-aware user embedding representations depict discriminative semantic information from various attribute-induced neighbors, which provides collaborative signals of neighbors comprehensively for target user's decisions. The global collaborative embedding $\mathbf{e}_u^{Att\_g}$ of various neighbors can improve the semantic ability for recommender systems. Similarly, we can obtain the global collaborative embedding of interaction view spaces as $\mathbf{e}_u^{Int\_g}$.

Given the item v, we can also obtain global attribute-neighbor-aware item embedding and interaction-neighbor-aware item embedding as $\mathbf{e}_v^{Att\_g}$ and $\mathbf{e}_v^{Int\_g}$.

Further, taking into account the collaborative signals matched by different neighbor embedding from the attribute and interactive views, we utilize cross-attention to model matched neighbor embedding. Based on the global collaborative embedding of interaction view, we can compute matched neighbor embedding of attribute view for user u as follows.

$$\beta_i = \mathbf{e}_u^{Int\_g} \mathbf{e}_u^{Att\_i} \tag{7}$$

Here, Equation (7) defines the preferential influence of user embedding in attribute i's view on the user's rating decision. Then, we normalize this influence using Equation (8).

$$\gamma_i = \frac{e^{\beta_i}}{\sum\limits_{s \in \{1,\ldots,m\}} e^{\beta_s}} \tag{8}$$

The normalized weight γ_i reflects the influence of multiscale attribute neighbor embedding for the user's interaction rating. Furthermore, the attribute embedding by incorporating user rating behavior preference can be updated as in Equation (9).

$$\mathbf{e}_u^{Att} = \sum_{i \in \{1,\ldots,m\}} \gamma_i \mathbf{e}_u^{Att\_i} \tag{9}$$

The multi-type matched signals based on attribute neighbors and interactive neighbors can enhance the embedding representation of nodes. Similarly, we can calculate the matched neighbor embedding of the interactive view as below.

$$w_j = \mathbf{e}_u^{Att\_g} \mathbf{e}_u^{Int\_j} \tag{10}$$

Here, w_j is the dependency influence of user embedding in rating tag j's view on user's attributes. In terms of different rating levels, this dependence effect can be normalized using Equation (11).

$$g_j = \frac{e^{w_j}}{\sum\limits_{p \in \{1,\ldots,t\}} e^{w_p}} \tag{11}$$

$$\mathbf{e}_u^{Int} = \sum_{p \in \{1,\ldots,t\}} g_j \mathbf{e}_u^{Int\_j} \tag{12}$$

In Equation (12), $\mathbf{e}_u^{Int}$ represents the user's interaction embedding incorporating the dependency of user attributes. Considering the matched neighbor embedding in the attribute view and neighbor embedding in the interactive view, we can model fused multiscale neighbor embedding with the concatenation operator as follows.

$$\mathbf{e}_u = \mathbf{e}_u^{Att} || \mathbf{e}_u^{Int} \tag{13}$$

4.3. Rating Prediction

Based on the fused multiscale neighbor embedding of users and items, the dot product is used to predict the rating of the user u on an item v, which is shown as follows:

$$\hat{y} = \mathbf{e}_u \cdot \mathbf{e}_v + b_g + b_u + b_v \tag{14}$$

Here, the parameters b_g, b_u, and b_v are global bias, user bias, and item bias. To preserve the preference information of user attributes over item attributes, we define rating prediction of user u for the item v with their attribute-view neighbor embeddings as follows:

$$\hat{y}_{Att} = \mathbf{e}_u^{Att} \cdot \mathbf{e}_v^{Att} + b_g + b_u + b_v \tag{15}$$

Similarly, we can compute the rating with the interaction-view neighbor embeddings in interactive space, which is shown as below.

$$\hat{y}_{Int} = \mathbf{e}_u^{Int} \cdot \mathbf{e}_v^{Int} + b_g + b_u + b_v \tag{16}$$

In the process of model optimization, to observe the influence of attribute neighbors and interaction neighbors on rating prediction, we define the joint root mean squared error (RMSE) loss function as follows.

$$L = \lambda RMSE(\hat{y}, y) + \lambda_{Att} RMSE(\hat{y}_{Att}, y) + \lambda_{Int} RMSE(\hat{y}_{Int}, y) \tag{17}$$

The joint loss function considers the global and local important neighbors to predict the user's rating of the item, which not only captures the user's attribute preference for the item from various attribute-view neighbors, but also retains the behavioral preference of collaborative fine-grained interaction neighbors.

5. Experiments

In this section, we verify the performance of the proposed model with the aim of answering the following three questions:

- RQ1: How does MSNAN perform compared with state-of-the-art neighbor-based collaborative filtering methods?
- RQ2: How does the multiscale neighbor node embedding perform for sparsity recommendation?
- RQ3: How do different types of neighbor embedding affect the performance of the model?

5.1. Dataset

We ran the proposed MSNAN model and baselines on three public datasets: Movielens-100kr (ML-100kr) (https://grouplens.org/datasets/movielens/ (accessed on 24 September 2023)), Book-Crossing-10croe (BK-10C) (http://bookcrossing.com (accessed on 24 September 2023)), and Douban (https://movie.douban.com/ (accessed on 24 September 2023)) datasets to verify their effectiveness. The ML-100kr contains the interaction ratings of 943 users on 1682 movies. Users have 5 attributes, and movies have 19 attributes. The rating score is on the scale 1–5. For the BK-10C dataset, we selected users who had rated at least 10 books and books that have been rated by at least 10 users, which involved 1820 users and 2030 books. The rating score uses the range of 1–10. The Douban dataset contains ratings of 6971 movies from 3022 users with rating values of 1–5 [39]. On all datasets, the higher the rating is, the more the user likes the movie/book. Statistical information of experimental datasets is shown in Table 1. During the experiment, we use the MAE and RMSE indexes for performance evaluation. The smaller MAE and RMSE values indicate better performance. All the datasets were divided into training set, validation set, and testing set with a proportion of 8:1:1. The rating prediction performance of the model is evaluated on the testing set.

Table 1. Statistical information of experimental datasets.

Datasets	Users	Items	Interactions	Rating	Sparsity
ML-100kr	943	1682	100,000	1–5	94.12%
BK-10C	1820	2030	41,456	1–10	98.87%
Douban	3022	6971	195,493	1–5	99.07%

5.2. Baseline Models

To test the contribution of MSNAN method, we employ traditional methods, deep learning-based recommendations, and GCN-based models to compare the performance. Several baselines are as follows.

(1) NCF [17]. A neural network method is used to learn the interaction information between users and items for collaborative filtering.

(2) Wide and Deep [20]. A method combining a generalized linear model and deep neural network is designed to improve the performance of recommender systems.

(3) NGCF [33]. A collaborative filtering method based on a graph neural network learns the embedding representations of users and items with a user–item interaction graph.
(4) GCN [40]. A graph convolutional neural network leverages the information of multi-order neighbors by superimposing several convolutional layers for recommendation.
(5) LightGCN [38]. The embedding representations of user and item are learned by aggregating linear neighbor information of nodes.
(6) GAT [41]. A graph convolutional neural network method together with attention mechanism learns weighted node embedding representation for recommendation.
(7) ACCM [3]. The attention mechanism is used to integrate a content-based method with collaborative filtering for rating prediction.
(8) AFM [5]. An attentional network factorization method learns the interactive importance of different features for prediction.
(9) TANP [42] A task-adaptive neural network is constructed to learn the relevance of different tasks for user cold-start recommendations.

5.3. Parameter Settings

In the experiment, we adopt a grid search to obtain the parameters for optimizing the performance of model. For the attention mechanism, we tune its dimension with the values {32, 64, 128, 256}. To prevent the model overfitting, we adjust the dropout values {0.1, 0.2, 0.3, 0.4, 0.5}. We utilize stochastic gradient descent to optimize the model with a learning rate of 0.01. In order to obtain a better comparison, we use inner product in the prediction layer for NGCF, GCN, LightGCN, and GAT methods to make rating predictions.

5.4. Experiment Settings

For the proposed multiscale neighbor method, the NGCF graph convolution model is used in Equations (2) and (3) to capture the cooperative signal propagation between nodes. We can also use other signal propagation methods to model the initial representation of nodes in a certain view, and then observe the effect of multiscale neighbors. During the experiment, for the multiscale neighbor graph, we adopted different graph neural network models, such as NGCF, GCN, LightGCN, and GAT, to learn various neighbor embedding and observe the robust effect of multiscale neighbor signals. In the experiment, the hyperparameters λ, λ_{Att}, and λ_{Int} are set to 0.01.

6. Experimental Results

In this subsection, we compare the proposed MSNAN model with the benchmark models in terms of MAE and RMSE metrics on three datasets. The experimental results are shown in Table 2. In Table 2, we have the following observations from the experimental results.

Table 2. MAE and RMSE results of different methods on three datasets.

Method	ML-100kr		BK-10C		Douban	
	MAE	RMSE	MAE	RMSE	MAE	RMSE
NCF	0.7457	0.9342	1.1611	1.5288	0.5781	0.7304
NGCF	0.7298	0.9195	1.1241	1.4776	0.5768	0.7271
GCN	0.7253	0.9178	1.1333	<u>1.4772</u>	0.5766	0.7259
LightGCN	0.7260	0.9182	1.1166	1.4794	0.5709	0.7213
GAT	0.7257	0.9186	1.1250	1.4813	0.5770	0.7238
AFM	0.7319	0.9238	1.1170	1.4786	<u>0.5643</u>	<u>0.7136</u>
Wide&Deep	0.7204	0.9152	<u>1.1151</u>	1.4807	0.5654	0.7141
ACCM	<u>0.7145</u>	<u>0.9027</u>	1.1983	1.5373	0.5789	0.7301
TANP	0.7881	0.9757	1.2081	1.5336	0.5767	0.7274
MSNAN + NGCF	**0.6910**	0.8842	1.1029	1.4586	0.5536	0.7069
MSNAN + GCN	0.7008	0.8890	1.1038	1.4598	0.5643	0.7091
MSNAN + LightGCN	0.6949	**0.8815**	**1.1003**	**1.4563**	**0.5517**	**0.7003**
MSNAN + GAT	0.6973	0.8901	1.1047	1.4607	0.5617	0.7034

Through the comparison results, the performance of MSNAN shows excellent improvements on three datasets. For example, compared with the ACCM method of collaborative filtering, the MAE and RMSE values of the MSNAN + LightGCN method increase by 2.74%, 2.35%, 8.18%, 5.27%, 4.70%, and 4.08% on three datasets, respectively. In addition, the MAE and RMSE of the MSNAN + NGCF model have achievements of 3.29%, 2.05%, 1.09%, 1.26%, 1.90%, and 0.94% over the best baseline on three datasets, respectively. Meanwhile, the improvements of the MSNAN + LightGCN model to the best baseline are 2.74% and 2.35% for MAE and RMSE on the ML-100kr dataset, 1.33% and 1.41% on the BK-10C dataset, and 2.23% and 1.86% on the Douban dataset, respectively. This shows that the proposed MSNAN model can better learn the embedding representation of users and items and effectively improve the accuracy of rating prediction, which verifies the significance of collaborative semantics of multiscale neighbors.

Compared with these graph neural network baselines, the proposed MSNAN based on multiscale neighbors achieves competitive improvements on all datasets. For example, on the ML-100kr dataset, the MSNAN + NGCF, MSNAN + GCN, MSNAN + LightGCN, and MSNAN + GAT methods improve by 5.32%, 3.83%; 3.38%, 3.14%; 4.28%, 4.00%; and 3.91%, 3.10% over the NGCF, GCN, LightGCN, and GAT baselines on MAE and RMSE metrics. In addition, on the BK-10C dataset, the MSNAN + NGCF, MSNAN + GCN, MSNAN + LightGCN, and MSNAN + GAT methods also improve MAE and RMSE values by 1.89%, 1.29%; 2.60%, 1.18%; 1.46%, 1.56%; and 1.80%, 1.39% over the corresponding graph model baselines, respectively. This is because the node embedding of MSNAN combines collaborative semantics of multiscale neighbors, which purifies the important information of similar neighbors in rating prediction. However, for different graph neural network models, the representation quality of nodes is different, and the superposition effect of MSNAN is also different. As is known, LightGCN simplifies the transformation matrix and activation function, which obtains the highest embedding quality of node representation, and can also improve the performance of this method. The performance of the general GCN method is relatively backward. In addition, compared with ACCM and AFM methods, the MSNAN approach can achieve a smaller error in the rating prediction scenario, which indicates that the proposed model can better learn the embedding representation of users and items with the collaborative signal of multiscale neighbors.

7. Discussion and Analysis

7.1. Sparsity Analysis

Sparse recommendation is a challenging problem in recommender systems. The sparsity problem of the user–item interaction matrix makes it difficult to learn the semantic representation of users and items, and reduces the accuracy of rating prediction results. In order to observe the robust advantage of multiscale neighbor signals for sparse recommendation, we conduct the comparative experiments by randomly shielding the rating labels and setting different sparsity proportions for three datasets. In the experiment, we compare the performance of proposed MSNAN model with several graph neural network methods on MAE and RMSE metrics. The experimental results are shown in Figures 2–7.

In the above figures, with the increase in data sparsity, the MAE and RMSE values of all methods show an increasing trend, which indicates that the error of rating prediction is gradually increasing. This demonstrates that high data sparsity can affect the quality of embedding representation of users and items and reduces the accuracy of recommendation results. In addition, with the improvement in data sparsity, the MAE and RMSE values of the graph neural network methods fluctuate. This is because the benchmark models of graph neural networks only consider one type of neighbor information of the user–item interaction matrix. Insufficient interaction records lead to the difficulties of higher-order information propagation of single-type neighbors, which limits the stability of node embedding representation.

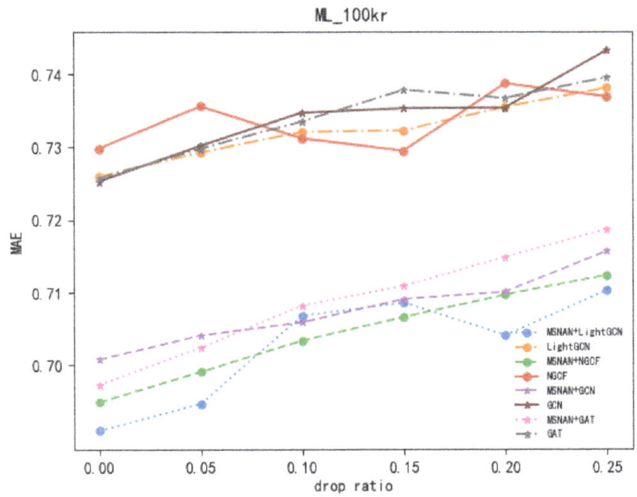

Figure 2. MAE results of different methods on the ML-100kr dataset with different sparsity proportions.

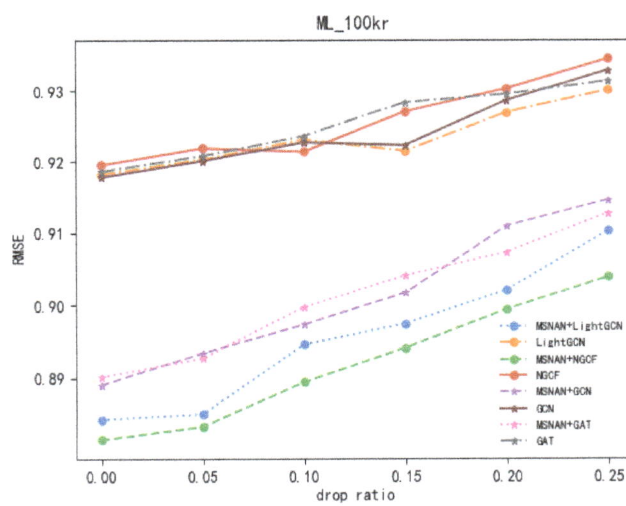

Figure 3. RMSE results of different methods on the ML-100kr dataset with different sparsity proportions.

Compared with graph neural network methods, the proposed MSNAN model consistently yields the best performance for MAE and RMSE on three datasets to adapt the sparsity ratio of different scales. For different graph neural network models, the result of blending the MSNAN model is more generalized than the original model. As shown in Figures 4–7, when there is a low drop ratio, LightGCN can obtain better performance compared with other graph neural network methods due to its better node representation quality. This is because LightGCN itself effectively learns the embedding representation of nodes by simplifying the nonlinear structure and reducing complexity. In large-scale e-commerce platforms, the sparsity of user–item interactions can be high. It is difficult to find collaborative neighbors based on the similarity of interaction behaviors. Although the user–item matrix loses a part of interaction records, the proposed model fuses various neighbor information from attribute views and rating views, which fully learns the representations of users and items. According to the attributes of users or items, we can find neighbors of different scales and conduct collaborative filtering recommendation.

Moreover, the model takes advantage of high-quality collaborative signals from multiscale neighbors for improving the quality of embedding representation, which is suitable for the practice of large-scale e-commerce and alleviates the performance impact of data sparsity to some extent.

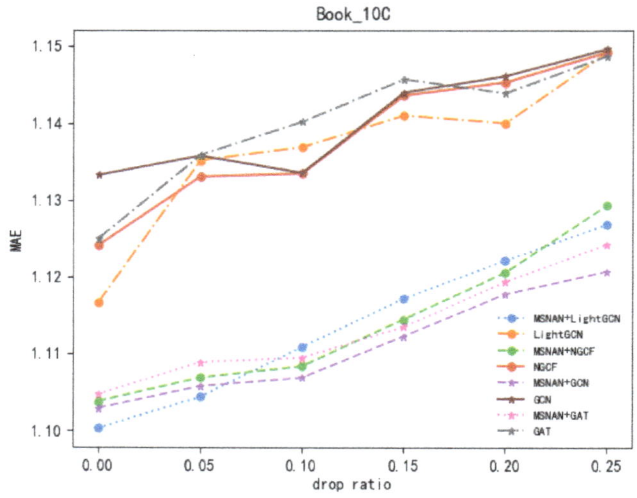

Figure 4. MAE results of different methods on the BK-10C dataset with different sparsity proportions.

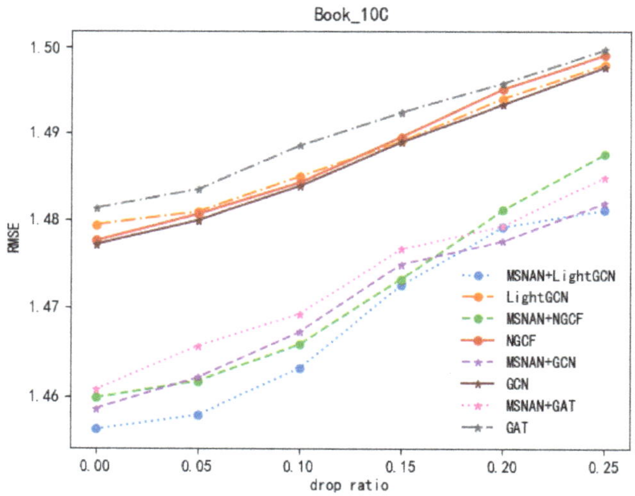

Figure 5. RMSE results of different methods on the BK-10C dataset with different sparsity proportions.

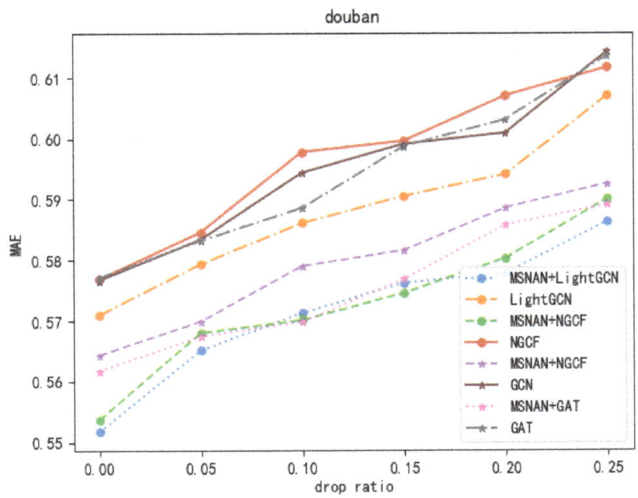

Figure 6. MAE results of different methods on the Douban dataset with different sparsity proportions.

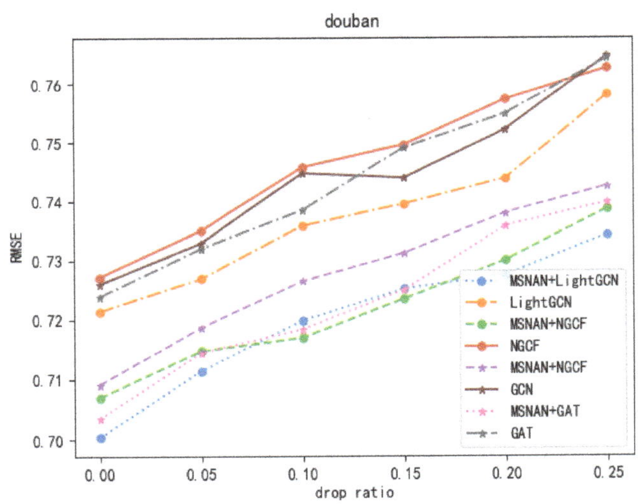

Figure 7. RMSE results of different methods on the Douban dataset with different sparsity proportions.

7.2. Impact of Neighbor Embedding

For the multiscale neighbor embedding, the neighbor embedding of different views contributes differently, which affects the result of rating prediction. To observe the roles of different types of neighbor embedding, Figures 8–10 give the performance of methods by removing different attribute-view neighbor embedding. In Figure 8, for the ML-100kr dataset, the users have three types of neighbor embedding. We can see that the MAE and RMSE of three att-view neighbor embedding change greatly compared with the other two types, which states that three att-view neighbor embedding plays an important role in predicting the interest rating of the target user. That is, through the neighbor interaction graphs with multiple similar attributes, the multiscale neighbor embedding achieves a stronger ability to express the preference of target user, which also demonstrates that neighbors with fine-grained interests play a greater role in collaborative recommendation. Similarly, for the BK-10C dataset, after removing the four att-view neighbor embedding, the MAE and RMSE values change the most, which indicates that four att-view neighbor embedding makes the greatest contribution on modeling the preference of the user. For the Douban dataset, two att-view neighbor embedding is important. These results indicate that multiscale neighbors on different views have a collaborative effect on the user's preference decisions.

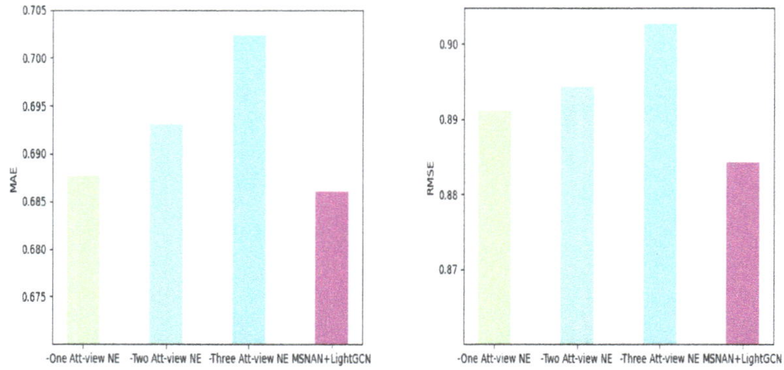

(a) MAE (b) RMSE

Figure 8. Ablation study of multiscale neighbor embedding on the ML-100kr dataset.

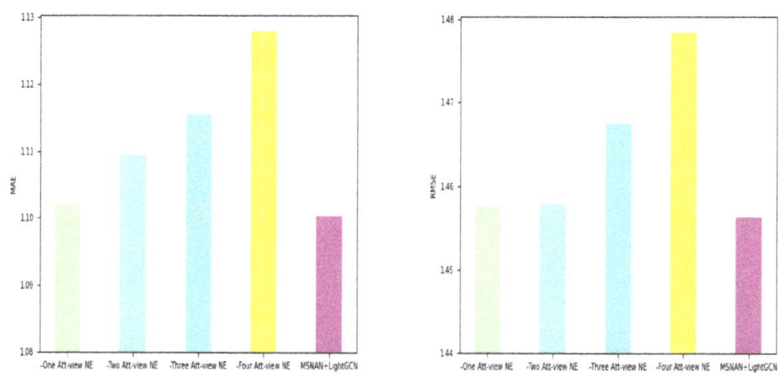

(a) MAE (b) RMSE

Figure 9. Ablation study of multiscale neighbor embedding on the BK-10C dataset.

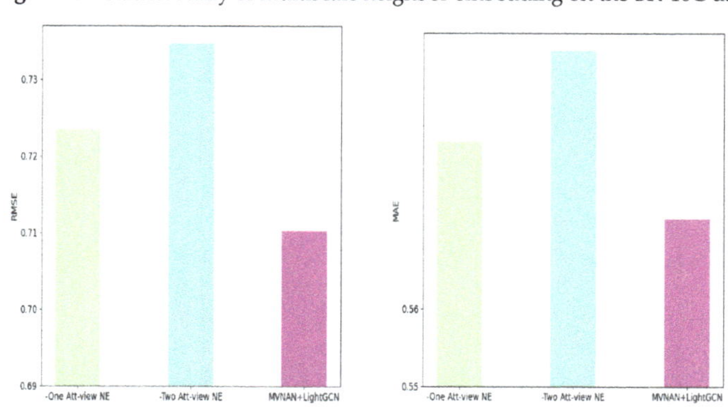

(a) MAE (b) RMSE

Figure 10. Ablation study of multiscale neighbor embedding on the Douban dataset.

7.3. Visual Explanation Study

In order to further observe the influence of multiscale neighbor embedding on user decision-making, Figures 11–13 report the attention weight of attribute-view neighbor embedding and interaction-view neighbor embedding of 10 users for three datasets. In Figure 11a, the weight of three att-view is the largest, which indicates that similar neighbors with multiple attributes can have a major contribution on capturing collaborative signals. One att-view neighbor embedding can also capture coarse-grained semantic interests for collaborative filtering. This important insight allows us to select neighbors for users and items with different numbers of attributes, and capture rich signals for collaborative filtering recommendation tasks. In addition, in Figure 11, the user's final embedding representation is most affected by the four att-view embedding and eight sco-view embedding. By observing the dataset of BK-10Core, we find that most users are used to evaluating the items with high score values, which leads to a large contribution of the high-score-view embedding. Meanwhile, the rating neighbor matrix is relatively sparse, which makes the difference between the weight values of various types of high-score-view embeddings small.

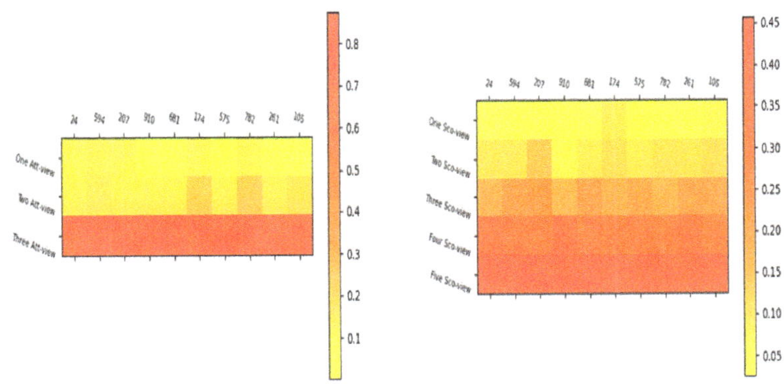

(**a**) Attribute-view (**b**) Interaction-view

Figure 11. Attention weight of multiscale neighbor embedding for 10 users on the ML-100kr dataset.

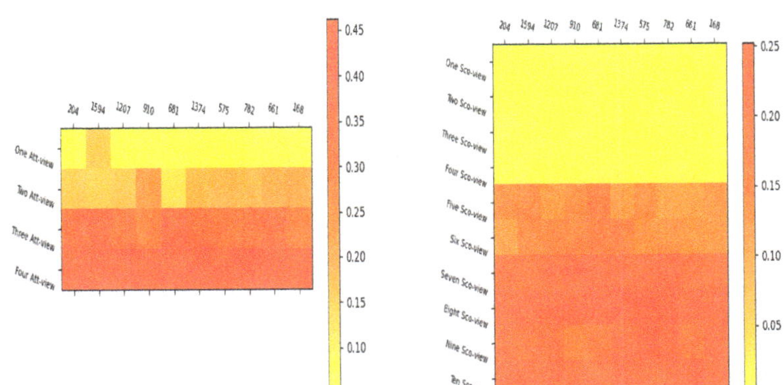

(**a**) Attribute-view (**b**) Interaction-view

Figure 12. Attention weight of multiscale neighbor embedding for 10 users on the BK-10C dataset.

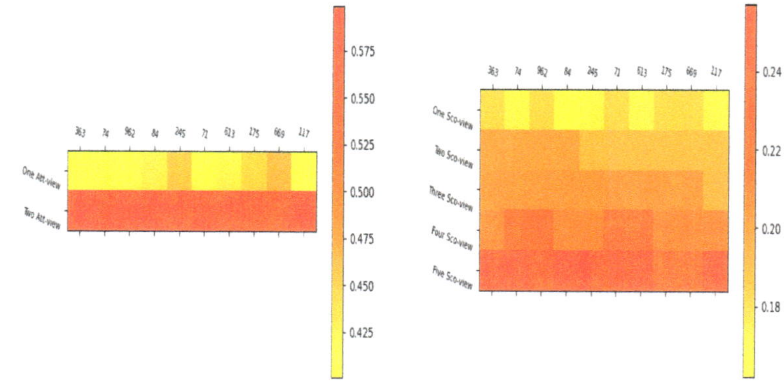

(**a**) Attribute-view (**b**) Interaction-view

Figure 13. Attention weight of multiscale neighbor embedding for 10 users on the Douban dataset.

8. Conclusions

In this paper, we propose a multiscale neighbor-aware attention network for collaborative filtering recommendation. The proposed strategy fuses the global semantics of various types of neighbors and important local embedding of multiscale neighbors. Multiple attribute-view neighbors and interaction-view neighbors provide collaborative signals to predict the user's rating of items. Experiments verify the effectiveness of collaborative contributions of multiscale neighbors for learning user and item representation. The key finding is that the combination of multiscale attribute neighbors and interactive neighbors can improve the accuracy of recommendation, and alleviate the poor performance of recommendation in the case of sparse data. The disadvantage is that the computation of multiscale neighbors requires different graph structures to learn the representation of nodes. The platform can construct the graph structure offline and calculate multiscale neighbors, which can reduce the online resource requirements. However, in the e-commerce scenario, the proposed method can realize targeted personalized recommendation according to different attribute neighbors of users. In particular, for some cold-start users who do not have interactive behaviors, the method can select neighbors with similar attributes for the target user according to one's social attribute set, and then conduct collaborative filtering recommendation. Some products and services can be recommended based on similar attribute preferences of neighbors for target users. In addition, by combining with the behavioral preferences of group users, we can make rating predictions and recommend popular products for target users.

In future work, by investigating the semantic difference of various attribute and interactive behavior views, we can focus on a consistency study of node representations on different behavior views and improve the accuracy and interpretability of the recommendation system. In addition, heterogeneous types of semantic information from different types of user behaviors such as evaluation, clicking, and buying can describe the ordered semantic interests of the user. We will distinguish the types of multiple interaction behaviors to learn heterogeneous semantic representations and model the sequential relations between different behaviors.

Author Contributions: Methodology, J.Z., F.C. and Q.C.; Software, T.J.; Investigation, J.Z. and Y.L.; Writing—original draft, J.Z.; Writing—review & editing, Y.K. and Q.C. All authors have read and agreed to the published version of the manuscript.

Funding: This work was partially supported by the National Natural Science Foundation of China (nos. 62272286, 62072291), the Natural Science Foundation of Shanxi Province (nos. 20210302123468, 202203021221021, 202203021221001).

Data Availability Statement: Data used in this manuscript consist of publicly available standard benchmark datasets.

Conflicts of Interest: The authors declare no conflict of interest.

References

1. Li, Z.; Cui, Z.; Wu, S.; Zhang, X.; Wang, L. Fi-gnn: Modeling feature interactions via graph neural networks for ctr prediction. In Proceedings of the 28th ACM International Conference on Information and Knowledge Management, Beijing, China, 3–10 November 2019; pp. 539–548.
2. Naghiaei, M.; Rahmani, H.; Deldjoo, Y. CPFair: Personalized Consumer and Producer Fairness Re-ranking for Recommender Systems. In Proceedings of the 45th International ACM SIGIR Conference on Research and Development in Information Retrieval, Madrid, Spain, 11–15 July 2022; pp. 770–779.
3. Shi, S.; Zhang, M.; Liu, Y.; Ma, S. Attention-based adaptive model to unify warm and cold starts recommendation. In Proceedings of the 27th ACM International Conference on Information and Knowledge Management, Turin, Italy, 22–26 October 2018; pp. 127–136.
4. Ge, Y.; Tan, J.; Zhu, Y.; Xia, Y.; Luo, J.; Liu, S.; Fu, Z.; Geng, S.; Li, Z.; Zhang, Y. Explainable Fairness for Feature-aware Recommender Systems. In Proceedings of the 45th International ACM SIGIR Conference on Research and Development in Information Retrieval, Madrid, Spain, 11–15 July 2022; pp. 1–11.
5. Xiao, J.; Ye, H.; He, X.; Zhang, H.; Wu, F.; Chua, T. Attentional factorization machines: Learning the weight of feature interactions via attention networks. In Proceedings of the 26th International Joint Conference on Artificial Intelligence, Beijing, China, 21–22 May 2017; pp. 3119–3125.
6. Wang, X.; He, X.; Cao, Y.; Liu, M.; Chua, T. Kgat: Knowledge graph attention network for recommendation. In Proceedings of the 25th ACM SIGKDD International Conference on Knowledge Discovery & Data Mining, Anchorage, AK, USA, 4–8 August 2019; pp. 950–958.
7. Koren, Y.; Bell, R.; Volinsky, C. Matrix Factorization Techniques for Recommender Systems. *Computer* **2009**, *42*, 30–37. [CrossRef]
8. Salakhutdinov, R.; Mnih, A. Probabilistic matrix factorization. In Proceedings of the 20th International Conference on Neural Information Processing Systems, Vancouver, WA, USA, 3–6 December 2007; pp. 849–858.
9. Koren, Y. Factorization meets the neighborhood: A multifaceted collaborative filtering model. In Proceedings of the 14th ACM SIGKDD International Conference on Knowledge Discovery and Data Mining, Las Vegas, NV, USA, 24–27 August 2008; pp. 426–434.
10. Lee, J.; Kim, S.; Lebanon, G.; Singer, Y. Local low-rank matrix approximation. In Proceedings of the 30th International Conference on International Conference on Machine Learning, Atlanta, GA, USA, 16–21 June 2013; pp. 82–90.
11. Ling, G.; Lyu, M.; King, I. Ratings meet reviews, a combined approach to recommend. In Proceedings of the 8th ACM Conference on Recommender Systems, Foster City, CA, USA, 6–10 October 2014; pp. 105–112.
12. Rendle, S.; Freudenthaler, C. Improving pairwise learning for item recommendation from implicit feedback. In Proceedings of the 7th ACM International Conference on Web Search and Data Mining, New York, NY, USA, 24–28 February 2014; pp. 273–282.
13. Ning, X.; Karypis, G. Sparse linear methods with side information for top-n recommendations. In Proceedings of the 6th ACM Conference on Recommender Systems, Dublin, Ireland, 9–13 September 2012; pp. 155–162.
14. Sedhain, S.; Menon, A.; Sanner, S.; Xie, L. Autorec: Autoencoders meet collaborative filtering. In Proceedings of the 24th International Conference on World Wide Web, Florence, Italy, 18–22 May 2015; pp. 111–112.
15. Park, J.; Nam, K. Group recommender system for store product placement. *Data Min. Knowl. Disc.* **2019**, *33*, 204–229. [CrossRef]
16. Zheng, J.; Liu, J.; Shi, C.; Zhuang, F.; Li, J.; Wu, B. Dual Similarity Regularization for Recommendation. In Proceedings of the 2016 Pacific-Asia Conference on Knowledge Discovery and Data Mining, Auckland, New Zealand, 19–22 April 2016; pp. 542–554.
17. He, X.; Liao, L.; Zhang, H.; Nie, L.; Hu, X.; Chua, T. Neural collaborative filtering. In Proceedings of the 26th International Conference on World Wide Web, Perth, Australia, 3–7 April 2017; pp. 173–182.
18. Lian, J.; Zhou, X.; Zhang, F.; Chen, Z.; Xie, X. DeepFM: Combining explicit and implicit feature interactions for recommender systems. In Proceedings of the 24th ACM SIGKDD International Conference on Knowledge Discovery and Data Mining, London, UK, 19–23 August 2018; pp. 1754–1763.
19. Cai, Y.; Ke, W.; Cui, E.; Yu, F. A deep recommendation model of cross-grained sentiments of user reviews and ratings. *Inf. Process Manag.* **2022**, *59*, 102842. [CrossRef]
20. Cheng, H.; Koc, L.; Harmsen, J.; Shaked, T.; Chandra, T.; Aradhye, H.; Anderson, G.; Corrado, G.; Chai, W.; Ispir, M.; et al. Wide & deep learning for recommender systems. In Proceedings of the 1st Workshop on Deep Learning for Recommender Systems, Boston, MA, USA, 15–19 September 2016; pp. 7–10.
21. He, X.; Chua, T. Neural factorization machines for sparse predictive analytics. In Proceedings of the 40th International ACM SIGIR conference on Research and Development in Information Retrieval, New York, NY, USA, 7–12 February 2017; pp. 355–364.
22. Wang, X.; Wang, R.; Shi, C.; Song, G.; Li, Q. Multi-component graph convolutional collaborative filtering. In Proceedings of the AAAI Conference on Artificial Intelligence, New York, NY, USA, 7–12 February 2020; pp. 6267–6274.
23. Magron, P.; Fevotte, C. Neural content-aware collaborative filtering for cold-start music recommendation. *Data Min. Knowl. Disc.* **2022**, *36*, 1971–2005. [CrossRef]

24. Jin, B.; Gao, C.; He, X.; Jin, D.; Li, Y. Multi-behavior recommendation with graph convolutional networks. In Proceedings of the 43rd International ACM SIGIR Conference on Research and Development in Information Retrieval, Xi'an, China, 25–30 July 2020; pp. 659–668.
25. Su, Z.; Dou, Z.; Zhu, Y.; Qin, X.; Wen, J. Modeling Intent Graph for Search Result Diversification. In Proceedings of the 44th International ACM SIGIR Conference on Research and Development in Information Retrieval, Online, 11–15 July 2021; pp. 736–746.
26. Li, H.; Chen, Z.; Li, C.; Xiao, R.; Deng, H.; Zhang, P.; Liu, Y.; Tang, H. Path-based Deep Network for Candidate Item Matching in Recommenders. In Proceedings of the 44th International ACM SIGIR Conference on Research and Development in Information Retrieval, Online, 11–15 July 2021; pp. 1493–1502.
27. Tai, C.; Huang, L.; Huang, C.; Ku, L. User-Centric Path Reasoning towards Explainable Recommendation. In Proceedings of the 44th International ACM SIGIR Conference on Research and Development in Information Retrieval, Online, 11–15 July 2021; pp. 879–889.
28. Duan, H.; Zhu, Y.; Liang, X.; Zhu, Z.; Liu, P. Multi-feature fused collaborative attention network for sequential recommendation with semantic-enriched contrastive learning. *Inf. Process Manag.* **2023**, *60*, 103416. [CrossRef]
29. Vaswani, A.; Shazeer, N.; Parmar, N.; Uszkoreit, J.; Jones, L.; Gomez, A.; Kaiser, L.; Polosukhin, I. Attention is all you need. In Proceedings of the 31st Advances in Neural Information Processing Systems, Long Beach, CA, USA, 4–9 December 2017; pp. 5998–6008.
30. Wang, X.; Jin, H.; Zhang, A.; He, X.; Xu, T.; Chua, T. Disentangled graph collaborative filtering. In Proceedings of the 43rd International ACM SIGIR Conference on Research and Development in Information Retrieval, Xi'an, China, 25–30 July 2020; pp. 1001–1010.
31. Chen, J.; Zhang, H.; He, X.; Nie, L.; Liu, W.; Chua, T. Attentive collaborative filtering: Multimedia recommendation with item-and component-level attention. In Proceedings of the 40th International ACM SIGIR Conference on Research and Development in Information Retrieval, Tokyo, Japan, 7–11 August 2017; pp. 335–344.
32. Tang, X.; Wang, T.; Yang, H.; Song, H. AKUPM: Attention-enhanced knowledge-aware user preference model for recommendation. In Proceedings of the 25th ACM SIGKDD International Conference on Knowledge Discovery & Data Mining, Anchorage, AK, USA, 4–8 August 2019; pp. 1891–1899.
33. Wang, X.; He, X.; Wang, M.; Feng, F.; Chua, T. Neural graph collaborative filtering. In Proceedings of the 42nd International ACM SIGIR Conference on Research and Development in Information Retrieval, Paris, France, 21–25 July 2019; pp. 165–174.
34. Wang, Z.; Lin, G.; Tan, H.; Chen, Q.; Liu, X. CKAN: Collaborative knowledge-aware attentive network for recommender systems. In Proceedings of the 43rd International ACM SIGIR Conference on Research and Development in Information Retrieval, Xi'an, China, 25–30 July 2020; pp. 219–228.
35. Niu, G.; Li, Y.; Tang, C.; Geng, R.; Dai, J.; Liu, Q.; Wang, H.; Sun, J.; Huang, F.; Si, L. Relational Learning with Gated and Attentive Neighbor Aggregator for Few-Shot Knowledge Graph Completion. In Proceedings of the 44th International ACM SIGIR Conference on Research and Development in Information Retrieval, Online, 11–15 July 2021; pp. 213–222.
36. Xiao, W.; Zhao, H.; Pan, H.; Song, Y.; Zheng, V.; Yang, Q. Social explorative attention based recommendation for content distribution platforms. *Data Min. Knowl. Disc.* **2021**, *35*, 533–567. [CrossRef]
37. Ye, H.; Song, Y.; Li, M.; Cao, F. A new deep graph attention approach with influence and preference relationship reconstruction for rate prediction recommendation. *Inf. Process Manag.* **2023**, *60*, 103439. [CrossRef]
38. He, X.; Deng, K.; Wang, X.; Li, Y.; Zhang, Y.; Wang, M. Lightgcn: Simplifying and powering graph convolution network for recommendation. In Proceedings of the 43rd International ACM SIGIR Conference on Research and Development in Information Retrieval, Xi'an, China, 25–30 July 2020; pp. 639–648.
39. Zheng, Y.; Tang, B.; Ding, W.; Zhou, H. A Neural Autoregressive Approach to Collaborative Filtering. In Proceedings of the 33rd International Conference on Machine Learning, New York, NY, USA, 19–24 June 2016; pp. 764–773.
40. Kipf, T.; Welling, M. Semi-supervised classification with graph convolutional networks. *Commun. ACM* **2007**, *50*, 36–44.
41. Veličković, P.; Cucurull, G.; Casanova, A.; Romero, A.; Liò, P.; Bengio, Y. Graph Attention Networks. In Proceedings of the 6th International Conference on Learning Representations, Vancouver, BC, Canada, 30 April–3 May 2018.
42. Liu, X.; Wu, J.; Zhou, C.; Pan, S.; Cao, Y.; Wang, B. Task-adaptive neural process for user cold-start recommendation. In Proceedings of International World Wide Web Conference, Ljubljana, Slovenia, 19–23 April 2021; pp. 1306–1316.

electronics MDPI

Article

Machine Learning for Energy-Efficient Fluid Bed Dryer Pharmaceutical Machines

Roberto Barriga [1], Miquel Romero [2] and Houcine Hassan [1,*]

[1] Departamento de Informática de Sistemas y Computadores, Universitat Politècnica de València, Camino de Vera, n°14, 46022 Valencia, Spain
[2] Industrias Farmacéuticas Almirall, Ctra. N-II, km. 593, 08740 Sant Andreu de la Barca, Spain
* Correspondence: husein@disca.upv.es

Abstract: The pharmaceutical industry is facing significant economic challenges due to measures aimed at containing healthcare costs and evolving healthcare regulations. In this context, pharmaceutical laboratories seek to extend the lifespan of their machinery, particularly fluid bed dryers, which play a crucial role in the drug production process. Older fluid bed dryers, lacking advanced sensors for real-time temperature optimization, rely on fixed-time deterministic approaches controlled by operators. To address these limitations, a groundbreaking approach taking into account Exploration Data Analysis (EDA) and a Catboost machine-learning model is presented. This research aims to analyze and enhance a drug production process on a large scale, showcasing how AI algorithms can revolutionize the manufacturing industry. The Catboost model effectively reduces preheating phase time, resulting in significant energy savings. By continuously monitoring critical parameters, a paradigm shift from the conventional fixed-time models is achieved. It has been shown that the model is able to predict on average a reduction of 50.45% of the preheating process duration and up to 59.68% in some cases. Likewise, the energy consumption of the fluid bed dryer for the preheating process could be reduced on average by 50.48% and up to 59.76%, which would result on average in around 3.120 kWh energy consumption savings per year.

Keywords: energy consumption; IoT-based power control systems; machine learning; optimization using sensor data; predictive control; pharmaceutical technology; process modeling; exploratory data analysis

Citation: Barriga, R.; Romero, M.; Hassan, H. Machine Learning for Energy-Efficient Fluid Bed Dryer Pharmaceutical Machines. *Electronics* **2023**, *12*, 4325. https://doi.org/10.3390/electronics12204325

Academic Editor: Adel M. Sharaf

Received: 20 September 2023
Revised: 11 October 2023
Accepted: 17 October 2023
Published: 18 October 2023

1. Introduction

The entire pharmaceutical manufacturing process comprises multiple stages, including dispensing, granulation, drying, compression, and coating [1], as depicted in the diagram below in Figure 1.

Fluid bed drying technology is widely employed in pharmaceutical manufacturing due to its high efficiency in drying granules obtained through wet granulation [2]. However, the primary challenge associated with using a fluid bed dryer lies in the time and energy it consumes to complete the process. The drying process entails three phases: (i) preheating the machine without introducing any product, (ii) drying the product, and (iii) cooling the machine for product cooling. Costs are incurred in all three phases, encompassing the time taken by the machines and the energy required for heating and air circulation. Additionally, the budget is impacted by the number of operators involved in handling the machine [3]. The fluid bed drying of wet granules obtained through high shear granulation involves a combination of moisture diffusion from the solid material, facilitated by hot air, and the entrainment of this moisture through forced convection. The success of this process relies on the uniform fluidization of the granules by hot air, ensuring efficient mass and energy transfer. The drying time can be reduced by increasing the temperature and intake airflow. However, each parameter must be carefully tailored for the specific granule type. The inlet air temperature is adjusted based on the temperature signal recorded by the air sensor in contact with the fluidized product, ensuring it does not exceed the critical temperature for pharmaceutical stability. Inlet air humidity is kept within a narrow dew-point range to achieve batch-to-batch reproducibility. Thus, under optimized conditions of temperature, humidity, and airflow entering the machine, drying takes less time and generates a high-quality product. Temperature, pressure, and flow sensors monitor the changes throughout the process [4,5].

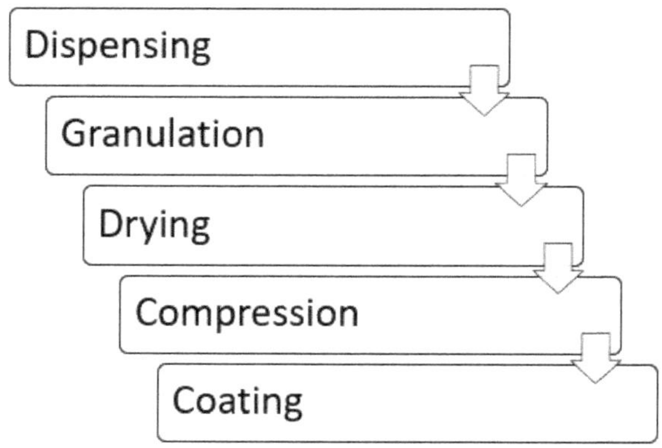

Figure 1. Pharmaceutical manufacturing process.

The financial landscape, ongoing measures implemented by authorities to control healthcare expenses, and recent changes in healthcare regulations significantly impact pharmaceutical laboratories and manufacturers of medical products. Due to the considerable cost of fluid bed dryers and other machinery used in medicine production, there is a concerted effort to maximize the lifespan of these machines [6]. In particular, older fluid bed dryers lack sensors that can indicate when the machine has reached the optimal temperature for any of the three phases (preheating, drying, and cooling). Deterministic methods are usually employed, meaning fixed times are used for each process phase, and the machine's operator is responsible for managing these times. Moreover, during the drying process, the operator halts the machine after a specific duration to obtain a product sample and measure humidity levels, thereby checking whether any critical machine parameters need adjustments (such as inlet air temperature or airflow). The primary aim of this study is to propose a Catboost machine-learning model that can reduce the time needed for the preheating phase, therefore reducing overall energy consumption, and to demonstrate a methodology for utilizing exploratory data analysis in the analysis and optimization of a drug production process on a large scale. The experiments were performed on a fluid bed dyer located in a pharmaceutical manufacturing plant in Spain. The methodology used to develop the model can be implemented in a wide range of equipment that does not possess state-of-the-art sensor technology. Our study embraces a groundbreaking approach that involves real-time monitoring of crucial manufacturing equipment parameters, representing a paradigm shift from the conventional model. The paper is organized as follows: Section 2 presents the related work on applying artificial intelligence algorithms to improve methods and processes in the manufacturing industry; Section 3 details the proposed methodology; Section 4 presents the experiment set up, including a description of the fluid bed dryer and the data collection; Section 5 presents the results in terms of energy savings for the fluid bed dryer preheating process after applying EDA and Catboost machine learning model; finally, Section 6 gives the main conclusions.

2. Related Work

The most significant hurdle in employing a fluid bed dryer lies in mitigating the substantial time and energy consumption associated with completing the process. Following the electric energy crises of the 1970s [7], electricity consumption became a topic of discussion. Furthermore, it has been established that global electric energy use is quickly expanding [8], specifically in the pharmaceutical industry, which is a growing field nowadays. As a result, every pharmaceutical company seeks to utilize as little electric energy as possible in many sectors, such as manufacturing fields, packing industrial processes, and transportation to different hospitals or medical stores [9]. Utilizing advanced analytics techniques, such as machine learning, enables us to anticipate the electricity consumption in diverse pharmaceutical manufacturing processes, allowing us to tailor strategies to specific domains [10]. The accurate prediction of electricity usage holds paramount importance for decision makers and policymakers within the pharmaceutical industry, given the energy-intensive nature of its machinery. In the context of increasingly dynamic electricity markets, where prices are subject to fluctuation, understanding and forecasting electricity usage becomes even more critical. The ability to predict electricity costs can significantly impact the bottom line for pharmaceutical manufacturers. Comprehending the expected electric energy consumption empowers us to envision enhancements in pharmaceutical manufacturing processes, aiming to reduce electricity usage. This predictive capability, whether in the short or long term, equips us with insights into energy-saving opportunities and strategies for optimizing current energy consumption, thus mitigating the potential impacts of rising electricity prices. With many variables, estimating energy usage is a problematic manufacturing task [11]. Machine learning models are currently employed in various fields, since they are beneficial. Machine learning operates similarly to a function that nicely maps the input data to the output. Machine-learning models can give high-accuracy predictions for energy usage in the pharmaceutical process or the heating process in the manufacturing process. As a result, pharmaceutical companies can use them to enact energy-saving initiatives in different manufacturing domains. For example, machine learning algorithms can forecast how much electric energy is utilized in a dryer machine in manufacturing [12]. They can also be used to forecast the future-energy consumption, such as power or organic gas [13]. Numerous studies have showcased the wide applicability of machine learning techniques in the pharmaceutical industry [14–18]. For instance, [19] conducted a comprehensive investigation into the implementation of Artificial Neural Networks (ANNs) for the development and formulation of pharmaceutical products using a Quality by Design approach for tablet formulations. By leveraging historical data, the researchers were able to gain valuable insights into the intricate interactions between formulation variables and drug specifications. The study's conclusions emphasized the efficiency of neural networks and genetic algorithms in optimizing formulations, ultimately leading to reduced energy consumption.

3. Proposed Methodology

Figure 2, from left to right, shows the overall approach for data modeling and simulating. First, a business need and objective have to be clearly agreed—in the present work, the modeling and optimization of the drying process—due to the high energetic cost and the evaluation that significant savings can be obtained. Next, the right data have to be captured in order to satisfy the business objective. This is followed by data exploration/processing, modeling and finally evaluation of the results [20]. Note that this can, in practice, become a cyclic processing iterating back from the result evaluation phase to the data collection phase, or even back to re-evaluate the business need.

- **Define business problem:** The initial phase of the machine learning workflow involves defining the business problem. The duration of this step varies, ranging from several days to a few weeks, depending on the complexity of the problem and its specific application. During this stage, data scientists collaborate with subject matter experts (SMEs) to gain a comprehensive understanding of the problem. This involves conducting interviews with key stakeholders, gathering pertinent information, and establishing overall project goals. In the case at hand, our objective is to minimize the energy consumption in the fluid bed dryer.

- **Obtain the data:** Once the understanding of the problem is achieved, it is about obtaining the information identified and available for solving the business problem. In our case, the data obtained from the fluid bed dryer will be used directly.
- **Explore the data:** The next step in the process is exploration data analysis (EDA), which involves analyzing the raw data. The primary objective of EDA is to delve into the data, evaluate its quality, identify any missing values, examine feature distributions, assess correlations, and so on.
- **Create the model:** Model creation encompasses various tasks, including dividing the data into training and testing sets, handling missing values, training multiple models, fine-tuning hyperparameters, consolidating models, evaluating performance metrics, and ultimately selecting the optimal model for deployment to forecast our target variable. In our specific scenario, we aimed to predict the duration required for the preheating process in order to minimize energy consumption. In this paper, Catboost machine learning model for optimizing fluid bed dryer energy consumption is used.

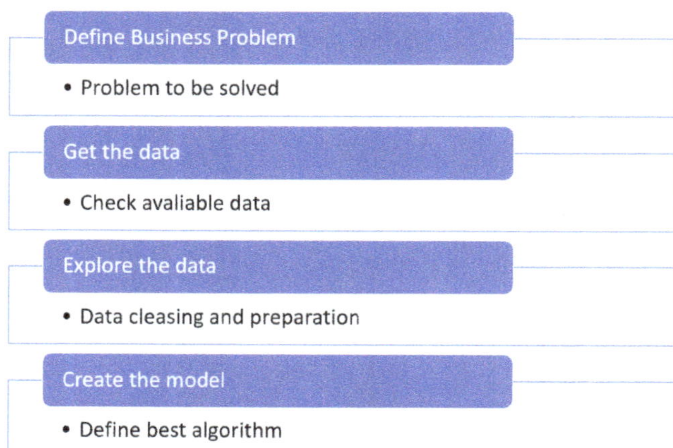

Figure 2. Overall procedure for data analysis and modeling.

Catboost Algorithm Application

Catboost Regression represents a relatively recent and purportedly potent machine learning algorithm, offering several advantages [21]. In essence, machine learning algorithms are commonly utilized to discern intricate patterns within extensive datasets, enabling predictions of future behaviors. Catboost specifically leverages gradient boosting for decision trees. In both regression and classification scenarios, gradient boosting serves as a machine learning technique that constructs a prediction model by combining multiple "weak prediction models", typically decision trees [22]. The fundamental concept revolves around applying steepest descent steps to a minimization problem, known as functional gradient descent. The gradient boosting process progressively generates a series of approximations Ft: $Rm \rightarrow R$, with $t = 0, 1, \ldots$, in a step-by-step manner. Each Ft is derived additively from the previous approximation Ft^{-1}, following the formula: $Ft = Ft^{-1} + \alpha ht$, where α represents a step size and function ht: $Rm \rightarrow R$, referred to as a base predictor, is selected from a family of functions H to minimize the expected loss ht. Catboost, in particular, implements gradient boosting using binary decision trees as the function $h(x)$, defined as

$$h(x) = \sum_{j=1}^{J} b_j l_{\{x \in R_j\}}$$

In Catboost's implementation, the regions R_j represent the disjoint leaves of the decision tree, and $b_j l_{\{x \in R_j\}}$ denotes the jth binary variable corresponding to attribute x. One notable advancement of Catboost is its ability to process mixed data types simultaneously for model construction. It can handle both categorical inputs (converted to numbers) and numerical inputs effectively. Additionally, two of its strong features are (i) the default hyper-parameters, which require minimal tuning and perform well across various data scenarios, and (ii) its built-in mechanism for auto-correction, which helps prevent overfitting. When applying Catboost to the data, certain measures were taken to address concerns about model size and memory consumption by setting specific meta-parameters:

- RAM limit—a limit value was set to restrict memory usage.
- Max_ctr_complexity—it was assigned a value of 1 or 2 to control the complexity of interactions. The default value is 4.
- Model_size_reg—a larger value was assigned to penalize heavy combinations.

It is worth noting that memory usage currently remains a significant limitation of Catboost. Catboost demands that all data be immediately accessible in memory for quick random sampling, unlike stochastic gradient and neural network models. An additional critical concern is the sensitivity of Catboost to hyper-parameters and the significance of conducting hyper-parameter tuning. These factors can be influenced by the Big Data environment, such as the Apache Spark distributed framework [23]. Further details regarding hyper-parameter tuning will be provided later in the study. When dealing with extremely large datasets, an approach to address this challenge involves fitting the Catboost model to a representative sample using the Catboost Python API. Subsequently, the model can be applied to the larger dataset using Apache Spark or Hadoop with the aid of Catboost's Java API. This methodology enables the efficient processing of massive datasets within the distributed computing environment.

4. Methodology Applied to Fluid Bed Dryer

4.1. Fluid Bed Dryer

The fluid bed drying machine utilized in this study is the Fielder Aeromatic MP, located within a pharmaceutical manufacturing plant in Spain, as depicted in Figure 3. This machine is equipped with 56 sensors controlled by Programmable Logic Controller (PLC), and the SCADA (Supervisory Control And Data Acquisition) system, which enables operators to monitor and adjust essential parameters, such as the inlet air temperature and air flow. Three critical parameters significantly impact the efficiency of the drying process and, consequently, can influence the final product's quality. These parameters are temperature, humidity, and air flow. In theory, a higher inlet air temperature and flow rate lead to a shorter drying time. However, it is essential to configure each of these three parameters correctly, depending on the specific product type, to prevent quality issues and degradation of the final product post-drying. Notably, it is crucial to ensure that the inlet air temperature does not exceed the critical temperature of the product to be dried, as surpassing this threshold could jeopardize its quality and pharmaceutical properties. Careful monitoring and regulation of these parameters are vital to maintaining product integrity and achieving desired outcomes during the drying process.

In this process, the operator utilizes SCADA to monitor the increase in outlet air temperature during the product drying phase. It is essential to note that when the product is completely dried, the outlet air temperature aligns closely with the inlet air temperature. At this critical point, the operation must be halted promptly to avoid jeopardizing the product's quality and prevent the unnecessary consumption of time and energy, leading to increased process costs. As the fluid bed drying machine lacks sensors to indicate the optimal temperatures for different drying phases (preheating, drying, and cooling), human operators typically rely on fixed time durations for these phases. However, the preheating phase may vary in time depending on the operator's experience with the machine. During the preheating phase, the fluid bed dryer contains no drug product; instead, it receives hot air for machine preheating. Once preheating is complete, operators introduce the drug product to initiate the drying phase. During drying, operators take samples to analyze various chemical parameters. After drying, the cooling process commences. Once all three phases are finished, the fluid bed dryer undergoes cleaning before a new batch is processed. Overall, continuous monitoring of the outlet air temperature through SCADA is crucial to ensure the preheating process is controlled effectively and prevent unnecessary energy consumption.

Figure 3. Fluid bed dryer Fielder Aeromatic.

4.2. Data Collection

For this study, a fluid bed dryer machine was used, which is presently operational in a real pharmaceutical plant belonging to a multinational company in Spain. This machine typically handles one to two batches of pharmaceutical drug granules each day, with each batch comprising approximately 150 kg of drug mixed with 25 kg of alcohol and 10 kg of another excipient before entering the fluid bed dryer. The fluid bed dryer is equipped with 56 sensors that measure several parameters, including inlet/outlet air temperature, air flow (m^3/h), motor rotation speed (rpm), air pressure (Pa), and others. Each sensor records data at a minute-by-minute interval. A dataset covering a year and a half of data has been accumulated, comprising more than 700,000 readings for each of the 56 signals. Data collection was accomplished through a Programmable Logic Controller (PLC) and stored in a SCADA system. These datasets served as the foundation for our subsequent analysis and optimization of the fluid bed drying process. Table 1 shows a sample from the fluid bed dryer sensors, including a description for each signal, the minimum and maximum value and their units of measure.

Some of these sensors are involved in different processes, such as granulation (column PMA), drying (column TSG) or cleaning (column CIP). For the exploration phase, be the sensors involved just in the drying process were selected (column TSG), but, as it will be explained in next section, for the data modeling, all of these were selected, to simulate a real situation where it would not able to differentiate which sensor belongs to which phase. Figure 4 illustrates the SCADA interface utilized by operators to interact with the machine, providing functionalities such as starting/stopping the controller and displaying indicators for the inlet air temperature, inlet air flow, and more. To conduct our analysis, the data from SCADA was exported into a tabular format comprising over 700,000 rows and

56 columns, containing the recorded information from the various sensors and parameters of the fluid bed dryer. This comprehensive dataset was the basis for our further exploration and optimization of the fluid bed drying process.

Table 1. Fluid Bed Dryer sensors.

Item	TagName (Symbol)	Description	Min	Max	Units	PMA	TSG	CIP
1	FS3_GEA_EIS1200_ME	Impeller power [Kw]	0	300	Kw	X		
2	FS3_GEA_EOP_GP	Current EOP in GP	0	1000	None	X		
3	FS3_GEA_EOP_MP	Current EOP in MP	0	1000	None		X	
4	FS3_GEA_FIC1217_ME	Liquid flow rate in GP [cl/min]	0	833	cl/min	X		
5	FS3_GEA_FIC1217_XS	Liquid flow setpoint in GP [cl/min]	0	833	cl/min	X		
6	FS3_GEA_FIC200_ME	Air flow [m^3/h]	0	4500	m^3/h		X	
7	FS3_GEA_FIC200_XS	Air flow setpoint [m^3/h]	0	4500	m^3/h		X	
8	FS3_GEA_FIC701_ME	Spray liquid flow in MP [cl/min]	0	667	cl/min		X	
9	FS3_GEA_FIC701_XS	Spray liquid flow rate setpoint in MP [cl/min]	0	667	cl/min		X	
10	FS3_GEA_LI940_ME	Cleaning water tank level [L]	0	500	L			X
11	FS3_GEA_MIS213_ME	Inlet air humidity [g/Kg]	0	250	g/Kg		X	
12	FS3_GEA_NFGP	No. Current phase in execution in GP	0	1000	None	X		
13	FS3_GEA_NFMP	No. Current phase in execution in MP	0	1000	None	X		
14	FS3_GEA_NFW	No. Current cleaning phase in execution	0	1000	None			X
15	FS3_GEA_NW	No. Current cleaning in execution	0	1000	None			X

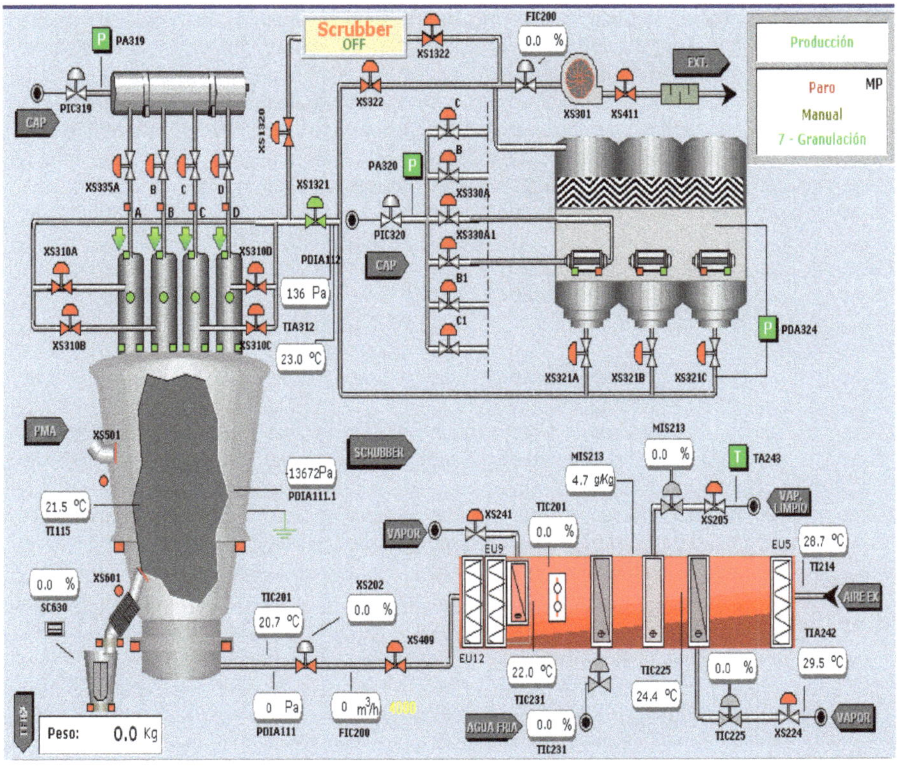

Figure 4. Fluid bed dryer SCADA.

On the SCADA screen, the status of the station in detail can be seen, including the values of the sensors and valves, for example the temperature or pressure, and in the upper right, it shows the state of the fluid bed dryer, what process it is carrying out and what state each of them is in (granulating, drying or cleaning). For example, when steam is added to the fluid bed dryer to control the humidity of the air that is introduced into the dryer, if the humidity is very low, more steam is added to increase it. The air that is introduced into the dryer allows us to control both the temperature and its humidity. The pressure of the dryer is indicative of the clogging of the filters: if there is a big difference between the internal pressure and the output pressure, it means that it has dirty filters, and you need to clean them. The SCADA records and monitors the operating status of the fluid bed dryer in its operating modes and states and the duration of these and the registers of the analog parameters involved. Taking into account the drying process and how the fluid bed machine works, four sensors were selected for the exploration analysis. The signals recorded by the different sensors in the fluid bed dryer were as follows:

- Fan motor: this signal indicates whether the fluid bed dryer is currently running (ON) or turned off (OFF).
- Air flow: This signal represents the quantity of air flowing into the fluid bed dryer, measured in cubic meters per hour (m^3/h). The machine operator configures this parameter. Monitoring the air flow helps distinguish between the preheating and drying phases, as both processes require air to be completed.
- Inlet air temperature: this signal indicates the initial temperature of the air entering the fluid bed dryer, and it is set by the machine operator at the start of the process.
- Outlet air temperature: this signal indicates the temperature of the air leaving the fluid bed dryer.

During the operation of the fluid bed dryer and the commencement of the hot air inlet process, it is essential to consider the heat absorbed by the machine to reach the preheating temperature. The temperature difference between the outlet air temperature and the inlet air temperature helps determine the amount of heat absorbed by the fluid bed dryer. When the machine reaches a point where it cannot absorb more heat, the inlet air temperature will become similar to the outlet air temperature. To better understand the behavior of the process, the temperature difference of the air inlet and outlet of the machine was utilized, denoted as TA_D, which is defined in Equation (1):

$$TA_D = TA_s - TA_e. \tag{1}$$

where TA_s represents the outlet air temperature, TA_e represents the inlet air temperature, and TA_D represents the temperature difference.

5. Experimental Results

5.1. Exploratory Data Analysis

The dataset used for machine learning analyses is the same that was used for the exploratory data analysis. It includes various parameters related to the fluid bed dryer's operation, such as the inlet and outlet air temperatures, airflow rate, and the phase number the machine is in (preheating, drying, or cooling). Information from over 200 batches of dried drug product, covering a span of 18 months of production, was also accessible for analysis. The variables used in the current study were the following:

- The phase indicator takes values 1, 2, or 3, representing the current phase of the fluid bed dryer. Phase 1 indicates preheating, Phase 2 is the drying phase, and Phase 3 indicates cooling after the drying process.
- The inlet air temperature sensor represents the temperature at which the air enters the machine during any of the three phases (preheating, drying, or cooling).
- The outlet air temperature signal corresponds to the temperature at which the air leaves the machine.
- The inlet airflow sensor indicates the volume of air supplied by the machine's fan.

- The fan motor signal is useful for determining when the machine is active during any of the three phases, indicating the fan motor's movement.

In the next step of the analysis, random days will be selected to observe the behavior of the machine signals during the preheating, drying, and cooling processes for each batch of pharmaceutical product processed. The primary goal of this exploration is to identify trends and gain a better understanding of fluid bed dryer processes, with the objective of identifying opportunities for improvement. Figure 5 visually depicts the behavior of the signals on different days, representing a full day of fluid bed dryer operation. The x-axis represents the elapsed time for one day of fluid bed dryer operation (1440 min, corresponding to 24 h), while the y-axis indicates the difference in temperature between the machine's inlet and outlet air. The blue dots indicate the preheating process, the orange dots represent the drying process, and the green dots represent the cooling process.

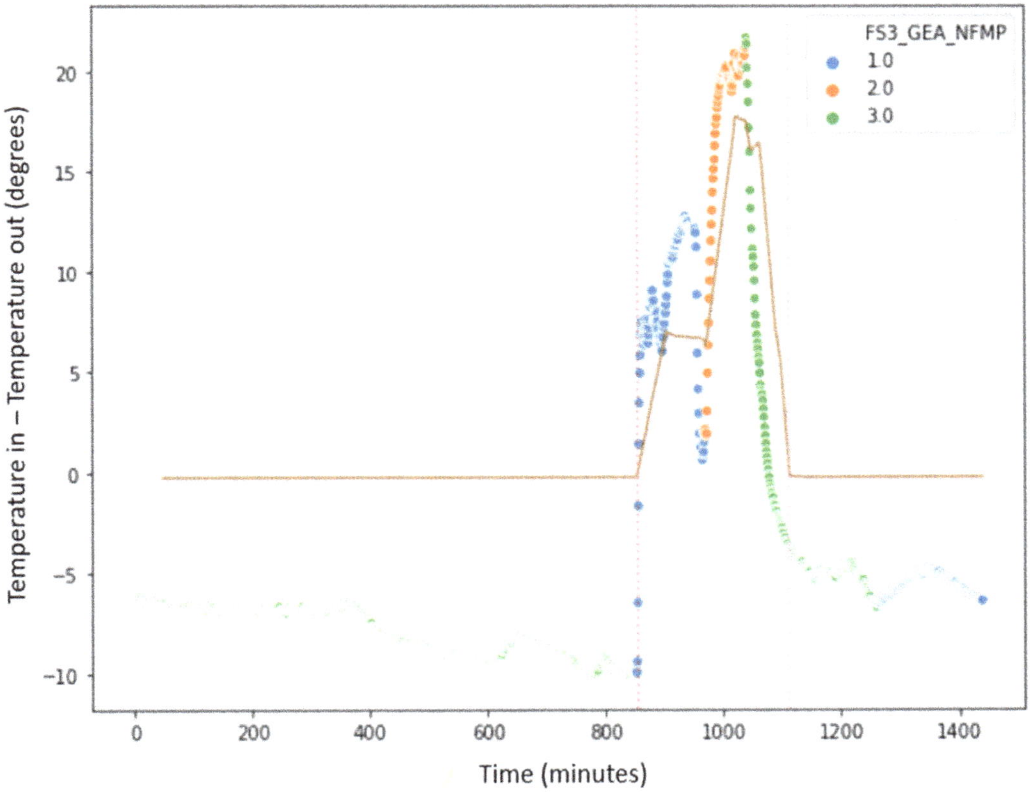

Figure 5. Plot of the drying steps.

Figure 6 shows a sample of 4 different days taken randomly, where it can be observed that some days, the fluid bed dryer processed one batch, and other days two batches, with an average of around 350 min per batch. In the figure dated 2 December 2019, there are two batches that were processed, and if looking at the blue dots, it can be seen that the preheating process lasted much longer in the two batches, compared to the duration of the preheating process; for example, on 7 October 2018, the blue dots were much smaller and the temperature difference, the y-axis, did not exceed 10 degrees. It can also be observed that the duration of the drying process, the orange dots, was more or less homogeneous, and lasted approximately the same for all the days and all the batches (x-axis); as well as this, the temperature differences were approximately similar (y-axis). As a conclusion, it is evident from the data that the duration of the preheating process exhibits variability. To preheat the fluid bed dryer, some batches take a longer time preheating the machine than others, with the consequent unnecessary consumption of energy.

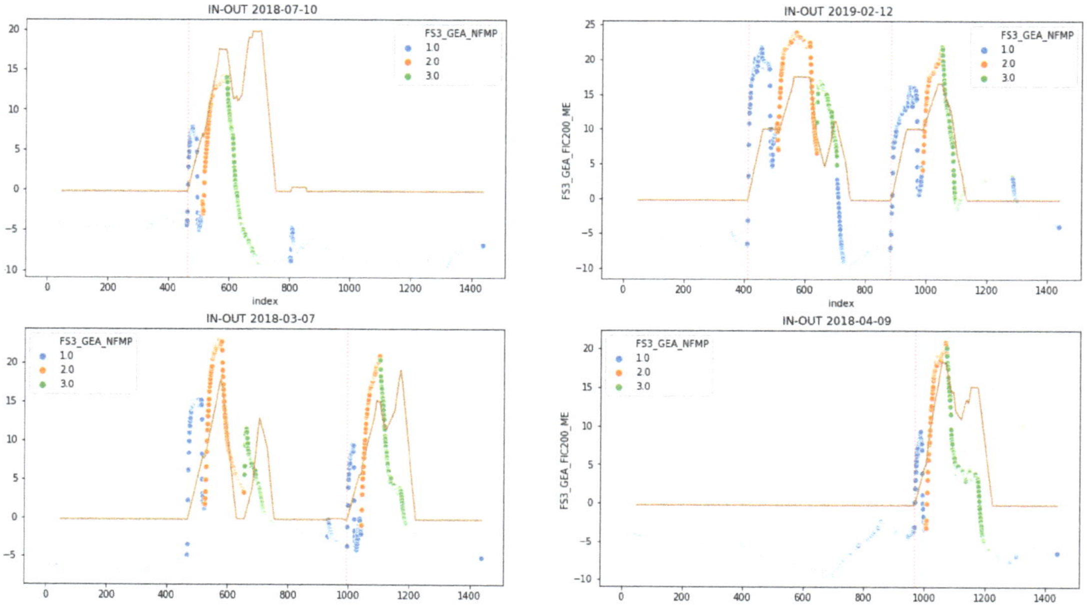

Figure 6. Example of 4 different days of batch drying. Above each figure is plotted the date of the batch (1.0 Preheating, 2.0 Drying, 3.0 Cooling).

5.1.1. EDA Timing Savings Analysis

In Figure 7, it can be observed for the 200 batches analyzed how many minutes on average the fluid bed dryer was used to perform the preheating process. Each blue line indicates, for each individual batch, the time taken to complete the preheating process in the fluid bed dryer. The variability in the preheating duration is attributed to the manual operation of the process, as it relies on the operator's discretion to start and finish the preheating phase; considering the age of the machine, the machine is kept for preheating less than 50.1 min, whereas other times, the machine is kept preheating for up to 180.3 min. The fluid bed dryer is initially set up with hot air inlet at 45 degrees and airflow 2000 m^3/h. However, the fluid bed dryer does not have any sensor notifying when the machine is warm enough to introduce the drug product and start the drying process. The brown line in the graph indicates the average that was 99.7 min to complete the preheating process. To summarize, this indicates again the opportunity to harmonize the preheating process by establishing an optimum preheating time, and potentially, to be able to reduce the preheating process time, and consequently reduce the fluid bed dryer energy consumption.

5.1.2. EDA Energy Savings Analysis

Figure 8 shows the variability of analyzing the energy consumption used to complete the preheating process for each batch in the fluid bed dryer. The energy consumption EC_b was calculated using Equation (2):

$$EC_b = Batch_t * \text{Cpm} \tag{2}$$

where $Batch_t$ is the time consumed by the fluid bed dryer for preheating the batch, and Cpm corresponds to the fluid bed dryer energy consumption per minute. The fluid bed dryer currently consumes 18.5 kWh during the preheating process; this means that for each minute it consumes 0.31 kWh (18.5 kWh/60 min = 0.31 kWh). If the preheating process may take between 50.1 and 180.3 min, the fluid bed dryer consumes between 15.5 kWh and 55.8 kWh for preheating the machine to dry one batch of drug product.

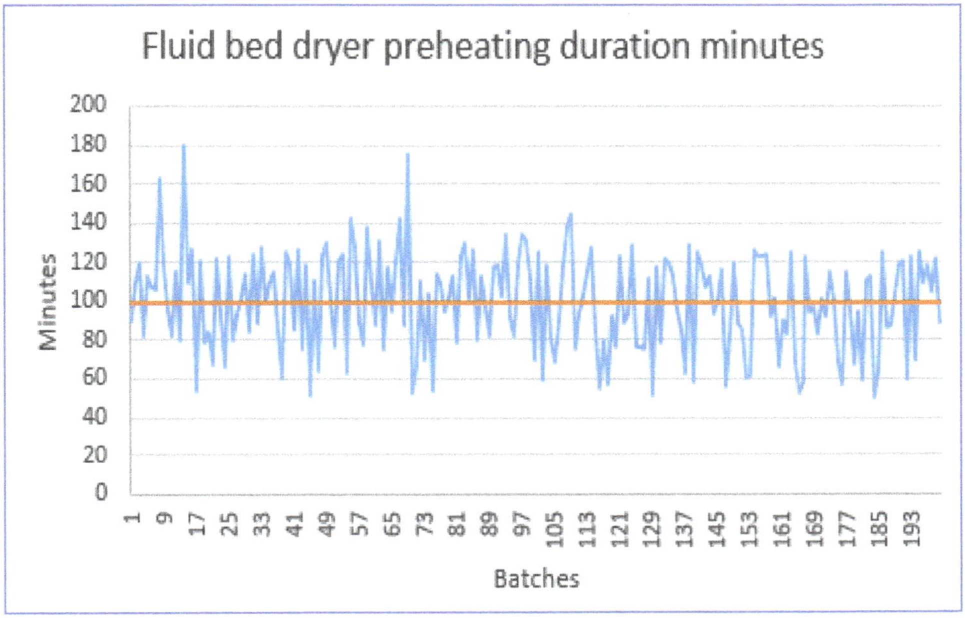

Figure 7. Fluid bed dryer preheating duration in minutes. Brown line indicates average duration.

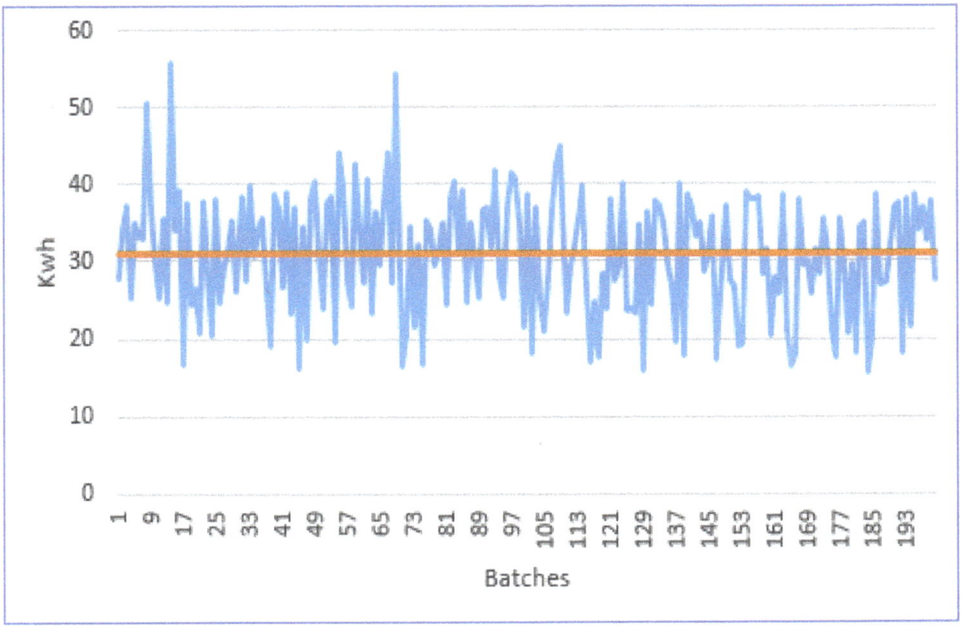

Figure 8. Fluid bed dryer preheating energy consumption (kWh). Brown line indicates average duration.

It can be observed that some batches needed 55.8 kWh; however, other batches needed less than 15.5 kWh, which means in some cases around 72.2% less energy consumption for some batches. The brown line indicates the average consumption for the 200 batches, around 30.9 kWh. This indicates important potential energy savings if the preheating process in the fluid bed dryer is optimized. To calculate the potential energy savings of the fluid bed dryer during the preheating process for each batch, a machine learning model was implemented, as discussed in the next chapter, to predict when the right time was to stop the process, and therefore, consume just the energy needed for preheating the fluid bed dryer.

5.2. Catboost Machine Learning Model Analysis

As described in this section, a Catboost model was selected and executed using the historical data obtained from the activity of the fluid bed dryer process in the production plant. Due to the fluid bed dryer's age, one of the primary problems is that it does not have sensors that can detect whether the air within is at the right temperature to conclude the preheating process. From the perspective of data modeling, the issue has a number of intriguing characteristics:

- Due to the inclusion of 56 sensors, there are a large number of possible inputs (700.000 rows and 3 GB data).
- There are a lot of manufacturing batches, more than 200, but the machine does not keep track of when the preheating operation starts or ends. As a result, the deduction is carried out using the machine's air inlet and output as well as temperature differences.
- The goal is to interpret the estimated model in a way that can reveal the factors that influence the air inlet- and outlet-temperature differential curves. Using this method, it is possible to estimate how long the preheating process will take.

To estimate the preheating time conceptually, a function model f (i) was constructed using the data of a matrix X, which contained the data taken from the fluid bed dryer. In order for the operators to know when it is best to cease the machine's preheating operation and so save energy, the information present in the machine learning model was used to anticipate the estimated preheating time for each batch. The time remaining for the fluid bed dryer to finish preheating (based on the inlet–outlet temperature disparities) was the predicted output from our model. Data preprocessing techniques were used on the input dataset to remove unnecessary information, such as missing values, in preparation for future analysis. The first step to select the most suitable model was to split the dataset into training and testing data. This technique is used for evaluating the performance of a machine learning algorithm. It can be used for classification or regression problems and can be used for any supervised learning algorithm. The process consists of taking a dataset and dividing it into two subsets. The first subset is used to fit the model and is referred to as the training dataset. The second subset is not used to train the model; instead, the input element of the dataset is provided to the model, then predictions are made and compared to the expected values. This second dataset is referred to as the test dataset. The objective of splitting the dataset into train and test is to estimate the performance of the machine learning model based on new data that will be captured directly from the fluid bed dryer, namely, to fit it on available data with known inputs and outputs, then make predictions on new examples in the future where there are not the expected output or target values. The train–test procedure is appropriate when there is a sufficiently large dataset available, which means that there are enough data to split the dataset into train and test datasets and each of the train and test datasets are suitable representations of the problem domain. To perform the evaluation and selection of the best-fit algorithm for the fluid bed dryer process, the Python libraries were used. The same dataset was injected in the different algorithms. The dataset contained 18 months data coming from the 56 sensors of the fluid bed dryer, and the values represented the average of 10-fold cross validation (partitioning of the dataset into 10 parts, 9 for train and one for test, then rotating 10 times to obtain different combinations of partitions). The results of the most relevant algorithm's evaluation are shown in Table 2.

Table 2. Benchmarking outcomes of various machine learning algorithms on the dataset.

Model	MAE	MSE	RMSE	R2
CatBoost Regressor	8.3507	144.1894	11.8781	0.6806
Light Gradient Boosting Machine	9.3655	168.9438	12.8561	0.6285
Extreme Gradient Boosting	9.3243	188.3094	13.3581	0.5845
Random Forest Regressor	10.6074	195.4555	13.8601	0.5616
Gradient Boosting Regressor	10.6236	211.7174	14.3607	0.5365
K Neighbors Regressor	13.3827	301.5125	17.1741	0.3466
AdaBoost Regressor	14.2045	297.2756	17.1463	0.3375
Orthogonal Matching Pursuit	16.1283	402.0394	19.9813	0.1090
Lasso Regression	16.3557	407.1744	20.0864	0.1020
Elastic Net	16.6512	412.9383	20.2066	0.0947
Bayesian Ridge	16.6316	420.1060	20.3491	0.0779
Decision Tree Regressor	14.0376	411.2750	19.9699	0.0285
Ridge Regression	16.4399	450.6532	21.0216	0.0156

Based on Table 2, the Catboost Regressor has the lowest MAE of 83.507 and the lowest RMSE of 118.781, indicating that it has the best predictive accuracy compared to the other models. It also has the highest R2 value of 0.6806, indicating that it can explain about 68.06% of the variance in the target variable. The Light Gradient Boosting Machine has the second-best performance, with slightly higher MAE and RMSE values than the Catboost model, and an R2 value of 0.67. The Extreme Gradient Boosting, Random Forest Regressor, and Gradient Boosting Regressor models have higher MAE, MSE, and RMSE values and lower R2 values than the Catboost and Light Gradient Boosting models, indicating that they may not perform as well on this specific dataset, the same as the rest of the models. To select the best metric for the Catboost algorithm, the nature of the problem and the evaluation criteria was considered. To measure the proportion of variance in the target variable that can be explained by the model, R2 was the most suitable metric. MAE was discarded because it focuses on minimizing the average absolute difference between predicted and actual values, and MSE or RMSE penalize larger errors more than smaller errors.

5.2.1. Time Duration Analysis

In Figure 9, the real duration of the preheating process per month from the historical dataset can be seen in blue. This duration is measured in minutes and represents the average of the time spent by the process for the whole month. This measure was performed for the 200 batches evaluated during 18 months. The results show that the preheating process duration varied from one month to another and fell between 88.5 and 110.6 min, depending on when the optimal temperature difference in–out was reached. The average duration of the 200 batches during the 18 months was around 99.7 min. This key information allowed us to calculate the real consumption of the preheating process. Figure 9 also shows the Catboost prediction duration of the preheating process. It can be observed how for the 200 batches, during 18 months of evaluation, the predicted time was always lower than the real time. The reduction in the predicted time was significant, ranging from 34.7 min (39.2% time reduction) in the month of December 2018 to 66.0 min (59.68.2% reduction) in the month of October 2018. The optimal time predicted by the algorithm corresponded to an average per month between 42.5 and 59.5 min, with an average of 49.4 min. The average predicted time reduction was 50.3 min. Therefore, the duration of the process can be reduced on average by 50.45%.

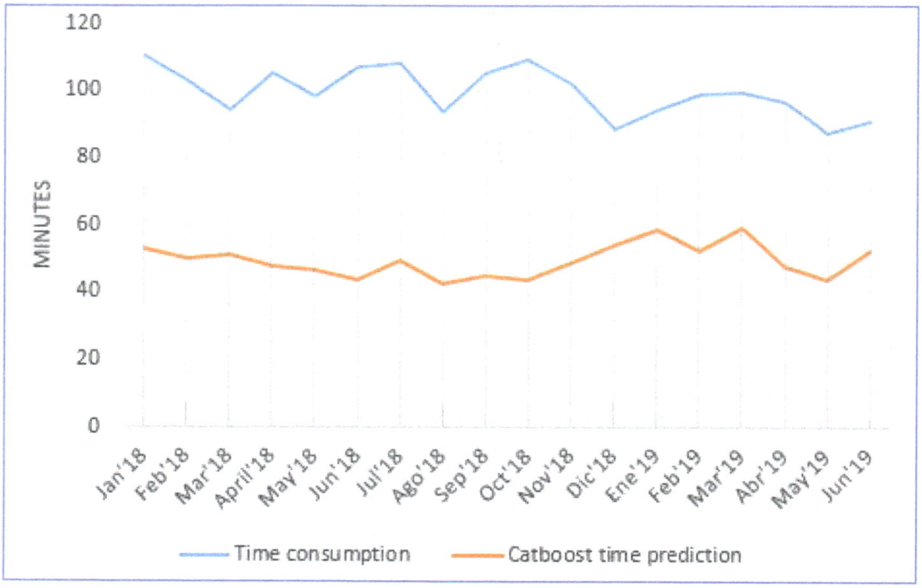

Figure 9. Time duration for preheating process comparing real duration with Catboost prediction.

5.2.2. Energy Saving Analysis

In Figure 10, the real energy consumption of the preheating process per month from the historical dataset can be seen in blue. This energy consumption is measured in kWh and represents the average of the energy spent by the process for the whole month per batch. This measure was performed for the 200 batches, evaluated over 18 months. The results show that the real preheating-process energy consumption varied from one month to another and fell between 27.1 kWh and 34.3 kWh per batch every month, depending on when the optimal temperature difference in–out was reached. The average energy consumption of the 200 batches as 30.9 kWh per batch.

Figure 10 shows also the Catboost prediction energy consumption of the preheating process. It can be observed how for the 200 batches, during the 18 months of evaluation, the predicted energy consumption was always lower than the real energy consumption. The reduction in the predicted energy consumption was significant, ranging from 10.8 kWh (39.8% energy reduction) in the month of December 2018 to 20.5 kWh (59.76% energy reduction) in the month of October 2018.

The optimal energy consumption predicted by the algorithm per batch corresponded to on average between 13.2 kWh and 18.4 kWh. The average predicted energy reduction was 15.6 kWh. Consequently, the reduction in energy consumption predicted by the algorithm, to complete the prehearing process, represented 50.48% less energy. The total energy saving was calculated using Equation (3), with $Nbatches$ being the number of batches and ES_b the energy saved per batch.

$$ES_t = Nbatches * ES_b \quad (3)$$

Based on Figure 9, there is a potential saving of 50.3 min per batch each time the fluid bed dryer is preheated. This means a saving of around 15.6 kWh per batch (50.3 min $\times$ 0.31 kWh). If the fluid bed dryer processes approximately 200 batches per year, based on the current estimation, then the annual potential energy savings could be approximately 3.120 kWh when applying Equation (3).

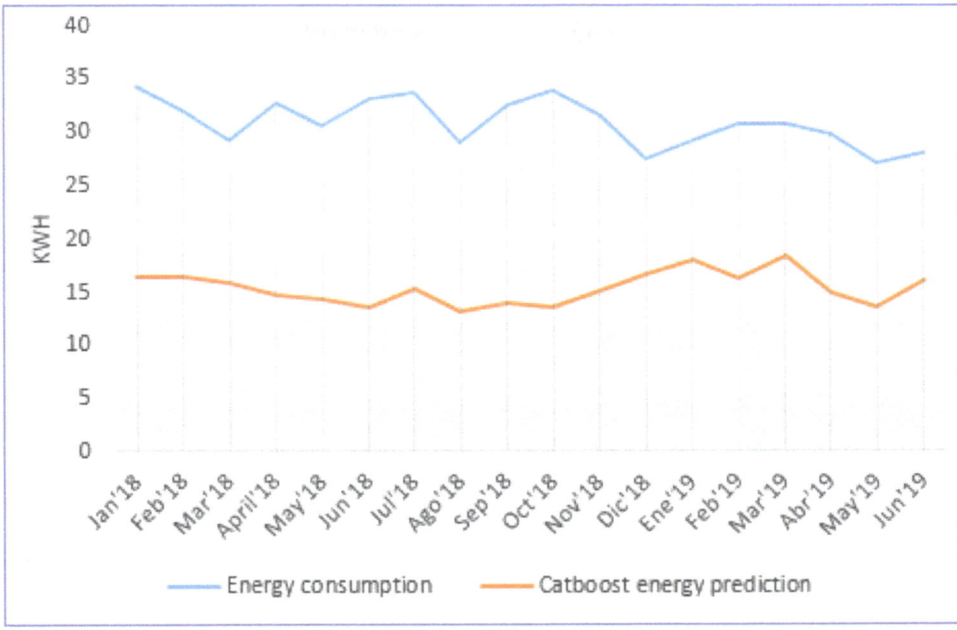

Figure 10. Energy consumption for preheating process comparing real energy with Catboost prediction.

6. Conclusions

This paper introduced an exploration data analysis methodology tailored for the analysis and optimization of a large-scale drug production process, and a Catboost machine learning model implementation, specifically focusing on the preheating stage of pharmaceutical granules using a fluid bed dryer. As a conclusion drawn from the exploratory data analysis of the signals, it can be stated that the preheating phase lasts longer than necessary. Some batches need less than 50.1 min to complete the preheating process; however, there are batches that take up to 180.3 min. In terms of energy consumption, this means that for some batches, the fluid bed dryer consumes 15.5 kWh, and for others it consumes 55.8 kWh, which could represent savings, in some cases, of 72.2% of energy. In addition, the most suitable model for the fluid bed dryer prediction process was selected based on the current dataset obtained from the activity of the fluid bed dryer process in the production plant. First, several models, including Catboost, Elastic net, Random Forest or Linear Regression, were compared. Catboost was selected because it provided the lowest error and, at the same time, the highest R2, as it has been described in previous sections. Once the model was selected, the analysis of the historical dataset, with 200 batches from 18 months of production, was performed. It has been shown that the model is able to predict on average a reduction of 50.45% of the preheating process duration and up to 59.68% in some cases. Likewise, the energy consumption of the fluid bed dryer for the preheating process could be reduced on average by 50.48% and up to 59.76%, which results on average in around 3.120 kWh of energy consumption savings per year.

Author Contributions: Conceptualization, R.B. and M.R.; methodology, R.B.; software, R.B.; validation, R.B. and M.R.; formal analysis, R.B.; investigation, R.B.; resources, R.B.; data curation, R.B.; writing—original draft preparation, R.B.; writing—review and editing all authors; visualization, all authors; supervision H.H. All authors have read and agreed to the published version of the manuscript.

Funding: This research received no external funding.

Data Availability Statement: Data is unavailable due to privacy restrictions.

Conflicts of Interest: The authors declare no conflict of interest.

References

1. Pharmaguide. Available online: https://www.pharmaguideline.com/2021/10/tablet-manufacturing-process-overview.html (accessed on 1 September 2023).
2. Parikh, D. *How to Optimize Fluid Bed Processing Technology: Part of the Expertise in Pharmaceutical Process Technology Series*; Academic Press: Cambridge, MA, USA, 2017.
3. Lourenço, V.; Lochmann, D.; Reich, G.; Menezes, J.; Herdling, T.; Schewitz, J. A quality by design study applied to an industrial pharmaceutical fluid bed granulation. *Eur. J. Pharm. Biopharm.* **2012**, *81*, 438–447. [CrossRef] [PubMed]
4. Burggraeve, A.; Monteyne, T.; Vervaet, C.; Remon, J.P.; De Beer, T. Process analytical tools for monitoring, understanding, and control of pharmaceutical fluidized bed granulation: A review. *Eur. J. Pharm. Biopharm.* **2013**, *83*, 2–15. [CrossRef] [PubMed]
5. Yüzgeç, U.; Becerikli, Y.; Türker, M. Dynamic neural-network-based model-predictive control of an industrial baker's yeast drying process. *IEEE Trans. Neural Netw.* **2008**, *19*, 1231–1242. [CrossRef]
6. Price, W.N. Making do in making drugs: Innovation policy and pharmaceutical manufacturing. *Boston Coll. Law Rev.* **2013**, *55*, 2013. [CrossRef]
7. Lifset, R.D. A new understanding of the American energy crisis of the 1970s. *Hist. Soc. Res. Hist. Sozialforschung* **2014**, *39*, 22–42.
8. Boyd, G.A. *Development of a Performance-based Industrial Energy Efficiency Indicator for Pharmaceutical Manufacturing Plants*; Duke University: Durham, NC, USA, 2013. [CrossRef]
9. Thomas, P. Will Pharma Wear the Energy Star. *Pharma Manufacturing*, 6 March 2006.
10. Pazhayattil; Babu, A.; Konyu-Fogel, G. An empirical study to accelerate machine learning and artificial intelligence adoption in pharmaceutical manufacturing organizations. *J. Generic Med.* **2023**, *19*, 17411343221151109.
11. Mujumdar, A.S. Research and development in drying: Recent trends and future prospects. *Dry. Technol.* **2014**, *22*, 1–26. [CrossRef]
12. Aghbashlo, M.; Mobli, H.; Rafiee, S.; Madadlou, A. The use of artificial neural network to predict exergetic performance of spray drying process: A preliminary study. *Dry. Technol.* **2012**, *88*, 32–43. [CrossRef]
13. Lai, J.-P.; Chang, Y.-M.; Chen, C.-H.; Pai, P.-F. A survey of machine learning models in renewable energy predictions. *Appl. Sci.* **2020**, *10*, 5975. [CrossRef]
14. Diaz, L.P.; Brown, C.J.; Ojo, E.; Mustoe, C.; Florence, A.J. Machine learning approaches to the prediction of powder flow behaviour of pharmaceutical materials from physical properties. *Digit. Discov.* **2023**, *2*, 692–701. [CrossRef]
15. Sciuto, G.L.; Susi, G.; Cammarata, G.; Capizzi, G. A spiking neural network-based model for anaerobic digestion process. In Proceedings of the 2016 International Symposium on Power Electronics, Electrical Drives, Automation and Motion (SPEEDAM), Capri, Italy, 22–24 June 2016; pp. 996–1003.
16. Kim, D.; Kim, M.; Kim, W. Wafer Edge Yield Prediction Using a Combined Long Short-Term Memory and Feed- Forward Neural Network Model for Semiconductor Manufacturing. *IEEE Access* **2020**, *8*, 215125–215132. [CrossRef]
17. Wang, J.; Zhang, J.; Wang, X. A Data Driven Cycle Time Prediction with Feature Selection in a Semiconductor Wafer Fabrication System. *IEEE Trans. Semicond. Manuf.* **2018**, *31*, 173–182. [CrossRef]
18. Aksu, B.; Matas, M.D.; Cevher, E.; Özsoy, Y.; Güneri, T.; York, P. Quality by design approach for tablet formulations containing spray coated ramipril by using artificial intelligence techniques. *Int. J. Drug Deliv.* **2012**, *4*, 59.
19. Peterson, J.J.; Snee, R.D.; McAllister, P.R.; Schoeld, T.L.; Carella, A.J. Statistics in pharmaceutical development and manufacturing. *J. Qual. Technol.* **2019**, *41*, 111–134. [CrossRef]
20. Prokhorenkova, L.; Gusev, G.; Vorobev, A.; Dorogush, A.V.; Gulin, A. CatBoost: Unbiased boosting with categorical features. In Proceedings of the 32nd International Conference on Neural Information Processing Systems, Red Hook, NY, USA, 3–8 December 2017.
21. Liu, Z. Using neural network to establish manufacture production performance forecasting in IOT environment. *J. Supercomput.* **2022**, *78*, 9595–9618. [CrossRef]
22. Markarian, J. Modernizing pharma manufacturing. *Pharm. Technol.* **2018**, *42*, 20–25.
23. Nettleton, D.F.; Wasiak, C.; Dorissen, J.; Gillen, D.; Tretyak, A.; Bugnicourt, E.; Rosales, A. Data Modeling and Calibration of In-Line Pultrusion and Laser Ablation Machine Processes. In Proceedings of the International Conference on Advanced Data Mining and Applications (ICADMA), Barcelona, Spain, 20–21 August 2018.

electronics

MDPI

Article

A Network Intrusion Detection Model Based on BiLSTM with Multi-Head Attention Mechanism

Jingqi Zhang [1], Xin Zhang [1], Zhaojun Liu [1], Fa Fu [1,*], Yihan Jiao [1] and Fei Xu [2]

1 College of Computer Science and Technology, Hainan University, Haikou 570228, China; 20203102578@hainanu.edu.cn (J.Z.); 20203101194@hainanu.edu.cn (X.Z.); 20203101299@hainanu.edu.cn (Z.L.); 20203101191@hainanu.edu.cn (Y.J.)

2 College of Civil and Architecture Engineering, Hainan University, Haikou 570228, China; 20203102564@hainanu.edu.cn

* Correspondence: fufa@hainanu.edu.cn

Abstract: A network intrusion detection tool can identify and detect potential malicious activities or attacks by monitoring network traffic and system logs. The data within intrusion detection networks possesses characteristics that include a high degree of feature dimension and an unbalanced distribution across categories. Currently, the actual detection accuracy of some detection models is relatively low. To solve these problems, we propose a network intrusion detection model based on multi-head attention and BiLSTM (Bidirectional Long Short-Term Memory), which can introduce different attention weights for each vector in the feature vector that strengthen the relationship between some vectors and the detection attack type. The model also utilizes the advantage that BiLSTM can capture long-distance dependency relationships to obtain a higher detection accuracy. This model combined the advantages of the two models, adding a dropout layer between the two models to improve the detection accuracy while preventing training overfitting. Through experimental analysis, the network intrusion detection model that utilizes multi-head attention and BilSTM achieved an accuracy of 98.29%, 95.19%, and 99.08% on the KDDCUP99, NSLKDD, and CICIDS2017 datasets, respectively.

Keywords: intrusion detection; deep learning; multi-head attention; BiLSTM

Citation: Zhang, J.; Zhang, X.; Liu, Z.; Fu, F.; Jiao, Y.; Xu, F. A Network Intrusion Detection Model Based on BiLSTM with Multi-Head Attention Mechanism. *Electronics* **2023**, *12*, 4170. https://doi.org/10.3390/electronics12194170

Academic Editors: Chao Zhang, Wentao Li, Huiyan Zhang and Tao Zhan

Received: 31 August 2023
Revised: 29 September 2023
Accepted: 6 October 2023
Published: 8 October 2023

1. Introduction

In recent years, network intrusion has gradually expanded, resulting in the theft of personal privacy and becoming the main attack platform [1]. Intrusion detection, as one of the most important network security protection tools after firewalls, plays a more important role in network security defense systems [2]. It can be defined as "network security devices that monitor network traffic to find unexpected patterns" [3]. Intrusion detection is the process of monitoring network traffic or system activity for unauthorized access, policy violations, and other malicious activities. The intrusion detection aims to identify potential security breaches and alert security personnel so they can take appropriate action to prevent further damage. There are various IDS (intrusion detection systems) to be used, including host-based IDS and network-based IDS. Host-based IDS monitor activity on a single device or server, while network-based IDS monitor all traffic on a network segment. Intrusion detection is a significant component of a comprehensive cybersecurity strategy. It can detect intrusions effectively by monitoring the status and activities of the protection system. Therefore, it has the ability to discover unauthorized or abnormal network behaviors.

As mentioned above, intrusion detection has three types, which are host-based detection, network-based detection, and collaborative detection [4]. HIDS (Host-based intrusion detection) is located in the software component of the monitored system, which mainly monitors activities within the host, such as system or shell program logs. Based on the detection techniques, there are two types of intrusion detection, including misuse detection [5] and anomaly detection [6]. Misuse detection, also known as signature-based detection, is an application of signature matching to identify intrusions. It can effectively detect known attacks and has a low rate of false alarms.

However, some machine learning techniques have some shortcomings, such as training time being too long in large training sets, too sensitive to irrelevant attributes, and so on [7], researchers try hard to adopt deep learning technology to solve these problems. Currently, artificial intelligence technology is constantly developing, and multitudinous methods of machine learning or deep learning have been applied to intrusion detection systems [8]. The methods that use machine learning have better performance than classical intrusion detection methods. These methods have the ability to learn from a quantity of intrusion data to build an intrusion detection model for distinguishing whether there is an intrusion or not. But it still has some problems, such as the need for plentiful training samples, taking a long time, and relying on feature selection.

Deep learning is usually a modification of artificial neural networks for feature extraction, perception, and learning. Its applications are now used in scores of fields, such as speech recognition, unmanned vehicles, image recognition and classification, natural language processing, bioinformatics, etc. There are various neural network models using deep learning technology. In this paper, we propose an intrusion detection model based on multi-head attention with BiLSTM. The model completes the selection and extraction of features based on the multi-head attention mechanism, and captures the dependencies of vectors over longer distances through the BiLSTM model, thus improving the accuracy and efficiency of identifying network intrusions.

2. Related Research

The first part of the related literature that uses BiLSTM for intrusion detection: S.Siviamohan et al. [9] proposed a local university intrusion detection method based on RNN-BiLSTM (Bidirectional Long Short-Term Memory Recurrent Neural Network), which uses a two-step mechanism to solve the network problem. The experimental results show that BiLSTM is better than all other RNN (Recurrent Neural Network) architectures in terms of classification accuracy, and the prediction accuracy on the CICIDS2017 dataset reaches 98.48%, but it does not specify whether it is dual classification or multi-classification.

Nelly Elsayed et al. [10] produced an intrusion detection model by using BiLSTM and CNN (Convolutional Neural Network). The BiLSTM recursive behavior is used to save the information used for intrusion detection, while the CNN perfectly extracts the data features. It can be implemented and applied to lots of smart home network gateways.

Huang Chi et al. [11] created a network intrusion detection method, which uses CNN and BiLSTM. The former extracts local parallel features, solving the problem of incomplete local feature extraction. The latter is used to extract long-distance related features, taking into account the influence of attributes before and after each data point in the sequence data, which can improve accuracy.

Liangkang Zhang et al. [12] produced a new model based on mean control, CNN and BiLSTM. During data preprocessing, the data standardization of mean control is used to standardize the original data, and then the CNN-BiLSTM algorithm is combined to predict.

The following are the related literature of the works that use an attention mechanism for intrusion detection: Jingyi Wang et al. [13] proposed an intrusion detection model that uses an attention mechanism. A SSAE (Stacked Sparse Autoencoder) was constructed to extract the high-level feature representation of related information. Furthermore, the double-layer BiGRU (Bidirectional Gated Recurrent Unit) network with an attention mechanism was used to classify data.

Haixia Hou et al. [14] proposed a method that uses HLSTM (Hierarchical LSTM) and an attention mechanism. First of all, in order to extract sequence features across multiple hierarchical structures on network record sequences, researchers used HLSTM. Then, the attention layer's function is to capture the correlation between features, redistribute the weight of features, and adaptively map the importance of each feature to different network attack categories.

Yalong Song et al. [15] proposed a mechanism using BiGRU and a multi-head attention mechanism. It can manage the data and capture the correlation between data and features.

The above articles all use artificial intelligence to predict intrusion detection but do not make a detailed classification prediction of the type of intrusion. Some articles [13] use binary classification, some articles [11,12,14] divide the main categories of network intrusion, and some articles [9,10,15] do not explain the classification clearly. The above-mentioned articles do not make a detailed prediction of intrusion detection. In order to resolve this situation and improve our classification and precision, we propose an intrusion detection model based on BiLSTM and a multi-head attention mechanism.

3. Model Methodology

In order to be suitable for the current NIDS (Network Intrusion Detection Systems) methods and their characteristics, we propose a new detection method to complete intrusion detection, which uses multi-head Attention and BiLSTM. The whole model consists of two phases, including the training phase and the prediction phase. In the first phase, the model's goal is mainly to learn the original vector features of the network intrusion data, training the network to adjust the weight parameters. Then, it can realize the training of the proposed model through true value comparison and loss function calculation. In the prediction phase, the predicted data was put into our model to obtain the final prediction results, this model can also calculate the relevant performance metrics. The overall training and evaluation structure is shown in Figure 1. A more detailed network model structure is shown in Figure 2.

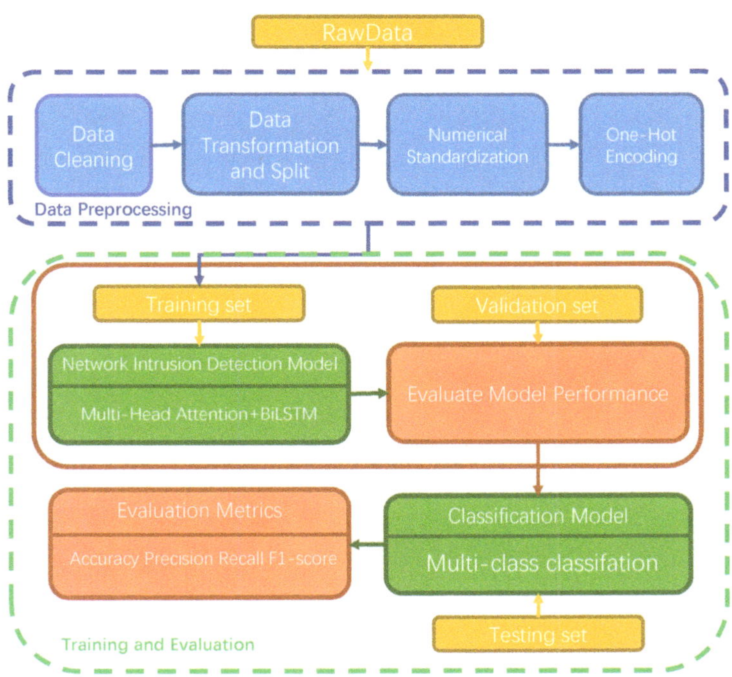

Figure 1. Overall structure of the model based on multi-head attention and BiLSTM.

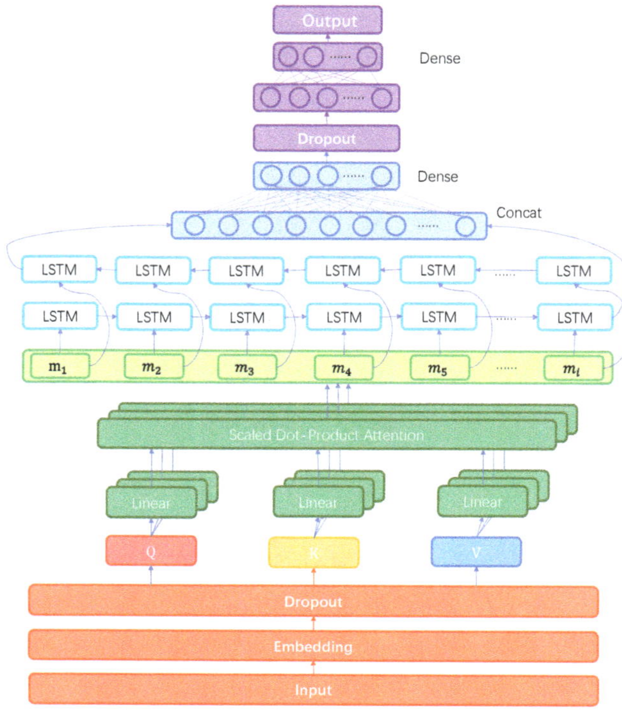

Figure 2. Intrusion detection model network structure.

3.1. Embedding

In Figure 2, we detail the model structure. First of all, we take the processed data and use the embedding layer to transform each feature of intrusion detection into the form of a vector.

The embedding layer is aimed to raise the dimension, and the input data is used to represent each feature value in the original vector (i ranges from 1 to 41 for KDDCUP99 and NSLKDD datasets, i ranges from 1 to 15 for CICIDS2017 datasets). Furthermore, $ai = Embedding(xi)$, where ai is the one-dimensional vector corresponding to each feature with a length of 32. At this time, the original vector is transformed into a two-dimensional vector (taking the NSLKDD dataset as an example, the data is transformed from one-dimensional data with a length of 41 to two-dimensional data with a length of [41, 32], where 41 is the number of features, 32 represents the embedding dimension). We hope that the features can be enlarged so that the model is capable of learning more characteristics of network intrusion activities with embedding. After the embedding layer is a dropout layer, which is used to improve the generalization ability of our proposed model. Without the dropout layer, the model we designed is prone to overfitting which leads to the low prediction accuracy of the test set. Therefore, the addition of the dropout layer can prevent overfitting to a certain extent, and that is, $ai = Dropout(ai)$.

3.2. Multi-Head Attention

Then, the processed data is passed to the multi-head attention [16] mechanism model. The reason why we use the mechanism is that it can pay attention to the important features in the vector, which is designed by imitating human vision [17]. It can give higher weights to important features and lower weights to other features. The data entering the multi-head attention mechanism model part is represented as $S = (a_1, a_2, \cdots, a_i)$, S will be multiplied with the weight data of the attention mechanism to obtain Q, K and V, that is, $q_i = SW_{qj}$,

$k_i = SW_{kj}$, and $v_i = SW_{vj}$, the range of j is from 1 to 3. The weight matrices W_q, W_k, and W_v can be continuously trained through learning, so the model's fitting ability can be further improved. The similarity matrix of different features is obtained by multiplying Q and K^T, and the similarity relationship between different features is obtained. After that, the similarity is normalized by the Softmax function, which reduces the amount of calculation to a certain extent. Finally, the obtained result is multiplied by V to obtain the data with the same dimension as the input according to the Equations (1) and (2). In Equation (2), d_K represents the dimension of the K matrix. Finally, we can obtain the final result according to Equation (3). Its structure is shown in Figure 3 schematically.

$$head_j(Q,K,V) = Softmax\left(\frac{QK^T}{\sqrt{d_k}}\right)V \tag{1}$$

$$Softmax(z_i) = \frac{\exp(z_i)}{\sum_j \exp(z_j)} \tag{2}$$

$$T = Concat(head_1, head_2, head_3) \tag{3}$$

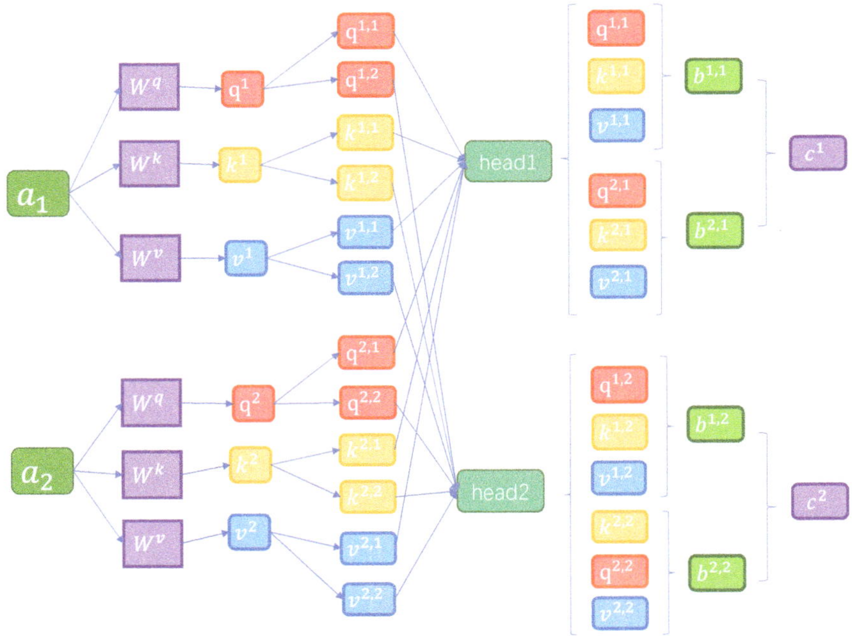

Figure 3. Schematic diagram of the multi-headed attention mechanism.

3.3. BiLSTM

After that, the data weighted by the attention mechanism is fed into the BiLSTM model. LSTM is a kind of RNN, that can learn and remember long-term dependencies, capture the relationship between different features in the feature vector, and will not encounter the problem of gradient disappearance or gradient explosion [18]. Graves et al. [19] reported an important improvement in classification accuracy when using LSTM in a bidirectional architecture. The feature vector of intrusion detection is not a time series data, but it can analyze the relationship between distant features, associate different features, and then make predictions. In this case, the output is a one-dimensional vector of length 128.

In the previous section, we were given the data T generated by the multi-head attention mechanism. Data T is a two-dimensional vector composed of multiple one-dimensional vectors, which we denote as $T = (m_1, m_2, \ldots, m_i)$.

We believe that for the detection of a certain relationship in the data, LSTM has the ability to capture this longer distance dependence while being able to avoid gradient disappearance, gradient explosion, and other problems. However, only using LSTM cannot encode back-to-back information. Therefore, we use BiLSTM to improve the ability to capture bidirectional features.

BiLSTM is composed of several small structures, with one basic unit consisting of four layers, which are the input layer, forward propagation layer, backward propagation layer, and output layer. The forward propagation layer is in charge of extracting the forward features of the vector from front to back, while the backward transmission layer is in charge of extracting the reverse features of the input sequence from back to front. The output layer integrates the data output from the forward propagation layer and the backward propagation layer. We want to extract the forward-backward correlation of the vectors, so the output formula of BiLSTM is shown in Equation (4).

$$h_i = [\overrightarrow{h_i} \oplus \overleftarrow{h_i}] \tag{4}$$

where $\oplus$ denotes the summation calculation of the corresponding elements. $\overrightarrow{h_i}$ denotes the forward output, $\overleftarrow{h_i}$ denotes the backward output. Finally, h_i denotes the result of the summation of the corresponding elements.

Among them, the BiLSTM network structure has lots of single LSTM structures, and the individual LSTM structure is shown in Figure 4.

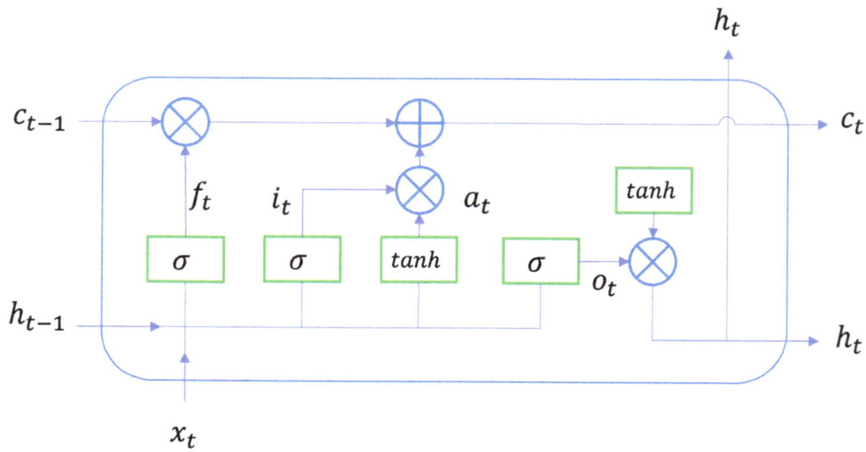

Figure 4. LSTM network structure.

LSTM adds three gating structures in the hidden layer, namely the forget gate, input gate, and output gate, and it also adds a new hidden cell state. In Figure 4, $f(t)$, $i(t)$, and $o(t)$ represent the forget gate, input gate, and output gate at time t, and $a(t)$ represents the initial feature extraction of $h(t-1)$ and $c(t)$ at time t. All formulas are shown in Equations (5)–(8).

$$f(t) = \sigma(W_f h_{t-1} + U_f m_t + b_f) \tag{5}$$

$$i(t) = \sigma(W_i h_{t-1} + U_i m_t + b_i) \tag{6}$$

$$a(t) = \tanh(W_a h_{t-1} + U_a m_t + b_a) \tag{7}$$

$$o(t) = \sigma(W_O h_{t-1} + U_O m_t + b_O) \tag{8}$$

where x_t represents the input at time t. h_{t-1} represents the hidden layer status value at time $t-1$. W_f, W_i, W_o represent the weight parameter of h_{t-1} in the feature extraction process of the forget gate, input gate, and output gate. U_f, U_i, U_o represent the weight parameter of x_t in the feature extraction process of forget gate, input gate, and output gate. b_f, b_i, and b_o represent the forget gate, input gate, output gate, and offset value in the process of feature extraction, respectively. The related functions are shown in Equation (9) [20] and Equation (10) [21]:

$$\tanh(x) = \frac{1-e^{-2x}}{1+e^{-2x}} \tag{9}$$

$$\sigma(x) = \frac{1}{1+e^{-x}} \tag{10}$$

The results of the forget gate and output gate calculations act on $c(t-1)$, which constitutes the cell state $c(t)$ at moment t, denoted as Equation (11). The final hidden state $h(t)$ at moment t is derived from the output gate o(t) as well as the cell state $c(t)$, denoted as Equation (12). where $\odot$ represents Hadamard product.

$$c(t) = c(t-1) \odot f(t) + i(t) \odot a(t) \tag{11}$$

$$h(t) = o(t) \odot \tanh(c(t)) \tag{12}$$

3.4. *Dense Layer*

Dense layers are a type of layer that helps the neural network to better understand the data and improve the accuracy of the model's predictions. We add two dense layers after BiLSTM, but each uses different activation functions. The first dense layer includes the ReLU activation function [22] (shown in Equation (13)). It aims to learn the features in the data and distinguish different features that improve the accuracy of the model prediction. The ReLU function is very simple and fast to calculate, which is very suitable for large-scale deep neural networks. Meanwhile, the ReLU function can effectively prevent the case of gradient disappearance, and the gradient is constant to 1 when the input is bigger than 0.

$$ReLU(x) = max(0, x) \tag{13}$$

The second dense layer uses the Softmax activation function (shown in Equation (2)), which is often used in multi-class classification problems, so it is very suitable for our network detection model. Then, the output of the dense layer also corresponds to the number of attack types. In the end, our model predicts all the mentioned attack types in the dataset, so it is of great importance to use the Softmax activation function.

Combining multi-head attention with BiLSTM has several advantages, including:

- Improved sequence modeling: BiLSTM is a type of RNN that can effectively model sequential data in both forward and backward directions. However, when combined with multi-head attention, they can further enhance the model's ability to capture long-range dependencies and improve the quality of sequence modeling.
- Increased interpretability: Multi-head attention mechanism allows the model to attend to distinct parts of the input sequence selectively, providing more transparency and interpretability to the model's decision-making process. This is particularly useful in detection tasks such as network intrusion detection.
- Robustness to noise and variations: By attending to multiple parts of the input sequence, the model becomes more robust to variations and noise in the data.
- Scalability: The combination of multi-head attention with BiLSTM allows the model to scale well to larger datasets and more complex tasks without compromising performance or accuracy. This makes it an effective approach for handling large-scale network intrusion detection tasks.

4. Implementation Details

In our experiments, the hardware environment is as follows: the CPU model is Intel Core i7-10750H, the GPU model is NVIDIA GeForce RTX2060 with Max-Q Design, the memory on the GPU card is 6 GB and the RAM on the computer is 32 GB.

The language and platform (software) environment are as follows: the operating system used in the experiment is Windows 11, and the programming environment is Python 3.9. The Keras deep learning framework and Scikit-learn framework are used to help us build the model and process the data.

We divided the data set into three parts, each part has a different function. The training set accounted for 64%, the validation set accounted for 16%, and the test set accounted for 20%. A total of 60 rounds of model training were performed, the batch size was 512, the random number seed was 0, and the number of heads of the multi-head attention mechanism was 3. The Adam algorithm [23] is used as the optimizer of the model, and the value of the learning rate is 0.0003. Adam is able to adaptively adjust the learning rate based on the gradient information and adjust the momentum to avoid falling into a local minimum too early. We add two dense layers after BiLSTM and in the output part, but they have distinct activation functions, one is the ReLU activation function, and the other is the Softmax activation function. Meanwhile, a dropout layer is added after the embedding layer and between the two dense layers with parameters of 0.8 and 0.3, respectively. The whole training time of the model is 55 min on the KDDCUP99 dataset, 4 min 45 s on the NSLKDD dataset, and 33 min on the CICIDS2017 dataset.

The core code of the model we designed is shown in Figure 5. The code is written in Python and built under the Keras deep learning framework.

```
inputs = Input(name='inputs', shape=[max_words, ], dtype='float64')
x = Embedding(top_words, input_length=max_words, output_dim=embed_dim)(inputs)
x = Dropout(0.8)(x)
x = MultiHeadAttention(num_heads=3, key_dim=embed_dim)(x, x, x)
x = Bidirectional(LSTM(hidden_dim[0]))(x)
x = Dense(64, activation='relu')(x)
x = Dropout(0.3)(x)
output = Dense(num_labels, activation='softmax')(x)
model = Model(inputs=inputs, outputs=output)
```

Figure 5. The core code of the model.

5. Experiment and Results

In the model experiment, in order to ensure the accuracy of the experiment, we use three data sets, namely the KDDCUP99 data set, the NSLKDD data set, and the CICIDS2017 data set, which can carry out a more comprehensive evaluation of our model.

5.1. Introduction to the Data Set

The KDDCUP99 dataset [24] is a widely used benchmark dataset in the field of network intrusion detection. It was created for not only the KDD Cup 1999 Data Mining but also the Knowledge Discovery competition. It aims to develop efficient algorithms for detecting network intrusions from TCP/IP network traffic data. This dataset consists of a large collection of network traffic data captured from a simulated environment. It also contains a variety of features extracted from network packets, such as protocol types, service types, and so on.

The NSLKDD dataset [25] is an improved version of the KDD Cup 1999 dataset. This dataset is widely used for network intrusion detection research. NSLKDD stands for "NSL-KDD Intrusion Detection Dataset". It was developed to address some limitations and drawbacks of the original KDD Cup dataset. The types of attacks are shown in Table 1.

Table 1. KDDCUP99 and NSLKDD dataset attack types.

Category	Attack	Interpretation
Normal	Normal	Normal network activity
DOS	back, land, neptune, pod, smurf, teardrop	Denial-of-Service (DoS) attack is a type of cyber attack where a perpetrator attempts to make a website or network resource unavailable to its intended users by overwhelming it with traffic or other types of data.
Probing	ipsweep, nmap, portsweep, satan	Surveillance and other detection activities.
R2L	ftp_write, guess_passwd, imap, multihop, phf, spy, warezclient, warezmaster	Remote-to-Local (R2L) attack is a type of cyber attack where an attacker tries to gain unauthorized access to a target system by exploiting vulnerabilities in remote services or applications.
U2R	buffer overflow, loadmodule, perl, rootkit	User-to-Root (U2R) attackis a type of cyber attack where an attacker with limited privileges on a system attempts to gain root-level access.

The CICIDS2017 dataset [26], also known as the Canadian Institute for Cybersecurity Intrusion Detection System (CIC-IDS2017), is a comprehensive dataset designed for evaluating NIDS. Its authors are researchers at the University of New Brunswick in Canada. This dataset consists of various network traffic features extracted from different types of network traffic, including normal traffic and several types of attacks. It also simulates a real-world network environment to provide a realistic representation of network traffic. A short description of its files is shown in Table 2. Descriptions of all the datasets are shown in Table 3 (Table 3 is the data set size that has been balanced by using the algorithm SMOTE. The relevant content is presented in Section 5.2.4).

Table 2. Description of files containing CICIDS2017 dataset.

Name of Files	Day Activity	Attacks Found	Advantage	Goal
Monday WorkingHours.pcap_ISCX.csv	Monday	Benign (Normal human activities)	This is a dataset that can further meet real-world standards, covering attack standards from 11 countries, making it more reliable and available [26].	Using this dataset can help improve our model's generalization ability and improve its accuracy in modern intrusion detection predictions, rather than just being applicable to the past.
Tuesday WorkingHours.pcap_ISCX.csv	Tuesday	Benign, FTP-Patator, SSH-Patator		
Wednesday workingHours.pcap_ISCX.csv	Wednesday	Benign, DoS GoldenEye, DoS Hulk, DoS Slowhttptest, DoS slowloris, Heartbleed		
Thursday-WorkingHours Morning-WebAttacks.pcap_ISCX.csv	Thursday	Benign, Web Attack-Brute Force, Web Attack-Sql Injection, Web Attack-XSS		
Thursday-WorkingHours-Afternoon-Infilteration.pcap_ISCX.csv	Thursday	Benign, Infiltration		
Friday-WorkingHours Morning.pcap_ISCX.csv	Friday	Benign, Bot		
Friday-WorkingHours-Afternoon PortScan.pcap_ISCX.csv	Friday	Benign, PortScan		
Friday-WorkingHours-Afternoon DDos.pcap_ISCX.csv	Friday	Benign, DDoS		

Table 3. Description of all datasets.

Dataset	Training Set	Validation Set	Test Set	Total	Input Vector Features	Number of Labels
KDDCUP99	3,108,950	777,237	971,547	4,857,734	41	40
NSLKDD	95,050	23,762	29,704	148,516	41	40
CICIDS2017	498,741	124,685	155,857	779283	78	15

5.2. Data Processing

We used numerical division, data normalization, one-hot encoding, and data balance to process the data set. For KDDCUP99 and NSLKDD data sets, the dimension after data processing is 41×1, and the data set has 40 classification tags, one of which is the normal category, and the other 39 represent various attack types. After one-hot encoding, they are converted into 40-dimensional vectors, so the dimension of the output layer is 40. For the CICIDS2017 dataset, the dimension after data processing is 78×1, and the dataset has 15 classification labels in total, one of which is the normal category, and the other 14 labels represent various attack types. After one-hot encoding, it is converted to a 15-dimensional vector, so the output dimension is 15.

5.2.1. Data Conversion

For the KDDCUP99 dataset, we convert the text type data into numeric types to facilitate our training and testing of the model, such as character type data protocal_type, service, and state, where protocal_type has 3 protocol types, service has 70 network service types, while flag haves 11 network connection types. For instance, the three protocol types characterized by protocol type are TCP, UDP, and ICMP, which we convert to 0, 1, and 2.

5.2.2. Data Normalization

The variation in individual features of the numerically processed data is large, and normalizing this data can avoid causing gradient dispersion when using the backpropagation algorithm. The problem with not normalizing the data is that the magnitude of the gradient decreases with backpropagation, which slows down the growth of the update weights of the intrusion detection model, resulting in a complex dataset of features that are not well extracted by deep learning. We use the normalization method of z-score to convert all the data of KDDCUP99 to $[-1, 1]$, as shown in Equation (14).

$$m'_i = \frac{m_i - \overline{m}}{x} \tag{14}$$

We use m_i and m'_i to represent the value of the data sample before and after normalization. Meanwhile, $\overline{m}$ is used to represent the average data value of the feature before normalization.

5.2.3. One-Hot Encoding

In the feature vectors of the three datasets, we did not use one-hot encoding for processing, because the original data has too many features; thus, using it may lead to the curse of dimensionality and reduce the accuracy. We perform one-hot encoding for the label values (predicted attack types) of the datasets. For the KDDCUP99 dataset and NSLKDD dataset, we convert 40 different categories into one-dimensional vectors with a length of 40 by one-hot encoding. For the CICIDS2017 dataset, we convert 15 different categories into one-dimensional vectors with a length of 15 by one-hot encoding. This encoding is more conducive to model training.

5.2.4. Dataset Balanced

For the KDDCUP99 and CICIDS2017 datasets, there is an imbalance between normal samples and attack samples. To make the dataset more balanced, we use the SMOTE (Synthetic Minority Over-sampling Technique) [27] algorithm to over-sample a small number of samples. To ensure the balance of the dataset and the generalization of our model, we randomly under-sample the samples with a large number of samples.

SMOTE works by creating synthetic examples of the minority class that are strategically placed between existing instances of the minority class. The algorithm randomly selects a minority instance and looks for its k nearest neighbors. It then selects one of these neighbors and calculates the difference between the feature values of the two instances. It multiplies this difference by a random value between 0 and 1 and adds it to the feature values of the selected minority instance. This generates a synthetic instance that is similar to the original minority instance but slightly different.

By repeating this process for multiple instances of the minority samples, SMOTE increases the number of these samples in the dataset. This operation helps to balance the class distribution and provides more training examples for the minority class, and it also improves the performance of machine learning algorithms. The size of the dataset after processing is shown in Table 3.

5.3. Evaluation Criteria

We use Accuracy, Precision, Recall, and $F1$-Score as the metrics for model evaluation, and the formulas for all metrics are shown in Table 4. Where TN indicates the number of correctly predicted negative samples, TP indicates the number of correctly predicted positive samples, FN indicates the number of incorrectly predicted negative samples, and FP indicates the number of incorrectly predicted positive samples.

Table 4. Model evaluation metrics.

Metric	Mathematical Formulae
Accuracy	$Accuracy = \frac{TP+TN}{TP+TN+FP+FN}$
Precision	$Precison = \frac{TP}{TP+FP}$
Recall	$Recall = \frac{TP}{TP+FN}$
$F1$-Score	$F1 - Score = 2 \times \frac{Precision \times Recall}{Precision \times Recall}$

5.4. Model Review

We have tested and evaluated our model using metrics in Table 4. The result is shown in Table 5. Through the test data, we obtain the information that the accuracy of our model on the KDDCUP99 dataset is 98.29%, the Precision against normal samples (unattacked samples) is 0.97, Recall is 1, and $F1$ is 0.98. The accuracy of our model on the NSLKDD dataset is 95.19%, the Precision against normal samples (unattacked samples) is 0.95, Recall is 0.98, and $F1$ is 0.97. For the CICIDS2017 dataset, the accuracy is 99.08%, the Precision against normal samples (unattacked samples) is 1, Recall is 0.99, and $F1$ is 0.99. It was found that changing the ratio of the training set and test set size did not have an impact on the detection results of the samples. The data results are shown in Figure 6.

Table 5. Performance of different network structures on the dataset.

Dataset	Structure	Accuracy(%)	Precision	Recall	$F1$-Score
KDDCUP99	Transformer	85.71	0.88	0.82	0.85
	BiLSTM	98.25	0.97	1	0.98
	Attention	71.65	0.63	0.99	0.77
	Multi-Head Attention	71.54	0.63	0.98	0.77
	Attention + BiLSTM	97.96	0.97	1	0.98

Table 5. *Cont.*

Dataset	Structure	Accuracy(%)	Precision	Recall	*F1*-Score
	Multi-Head Attention + BiLSTM	98.29	0.97	1	0.98
NSLKDD	Transformer	73.26	0.75	0.81	0.78
	BiLSTM	95.13	0.96	0.97	0.97
	Attention	65.01	0.76	0.62	0.68
	Multi-Head Attention	65.01	0.76	0.62	0.68
	Attention + BiLSTM	94.7	0.95	0.98	0.96
	Multi-Head Attention + BiLSTM	95.19	0.95	0.98	0.97
CICID17	Transformer	97.94	0.98	0.97	0.97
	BiLSTM	98.51	0.99	0.98	0.99
	Attention	97.75	0.98	0.97	0.98
	Multi-Head Attention	97.88	0.99	0.97	0.98
	Attention + BiLSTM	97.24	0.97	0.97	0.97
	Multi-Head Attention + BiLSTM	99.08	1	0.99	0.99

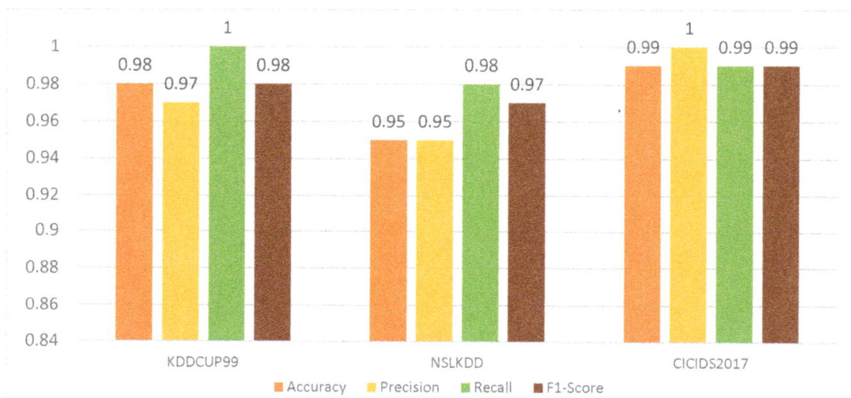

Figure 6. Performance evaluation of three datasets.

5.5. Model Accuracy and Loss Variation

Next, Figures 7–9 show the loss and accuracy during the training of the KDDCUP99 dataset, NSLKDD dataset, and CICIDS2017 dataset, respectively. The experiment carried out 60 rounds of training. For the KDDCUP99 dataset, the curve became smoother after 10 rounds of training, and finally, the value of loss dropped to 0.0049. For the NSLKDD dataset, the curve became smoother after 30 rounds of training, and the value of loss dropped to 0.1481. For the CICIDS2017 dataset, the curve became smoother after 20 rounds of training, and the value of loss dropped to 0.0209. It can be seen from the training process that the model proposed by us has better learning ability and fitting speed. The normalized confusion matrix of the CICIDS2017 dataset is shown in Figure 10.

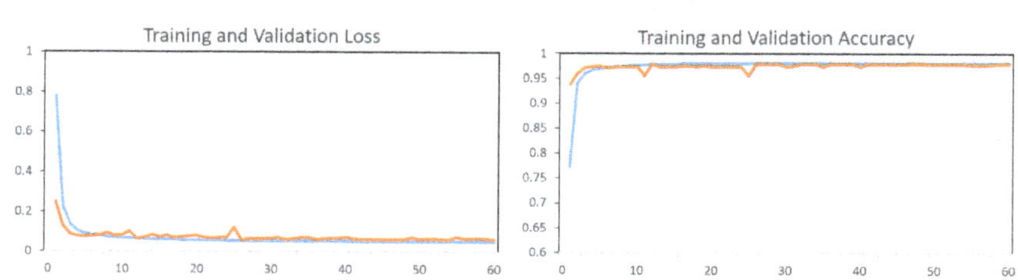

Figure 7. Accuracy and loss variation of model training on the KDDCUP99 dataset.

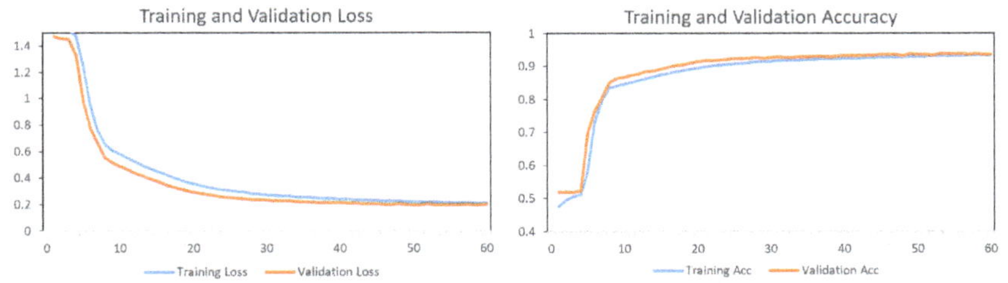

Figure 8. Accuracy and loss variation of model training on the NSLKDD dataset.

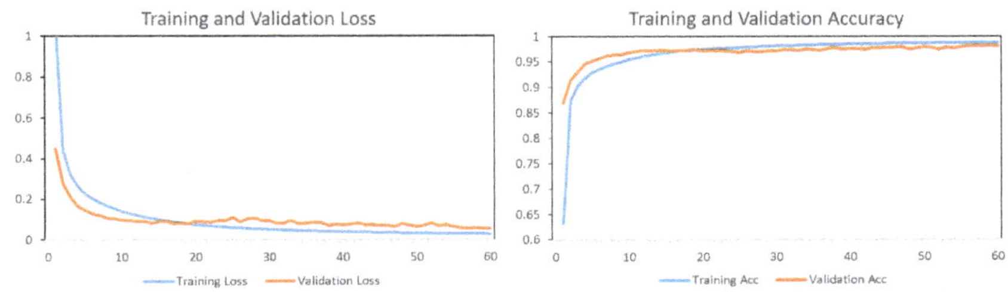

Figure 9. Accuracy and loss variation of model training on the CICIDS2017 dataset.

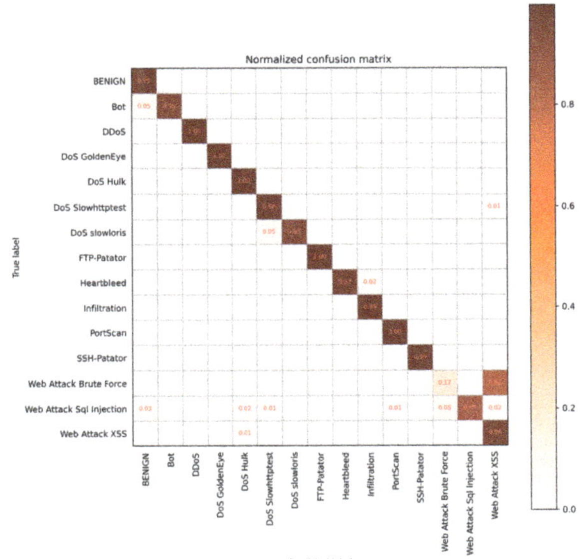

Figure 10. The normalized confusion matrix of CICIDS2017 dataset.

5.6. Ablation Experiments

We believe that the reasons for the high accuracy as well as the better performance of the proposed model are:

- We use BiLSTM to build our model. On the one hand, it can capture the bidirectional features better, on the other hand, it has the ability to avoid situations such as gradient disappearance and gradient explosion, which are very suitable for network intrusion detection.

- The addition of the multi-headed attention mechanism allows different attention weights for each vector in the feature vector to strengthen the relationship between certain vectors and the type of detected attacks, which improves the accuracy of detection. It also avoids the problem of over-focusing attention on its position.

We have verified the performance and accuracy of the Transformer, BiLSTM, Attention, Multi-Head Attention, Attention + BiLSTM, and Multi-Head attention + BiLSTM structures on three data sets. For Multi-Head Attention, we obtain the best head parameter of 3 according to pre-training. In Table 5, we show the performance indicators of different structures on three datasets by using four metrics shown in Table 4 for normal samples (samples that are not under attack). It can be seen from the results that the Multi-Head Attention + BiLSTM structure has higher accuracy and higher F1-Score.

From the data in Table 5, we can obtain the information that the classification effect of the single attention mechanism model or the multi-head attention mechanism model is not good. Furthermore, the performance on the KDDCUP99 dataset and NSLKDD dataset is even worse. The Transformer model alone is not fully competent for the task of intrusion detection, and the accuracy is relatively low. However, when the multi-head attention mechanism is combined with the BiLSTM model, the accuracy of intrusion detection recognition can be increased, and its detection accuracy is also better than that of the BiLSTM model alone, indicating that the combination of the two can show the advantages of both, it also can adapt well to the task of intrusion detection.

5.7. Comparison with Other Models

The experiments of the model were trained and predicted on the KDDCUP99, NSLKDD, and CICIDS2017 datasets, and we compared our model with other models, comparing metrics including accuracy as well as F1-score. the final results are shown in Table 6. The accuracy is shown in Figure 11.

In Table 6 and Figure 11, the information we have obtained indicates that our proposed model has better predictive performance, which means that more refined intrusion detection can be performed. At the same time, we also achieve the most refined classification and inform the relevant people with more specific information.

Table 6. Comparison with other models on the dataset.

Dataset	Algorithm	Accuracy (%)	*F1*-Score (%)
KDDCUP99	CLAIRE [28]	93.58	95.9
	MCLDM [29]	93.94	96.06
	Improved LSTM [30]	97.79	96.95
	Our Method	98.29	98
NSLKDD	BAT [31]	84.25	-
	ICVAE-DNN [32]	85.97	86.27
	Deep AE [33]	87	81.21
	Our Method	95.19	97
CICIDS2017	KELM [34]	97.15	-
	CLAIRE [28]	98	-
	CNN [35]	98	98
	Our Method	99.08	99

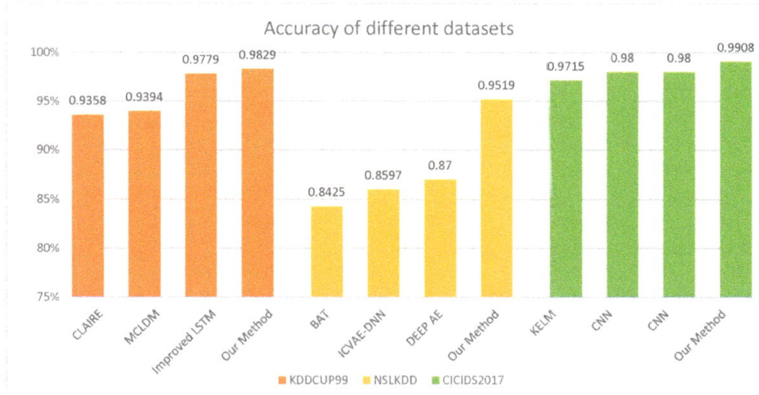

Figure 11. Comparison of different models on three datasets.

6. Conclusions

In this paper, we propose an intrusion detection model based on a multi-head attention mechanism and BiLSTM. The embedding layers can convert sparse high-dimensional feature vectors into low-dimensional feature vectors. This operation can fuse a large amount of valuable information. Then, we try to use the attention mechanism to introduce different attention weights for each vector in the feature vector, not only strengthening the relationship between certain vectors and the type of detected attacks but also improving the accuracy of detection. We also improve the use of a multi-head attention mechanism to avoid focusing too much attention on certain elements in the vector. Finally, we apply the BiLSTM network to detect some kind of relationship that exists in the data, while LSTM aims to capture long-distance dependencies and also can avoid lots of situations like gradient disappearance and gradient explosion. The experimental comparison shows that our proposed model has better accuracy and $F1$-score on the KDDCUP99, NSLKDD, and CICIDS2017 datasets than other models, and it is more accurate for multiple types of intrusion detection than other binary intrusion detection models.

Of course, our model still has some shortcomings. For example, our model is a multi-classification model, hoping to make more detailed and accurate predictions for different network intrusions. If the intrusion is a new type, our model can not give its definition or report what kind of specific intrusion method it is, but it can still be classified as an intrusion type rather than a normal network activity for professionals to study. In addition, the normal samples of the KDDCUP99 dataset and the normal samples of the CICIDS2017 dataset are too large. To ensure the availability of the detection model, we use oversampling and undersampling to ensure the balance of the data, but the sampling process has great randomness, which may delete some important information in the majority sample. In future improvements, we will try to solve these problems.

Author Contributions: Conceptualization, J.Z., X.Z., Z.L. and F.F.; methodology, J.Z.; software, J.Z. and Y.J.; validation, J.Z. and Y.J.; formal analysis, J.Z.; investigation, X.Z.; resources, X.Z.; data curation, J.Z.; writing—Original draft preparation, J.Z. and X.Z.; writing—Review and editing, Z.L., F.F. and F.X.; supervision, F.F.; project administration, J.Z. and X.Z. All authors have read and agreed to the published version of the manuscript.

Funding: This work was funded by Hainan Province Science and Technology Special Fund (Grant No. ZDYF2021GXJS006), Haikou Science and Technology Plan Project (Grant No. 2022-007) and Key Laboratory of PK System Technologies Research of Hainan, China.

Data Availability Statement: The datasets analyzed during this study are available from the corresponding author upon reasonable request.

Conflicts of Interest: The authors declare no conflict of interest.

References

1. Manzoor, I.; Kumar, N. A feature reduced intrusion detection system using ANN classifier. *Expert Syst. Appl.* **2017**, *88*, 249–257.
2. Thapa, S.; Mailewa, A. The role of intrusion detection/prevention systems in modern computer networks: A review. In Proceedings of the Conference: Midwest Instruction and Computing Symposium (MICS), Online, 3–4 April 2020 ; Volume 53, pp. 1–14.
3. Patgiri, R.; Varshney, U.; Akutota, T.; Kunde, R. An investigation on intrusion detection system using machine learning. In Proceedings of the 2018 IEEE Symposium Series on Computational Intelligence (SSCI), Bangalore, India, 18–21 November 2018; pp. 1684–1691.
4. Liu, M.; Xue, Z.; Xu, X.; Zhong, C.; Chen, J. Host-based intrusion detection system with system calls: Review and future trends. *ACM Comput. Surv. (CSUR)* **2018**, *51*, 1–36. [CrossRef]
5. Pu, G.; Wang, L.; Shen, J.; Dong, F. A hybrid unsupervised clustering-based anomaly detection method. *Tsinghua Sci. Technol.* **2020**, *26*, 146–153. [CrossRef]
6. Buczak, A.L.; Guven, E. A survey of data mining and machine learning methods for cyber security intrusion detection. *IEEE Commun. Surv. Tutor.* **2015**, *18*, 1153–1176. [CrossRef]
7. Momand, A.; Jan, S.U.; Ramzan, N. A Systematic and Comprehensive Survey of Recent Advances in Intrusion Detection Systems Using Machine Learning: Deep Learning, Datasets, and Attack Taxonomy. *J. Sens.* **2023**, *2023*, 6048087. [CrossRef]
8. Liu, H.; Lang, B. Machine learning and deep learning methods for intrusion detection systems: A survey. *Appl. Sci.* **2019**, *9*, 4396. [CrossRef]
9. Sivamohan, S.; Sridhar, S.; Krishnaveni, S. An effective recurrent neural network (RNN) based intrusion detection via bi-directional long short-term memory. In Proceedings of the 2021 international conference on intelligent technologies (CONIT), Hubli, India, 25–27 June 2021; pp. 1–5.
10. Elsayed, N.; Zaghloul, Z.S.; Azumah, S.W.; Li, C. Intrusion detection system in smart home network using bidirectional LSTM and convolutional neural networks hybrid model. In Proceedings of the 2021 IEEE International Midwest Symposium on Circuits and Systems (MWSCAS), Lansing, MI, USA, 9–11 August 2021; pp. 55–58.
11. Chi, H.; Lin, C. Industrial Intrusion Detection System Based on CNN-Attention-BILSTM Network. In Proceedings of the 2022 International Conference on Blockchain Technology and Information Security (ICBCTIS), Huaihua City, China, 15–17 July 2022; pp. 32–39.
12. Zhang, L.; Huang, J.; Zhang, Y.; Zhang, G. Intrusion detection model of CNN-BiLSTM algorithm based on mean control. In Proceedings of the 2020 IEEE 11th International Conference on Software Engineering and Service Science (ICSESS), Beijing, China, 16–18 October 2020; pp. 22–27.
13. Wang, J.; Chen, N.; Yu, J.; Jin, Y.; Li, Y. An efficient intrusion detection model combined bidirectional gated recurrent units with attention mechanism. In Proceedings of the 2020 7th International Conference on Behavioural and Social Computing (BESC), Bournemouth, UK, 5–7 November 2020; pp. 1–6.
14. Hou, H.; Di, Z.; Zhang, M.; Yuan, D. An Intrusion Detection Method for Cyber Monintoring Using Attention based Hierarchical LSTM. In Proceedings of the 2022 IEEE 8th Intl Conference on Big Data Security on Cloud (BigDataSecurity), IEEE Intl Conference on High Performance and Smart Computing,(HPSC) and IEEE Intl Conference on Intelligent Data and Security (IDS), Jinan, China, 6–8 May 2022; pp. 125–130.
15. Song, Y.; Zhang, D.; Li, Y.; Shi, S.; Duan, P.; Wei, J. Intrusion Detection for Internet of Things Networks using Attention Mechanism and BiGRU. In Proceedings of the 2023 5th International Conference on Electronic Engineering and Informatics (EEI), Wuhan, China, 30 June–2 July 2023; pp. 227–230.
16. Vaswani, A.; Shazeer, N.; Parmar, N.; Uszkoreit, J.; Jones, L.; Gomez, A.N.; Kaiser, Ł.; Polosukhin, I. Attention is all you need. In Proceedings of the NIPS'17: Proceedings of the 31st International Conference on Neural Information Processing Systems, Long Beach, CA, USA, 4–9 December 2017.
17. Liu, C.; Liu, Y.; Yan, Y.; Wang, J. An intrusion detection model with hierarchical attention mechanism. *IEEE Access* **2020**, *8*, 67542–67554. [CrossRef]
18. Hochreiter, S.; Schmidhuber, J. Long short-term memory. *Neural Comput.* **1997**, *9*, 1735–1780. [CrossRef]
19. Graves, A.; Schmidhuber, J. Framewise phoneme classification with bidirectional LSTM and other neural network architectures. *Neural Netw.* **2005**, *18*, 602–610. [CrossRef]
20. Medsker, L.R.; Jain, L. Recurrent neural networks. *Des. Appl.* **2001**, *5*, 2.
21. Schtickzelle, M. Pierre-François Verhulst (1804–1849). La première découverte de la fonction logistique. *Population* **1981**, *3*, 541–556. [CrossRef]
22. Sudjianto, A.; Knauth, W.; Singh, R.; Yang, Z.; Zhang, A. Unwrapping the black box of deep relu networks: Interpretability, diagnostics, and simplification. *arXiv* **2020**, arXiv:2011.04041.
23. Kingma, D.P.; Ba, J. Adam: A method for stochastic optimization. *arXiv* **2014**, arXiv:1412.6980.
24. Tavallaee, M.; Bagheri, E.; Lu, W.; Ghorbani, A.A. A detailed analysis of the KDD CUP 99 data set. In Proceedings of the 2009 IEEE symposium on computational intelligence for security and defense applications, Ottawa, ON, Canada, 8–10 July 2009; pp. 1–6.
25. Revathi, S.; Malathi, A. A detailed analysis on NSL-KDD dataset using various machine learning techniques for intrusion detection. *Int. J. Eng. Res. Technol.* **2013**, *2*, 1848–1853.

26. Sharafaldin, I.; Lashkari, A.H.; Ghorbani, A.A. Toward generating a new intrusion detection dataset and intrusion traffic characterization. *ICISSp* **2018**, *1*, 108–116.
27. Chawla, N.V.; Bowyer, K.W.; Hall, L.O.; Kegelmeyer, W.P. SMOTE: Synthetic minority over-sampling technique. *J. Artif. Intell. Res.* **2002**, *16*, 321–357. [CrossRef]
28. Andresini, G.; Appice, A.; Malerba, D. Nearest cluster-based intrusion detection through convolutional neural networks. *Knowl.-Based Syst.* **2021**, *216*, 106798. [CrossRef]
29. Luo, J.; Zhang, Y.; Wu, Y.; Xu, Y.; Guo, X.; Shang, B. A Multi-Channel Contrastive Learning Network Based Intrusion Detection Method. *Electronics* **2023**, *12*, 949. [CrossRef]
30. Zhang, L.; Yan, H.; Zhu, Q. An Improved LSTM Network Intrusion Detection Method. In Proceedings of the 2020 IEEE 6th International conference on Computer and Communications (ICCC), Chengdu, China, 11–14 December 2020; pp. 1765–1769.
31. Su, T.; Sun, H.; Zhu, J.; Wang, S.; Li, Y. BAT: Deep Learning Methods on Network Intrusion Detection Using NSL-KDD Dataset. *IEEE Access* **2020**, *8*, 29575–29585. [CrossRef]
32. Yang, Y.; Zheng, K.; Wu, C.; Yang, Y. Improving the Classification Effectiveness of Intrusion Detection by Using Improved Conditional Variational AutoEncoder and Deep Neural Network. *Sensors* **2019**, *19*, 2528. [CrossRef]
33. Ieracitano, C.; Adeel, A.; Morabito, F.C.; Hussain, A. A novel statistical analysis and autoencoder driven intelligent intrusion detection approach. *Neurocomputing* **2020**, *387*, 51–62. [CrossRef]
34. Wang, Z.; Zeng, Y.; Liu, Y.; Li, D. Deep belief network integrating improved kernel-based extreme learning machine for network intrusion detection. *IEEE Access* **2021**, *9*, 16062–16091. [CrossRef]
35. Mendonça, R.V.; Teodoro, A.A.; Rosa, R.L.; Saadi, M.; Melgarejo, D.C.; Nardelli, P.H.; Rodríguez, D.Z. Intrusion detection system based on fast hierarchical deep convolutional neural network. *IEEE Access* **2021**, *9*, 61024–61034. [CrossRef]

electronics

MDPI

Article

ADQE: Obtain Better Deep Learning Models by Evaluating the Augmented Data Quality Using Information Entropy

Xiaohui Cui [1,2], Yu Li [1,2], Zheng Xie [1,2], Hanzhang Liu [1], Shijie Yang [1] and Chao Mou [1,2,*]

1 School of Information Science and Technology, Beijing Forestry University, Beijing 100083, China; cuixiaohui@bjfu.edu.cn (X.C.)
2 Engineering Research Center for Forestry-Oriented Intelligent Information Processing of National Forestry and Grassland Administration, Beijing 100083, China
* Correspondence: chao_m@bjfu.edu.cn

Abstract: Data augmentation, as a common technique in deep learning training, is primarily used to mitigate overfitting problems, especially with small-scale datasets. However, it is difficult for us to evaluate whether the augmented dataset truly benefits the performance of the model. If the training model is relied upon in each case to validate the quality of the data augmentation and the dataset, it will take a lot of time and resources. This article proposes a simple and practical approach to evaluate the quality of data augmentation for image classification tasks, enriching the theoretical research on data augmentation quality evaluation. Based on the information entropy, multiple dimensional metrics for data quality augmentation are established, including diversity, class balance, and task relevance. Additionally, a comprehensive data augmentation quality fusion metric is proposed. Experimental results on the CIFAR-10 and CUB-200 datasets show that our method maintains optimal performance in a variety of scenarios. The cosine similarity between the score of our method and the precision of model is up to 99.9%. A rigorous evaluation of data augmentation quality is necessary to guide the improvement of DL model performance. The quality standards and evaluation defined in this article can be utilized by researchers to train high-performance DL models in situations where data are limited.

Keywords: data augmentation; deep learning; data quality; big data; data mining

Citation: Cui, X.; Li, Y.; Xie, Z.; Liu, H.; Yang, S.; Mou, C. ADQE: Obtain Better Deep Learning Models by Evaluating the Augmented Data Quality Using Information Entropy. *Electronics* **2023**, *12*, 4077. https://doi.org/10.3390/electronics12194077

Academic Editor: Byung-Gyu Kim

Received: 21 August 2023
Revised: 16 September 2023
Accepted: 27 September 2023
Published: 28 September 2023

1. Introduction

Data-driven deep learning (DL) has achieved many significant achievements in the past few years [1–3], and data augmentation has played an important role in it [4–6]. In practical applications, often the scale of the available data is insufficient for model training, which is the problem that data augmentation aims to solve [7]. Data augmentation is a regularization technique that increases the size and diversity of datasets by transforming and augmenting the original data [8]. Typically, data augmentation has a positive effect on the training and performance of DL models, but in practice, phenomena such as decreased precision or overfitting can occur after data augmentation [9]. Therefore, we need an evaluation criterion for the quality of data augmentation to assess the effectiveness of the employed data augmentation methods and their ability to improve model performance and generalization. Most of the research on data augmentation focuses on enhancing the generalization characteristics of the models, such as precision and F1 score [10,11]. Only a few papers have proposed a general framework for explaining data augmentation by studying its regularization effects [12–14], its influence on feature selection [15–17], rough set [18–20], and invariance perspectives [21,22]. However, the evaluation methods based on model performance cannot explain in which aspect data augmentation improves data quality. Neither can it guide researchers to choose a more optimal data enhancement strategy, even if a lot of time is spent on training the model. In order to determine to what extent data augmentation can improve data quality, evaluation standards need to be more theoretical and comprehensive. By visualizing and analyzing the augmented data, we can evaluate the effectiveness of data augmentation, observe whether the changes in data are reasonable, and cover different categories and tasks. This intuitive and simple method can efficiently evaluate the effects of data augmentation and quickly find suitable data augmentation methods, which can help obtain the most suitable dataset for model training in advance.

Studies [23,24] have shown that data quality is a multidimensional concept. Data quality has different meanings in different contexts. For example, data quality can be about measuring defective or outlier data in a general context [25–27], or describing whether the data meet the expected purpose in a specific context [28]. In this paper, we define data quality as a measure of data suitability for constructing a DL training set. Existing data quality assessments consider both intrinsic data quality and contextual quality [29], but the definitions of contextual quality vary. The most common idea is to divide contextual quality into two parts based on the process of DL: diversity within the training set and similarity between the training and testing sets. The main idea is to make the training set complex enough to encompass all the features and be similar to the real distribution represented by the testing set so that the DL model can learn adequately from this dataset. However, they overlook the fact that the performance of deep learning models is not only influenced by the problem space covered by the data. For instance, imbalanced classes in the dataset may lead to model bias [30,31], and these imbalances can occur in terms of quantity, features, or colors. Creating more dimensions based on the task and data features can better describe the quality of the data and its value for deep learning models. We hope to construct a universal, robust, and highly generalizable multidimensional quality evaluation method by refining and differentiating the definition of quality metrics, which can provide strong support for the quality evaluation of data augmentation.

In addition, due to the curse of dimensionality, such as in the case of image and text data, there arise computational and statistical challenges, with computational complexity growing exponentially. Hence, many works have used average similarity and minmax similarity between samples to calculate these two dimensions [29,32]. Although average or minmax similarity between samples can quickly assess the quality of a dataset, they cannot accurately approximate the precision of models trained on that dataset. Information entropy [33] can provide a comprehensive evaluation of data distribution, considering global characteristics such as sample diversity, rather than just focusing on average differences in the data [34]. Its feature as a non-linear measure based on probability distribution can better capture non-linear relationships in data distribution, with less sensitivity to noise and stronger interpretability [35]. Because of the low noise sensitivity, it is suitable to improve the computational efficiency with dimension reduction technology. The computation problem caused by a dimension disaster can be avoided. In summary, information entropy, as a metric for evaluating the quality of data augmentation, possesses more comprehensive, robust, and interpretable characteristics, making it more suitable for approximating the precision of models.

Therefore, this paper proposes an information entropy-based method for evaluating the quality of data augmentation. By attempting to deconstruct the dimensions of the data, we assess the quality of the dataset and data augmentation. In our approach, the augmented dataset is initially broken down into three dimensions, including diversity, class balance, and task relevance. Furthermore, taking image data as an example, for each dimension, numerous sub-dimensions are derived based on the task and data characteristics. Finally, by considering the correlations between the metrics, we calculate the ultimate composite metric score, providing insights into the impact of the current augmentation strategy on model performance.

- In this paper, we design and implement a data enhancement quality evaluation method, which can optimize and generate large-scale, high-quality data sets by disassembling and balancing the quality dimensions of data sets.

- This paper discusses the choice of mathematical tools for statistical analysis of data dimensions, and determines that information entropy is more suitable than other methods for evaluating the information content of data.
- This paper extensively evaluates the proposed method on various data augmentation techniques, datasets, and models. There is a strong correlation between the experimental results of the deep learning model and the evaluation results of the method, which shows that the method can improve the performance of the model on related tasks by evaluating the data enhancement quality.

2. Methods

In this work, we aim to explore the effectiveness of data augmentation in enhancing datasets, with the hope of replacing expensive model training with more comprehensive statistical metrics to evaluate the quality of augmented datasets. The primary goal of data augmentation is to generate a diverse and balanced dataset that is highly relevant to the task.

2.1. Preliminaries

Before presenting the details of our method, we give a brief overview of deep learning and data augmentation, which provides the theoretical basis of our algorithm design. For better illustration, some notations are summarized in Table 1.

Table 1. Some important mathematical notation.

Mathematical Notation	Description
X, Y	The data and labels of the dataset, as well as the input and output space of the model. X and Y represent the original training dataset. X' and Y' represent the augmented training dataset. X_t and Y_t represent the test dataset.
x, y	The x denotes an input data and the y denotes a label for the data. The x and y represent the original training dataset. The x' and y' represent the augmented training dataset. The x_t and y_t represent the test dataset.
P, p	Both represent a probability function that describes the distribution of the sample space.
R	It represents the risk function. Its subscripts represent the computational ideas used, empirical and expectation, respectively.
Q	It represents a collection of data augmentation quality metrics. Each $Q_i \in Q$ counts the different dimensions of the data.
$pixel$	It represents the pixel of the image data.
D	D denotes the dataset. D represents the original training dataset. D' represents the augmented training dataset. D_t represents the test dataset.
C, c_i	C represents the number of classes in the dataset and c_i represents the number of samples in the ith classes of the training dataset.
N	It represents the number of samples in the dataset. N represents the original training dataset. N' represents the augmented training dataset. N_t represents the test dataset.

2.1.1. Deep Learning

In machine learning, we formally refer to the sets of all possible values for the input and output of models as the input space X and the output space Y, respectively. Each specific input is an instance x, usually represented by a feature vector. In this case, the space where all feature vectors exist is called the feature space. The specific output is denoted as y, typically representing the label of the input X. At this point, the input and output variables X and Y follow a joint probability distribution $P(X, Y)$. The input space X and the output space Y together form a sample space. For a sample $(x, y) \in (X, Y)$ in the sample space, it is assumed that there exists an unknown true mapping function $f : X \rightarrow Y$, such that $y = f(x, \theta)$, where $\theta \in R^m$ represents the parameters in the function space and m is the number of parameters. In this case, we can measure the distance between the model and the data distribution $P(X,Y)$ to train the model. Therefore, the expected loss of the model $f(x,\theta)$ with respect to the joint distribution $P(X,Y)$ is expressed as

$$R_{exp}(\theta) = E_P[L(y,f(x,\theta))] = \int_{X\times Y} L(y,f(x,\theta))P(x,y)\,dx\,dy, \tag{1}$$

since the $L(y,f(x,\theta))$ represents the loss function, which quantifies the difference between individual input and output instances. However, in reality, the data distribution $P(X,Y)$ is often unknown, and we only have knowledge of the distribution of samples in the training set. Therefore, to deal with this situation, in DL, the approach is to minimize the expected loss on the training set. As shown in Equation (2), the empirical distribution $\hat{P}(X,Y)$ based on the training set is used instead of the true distribution $P(X,Y)$ to calculate the empirical loss R_{emp}. This way, during the training process, the model performs parameter optimization based on the sample distribution in the training set, aiming to approximate the performance of the true distribution as closely as possible.

$$R_{emp}(\theta) = \frac{1}{N}\sum_{n=1}^{N} L(y,f(x,\theta)). \tag{2}$$

2.1.2. Data Augmentation

Data augmentation refers to any method that uses artificial transformations of data and labels to expand the original training set. It can be represented as a mapping of the set. This function is defined as

$$P(X',Y') = f((X,Y)|\beta)\bigcup P(X,Y), \tag{3}$$

where β represents the data augmentation strategy, $f((X,Y)|\beta)$ represents the augmented part, and $P(X',Y')$ represents the final augmented training dataset.

2.1.3. Benefits of Data Augmentation

To address the issue of unreliable empirical risk in situations with limited data, we need to introduce prior knowledge. Prior knowledge is used to augment the dataset, even in cases where data are scarce.

Lemma 1 (Chebyshev's inequality). *Let t be a random variable with finite expected value μ. And there are n variables in total. Then, for any small positive number ϵ,*

$$\lim_{n\to\infty} P(|\frac{\sum t_i}{n} - \mu < \epsilon|) = 1. \tag{4}$$

According to the law of large numbers, Equations (2) and (4), as the sample size N becomes sufficiently large, the empirical risk R_{emp} tends to the expected risk R_{exp}. However, when it comes to DL datasets, considering only the distribution of samples and labels is insufficient. For instance, simply resampling data can lead to more severe overfitting of the model. We also need to ensure that the features in the data align closely with the true distribution, which is a key problem addressed by data augmentation. In general, data augmentation is achieved by modifying the original data based on prior knowledge to expand the dataset. The generated data may have the same labels as the original data, but the features extracted by the model are different. This enables the model to more easily recognize the critical features relevant to the task at hand. The parameters for data enhancement need to satisfy the following expression:

$$\arg\min_{\beta} distance(\arg\max_{P()} quality(P(X',Y')), P(X_t,Y_t)), \tag{5}$$

where the *distance* is defined as a function of measuring distribution distance, such as similarity, and *quality* is defined as a function that measures the distribution of data sets.

The ideal scenario is that the dataset remains consistent with the true distribution for all features $P(X_t, Y_t)$. However, this is an ideal situation and the true distribution is still unknown. Therefore, we use the test set instead of the true distribution for the estimation.

Theorem 1. *The expectation and variance of the original dataset are μ and σ, the expectation and variance of the augmented dataset are μ' and σ'. Assuming that the expectation and variance of true distribution are μ_t and σ_t. Equation (5) can be expressed as*

$$fusion(distance(\mu', \mu_t), distance(\sigma', \sigma)). \tag{6}$$

However, due to the randomness of data augmentation, the generated data may not necessarily be more in line with the true distribution compared to the original data. Therefore, we need to perform quality estimation on it.

Lemma 2. *Expectation and variance are not mutually independent, only when the distribution does not follow a normal distribution.*

Therefore, the quality evaluation metric for data augmentation is defined as

$$Q_{Augmentation} = Q_\sigma \times Q_\mu = Q_\sigma \times Q_{TaskRelevance}, \tag{7}$$

which is the product of the statistical values of expectation Q_σ and variance Q_μ . The expectation of the dataset is primarily influenced by the target task, and the value obtained from expectation is also referred to as task relevance.

Unlike expectation values, data can have different distributions based on different feature selections, leading to varying variances. These variances can be mainly divided into two categories. One is the distribution of semantic features that approximate a normal distribution, and the other is the distribution of categories, which is mostly a uniform distribution. The variances of these two distributions are independent of each other, so statistical values of variances can be obtained using addition.

Theorem 2. *Output space $Y = (y_1, y_2, \cdots, y_n) \sim U(\lambda)$, input space X is abstracted into $feature = (feature_1, feature_2, \cdots .feature_{k_1}) \sim N(\mu_1, \sigma_1)$. So $P(x, y)$ can be represented by $P(feature, label)$. By delineating the distribution of label-independent features, we have the $semantic_i = (semantic_1, semantic_2, \cdots, semantic_{k_2}) \sim N(\mu_2, \sigma_2), semantic_{k_2} \in C$, then*

$$P(label, semantic) = \sum_i^C P(label) \times P(semantic_i) = \sum_i^C P(semantic_i). \tag{8}$$

Features are classified according to whether they have a relationship with the category, and the same is true for data quality. So we have

$$Q_\sigma = (Q_{label} + Q_{(label,feature)}) + Q_{semantic} = Q_{ClassBalance} + Q_{Diversity}. \tag{9}$$

Finally, the augmented datasets quality evaluation formula is defined as

$$Q_{Augmentation} = (Q_{Diversity} + Q_{ClassBalance}) \times Q_{TaskRelevance}. \tag{10}$$

2.2. The Overall Framework

The overall framework of the method proposed in this paper is shown in Figure 1 and Algorithm 1. The methodology of this paper will consist of the following components: (1) Feature Extraction: The dataset will be processed using different data augmentation strategies. And each sample in the datasets will be mapped to a high-dimensional feature space. (2) Metric Scores Calculating: The quality metrics of the dataset will be divided into three dimensions: diversity, class balance, and task relevance. Information entropy is primarily used to evaluate the diversity and class balance metric. (3) Result Statistics: The scores of each augmented dataset will be ranked, and the datasets with higher quality will be selected for model training and validation. The implementation code is available at: https://github.com/ForestryIIP/ADQE (accessed on 20 August 2023).

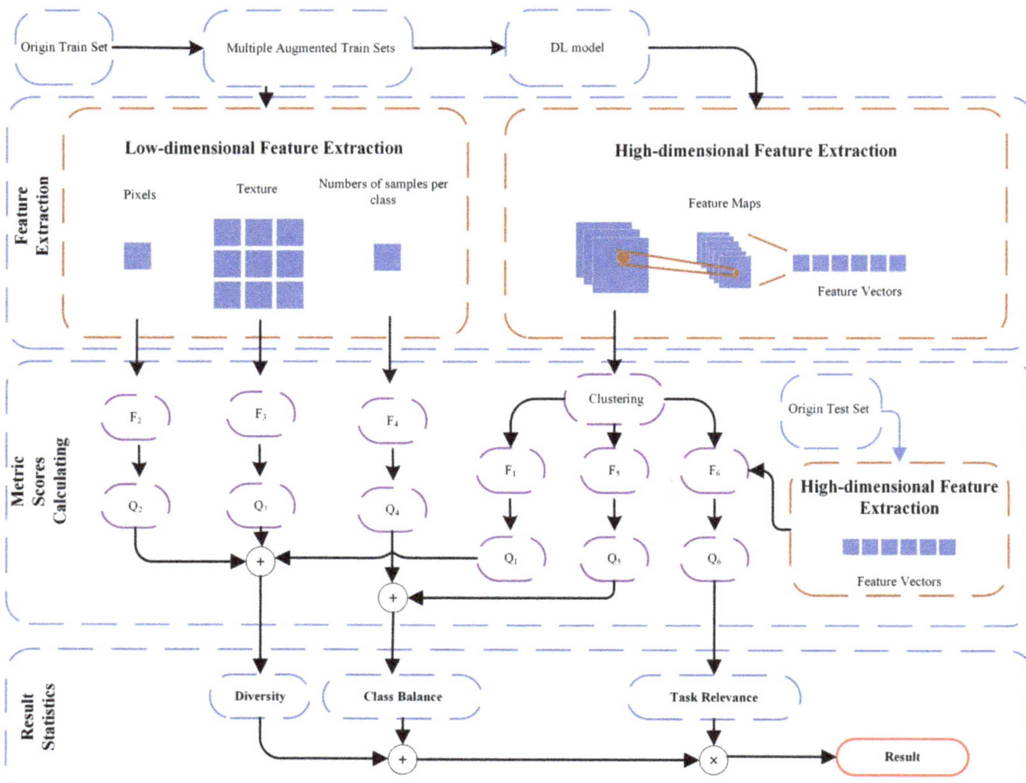

Figure 1. The framework of the proposed method.

Algorithm 1: Calculation of all augmented datasets evaluations.

Data: The collection of augmented datasets $D' = \{D'_0, D'_1, \cdots, D'_s\}$, A test dataset D_t, DL model *NN*

Result: the quality metric of the augmented dataset Q

$Q \leftarrow \emptyset$;
$cnt \leftarrow 0$;
for *each dataset* $D'_i \in D'$ **do**
  Update Q_4 *based on Algorithm* 4;
  Update Q_2 *and* Q_3 *based on Algorithm* 3;
  Extracting the feature vector V from D'_i with help of *NN*;
  Clustering feature vector V into k classes of V';
  Update Q_5 *based on Algorithm* 5;
  Update Q_1 *based on Algorithm* 2;
  Update Q_6 *based on Algorithm* 6;
  if $cnt \mathrel{!}= 0$ **then**
    $Q \leftarrow$ *result of Equation* (21)
  end
end

2.3. Feature Extraction

2.3.1. Feature Selection

The selection of features is crucial for evaluating the quality of data augmentation, as it impacts the effectiveness and comprehensiveness of metric calculations. In order to obtain more effective image representations, this paper utilizes the pre-trained model ResNet101 [36] to map images into high-dimensional feature vectors. Additionally, to achieve more comprehensive metrics, we supplement the quality indicators by incorporating features such as pixel brightness, texture, and class sample counts based on the characteristics of the images and the dataset itself.

2.3.2. Clustering

After data augmentation, we can proceed with the calculation of various metrics. For diversity computation, calculating low-dimensional feature diversity simply requires measuring the individual feature values of each image, which can be performed in $O(n)$ time complexity. However, in the case of high-dimensional features, the complexity increases as we need to calculate the similarity between each pair of samples and compute the eigenvalues of the nn similarity matrix, resulting in a time complexity of $O(n^3)$. Similarly, for task relevance, we need to compute the similarity between sample pairs from the training set and the test set, resulting in a high time complexity of $O(n \times m)$. To address these challenges, this study adopts the FINCH clustering algorithm [37], which employs pooling and sampling techniques to reduce the dimensionality and scale of the dataset, thereby mitigating the computational cost and time overhead. In the calculation of class balance, the clustering algorithm can directly compute the desired metrics.

2.4. Metric Scores Calculating

2.4.1. Diversity

The main benefit of data augmentation comes from increasing diversity. However, only comparing the average similarity of data in the augmented dataset in high-dimensional space has limitations. It just only indicates the pairwise distances between data points and does not reflect the overall distribution of data in the high-dimensional feature space. The Vendi score [38] introduces an ecological concept by using the entropy index of species distribution. This effectively measures the diversity of feature distributions in the training set samples. The Vendi score is defined as the exponential Shannon entropy of the eigenvalues of a similarity matrix K, such as

$$K(x,y) = \{similarity(x,y)|x,y \in D^{'}\}. \tag{11}$$

This matrix is derived from a user-defined similarity function applied to the samples under evaluation for diversity:

$$Q_1 = \frac{1}{C}\sum_{i=1}^{C} exp(-\sum_{j=1}^{c_i} \lambda_j log\lambda_j), \tag{12}$$

where the λ_j represents the eigenvalues of the similarity matrix K. The K is a positive-definite matrix obtained from a set of samples x and the similarity function. For all x, $similarity(x,x) = 1$. This method , as shown in Algorithm 2, primarily quantifies the effective number of distinct elements in the data. For example, after extracting features from an image using a neural network, you typically obtain a 2048-dimensional vector. This vector stores features across different dimensions of the image. Assuming the similarity function is a cosine similarity, this metric measures whether the directions of two vectors align. If the feature dimensions forming these directions are more similar, the metric value is higher. Consequently, a similarity matrix K can be computed. Eigenvalues generally represent inherent structural properties and patterns within the data [39]. Each eigenvalue corresponds to a mode of variation or structure in the data. In the context of image similarity, they can indicate different similarity patterns or clusters among images. The magnitude of eigenvalues also reflects the proportion of corresponding patterns in the data. Therefore, computing the entropy of eigenvalues is equivalent to quantifying the richness of patterns in the data. If a single pattern dominates the data, the style and content of the images are highly certain, resulting in low information uncertainty. Conversely, when multiple similar patterns share similar proportions, the style and content of the images become less certain, leading to higher information uncertainty. In summary, the eigenvalues of similarity matrix K and the entropy of these eigenvalues provide valuable insights into the data's structure and diversity. High entropy indicates complexity and diversity, whereas low entropy suggests simplicity or uniformity in similarity patterns within the data. They can guide overall data analysis in the context of diversity metrics.

Algorithm 2: Rapid calculation of diversity by clustering.

Data: A training dataset $D' = \{X_1, X_2, \cdots, X_{N'}\}$, dataset size N', DL model NN
Result: Diversity Metric Scores Q_1

$K \leftarrow 0_{[N' \times N']}$;
$Q_1 \leftarrow 0$;
Extracting the feature vector V from D' with help of NN;
Clustering feature vector V into k classes of V';
$D'_{clustered} \leftarrow \emptyset$;
for $i \leftarrow 0$ *to* k **do**
    $D'_{clustered} \leftarrow$ randomly select C data from different categories in V'
end
for *class index* $\leftarrow 0$ *to* C **do**
    for *each vector* v_i *and* $v_j \in$ *collection of class index in* V' **do**
        $K[i][j] \leftarrow similarity(v_i, v_j)$;
    end
    $\lambda \leftarrow eigenvalue\,of\,K$;
    $Q_1 \leftarrow Q_1 + exp(-\sum_{j=1}^{S_i} \lambda_j log \lambda_j)$;
end
$Q_1 \leftarrow Q_1 \div C$;

Ignoring the time of feature extraction and clustering part, the computation of Q_1 is mainly divided into two steps. They are the computation of similarity adjacency matrix and its eigenvalues, respectively. For each pair of vectors, similarity computation, such as cosine similarity, requires $O(K)$ time complexity. Where K is the dimension size of the image vector output by the model. Every element in the dataset needs to be computed, then computing the adjacency matrix requires $O(KN^2)$ time complexity. The next step is to solve for the eigenvalues, which are also usually $O(N^3)$. Generally, $K \ll N$, so the time complexity is $O(N^3)$. Although the time complexity is high, the data size is reduced by clustering in advance. The actual computation time is still within acceptable limits.

Furthermore, analyzing the diversity of data solely based on high-dimensional features provides an abstract understanding of diversity, but it may not provide an intuitive and clear understanding. Therefore, it is still necessary to define the diversity of data in low-dimensional features. For example, in image data, models often learn texture features extensively from the dataset [40]. To address this, we can calculate the occurrence probabilities of different textures in all the data:

$$Q_2 = exp(\sum_{i=1}^{N_{Texture}} p_{Texture_{i,j}} log p_{Texture_{i,j}}), \quad (13)$$

where $N_{Texture}$ is the number of different *Texture* and $p_{Texture_{i,j}}$ represents the probability that the value of $Texture_{i,j}$ will occur. The Texture is defined as a combination of $pixel_{i,j}$ and adjacent pixels:

$$Texture_{i,j} = \begin{pmatrix} pixel_{i-1,j+1} & pixel_{i,j+1} & pixel_{i+1,j+1} \\ pixel_{i-1,j} & pixel_{i,j} & pixel_{i+1,j} \\ pixel_{i-1,j-1} & pixel_{i,j-1} & pixel_{i+1,j-1} \end{pmatrix}. \tag{14}$$

Due to computational complexity and memory limitations, this paper calculates the information entropy of the average pixel values within a 3 × 3 window instead of the information entropy of pixel combinations. Lastly, considering that brightness can have an impact on the model [41], this paper also calculates the probability of brightness for each pixel in the entire dataset and computes its entropy value:

$$Q_3 = \frac{1}{3} \sum_{RGB=1}^{3} exp(\sum_{level=1}^{256} p_{level} log p_{level}), \tag{15}$$

where *RGB* represents the three color channels of an image and *level* represents the component of the channel, up to a maximum of 256. Texture and brightness are counting the pixels of the image, so the time complexity is $O(PN)$ and P is the average number of pixel points in the image. As shown in Algorithm 3, the variables for the Q_2 and Q_3 are similar and can be run together in the feature extraction phase.

Algorithm 3: The calculation of image datasets' color and brightness space.

Data: A training dataset D'
Result: Diversity Metric Scores Q_2 and Q_3
$P_1 \leftarrow 0_{[3\times256]}$;
$P_2 \leftarrow 0_{[3\times256]}$;
for *each image* $x \in D'$ **do**
  for $channel_{RGB} \leftarrow 0$ *to* 2 **do**
    for *each pixel* $\in x$ **do**
      $t \leftarrow Average\ of\ adjacent\ pixels$;
      $P_1[channel_{RGB}][pixel] \leftarrow P_1[channel_{RGB}][pixel] + 1$;
      $P_2[channel_{RGB}][t] \leftarrow P_2[channel_{RGB}][t] + 1$;
    end
  end
end
$Q_2 \leftarrow 0$;
$Q_3 \leftarrow 0$;
for $channel_{RGB} \leftarrow 0$ *to* 2 **do**
  $P_1[channel_{RGB}] \leftarrow sum(P_1[channel_{RGB}])$;
  $Q_2 \leftarrow Q_2 + exp(-\sum_{j=0}^{255} P_1[channel_{RGB}][j] log(P_1[channel_{RGB}][j]))$;
  $P_2[channel_{RGB}] \leftarrow sum(P_2[channel_{RGB}])$;
  $Q_3 \leftarrow Q_3 + exp(-\sum_{j=0}^{255} P_2[channel_{RGB}][j] log(P_2[channel_{RGB}][j]))$;
end
Calculate entropy;

2.4.2. Class Balance

From the perspective of data categories, real-world datasets often suffer from the long-tail problems [42]. Models tend to overlook minority classes due to the majority of samples being concentrated in a few categories, resulting in poorer predictive performance. To measure class balance, the first aspect to consider is whether the number of samples in each class is equal, which is the most fundamental metric. The formula for balance of number in class is as follows:

$$Q_4 = 1 - \frac{1}{C}\sum_{i=1}^{C}(c_i - \bar{c}), \tag{16}$$

where the $\bar{c}$ represents the average count of samples across classes. The variance of the classes is computed with a time complexity of $O(C)$.The algorithm is illustrated in Algorithm 4.

Algorithm 4: The calculation of quantity balance between classes.

Data: A training dataset D', The number of classes C
Result: Class Balance Metric Scores Q_4

$K \leftarrow 0_{[C\times 1]}$;
for $i \leftarrow 0$ *to* C **do**
  $K[i] \leftarrow c_i$
end
$Q_4 \leftarrow 1 - \frac{1}{C}(\sum_{i=1}^{C}(K[i] - average(K)))$;

Additionally, we need to assess the distribution differences between classes. When performing a classification task, it is generally easier to distinguish between animals and humans compared to distinguishing between males and females. If both labels are present in the same dataset, the labels for males and females can easily be confused. This results in poor data quality because the model cannot learn precise features to differentiate between males and females accurately [43]. Unsupervised clustering can group similar samples into the same class. In a dataset with balanced difficulty and granularity of classification, the clustering results should align with the original labels. By calculating the mutual information between the clustering results and the original labels, we can obtain the classification difficulty of the samples, which serves as an indicator for the class feature balance of the dataset. The basic definition of information entropy is

$$H(X) = -\sum_{X} p(x)logp(x), \tag{17}$$

where $H(X)$ represents the entropy of the classes. The basic definition of mutual information is

$$I(X;Y) = \sum_{X}\sum_{Y} p(x,y)log\frac{p(x,y)}{p(x)p(y)}, \tag{18}$$

where $I(X;Y)$ represents the mutual information and $p(x,y)$ is the joint distribution between the clustering results and the ground truth. The $p(x)$ and $p(y)$ are the marginal distributions of the ground truth labels and clustering results, respectively. Although mutual information can also measure the degree of similarity between two clustering results, its value is strongly influenced by the sample size. Normalized mutual information (NMI), on the other hand, can better measure the degree of similarity between two clustering results by normalizing the mutual information values to the same range of values. The NMI is defined as

$$Q_5 = 2\frac{I(X;Y)}{H(x) + H(y)}. \tag{19}$$

The details are presented in Algorithm 5. Ignoring the time of feature extraction and clustering part, the computational time complexity of the mutual information of the label distributions before and after clustering is only $O(N)$.

Algorithm 5: The calculation of features balance between classes.

Data: A training dataset $D^{'}$
Result: Class Balance Metric Scores Q_5
Extracting the feature vector V from $D^{'}$ with help of NN;
Clustering feature vector V into k classes of $V^{'}$;
$P_{D'} \leftarrow distribution\,of\,V;$
$P_{D'_{clustered}} \leftarrow distribution\,of\,V^{'};$
$P_{D',D'_{clustered}} \leftarrow jointdistribution\ between\ Vand\ V^{'};$
Update Q_5 based on Equations (18) and (19);

2.4.3. Task Relevance

The most common issue that can occur after image augmentation is the loss of semantic information, where important features, such as the pixels representing a dog in a dog image, are completely erased. In such cases, even humans would struggle to determine the correct label for the modified image. Therefore, assessing the relevance between the augmented training set and the task at hand can provide insights into the quality of the dataset. Considering that models are typically evaluated on a separate test set for various tasks, this study compares the average similarity between the test set and the augmented training set in the high-dimensional feature space:

$$Q_6 = \frac{1}{N_t * N'} \sum_{i=1}^{N_t} \sum_{j=1}^{N'} similarity(i, j), \tag{20}$$

where $similarity(i, j)$ denotes the similarity function used to measure the similarity between samples. Based on the Equation (20), we can get the Algorithm 6.

Algorithm 6: Rapid calculation of task relevance by clustering.

Data: A training dataset $D^{'} = \{x_1, x_2, \cdots, x_{N'}\}$, dataset size $N^{'}$, The number of classes C, A test dataset D_t, test dataset size N_t, DL model NN
Result: Task Relevance Metric Scores Q_6
$Q_6 \leftarrow 0;$
Extracting the feature vector V from $D^{'}$ and V_t from D_t with help of NN;
Clustering feature vector V into k classes of $V^{'}$;
$D^{'}_{clustered} \leftarrow \emptyset;$
for $i \leftarrow 0\ to\ k$ **do**
  $D^{'}_{clustered} \leftarrow$ randomly select C data from different categories in $V^{'}$
end
for $class\ index \leftarrow 0\ to\ C$ **do**
  for $each\ vector\ v_i\ and\ v_j \in collection\ of\ class\ index\ in\ V^{'}\ and\ V_t$ **do**
    $Q_6 \leftarrow Q_6 + similarity(v_i, v_j);$
  end
end
$Q_6 \leftarrow average(Q_6);$

Ignoring the time of feature extraction and clustering part, the computation of Q_6 is mainly divided into two steps. They are the computation of the similarity matrix and its average, respectively. The similarity of all unordered pairs of the dataset and test set needs to be computed, assuming that the dataset size is N and the test set size is M, then the total number is $N \times M$. The similarity matrix time complexity is $O(KMN)$. Where K is the dimension size of the image vector output by the model. And calculating the mean only requires $O(NM)$. So the total time complexity is $O(KMN)$.

2.5. Result Statistics

To combine various metrics, existing approaches often employ weighted fusion [44] or rank fusion [45] to directly rank the augmented datasets. However, these methods are not suitable in this case because the metrics from different dimensions cannot be directly compared, and a comparison between the augmented training set and the original training set is required. In this paper, a fusion approach is proposed that calculates the quality of data augmentation by considering the ratio of quality before and after augmentation. All metrics are shown in Table 2 and the fusion equation is defined as

$$Q = \frac{Q_6'}{Q_6} \times \sum_{i=1}^{5} w_i \frac{Q_i'}{Q_i} / \sum_{i=1}^{5} w_i, \tag{21}$$

where the Q_i' represents the quality metric of the augmented dataset, Q_i represents the quality metric of the original dataset and w_i represents the weight of each metric to the final fusion metric Q. By assigning appropriate weight coefficients to the metrics of different parts, the balance of influence of different factors in the metrics can be ensured. For example, when you encounter high precision requirements of the task, task relevance metrics are more critical and need to be assigned a higher weight.

Table 2. Data augmentation quality metrics.

Quality Metrics	Full Name	Category
Q_1	diversity of features	diversity
Q_2	diversity of textures	
Q_3	diversity of brightness	
Q_4	balance of number between class	class balance
Q_5	balance of features between class	
Q_6	task relevance	task relevance

3. Experimental

3.1. Datasets and Data Augmentation

In order to validate the effective evaluation of data augmentation quality, this study selected different augmentation strategies, datasets at different granularities, and several specific tasks as the objects of method evaluation. The CIFAR-10 and CUB-200 [46] datasets were chosen as the experimental datasets for image classification tasks. The CIFAR-10 and CUB-200 represent different areas of computer vision problems. CIFAR-10 is an image classification dataset containing 10 different categories of common objects such as aircraft, dogs, cars, and so on. The CUB-200, which focuses on bird identification, contains images of 200 different species of birds. The multi-domain coverage of these two datasets allowed us to explore the impact of diversity and task relevance in different application contexts. They also exhibit varying levels of image diversity and class balance. CIFAR-10 includes diverse scenes, lighting conditions, angles, and variations. In contrast, CUB-200 exhibits limited image diversity, with predominantly consistent backgrounds. Furthermore, both datasets represent numerous practical application scenarios, such as image classification, object detection, and object recognition. By conducting experiments on CIFAR-10 and CUB-200, we gain a better understanding of the performance and applicability of our methods.

In the context of image data applications, data augmentation methods add noise to the original data to simulate other real-world scenarios, thus creating augmented images for model training. To evaluate the improvement in data quality brought by different data augmentation methods, this study employs RandAugment's data augmentation search strategy [47]. Unlike RandAugment, which applies random transformations to images during training, this study scales up the dataset by employing the RandAugment strategy prior to training. To generate data diversity, this study selects n transformations from a set of $k = 16$ data augmentation transformations with uniform probability, where the augmentation magnitude for each transformation is set to m. By varying these two parameters, the strategy can express a total of $m \times 2^n$ potential augmentation policies, where n represents the strategy for selecting data enhancement and m represents the intensity of data augmentation. Each parameter of the transformations is scaled using the same linear scale, ranging from 0 to 30, where 30 represents the maximum scale for a given transformation, and then mapped to the parameter range of each transformation. Subsequently, during the expansion and generation of the augmented dataset, we uniformly sample dataset samples with probability c for transformation. All generated images are then merged with the original dataset to create the augmented dataset. However, image data alone is insufficient for calculating quality metrics. Therefore, the image data needs to be abstracted into 2048-dimensional feature vectors. In this study, a pre-trained model, ResNet101, is utilized to extract features from the generated augmented dataset.

3.2. Baseline

In this paper, we will present the results of our proposed method on CIFAR-10 and CUB-200 datasets to demonstrate how our approach captures the intuitive notion of data augmentation and can be applied to assess the quality of data augmentation in DL. The existing data augmentation evaluation work can be divided into two categories: model validation for assessing effectiveness and statistical analysis for improving data quality. Model validation aims for the highest level of accuracy, as seen in approaches like AutoAugment. However, due to the immense computational demands, it may not be practically feasible in real-world applications. On the other hand, work in the field of statistical analysis for enhancing data quality tends to focus on calculating dataset quality at a finer granularity within a specific dimension [38,48]. Quality is multidimensional, and solely relying on a single dimension for analysis can provide a limited perspective. These two articles are both based on the intrinsic attributes of the data and the contextual tasks for analysis [29,32]. However, the mathematical tools chosen in statistical analysis cannot achieve the goal of correctly assessing data quality. This paper divides data quality into multiple dimensions, allowing researchers to comprehensively assess the quality of data and identify areas in need of improvement. Since it focuses on enhancing data to improve model performance, it is essential to clearly define their definitions and relationships, while having effective methods for utilizing these dimensions to evaluate the effectiveness of data augmentation. So, we will compare our method with two baseline approaches: diversity and task independence calculated using the mean and min-max criteria. Since the criteria for quality fusion differ among these methods, we will employ our proposed quality fusion method to calculate the final scores for all the approaches. Baseline's formula is shown in Table 3.

Table 3. The formula of baseline.

Baseline	Metrics	Formula
mean criteria	$mean\_Q_1$	$1 - \frac{1}{N'^2} \sum_i^{N'} \sum_j^{N'} similarity(i,j)$
	$mean\_Q_6$	$\frac{1}{N' * N_t} \sum_i^{N'} \sum_j^{N_t} similarity(i,j)$
	$mean\_Q$	$mean\_Q_1 * mean\_Q_6$
minmax criteria	$minmax\_Q_1$	$1 - \frac{1}{N'} \sum_i^{N'} \min_{j \in D'} similarity(i,j)$
	$minmax\_Q_6$	$\frac{1}{N'} \sum_i^{N'} \max_{j \in D_t} similarity(i,j)$
	$minmax\_Q$	$minmax\_Q_1 * minmax\_Q_6$

In this section, this study conducted two sets of experiments to validate the effectiveness of the proposed method. The first set of experiments evaluated the correlation between the model precision on the test set and the quality evaluation results of the generated augmented dataset using different parameters for data augmentation. Due to the enormous search space of data augmentation strategies, it was challenging to define it precisely. Therefore, this study randomly selected 7 sets of parameters as the comparative parameters for the experiment. The second set of experiments involved generating augmented datasets of different sizes using the same set of parameters for data augmentation. The performance of the model on these datasets was observed to see if it aligns with the algorithm results.

3.3. Evaluation Index

For the purpose of achieving higher model precision, this study adopts model evaluation metrics as the evaluation criteria for algorithm performance. After training on the augmented dataset, the model's precision is tested on the original test set. The model training parameter settings are shown in Table 4.

Table 4. The hyperparameters of DL model.

Dataset	Model	Hyperparameters
CIFAR10	densenet161	Epochs = 90, Init lr = 0.1 divide by 5 at 40th, 60th, 80th, Batchsize = 256, Weight decay = 5×10^{-4}, momentum = 0.9
CUB200	NtsNet	Epochs = 50, Lr = 0.001, Batchsize = 16, Weight decay = 1×10^{-4}, Momentum = 0.9

Due to the different dimensions between the scores of the method and the baseline and the accuracy of the model, we first divide all the model accuracies by the accuracy of the model on the original training set, similar to how the scores are calculated. We refer to this as the actual score of the augmented training set. Then, we consider using Cosine Similarity (CS) to calculate the similarity between the estimated scores and the actual scores. We sort the scores into a vector according to certain rules, such as enhancement magnitude or scale, and then calculate the cosine value of the angle between them. This cosine value reflects the similarity of the changing trends between the estimated scores and the actual scores. However, we still need to know the absolute distance between the two scores. Therefore, we also select Mean Squared Error (MSE) as our second metric. By combining these two metrics, we can comprehensively analyze the strengths and weaknesses of the algorithm. A larger CS and a smaller MSE indicate better results.

3.4. Ablation Study

This paper introduces an evaluation method for assessing the quality of data augmentation in an image dataset, aiming to better select augmentation techniques. In order to demonstrate the effectiveness of our proposed method in evaluating data augmentation based on local independence and global correlation information, as well as to validate our pipeline design, we conducted two ablation experiments. In Experiment 1, we focused on metric adjustment to confirm the observed differences between method results and estimation results from a statistical perspective. The Spearman's rank correlation coefficient contains a p-value to show the level of significance. The technique was statistically different when and only when the P value was below 0.05. In Experiment 2, we replaced the components in the framework to evaluate their impact on the results. Both ablation experiments were conducted on the CIFAR-10 dataset, and the experimental setup described earlier was used to train the network from scratch.

3.5. Case Study

In general, because different data types and different tasks require the machine learning model to learn different content, there is no absolute universal method for evaluating data augmentation quality. This article sets up experimental datasets from the perspectives of data types and downstream tasks. We will apply the methods described in this paper to two different datasets selected from various domains and tasks: EuroSAT and IMDB.

We have chosen datasets from significantly different domains—images and text, which represent two major data types. We also ensure that the datasets are large enough to facilitate meaningful augmentation and analysis. By calculating the score differences after augmentation using ADQE, we evaluate and visualize the differences between the two datasets with the largest score differences. EuroSAT, being an image dataset, undergoes the same augmentation and evaluation methods as described in this paper. For the text dataset IMDB, we apply data augmentation methods such as Optical Character Recognition (OCR), semantic augmentation, and summarization. Since the data type is different, we cannot directly use ADQE to calculate evaluation scores, and adjustments need to be made to the methods. The text dataset also involves extracting feature vectors to calculate diversity and task relevance, but text is composed of words rather than pixels. Q_2 and Q_3 need to be recalculated using words and characters. We combine these two metrics and redefine them as the information entropy calculated from frequency of the top 10,000 most frequent words.

4. Results and Discussion

4.1. Results of Comparative Experiments

The experimental results of the first group are presented in Figure 2. Figure 2 primarily illustrates the performance of our method compared to state-of-the-art methods on two different datasets, including the visual comparison and statistical analysis between the evaluation scores and model accuracy, as well as the variation trend of scores under different data augmentations. Figure 2a,b show the experimental results based on the CIFAR-10 dataset, while Figure 2c,d display the experimental results based on the CUB-200 dataset. From the experimental results, it can be observed that the entropy-based method provides the most similar quality results for all augmented datasets, which is further validated by the model's accuracy.

In Figure 2a,b, the entropy-based method shows the best fit to the model accuracy curve. Meanwhile, in Figure 2e,f, our results are validated using statistical analysis. We achieved excellent scores in both indicators, with CS approximately equal to 1 and MSE close to 0, indicating that entropy effectively captures the data characteristics and estimates corresponding evaluation scores. On the other hand, the evaluation results of methods like the minmax criteria continuously increase with the magnitude of data augmentation, deviating significantly from the trend of model accuracy. This is likely due to the noise introduced by data augmentation, leading to an overestimation of the quality of augmented datasets using the minmax criteria. This principle overlooks the overall distribution of the dataset and approximates its quality by statistical extreme values, resulting in a significant decrease in evaluation accuracy. Among all the figures, the minmax criteria consistently performs the worst. This observation is also reflected in the evaluation results based on the mean criteria in Figure 2b. The evaluation under overly strong data augmentation also overestimates the quality of the augmented dataset. In CUB-200, where the original data are scarce and very similar, with backgrounds mostly consisting of single skies, oceans, and forests, and birds differing slightly in feather color and shape, data augmentation significantly increases the diversity of the dataset. Changing a large portion of data distribution without perfectly matching the actual data distribution amplifies the diversity evaluation scores. The CS drops by one percentage point, and the MSE increases from below 0.1 to 1.1.

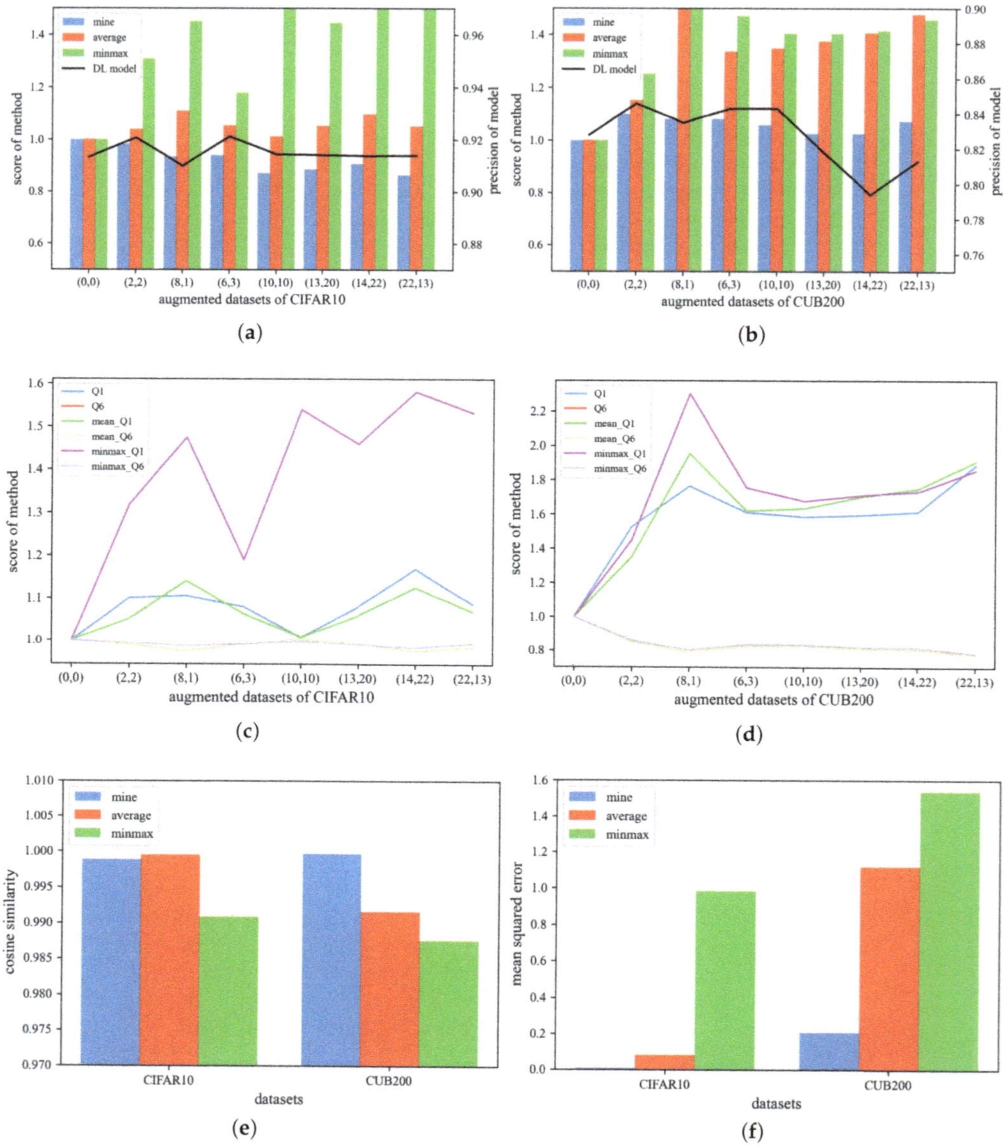

Figure 2. Under the same scale and different data augmentation scenarios, our method achieves the best quality evaluation results. (**a**) Evaluation of three algorithms and the model on the augmented dataset using the CIFAR-10 dataset. (**b**) Evaluation of three algorithms and the model on the augmented dataset using the CUB-200 dataset. (**c**) Partial indicator scores of the augmented dataset by three algorithms on the CIFAR-10 dataset. (**d**) Performance evaluation of the three algorithms using CS. (**e**) Performance evaluation of the three algorithms using MSE. (**f**) Partial indicator scores of the augmented dataset by three algorithms on the CUB-200 dataset. In (**a–c**,**f**), the x-axis represents the data augmentation strategies, including the selection probability of data augmentation and the magnitude of data augmentation. The selection of these strategies is randomly chosen within the given range. The word "mine" represents the methodology of this paper.

In our data augmentation quality evaluation metric calculation, we partition the overall evaluation metric based on the mean and variance of the data augmentation quality. The mean measures the distance between the augmented data and the target data distribution. The variance measures the distribution uniformity and diversity of the augmented data. From the q_6 curves in Figure 2c,d, for all methods, their mean calculation results are relatively consistent. They accurately calculate the distance between all augmented training data and the test set distributions. However, only the entropy method meets the criteria for variance estimation. Since semantic information has multiple dimensions and is not entirely related to labels, it is divided into class balance and diversity. Even through augmentation, class-related semantics will not be changed or lost. By using clustering dimensionality reduction, we can clearly understand the spatial distribution of the data, which remains relatively unchanged before and after augmentation. However, within the class, due to color changes, deformations, inversions, and other operations, these noises are substantially supplemented. We need to assess the distribution balance of data label-related features, and so on. The minmax criteria emphasizes the farthest distance of the data distribution. Although the mean criteria is more balanced, it is also affected by extreme values. Moreover, data augmentation methods are likely to inject many extreme values due to their randomness, thus affecting the evaluation. Entropy can smooth these extreme values, categorizing them together, and calculate the overall diversity evaluation value by statistically counting the effective number of categories under different dimensions. Experimental results demonstrate that our framework can effectively evaluate data augmentation quality by incorporating entropy, visually and comprehensively showcasing the improvements brought by data augmentation to the dataset.

From another perspective, Figure 2a–d illustrate that the model accuracy is significantly enhanced primarily through the increased diversity of the data. However, as the diversity increases, there is a bottleneck, and the task relevance decreases, resulting in poorer performance than the initial state. This is primarily related to the data augmentation techniques chosen in this paper. Most of the changes involve alterations in color space, shape, and orientation, which enhance data diversity and model robustness. Fewer enhancements focus on image quality or denoising, reducing the degree of association between samples in the dataset and the task objectives, which increases the dimensions of the data that need to be analyzed. The target variables become less understandable and predictable. Therefore, the experimental results inevitably show a decrease in task relevance and an increase in diversity. The intensity of data enhancement up to a certain point reduces the benefits and loses more data quality. By clearly defining dimensions, researchers can better guide the selection and implementation of data augmentation strategies. Using data quality dimensions to evaluate the effectiveness of data augmentation in experiments helps provide empirical evidence. This allows researchers to quantify improvements and demonstrate that the measures taken have indeed enhanced model performance.

The results of the second group of experiments are shown in Figure 3. The evaluation trends of the three methods are mostly consistent, showing an improvement in the quality fusion score as the scale increases, although the rate of improvement decreases. The minmax principle still exaggerates the quality of data augmentation. The model precision also increases as the scale expands, but gradually starts to decline after reaching a four-fold scale. This indicates that the repetition of data augmentation-generated images can have a negative impact on model performance, leading to overfitting. It demonstrates that more data are not always better. According to Figure 3b, our algorithm and the averaging method have similar performance, which is superior to the minmax values.

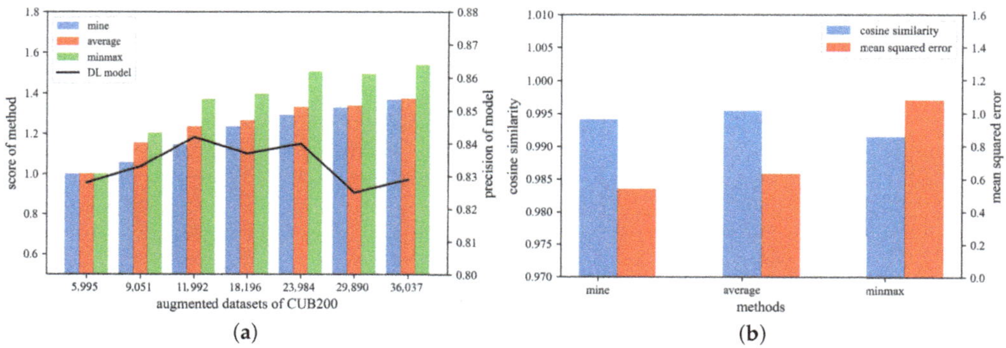

Figure 3. Under the same data augmentation with different scales, our algorithm and the averaging method exhibit similar performance. (**a**) Evaluation of the three algorithms and the model on the augmented dataset of CUB-200 dataset. (**b**) Performance evaluation of the three algorithms using CS and MSE. The word "mine" represents the methodology of this paper.

4.2. Results of Ablation Experiments

Table 5 presents four methods for metrics fusion. In the case of calculating metrics using multiplication, if we continue to use multiplication for index weights, it would lose the balancing effect of importance. Therefore, we chose to use power operations to further enhance the balancing role of parameters. For each fusion method, we evaluate the accuracy of the evaluation results using CS and MSE. The fusion method used in this study achieved the best scores in all four indicators. The results in the second and third rows are one order of magnitude higher than those in the first and fourth rows, with MSE increasing from 0.1357 and 0.5363 to 5.6027 and 9.7748. This also confirms that in the estimation of variance, the labels and semantic cannot be simply regarded as two sets of linearly related variables. The semantic information needs further decomposition, and the final scores need to be fused using addition for the unrelated parts. The first and fourth rows achieved approximately the same excellent scores, with CS differing by less than 1 percentage point. However, in the CUB dataset, the MSE differed by three times. The reason is the neglect of the correlation between variance and mean. In large and relatively balanced datasets like CIFAR-10, this effect is not evident. But for small datasets, data augmentation can easily disrupt data semantics and introduce irrelevant noise, affecting model training. This is reflected in the increase in variance evaluation results while the mean evaluation results decrease. Therefore, using addition to fuse variance and mean cannot well reflect the relationship between the two indicators.

In Tables 6 and 7, we can visually observe the correlations between indicators. One notable observation is that the correlation between Q_1 and Q_6 is approximately -1. This confirms that there is indeed a strong negative correlation between them, consistent with the previous discussions. Surprisingly, we also found high correlations between Q_1 and Q_4, as well as between Q_2, Q_3, and Q_5. The high correlation between Q_1 and Q_4 may be due to the random generation of augmented datasets. It results in different numbers of samples for each class, causing Q_4 and Q_1 to increase synchronously and be considered highly correlated. If the size of tests is larger and more diverse, this correlation will gradually decrease. The high correlation between Q_2, Q_3, and Q_5 may be due to manual rule intervention, which leads to a more uniform color space distribution in the dataset and reduces the use of color as a criterion for unsupervised clustering. This improves the accuracy of clustering results. This also indirectly demonstrates that data augmentation indeed helps the model learn from data.

Table 5. Results of ablation experiments.

Fusion of Quality Metrics	Formula	CS_{cub}	MSE_{cub}	CS_{cifar}	MSE_{cifar}
$(Q_{Diversity} + Q_{ClassBalance}) \times Q_{TaskRelevance}$	$\frac{Q_6'}{Q_6} \times \sum_{i=1}^{5} w_i \frac{Q_i'}{Q_i} / \sum_{i=1}^{5} w_i$	**0.9999** *	**0.1357**	**0.9997**	**1.4841**
$Q_{Diversity} \times Q_{ClassBalance} \times Q_{TaskRelevance}$	$\prod_{i=1}^{6} (\frac{Q_i'}{Q_i})^{r_i}$	0.9805	5.6027	0.9721	16.5365
$Q_{Diversity} \times (Q_{ClassBalance} + Q_{TaskRelevance})$	$\prod_{i=1}^{3} (\frac{Q_i'}{Q_i})^{r_i} \times \sum_{i=4}^{6} w_i \frac{Q_i'}{Q_i} / \sum_{i=4}^{6} w_i$	0.9727	9.7748	0.9916	98.9535
$Q_{Diversity} + Q_{ClassBalance}) + Q_{TaskRelevance}$	$\sum_{i=1}^{6} w_i \frac{Q_i'}{Q_i} / \sum_{i=1}^{6} w_i$	0.9977	0.5363	0.9995	1.8736

* The bold values are the best.

Table 6. The correlation coefficient of each metric in CIFAR-10.

Metrics	Q_1	Q_2	Q_3	Q_4	Q_5	Q_6
Q_1	1					
Q_2	−0.05	1				
Q_3	−0.29	0.76 *	1			
Q_4	0.81 *	0.10	−0.10	1		
Q_5	−0.19	0.62 *	0.81 *	−0.14	1	
Q_6	−0.92 **	0.02	0.24	−0.57	0.07	1

* $p < 0.05$, ** $p < 0.01$.

Table 7. The correlation coefficient of each metric in CUB-200.

Metrics	Q_1	Q_2	Q_3	Q_4	Q_5	Q_6
Q_1	1					
Q_2	−0.23	1				
Q_3	−0.71 *	0.73 *	1			
Q_4	0.76 *	0.11	−0.24	1		
Q_5	−0.71 *	0.79 *	0.95 **	−0.19	1	
Q_6	−0.97 **	0.36	0.79 *	−0.66	0.8 *	1

* $p < 0.05$, ** $p < 0.01$.

However, there are slight differences in the experimental results between the two datasets. For example, in CIFAR-10, Q_6 is strongly correlated only with Q_1. But in CUB-200, Q_6 is strongly correlated with not only Q_1 but also Q_3 and Q_5. In particular, the correlation with Q_5 increased from 0.07 to 0.8, showing two extremes. From the perspective of dataset properties, CIFAR-10 is a large-scale dataset with data from various scenarios and weather conditions, and data augmentation has a limited impact on its basic image feature distribution, and the distribution of each class changes little. Data augmentation mainly affects semantic information in this dataset, such as object inversion not affecting recognition. Therefore, only Q_1 and Q_6 show correlation. However, in the small dataset CUB-200, most images have a single color and similar backgrounds. Image features have a greater impact on the final results, which is also reflected in the correlation of indicator results. Therefore, when using our framework to evaluate data augmentation quality, analyzing the correlations of various indicators can help understand the strengths and weaknesses of the dataset.

Figure 4 illustrates the impact of clustering algorithms on the framework results, where only Q_1 and Q_6 are accelerated by clustering in the framework. Figure 4a,b show the accuracy of the results before and after clustering. The original algorithm uses brute force to compute the distance between pairwise image vectors to obtain evaluation results. In Figure 4a, the clustered results show some differences compared to the original results, with a similar trend but a decrease in evaluation scores after clustering. However, as Figure 4b indicates, both indicators improve after clustering, proving that clustering actually helps improve the accuracy of the evaluation results. Figure 4c displays the saved running time through clustering. The x-axis represents the product of the number of classes and the number of samples in each class in the dataset. It is evident that when the number of classes is large and the number of samples per class is small, clustering only saves about half of the time. However, when the number of samples per class is much larger than the number of classes, clustering can save over 90% of the time. This is because Q_1 separates the dataset based on classes, and the time complexity within each class is $O(n^2)$. After fast clustering, since the number of classes is roughly the same, it can be considered a constant factor. Therefore, the fewer the number of classes and the more samples per class, the more time can be saved. In Figure 5, it can be observed that replacing different pre-trained models does not significantly affect the evaluation results of this method. The more sufficient the pre-training, the more accurate the grasp of image features, and the generated image feature vectors also have certain discriminative power. The two pre-trained models selected in this study, both trained on ImageNet, did not show significant differences.

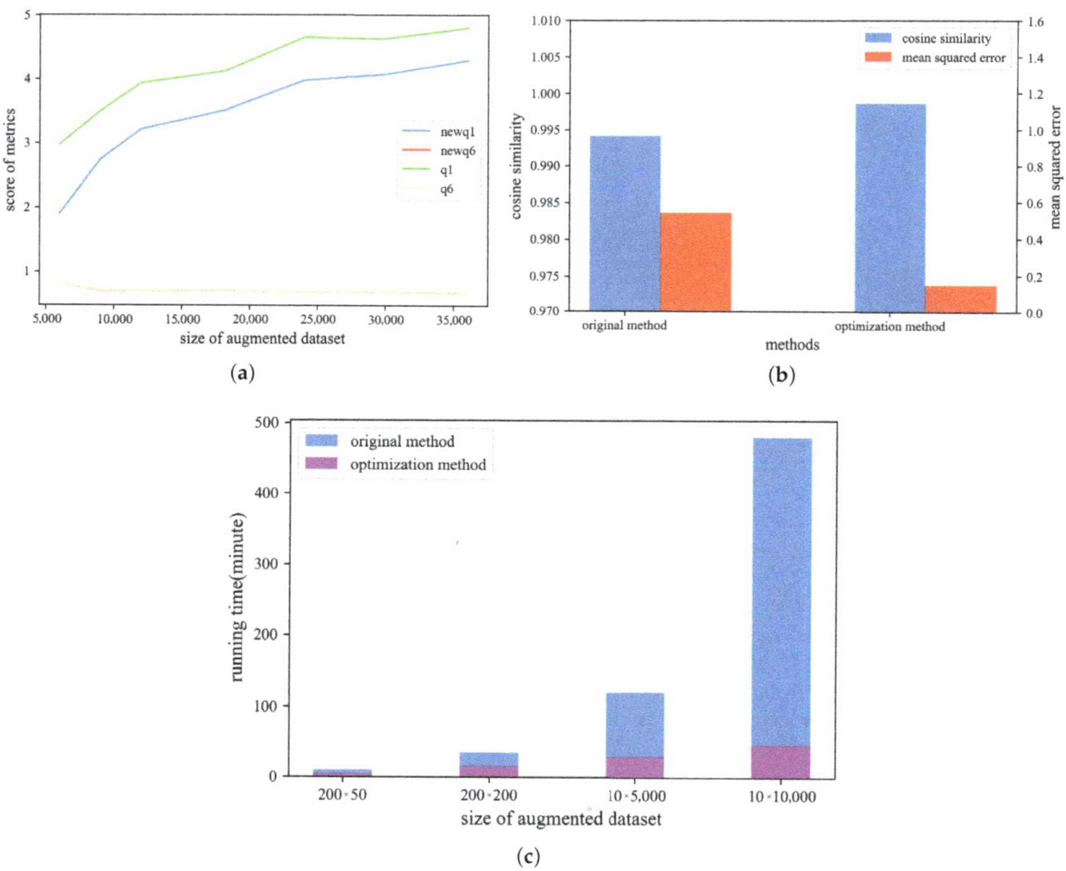

Figure 4. The clustering results have not only optimized the computation of the algorithm but also improved its performance. (**a**) The running times of the unoptimized and optimized algorithms are shown for different numbers of categories and samples within each category. The x-axis represents the product of the number of categories and the number of samples within each category in the dataset. (**b**) Partial scores of the original algorithm and the optimized algorithm are presented for the CUB-200 dataset. The x-axis represents the total number of samples in the augmented training set. (**c**) Performance evaluation of the original algorithm and the optimized algorithm in the CUB-200 dataset.

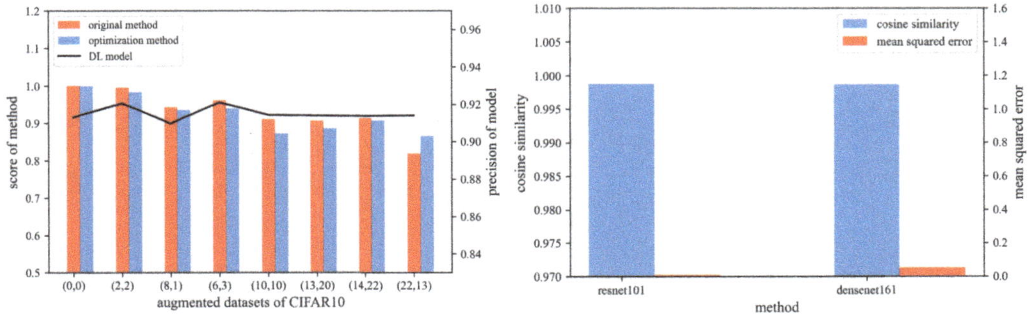

(**a**) (**b**)

Figure 5. Under the same data augmentation with different scales, our algorithm and the averaging method exhibit similar performance. (**a**) Evaluation of the three algorithms and the model on the augmented dataset of CUB-200 dataset. (**b**) Performance evaluation of the three algorithms using CS and MSE.

4.3. Applicability and Potential Applications

In Figures 6 and 7, we assessed the quality of each dataset and visualized the scores for the original dataset and the highest dataset. It can be seen that from a visual perspective, there are significant differences between the two. Figure 6 presents a comparison of the EuroSAT dataset. On the left, the original images within the same category mostly share a similar color and content. On the right, due to the effects of data augmentation, there is increased diversity in color and brightness. However, in some images, it becomes difficult to discern their content with the naked eye. Figure 7 illustrates the dimensionality reduction visualization of the IMDB dataset using the t-SNE algorithm. It is evident that the dataset on the right, which has higher scores, exhibits better data separation. Most of the positive data points are concentrated on the right. This indicates that the classes have good separability, which can effectively enhance the model's performance.

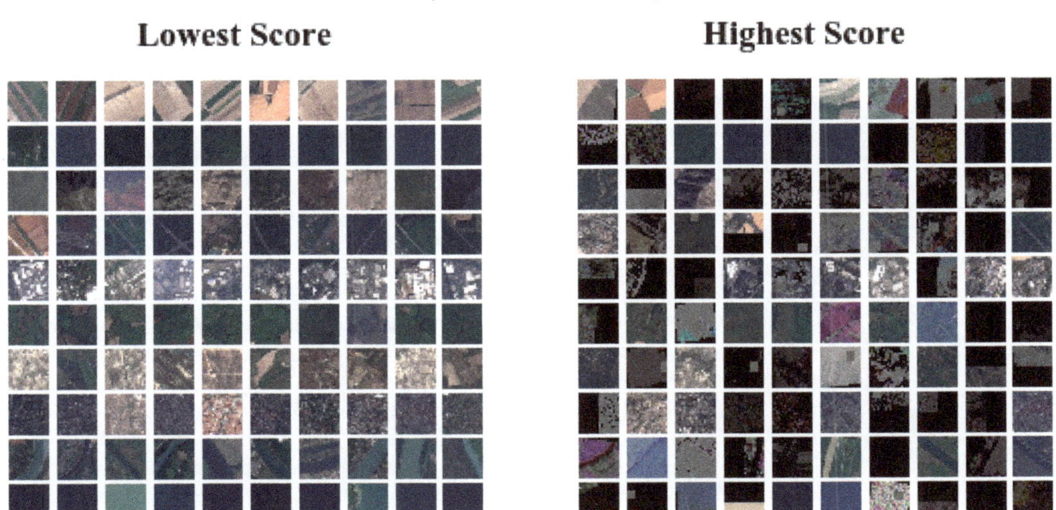

(**a**) (**b**)

Figure 6. This is a display of the EuroSAT dataset, each row is a different class. (**a**) Origin dataset. (**b**) The augmented dataset with the highest score.

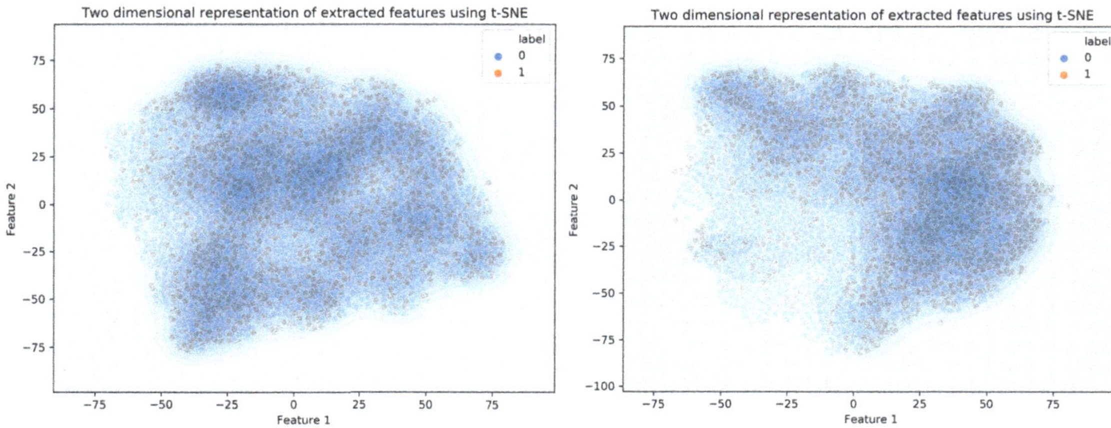

(**a**) (**b**)

Figure 7. The t-SNE algorithm reduces dimensionality of all features in the IMDB dataset and visualizes them. (**a**) Origin dataset. (**b**) The augmented dataset with the highest score.

The context of intrinsic data attributes and the intended use of the data in the evaluation method cannot be balanced well, which limits its applicability. For example, remote sensing images inherently have issues like low contrast, noise, and blurriness. Additionally, remote sensing images often contain imprecisely shaped objects, unlike medical imaging, which require precise target recognition. Therefore, remote sensing images are not highly sensitive to task relevance, and optimizing intrinsic data attributes for quality improvement results in more significant quality loss compared to the decrease in task relevance. In contrast, medical imaging has sufficiently high image quality. In an attempt to enhance diversity, it leads to a loss in data quality and task relevance, ultimately resulting in an overall decline in quality. Different tasks have varying quality requirements for datasets, and during the calculation, parameter variations in three areas—diversity, class balance, and task relevance—need to be considered. This is also a direction we need to focus on in the future. When our method is not tuned to the correct parameter resulting in poor performance, a more well-performing parameter can be fitted by combining the evaluation results of multiple augmented datasets with a small number of training results. Then try evaluating a new augmented dataset again until the parameter is similar to most of the training results.

Different downstream tasks have different priorities. The evaluation method may succeed for the assumed task but fail for the target task. For example, in the case of object recognition tasks, the optimal augmented dataset obtained using the method described in this paper may be suboptimal for this task. This is because image similarity metrics are better suited for classification tasks rather than object recognition, where the goal is to detect similarity between small regions. Task relevance metrics, on the other hand, can replace this by calculating the similarity between annotated regions and other regions.

Data can exhibit significant differences in terms of structure, format, complexity, size, noise levels, and more. Evaluation methods tailored to one data type may fail when applied to another data type. For image datasets, we calculate statistics on pixels and textures. However, text datasets are composed of characters, words, and sentences, so the intrinsic data quality of different data types needs rules established by domain experts for measurement. In addition to this, this paper's method reflects high applicability and robustness in data diversity. The main reason is that no matter what type of deep learning data can use neural networks to extract multi-dimensional feature vectors, which constitutes most of the indicators of this paper's method based on the calculation of feature vectors, so to a certain extent this paper has the ability to adapt to most of the fields.

In most real-world datasets, a certain degree of anomalies, noise, errors, and outliers can be expected. Truly clean and pristine data are a rare find. This is particularly true for sensor data, remote sensing data, and internet-derived data, which often exhibit higher variability and a higher incidence of anomalies. The process of data augmentation can introduce new error patterns. In this paper, we split the dataset into three attributes and utilize techniques such as information entropy to effectively quantify the number of patterns and task relevance. Excessive errors can lead to a decrease in task relevance, which can manifest in the final results. However, because this paper employs unsupervised algorithms, it is challenging to distinguish and optimize true anomalies from acceptable variations.

The diversity and complexity of real-world data make the development of universally applicable data quality assessment methods inherently challenging. Thoughtful method design, extensive evaluation, avoiding overfitting, and relaxing assumptions can enhance applicability. In practical scenarios, the methods outlined in this paper can guide the construction of datasets. If the collected sample dataset is insufficient for training deep learning models, you can evaluate the effectiveness of data augmentation using quantitative metrics defined based on the three dataset attributes and the ultimate goal, such as improving classification accuracy, reducing error rates, or enhancing signal-to-noise ratios. Utilizing data augmentation and assessment methods can address the long-tail problem in data, improve data consistency, and reduce the labeling workload. As more real-world data become available, it is essential to continuously reassess and enhance data augmentation. The demand for augmentation may evolve over time. In summary, the methods described in this paper contribute to the development of well-generalized models from limited and imperfect real-world data.

Furthermore, the methods outlined in this paper are easy to implement and can be seamlessly integrated into the deep learning workflow using popular frameworks like TensorFlow or PyTorch. Leveraging the data pipelines within these frameworks and the data monitoring integrated into them, it becomes straightforward to quickly compute intrinsic data attribute metrics and feature vectors generated by pre-trained models after generating multiple augmented datasets from the input data. Once the optimal dataset for evaluation is obtained, you can proceed directly to training your model.

5. Conclusions

This study aims to enhance model performance through the improvement of data quality. By categorizing data quality into three key dimensions, diversity, class balance, and task relevance, and evaluating the effectiveness of data augmentation within these dimensions, we have achieved the following key outcomes and conclusions.

Firstly, we have successfully deconstructed the complexity of data quality into three essential dimensions. This aids in providing a more comprehensive understanding of data quality. Diversity ensures the inclusion of various sample types in the dataset, class balance helps address imbalances in class distribution, and task relevance ensures the alignment of data with the actual task at hand.

Secondly, by assessing the impact of data augmentation methods across these three dimensions, we can quantitatively measure the influence of different enhancement strategies on data quality. Our experimental results demonstrate that, with reasonable selection and adjustment of augmentation strategies, significant improvements can be made in data diversity and class balance while maintaining a high degree of relevance to the task.

Most importantly, our work holds significant practical implications. In the modern fields of machine learning and artificial intelligence, data serves as the foundation of successful models. By elevating data quality, we can enhance model generalization, mitigate overfitting risks, improve model robustness in real-world scenarios, and provide more accurate predictive and decision-making support across various application domains. This impact extends to critical areas such as medical diagnostics, financial risk analysis, autonomous driving, and beyond.

In the future, we aim to further explore the relationships among data quality dimensions and strive towards the automatic selection of parameters tailored to specific domains and tasks. The most notable improvement will be in terms of efficiency, as manual parameter selection and adjustment will no longer be necessary. Additionally, this approach will reduce subjectivity and bias, ensuring the replicability and comparability of experimental results. Most importantly, it will simplify experimentation, making this method accessible to a broader range of researchers interested in understanding the principles of data construction.

In summary, our research provides a systematic approach to enhancing data quality, enabling researchers to better comprehend, evaluate, and enhance data for improved machine learning model performance. This work offers robust guidance for future research and applications, with the potential to make a positive impact in data-driven fields.

Author Contributions: Conceptualization, X.C., Y.L. and C.M.; methodology, X.C. and Y.L.; software, X.C., Y.L. and C.M.; validation, X.C., Y.L. and C.M.; formal analysis, X.C. and Y.L.; investigation, Y.L.; resources, X.C., Y.L., Z.X. and C.M.; data curation, X.C., Y.L., H.L. and S.Y.; writing—original draft preparation, X.C. and Y.L.; visualization, X.C. and Y.L.; supervision, X.C., Z.X. and C.M.; project administration, X.C. and C.M.; funding acquisition, X.C. and C.M.; X.C., Y.L. and C.M.: significance contributions to the manuscript. All authors have read and agreed to the published version of the manuscript.

Funding: This research was supported by Outstanding Youth Team Project of Central Universities(QNTD202308) and National Key R&D Program of China (2022YFF1302700).

Data Availability Statement: Not applicable.

Acknowledgments: The authors thank the anonymous reviewers for their valuable comments.

Conflicts of Interest: The authors declare no conflict of interest.

References

1. Zhang, T.; Chen, J.; Li, F.; Zhang, K.; Lv, H.; He, S.; Xu, E. Intelligent fault diagnosis of machines with small & imbalanced data: A state-of-the-art review and possible extensions. *ISA Trans.* **2022**, *119*, 152–171. [PubMed]
2. Chlap, P.; Min, H.; Vandenberg, N.; Dowling, J.; Holloway, L.; Haworth, A. A review of medical image data augmentation techniques for deep learning applications. *J. Med. Imaging Radiat. Oncol.* **2021**, *65*, 545–563. [CrossRef] [PubMed]
3. Silver, D.; Hubert, T.; Schrittwieser, J.; Antonoglou, I.; Lai, M.; Guez, A.; Lanctot, M.; Sifre, L.; Kumaran, D.; Graepel, T.; et al. A general reinforcement learning algorithm that masters chess, shogi, and Go through self-play. *Science* **2018**, *362*, 1140–1144. [CrossRef] [PubMed]
4. Hao, X.; Liu, L.; Yang, R.; Yin, L.; Zhang, L.; Li, X. A Review of Data Augmentation Methods of Remote Sensing Image Target Recognition. *Remote Sens.* **2023**, *15*, 827. [CrossRef]
5. Chen, Y.; Yang, X.H.; Wei, Z.; Heidari, A.A.; Zheng, N.; Li, Z.; Chen, H.; Hu, H.; Zhou, Q.; Guan, Q. Generative adversarial networks in medical image augmentation: A review. *Comput. Biol. Med.* **2022**, *144*, 105382. [CrossRef]
6. Yang, J.; Guo, X.; Li, Y.; Marinello, F.; Ercisli, S.; Zhang, Z. A survey of few-shot learning in smart agriculture: Developments, applications, and challenges. *Plant Methods* **2022**, *18*, 28. [CrossRef]
7. Maslej-Krešňáková, V.; Sarnovský, M.; Jacková, J. Use of Data Augmentation Techniques in Detection of Antisocial Behavior Using Deep Learning Methods. *Future Internet* **2022**, *14*, 260. [CrossRef]
8. Shorten, C.; Khoshgoftaar, T.M.; Furht, B. Text data augmentation for deep learning. *J. Big Data* **2021**, *8*, 101. [CrossRef]
9. Gong, C.; Wang, D.; Li, M.; Chandra, V.; Liu, Q. Keepaugment: A simple information-preserving data augmentation approach. In Proceedings of the IEEE/CVF Conference on Computer Vision and Pattern Recognition, Nashville, TN, USA, 20–25 June 2021; pp. 1055–1064.
10. Iwana, B.K.; Uchida, S. An empirical survey of data augmentation for time series classification with neural networks. *PLoS ONE* **2021**, *16*, e0254841. [CrossRef]
11. Zhou, X.; Hu, Y.; Wu, J.; Liang, W.; Ma, J.; Jin, Q. Distribution bias aware collaborative generative adversarial network for imbalanced deep learning in industrial IoT. *IEEE Trans. Ind. Inform.* **2022**, *19*, 570–580. [CrossRef]
12. Bishop, C.M. Training with noise is equivalent to Tikhonov regularization. *Neural Comput.* **1995**, *7*, 108–116. [CrossRef]
13. Hernández-García, A.; König, P. Data augmentation instead of explicit regularization. *arXiv* **2018**, arXiv:1806.03852.
14. Carratino, L.; Cissé, M.; Jenatton, R.; Vert, J.P. On mixup regularization. *arXiv* **2020**, arXiv:2006.06049.
15. Shen, R.; Bubeck, S.; Gunasekar, S. Data augmentation as feature manipulation. In Proceedings of the International Conference on Machine Learning, Baltimore, MD, USA, 17–23 July 2022; pp. 19773–19808.

16. Ilse, M.; Tomczak, J.M.; Forré, P. Selecting data augmentation for simulating interventions. In Proceedings of the International Conference on Machine Learning, Virtual Event, 18–24 July 2021; pp. 4555–4562.
17. Allen-Zhu, Z.; Li, Y. Feature purification: How adversarial training performs robust deep learning. In Proceedings of the 2021 IEEE 62nd Annual Symposium on Foundations of Computer Science (FOCS), Denver, CO, USA, 7–10 February 2022; pp. 977–988.
18. Kong, Q.; Chang, X. Rough set model based on variable universe. *CAAI Trans. Intell. Technol.* **2022**, *7*, 503–511. [CrossRef]
19. Zhao, H.; Ma, L. Several rough set models in quotient space. *CAAI Trans. Intell. Technol.* **2022**, *7*, 69–80. [CrossRef]
20. Kusunoki, Y.; Błaszczyński, J.; Inuiguchi, M.; Słowiński, R. Empirical risk minimization for dominance-based rough set approaches. *Inf. Sci.* **2021**, *567*, 395–417. [CrossRef]
21. Chen, S.; Dobriban, E.; Lee, J.H. A group-theoretic framework for data augmentation. *J. Mach. Learn. Res.* **2020**, *21*, 9885–9955.
22. Mei, S.; Misiakiewicz, T.; Montanari, A. Learning with invariances in random features and kernel models. In Proceedings of the Conference on Learning Theory, Boulder, CO, USA, 15–19 August 2021; pp. 3351–3418.
23. Wand, Y.; Wang, R.Y. Anchoring data quality dimensions in ontological foundations. *Commun. ACM* **1996**, *39*, 86–95. [CrossRef]
24. Abdullah, M.Z.; Arshah, R.A. A review of data quality assessment: Data quality dimensions from user's perspective. *Adv. Sci. Lett.* **2018**, *24*, 7824–7829. [CrossRef]
25. Firmani, D.; Mecella, M.; Scannapieco, M.; Batini, C. On the meaningfulness of "big data quality". *Data Sci. Eng.* **2016**, *1*, 6–20. [CrossRef]
26. Jarwar, M.A.; Chong, I. Web objects based contextual data quality assessment model for semantic data application. *Appl. Sci.* **2020**, *10*, 2181. [CrossRef]
27. Sim, K.; Yang, J.; Lu, W.; Gao, X. MaD-DLS: Mean and deviation of deep and local similarity for image quality assessment. *IEEE Trans. Multimed.* **2020**, *23*, 4037–4048. [CrossRef]
28. Senaratne, H.; Mobasheri, A.; Ali, A.L.; Capineri, C.; Haklay, M. A review of volunteered geographic information quality assessment methods. *Int. J. Geogr. Inf. Sci.* **2017**, *31*, 139–167. [CrossRef]
29. Chen, H.; Chen, J.; Ding, J. Data evaluation and enhancement for quality improvement of machine learning. *IEEE Trans. Reliab.* **2021**, *70*, 831–847. [CrossRef]
30. Gosain, A.; Saha, A.; Singh, D. Measuring harmfulness of class imbalance by data complexity measures in oversampling methods. *Int. J. Intell. Eng. Inform.* **2019**, *7*, 203–230. [CrossRef]
31. Bellinger, C.; Sharma, S.; Japkowicz, N.; Zaïane, O.R. Framework for extreme imbalance classification: SWIM—Sampling with the majority class. *Knowl. Inf. Syst.* **2020**, *62*, 841–866. [CrossRef]
32. Li, A.; Zhang, L.; Qian, J.; Xiao, X.; Li, X.Y.; Xie, Y. TODQA: Efficient task-oriented data quality assessment. In Proceedings of the 2019 15th International Conference on Mobile Ad-Hoc and Sensor Networks (MSN), Shenzhen, China, 11–13 December 2019; pp. 81–88.
33. Delgado-Bonal, A.; Marshak, A. Approximate entropy and sample entropy: A comprehensive tutorial. *Entropy* **2019**, *21*, 541. [CrossRef]
34. Li, Y.; Chao, X.; Ercisli, S. Disturbed-entropy: A simple data quality assessment approach. *ICT Express* **2022**, *8*, 309–312. [CrossRef]
35. Liu, L.; Miao, S.; Liu, B. On nonlinear complexity and Shannon's entropy of finite length random sequences. *Entropy* **2015**, *17*, 1936–1945. [CrossRef]
36. He, K.; Zhang, X.; Ren, S.; Sun, J. Deep residual learning for image recognition. In Proceedings of the IEEE Conference on Computer Vision and Pattern Recognition, Las Vegas, NV, USA, 27–30 June 2016; pp. 770–778.
37. Sarfraz, S.; Sharma, V.; Stiefelhagen, R. Efficient parameter-free clustering using first neighbor relations. In Proceedings of the IEEE/CVF Conference on Computer Vision and Pattern Recognition, Long Beach, CA, USA, 15–20 June 2019; pp. 8934–8943.
38. Friedman, D.; Dieng, A.B. The Vendi Score: A Diversity Evaluation Metric for Machine Learning. *arXiv* **2022**, arXiv:2210.02410.
39. Mishra, S.P.; Sarkar, U.; Taraphder, S.; Datta, S.; Swain, D.P.; Saikhom, R.; Panda, S.; Laishram, M. Multivariate Statistical Data Analysis- Principal Component Analysis (PCA). *Int. J. Livest. Res.* **2017**, *7*, 60–78.
40. Geirhos, R.; Rubisch, P.; Michaelis, C.; Bethge, M.; Wichmann, F.A.; Brendel, W. ImageNet-trained CNNs are biased towards texture; increasing shape bias improves accuracy and robustness. *arXiv* **2018**, arXiv:1811.12231.
41. Lore, K.G.; Akintayo, A.; Sarkar, S. LLNet: A deep autoencoder approach to natural low-light image enhancement. *Pattern Recognit.* **2017**, *61*, 650–662. [CrossRef]
42. Yang, Y.; Xu, Z. Rethinking the value of labels for improving class-imbalanced learning. *Adv. Neural Inf. Process. Syst.* **2020**, *33*, 19290–19301.
43. Lin, T.Y.; Goyal, P.; Girshick, R.; He, K.; Dollár, P. Focal loss for dense object detection. In Proceedings of the IEEE International Conference on Computer Vision, Venice, Italy, 22–29 October 2017; pp. 2980–2988.
44. Xu, Y.; Lu, Y. Adaptive weighted fusion: A novel fusion approach for image classification. *Neurocomputing* **2015**, *168*, 566–574. [CrossRef]
45. Ahmad, S.; Pal, R.; Ganivada, A. Rank level fusion of multimodal biometrics using genetic algorithm. *Multimed. Tools Appl.* **2022**, *81*, 40931–40958. [CrossRef]
46. Nawaz, S.; Calefati, A.; Caraffini, M.; Landro, N.; Gallo, I. Are these birds similar: Learning branched networks for fine-grained representations. In Proceedings of the 2019 International Conference on Image and Vision Computing New Zealand (IVCNZ), Dunedin, New Zealand, 2–4 December 2019; pp. 1–5.

47. Cubuk, E.D.; Zoph, B.; Shlens, J.; Le, Q.V. Randaugment: Practical automated data augmentation with a reduced search space. In Proceedings of the IEEE/CVF Conference on Computer Vision and Pattern Recognition Workshops, Seattle, WA, USA, 14–19 June 2020; pp. 702–703.
48. Salimans, T.; Goodfellow, I.; Zaremba, W.; Cheung, V.; Radford, A.; Chen, X. Improved techniques for training gans. In Proceedings of the 30th International Conference on Neural Information Processing Systems, Barcelona, Spain, 5–10 December 2016; pp. 2234–2242.

electronics

MDPI

Article

An Improved Spatio-Temporally Smoothed Coherence Factor Combined with Delay Multiply and Sum Beamformer

Ziyang Guo [1,2], Xingguang Geng [1], Fei Yao [1], Liyuan Liu [1,2], Chaohong Zhang [1,2], Yitao Zhang [1,2,*] and Yunfeng Wang [1,2]

1 Institute of Microelectronics of Chinese Academy of Sciences, Beijing 100029, China
2 University of Chinese Academy of Sciences, Beijing 100049, China
* Correspondence: zhangyitao@ime.ac.cn

Abstract: Delay multiply and sum beamforming (DMAS) is a non-linear method used in ultrasound imaging which offers superior performance to conventional delay and sum beamforming (DAS). While the combination of DMAS and coherence factor (CF) can further improve single plane-wave imaging lateral resolution, by using CF to weight the DMAS output, the main lobe width and aberration effects can be suppressed, which will improve the disadvantage of low lateral resolution when imaging with a single plane-wave. However, in low signal-to-noise ratio (SNR) environments, the speckle variance of the image increases, and there are black area artifacts around high echo objects. To improve the quality of the scatter without significantly reducing the lateral resolution of the DMAS-CF, this paper proposes an adaptive spatio-temporally smoothed coherence factor (GSTS-CF) combined with delay multiply and sum beamformer (DMAS + GSTS-CF), which uses the generalized coherence factor (GCF) as a local coherence detection tool to adaptively determine the subarray length to obtain an improved adaptive spatio-temporally smoothed factor, and uses this factor to weight the output of DMAS. The simulation and experimental data show that the proposed method improves lateral resolution (20 mm depth) by 86.87% compared to DAS, 52.13% compared to DMAS, 15.84% compared to DMAS + STS-CF, and has a full width at half maxima (FWHM) similar to DMAS-CF. The proposed method improves the speckle signal-to-noise ratio (sSNR) by 87.85% (simulation) and 77.84% (in carotid) compared to DMAS-CF, 20.37% (simulation) and 40.74% (in carotid) compared to DMAS, 15.03% (simulation) and 13.46% (in carotid) compared to DMAS + STS-CF, and has sSNR and scatter variance similar to DAS. This indicates that the method improves scatter quality (lower scatter variance and higher sSNR) without significantly reducing lateral resolution.

Keywords: ultrasound imaging; plane-wave; beamforming; coherence factor; adaptive; spatio-temporally smoothed; delay multiply and sum beamforming

Citation: Guo, Z.; Geng, X.; Yao, F.; Liu, L.; Zhang, C.; Zhang, Y.; Wang, Y. An Improved Spatio-Temporally Smoothed Coherence Factor Combined with Delay Multiply and Sum Beamformer. *Electronics* **2023**, *12*, 3902. https://doi.org/10.3390/electronics12183902

Academic Editor: Stefano Ricci

Received: 16 August 2023
Revised: 4 September 2023
Accepted: 7 September 2023
Published: 15 September 2023

1. Introduction

In ultrasound imaging, we generally use a fixed focus at emission and a dynamic focus at reception, which allows imaging at frame rates in the tens of Hertz to be obtained [1]; however, this is not suitable for high frame rate imaging such as echocardiography, 3D imaging, and elastography [2,3]. The plane-wave scanning approach achieves high frame rates, but the lack of emission focus severely degrades the imaging quality. Although some traditional coherence factor methods can suppress clutter and improve image contrast with low computational complexity, the ability to reduce scattering variance and remove black-area artefacts is still insufficient, which makes the image background inhomogeneous. Therefore, simultaneously improving the lateral resolution and background scattering quality has become the focus of this research. Based on the spatio-temporally smoothed coherence factor (STS-CF) [4], we propose a new coherence factor method that can adaptively change the length of the smoothing subarray according to the detection target. It provides better side and/or grating lobe suppression, clutter reduction, and aberration correction. To achieve a trade-off between lateral resolution and scattering retention performance, we combined the new coherence factor with the DMAS method to further improve lateral resolution.

One of the most common beamformers is the delay and sum beamforming (DAS), but its ability to improve image resolution and suppress clutter interference is limited. Giulia Matrone et al. introduced DMAS based on receive aperture autocorrelation, which, unlike DAS, is a non-linear algorithm in which signals are combined and coupled and then multiplied before summing [5]. This means that a correlation operation is conducted on the echoes. Since DMAS multiplies echoes of almost the same frequency, DC and second harmonic components appear in the output spectrum. Therefore, a band-pass filter is added after the DMAS output to filter out DC components and higher harmonic components, while the signal centered at $2f_0$ remains unchanged (f_0 is the central frequency of the echo), and finally the output of the filtered delayed multiplication sum (F-DMAS) is obtained [6]. Compared to DAS, DMAS better suppresses spurious and noise via a correlation operation, brings the measure of backward scattering signal coherence into the beamforming process, and increases the number of new "artificial apertures" (because the coefficient of the autocorrelation function is $2N - 1$ and N is the number of received apertures), thus reducing $f\#$ and resulting in an improvement in lateral resolution [7]. However, this imaging method requires further suppression of the side lobes to improve the imaging quality.

A number of scholars have proposed correcting the output of DMAS with a CF-like method, which can develop for side lobe suppression, clutter reduction, and aberration correction. By adaptively weighting the beamsum, they can enhance image contrast without sacrificing spatial resolution. In addition, they have low computational complexity and are easy to implement. The most representative CF-like method is the coherence factor (CF) [8,9], which is defined as the ratio between the coherent energy of the aperture received signal and the total energy (non-coherent). By using CF to weigh the output of beamforming, the side lobes and aberration effects can be suppressed, but this darkens the image and even reconstructs the image with errors. p.-C. Li and M.-L. Li proposed the generalized coherence factor (GCF), a spatial frequency domain version of CF, which adds a low frequency signal that is not very different from the axial fundamental frequency signal to the molecule of CF and improves the preservation of scatter [10]. Camacho et al. [11] designed the phase coherence factor (PCF) and the sign coherence factor (SCF). The principle is to replace the amplitude information with the phase information, and a linear or exponential relationship curve was added to regulate the suppression of off-axis signals and the retention of background scatter. This suppresses the side lobe and improves the lateral resolution. The implementation of this technique is simple and practical.

Although the CF-like method combined with the DMAS beamformer has advantages, the images may suffer some undesirable effects in a low-SNR environment. They include overall image brightness reduction, increased speckle variance, underestimation of the size of the point target, black area artifacts in the region around the high echo reflector, and even removal of the speckle pattern. To solve the above problems, we introduced and improved the spatio-temporally smoothed coherence factor (STS-CF) proposed by MengLing Xu et al. [12], the essence of which is to measure the coherence between the split subarrays. It uses spatial smoothing (i.e., sub-aperture averaging) to create overlapping subarrays and temporal smoothing across multiple time samples to calculate the energy of the coherence sum [13]. This method introduces a tunable factor, achieving a balance between image quality and algorithmic robustness. However, the value of the subarray length L is determined empirically, and the most appropriate L is different in different environments, so the method performs generally in clinical applications.

To solve the above problem, we propose the GSTS-CF. In this study, we use GCF to detect local coherence and adaptively determine the subarray length for spatial smoothing [14]. Using this factor to weigh the DMAS output can improve the scatter quality without significantly reducing the lateral resolution. This is more applicable in complex clinical settings. Section 2 briefly introduces the framework background of GCF, STS-CF, and DMAS, and then we describe the proposed method. Section 3 describes the simulation setup and experimental steps and provides some metrics for evaluating a different beamformer. Section 4 shows the obtained images and discusses the results. The performance of these methods and the possibilities for further improvements are discussed in Section 5. Section 6 provides a conclusion of the proposed method of this paper.

2. Materials and Methods

2.1. Spatio-Temporally Smoothed Coherence Factor

In CF, coherence is measured by a single element at a single time metric [15]; the formula of the CF is:

$$CF(p) = \frac{\left|\sum_{m=0}^{N} x_m(p)\right|^2}{N\sum_{m=0}^{N}|x_m(p)|^2} \tag{1}$$

where N is the number of elements, and $x_m(p)$ is the delayed signal received by the m-th element at point p. These signals are susceptible to noise and side lobes interference. To make the background region scatter more uniform, the STS-CF is introduced [11]. It divides the array into $N-L+1$ mutually overlapping subarrays and measures the coherence of the array signal at $2K+1$ sampling points. It uses the beamsum of the subarrays instead of the single element signal [16]. The mathematical expression is defined as:

$$\text{STS-CF(p)} = \frac{\sum_{k=-K}^{K}\left|\sum_{l=1}^{N-L+1}\sum_{m=l}^{L+l-1} x_m(p+k)\right|^2}{(\text{N}-\text{L}+1)\sum_{k=-K}^{K}\sum_{l=1}^{N-L+1}\left|\sum_{m=l}^{L+l-1} x_m(p+k)\right|^2} \tag{2}$$

where L is the length of the subarrays, and $x_m(p+k)$ is the delayed signal received by the m-th element at time index $p+k$. The temporal smoothed technique divides the received array into $N-L+1$ overlapping subarrays containing L elements and uses the subarray instead of a single element to measure the coherence of the signal. L as an adjustable parameter is able to balance between performance and algorithmic robustness, when $K=0$ and $L=0$, STS-CF is CF; when $K=0$ and $L=M$, $STS-CF\equiv 1$, which means no correction to the beamformer output.

2.2. Generalized Coherence Factor

CF not only suppresses off-axis interference but also filters out useful off-axis signals, which is unstable in a low signal-to-noise ratio environment [14]. The GCF method is based on an improved form of CF. To enhance the robustness, GCF considers the energy of non-coherent scattering signals. It adds low-frequency signals that do not differ much from the axial orientation to the numerator of Equation (1) [15]. It is defined as the ratio of the spectral energy of a particular low-frequency region to the total energy, and its mathematical expression is:

$$GCF(p) = \frac{\sum_{f\in LFR}|h(f,p)|^2}{\sum_{f=0}^{N-1}|h(f,p)|^2} \tag{3}$$

where $h(f,p)$ is the discrete Fourier transform of the delay-compensated aperture data for imaging point p, and the low-frequency region (LFR) is determined by the cutoff frequency M_0, which determines the performance of the GCF. When $M_C=0$, GCF becomes CF.

2.3. Delay Multiply and Sum Beamforming

In DMAS, the signal is delayed, coupled, and multiplied, and then the absolute value of the multiplied couple is square-rooted while preserving the sign, and the resulting signal is summed and band-pass filtered (BP). If the receiving aperture is N transducers, then there will be $N(N-1)/2$ combinations. The expression of DMAS is:

$$y_{DMAS}(t) = \sum_{i=1}^{N-1} \sum_{j=i+1}^{N} \hat{s}_{ij}[t] = \sum_{i=1}^{N-1} \sum_{j=i+1}^{N} sign(s_i(t)s_j(t)) \cdot \sqrt{|s_i(t)s_j(t)|} \tag{4}$$

where $s_i(t)$ and $s_j(t)$ are the delayed RF signals received by the i-th and j-th transducer elements, and $sign(x)$ is the sign function. Due to the multiplication of signals with similar frequencies, a DC and a second harmonic component appear in the spectrum of the DMAS. Therefore, band-pass filtering is further introduced to filter the DC and higher frequency components, and the filtered DMAS is called F-DMAS.

2.4. Proposed Method

We proposed an adaptive spatio-temporally smoothed coherence factor based on GCF, called GSTS-CF, which uses GCF as a local coherence detection tool to determine the subarray length adaptively [13]. To further improve the image sharpness of GSTS-CF, it is combined with the DMAS, which is DMAS + GSTS-CF. The algorithm flowchart is shown in Figure 1.

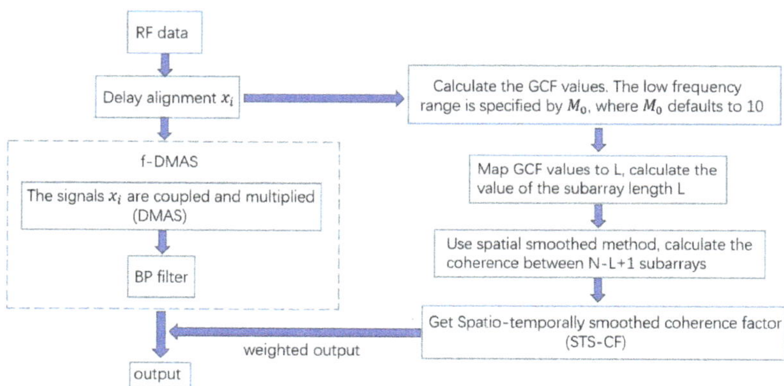

Figure 1. Flow chart of the proposed methods.

The flow of the algorithm is to weigh the output of the DMAS with the GSTS-CF factor. The key point is to calculate the GSTS-CF factor. The calculation process of GSTS-CF is divided into two steps. The first step is to improve the CF using a spatio-temporal smoothing method to obtain the STS-CF factor. The second step is to detect local coherence with the GCF and map the GCF onto the subarray length L so that its value varies adaptively. So first, we divide the array into $N-L+1$ subarrays, each containing L elements. This is called the spatial smoothed method, and the diagram is shown in Figure 2.

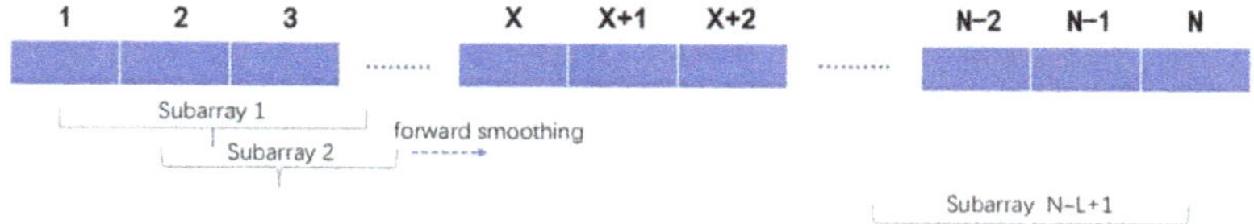

Figure 2. Diagram of the spatial smoothing method.

Figure 2 shows the ultrasonic probe consisting of N array elements. The spatial smoothing method is used to process the data by dividing the N array elements into $N-L+1$ subarrays, the first subarray contains the 1st to the Lth array element, the second subarray contains the 2nd to the $L+1$th array element, and so on, and the $N-L+1$th subarray contains the $N-L+1$th to the Nth array element. The spatial smoothed method computes the coherence of the subarray beamsums instead of computing the coherence with a single element. Equation (1) calculates coherence using a single subarray, while Equation (2) is a spatial smoothed method based on Equation (1), which uses a subarray of L array elements instead of a single array element to calculate the coherence between the arrays. Further, we measure the coherence of the subarrays at $2k+1$ neighbouring time samples instead of a single time sample, which have improved array gain on SNR and lowered side lobe levels. This is called the spatio-temporally smoothed method. L, as an adjustable parameter, can only be determined empirically. In order to make L adaptively changeable, we use the GCF to detect local coherence and map the GCF to the subarray length L.

In general, as the subarray length increases, the STS-CF approaches 1, thus enhancing robustness at the expense of lateral resolution. Therefore, to maintain the scatter pattern, the L value should be larger, while for echo-free cysts and highly echogenic reflectors, the L value should be smaller to obtain satisfactory image resolution and contrast [16].

Considering the performance of the GCF, the GCF values are small in non-coherent scattering targets (i.e., in echo-free capsules), large in strongly coherent scattering targets (i.e., high-echo reflectors), and tend to be moderate in low-coherent scattering targets (i.e., scattering spots).

According to the analysis above, when the value of GCF is large or small, we want the corresponding L to be small, while when the value of GCF tends to be medium, we want the value of L to be large.

In order to further determine the mapping relationship between GCF and L, we select some points in the incoherent region, strongly coherent region, and low-coherent region, respectively, to calculate the GCF value, and determine the appropriate L value at that point. The evaluation criterion of L-optimal solutions is that in the low coherence region, the scattering variance is the smallest, and in the strongly coherent region and the incoherent region, the lateral resolution is the best. Different echo targets correspond to different GCF values, and we selected 50 different targets whose GCF values were uniformly distributed between 0 and 1. We also calculated the optimal L values corresponding to these 50 targets and plotted a scatter plot, as shown in Figure 3a, using the GCF value of the point as the horizontal coordinate and the L value as the vertical coordinate. The scatter plot is plotted in matlab and then the geom_smooth() function is used to fit the scatter points; the optimal model is selected as the Gaussian function model by calculating the AIC value. Thus, the mapping of GCF to L is shown in Equation (5), and the fitting curve is shown in Figure 3b.

$$L(p) = fix\left[N \times e^{-\left(\frac{GCF(p)-0.5}{\alpha}\right)^2}\right] \tag{5}$$

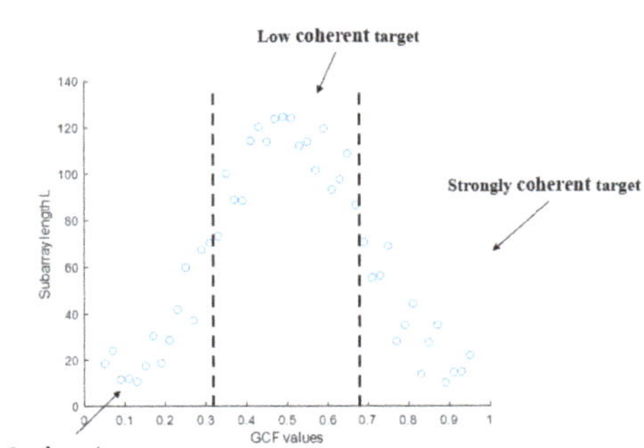

(a)

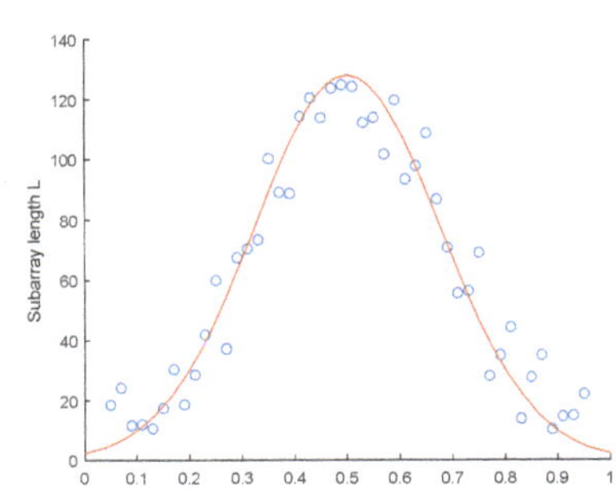

(b)

Figure 3. (**a**) The scatter plot of appropriate L for GCF. (**b**) Fitting curves for scattered points.

The fix is an integer operation, and $GCF(p)$ is derived from Equation (3), which takes values from 0 to 1, and the range of $L(p)$ is 0 to N. The value of parameter α ranges from 0 to 2. The smaller α is, the faster $L(p)$ changes, and the more sensitive the algorithm is to the detection of the target but the less robust [17]. In this paper, $\alpha = 0.2$. The cutoff frequency M_0 of the low-frequency region in GCF is selected empirically, and different selection will affect the results. In this paper, $M_0 = 10$.

Then, bring $L(p)$ into Equation (2) and replace fixed L with adaptive $L(p)$ to obtain the expression of GSTS-CF at point p:

$$\text{GSTS-CF(p)} = \frac{\sum_{k=-K}^{K}\left|\sum_{l=1}^{M-L(p)+1}\sum_{m=l}^{L(p)+l-1} x_m(p+k)\right|^2}{(M-L(p)+1)\sum_{k=-K}^{K}\sum_{l=1}^{M-L(p)+1}\left|\sum_{m=l}^{L(p)+l-1} x_m(p+k)\right|^2} \quad (6)$$

The method adaptively changes the subarray length, which gives it a stronger scatter retention capability compared with the conventional CF, but the noise reduction capability is not sufficient, thus combining it with DMAS. Weighting the output of DMAS with GSTS-CF to obtain the output expression of DMAS + GSTS-CF as:

$$y_{DMAS+GSTS-CF} = y_{DMAS} \times GSTS - CF(p) \quad (7)$$

y_{DMAS} is derived from Equation (4).

3. Simulation and Experimental Datasets

3.1. Simulated Data Set

A linear array probe with 128 elements (element width = 0.25 mm, spacing = 0.28 mm, gap = 0.03 mm) was simulated using field ii for the generation of single plane-waves. The excitation was a two-period Hanning-weighted sinusoidal excitation pulse with the central frequency set at 3 MHz. The sampling frequency was 40 MHz. Two experiments were set up in the sound field: the first was scattering points simulation and the second d was a cyst simulation. All are images displayed at a 60 dB dynamic range. The details are as follows.

In the point imaging simulation, four-point pairs are placed evenly at depths of 20 mm to 50 mm. The two points in a point pair are spaced 2 mm apart laterally. This experiment analyzes the lateral resolution of the algorithm by whether the two points in the point pair are clearly visible.

In the cyst simulation, the test model consists of 10,000 points, randomly distributed in a $40 \times 10 \times 5$ mm^3 box. Within this volume, the reflectivity has a Gaussian distribution. Embedded in this area is a cylindrical cyst of 3 mm diameter, centrally located at (x, y, z) = (0, 0, 45) mm. It is assumed that the cyst is echo-free and that the reflection coefficient of its internal scattering points is zero.

3.2. Experimental Data Set

Vivo carotid scanning experiments were carried out using a Heskell 256-channel ultrasound signal collector and RF data before beamforming was acquired using an L-15 (40 MHz) line array probe. The probe emits a single plane-wave during the experiment. The dynamic range of the images was 60 dB [17]. All subjects gave their informed consent for inclusion before they participated in the study. The study was conducted in accordance with the Declaration of Helsinki.

3.3. Image Quality Metrics

In order to quantitatively assess the performance of different beamforming methods, the following metrics are quoted in Section 4: the lateral resolution, contrast ratio (CR), contrast-to-noise ratio (CNR), and speckle SNR (sSNR) [18]. The lateral resolution is measured by calculating the full width at half maximum (FWHM, −6 dB beam width) [19] of the main lobe in a lateral direction. The CNR is used to evaluate whether the image detail is sufficiently sharp, and sSNR is used to evaluate the quality of background scatter [20]. The CR, CNR, and sSNR are defined as follows:

$$CR = 20log_{10}\left(\frac{\mu_i}{\mu_b}\right) \tag{8}$$

$$CNR = \frac{|\mu_i - \mu_b|}{\sqrt{\sigma_b^2 - \sigma_i^2}} \tag{9}$$

$$\text{sSNR} = \frac{\mu_b}{\sigma_i} \tag{10}$$

where μ_i and μ_b are the average image intensities (before logarithmic compression) of the cyst and background within a region, while σ_i^2 and σ_b^2 are the corresponding variances.

4. Results

4.1. Simulated Point Target Results

From Figure 3a,b, we can see that the artifacts of DMAS are much smaller than those of DAS, which is due to the fact that the correlation operation brings the metric of backscattered signal coherence into the beamforming process, achieving better spurious and noise rejection, but the image intensity decays as the depth increases [21–23]. The weighted image has a better contrast and resolution, which can be seen in Figure 4c–f. In the four images above, it can be found that the CF-weighted image has the sharpest resolution of point pairs and fewer surrounding artefacts. The method proposed in this paper (GSTS-CF) is the next best, and finally STS-CF is evaluated [14,24].

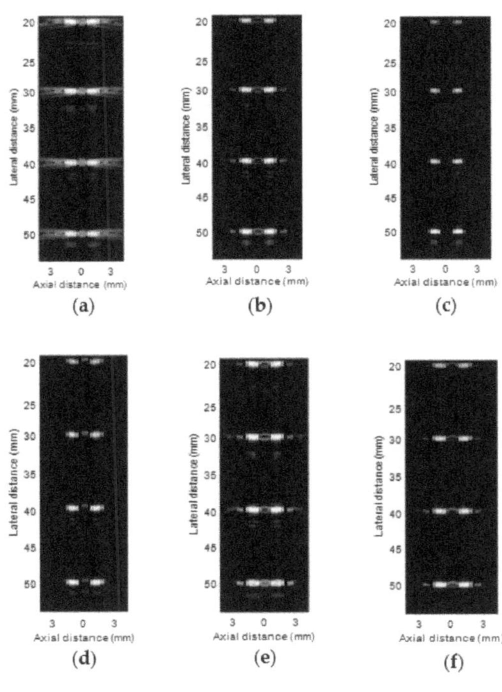

Figure 4. Simulated single plane-wave imaging with a dynamic range of 60 dB. (**a**) DAS image, (**b**) DMAS image, (**c**) DMAS weighted by the CF, (**d**) DAS weighted by the CF, (**e**) DMAS weighted by the STS-CF, (**f**) DMAS weighted by the GSTS-CF. All images are shown in a 60 dB dynamic range.

To further compare the lateral resolution of the different beamformer, we draw the lateral projections of Figure 4 in Figure 5, while calculating the FWHM of the different beamformer based on the lateral projections [25]. From Figure 5, it can be found that the combination of CF and DMAS has the narrowest main lobe width. Weighting DMAS with STS-CF also improves the lateral resolution, but compared to DMAS-CF, there is still a gap. In point imaging, the combination of DMAS and CF gives the best resolution because of less clutter interference and a high echo SNR, but in complex environments with a low SNR, the method will produce a large number of artifacts. The proposed method, which combines the improved GSTS-CF with DMAS, has a slightly worse main lobe width and side lobe amplitude than DMAS-CF, which weakens the suppression effect of CF and makes a compromise between imaging clarity and algorithmic robustness [26]. It can also be seen from Figure 5 that the GSTS-CF has a much-improved lateral resolution compared to the STS-CF, and its FWHM converges to that of the CF-weighted beamformer. Although CF-weighted beamformers have the best lateral resolution, they produce artefacts and uneven background scatter in complex environments, as will be seen in the following experiments.

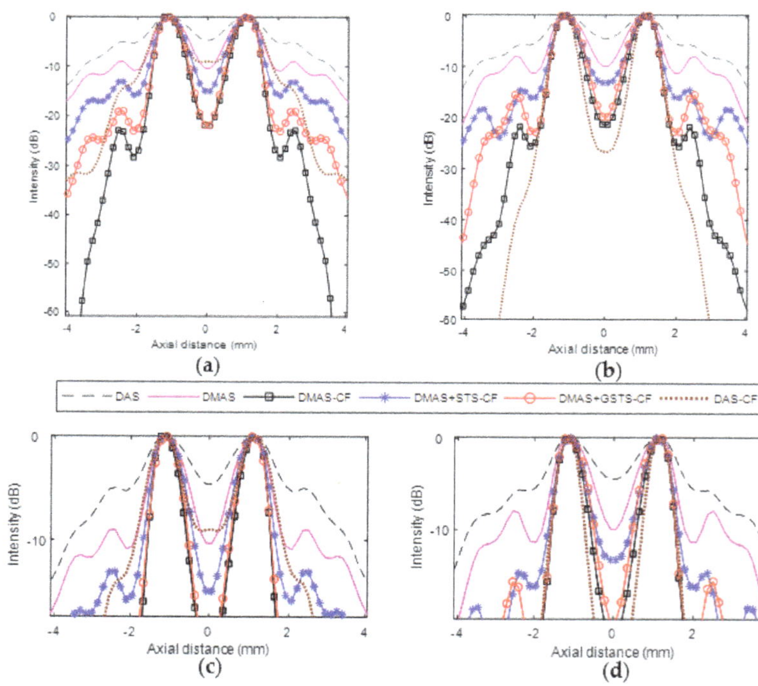

Figure 5. Lateral projections of the single plane-wave images in Figure 4. The point pairs were located at (**a**) 20 mm and (**b**) 40 mm. The corresponding zoomed-in figures are shown in (**c**,**d**).

From Table 1, we can see the variation in the lateral resolution of different beamformers. The CF-weighted beamformer has a narrower main lobe and higher lateral resolution due to the fact that CF works well in a simple scattering environment with a high signal-to-noise ratio [27]. The combination of GSTS-CF and DMAS achieves a lateral resolution similar to that of the CF-weighted beamformer, which is significantly better than DMAS as well as DMAS + STS-CF.

Table 1. The FWHM for different methods at 20 mm and 40 mm depth.

Method	FWHM (mm)	
	20 mm	40 mm
DAS	3.762	3.831
DMAS	1.032	1.157
DAS-CF	0.525	0.543
DMAS-CF	0.463	0.584
DMAS + STS-CF	0.587	1.034
DMAS + GSTS-CF	0.494	0.612

4.2. Simulated Cyst Target Result

A cyst is synthesized to compare the imaging quality of different algorithms [18]. Figure 6 shows that compared to DAS, DMAS has a clearer cyst demarcation line and lower overall gray value. This is because DMAS has a narrower main lobe and noise floor. Thus, we know that the performance of multiplying and summing is higher than summing, and using adaptive weighting for DMAS can further improve the imaging quality. In Figure 6c,d, the CF over-suppresses the signal. Although the clutter in the cyst was effectively removed, the background region was also over-suppressed, resulting in a lower average level of background intensity (see Table 1) and an increase in background scatter variance, along with a large number of artefacts [19]. It is also seen that the imaging reconstruction in Figure 6c is biased. Because the off-axis signal amplitude is much larger than the true signal amplitude when imaging in the region near the highlighted scatterer, using CF directly would only result in incorrect imaging reconstruction [28]. By making a comparison, we can see that DMAS + STS-CF and DMAS + GSTS-CF have clear cyst demarcation lines, effectively eliminating background region artifacts and suppressing clutter in the cyst region. The method proposed in this paper adaptively evaluates the subarray length with GCF, and it has a more uniform background and lower scatter variance than DMAS + STS-CF [29].

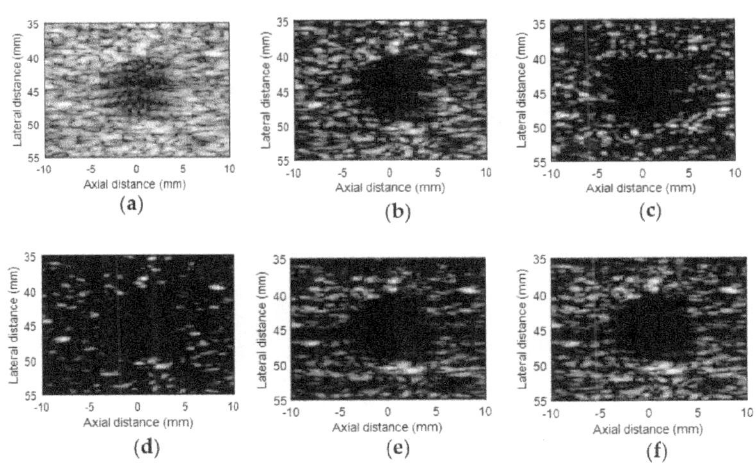

Figure 6. Single plane-wave images of the computer-generated cyst phantom reconstructed using (**a**) DAS, (**b**) DMAS, (**c**) DAS+CF, (**d**) DMAS + CF, (**e**) DMAS + STS-CF, (**f**) DMAS + GSTS-CF. All images are shown in a 60 dB dynamic range.

Lateral cross-sections through the cyst target in the simulated images are shown in Figure 7. It can be seen that DMAS-CF and DMAS + GSTS-CF have the lowest average grey values, indicating that the internal clutter of the cyst is effectively removed [30]. The greatest variation is seen at the cyst demarcation line, indicating that the two methods have the clearest boundaries. Table 2 presents the results of the evaluation of the parameters

σ_b^2, CR, CNR, and sSNR, and the areas used to calculate these indicators are shown in the rectangle in Figure 6 [31]. It can be seen that DMAS-CF has the highest CR and the lowest CNR and sSNR, while DAS has the lowest CR and the highest CNR and sSNR. This indicates that the normal DAS has the best ability to preserve scatter quality despite its low CR; however, neither method has the best CR, CNR, and sSNR at the same time. The DMAS + GSTS-CF is a compromise between the CR and CNR and sSNR. Its CR is comparable to the DMAS-CF, and CNR and sSNR are comparable to the DAS, while it can be seen that the GSTS-CF outperforms the STS-CF. For σ_b^2, DMAS + GCTS-CF is also low, just above DAS. This indicates a more uniform background scatter.

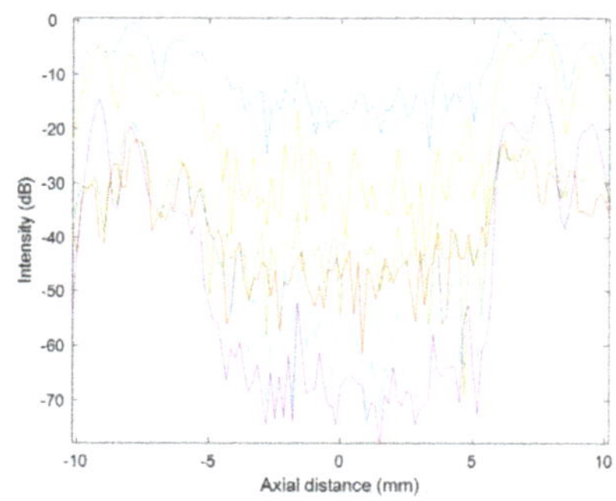

Figure 7. Lateral cross-sections through the cyst target in the simulated images.

Table 2. The cyst average intensity, background average intensity, CR, CNR, and sSNR of cysts for different methods.

Method	μ_i	μ_b	CR (dB)	CNR	sSNR	σ_b^2
DAS	2.3838×10^{-4}	3.60×10^{-3}	−12.7348	1.7998	1.9323	0.0173
DMAS	1.7782×10^{-4}	7.09×10^{-4}	−16.6694	1.2375	1.4648	0.0698
DAS-CF	3.2876×10^{-5}	6.03×10^{-4}	−20.2362	0.9213	1.1923	0.0876
DMAS-CF	4.2484×10^{-7}	3.03×10^{-5}	−32.9846	0.7694	0.9386	0.0945
DMAS + STS-CF	5.9576×10^{-5}	3.09×10^{-4}	−21.6455	1.3393	1.5328	0.0583
DMAS + GSTS-CF	1.1661×10^{-6}	5.03×10^{-4}	−29.3099	1.5420	1.7632	0.0352

4.3. Carotid Artery Experiment

Experiments were performed on human carotid artery scans, and the scanning site and ultrasound signal acquisition device are shown in Figure 8. The RF data were processed using a series of beamforming methods and represented on a 60 dB dynamic range. The model diagram of the imaging site and the final imaging results are shown in Figure 9.

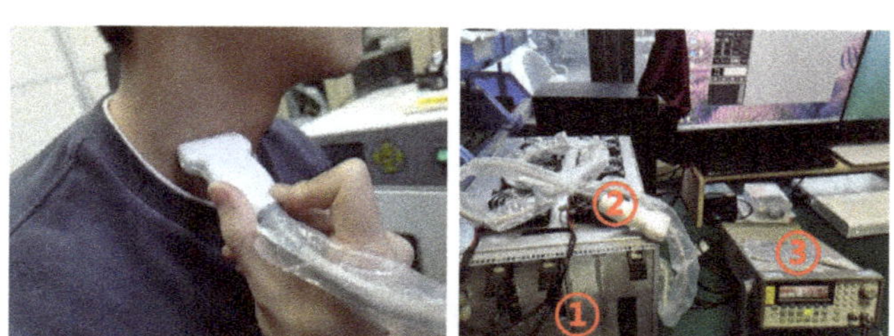

(a) **(b)**

Figure 8. (**a**) Location of the collected carotid artery data. (**b**) Ultrasound signal acquisition device. ① RF receiver and transmitter circuits, ② line array probe, ③ signal generator.

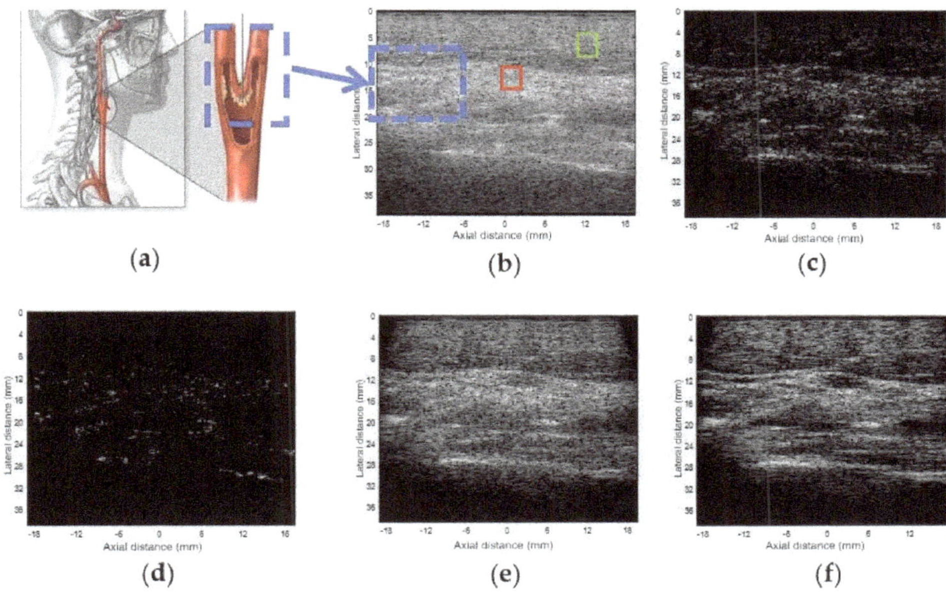

Figure 9. Carotid artery model maps and plane-wave imaging results. The green and red boxes indicate the areas where data μ_i and μ_b were collected (**a**) Carotid artery model maps, (**b**) DAS, (**c**) DMAS, (**d**) DMAS + CF, (**e**) DMAS + STS-CF, (**f**) DMAS + GSTS-CF.

Compared with the simulation, this experimental object has a more complex structure and more noise disturbances. In addition to the coherent noise of the echoes considered in the simulation, there are a series of conditions affecting the signal quality, such as phase distortion caused by the different transmission media and signal distortion caused by the limited performance of the acquisition device. Therefore, to improve the imaging quality of the experiment, we need a more precise signal acquisition device with the assistance of interpolation processing and ultrasonic image denoising technology [32]. Although all of this experimental imaging has some speckle noise, the analysis of the performance of different algorithms is not affected.

It can be seen from Figure 9 that DMAS brings about a higher contrast ratio, but at the same time it results in artifacts in the background, and the overall quality of the image becomes darker. DMAS-CF suppresses the signal excessively, leading to image reconstruction errors, while DMAS + GSTS-CF has a more uniform background area, and the demarcation line between the vessel wall and the lumen can be seen more clearly. At the same time, there are fewer artifacts in the echoless region inside the lumen. Compared to the over-suppression of DMAS + CF, DMAS + GSTS-CF provides higher algorithmic robustness, suppresses artifacts within the vessel lumen, and results in more uniform tissue in the perivascular region. In contrast, although DMAS + STS-CF had similar effects, its performance was lower than that of DMAS + GSTS-CF.

The μ_{lumen} CR, CNR, and sSNR obtained by the different methods are given in Table 3, and the regions used to estimate these metrics are marked with rectangular boxes in Figure 9 (μ_{lumen} is the average value of the vascular lumen region, which reflects the ability of clutter suppression in the vascular lumen). In a complex scattering environment, the CF-weighted beamformer has a very low CNR and sSNR, while the DAS has the lowest CR. The GSTS-CF well balances the CR, CNR, and sSNR, suggesting that the GSTS-CF can preserve scatter patterns well and achieve a high contrast.

Table 3. Average intensity of the carotid lumen and sSNR of the scattered area.

Method	μ_{lumen}	CR	CNR	sSNR	σ_b^2
DAS	7.8059×10^4	−11.324	2.0238	2.9260	1.58
DMAS	4.6285×10^4	−19.736	1.2876	2.0006	2.87
DMAS-CF	3.2523×10^3	−33.254	0.2375	1.5832	5.47
DMAS + STS-CF	3.3379×10^4	−25.232	1.3236	2.4815	2.53
DMAS + GSTS-CF	1.7438×10^4	−28.765	1.6342	2.8156	1.96

5. Discussion

In this paper, we use a coherence factor to improve the performance of DMAS, and our motivation is to achieve a trade-off between lateral resolution and scattering retention performance. Compared with DMAS + CF, the proposed method better preserves the scatter pattern without significantly reducing the lateral resolution. Compared with DAS, the proposed method greatly improves the lateral resolution and contrast while having an approximate background pattern. GSTS-CF is essentially an improved STS-CF who uses GCF as a local coherence detection tool and adaptively selects the appropriate subarray length to conduct spatial smoothing. The combination of GSTS-CF and DMAS further improves the image quality.

Tables 1–3 show that the proposed method improves lateral resolution (20 mm depth) by 86.87% compared to DAS, 52.13% compared to DMAS, 15.84% compared to DMAS + STS-CF, and has a full width at half maxima (FWHM), similar to DMAS-CF. The proposed method improves the speckle signal-to-noise ratio (sSNR) by 87.85% (simulation) and 77.84% (in carotid) compared to DMAS-CF, 20.37% (simulation) and 40.74% (in carotid) compared to DMAS, 15.03% (simulation) and 13.46% (in carotid) compared to DMAS + STS-CF, and has sSNR and scatter variance similar to DAS.

Because the subarray length L for GSTS-CF is estimated by GCF, the performance of the proposed method is influenced by the M_0 (cutoff frequency in the molecule of GCF). As can be seen from Equation (1), the GCF increases with the increase in M_0, leading to the change in $L(p)$ in Equation (4). From Figure 3, it can be seen that in incoherent target, $L(p)$ increases with the increase in GCF, which leads to a slight decrease in lateral resolution but a more uniform scattering area. In strongly coherent target, $L(p)$ decreases with the increase in GCF, leading to an improvement in the lateral resolution but a decrease in the scattering quality. In low-coherent targets, the change in GCF has little effect on the results because the curve of GCF values mapping to the length of subarray $L(p)$ changes slowly in this region. In clinical applications, we can change the value of M0 according to different environments, and the default of M_0 in this paper is 10.

During carotid data acquisition, the SNR of the signal is much lower than during cyst simulation, especially when receiving echoes from deeper regions with greater signal attenuation. In Figure 9, the images generated by DMAS + CF may not be suitable for clinical applications because of the significantly lower amplitude levels in the background and the loss of texture information. This is due to the fact that each channel signal has different amounts of correlated noise and interference, and they have different SNRs. Therefore, the weighting factor varies widely, and these artifacts may appear.

DMAS + STS-CF uses the spatio-temporally smoothed method, which enhances the robustness to noise interference and side lobe interference in coherent measurements. As a result, there is a lower scattering variance and higher sSNR, but the lateral resolution is reduced. However, the artifacts of DMAS + STS-CF are more severe in the carotid experiment (Figure 9) than in the cyst simulation (Figure 6). Because the choice of L is fixed, it is necessary to choose a different L for different scenarios in clinical applications, which affects the performance of the algorithm.

In Figure 9f, DMAS + GSTS-CF has higher robustness in complex environments. Compared to DMAS + STS-CF, the method estimates a value of $L(p)$ with GCF at each imaging point, which significantly removes these artifacts by improving the scatter pattern (lower scatter variance and higher sSNR) while maintaining the clutter rejection capability (lower mean value (μ_{lumen}) in the vessel lumen).

Since the adaptive subarray length is estimated by the GCF, the performance of the proposed method is affected by the cut-off frequency M0, and therefore in clinical applications there may be drawbacks, such as noise, clutter, or other types of artefacts, if the parameters are not set correctly. Nevertheless, the proposed method also has potential in clinical applications. This is because the proposed method allows for a flexible selection of the subarray length L according to the echo target and thus improves the lateral resolution along with speckle protection. Thus, it may have potential for applications in the heart, carotid artery, thyroid, tumors, etc. [33–36]. Using the GCF to estimate the subarray length $L(p)$ leads to a significant increase in computational effort because of the large number of Fourier transforms in the GCF. However, graphics processing unit calculations have been used to accelerate these beamformers for real-time imaging and thus can be used in the method proposed in this paper to improve computational efficiency.

6. Conclusions

To improve the speckle quality (lower scatter variance and higher sSNR) without significantly reducing the lateral resolution, we propose an adaptive spatio-temporally smoothed coherence factor called GSTS-CF and combine it with DMAS. The simulation and experimental results show that the method can obtain better background scattering without affecting the lateral resolution. The algorithm is more robust and more suitable for clinical applications.

Author Contributions: In this paper, Z.G. is responsible for methodology and software. X.G. is responsible for data curation and investigation. F.Y. is responsible for visualization, writing—original draft preparation, and validation. Y.Z. is responsible for writing—reviewing and conceptualization. L.L., C.Z. and Y.W. are responsible for supervision and editing. All team members participated in the article work, and there were no conflicts of interest between the teams. All authors have read and agreed to the published version of the manuscript.

Funding: This research was funded by Sichuan Science and Technology Major Project (No.2022ZDZX0033) and the Key Research Pro-gram of the Chinese Academy of Sciences (No. ZDRW-ZS-2021-1).

Data Availability Statement: Data sharing is not applicable to this article.

Conflicts of Interest: There are no conflict of interest among the authors.

References

1. Wells, P.N.T. Ultrasonics in medicine and biology. *Phys. Med. Biol.* **1977**, *22*, 629–669. [CrossRef] [PubMed]
2. Tanter, M.; Fink, M. Ultrafast imaging in biomedical ultrasound. *IEEE Trans. Ultrason. Ferroelectr. Freq. Control* **2014**, *61*, 102–119. [CrossRef] [PubMed]
3. Montaldo, G.; Tanter, M.; Bercoff, J.; Benech, N.; Fink, M. Coherentplane-wave compounding for very high frame rate ultrasonography and transient elastography. *IEEE Trans. Ultrason. Ferroelectr. Freq. Control* **2009**, *56*, 489–506. [CrossRef] [PubMed]
4. Matrone, G.; Savoia, A.S.; Caliano, G.; Magenes, G. The delay multiply and sum beamforming algorithm in ultrasound B-mode medical imaging. *IEEE Trans. Med. Imaging* **2015**, *34*, 940–949. [CrossRef] [PubMed]
5. Matrone, G.; Ramalli, A.; Tortoli, P.; Magenes, G. Experimental evaluation of ultrasound higher order harmonic imaging with filtered delay multiply and sum (F-DMAS) non-linear beamforming. *Ultrasonics* **2018**, *86*, 59–68. [CrossRef] [PubMed]
6. Synnevåg, J.F.; Austeng, A.; Holm, S. Adaptive beamforming applied to medical ultrasound imaging. *IEEE Trans. Ultrason. Ferroelectr. Freq. Control* **2007**, *54*, 1606–1613. [CrossRef]
7. Synnevåg, J.F.; Austeng, A.; Holm, S. A low-complexity data-dependent beamformer. *IEEE Trans. Ultrason. Ferroelectr. Freq. Control* **2010**, *57*, 281–289.
8. Hollman, K.W.; Rigby, K.W.; O'donnell, M. Coherence factor of speckle from a multi-row probe. In Proceedings of the 1999 IEEE Ultrasonics Symposium, Tahoe, NV, USA, 17–20 October 1999; pp. 1257–1260.
9. Nilsen, C.I.C.; Holm, S. Wiener beamforming and the coher-ence factor in ultrasound imaging. *IEEE Trans. Ultrason. Ferroelectr. Freq. Control* **2010**, *57*, 1329–1346. [CrossRef]

10. Li, P.C.; Li, M.L. Adaptive imaging using the generalized co-herence factor. *IEEE Trans. Ultrason. Ferroelectr. Freq. Control* **2003**, *50*, 128–141.
11. Xu, M.; Yang, X.; Ding, M.; Yuchi, M. Spatio-temporally Smoothed Coherence Factor for Ultrasound Imaging. *IEEE Trans. Ultrason. Ferroelectr. Freq. Control* **2014**, *61*, 182–190. [CrossRef]
12. Shan, T.J.; Wax, M.; Kailath, T. On spatial smoothing for direction-of-arrival estimation of coherent signals. *IEEE Trans. Acoust. Speech Signal Process.* **1985**, *33*, 806–811. [CrossRef]
13. Lan, Z.; Jin, L.; Feng, S. Joint Generalized Coherence Factor and Minimum Variance Beamformer for Synthetic Aperture. *IEEE Trans. Ultrason. Ferroelectr. Freq. Control* **2021**, *68*, 1167–1183
14. Varray, F.; Kalkhoran, M.A.; Vray, D. Adaptive minimum variance coupled with sign and phase coherence factors in IQ domain for plane wave beamforming. In Proceedings of the International Ultrasonic Symposium (IUS), Tours, France, 18–21 September 2016; pp. 1–4.
15. Behar, V.; Adam, D.; Friedman, Z. A new method of spatial compounding imaging. *Ultrasonics* **2003**, *41*, 377–384. [CrossRef] [PubMed]
16. Wang, Y.; Li, P. SNR-Dependent Coherence-Based Adaptive Imaging for High-Frame-Rate Ultrasonic and Photoacoustic Imaging. *IEEE Trans. Ultrason. Ferroelectr. Freq. Control* **2014**, *61*, 1419–1432. [CrossRef] [PubMed]
17. Wu, X.; Gao, Q.; Lu, M. An improved spatio-temporally smoothed coherence factor combined with eigenspace-based minimun variance beamformer for plane-wave imaging in medical ultrasound. In Proceedings of the 2017 IEEE International Ultrasonics Symposium (IUS), Washington, DC, USA, 6–9 September 2017.
18. Wagner, R.F.; Insana, M.F.; Smith, S.W. Fundamental correlation lengths of coherent speckle in medical ultrasonic images. *IEEE Trans. Ultrason. Ferroelectr. Freq. Control* **1988**, *35*, 34–44. [CrossRef] [PubMed]
19. Synnevåg, J.F.; Nilsen, C.I.C.; Holm, S. P2B-13 Speckle statistics in adaptive beamforming. In Proceedings of the 2007 IEEE Ultrasonics Symposium, New York, NY, USA, 28–31 October 2007; pp. 1545–1548.
20. Matone, G.; Ramalli, A.; D'hooge, J. Spatial Coherence Based Beamforming in Multi-Line Transmit Echocardiography. In Proceedings of the 2018 IEEE International Ultrasonics Symposium (IUS), Kobe, Japan, 22–25 October 2018. [CrossRef]
21. Synnevåg, J.-F.; Austeng, A.; Holm, S. A low-complexity data dependent beamformer. *IEEE Trans. Ultrason. Ferroelectr. Freq. Control* **2011**, *58*, 281–289. [CrossRef] [PubMed]
22. Synnevag, J.-F.; Austeng, A.; Holm, S. Benefits of minimumvariance beamforming in medical ultrasound imaging. *IEEE Trans. Ultrason. Ferroelectr. Freq. Control* **2009**, *56*, 1868–1879. [CrossRef]
23. Zimbico, J.; Granado, D.W.; Schneider, F.K.; Maia, J.M.; Assef, A.A.; Schiefler, N., Jr.; Costa, E.T. Eigenspace generalized sidelobe canceller combined with SNR dependent coherence factor for plane wave imaging. *Biomed. Eng. Online* **2018**, *17*, 109. [CrossRef]
24. Asl, B.M.; Mahloojifar, A. Minimum variance beamforming combined with adaptive coherence weighting applied to medical ultrasound imaging. *IEEE Trans. Ultrason. Ferroelectr. Freq. Control* **2009**, *56*, 1923–1931. [CrossRef]
25. Zhao, J.; Wang, Y.; Yu, J.; Guo, W.; Li, T.; Zheng, Y.-P. Subarray coherence based postfilter for eigenspace based minimum variance beamformer in ultrasound plane-wave imaging. *Ultrasonics* **2016**, *65*, 23–33. [CrossRef]
26. Deylami, A.M.; Jensen, J.A.; Asl, B.M. An improved minimum variance beamforming applied to plane-wave imaging in medical ultrasound. In Proceedings of the 2016 IEEE International Ultrasonics Symposium (IUS), Tours, France, 18–21 September 2016; pp. 1–4.
27. Qi, Y.; Wang, Y.; Yu, J.; Guo, Y. 2-D Minimum Variance Based Plane Wave Compounding with Generalized Coherence Factor in Ultrafast Ultrasound Imaging. *Sensors* **2018**, *18*, 4099. [CrossRef] [PubMed]
28. Asl, B.M.; Mahloojifar, A. Contrast enhancement and robustness improvement of adaptive ultrasound imaging using forward-backward minimum variance beamforming. *IEEE Trans. Ultrason. Ferroelectr. Freq. Control* **2011**, *58*, 858–867. [CrossRef] [PubMed]
29. Zhang, C.; Geng, X.; Yao, F.; Liu, L.; Guo, Z.; Zhang, Y.; Wang, Y. The Ultrasound Signal Processing Based on High-Performance CORDIC Algorithm and Radial Artery Imaging Implementation. *Appl. Sci.* **2023**, *13*, 5664. [CrossRef]
30. Ali, I.; Saleem, M.T. Spatiotemporal Dynamics of Reaction–Diffusion System and Its Application to Turing Pattern Formation in a Gray–Scott Model. *Mathematics* **2023**, *11*, 1459. [CrossRef]
31. Kaddoura, T.; Zemp, R.J. Hadamard Aperiodic Interval Codes for Parallel-Transmission 2D and 3D Synthetic Aperture Ultrasound Imaging. *Appl. Sci.* **2022**, *12*, 4917. [CrossRef]
32. Khan, S.U.; Ali, I. Application of Legendre spectral-collocation method to delay differential and stochastic delay differential equation. *AIP Adv.* **2018**, *8*, 035301. [CrossRef]
33. Rindal, O.M.H.; Aakhus, S.; Holm, S.; Austeng, A. Hypothesis of improved visualization of microstructures in the interventricular septum with ultrasound and adaptive beamforming. *Ultrasound Med. Biol.* **2017**, *43*, 2494–2499. Available online: http://www.sciencedirect.com/science/article/pii/S0301562917302466 (accessed on 5 May 2018). [CrossRef]
34. Nguyen, N.Q.; Prager, R.W. Minimum variance approaches to ultrasound pixel-based beamforming. *IEEE Trans. Med. Imaging* **2017**, *36*, 374–384. [CrossRef]

35. Qi, Y.; Wang, Y.; Guo, W. Joint subarray coherence and minimum variance beamformer for multitransmission ultrasound imaging modalities. *IEEE Trans. Ultrason. Ferroelectr. Freq. Control* **2018**, *65*, 1600–1617. [CrossRef]
36. Szasz, T.; Basarab, A.; Kouame, D. Beamforming through regularized inverse problems in ultrasound medical imaging. *IEEE Trans. Ultrason. Ferroelectr. Freq. Control* **2016**, *63*, 2031–2044. [CrossRef]

electronics

MDPI

Article

Underwater AUV Navigation Dataset in Natural Scenarios

Can Wang, Chensheng Cheng, Dianyu Yang, Guang Pan and Feihu Zhang *

School of Marine Science and Technology, Northwestern Polytechnical University, Xi'an 710072, China; can.wang@mail.nwpu.edu.cn (C.W.); chensheng.cheng@mail.nwpu.edu.cn (C.C.); 2019100496@mail.nwpu.edu.cn (D.Y.); panguang@nwpu.edu.cn (G.P.)
* Correspondence: feihu.zhang@nwpu.edu.cn; Tel.: +86-029-88492611

Abstract: Autonomous underwater vehicles (AUVs) are extensively utilized in various autonomous underwater missions, encompassing ocean environment monitoring, underwater searching, and geological exploration. Owing to their profound underwater capabilities and robust autonomy, AUVs have emerged as indispensable instruments. Nevertheless, AUVs encounter several constraints in the domain of underwater navigation, primarily stemming from the cost-intensive nature of inertial navigation devices and Doppler velocity logs, which impede the acquisition of navigation data. Underwater simultaneous localization and mapping (SLAM) techniques, along with other navigation approaches reliant on perceptual sensors like vision and sonar, are employed to augment the precision of self-positioning. Particularly within the realm of machine learning, the utilization of extensive datasets for training purposes plays a pivotal role in enhancing algorithmic performance. However, it is common for data obtained exclusively from inertial sensors, a Doppler Velocity Log (DVL), and depth sensors in underwater environments to not be publicly accessible. This research paper introduces an underwater navigation dataset derived from a controllable AUV that is equipped with high-precision fiber-optic inertial sensors, a DVL, and depth sensors. The dataset underwent rigorous testing through numerical calculations and optimization-based algorithms, with the evaluation of various algorithms being based on both the actual surfacing position and the calculated position.

Keywords: AUV; underwater navigation; dataset; inertial navigation; DVL

Citation: Wang, C.; Cheng, C.; Yang, D.; Pan, G.; Zhang, F. Underwater AUV Navigation Dataset in Natural Scenarios. *Electronics* **2023**, *12*, 3788. https://doi.org/10.3390/electronics12183788

Academic Editors: Wentao Li, Huiyan Zhang, Tao Zhan and Chao Zhang

Received: 8 August 2023
Revised: 29 August 2023
Accepted: 1 September 2023
Published: 7 September 2023

1. Introduction

In recent years, AUVs have garnered increasing attention in underwater exploration and military missions [1]. Due to the attenuation of underwater electromagnetic signals, AUVs rely on inertial sensors and acoustic devices for navigation instead of GPS [2,3]. Specifically, for stand-alone strapdown inertial navigation systems (SINS), the estimation of relative velocity and position involves the integration of accelerometer and gyroscope sensor data, which introduces errors and leads to significant drift in the estimated position and velocity [4]. The introduction of a DVL assists navigation by measuring bottom-to-water relative velocity to improve positioning accuracy, but prolonged error accumulation remains a challenge [5]. Navigation methods based on other perceptual sensors, such as vision and sonar, have been applied in specific scenarios but mostly remain in the laboratory stage [6,7]. Limitations in sensing conditions, such as vision being unsuitable in turbid environments and limited acoustic sensing resolution, make perception-based navigation non-universal for underwater applications. In contrast, inertial navigation combined with DVLs and depth sounders has become a mature technology widely used in underwater navigation solutions.

In contrast to indoor settings, the navigation of underwater environments poses significant challenges on a large scale. Consequently, the integration of inertial and Doppler Velocity Log (DVL) navigation techniques can be effectively employed in various scenarios. However, the utilization of optical devices and navigation applications based on structured environments is restricted due to the turbidity and intricacy inherent in natural underwater environments [8]. Concurrently, the high cost of high-precision inertial sensors and sound velocity measurement devices restricts the application and data collection of small AUVs [9]. In particular, the cost of common fiber-optic inertial navigation systems can reach tens of thousands of dollars, making them impractical for small teams. In such cases, the availability of underwater high-precision navigation data allows researchers to analyze the generation and propagation of AUV navigation errors more profoundly and devise strategies to mitigate any potential errors or limitations. This will provide a research foundation for various fields, including marine biology, geology, environmental monitoring, and defense operations [10].

Currently, the collection of AUV navigation data in natural underwater environments is concentrated in authoritative experimental institutions, such as the Naval Surface Warfare Center (NSWC) [11], the Naval Undersea Warfare Center (NUWC) [12], the European Research Agency [13], and the Australian military, etc. [14]. The collection of pertinent data necessitates the utilization of shore-based platforms or motherships, and in intricate environments, human divers are also employed to facilitate navigation, thereby resulting in substantial costs. The range of underwater sensors is limited by the environment, especially in visible light, where active light sources are subject to forward and backward scattering [15]. Acoustic-based forward-looking sonar and side-scan sonar (SSS) are widely used for underwater environment sensing and terrain-matching navigation [16]. Therefore, the collection and organization of underwater datasets will be a high-cost, complex task, and the number of existing publicly available datasets is very small, which reflects this paper's significance in this work.

In this paper, we present a novel dataset to expand research in underwater navigation. The uniqueness lies in the data sourced from multiple sensors, including high-quality inertial navigation systems (INS) and Differential Global Positioning Systems (DGPS), as well as synchronized DVLs and depth sounder data. In particular, all data are collected in the natural environment, including lakes, reservoirs, and offshore areas. Moreover, the data are generated during the autonomous navigation of the AUV, meaning that the navigation data conform to the kinematics of the vehicle. To the best of our knowledge, this is the first publicly released AUV lake/ocean navigation dataset based on high-precision sensors. The dataset is accessible via the following link: https://github.com/nature1949/AUV_navigation_dataset (accessed on 8 June 2023).

In summary, the main contributions of this article are as follows:

- Presentation of a substantial amount of underwater high-precision navigation data, covering approximately 147 km;
- Collection of data from real scenarios in three different regions, encompassing diverse trajectories and time spans;
- Introduction of navigation challenges in underwater environments and the proposed methods based on dead reckoning and collaborative localization, evaluated against our benchmark.

The paper is structured as follows: Section 2 describes the research foundation and current status of underwater navigation, as well as the characteristics and limitations of publicly available datasets for underwater navigation. Section 3 describes the platforms and sensors used for data acquisition, as well as the acquisition process. Section 4 describes the dataset structure and typical trajectories and tests the dataset by common methods. A discussion of the results and data is carried out in Section 5 and finally summarized in Section 6.

2. Related Work

2.1. Underwater Navigation Methods

Typically, AUVs employ inertial navigation combined with acoustics for collaborative navigation, while ROVs, due to their limited mobility, additionally use visually and acoustically aided navigation. Positioning algorithms often apply filtering or optimization methods, including traditional EKF, UKF, and the latest SLAM techniques, among others.

For instance, Harris et al. developed an AUV position estimation algorithm using the ensemble Kalman filter (EnKF) and fuzzy Kalman filter (FKF), which avoids linearization of the AUV's dynamics model [5]. Jin et al. proposed a single-source assisted passive localization method that combines acoustic positioning with inertial navigation and concluded that time difference of arrival (TDOA) + AOA yields better results [17]. This method utilizes fixed sound sources to periodically emit sound pulses underwater and locate the source using a TDOA positioning technique.

Jorgensen et al. based their approach on the XKF principle and constructed an observer for estimating position, velocity, attitude, underwater wave speed, rate sensor, and accelerometer biases, which demonstrated stability and achieved near-noise optimal performance [18]. Wang et al. integrated depth information into two-dimensional visual images and proposed an online fusion method based on tightly coupled nonlinear optimization to achieve continuous and robust localization in rapidly changing underwater environments [19]. Manderson et al. presented an efficient end-to-end learning approach for training navigation strategies using visual data and demonstrated autonomous visual navigation over a distance of more than one kilometer [20]. Machine learning-based approaches require massive amounts of training data, which highlights the importance of collecting underwater navigation data to enhance the performance of navigation algorithms.

2.2. Underwater Natural Scene Datasets

Publicly available underwater natural scene datasets are continuously released and used for underwater navigation, 3D reconstruction, underwater target recognition, and scene perception. Singh et al. released a marine debris dataset for forward-looking sonar semantic segmentation, which contains typical marine debris segmentation grayscale maps [21]. Zhou et al. constructed a common target detection dataset for sonar image detection and classification, which contained targets such as underwater shipwrecks, wrecks of crashed airplanes, victims, etc. [22]. Zhang et al. disclosed a homemade sonar common target detection dataset and evaluated the performance of a self-trained AutoDL detector [23]. Hou et al. published real side-scan sonar image datasets containing images of different classes of undersea targets and proposed a semi-synthetic data generation method that combines image segmentation with simulation of intensity distributions in different regions using optical images as input [24].

Given the difficulty of acquiring underwater datasets, there are also many researchers investigating deep neural network-based enhancement methods for underwater data. Chang et al. proposed a real-world underwater image dataset (UIDEF) containing multiple degradation types and different shooting perspectives and a color-contrast-complementary image enhancement framework consisting of adaptive chromatic balancing and multiscale weighted fusion [25]. Yin et al. proposed an underwater image restoration (UIR) method based on a convolutional neural network (CNN) architecture and a synthetic underwater dataset that can realize the direct restoration of degraded underwater images [26]. Chen et al. created class-balanced underwater datasets capable of generating underwater datasets with various color distortions and haze effects, and generated class-balanced underwater datasets from the open competition underwater dataset URPC18 via the class-based style enhancement (CWSA) algorithm [27]. Polymenis et al. used object images taken in a controlled laboratory environment to generate underwater images by generative adversarial networks (GANs) in combination with images featuring the underwater environment [28]. Boittiaux et al. provided image datasets from multiple visits to the same hydrothermal vent edifice and estimated camera poses and scenes from navigation data and motion structures [29]. Extending underwater data through data augmentation methods is a future research direction, but original underwater datasets of natural scenes are still indispensable.

2.3. Underwater Navigation Datasets

The collection of underwater data presents challenges due to costs, technical requirements, and limitations imposed by the underwater environment. Nevertheless, an increasing number of research teams have released datasets related to AUV autonomous navigation, with a focus on easily obtainable visual information. The establishment of these datasets has facilitated the development of AUV technologies, particularly in underwater target recognition and underwater SLAM techniques. For instance, Cheng et al. provided data collected in inland waterways using a stereo camera, LiDAR system, global positioning system antenna, and inertial measurement unit [30]. Song et al. obtained a millimeter-precision underwater visual-inertial dataset through a motion capture system, but the data were acquired in a laboratory setting [31]. Tomasz et al. introduced an underwater visual navigation SLAM dataset that includes ground truth tracking of vehicle positions obtained through underwater motion capture [32].

Martin et al. offered canoe attitude and stereo camera data collected in natural river environments [33]. Panetta et al. presented the Underwater Object Tracking (UOT100) benchmark dataset, which comprises 104 underwater video sequences and over 74,000 annotated frames from natural and artificial underwater videos with various distortions [34]. Angelos et al. provided data collected by AUVs in complex underwater cave systems, particularly equipped with two mechanically scanned imaging sonars [35]. Notably, Kristopher et al. simulated AUV data with advanced sensors by equipping a ground vehicle with two multibeam sonars and a set of navigation sensors [36]. More recently, Maxime et al. collected ROS data for underwater SLAM using a monocular camera, an inertial measurement unit (IMU), and other sensors in harbors and archaeological sites [37]. Li et al. presented an adaptive AUV-assisted ocean current data collection strategy, formulating an optimization problem to maximize the VoI energy ratio, thereby reducing AUV energy consumption and ensuring timely data acquisition [38].

However, the existing datasets primarily concentrate on underwater vision and are obtained from natural environments or created through data augmentation. These datasets are primarily utilized for various applications, including underwater image recognition, underwater 3D reconstruction, and visual/visual-inertial SLAM. However, the availability of independent datasets specifically focused on underwater inertial/DVL navigation remains limited. Therefore, this paper aims to address this gap by compiling AUV navigation data gathered from diverse natural scenarios and presenting it in an enhanced KITTI format, facilitating the extraction of algorithmic data for general purposes.

3. Data Acquisition

3.1. Platform

We used a 325 mm diameter AUV as the acquisition platform and collected data through different trajectories at different times and locations to achieve a diversified data type and a more representative sample set. The platform was equipped with high-precision inertial navigation, differential RTK, DVL, depth finder, and other sensors. The computing platform used a customized motherboard, which allowed different devices access and provided high-speed computing power. The platform structure and sensor layout are shown in Figure 1. The perception and navigation sensors were fixed on the vehicle and could be associated with rigid body transformations, but the provided data were obtained through rotational transformations based on their own sensors.

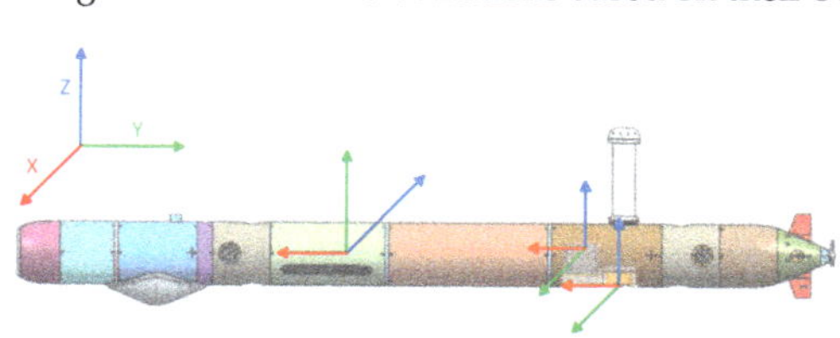

Figure 1. Schematic diagram of AUV structure and sensor layout.

3.2. Sensor

This section introduces the hardware and software used for data collection, including navigation sensors, DVL, depth sensors, and other payloads. These components work in harmony to capture comprehensive and accurate underwater navigation data. The high-precision fiber-optic inertial navigation system performs inertial measurements, provides six-axis angular velocity and linear acceleration, and has the internal potential for satellite and Doppler fusion. The DVL features a four-phase beam array, allowing it to calculate the vehicle's velocity relative to the water independently. This additional velocity information contributes to a more comprehensive understanding of the AUV's motion and aids in precise positioning during underwater navigation. Table 1 lists the complete hardware.

Table 1. Overview of platform and sensor specifications and performance.

System	Parameter	Performance	Coordinate System
AUV	Rated depth Weight Size Maximum speed	100 m 200 kg 10.5×1.06 ft 8 Kn	Forward, up, right
Inertial sensor	Heading alignment accuracy Gyro zero deviation stability Accelerometer zero offset stability	$0.05°(1\sigma)$ $\leq 0.01°/h$ ≤ 30 μg (1σ)	Forward, left, up
DVL	Frequency Velocity accuracy Altitude	300 kHz $0.5\% \pm 0.3$ cm/s 3–200 m	Forward, left, up
Depth sensor	Pressure range Output range Output zero point Repeatability	0–5 Pa 4 V $\pm$ 1% 1 V $\pm$ 1% of span $\pm$0.25 range percent	N.A.

Through the utilization of this sophisticated hardware configuration, we guarantee the attainment of sensor data of exceptional quality, thereby enabling a precise and comprehensive examination of underwater navigation efficacy. The amalgamation of this dataset with the robust and accurate hardware employed presents researchers with a significant resource for the purpose of benchmarking algorithms and assessing navigation performance.

3.3. Data Collection

This dataset was collected from April 2021 to August 2022, spanning 12.63° of longitude and 11.22° of latitude. The data mainly include underwater tracking tasks, surface tracking tasks, and surface manual remote control tasks. Among them, underwater tasks were generally performed at a depth of 10 m to ensure the safety of the equipment. The trajectory estimated from the navigational position does not precisely align with the actual point of departure from the water, indicating a navigation error. Conversely, in surface missions, the presence of GPS signals results in the trajectory error being solely attributed to control error. It should be noted that underwater tasks may be affected by water flow, which limits the accuracy of navigation. However, such data are not allowed to be disclosed at this time. When the surface is moored, drift is generated due to interference from surface factors such as water flow and wind.

3.4. Synchronization

Various sensors first record the timestamp of the captured frames. Through the central processing platform, they are fused with the GPS time and computer time to record events with a time accuracy of 100 ms. For asynchronous inertial and DVL measurements, the lower frequency is mainly recorded. The period is not fixed due to the inconsistent frequencies and times of different sensors.

4. Dataset

4.1. Data Structures

Existing public underwater datasets adopt non-standard data structures based on different types of sensors, which also causes difficulties for researchers to interpret. To unify, this dataset is based on the data structure of the commonly used KITTI car dataset [39] and adds additional data including DVL velocity and depth information. It is finally provided in CSV format, together with Python tools that can be directly transferred to ROS. The file directory structure is shown in Figure 2.

Several typical vehicle trajectories are shown in Figure 3. Note that the dataset contains structured underwater and surface maneuvering trajectories and that switching between surface and underwater sailing also occurs in one segment of the trajectory. We labeled the surface and underwater navigation sections to differentiate between them, while the most significant feature of underwater navigation is the absence of GPS signals, resulting in constant latitude and longitude received from GPS.

Compared with other underwater datasets, most of this dataset is driven autonomously according to the AUV's own driving method rather than with manual assistance. This is conducive to analyzing the kinematic rules of the AUV. At the same time, underwater navigation based on inertial units inevitably leads to error accumulation, which is fatal to the task. In the dataset, precise global positioning results measured by differential RTK are provided, with RTK base stations deployed on land. During underwater cruising, GPS measurements are unavailable until the AUV surfaces. This helps to determine the accuracy of underwater navigation. The latitude and longitude derived based on waypoints are the initial results of onboard calculations and do not indicate navigation error performance.

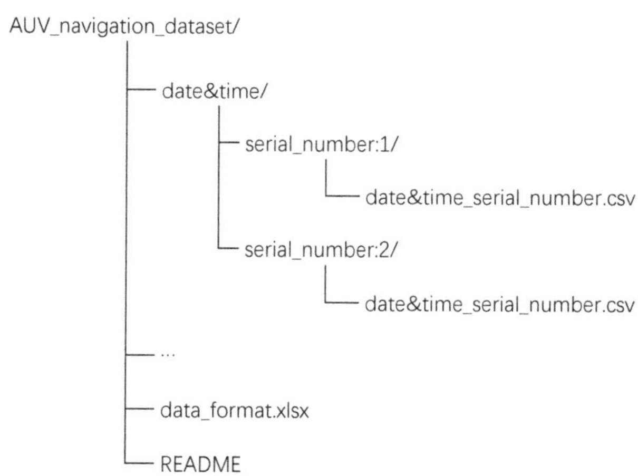

Figure 2. Directory structure of the file set.

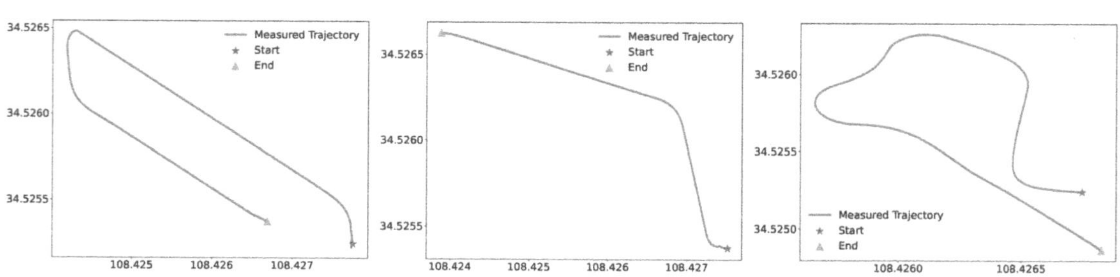

(a) (b) (c)

Figure 3. *Cont.*

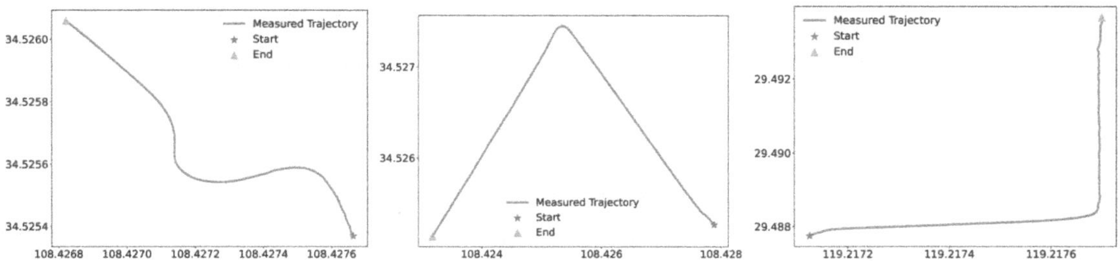

(d) (e) (f)

Figure 3. Various types of AUV tracks in different regions. (**a**) Scenario 1. (**b**) Scenario 2. (**c**) Scenario 3. (**d**) Scenario 4. (**e**) Scenario 5. (**f**) Scenario 6.

4.2. Testing

The mathematical model and engineering implementation of the fiber-optic inertial navigation system are mature cases. In this paper, the performance of the data is initially evaluated by solving the inertial navigation part underwater and comparing it with the initial navigation estimation results. The evaluation uses data acquired from gyroscopes, accelerometers, depth sensors, magnetometers, etc., and performs position estimation after calibration, filtering, and time synchronization. Methods for compensating errors due to underwater welfare, flow, etc., are not the focus of this paper and are to be further investigated.

To begin the evaluation process, we utilize the position and attitude of the AUV at the entry point as the initial state. The initial longitude, latitude, and height values are recorded as λ_0, L_0, and h_0, respectively. The initial velocity is obtained from either the DVL's effective state or the inertial navigation system. Additionally, the initial strapdown attitude matrix is expressed in (1), providing a reference for the AUV's initial orientation.

By comparing the results of the inertial navigation solution with those of the initial navigation estimate, we can assess the accuracy and reliability of the fiber-optic inertial navigation system in the underwater environment and verify its ability to provide accurate navigation information to the AUV throughout the underwater mission. Accumulated errors will occur during the solving process, and can be resolved through periodic error correction and position calibration using known landmarks or reference points for accurate navigation [40].

$$\begin{aligned} T_b^n(0) &= \begin{bmatrix} T_{11} & T_{12} & T_{13} \\ T_{21} & T_{22} & T_{23} \\ T_{31} & T_{32} & T_{33} \end{bmatrix} \\ &= \begin{bmatrix} \cos\gamma_0\cos\psi_{g0} - \sin\gamma_0\sin\theta_0\sin\psi_{g0} & -\cos\theta_0\sin\psi_{g0} & \sin\gamma_0\cos\psi_{g0} + \cos\gamma_0\sin\theta_0\sin\psi_{g0} \\ \cos\gamma_0\sin\psi_{g0} + \sin\gamma_0\sin\theta_0\cos\psi_{g0} & \cos\theta_0\cos\psi_{g0} & \sin\gamma_0\sin\psi_{g0} - \cos\gamma_0\sin\theta_0\cos\psi_{g0} \\ -\sin\gamma_0\cos\theta_0 & \sin\theta_0 & \cos\gamma_0\cos\theta_0 \end{bmatrix}, \end{aligned} \tag{1}$$

The initial position matrix is expressed as:

$$\begin{aligned} C_e^n(0) &= \begin{bmatrix} C_{11} & C_{12} & C_{13} \\ C_{21} & C_{22} & C_{23} \\ C_{31} & C_{32} & C_{33} \end{bmatrix} \\ &= \begin{bmatrix} -\sin\lambda_0 & \cos\lambda_0 & 0 \\ -\sin L_0\cos\lambda_0 & -\sin L_0\sin\lambda_0 & \cos L_0 \\ \cos L_0\cos\lambda_0 & \cos L_0\sin\lambda_0 & \sin L_0 \end{bmatrix}, \end{aligned} \tag{2}$$

The initial angular velocity of the Earth's rotation is expressed as:

$$\omega_{ie}^{n}(0) = C_{e}^{n}(0)\omega_{ie}^{e}, \tag{3}$$

The angular velocity of the AUV, ω_{nb}^{b}, is calculated based on the output value ω_{ib}^{b} of the gyroscope. The calculation process is as follows. Compared with the original text, the following optimizations are made:

$$\omega_{nb}^{b} = \omega_{ib}^{b} - T_{n}^{b}\omega_{in}^{n} = \omega_{ib}^{b} - T_{n}^{b}(\omega_{ie}^{n} + \omega_{en}^{n}), \tag{4}$$

The quaternion is instantaneously updated using the fourth-order Runge–Kutta method and finally normalized. Over three consecutive sampling periods, Δt, the update equation is expressed as:

$$\begin{aligned} K_1 &= \frac{1}{2}Q(k)\omega_{nb}^{b}(k) \\ K_2 &= \frac{1}{2}(Q(k) + \frac{\Delta t}{2}K_1)\omega_{nb}^{b}(k + \frac{\Delta t}{2}) \\ K_3 &= \frac{1}{2}(Q(k) + \frac{\Delta t}{2}K_2)\omega_{nb}^{b}(k + \frac{\Delta t}{2}) \\ K_4 &= \frac{1}{2}(Q(k) + \frac{\Delta t}{2}K_3)\omega_{nb}^{b}(k + \frac{\Delta t}{2}) \\ Q_{k+1} &= Q(k) + \frac{\Delta t}{6}(K_1 + 2K_2 + 2K_3 + K_4) \end{aligned}, \tag{5}$$

At this point, the updated strapdown matrix is shown in (6).

$$T_{b}^{n} = \begin{bmatrix} q_0^2 + q_1^2 - q_2^2 - q_3^2 & 2(q_1q_2 - q_0q_3) & 2(q_1q_3 + q_0q_2) \\ 2(q_1q_2 + q_0q_3) & q_0^2 - q_1^2 + q_2^2 - q_3^2 & 2(q_2q_3 - q_0q_1) \\ 2(q_1q_3 - q_0q_2) & 2(q_2q_3 + q_0q_1) & q_0^2 - q_1^2 - q_2^2 + q_3^2 \end{bmatrix}, \tag{6}$$

The acceleration output of the accelerometer needs to be transformed from the carrier coordinate system to the navigation coordinate system, that is, $f^n = T_b^n f^b$. At this point, the acceleration with respect to the ground is shown in (7).

$$\dot{V}^n = f^n - \begin{bmatrix} 0 & -(2\omega_{iez}^{n} + \omega_{enz}^{n}) & 2\omega_{iey}^{n} + \omega_{eny}^{n} \\ 2\omega_{iez}^{n} + \omega_{enz}^{n} & 0 & -(2\omega_{iex}^{n} + \omega_{enx}^{n}) \\ -(2\omega_{iey}^{n} + \omega_{eny}^{n}) & 2\omega_{iex}^{n} + \omega_{enx}^{n} & 0 \end{bmatrix} V^n + g, \tag{7}$$

The velocity update is expressed as:

$$V^n(k) = V^n(k-1) + \frac{\Delta t}{2}(\dot{V}^n(k-1) + \dot{V}^n(k)), \tag{8}$$

The position angular velocity update equation is:

$$\omega_{ie}^{n} = \begin{bmatrix} -\frac{V_y^n}{R_M} \\ \frac{V_x^n}{R_N} \\ \frac{V_x^n}{R_N}\tan L \end{bmatrix}, \tag{9}$$

Due to the slow changes in position during the navigation process, the update of the position matrix can be represented as follows:

$$C_{e}^{n}(k) = C_{e}^{n}(k-1) + \Delta t \dot{C}_{e}^{n}(k), \tag{10}$$

At this moment, the position of the AUV is calculated using the following formula:

$$\begin{aligned} \lambda &= \arctan \frac{C_{32}}{C_{31}} \\ L &= \arctan \frac{C_{33}}{\sqrt{C_{31}^2 + C_{32}^2}} \end{aligned} . \tag{11}$$

Here, we assess three representative sequences from the dataset, encompassing distinct movement attributes and geographic regions, as illustrated in Figures 4–6. Observing the results, it is evident that the computed velocities and attitude angles align well with the initial data. While the altitude direction remains linked to the resolved velocity, depth gauge measurements offer increased reliability. The deviation between the updated trajectory from the odometer and the measured trajectory, which does not accurately represent the true values, results from the accumulation of measurement errors.

For independent strapdown inertial navigation systems (SINS), the estimation of relative velocity and position involves the integration of accelerometer and gyro sensor data, which can introduce errors and result in significant drift in the estimated position and velocity [4]. Integrating the DVL and depth gauge measurements would notably enhance underwater navigation precision, even though the challenge of mitigating errors persists.

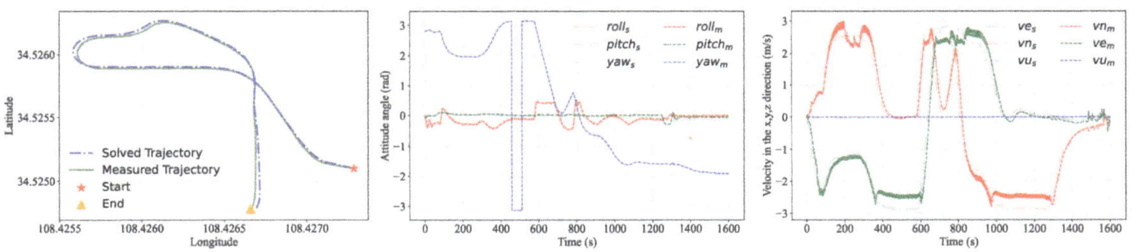

(a) (b) (c)

Figure 4. A comparison between the solved and measured values for Case 1 is presented. The subscript "s" denotes the solved result, while the subscript "m" indicates the measured result. (**a**) entails a comparison between the initial heading projection position and the solved position. (**b**) involves a comparison between the measured attitude and the solved attitude. Lastly, (**c**) examines the contrast between the measured and solved velocity values in the northeast sky direction.

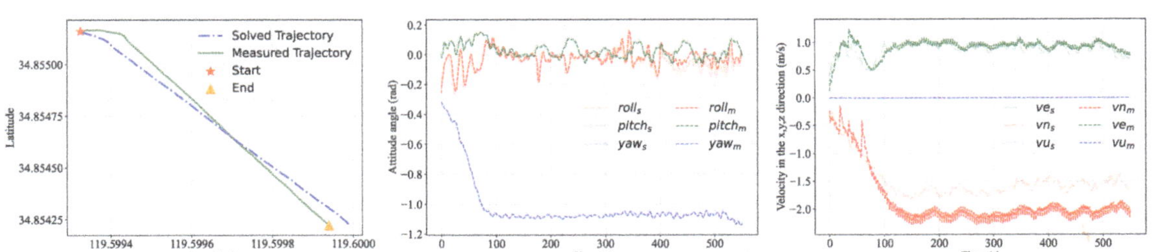

(a) (b) (c)

Figure 5. A comparison between the solved and measured values for Case 2 is presented. The subscript "s" denotes the solved result, while the subscript "m" indicates the measured result. (**a**) entails a comparison between the initial heading projection position and the solved position. (**b**) involves a comparison between the measured attitude and the solved attitude. Lastly, (**c**) examines the contrast between the measured and solved velocity values in the northeast sky direction.

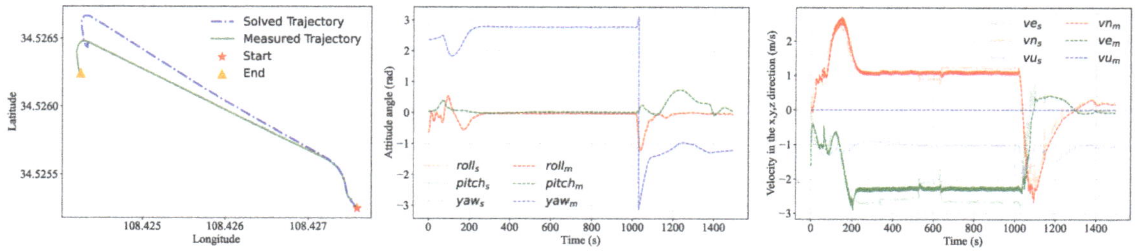

(a) (b) (c)

Figure 6. A comparison between the calculated and measured values for Case 3 is presented. The subscript "s" denotes the solved result, while the subscript "m" indicates the measured result. (**a**) entails a comparison between the initial heading projection position and the calculated position. (**b**) involves a comparison between the measured attitude and the calculated attitude. Lastly, (**c**) examines the contrast between the measured and calculated velocity values in the northeast sky direction.

5. Discussion

Navigation data on lakes and oceans were gathered by employing autonomous underwater vehicles (AUVs) equipped with rudimentary sensors. The trajectories and performance of the navigation data were acquired via dead-reckoning. By considering diving points and upper floating points, the underwater state of the vehicle can be ascertained and examined for diverse trajectories. In particular, the kinematic model of the AUV enables a meticulous analysis of its navigation trajectory attributes, thereby facilitating the augmentation of precision in underwater navigation. Extensive scholarly investigations have been conducted on fusion navigation algorithms for IMU/DVL. However, it is of utmost importance to conduct an integrated evaluation that incorporates openly accessible datasets. The profusion of underwater navigation data presents a valuable resource for comprehensively analyzing the interdependent connection between navigation strategies and devices, thereby revealing possibilities for attaining high-precision navigation through cost-effective sensor solutions. Additionally, future endeavors will explore applications of the navigation dataset.

Until new and efficient means of underwater navigation are developed, the capacity of AUVs to achieve high-precision navigation remains constrained by cost and technological limitations. The predominant approach to AUV navigation is centered on aided navigation techniques based on inertial navigation principles and amalgamating diverse measurements [41]. However, the complex interaction of practical environmental limitations, hydroacoustic channel multipath effects, and submerged ambient noise interference often leads to significant irregularities [42]. Consequently, addressing these challenges, such as mitigating cumulative errors arising from inertial navigation and rectifying measurement inaccuracies from various sensors, becomes a crucial focus for future research efforts.

In future work, we plan to expand our dataset further by incorporating additional sensing modalities, such as perception and acoustic data, to extend its usability. Specifically, we are interested in exploring underwater SLAM techniques based on forward-looking and side-scan sonar data, which will open up new avenues in underwater navigation. Moreover, data-driven pedestrian dead reckoning (PDR) research has already shown promising results with extensive datasets, inspiring us to further improve underwater navigation accuracy through large-scale learning approaches.

6. Conclusions

We have compiled a navigation dataset of AUVs operating in various regions, collected using high-precision inertial navigation, DVL, and depth sensors. The dataset encapsulates a myriad of natural scenarios involving AUVs navigating in both underwater and surface environments and spanning diverse latitudes, longitudes, and timelines. This dataset represents a pioneering collection of underwater navigation data obtained through the combination of high-cost fiber-optic gyroscopes. Drawing upon our dataset, we offer significant data support for the enhancement of underwater navigation algorithms. The assessment of typical algorithms has substantiated the practicality and effectiveness of our dataset. We hope that this dataset will be beneficial to other researchers in the field of autonomous exploration in constrained underwater environments.

Author Contributions: Conceptualization, C.W. and F.Z.; methodology, C.W.; software, C.W. and C.C.; validation, D.Y.; investigation, F.Z.; resources, F.Z. and G.P.; formal analysis, C.C. and D.Y.; writing—original draft preparation, C.W. and C.C.; writing—review and editing, C.W. and F.Z.; visualization, C.W. and C.C.; supervision, F.Z.; project administration, F.Z. and G.P.; funding acquisition, F.Z. All authors have read and agreed to the published version of the manuscript.

Funding: This research was funded by the National Natural Science Foundation of China (52171322), the National Key Research and Development Program (2020YFB1313200), and the Fundamental Research Funds for the Central Universities (D5000210944).

Data Availability Statement: Data available in a publicly accessible repository. The data presented in this study are openly available in the AUV_navigation_dataset at https://github.com/nature1949/AUV_navigation_dataset (accessed on 8 June 2023).

Acknowledgments: The authors gratefully acknowledge the support provided by the Key Laboratory of Unmanned Underwater Transport Technology during the data collection process, as well as the assistance of the research team members.

Conflicts of Interest: The authors declare no conflict of interest.

Abbreviations

The following abbreviations are used in this manuscript:

AUV	Autonomous underwater vehicle
DVL	Doppler Velocity Log
SLAM	Simultaneous localization and mapping
DGPS	Differential Global Positioning Systems
SINS	Strapdown inertial navigation systems
INS	Inertial navigation systems
IMU	Inertial measurement unit
TDOA	Time difference of arrival

References

1. Lapierre, L.; Zapata, R.; Lepinay, P.; Ropars, B. Karst exploration: Unconstrained attitude dynamic control for an AUV. *Ocean Eng.* **2021**, *219*, 108321. [CrossRef]
2. Yan, J.; Ban, H.; Luo, X.; Zhao, H.; Guan, X. Joint Localization and Tracking Design for AUV With Asynchronous Clocks and State Disturbances. *IEEE Trans. Veh. Technol.* **2019**, *68*, 4707–4720. [CrossRef]
3. Liu, R.; Liu, F.; Liu, C.; Zhang, P. Modified Sage-Husa Adaptive Kalman Filter-Based SINS/DVL Integrated Navigation System for AUV. *J. Sens.* **2021**, *2021*, 9992041. [CrossRef]
4. Sahoo, A.; Dwivedy, S.K.; Robi, P. Advancements in the field of autonomous underwater vehicle. *Ocean Eng.* **2019**, *181*, 145–160. [CrossRef]
5. Harris, Z.J.; Whitcomb, L.L. Cooperative acoustic navigation of underwater vehicles without a DVL utilizing a dynamic process model: Theory and field evaluation. *J. Field Robot.* **2021**, *38*, 700–726. [CrossRef]
6. Bucci, A.; Zacchini, L.; Franchi, M.; Ridolfi, A.; Allotta, B. Comparison of feature detection and outlier removal strategies in a mono visual odometry algorithm for underwater navigation. *Appl. Ocean Res.* **2022**, *118*, 102961. [CrossRef]
7. Franchi, M.; Ridolfi, A.; Zacchini, L. 2D Forward Looking SONAR in Underwater Navigation Aiding: An AUKF-based strategy for AUVs*. *IFAC-Papersonline* **2020**, *53*, 14570–14575. [CrossRef]
8. Zhou, W.H.; Zhu, D.M.; Shi, M.; Li, Z.X.; Duan, M.; Wang, Z.Q.; Zhao, G.L.; Zheng, C.D. Deep images enhancement for turbid underwater images based on unsupervised learning. *Comput. Electron. Agric.* **2022**, *202*, 107372. [CrossRef]
9. Su, R.; Zhang, D.; Li, C.; Gong, Z.; Venkatesan, R.; Jiang, F. Localization and Data Collection in AUV-Aided Underwater Sensor Networks: Challenges and Opportunities. *IEEE Netw.* **2019**, *33*, 86–93. [CrossRef]
10. Howe, J.A.; Husum, K.; Inall, M.E.; Coogan, J.; Luckman, A.; Arosio, R.; Abernethy, C.; Verchili, D. Autonomous underwater vehicle (AUV) observations of recent tidewater glacier retreat, western Svalbard. *Mar. Geol.* **2019**, *417*, 106009. [CrossRef]

11. Gallagher, D.G.; Manley, R.J.; Hughes, W.W.; Pilcher, A.M. Development of an enhanced underwater navigation capability for military combat divers. In Proceedings of the OCEANS 2016 MTS/IEEE Monterey, Monterey, CA, USA, 19–23 September 2016; pp. 1–4. [CrossRef]
12. Dzikowicz, B.R.; Yoritomo, J.Y.; Heddings, J.T.; Hefner, B.T.; Brown, D.A.; Bachand, C.L. Demonstration of Spiral Wavefront Navigation on an Unmanned Underwater Vehicle. *IEEE J. Ocean. Eng.* **2023**, *48*, 297–306. [CrossRef]
13. Huet, C.; Mastroddi, F. Autonomy for Underwater Robots—A European Perspective. *Auton. Robot.* **2016**, *40*, 1113–1118. [CrossRef]
14. Bil, C. Concept Evaluation of a Bi-Modal Autonomous System. In Proceedings of the AIAA AVIATION 2023 Forum, San Diego, CA, USA, 12–16 June 2023. [CrossRef]
15. Li, H.; Zhu, J.; Deng, J.; Guo, F.; Zhang, N.; Sun, J.; Hou, X. Underwater active polarization descattering based on a single polarized image. *Opt. Express* **2023**, *31*, 21988–22000. [CrossRef] [PubMed]
16. Franchi, M.; Ridolfi, A.; Pagliai, M. A forward-looking SONAR and dynamic model-based AUV navigation strategy: Preliminary validation with FeelHippo AUV. *Ocean Eng.* **2020**, *196*, 106770. [CrossRef]
17. Jin, B.; Xu, X.; Zhu, Y.; Zhang, T.; Fei, Q. Single-Source Aided Semi-Autonomous Passive Location for Correcting the Position of an Underwater Vehicle. *IEEE Sens. J.* **2019**, *19*, 3267–3275. [CrossRef]
18. Jorgensen, E.K.; Fossen, T.I.; Bryne, T.H.; Schjolberg, I. Underwater Position and Attitude Estimation Using Acoustic, Inertial, and Depth Measurements. *IEEE J. Ocean. Eng.* **2020**, *45*, 1450–1465. [CrossRef]
19. Wang, Y.; Ma, X.; Wang, J.; Wang, H. Pseudo-3D Vision-Inertia Based Underwater Self-Localization for AUVs. *IEEE Trans. Veh. Technol.* **2020**, *69*, 7895–7907. [CrossRef]
20. Manderson, T.; Gamboa Higuera, J.C.; Wapnick, S.; Tremblay, J.F.; Shkurti, F.; Meger, D.; Dudek, G. Vision-Based Goal-Conditioned Policies for Underwater Navigation in the Presence of Obstacles. *arXiv* **2020**, arXiv:2006.16235.
21. Singh, D.; Valdenegro-Toro, M. The Marine Debris Dataset for Forward-Looking Sonar Semantic Segmentation.In Proceedings of the 2021 IEEE/CVF International Conference on Computer Vision Workshops (ICCVW), Virtual, 11–17 October 2021; pp. 3734–3742.
22. Zhou, Y.; Chen, S.; Wu, K.; Ning, M.; Chen, H.; Zhang, P. SCTD 1.0:Sonar Common Target Detection Dataset. *Comput. Sci.* **2021**, *48*, 334–339. [CrossRef]
23. Zhang, P.; Tang, J.; Zhong, H.; Ning, M.; Liu, D.; Wu, K. Self-Trained Target Detection of Radar and Sonar Images Using Automatic Deep Learning. *IEEE Trans. Geosci. Remote Sens.* **2022**, *60*, 1–14. [CrossRef]
24. Huo, G.; Wu, Z.; Li, J. Underwater Object Classification in Sidescan Sonar Images Using Deep Transfer Learning and Semisynthetic Training Data. *IEEE Access* **2020**, *8*, 47407–47418. [CrossRef]
25. Chang, L.; Song, H.; Li, M.; Xiang, M. UIDEF: A real-world underwater image dataset and a color-contrast complementary image enhancement framework. *ISPRS J. Photogramm. Remote Sens.* **2023**, *196*, 415–428. . [CrossRef]
26. Yin, X.; Liu, X.; Liu, H. FMSNet: Underwater Image Restoration by Learning from a Synthesized Dataset. In Proceedings of the Artificial Neural Networks and Machine Learning—ICANN 2021, Bratislava, Slovakia, 14–17 September 2021; Farkaš, I., Masulli, P., Otte, S., Wermter, S., Eds.; Springer: Cham, Switzerland, 2021; pp. 421–432.
27. Chen, L.; Dong, J.; Zhou, H. Class balanced underwater object detection dataset generated by class-wise style augmentation. *arXiv* **2021**, arXiv:2101.07959.
28. Polymenis, I.; Haroutunian, M.; Norman, R.; Trodden, D. Artificial Underwater Dataset: Generating Custom Images Using Deep Learning Models. In Proceedings of the ASME 2022 41st International Conference on Ocean, Offshore and Arctic Engineering, Hamburg, Germany, 5–10 June 2022. [CrossRef]
29. Boittiaux, C.; Dune, C.; Ferrera, M.; Arnaubec, A.; Marxer, R.; Matabos, M.; Audenhaege, L.V.; Hugel, V. Eiffel Tower: A deep-sea underwater dataset for long-term visual localization. *Int. J. Robot. Res.* **2023**, 02783649231177322. [CrossRef]
30. Cheng, Y.; Jiang, M.; Zhu, J.; Liu, Y. Are We Ready for Unmanned Surface Vehicles in Inland Waterways? The USVInland Multisensor Dataset and Benchmark. *IEEE Robot. Autom. Lett.* **2021**, *6*, 3964–3970. [CrossRef]
31. Song, Y.; Qian, J.; Miao, R.; Xue, W.; Ying, R.; Liu, P. HAUD: A High-Accuracy Underwater Dataset for Visual-Inertial Odometry. In Proceedings of the 2021 IEEE Sensors, 31 October–3 November 2021; pp. 1–4. [CrossRef]
32. Luczynski, T.; Scharff Willners, J.; Vargas, E.; Roe, J.; Xu, S.; Cao, Y.; Petillot, Y.; Wang, S. Underwater inspection and intervention dataset. *arXiv* **2021**, arXiv:2107.13628. [CrossRef]
33. Miller, M.; Chung, S.J.; Hutchinson, S. The Visual–Inertial Canoe Dataset. *Int. J. Robot. Res.* **2018**, *37*, 13–20. [CrossRef]
34. Panetta, K.; Kezebou, L.; Oludare, V.; Agaian, S. Comprehensive Underwater Object Tracking Benchmark Dataset and Underwater Image Enhancement With GAN. *IEEE J. Ocean. Eng.* **2022**, *47*, 59–75. [CrossRef]
35. Mallios, A.; Vidal, E.; Campos, R.; Carreras, M. Underwater caves sonar data set. *Int. J. Robot. Res.* **2017**, *36*, 1247–1251. [CrossRef]
36. Krasnosky, K.; Roman, C.; Casagrande, D. A bathymetric mapping and SLAM dataset with high-precision ground truth for marine robotics. *Int. J. Robot. Res.* **2022**, *41*, 12–19. [CrossRef]
37. Ferrera, M.; Creuze, V.; Moras, J.; Trouvé-Peloux, P. AQUALOC: An underwater dataset for visual–inertial–pressure localization. *Int. J. Robot. Res.* **2019**, *38*, 1549–1559. [CrossRef]
38. Li, Y.; Sun, Y.; Ren, Q.; Li, S. AUV-Aided Data Collection Considering Adaptive Ocean Currents for Underwater Wireless Sensor Networks. *China Commun.* **2023**, *20*, 356–367. [CrossRef]

39. Geiger, A.; Lenz, P.; Stiller, C.; Urtasun, R. Vision meets robotics: The KITTI dataset. *Int. J. Robot. Res.* **2013**, *32*, 1231–1237. [CrossRef]
40. Wang, C.; Cheng, C.; Yang, D.; Pan, G.; Zhang, F. AUV planning and calibration method considering concealment in uncertain environments. *Front. Mar. Sci.* **2023**, *10*, 1228306. [CrossRef]
41. Zhai, W.; Wu, J.; Chen, Y.; Jing, Z.; Sun, G.; Hong, Y.; Fan, Y.; Fan, S. Research on Underwater Navigation and Positioning Method Based on Sea Surface Buoys and Undersea Beacons. In Proceedings of the China Satellite Navigation Conference (CSNC) 2020 Proceedings, Chengdu, China, 22–25 November 2020; Sun, J., Yang, C., Xie, J., Eds.; Springer: Singapore, 2020; Volume III, pp. 390–404.
42. Wang, J.; Zhang, T.; Jin, B.; Zhu, Y.; Tong, J. Student's t-Based Robust Kalman Filter for a SINS/USBL Integration Navigation Strategy. *IEEE Sens. J.* **2020**, *20*, 5540–5553. [CrossRef]

electronics

Article

Local-Aware Hierarchical Attention for Sequential Recommendation

Jiahao Hu *, Qinxiao Liu and Fen Zhao

School of Artificial Intelligence, Chongqing University of Technology, Chongqing 401135, China; liuqinxiao@cqut.edu.cn (Q.L.); zhaof@cqut.edu.cn (F.Z.)
* Correspondence: jhaohu@stu.cqut.edu.cn

Abstract: Modeling the dynamic preferences of users is a challenging and essential task in a recommendation system. Taking inspiration from the successful use of self-attention mechanisms in tasks within natural language processing, several approaches have initially explored integrating self-attention into sequential recommendation, demonstrating promising results. However, existing methods have overlooked the intrinsic structure of sequences, failed to simultaneously consider the local fluctuation and global stability of users' interests, and lacked user information. To address these limitations, we propose LHASRec (Local-Aware Hierarchical Attention for Sequential Recommendation), a model that divides a user's historical interaction sequences into multiple sessions based on a certain time interval and computes the weight values for each session. Subsequently, the calculated weight values are combined with the user's historical interaction sequences to obtain a weighted user interaction sequence. This approach can effectively reflect the local fluctuation of the user's interest, capture the user's particular preference, and at the same time, consider the user's general preference to achieve global stability. Additionally, we employ Stochastic Shared Embeddings (SSE) as a regularization technique to mitigate the overfitting issue resulting from the incorporation of user information. We conduct extensive experiments, showing that our method outperforms other competitive baselines on sparse and dense datasets and different evaluation metrics.

Keywords: sequential recommendation; local fluctuation; global stability; Stochastic Shared Embeddings

Citation: Hu, J.; Liu, Q.; Zhao, F. Local-Aware Hierarchical Attention for Sequential Recommendation. *Electronics* **2023**, *12*, 3742. https://doi.org/10.3390/electronics12183742

Academic Editors: Domenico Ursino, Chao Zhang, Wentao Li, Huiyan Zhang and Tao Zhan

Received: 7 August 2023
Revised: 25 August 2023
Accepted: 26 August 2023
Published: 5 September 2023

1. Introduction

In recent years, personalized recommendation tasks have become increasingly important. As the volume of information grows on the Internet, providing accurate recommendations based on changes in the user's interests has become a challenge. To address this challenge, researchers have regarded the user's historical interaction behaviors as ordered sequences, aiming to capture the dynamic changes in the user's interests from these sequences and predict the next interactive item they may be interested in. This prediction is crucial for providing personalized recommendations, which helps the platform better meet user needs and improve user experience.

To model the user's dynamic interests, researchers have explored various modeling strategies and algorithms [1–3] for the sequential features of the user's historical interaction behavior. In the early stages, Markov chain models [1,4] were commonly used to capture the transition of the user's preferences from their interaction history with items. FPMC [5] integrated the idea of matrix factorization with Markov chains, storing information about user transition matrices in a three-dimensional matrix. With the popularity of deep learning, Recurrent Neural Networks (RNNs) [6] and Convolutional Neural Networks (CNNs) [2] have gradually been applied to the sequential recommendation. Compared to Markov chains, RNNs can more effectively capture temporal relationships in the user's behavior sequences due to their inherent structural characteristics. CNNs can capture the user's interests from sequences with complex relationships. Additionally, some methods based on self-attention mechanisms [3,7,8] introduced weight adaptive adjustment mechanisms to model the importance of different elements in the sequence dynamically. These methods can more accurately capture changes and associations in the user's interests, thereby improving the accuracy and personalization of recommendation results.

In sequential recommendation, the user's preferences are typically influenced by a combination of long-term and short-term factors. Long-term preferences reflect the user's general interests, which are relatively stable and less prone to change within the sequence. On the contrary, short-term preferences reflect the user's transient "special" interests that may deviate from their general interests and exhibit the relative fluctuation within the sequence. For example, a user may prefer comedy movies as his/her favorite genre. However, due to the influence of his/her friends, he/she may develop a temporary fondness for art films for a certain period. However, traditional sequential recommendation models often treat the user's historical interaction sequence as a homogeneous entity, lacking simultaneous consideration of the local fluctuation and global stability of the user's interests. This can potentially impact the model's ability to learn the user's preferences and subsequently affect the effectiveness of recommendations.

Regarding the above issues, in this paper, we propose a Local-Aware Hierarchical Attention recommendation system that combines the local fluctuation and global stability of the user's interests. This model enhances the model's ability to model the user's behavior preferences, enabling the model to more accurately reflect the user's personalized interests and preferences. Consequently, it can more comprehensively understand the user's behavior patterns and provides users with more targeted recommendation results. Furthermore, we also consider the user information from SSE-PT [7] and employ the Stochastic Shared Embedding regularization technique to handle user and item embeddings in the input and prediction parts, alleviating overfitting issues. In summary, our main contributions are as follows:

- We comprehensively consider the local fluctuation and global stability of the user's interests to better capture users' long-term and short-term preferences.
- We employ the Stochastic Shared Embedding regularization technique to handle user embeddings and item embeddings in the input and prediction parts to alleviate the overfitting problem.
- We conducted extensive experiments on the MovieLens, Steam, and Beauty datasets, and the experimental results demonstrate that our model outperforms other competitive baselines.

2. Related Work

The user's behavior is a time-ordered behavior sequence, and their interests also dynamically change over time. Therefore, extracting temporal information from sequential data can provide valuable information. Early sequential recommendation models utilized Markov chains (MCs) [9] to capture the correlations within the sequential data. Shani et al. used the Markov chain [1] to mine the correlation between users' short-term behaviors, thus achieving a good recommendation effect. Rendle et al. combined the idea of Matrix Factorization (MF) with Markov chains [5,10] by storing user transition matrices in a three-dimensional matrix and explored the temporal information in the user's short-term behavior sequences by predicting the user's interests in other items. However, due to the scalability issue of Markov chains, the time and space complexity of the models significantly increase when dealing with longer sequences, leading to suboptimal recommendation performance.

Compared to Markov chains, Recurrent Neural Networks (RNNs) [11,12], benefiting from their distinctive structure, are more effective in handling sequential data. B. Hidasi et al. first applied RNNs to sequential recommendation [6] and proposed the session-based sequential recommendation model. It divided the user's behavior into multiple sessions based on a certain time interval, modeled each session's behavior using an RNN, and predicted the next item the user interacted with. To further improve sequential recommendation performance, Hidasi et al. introduced the parallel RNN session-based recommendation model [13], which used RNN to find the dependencies between items in the session and parallel RNN to model other attribute characteristics of items in the session, which improved the effect of sequential recommendation. Zhang et al. proposed an RNN-based sequential search click prediction model [14] that not only modeled the user's click events but also incorporated features of users and items and the information of the user's dwell time after clicking an item, resulting in the improved predictive capability of the model.

However, RNN models assume that any adjacent interactions in a sequence are mutually dependent, while in reality, there exist intricate and complex relationships among the user's consecutive actions. Linearly modeling the user's historical behavior makes it difficult to capture their true interests within a sequence with complex relationships. As a result, researchers have begun exploring the domains of CNN and self-attention mechanisms. Tang et al. proposed Caser [2], a sequential recommendation model based on convolutional embeddings, which applies CNN concepts to sequential recommendation. Caser employed multiple convolutional filters with different weights to extract various sequential pattern information from the sequence, thereby improving the accuracy of personalized recommendations. Similarly, Yuan et al. extended the Caser model with a future-oriented recommendation framework called NextItNet [15], utilizing dilated convolutions to capture more complex dependencies in the user's behavior sequence. Kang et al. introduced SASRec [3], a sequential recommendation model based on self-attention mechanisms, to address the dependency issue of RNNs. Additionally, Li et al. proposed MIND [16], which reflected a user's multidimensional interests by using multiple vectors to represent each user and predicted the match between the candidate items and the user's interests in each dimension.

Nevertheless, some existing models ignore the internal structure of the sequence, do not consider the local fluctuation and global stability of the user's interest simultaneously, and lack user information. In our study, we first divide the user's historical interaction sequence into multiple sessions based on a certain time interval. By comprehensively considering the weak correlation between sessions and the strong correlation among items within each session, we accurately reflect the local fluctuation of the user's interests. Next, we combine the calculated weights of each session with the user's historical interaction sequence to achieve the influence of the local fluctuation on the global stability of the user's interests. Additionally, we introduce user information into the model and employ the Stochastic Shared Embedding regularization technique to mitigate the overfitting problem that may arise from incorporating user information.

3. Method

We propose a hierarchical attention-based sequential recommendation model, LHASRec. The model consists of an embedding layer, a local-aware layer, a global attention layer, and a prediction layer. This section will describe how to construct this sequential recommendation model. The architecture of LHASRec is illustrated in Figure 1.

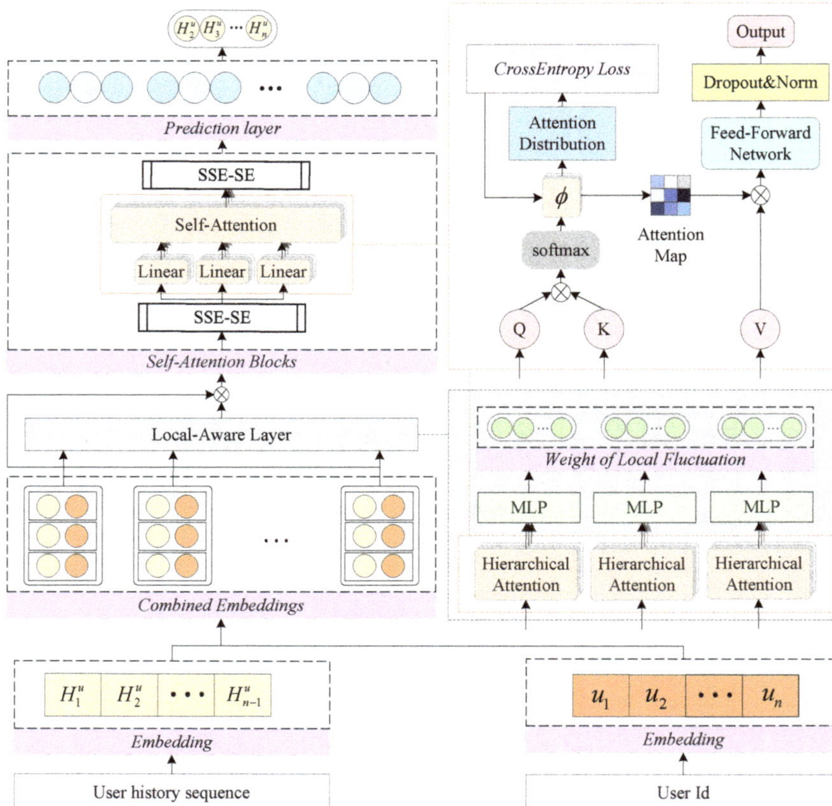

Figure 1. The overall framework of LHASRec. The model primarily consists of an embedding layer, a local-aware layer, a global attention layer, and a prediction layer, and the input and output of the global attention layer are handled using SSE regularization. (a) The target user's historical behavior sequence is combined with user information and divided into multiple sessions based on time intervals. (b) Each session is individually processed by the local-aware layer to generate local attention weights, which are then combined with the sequences containing item information and user information to serve as the input for the global attention layer. (c) The SSE regularization technique is applied to the input matrix. (d) The global attention layer captures the representation of the user's local and global preferences. (e) The output matrix is regularized using SSE. (f) Recommendations are made based on the target user's local and global preferences.

3.1. Sequential Recommendation Target

In the sequential recommendation task of this section, we define the user's historical behavior sequence as $H^u = (H^u_1, H^u_2, \cdots, H^u_{|H^u|})$, where $u \in U, H^u_i \in I$. During the model training process, at time step t, the model predicts the next item the target user will likely interact with based on the preceding t items. In other words, we use the user's historical behavior sequence $(H^u_1, H^u_2, \cdots, H^u_{|H^u|-1})$ and the user's information as inputs to the model, with the expected output denoted as $(H^u_2, H^u_3, \cdots, H^u_{|H^u|})$. The symbols used in our study are summarized in Table 1.

Table 1. Notation.

Notation	Description
U, I	user and item set
H^u	historical interaction sequence for the user u
$t_d \in \mathbb{N}$	division time interval
$n \in \mathbb{N}$	maximum sequence length
$n_s \in \mathbb{N}$	length of each session
$k \in \mathbb{N}$	number of sessions
$b \in \mathbb{N}$	number of stacked temporal attention blocks
$d \in \mathbb{N}$	latent vector dimension of the model
$d_i, d_u \in \mathbb{N}$	latent dimension of item and user
$M^I \in \mathbb{R}^{\|I\| \times d_i}$	item embedding matrix
$M^U \in \mathbb{R}^{\|U\| \times d_u}$	user embedding matrix
$\hat{S}_1, \hat{S}_2, \cdots, \hat{S}_k \in \mathbb{R}^{n_s \times d}$	input embedding matrix of local-aware layer
$\hat{E} \in \mathbb{R}^{n \times d}$	input embedding matrix of global attention layer
$c_1, c_2, \cdots, c_k \in \mathbb{R}$	fluctuation coefficient of each session
$A \in \mathbb{R}^{n \times d}$	output of the self-attention layer
$F \in \mathbb{R}^{n \times d}$	output of the point-wise feed-forward network

3.2. Embedding Layer

For the historical behavior sequence of the target user, denoted as $(H_1^u, H_2^u, \cdots, H_{|H^u|-1}^u)$, we fix its length to a specific value $n \in \mathbb{N}$ to obtain the sequence, denoted as $(h_1, h_2, \cdots, h_n)$, where n represents the maximum sequence length. The rule for fixing the length sequence is as follows:

$$\begin{cases} padding & H_{|H^u|-1}^u < n \\ unchange & H_{|H^u|-1}^u = n \\ cutting & H_{|H^u|-1}^u > n \end{cases} \tag{1}$$

It is worth noting that when the length of the original sequence is smaller than n, we pad the left side of the sequence with zeros. When the length of the original sequence is greater than n, we only consider the most recent n interactions. We construct the item embedding matrix $M^I \in \mathbb{R}^{|I| \times d_i}$ and the user embedding matrix $M^U \in \mathbb{R}^{|U| \times d_u}$, where $d_i, d_u \in \mathbb{N}$ represent the latent embedding dimensions for items and users, respectively. From these two embedding matrices, we retrieve the user information embedding for the target user and the embeddings of each item in the user's input sequence. These embeddings are combined to obtain the input embedding $E \in \mathbb{R}^{n \times d}$, where $d = d_i + d_u$:

$$E = \begin{bmatrix} \left[M_{h_1}^I; M_u^U\right] \\ \left[M_{h_2}^I; M_u^U\right] \\ \cdots \\ \left[M_{h_n}^I; M_u^U\right] \end{bmatrix} \tag{2}$$

where $\left[M_{h_i}^I; M_u^U\right]$ is the concatenation of the embedding vectors for item h_i and user u. We believe that when the time intervals between several user interactions are close, it indicates that the user is selecting items of the same kind of interest, indicating a strong correlation between these items. On the other hand, when the time interval between two adjacent items is large, it may suggest a change in the user's interest, resulting in a change in the selected items' category and indicating a weak correlation between these two items. In the user's historical behavior, the local fluctuation is generated with the change of the user's interest. To capture these fluctuations, we introduce a time interval threshold, denoted as $t_d \in \mathbb{N}$, and examine the time intervals between every two adjacent items. If the interval exceeds t_d, we split the input sequence accordingly. Following this approach, we divide the input sequence E into k sessions, i.e., local interaction sequences. Within each session, the items exhibit strong correlations, while there are weak correlations between items from different sessions. As the divided sessions have different lengths, we apply the same fixed-length rule for each session $S_i \in \mathbb{R}^{n_s \times d}$ to adjust its length to a specific value, denoted as $n_s \in \mathbb{N}$:

$$S_1 = \begin{bmatrix} \left[M^I_{h_1^{S_1}}; M^U_u\right] \\ \left[M^I_{h_2^{S_1}}; M^U_u\right] \\ \cdots \\ \left[M^I_{h_{n_s}^{S_1}}; M^U_u\right] \end{bmatrix}, \cdots, S_k = \begin{bmatrix} \left[M^I_{h_1^{S_k}}; M^U_u\right] \\ \left[M^I_{h_2^{S_k}}; M^U_u\right] \\ \cdots \\ \left[M^I_{h_{n_s}^{S_k}}; M^U_u\right] \end{bmatrix} \tag{3}$$

where $M^I_{h_j^{S_i}} \in \mathbb{R}^d$ represents the embedding of item h_j in the i-th session. Since the self-attention mechanism is unaware of the positional relationship of items in the sequence, we introduced learnable position embeddings for each session:

$$\hat{S}_1 = \begin{bmatrix} \left[M^I_{h_1^{S_1}}; M^U_u\right] + p_1^{S_1} \\ \left[M^I_{h_2^{S_1}}; M^U_u\right] + p_2^{S_1} \\ \cdots \\ \left[M^I_{h_{n_s}^{S_1}}; M^U_u\right] + p_{n_s}^{S_1} \end{bmatrix}, \cdots, \hat{S}_k = \begin{bmatrix} \left[M^I_{h_1^{S_k}}; M^U_u\right] + p_1^{S_k} \\ \left[M^I_{h_2^{S_k}}; M^U_u\right] + p_2^{S_k} \\ \cdots \\ \left[M^I_{h_{n_s}^{S_k}}; M^U_u\right] + p_{n_s}^{S_k} \end{bmatrix} \tag{4}$$

where $p_j^{S_i} \in \mathbb{R}^d$ represents the embedding of the j-th position in the i-th session.

3.3. Local-Aware Layer

We employ hierarchical attention to capture the strong correlations among items within each session, where each layer utilizes the self-attention mechanism [17] defined as follows:

$$Attention(Q, K, V) = softmax\left(\frac{QK^T}{\sqrt{d}}\right)V \tag{5}$$

where Q represents the query matrix, and K and V denote the key and value matrices, respectively. $\sqrt{d}$ is a scaling factor used to mitigate the problem of large inner products when the dimension is high.

We feed each session of the user separately into different attention layers to avoid mutual interference between items with weak correlations, ultimately obtaining the fluctuation coefficient corresponding to each session, thereby capturing the local fluctuations of the user's interests. Specifically, for the i-th session, it is linearly projected into three matrices, which are then fed into the i-th attention layer:

$$A_i^S = HA(\hat{S}_i) = Attention(\hat{S}_i W_{S_i}^Q, \hat{S}_i W_{S_i}^K, \hat{S}_i W_{S_i}^V) \tag{6}$$

where $W_{S_i}^Q, W_{S_i}^K, W_{S_i}^V \in \mathbb{R}^{d \times d}$ represent the projection matrices of Q, K, and V, respectively, for the matrix S_i. Due to the strong correlations among items within each session, we consider that their sequential order can be ambiguous, allowing subsequent keys to be connected to the current query to fully capture the representation power of the self-attention mechanism.

After the self-attention layer, we employ two MLP layers to model the non-linear relationships among items within the session to obtain the fluctuation coefficient c_i corresponding to the i-th session:

$$L_i = A_i^S W_i^M + b_i^M \tag{7}$$

$$c_i = GELU\Big((L_i)^T W_i^c + b_i^c\Big) \tag{8}$$

where $W_i^M \in \mathbb{R}^{d\times 1}, W_i^c \in \mathbb{R}^{n_s\times 1}, b_i^M \in \mathbb{R}^{n_s\times 1}, b_i^c \in \mathbb{R}$ are learnable parameters. Instead of the ReLU function, we utilize the smoother GELU [18] function for activation.

3.4. Global Attention Layer

For each session, we utilize its fluctuation coefficient to reflect the local interest fluctuation of the user, where there is a strong correlation among items within the same session. However, no single session can fully represent the user's global preferences, and there may also exist connections between different local interests. Therefore, we combine the local fluctuation coefficients with the overall input sequence to achieve global stability. Specifically, for the input sequence E, we extract the fluctuation coefficients of each item according to the fluctuation coefficient corresponding to the session in which the item belongs, resulting in a local fluctuation sequence $C = (c^{h_1}, c^{h_2}, \cdots, c^{h_n})$, where c^{h_i} is the fluctuation coefficient corresponding to the session where item h_i belongs. We introduce the fluctuation coefficients as weights to combine the input sequence that incorporates item and user information, along with learnable position embeddings, yielding a new input embedding $\hat{E} \in \mathbb{R}^{n\times d}$:

$$\hat{E} = \begin{bmatrix} c^{h_1}\left[M^I_{h_1}; M^U_u\right] + P_1 \\ c^{h_2}\left[M^I_{h_2}; M^U_u\right] + P_2 \\ \cdots \\ c^{h_n}\left[M^I_{h_n}; M^U_u\right] + P_n \end{bmatrix} \tag{9}$$

where $P_i \in \mathbb{R}^d$ represents the embedding for the i-th position.

Attention layer: We perform linear projections on the input embedding $\hat{E}$ to obtain three matrices, then feed into the attention layer:

$$A = HA(\hat{E}) = Attention(\hat{E}W^Q, \hat{E}W^K, \hat{E}W^V) \tag{10}$$

where $W^Q, W^K, W^V \in \mathbb{R}^{d\times d}$ is the projection matrix. Similar to [3], we introduce a mask to prevent any connection between Q_i and K_j ($j > i$) to prevent subsequent items from affecting the current item to be predicted.

Point-Wise Feed-Forward Network: To introduce non-linearity and consider interactions among different latent dimensions, similar to [3,8], we apply the same two-layer Point-Wise Feed-Forward Network (with shared weights) to A and use the ReLU activation function:

$$F = FFN(A) = ReLU(AW^{(1)} + b^{(1)})W^{(2)} + b^{(2)} \tag{11}$$

where $W^{(1)}, W^{(2)} \in \mathbb{R}^{d\times d}, b^{(1)}, b^{(2)} \in \mathbb{R}^d$ represents the learnable parameters. As the number of parameters in the network increases, several issues may arise, including overfitting, unstable training process (such as gradient vanishing), and longer training time. Similar to [3,8], we employ layer normalization, residual connections, and dropout regularization techniques after the attention layer and Point-Wise Feed-Forward Network to alleviate these issues:

$$f(x) = x + Dropout(f(LayerNorm(x))) \tag{12}$$

where $f(x)$ represents the self-attention layer or Point-Wise Feed-Forward Network. The definition of layer normalization is as follows:

$$LayerNorm(x) = \alpha \odot \frac{x-\mu}{\sqrt{\sigma^2+\epsilon}} + \beta \tag{13}$$

where x is a vector containing all the features of the samples, μ and σ denote the mean and variance, α is a learnable scale factor, and β is a bias term.

We merge one layer of the self-attention and one layer of the Feed-Forward Network into one attention module. To capture the user's preferences more accurately, we stack b attention modules to learn more complex item transformations, ultimately obtaining the representation of the user's preferences.

3.5. Prediction Layer

After b attention modules, the model obtains the global representation of the target user's preferences. Using this representation, at time step t, we predict the next item that the user may interact with:

$$r_{ti} = F_t\left[M^I_{h_i}; M^U_u\right] \tag{14}$$

where r_{ti} represents the score of item h_i given the previous t items, i.e., the possibility that the next item is h_i. $\left[M^I_{h_i}; M^U_u\right]$ represents the combined feature embedding that incorporates both item information and user information. At time step t, for each positive sample item $i = h_{t+1}$, we randomly sample a negative sample $m \notin H^u$. Due to the faster weight update rate of the binary cross-entropy loss function compared to the mean squared error loss function, we use binary cross-entropy as the loss function:

$$-\sum_{H^u \in H}\sum_{t\in[1,2,\cdots,n]}\left[log(\sigma(r_{ti})) + \sum_{m\notin H^u} log(1-\sigma(r_{tm}))\right] \tag{15}$$

where $\sigma(\cdot)$ is the sigmoid function. Because ADAM demonstrates greater robustness in handling noise and outliers compared to the stochastic gradient descent algorithm (SGD), we optimize the model using the ADAM optimizer [19]. The top-K recommendations for the target user at time step t can be obtained by sorting the scores of all items, and the top K items in the sorted list are the recommended items.

4. Experiments

In this section, we will present our experimental setup and show the results of our experiments. The experiments conducted aim to answer the following research questions:

RQ1: Can our proposed method outperform the state-of-the-art baselines?

RQ2: Does the choice of different time interval values for sequence dividing affect the model's ability to capture the local fluctuation of the user's interests?

RQ3: How do parameters such as maximum sequence length and the number of attention blocks impact the model's performance?

4.1. Datasets

We evaluated LHASRec on four datasets. These datasets cover different domains, sizes, and sparsity levels, and all of them are publicly available:

Movielens: https://grouplens.org/datasets/movielens/ (accessed on 25 August 2023) This dataset is sourced from the GroupLens Research project at the University of Minnesota. It is a widely used benchmark dataset. We utilized the Movielens-1M version, which consists of 1 million ratings from 6040 users on 3900 movies.

Amazon: http://jmcauley.ucsd.edu/data/amazon/ (accessed on 25 August 2023) We utilized the users' purchase and rating dataset from the e-commerce platform Amazon, which was collected by McAuley et al. [20]. To enhance the usability of the dataset, the researchers divided it based on high-ranking categories on Amazon. Specifically, we selected the **"Beauty"** and **"Video Games"** categories for our study.

Steam: https://cseweb.ucsd.edu/~jmcauley/datasets.html#steam_data (accessed on 25 August 2023) It originates from the popular digital game distribution platform, Steam.

The dataset captures users' behaviors, such as game purchases, game ratings, and game social interactions on the Steam platform.

These four datasets all include timestamps of users' interactions. We followed the methods described in [3,7] to preprocess the data. Firstly, we sorted the user–item interactions in ascending order based on the timestamps. To ensure the validity of the data, we excluded those cold-start users and those with less than three user–item interactions. Similar to the approach in [3], we used the last item in the interaction sequence (i.e., the most recent item interacted with by the user) as the test set, the second-to-last item as the validation set, and the remaining items as the training set. Through these preprocessing steps, we reduced redundant information while preserving the data's original meaning, facilitating further research and algorithm evaluation in the recommendation system domain. Table 2 provides an overview of these datasets, highlighting their characteristics. Among them, Movielens-1M is the densest dataset, with fewer users and items. On the other hand, the Steam dataset is the sparsest, containing relatively fewer interactions.

Table 2. Dataset statistics.

Dataset	Users	Items	Avg. Sequence Length	Sparsity
MovieLens-1M	6040	3706	163.6	95.58%
Beauty	22,363	12,101	6.88	99.94%
Games	24,303	10,672	7.54	99.92%
Steam	144,051	11,153	3.49	99.97%

4.2. Compared Methods

We compared LHASRec with various methods, including the classic recommendation approach (BPR) and recommendation models based on different techniques. Among them, we considered methods based on first-order Markov chains (such as FMC, FPMC, TransRec), transformer-based methods (such as SASRec, SSE-PT, TiSASRec), convolutional neural network-based methods (Caser, TARN), fusion model-based methods (BAR), and multilayer perceptron-based methods (FMLP-Rec).

BPR [21]: Bayesian personalized ranking model (BPR) is a traditional recommendation method that employs matrix factorization for the recommendation.

FPMC [5]: Factorizing personalized Markov chains model (FPMC) amalgamates matrix factorization with the initial-order Markov chain technique, enabling the model to encompass both users' long-term preferences and the dynamic transitions of items.

TransRec [22]: Translation-based recommendation model (TransRec) represents a first-order sequential recommendation approach, where items undergo embedding within a transformational domain, while users are depicted as translation vectors that encapsulate shifts from the present item to the subsequent one.

SASRec [3]: Self-attentive sequential recommendation model (SASRec) is the first transformer-based model that extracts context from all past interactions like recurrent neural networks while making predictions based on a limited number of interactions, similar to Markov chains.

SSE-PT [7]: Sequential recommendation via personalized transformer model (SSE-PT) integrates the embedding vector of the user ID and employs a novel regularization approach.

TiSASRec [8]: Time interval aware self-attention for sequential recommendation model (TiSASRec) leverages the advantage of attention mechanisms to handle items at different ranges in different datasets adaptively and adjusts the weights based on different items, absolute positions, and time intervals.

TARN [23]: Neural time-aware recommendation network (TARN) simultaneously captures users' static and dynamic preferences by fusing a feature interaction network with a convolutional neural network.

BAR [24]: Behavior-aware recommendation model (BAR) integrates behavioral information into the representation module and employs the innovative module across diverse backbone models.

FMLP-Rec [25]: Filter-enhanced MLP model (FMLP-Rec) is a pure MLP architecture model that encodes user sequences using learnable filters.

4.3. Implementation Details

We implemented the LHASRec using PyTorch, with the same number of transformer encoding blocks as SASRec and SSE-PT (i.e., $b = 2$). To optimize the model, we chose ADAM as the optimizer with a learning rate of 0.001 and a momentum decay rate of $\beta_1 = 0.9, \beta_2 = 0.98$. The batch size was set to 128. For the Movielens-1M dataset, we set the dropout rate to 0.2, while for the other three datasets, it was set to 0.5. Regarding the maximum length of the sequences, we set it to 190 for the Movielens-1M dataset and 50 for the other three datasets. Additionally, to further enhance the effectiveness of personalized recommendations, we fine-tuned two parameters of the SSE to improve its performance.

4.4. Evaluation Metrics

To assess the effectiveness of all the models, we employed $HR@N$ and $NDCG@N$ as evaluation metrics [26], defined as follows:

$$HR@N = \frac{1}{M}\sum_{i=1}^{M} hits(i) \tag{16}$$

$$NDCG@N = \frac{1}{M}\sum_{i=1}^{M} \frac{1}{log_2(p_i + 1)} \tag{17}$$

where M is the number of users, $hits(i)$ indicates whether the item interacted with by the i-th user is present in the recommendation list of length N, and p_i represents the position of the item interacted with by the i-th user in the recommendation list. In our experiments, we set the length N of the recommendation list to 10. To evaluate the performance of the recommendation algorithms, we employed $HR@10$ and $NDCG@10$ as the two metrics. Specifically, we appended 100 negative samples [27] randomly after each user's actual items and calculated the metric values based on the rankings of these 101 items. It is worth noting that higher values of $HR@10$ and $NDCG@10$ indicate better model performance.

4.5. Recommendation Performance (RQ1)

Table 3 presents the recommendation performance of various methods on the four datasets **(RQ1)**. For the dense dataset, TiSASRec outperforms other baseline methods. Its advantage lies in the effective utilization of attention mechanisms, and it can dynamically adjust the weights according to different items, absolute positions, and time intervals to adapt to variations in dataset ranges. For the sparse dataset, FMLP-Rec demonstrates superior recommendation performance compared to other baseline methods. Replacing the complex Transformer architecture with MLP layers in the frequency domain effectively addresses the overfitting issue caused by insufficient available information in sparse datasets. Neural network-based methods (Caser) excel at capturing long-term sequential patterns, thus performing well on dense datasets. In contrast, methods based on Markov chains (such as FMC, FPMC, and TransRec) focus more on item transitions, resulting in better performance on sparse datasets. Furthermore, the TARN approach, which concurrently captures both users' dynamic and static preferences, achieves superior performance compared to the SASRec technique, which focuses solely on a single type of preference, across all the datasets. Moreover, the BAR technique demonstrates superior performance compared to its underlying model, SASRec, across all datasets, highlighting the effectiveness of segregating the user's historical interaction sequence into item sequence and behavior sequence as a productive modeling strategy.

Table 3. Recommended performance. We have bolded the best-recommended method in each row and underlined the second-best-performing approach in each row.

Methods	Beauty		Games		ML-1M		Steam	
	Hit@10	**NDCG@10**	**Hit@10**	**NDCG@10**	**Hit@10**	**NDCG@10**	**Hit@10**	**NDCG@10**
BPR	0.3775	0.2183	0.4853	0.2875	0.5781	0.3287	0.7061	0.4436
FMC	0.3771	0.2477	0.6358	0.4456	0.6983	0.4676	0.7731	0.5193
FPMC	0.4310	0.2891	0.6082	0.4680	0.7599	0.5176	0.7710	0.5011
TransRec	0.4607	0.3020	0.6838	0.4557	0.6413	0.3969	0.7624	0.4852
Caser	0.4264	0.2547	0.5282	0.3214	0.7886	0.5538	0.7874	0.5381
SASRec	0.4663	0.3080	0.6843	0.4602	0.8285	0.5982	0.7867	0.5108
SSE-PT	0.4963	0.3159	0.6955	0.4677	0.8346	0.6163	0.7885	0.5369
TiSASRec	0.4981	0.3329	0.7080	0.467	<u>0.8359</u>	0.6156	<u>0.8053</u>	<u>0.5523</u>
TARN	0.4979	0.3324	0.6996	0.4698	0.8325	0.6139	0.7985	0.5476
BAR	0.4995	0.3336	0.7073	0.4704	0.8351	<u>0.6192</u>	0.8039	0.5492
FMLP-Rec	<u>0.5029</u>	<u>0.3351</u>	<u>0.7091</u>	<u>0.4773</u>	0.8291	0.5333	0.8031	0.5470
LHASRec	**0.5150**	**0.3402**	**0.7359**	**0.5072**	**0.8396**	**0.6197**	**0.8218**	**0.5611**

LHASRec outperforms the leading benchmark techniques in recommendation performance across all the datasets. This achievement can be attributed to two key factors. Firstly, in the case of sparse data, the introduction of user information embedding enhances the correlation between users and items, thereby improving data representation. This enables LHASRec to capture the user's preferences better and achieve more accurate personalized recommendations. Secondly, the model considers both the local fluctuation and global stability of the user's interests, demonstrating the ability to model the user's behavior accurately. This comprehensive modeling approach helps capture users' long-term and short-term preferences more accurately.

4.6. Local-Aware Ability (RQ2)

When utilizing sequential models for handling the user's historical interaction sequences, it is often prone to overlooking the impact of the local fluctuation of the user's interests on global stability. To delve deeper into this issue, we conducted a series of experiments. We divided the user's historical interaction sequences into multiple sessions based on different division time interval values (t_d) between adjacent items and compared the performance across four datasets. As shown in Table 4, selecting excessively small t_d (resulting in numerous sessions) or tremendous t_d (resulting in too few sessions) led to a certain degree of decline in the model's recommendation capability. Specifically, when t_d is too small, the model's local-aware ability becomes excessively strong, focusing excessively on the user's short-term interests and disregarding the strong correlations among items in the sequence, thus affecting the accuracy of recommendations. Conversely, when t_d is too large, the model's local-aware ability becomes weak, making it challenging to capture the user's short-term specific interests and failing to promptly reflect changes in the user's interests. We found that the model performed best on the Movielens-1M dataset when the value of t_d was 30 min. For the Games, Beauty, and Steam datasets, the model achieved the best recommendation results when the value of t_d was 15 min. In order to gain a deeper insight into the factors influencing the model's performance, we explored the possibility that it might be due to the additional information used by the model. Consequently, we modified the LHASRec by removing specific user attributes and compared the resulting model with the baseline model. As shown in Table 5, it is evident that even after removing specific user attributes, LHASRec maintains superior performance on the Beauty, Games, and Steam datasets. While its performance on the MovieLens-1M dataset is slightly below that of the optimal model, it still remains significant. This indicates that dividing the user's historical interaction sequence into a reasonable number of sessions is meaningful. It allows for a comprehensive consideration of the user's short-term special interests and the strong correlations among items within sessions, thus better reflecting the local fluctuation of the user's interests and laying a solid foundation for achieving global stability of the user's interests.

Table 4. Impact of different division time interval values on the recommendation performance of the models across four datasets. We have bolded the best-recommended method in each row.

t_d (min)	Beauty		Games		ML-1M		Steam	
	Hit@10	NDCG@10	Hit@10	NDCG@10	Hit@10	NDCG@10	Hit@10	NDCG@10
1	0.5052	0.3239	0.7214	0.4868	0.8214	0.5881	0.8141	0.5575
15	**0.5150**	**0.3402**	**0.7359**	**0.5072**	0.8242	0.5964	**0.8218**	**0.5611**
30	0.5137	0.3367	0.7310	0.5005	**0.8396**	**0.6198**	0.8116	0.5601
45	0.5086	0.3297	0.7258	0.4916	0.8235	0.5909	0.8075	0.5522
60	0.5059	0.3266	0.7218	0.4877	0.8225	0.5892	0.8001	0.5450

Table 5. Recommended performance. We removed specific user attributes from LHASRec and compared the resulting model with the baseline model for analysis. We have bolded the best-recommended method in each row and underlined the second-best-performing approach in each row.

Methods	Beauty		Games		ML-1M		Steam	
	Hit@10	NDCG@10	Hit@10	NDCG@10	Hit@10	NDCG@10	Hit@10	NDCG@10
BPR	0.3775	0.2183	0.4853	0.2875	0.5781	0.3287	0.7061	0.4436
FMC	0.3771	0.2477	0.6358	0.4456	0.6983	0.4676	0.7731	0.5193
FPMC	0.4310	0.2891	0.6082	0.4680	0.7599	0.5176	0.7710	0.5011
TransRec	0.4607	0.3020	0.6838	0.4557	0.6413	0.3969	0.7624	0.4852
Caser	0.4264	0.2547	0.5282	0.3214	0.7886	0.5538	0.7874	0.5381
SASRec	0.4663	0.3080	0.6843	0.4602	0.8285	0.5982	0.7867	0.5108
SSE-PT	0.4963	0.3159	0.6955	0.4677	0.8346	0.6163	0.7885	0.5369
TiSASRec	0.4981	0.3329	0.7080	0.467	**0.8359**	0.6156	<u>0.8053</u>	<u>0.5523</u>
TARN	0.4979	0.3324	0.6996	0.4698	0.8325	0.6139	0.7985	0.5476
BAR	0.4995	0.3336	0.7073	0.4704	0.8351	**0.6192**	0.8039	0.5492
FMLP-Rec	<u>0.5029</u>	<u>0.3351</u>	<u>0.7091</u>	<u>0.4773</u>	0.8291	0.5333	0.8031	0.5470
LHASRec	**0.5092**	**0.3387**	**0.7280**	**0.4996**	<u>0.8354</u>	<u>0.6167</u>	**0.8166**	**0.5556**

4.7. Stochastic Shared Embeddings

In the process of stacking self-attention modules and incorporating user information, the model is prone to overfitting. To alleviate this issue, we conducted a series of experiments with various regularization methods and compared their performance on the MovieLens-1M dataset (Table 6). Through the analysis of Table 6, we found that Stochastic Shared Embeddings (SSE) is a more effective regularization method compared to existing techniques such as Dropout and weight decay. Specifically, we investigated the recommendation performance when using Dropout and L2 regularization alone and their combination. The results showed that in the LHASRec, the overfitting problem was mitigated to some extent by adopting the SSE regularization method, which randomly replaces embedding matrices. Compared to Dropout or L2 regularization alone, the recommendation performance of the LHASRec on the MovieLens-1M dataset improved by approximately 3% and 8%, respectively. Overall, considering the results of our experiments, the combination of SSE, Dropout, and weight decay is the optimal choice for regularization. This comprehensive approach effectively reduces the risk of overfitting and improves the model's performance in recommendation tasks. Therefore, we recommend adopting this combined regularization strategy in practical applications for better recommendation results.

Table 6. Impact of different regularization methods on the recommendation effect on the MovieLens-1M. We have bolded the best-recommended method in each row and underlined the second-best-performing approach in each row.

Methods	Value	NDCG@10	Hit@10
L2	0.0005	0.5049	0.7447
	0.001	0.5083	0.7550
Dropout	0.4	0.5457	0.7990
	0.6	0.5192	0.7770
SSE-SE	-	0.5616	0.8035
L2+ Dropout	0.001 + 0.4	0.5523	0.8089
	0.001 + 0.6	0.5418	0.7952
	0.0005 + 0.4	0.5427	0.7980
	0.0005 + 0.6	0.5382	0.7892
L2 + SSE-SE	0.0005 + SSE-SE	0.5600	0.7982
	0.001 + SSE-SE	0.5641	0.8096
Dropout + SSE-SE	0.4 + SSE-SE	0.5512	0.8002
	0.6 + SSE-SSE	0.5649	0.8103
L2 + Dropout + SSE-SE	0.001 + 0.4 + SSE-SE	**0.5877**	**0.8243**

4.8. Ablation Study (RQ3)

The influence of the maximum sequence length on the model: Considering the different average sequence lengths of the datasets, we set different maximum sequence lengths based on the principle that each dataset's average sequence length is roughly proportional to the model's maximum sequence length. Through experimental observations, we found that the model's recommendation results are notably influenced by the maximum sequence length. Generally, a longer maximum sequence length leads to better recommendation performance. However, when the maximum sequence length exceeds a certain threshold, the recommendation performance of the model starts to decline. In the experimental data shown in Figure 2, we illustrate the variation in recommendation performance of the LHASRec under different maximum sequence lengths. The results indicate that the recommendation performance of the LHASRec improves as the maximum sequence length increases and reaches its optimum at a certain length (e.g., 190 for MovieLens-1M, 50 for Beauty and Games datasets, and 30 for the Steam dataset). However, the recommendation performance starts to deteriorate when the maximum sequence length surpasses this critical value. This is because an excessively long maximum sequence length may introduce irrelevant noise to the recommendation task, affecting the model's ability to utilize limited information for recommendations effectively. In conclusion, the model's recommendation performance is notably affected by the maximum sequence length.

The influence of the number of attention blocks on the model: The number of attention blocks in the model has a significant impact on the recommendation results. Generally, more self-attention blocks can improve the model's ability to fit the data. However, when the number of blocks is too low, the model tends to underfit, while an excessive number of blocks increases the model's complexity, resulting in a long time for fitting the data, and may lead to overfitting, thereby reducing the recommendation performance. In our experiments, we investigated the impact of using different numbers of self-attention blocks in the LHASRec on the recommendation results across four datasets. Based on the experimental data shown in Figure 3, we found that when $b = 1$, the model fails to fit the data well, resulting in poor recommendation performance. When $b = 2$, the model achieves the best performance and optimal recommendations. However, as the number of blocks exceeds 2, the model gradually starts to overfit, and the performance declines. Based on these experimental results, we select two self-attention blocks as the optimal setting across all datasets to balance the model's fitting ability and complexity, thereby obtaining better recommendation performance.

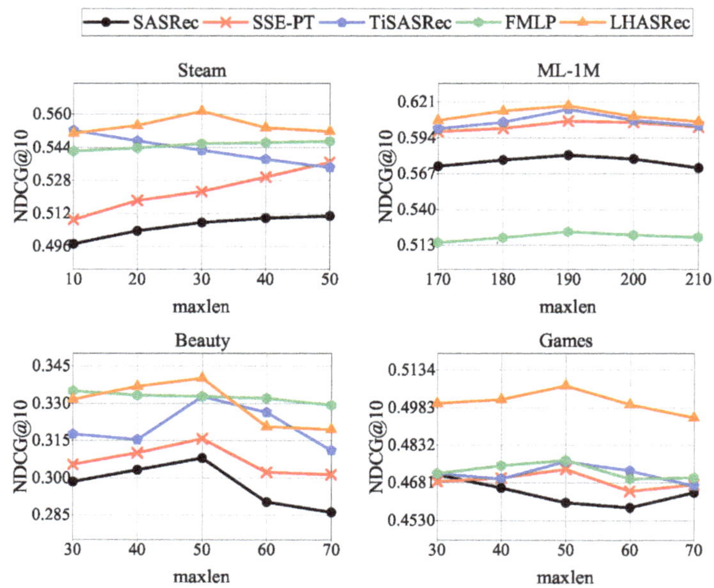

Figure 2. Influence of maximum sequence length on ranking performance (*NDCG*@10).

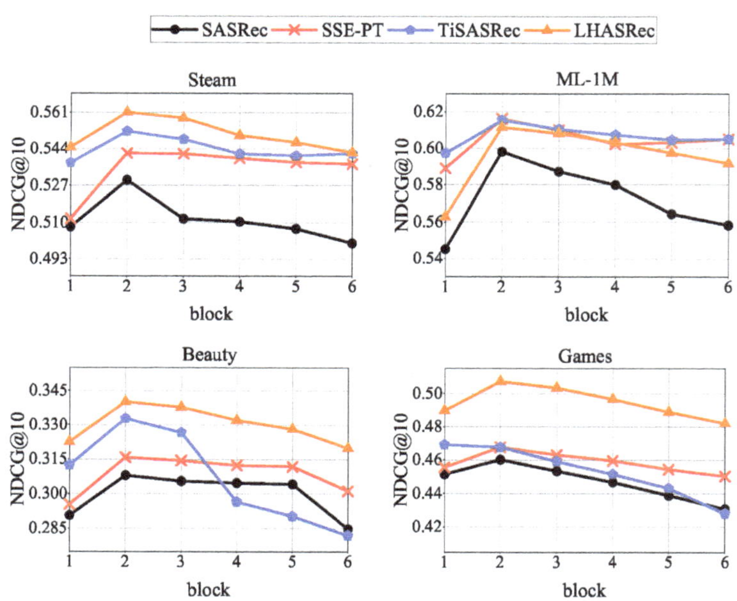

Figure 3. Influence of the number of attention blocks on ranking performance (*NDCG*@10).

5. Conclusions

In this work, we propose a sequential model with local-aware ability (LHASRec). The model comprehensively considers the local fluctuation and global stability of the user's interests to capture the long-term and short-term preferences more accurately. Meanwhile, we enhance the user's historical interaction sequences by embedding user information.

Additionally, we employ the Stochastic Shared Embeddings regularization technique to alleviate overfitting caused by embedding a large amount of user information in the model. Through experiments conducted on sparse and dense datasets, we demonstrate the superiority of LHASRec over various state-of-the-art baseline models. These results highlight the effectiveness and superiority of our proposed approach.

Author Contributions: Conceptualization, J.H.; methodology, J.H.; software, J.H. and Q.L.; validation, J.H., Q.L. and F.Z.; formal analysis, J.H.; investigation, J.H.; resources, J.H.; data curation, J.H.; writing—original draft preparation, J.H.; writing—review and editing, J.H. and F.Z.; supervision, J.H. and Q.L.; project administration, J.H. All authors have read and agreed to the published version of the manuscript.

Funding: This work is supported by the Action Plan for High-Quality Development of Graduate Education of Chongqing University of Technology (No. gzlcx20232102).

Data Availability Statement: Not applicable.

Conflicts of Interest: The authors declare no conflicts of interest.

References

1. Shani, G.; Heckerman, D.; Brafman, R.I.; Boutilier, C. An MDP-based recommender system. *J. Mach. Learn. Res.* **2005**, *6*, 1265–1295.
2. Tang, J.; Wang, K. Personalized top-n sequential recommendation via convolutional sequence embedding. In Proceedings of the Eleventh ACM International Conference on Web Search and Data Mining, Marina Del Rey, CA, USA, 5–9 February 2018; pp. 565–573.
3. Kang, W.C.; McAuley, J. Self-attentive sequential recommendation. In Proceedings of the 2018 IEEE International Conference on Data Mining (ICDM), Singapore, 17–20 November 2018; pp. 197–206.
4. Hosseinzadeh Aghdam, M.; Hariri, N.; Mobasher, B.; Burke, R. Adapting recommendations to contextual changes using hierarchical hidden markov models. In Proceedings of the 9th ACM Conference on Recommender Systems, Vienna, Austria, 16–20 September 2015; pp. 241–244.
5. Rendle, S.; Freudenthaler, C.; Schmidt-Thieme, L. Factorizing personalized markov chains for next-basket recommendation. In Proceedings of the 19th International Conference on World Wide Web, Raleigh, NC, USA, 26–30 April 2010; pp. 811–820.
6. Hidasi, B.; Karatzoglou, A.; Baltrunas, L.; Tikk, D. Session-based recommendations with recurrent neural networks. *arXiv* **2015**, arXiv:1511.06939.
7. Wu, L.; Li, S.; Hsieh, C.J.; Sharpnack, J. SSE-PT: Sequential recommendation via personalized transformer. In Proceedings of the 14th ACM Conference on Recommender Systems, Virtual Event, Brazil, 22–26 September 2020; pp. 328–337.
8. Li, J.; Wang, Y.; McAuley, J. Time interval aware self-attention for sequential recommendation. In Proceedings of the 13th International Conference on Web Search and Data Mining, Houston, TX, USA, 3–7 February 2020; pp. 322–330.
9. Brémaud, P. *Markov Chains: Gibbs Fields, Monte Carlo Simulation, and Queues*; Springer Science & Business Media: Berlin, Germany, 2001; Volume 31.
10. Xue, H.J.; Dai, X.; Zhang, J.; Huang, S.; Chen, J. Deep matrix factorization models for recommender systems. In Proceedings of the Twenty-Sixth International Joint Conference on Artificial Intelligence (IJCAI-17), Melbourne, Australia, 19–25 August 2017; Volume 17, pp. 3203–3209.
11. Hopfield, J.J. Neural networks and physical systems with emergent collective computational abilities. *Proc. Natl. Acad. Sci. USA* **1982**, *79*, 2554–2558. [CrossRef] [PubMed]
12. Elman, J.L. Finding structure in time. *Cogn. Sci.* **1990**, *14*, 179–211. [CrossRef]
13. Hidasi, B.; Quadrana, M.; Karatzoglou, A.; Tikk, D. Parallel recurrent neural network architectures for feature-rich session-based recommendations. In Proceedings of the 10th ACM Conference on Recommender Systems, Boston, MA, USA, 15–19 September 2016; pp. 241–248.
14. Zhang, Y.; Dai, H.; Xu, C.; Feng, J.; Wang, T.; Bian, J.; Wang, B.; Liu, T.Y. Sequential click prediction for sponsored search with recurrent neural networks. In Proceedings of the AAAI Conference on Artificial Intelligence, Québec City, QC, Canada, 27–31 July 2014; Volume 28.
15. Yuan, F.; Karatzoglou, A.; Arapakis, I.; Jose, J.M.; He, X. A simple convolutional generative network for next item recommendation. In Proceedings of the Twelfth ACM International Conference on Web Search and Data Mining, Melbourne, VIC, Australia, 11–15 February 2019; pp. 582–590.
16. Li, C.; Liu, Z.; Wu, M.; Xu, Y.; Zhao, H.; Huang, P.; Kang, G.; Chen, Q.; Li, W.; Lee, D.L. Multi-interest network with dynamic routing for recommendation at Tmall. In Proceedings of the 28th ACM International Conference on Information and Knowledge Management, Beijing, China, 3–7 November 2019; pp. 2615–2623.
17. Vaswani, A.; Shazeer, N.; Parmar, N.; Uszkoreit, J.; Jones, L.; Gomez, A.N.; Kaiser, Ł.; Polosukhin, I. Attention is all you need. *Adv. Neural Inf. Process. Syst.* **2017**, *30*.

18. Hendrycks, D.; Gimpel, K. Bridging Nonlinearities and Stochastic Regularizers with Gaussian Error Linear Units. *CoRR* **2016**, abs/1606.08415. Available online: https://www.bibsonomy.org/bibtex/9aaf203ef9c9e38569532ac88603af8e (accessed on 24 August 2023).
19. Kingma, D.P.; Ba, J. Adam: A method for stochastic optimization. *arXiv* **2014**, arXiv:1412.6980.
20. McAuley, J.; Targett, C.; Shi, Q.; Van Den Hengel, A. Image-based recommendations on styles and substitutes. In Proceedings of the 38th International ACM SIGIR Conference on Research and Development in Information Retrieval, Santiago, Chile, 9–13 August 2015; pp. 43–52.
21. Rendle, S.; Freudenthaler, C.; Gantner, Z.; Schmidt-Thieme, L. BPR: Bayesian personalized ranking from implicit feedback. *arXiv* **2012**, arXiv:1205.2618.
22. He, R.; Kang, W.C.; McAuley, J. Translation-based recommendation. In Proceedings of the Eleventh ACM Conference on Recommender Systems, Como, Italy, 27–31 August 2017; pp. 161–169.
23. Zhang, Q.; Cao, L.; Shi, C.; Niu, Z. Neural time-aware sequential recommendation by jointly modeling preference dynamics and explicit feature couplings. *IEEE Trans. Neural Netw. Learn. Syst.* **2021**, *33*, 5125–5137. [CrossRef] [PubMed]
24. He, M.; Pan, W.; Ming, Z. BAR: Behavior-aware recommendation for sequential heterogeneous one-class collaborative filtering. *Inf. Sci.* **2022**, *608*, 881–899. [CrossRef]
25. Zhou, K.; Yu, H.; Zhao, W.X.; Wen, J.R. Filter-enhanced MLP is all you need for sequential recommendation. In Proceedings of the ACM Web Conference 2022, Lyon, France, 25–29 April 2022; pp. 2388–2399.
26. He, X.; Liao, L.; Zhang, H.; Nie, L.; Hu, X.; Chua, T.S. Neural collaborative filtering. In Proceedings of the 26th International Conference on World Wide Web, Perth, Australia, 3–7 April 2017; pp. 173–182.
27. Koren, Y. Factorization meets the neighborhood: A multifaceted collaborative filtering model. In Proceedings of the 14th ACM SIGKDD International Conference on Knowledge Discovery and Data Mining, Las Vegas, NV, USA, 24–27 August 2008; pp. 426–434.

electronics MDPI

Article

An Off-Line Error Compensation Method for Absolute Positioning Accuracy of Industrial Robots Based on Differential Evolution and Deep Belief Networks

Yong Tao [1,2,*], Haitao Liu [1], Shuo Chen [3], Jiangbo Lan [3], Qi Qi [1] and Wenlei Xiao [1]

1 School of Mechanical Engineering and Automation, Beihang University, Beijing 100191, China; by2107062@buaa.edu.cn (H.L.); ghost921@buaa.edu.cn (Q.Q.); xiaowenlei@buaa.edu.cn (W.X.)
2 Research Institute of Aero-Engine, Beihang University, Beijing 102206, China
3 School of Large Aircraft Engineering, Beihang University, Beijing 100191, China; mechanics_chenshuo@163.com (S.C.); lanjiangbo@buaa.edu.cn (J.L.)
* Correspondence: taoy@buaa.edu.cn; Tel.: +86-010-8231-3905

Abstract: Industrial robots have been increasingly used in the field of intelligent manufacturing. The low absolute positioning accuracy of industrial robots is one of the difficulties in their application. In this paper, an accuracy compensation algorithm for the absolute positioning of industrial robots is proposed based on deep belief networks using an off-line compensation method. A differential evolution algorithm is presented to optimize the networks. Combined with the evidence theory, a position error mapping model is proposed to realize the absolute positioning accuracy compensation of industrial robots. Experiments were conducted using a laser tracker AT901-B on an industrial robot KR6_R700 sixx_CR. The absolute position error of the end of the robot was reduced from 0.469 mm to 0.084 mm, improving the accuracy by 82.14% after the compensation. Experimental results demonstrated that the proposed compensation algorithm could improve the absolute positioning accuracy of industrial robots, as well as its potential uses for precise operational tasks.

Keywords: absolute positioning accuracy; deep belief network; differential evolution algorithm; industrial robot; off-line error compensation

Citation: Tao, Y.; Liu, H.; Chen, S.; Lan, J.; Qi, Q.; Xiao, W. An Off-Line Error Compensation Method for Absolute Positioning Accuracy of Industrial Robots Based on Differential Evolution and Deep Belief Networks. *Electronics* **2023**, *12*, 3718. https://doi.org/10.3390/electronics12173718

Academic Editor: Christos Volos

Received: 7 August 2023
Revised: 29 August 2023
Accepted: 30 August 2023
Published: 2 September 2023

1. Introduction

Industry 4.0 technologies are critical and indispensable tools to propel social and technological innovation advancements. Previous research [1,2] noted that the use of Industry 4.0 technologies can better sustain current resources, reduce labor costs, be better sources of energy, and potentially produce higher-quality sustainable products. Examples of Industry 4.0 technologies include, but are not limited to, machine learning, virtual and augmented reality, IoT, artificial intelligence, big data, and robotics [3]. Ref. [4] found Industry 4.0 technologies assist manufacturing companies' sustainability and increase their economic potential. Scholars have examined Industry 4.0 technologies in diverse industries besides manufacturing. Ref. [5] implemented a systematic review to understand the use of Industry 4.0 technologies in managing pandemics. The use of Industry 4.0 technologies to meet the increasing demands of society is ubiquitous; applications include robotics [6], artificial intelligence [7], IoT [8], augmented reality [9], big data [10], and machine learning [11] in food and agricultural sciences, to assist in more efficient and enhanced production, which is needed to feed a growing world population. Ref. [12] investigated the use of Industry 4.0 technologies in the manufacturing sector, as examined in 380 papers prior to 2020. Ref. [13] sought to understand the manufacturing patterns implemented based on Industry 4.0 technologies.

Modern advanced manufacturing technology and key technologies demonstrate the fundamental competitiveness of a nation's manufacturing industry. Robotics is a significant Industry 4.0 innovation that offers immeasurable possibilities in manufacturing disciplines [14]. An industrial robot is a complex system that succeeds in situations involving cross-work environments, high repetition, and high-precision processing. The current manual-based processing methods cannot meet all the requirements of a short development cycle and high assembly precision [15], and the use of industrial robots for processing is an excellent solution to this problem. The repeat positioning accuracy of industrial robots during the actual work process is usually quite satisfactory [16], and is typically 0.1 mm. However, the absolute positioning accuracy is poor, with an accuracy range of only approximately 2–3 mm. The absolute positioning accuracy severely limits the promotion and application of industrial robots in the manufacturing industry.

To address the problem of poor absolute positioning accuracy at the end of industrial robots, scholars at home and abroad have proposed various solutions [17]. Kinematic model-based control of robot joints makes it possible to compensate for absolute positioning errors in industrial robots. However, the positioning accuracy of the robot is affected by the size of the error of each kinematic parameter. The kinematic parameter errors can be addressed using kinematic parameter identification. The resulting parameter errors can be applied to the kinematic model to realize the adjustment of the model. This increases the positioning accuracy of the robot in the actual working environment, and the generated kinematic errors can be used to upgrade the kinematic model. This improves the positioning accuracy requirements of robots in actual workplaces.

In addition to positioning errors caused by geometric factors, nongeometric factors, such as gear gap, joint deformation, and temperature change, also affect the end positioning accuracy of robots [18]. The error mechanisms affecting robot positioning accuracy are complex and interconnected [19]. It is difficult to establish an accurate kinematic model that can account for all sources of error. Researchers have begun to investigate the establishment of a mapping relationship between the theoretical and actual position values.

A co-kriging-based error compensation method [20] was suggested to improve the positioning accuracy of the aerial drilling robot. A compensation method based on error similarity and error correlation was proposed to increase the robot's positioning accuracy [21]. First, the maximum working stiffness of the robotic drilling system in a specific machining task was obtained. This was achieved by optimizing the mounting angle between the motor spindle and the robot end flange. This optimization laid the groundwork for achieving high hole-processing accuracy. Second, the method for calculating the corresponding compensation value was introduced. This was undertaken according to the position to be drilled. The method took into account the force deformation at the robot end and the absolute positioning error of the robot [22]. Combining error similarity and a radial basis function (RBF) neural network, Wang [23] developed a position error compensation approach. The robot joint angle and position error were used to fit the experimental semi-variance function. The bandwidth of the RBF neural network was modified using the parameters of the semi-variance function. The position error of the target position was also estimated using the RBF neural network. The estimated position error was used to modify the target position to achieve the compensation effect. In precision manufacturing, Li [24] introduced a synchronization estimation approach for the total inertia and load torque of spindle-tool systems. The synchronization method was based on a novel double extended sliding mode observer (DESMO), which synchronously tracked the total inertia and load torque. The robustness of DESMO was enhanced by inserting a robust activator to reduce the effect of coupling errors between the two expansion terms. This was critical to the precision control of the spindle tool and directly influenced the control performance.

Long-term research has been carried out by Tian Wei's team at the Nanjing University of Aeronautics and Astronautics to increase the absolute positioning accuracy of industrial robots. A robot positioning error compensation method based on a deep neural network [25] was proposed to perform Latin hypercube sampling in Cartesian space. A positioning error prediction model based on genetic particle swarm optimization and a deep neural network (GPSO-DNN) was developed to predict and compensate for positioning error. Then, a practical positioning error compensation scheme for mobile industrial robots was proposed [16]. A binocular vision measurement method for robot positioning was developed. A mapping model between theoretical and actual robot pose errors was proposed based on deep belief networks (DBN), and the pose error estimation was realized.

A method for optimizing neural networks using the genetic particle swarm algorithm was proposed. This was done to improve the positioning accuracy of robots [26]. The aim was to model and predict the positioning error of industrial robots and achieve the compensation of target points within the robot workspace. Tian [27] proposed an absolute positioning error compensation scheme based on the DBN and error similarity. Relevant scholars have conducted in-depth research. The accuracy and versatility of error prediction can be further studied to improve the absolute positioning accuracy of the robot [28]. In addition, there are many mathematical methods [29–32] that can also be used to study the positioning accuracy of robots. Analytic methods such as fractional order have gained more and more attention [33,34].

The DBN is simple in structure and is suitable for data training in industrial robots. The training time of the DBN is short, thereby helping to improve the efficiency of the robot. Meanwhile, the differential evolution (DE) algorithm is an optimization algorithm based on the theory of swarm intelligence. It has been widely used in many fields because of its simple principle, small number of controlled parameters, and strong robustness. Finally, evidence theory can make experimental results more reliable.

Min [35] proposed a stable and high-accuracy model-free calibration method for unopened robotic systems, which can significantly improve the robot positional accuracy. Ref. [36] proposed an adaptive hierarchical compensation method based on fixed-length memory window incremental learning and incremental model reconstruction. Real-time trajectory position error compensation technology that considers non-kinematic errors [37,38] has also been proposed.

An absolute positioning accuracy compensation algorithm is proposed for industrial robots based on the DBN. The DE algorithm is used to optimize the DBN. The number of layer nodes, learning rate, momentum factor, restricted Boltzmann machine (RBM) iterations, and DBN fine-tuning iterations improve the optimization effect based on six dimensions and nine parameters. Combined with the evidence theory, the position error mapping model of industrial robots is established to realize its absolute positioning accuracy compensation. The technical process is shown in Figure 1.

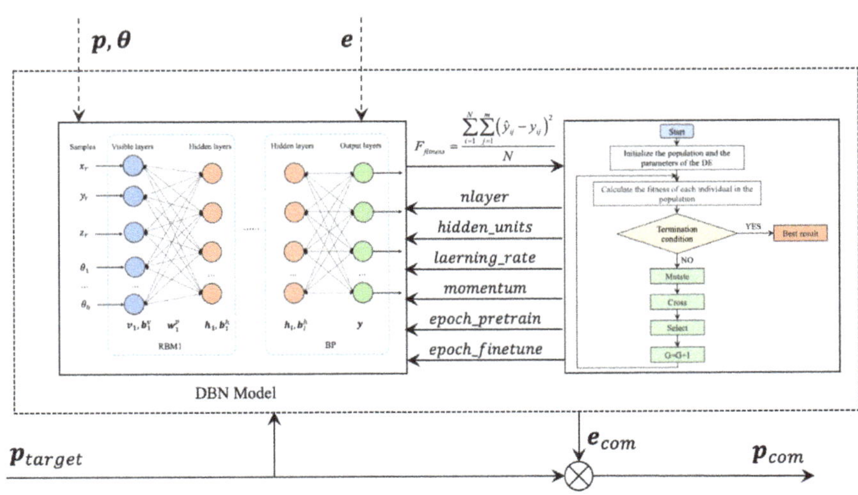

Figure 1. Accuracy compensation algorithm based on DE and DBN.

Combined with the off-line feed-forward compensation method, the prediction error of the theoretical pose coordinates of the robot target is superimposed on the robot control instructions. The validity and superiority of the scheme are verified using a AT901-B laser tracker on KUKA KR6_R700 sixx_CR. The absolute positioning error at the end of the robot was reduced by 82.14%, from 0.469 mm to 0.084 mm. Future work can further consider industrial robot load, motion speed, acceleration, ambient temperature, or other factors that affect the absolute positioning accuracy of the robot.

The chapter arrangement of this paper is as follows: Section 1 serves as the introduction, providing an overview of the absolute positioning accuracy of industrial robots and the method proposed in this paper. Section 2 focuses on the robot positioning error prediction algorithm based on DE and DBN. Section 3 presents supervised predictive optimization based on evidence theory. Section 4 includes experimental setup, data collection, model training, and result analysis. Finally, Section 5 presents the conclusion.

2. Robot Positioning Error Prediction Algorithm Based on DE and DBN

2.1. Principle of the DBN

The DBN is a probabilistic generative model proposed by Geoffrey Hinton [39] in 2006. It is composed of multiple RBM connections and a regression layer, and created by fine-tuning the resulting deep network through gradient descent and back propagation (BP) to form the best model.

RBM, as the basic component of the DBN [40], is a generative random artificial neural network, which can learn probability distribution from the inputs. The structure of RBM is shown in Figure 2. RBM consists of two layers: the visible layer $\boldsymbol{v}$ and the hidden layer $\boldsymbol{h}$. The visible layer is used to receive training data. In this study, the visible layer is used to accept the desired position of the end in the robot coordinate system and the current robot joint angle. The input of the hidden layer is the output of the visible layer, which is used to extract features. The neurons of the two layers have intra-layer no-connection and inter-layer full-connection relationships. Between the visible and hidden layers is the weight matrix w^p. $\boldsymbol{v}$ and $\boldsymbol{h}$ represent the visible-layer and hidden-layer vectors, respectively, and $\mathbf{b}^v$ and $\boldsymbol{b}^h$ represent the biases of the visible and hidden layers, respectively.

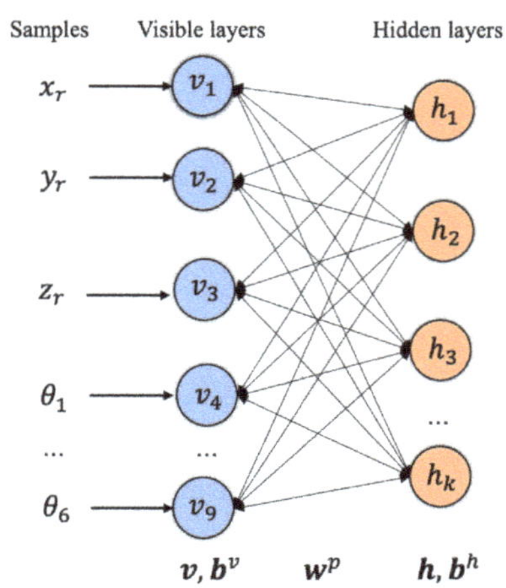

Figure 2. Architecture of RBM.

The multilayer RBM and the BP layer are stacked to form a DBN, as shown in Figure 3.

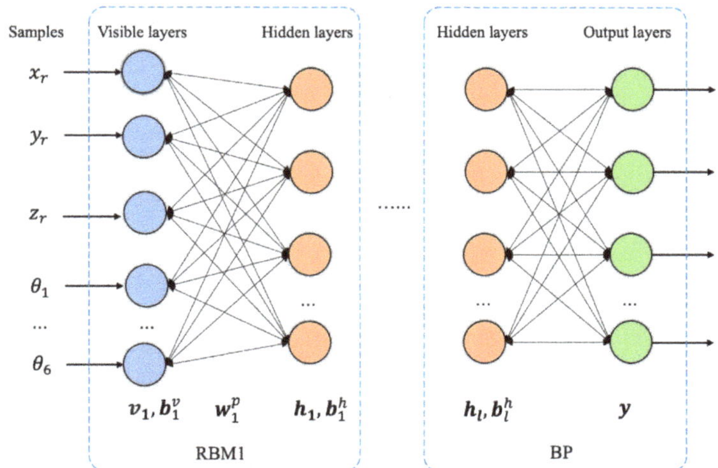

Figure 3. Architecture of DBN.

The first layer of RBM consists of a visible layer v_1 and a hidden layer h_1. The visible layer v_2 of the second layer of RBM is the hidden layer h_1 of the first layer of RBM, that is, $v_1 = h_1$, etc. The DBN realizes its layer-by-layer learning by stacking multiple RBMs, so as to extract the features of the data. The last layer of the DBN sets the BP network.

Unsupervised pretraining and fine-tuning are two processes of DBN training [41]. In the pretraining process, the greedy algorithm is used. The result obtained by the previous RBM training is used as the input of the next RBM until all RBMs are trained. The initial parameters of each RBM are obtained at the same time. The energy function of RBM is defined as follows:

$$E(v,h;\theta) = -\sum_{i=1}^{m} b_i v_i - \sum_{j=1}^{n} c_j h_j - \sum_{i,j=1}^{m,n} w_{ij} v_i h_j \tag{1}$$

$$\theta = \{w_{ij}, b_i, c_j\} \tag{2}$$

where m and n are the number of nodes in the visible and hidden layers, respectively. v_i,b_i are the biases between neurons in the visible layer. c_j, h_j are the biases between neurons in the hidden layer. w_{ij} is the connection weight value between the i-th neuron in the visible layer and the j-th neuron in the hidden layer. Based on the energy function, the probability distribution can be obtained as:

$$P(v,h;\theta) = \frac{1}{Z(\theta)} \exp(-E(v,h;\theta)) \tag{3}$$

where $Z(\theta)$ is the normalization factor expressed as:

$$Z(\theta) = \sum_{v,h} \exp(-E(v,h;\theta)) \tag{4}$$

The state probabilities of the hidden and visible layers are:

$$P(h_j = 1|v;\theta) = \varphi\left(b_j + \sum_{i=1}^{n} w_{ij} v_i\right) \tag{5}$$

$$P(v_j = 1|h;\theta) = \varphi\left(b_j + \sum_{j=1}^{m} w_{ij} h_i\right) \tag{6}$$

φ is the activation function, $\varphi = \frac{1}{1+\exp(-x)}$.

Fine-tuning refers to using the BP algorithm to train the entire network after pretraining [42], so that the entire DBN is in the best state, avoiding the disadvantages of critical points and long training time. Supposing $\boldsymbol{y}$ and $\hat{\boldsymbol{y}}$ are the actual output and desired output of the DBN, respectively, the loss function of the output layer is:

$$F(\tau) = \frac{1}{2}\sum_{1}^{N}(\hat{y} - y)^2 \tag{7}$$

where τ is the number of iterations and N is the number of training samples.

The weights between the hidden and output layers of the last layer of the network are iterated through an update function, and η is iteration rate of the DE algorithm:

$$w_{out}(\tau + 1) - w_{out}(\tau) = -\eta\frac{\partial F(\tau)}{\partial w_{out}(\tau)} \tag{8}$$

The DBN has strong robustness and fault tolerance because the information is distributed in the neurons in the network, and it can approximate any complex nonlinear system. Therefore, it is suitable for dealing with the nonlinear problem of error compensation.

2.2. DBN Optimization Based on DE Algorithm

The parameters such as number of hidden layers, number of nodes in the hidden layer, learning rate, momentum factor, number of RBM iterations, and the number of DBN fine-tuning iterations of DBN determine the complexity of the network. These parameters are important factors influencing the accuracy of the results of the prediction model. Other parameters such as speed of training and performance can also influence the accuracy of the results.

To achieve the best training effect, it is necessary to optimize the training parameters to obtain the optimal parameters. The DE algorithm is an optimization algorithm based on the theory of swarm intelligence. This algorithm was proposed by Rainer Storn and Kenneth Price [43] in 1995. It has been widely used in many fields because of its simple principle, small number of controlled parameters, and strong robustness.

The DE algorithm mainly includes four operations: initialization, mutation, crossover, and selection [44]. The population of DE is generated by random steps. The population size is N. The dimension of the search space is D. The dimension is the number of parameters to be optimized in this paper. The population initialization can be expressed as:

$$x_{i,j}(0) = x_{j,\min} + \textbf{\textit{rand}}[0,1] \cdot (x_{j,\max} - x_{j,\min}) \tag{9}$$

where $i \in [1, 2, \cdots, N]$, $j \in [1, 2, \cdots, D]$, $x(0)$ represents the 0th generation individual. $\textbf{\textit{rand}}[0,1]$ denotes uniformly distributed random numbers in $[0,1]$. $x_{j,\max}$ and $x_{j,\min}$ indicate the upper and lower bounds of the j-th chromosome, respectively.

After initialization, three different individuals are randomly selected from the population for mutation:

$$v_i^{child} = x_{ra} + F \cdot (x_{rb} - x_{rc}) \tag{10}$$

where F is the zoom factor, which can be set according to the actual situation.

A crossover operation is required to increase the diversity of the population. The new individual is generated by dimensionally crossing for each individual in the contemporary population and the new individual obtained from its mutation. The specific cross-operation process is defined as follows:

$$u_{i,G}^{j} = \begin{cases} v_{i,G}^{j}, rand[0,1] \leq CR \, or \, j = j_{rand} \\ x_{i,G}^{j}, otherwise \end{cases} \tag{11}$$

where ***CR*** is a crossover factor with a value range of $[0, 1]$. $u^j_{i,G}$ is a new individual generated by the crossover strategy.

After the crossover is completed, the DE algorithm selects each individual of the current population and the crossover individual. It keeps the best individual among the two as the next-generation population individual:

$$x^j_{i,G+1} = \begin{cases} u^j_{i,G}, if\, f(u_{i,j}) \leq f(x_{i,G}) \\ x^j_{i,G}, otherwise \end{cases} \tag{12}$$

The mean square error (MSE) between the expected output of the DBN and the actual output of the data is used as the fitness function of the DE algorithm:

$$F_{fitness} = \frac{\sum_{i=1}^{n}\sum_{j=1}^{m}\left(\hat{y}_{ij} - y_{ij}\right)^2}{N} \tag{13}$$

where ***N*** is the number of training sample data sets. m is the dimension of the network output. $\hat{y}_{ij}$ refers to the expected output of the sample. y_{ij} refers to the actual output of the network.

The DE algorithm is known as an efficient global optimizer with the advantages of convergence speed and high precision. The fitness function of the DE algorithm is related to the DBN, and the smaller the fitness function value, the better the optimization result.

The principle of the DE algorithm is shown in Figure 4.

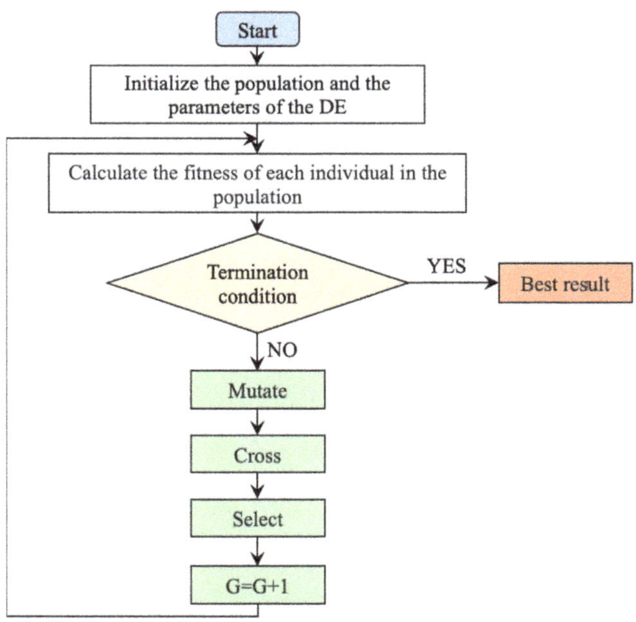

Figure 4. Flow chart of the DE algorithm.

The DE algorithm starts searching from a group, that is, multiple points instead of the same point [45]. This is the main reason why it can find the overall optimal solution with a greater probability. The evolution criterion of the DE algorithm is based on fitness information [46] with the help of other auxiliary information. It has inherent parallelism, which is suitable for large-scale parallel distributed processing [47]. The DE parameter settings are shown in Table 1.

Table 1. Parameters optimized by the DE algorithm.

Parameter	Symbol	Value
Dimensions of the objective function	n_dim	9
Maximum number of iterations	max_iter	200
Population size	size_pop	30
Mutation probability	prob_mut	0.8
Crossover probability	Crossp	0.7

The DBN input layer has nine channels. They are the theoretical position coordinates of the robot and the angles of the corresponding six joints $(x, y, z, \theta_1, \theta_2, \theta_3, \theta_4, \theta_5, \theta_6)$. The output layer of DBN is three channels, which are the position error (e_x, e_y, e_z) of the robot.

The maximum number of hidden layers of DBN nodes is set to four layers. The range of nodes in the hidden layer is (10, 101). The initial learning rate is 0.01. The momentum factor is 0.8. The activation function is Sigmoid, and the MSE is used as the loss function. The hyperparameters that need to be optimized are the number of hidden layers of the DBN, number of hidden-layer nodes, learning rate, momentum factor, RBM iterations, and DBN fine-tuning iterations. The hyperparameters are shown in Table 2.

Table 2. Parameters and optimization range of the DBN.

Parameter	Symbol	Range
Number of hidden layers of DBN	nlayer	(1, 4)
Number of nodes in DBN hidden layer 1	hidden_units [1]	(10, 101)
Number of nodes in DBN hidden layer 2	hidden_units [2]	(10, 101)
Number of nodes in DBN hidden layer 3	hidden_units [3]	(10, 101)
Number of nodes in DBN hidden layer 4	hidden_units [4]	(10, 101)
Learning rate of DBN	learning_rate	(0.01, 1)
Momentum factor of DBN	momentum	(0.8, 1)
Iterations of RBM	epoch_pretrain	(1, 100)
Fine-tuning iterations of DBN	epoch_finetune	(1, 200)

3. Supervised Predictive Optimization Based on Evidence Theory

The evidence theory is used as a supervised approach for assessing the uncertainty of DBN predictions. In order to assess the credibility of the DBN prediction, the uncertainty of the DBN prediction is evaluated [48]. This approach is taken to prevent the model from making mistakes due to overconfident performance and to improve the reliability of the prediction system. Evidence theory is a mathematical reasoning method with the characteristic of clearly expressing uncertainty. At present, in most cases, the basic probability assignment function is obtained by referring to expert experience and knowledge [49,50].

3.1. Evidence Theory

Evidence theory, also known as the Dempster–Shafer (DS) theory [51], is an uncertain reasoning theory widely used in the fields of information fusion and uncertain reasoning. The evidence theory can fuse evidence with comprehensible composite rules without prior probability. It can effectively deal with cognitive uncertainty in various engineering fields [52]. It can describe the corresponding fluctuation range of the system output through two boundary values: the belief function and the likelihood function. The fusion framework of the evidence theory usually consists of three parts: representation of evidence, fusion of evidence, and decision making of evidence [53].

The evidence theory is built on a framework of identification (FD) [54], usually represented by a nonempty set Θ, $\Theta = \{\theta_1, \theta_2, \cdots, \theta_n\}$. θ represents independent elements of a collection. The identification framework has 2^n subsets in total. Each subset A corresponds to a possible outcome of a proposition, that is, the proposition $\{\theta_i, \theta_j\}$ indicates that at least one of the two basic events is true. The confidence interval is usually used to describe events due to the lack of subjective knowledge.

The confidence interval is a closed interval composed of a belief function ($\boldsymbol{Bel}$) and a likelihood function ($\boldsymbol{Pl}$), which is used to indicate the degree of support for event $\boldsymbol{\theta}$ [55]. $\boldsymbol{Bel}(\boldsymbol{A})$ is the sum of the basic probability distributions of all subsets of $\boldsymbol{A}$, which indicates the degree of trust in $\boldsymbol{A}$. It is expressed as:

$$\begin{cases} \boldsymbol{Bel}(\boldsymbol{A}) = \sum\limits_{B \subseteq A} \boldsymbol{m}(\boldsymbol{B}) \\ A \subseteq \Theta \end{cases} \tag{14}$$

$\boldsymbol{Pl}(\boldsymbol{A})$ is the sum of the basic probability assignments of all subsets intersecting with $\boldsymbol{A}$. It indicates the degree of non-denial to $\boldsymbol{A}$. It is expressed as:

$$\begin{cases} \boldsymbol{Pl}(\boldsymbol{A}) = \sum\limits_{B \cap A \neq \varnothing} \boldsymbol{m}(\boldsymbol{B}) \\ A \subseteq \Theta \end{cases} \tag{15}$$

Let the finite nonempty set $\Theta = \{\boldsymbol{\theta}_1, \boldsymbol{\theta}_2, \cdots, \boldsymbol{\theta}_n\}$ be the identification framework, and the function $\boldsymbol{m}$:$2^n \to [0,1]$ be the basic probability assignment function on the identification framework Θ. In this study, $\boldsymbol{Bel}$ and $\boldsymbol{Pl}$ represent the upper and lower bounds of the reliability of the positioning accuracy of industrial robots. For the hypothetical conclusion $\boldsymbol{A}$ in the identification framework, $\boldsymbol{Bel}(\boldsymbol{A})$ and $\boldsymbol{Pl}(\boldsymbol{A})$ form a confidence interval denoted $\boldsymbol{Bel}(\boldsymbol{A}), \boldsymbol{Pl}(\boldsymbol{A})$.

This represents a propositional uncertainty, where the probability of the occurrence of proposition $\boldsymbol{A}$ lies somewhere between $\boldsymbol{Bel}$ and $\boldsymbol{Pl}$ bounds. It is shown in Figure 5.

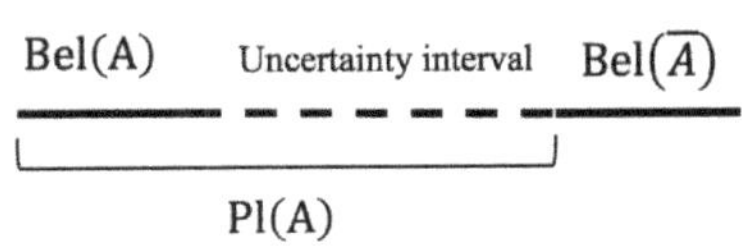

Figure 5. Propositional uncertainty representation.

3.2. *Optimization of Position Error Based on Evidence Theory*

This study evaluates the uncertainty of DBN prediction using the evidence theory. It combines the evidence theory with the DBN model and achieves the fusion of evidence through the basic probability assignment function and the Dempster combination rule. The evidence theory does not rely on prior information [56] and is relatively more suitable for situations where it is inconvenient to obtain prior probability and conditional probability [57]. In the DBN prediction task, *n* feature vectors of the input sample are represented as $\boldsymbol{\varphi}(\boldsymbol{x}) = (\boldsymbol{\varphi}_1(\boldsymbol{x}), \boldsymbol{\varphi}_2(\boldsymbol{x}), \cdots, \boldsymbol{\varphi}_n(\boldsymbol{x}))$. The identification frame $\Theta = \{\boldsymbol{\theta}_1, \boldsymbol{\theta}_2, \cdots, \boldsymbol{\theta}_k\}$ can be regarded as evidence corresponding to a subset in the identification frame. For each category $\boldsymbol{\theta}_k$, evidence $\boldsymbol{\varphi}_j(\mathbf{x})$ can be considered to support either $\{\boldsymbol{\theta}_\mathbf{k}\}$ or $\overline{\{\boldsymbol{\theta}_k\}}$, and the specific support depends on the weight coefficient of the evidence $\boldsymbol{\varphi}_j(\boldsymbol{x})$:

$$\omega = \boldsymbol{\beta}\boldsymbol{\varphi}_j(\boldsymbol{x}) + \boldsymbol{\alpha} \tag{16}$$

where $\boldsymbol{\beta}$ and $\boldsymbol{\alpha}$ are two weight parameters.

Assume that the evidence weights of $\{\boldsymbol{\theta}_k\}$ and $\overline{\{\boldsymbol{\theta}_k\}}$ are the positive and negative parts of ω, respectively. For each $\boldsymbol{\varphi}_j(\boldsymbol{x})$ and each category $\boldsymbol{\theta}_k$, two basic probability assignment functions exist:

$$\boldsymbol{m}^{+} = \{\boldsymbol{\theta}_k\}^{\omega^{+}} \tag{17}$$

$$\boldsymbol{m}^{-} = \{\boldsymbol{\theta}_k\}^{\omega^{-}} \tag{18}$$

where $\boldsymbol{m}^{+}$ represents $\boldsymbol{\varphi}_j(\boldsymbol{x})$ support for $\{\boldsymbol{\theta}_k\}$, and $\boldsymbol{m}^{-}$ represents $\boldsymbol{\varphi}_j(\boldsymbol{x})$ support for $\overline{\{\boldsymbol{\theta}_k\}}$.

The process of quantitative evaluation method for the uncertainty of DBN prediction is mainly divided into two stages. The first stage is the construction stage of the evidence classifier. The appropriate DBN model is built according to the needs, and the training samples are used to train the model to ensure that the model can meet the requirements of high classification accuracy. The model is saved. The model parameters are loaded. The parameters of the last hidden layer of the model are extracted. The extracted parameters are converted into the basic probability assignment function. The Dempster synthesis rule is used to combine and output the final basic probability assignment function.

The second stage is the uncertainty modeling of DBN prediction. The basic probability assignment function is used as the decision index to predict the classification results. At the same time, the basic probability assignment function is further calculated to obtain the conflict value and uncertainty. The uncertainty of DBN prediction is quantitatively evaluated using the uncertainty evaluation module. The uncertainty evaluation method for DBN prediction is shown in Figure 6.

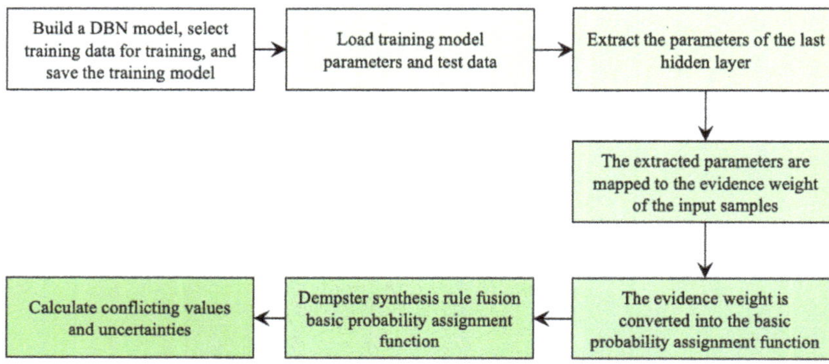

Figure 6. Uncertainty evaluation method for DBN prediction.

The evaluation method proposed earlier is based on the evidence classifier. The evidence classifier is modeled by extracting the parameters of the hidden layer of the trained DBN model. During the modeling process, the modification of the training loss function and the retraining of the DBN model are not required. Such characteristics mean that the evaluation method can be applied to any pre-trained DBN model and has strong scalability in the application of the DBN model.

4. Experiments

4.1. Experimental Setup and Data Collection

The experimental platform for compensating absolute positioning accuracy compensation of industrial robots is shown in Figure 7. The industrial robot used for precision compensation is KUKA's KR6_R700 sixx_CR. It has a load capacity of 6 kg and a working range of 700 mm radius. The volume of the working space is 1.36 m^3, the position repeatability is ± 0.03 mm, and the absolute positioning accuracy is ± 0.6 mm. A Leica AT901-B laser tracker is used to measure the position error and the error is ± 15 µm + 6 µm/m. The error of the laser tracker increases with the increase in distance.

AT901-B uses an angle encoder to measure the angle and an absolute interferometer to measure the distance. The absolute interferometer in the AT901 integrates a helium-neon laser interferometer and an absolute range finder. The two lasers can work independently. The laser beam emitted by the laser is directed to the target through the universal mirror. The interferometer laser beam also serves as the collimation axis for the tracker. The reflected laser light is measured using the tracker's built-in dual-axis position detector. The pulse generated by the position detector is processed by the processor of the tracker. The output is then fed back to the servo motor, which drives the motor to track the target mirror of the tracker in real time. Finally, the tracking distance measurement is realized, which is used to measure the actual pose of the end effector of the industrial robot.

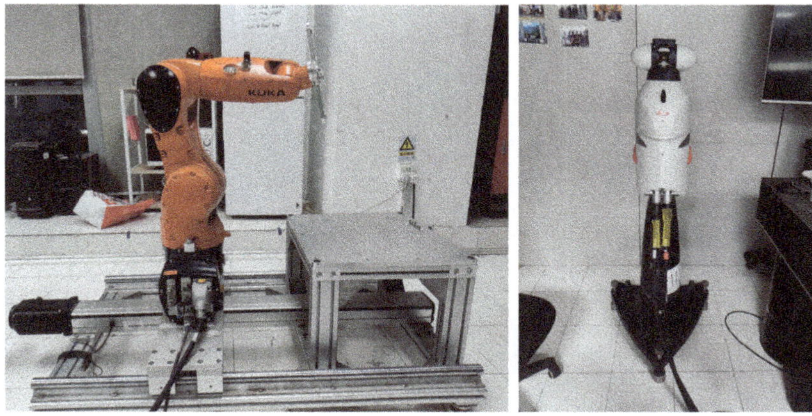

Figure 7. Error compensation platform for laser trackers and industrial robots.

The control and communication diagram of the error compensation platform is shown in Figure 8. The computer is used as the TwinCAT master, that is, the primary controller of the control system. The TwinCAT master uses industrial Ethernet EtherCAT to communicate with industrial robots. The laser tracker communicates with the TwinCAT master via Ethernet (TCP/IP protocol).

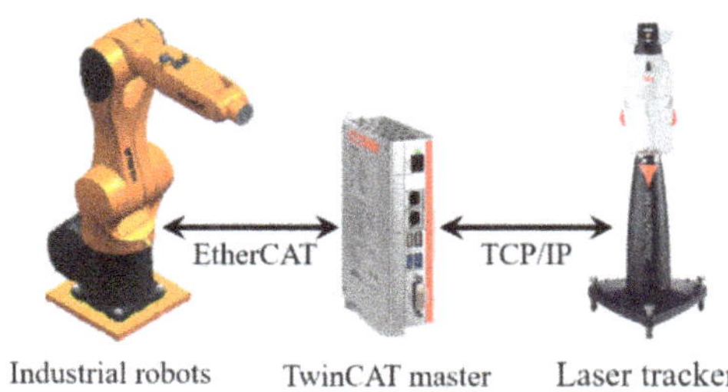

Figure 8. Communication of the experimental setup.

In this study, an off-line compensation method [58] is adopted, which uses a laser tracker to obtain the actual position of the manipulator. The DBN based on the DE algorithm is employed to perform the error compensation function.

Assuming that the measurement requirements of the laser tracker are met, the target ball of the fixed tooling of the industrial robot is installed in the 240 mm × 240 mm × 200 mm working space, and about 8000 sets of data are measured. For the universality and randomness of the experimental data, the random number module *drand* is used in TwinCAT3 to randomly generate specific sampling data within a predetermined sampling space. In order to obtain the real position of the robot and the laser tracker in steady state, each sampling is divided into three steps. First, the robot arrives at the sampling point and remains there for 2000 ms. Then, the laser tracker records the data for 1000 ms. Finally, the devices are delayed for another 1000 ms in order to reset them. The theoretical position coordinates and joint angles of the robot are the input of the model. The absolute position error of the robot end constitutes the output of the model.

The data set is divided into the training set and the test set in a ratio of 0.3. The 8000 sets of collected data are divided into 5600 sets of training data and 2400 sets of test data. As shown in Figure 9, the blue dots represent the training set and the red dots represent the test set.

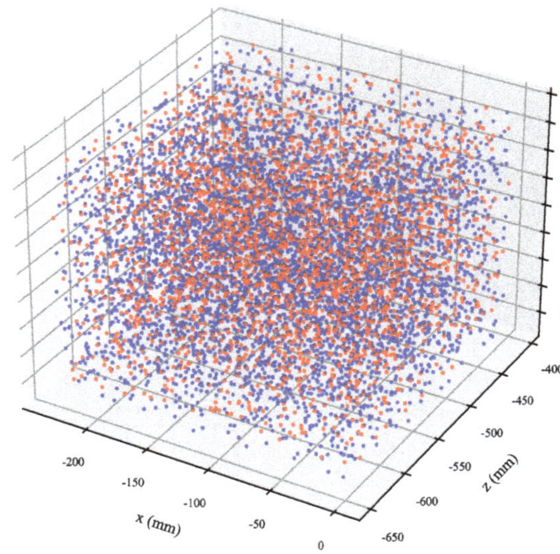

Figure 9. Sample data set.

4.2. Model Training and Result Analysis

According to the setting in Section 2.2, the input layer of DBN has nine channels, which are the theoretical position coordinates and joint angles of the robot $(x, y, z, \theta_1, \theta_2, \theta_3, \theta_4, \theta_5, \theta_6)$. The output layer of DBN has three channels, which are the position error (e_x, e_y, e_z) of the robot. Figure 10 shows the loss function decline curve when the DBN is trained alone.

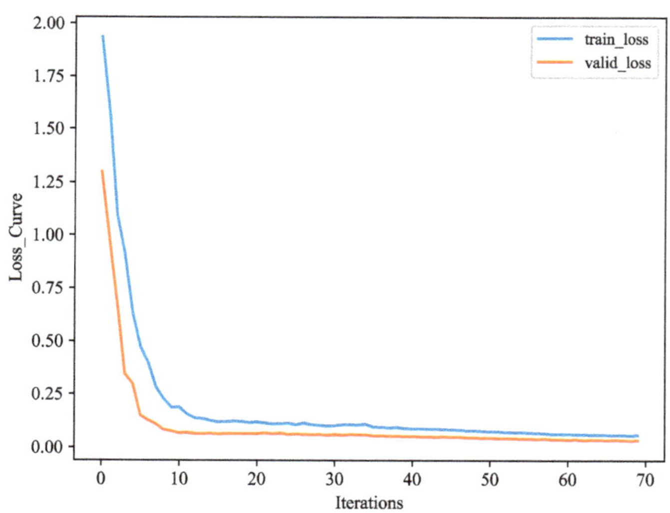

Figure 10. Loss curves.

The DE algorithm is utilized to determine the number of hidden layers of the DBN, number of nodes in the hidden layer, learning rate, momentum factor, number of iterations of RBM, and number of iterations of DBN fine-tuning. Figure 11 shows the fitness decline curve of the DE-optimized DBN, which is reduced by 60.7% from 0.387 to 0.152. After 150 iterations of training, the optimal fitness iteration number can be found at the end of the 92nd iteration, and the optimal parameters of the DBN can be output.

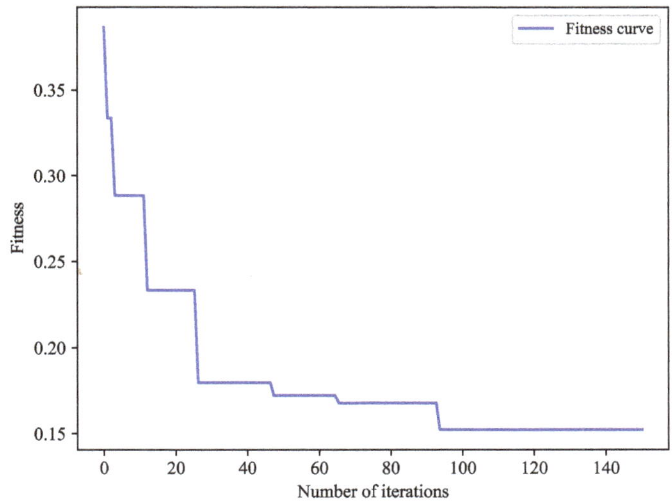

Figure 11. Fitness curve.

The DE algorithm is utilized to determine the number of hidden layers of the DBN, number of hidden-layer nodes, learning rate, momentum factor, number of iterations of RBM, and number of iterations of DBN fine-tuning. The fitness of each particle is calculated according to the fitness calculation conditions. When the training error reaches the allowable value or the number of iterations reaches the maximum value, the DE algorithm iteration terminates.

Finally, the DBN hyperparameters determined according to the DE algorithm are shown in Table 3.

Table 3. Optimal parameters of DBN.

Parameter	Symbol	Optimal Value
Number of hidden layers of DBN	nlayer	4
Number of nodes in DBN hidden layer 1	hidden_units [1]	7
Number of nodes in DBN hidden layer 2	hidden_units [2]	56
Number of nodes in DBN hidden layer 3	hidden_units [3]	62
Number of nodes in DBN hidden layer 4	hidden_units [4]	56
Learning rate of DBN	learning_rate	0.9360
Momentum factor of DBN	momentum	0.9618
Iterations of RBM	epoch_pretrain	100
Fine-tuning iterations of DBN	epoch_finetune	70

Supervised learning in machine learning essentially yields a series of training samples and establishes a mapping relationship so that the fitting result is as close as possible to the real output. The loss function is an important indicator for analyzing the quality of the training results. In this paper, the DBN is analyzed using five indicators: MSE, root mean square error (RMSE), mean absolute percentage error (MAPE), mean absolute error (MAE), and R^2. These five indicators are expressed as follows:

$$\mathrm{MSE} = \frac{1}{n}\sum_{i=1}^{n}(y_i - \hat{y}_i)^2 \tag{19}$$

$$\mathrm{RMSE} = \sqrt{\frac{1}{n}\sum_{i=1}^{n}(y_i - \hat{y}_i)^2} = \sqrt{\mathrm{MSE}} \tag{20}$$

$$\mathrm{MAPE} = \frac{1}{n}\sum_{i=1}^{n}\frac{|y_i - \hat{y}_i|}{y_i} \tag{21}$$

$$\mathrm{MAE} = \frac{1}{n}\sum_{i=1}^{n}|y_i - \hat{y}_i| \tag{22}$$

$$\mathrm{R}^2 = 1 - \frac{\sum\limits_{i=1}^{n}(y_i - \hat{y}_i)^2}{\sum\limits_{i=1}^{n}(y_i - \overline{y}_i)^2} = 1 - \frac{\mathrm{RMSE}}{\mathrm{Var}} \tag{23}$$

where y_i represents the true value of the data set. $\hat{y}_i$ represents the predicted value of the data set. $\overline{y}_i$ represents the average value of the predicted value of the data set. n represents the number of data sets. Var represents the data variance.

MSE is the mean of the sum of squares of the corresponding point errors between the predicted data and the original data. RMSE is the square root of MSE, also known as the fitting standard deviation of the regression system. MAPE is often used to measure the accuracy of predictions. However, when the real data are equal to zero, the denominator becomes zero and the formula is not available. The situation where the true value is zero does not appear in this study. The MAE refers to the average value of the absolute value of the deviation of each measurement value, which accurately reflects the size of the actual prediction error. The closer the four indicators approach 0, the closer the predicted value will be to the real value, indicating a better prediction effect.

R^2 represents the coefficient of determination of the model. The best score is 1, indicating that the model perfectly predicts the real value. It may also be negative because the model can be arbitrarily worse, that is, no mapping-fitting relationship exists between the predicted data and the real data.

Table 4 shows the calculated MSE, RMSE, MAPE, MAR, and R^2 from the DBN model.

Table 4. MSE, RMSE, MAPE, MAE, and R^2 values.

	x	y	z
MSE	0.0104	0.0017	0.0003
RMSE	0.1021	0.0412	0.0194
MAPE	0.1021	0.0412	0.0193
MAE	0.0824	0.0284	0.0900
R^2	0.8701	0.9038	0.9582

The MSE, RMSE, MAPE, and MAE of the position error prediction value of the robot end effector are all found to be close to 0 using the proposed position error prediction model for industrial robots. The coefficient of determination (R^2) of the predicted value of the position error is shown in Figures 12–14. The figures show that the predicted value (blue dot) is closely distributed around the real value (red line). In addition, R^2 is close to 1, indicating the correlation between the predicted value and the actual value. The higher the value, the higher the fitting accuracy. Therefore, the proposed machine learning model has good adaptability and robustness in the prediction of industrial robot position errors.

The R^2 of the robot end precision compensation error is around 0.87–0.95, and the overall effect is good. As the DBN is trained and iterated in three dimensions, the characteristics of the three dimensions interact and couple with each other. It is also disturbed by nonlinear factors such as the accuracy of data acquisition and environmental conditions, which makes some differences in the effect of robot end accuracy error compensation.

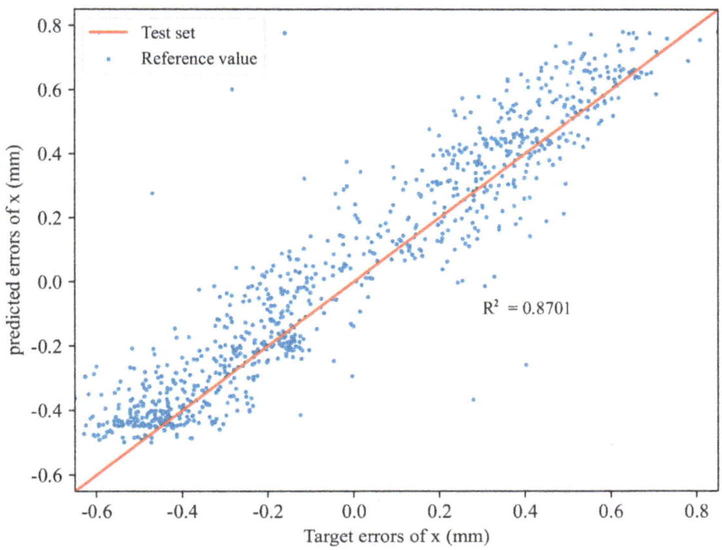

Figure 12. R^2 diagram of *x*.

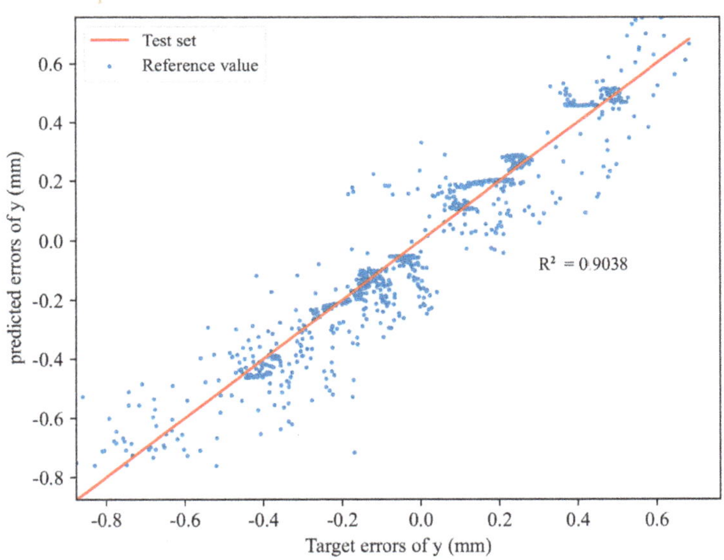

Figure 13. R^2 diagram of *y*.

A total of 50 random verification points were selected in the robot motion space with a measurement space of 240 × 240 × 200 mm^3. These are presented to verify the effectiveness and improvement effect of the DBN optimization based on the DE algorithm. The distribution of position errors before and after compensation in the *x*, *y*, and *z* directions are shown in Figures 15–17, respectively. The compensation results revealed the following. Before compensation, the errors in the *x* direction are basically evenly distributed above and below 0. The errors in the *y* direction are also distributed around 0, but are more negative. The errors in the *z* direction are basically negative. The error compensation technology proposed in this study is used to optimize the DBN with the DE algorithm. The errors in the three directions are basically distributed around 0 and fluctuate around ±0.2 mm, ±0.1 mm, and ±0.5 mm. The range of fluctuations is extremely small, indicating that the accuracy after compensation has high stability and that the accuracy of robot operation can be improved.

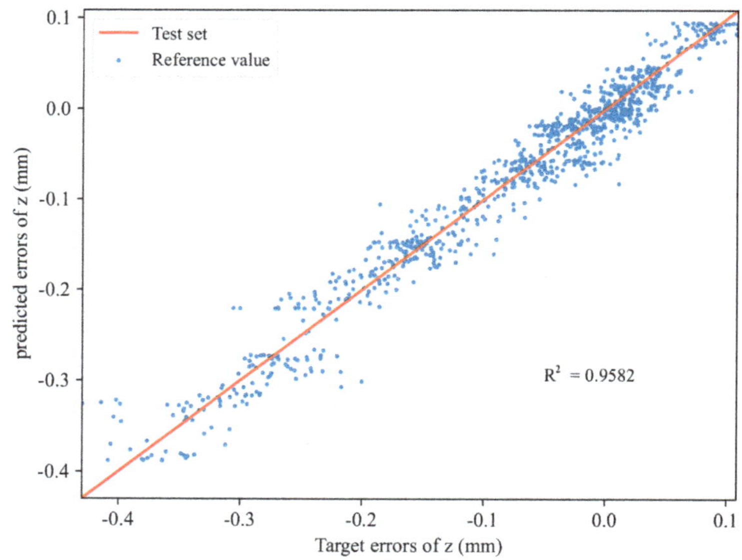

Figure 14. R^2 diagram of z.

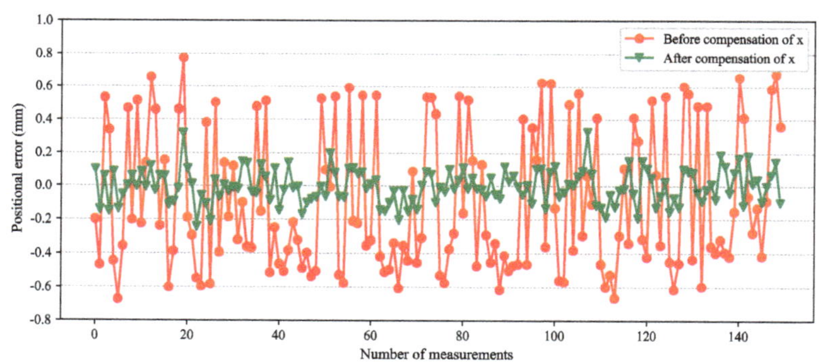

Figure 15. Position error on x before and after compensation of the robot.

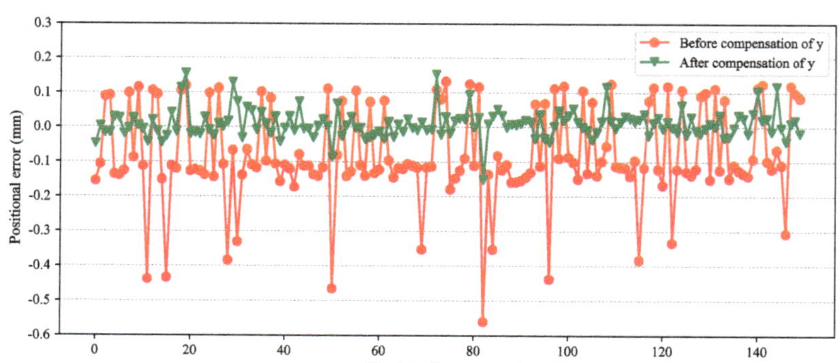

Figure 16. Position error on y before and after compensation of the robot.

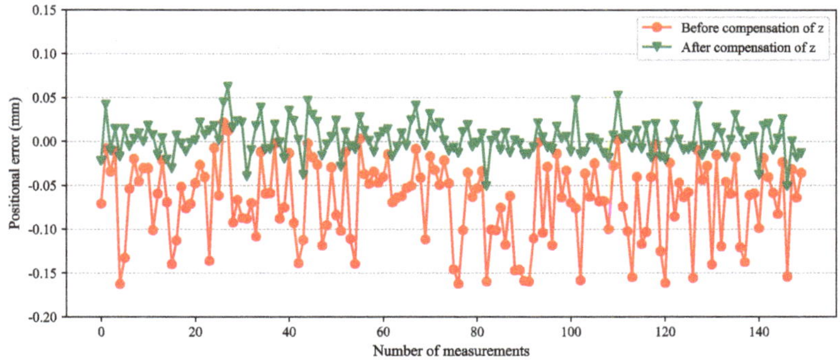

Figure 17. Position error on z before and after compensation of the robot.

Table 5 shows the static statistical analysis results before and after the robot end position error compensation. The x, y, and z directions are improved by 65.56%, 55.22%, and 49.12%, respectively.

Table 5. Statistical results of the positional errors.

		Error Range	Confidence	Percent Improvement
x error (mm)	Before	[−0.674, 0.773]	[−0.500, 0.345]	65.56%
	After	[−0.130, 0.368]	[−0.017, 0.017]	
y error (mm)	Before	[0.133, −0.559]	[−0.225, 0.063]	55.22%
	After	[0.108, −0.201]	[−0.047, 0.031]	
z error (mm)	Before	[0.003, −0.162]	[−0.115, −0.026]	49.12%
	After	[0.029, −0.054]	[−0.021, 0.009]	

The experimental platform for data acquisition and verification of the robot is the light industrial robot KR6_R700 sixx_CR. The error range is much smaller compared with that of the traditional heavy industrial robot. Hence, using the DBN for feature extraction, model training, and optimization is difficult. The network is optimized and combined with the evidence theory, and the position error mapping model of industrial robots is established. The comprehensive analysis of the compensation effect of the robot end accuracy in three directions is shown in Figure 18.

The method used in this study is compared with previous methods to verify the test results [25,27]. The results are shown in Table 6. After off-line compensation, the minimum value is reduced from 0.097 mm to 0.006 mm. The average value is reduced from 0.110 mm to 0.083 mm. Therefore, the proposed DE-DBN method was successful in improving the minimum and average values after the end error compensation of the robot.

Table 6. Comparison of data compensation on the results of previous models.

Error (mm)	Max	Min	Average
Uncompensated	1.529	0.124	0.754
GA-DNN	0.965	0.017	0.284
PSO-DNN	0.519	0.172	0.333
GPSO-DNN	0.364	0.097	0.249
PSO-DBN	0.244	Null	0.110
Uncompensated	0.701	0.139	0.469
DE-DBN	0.255	0.006	0.083

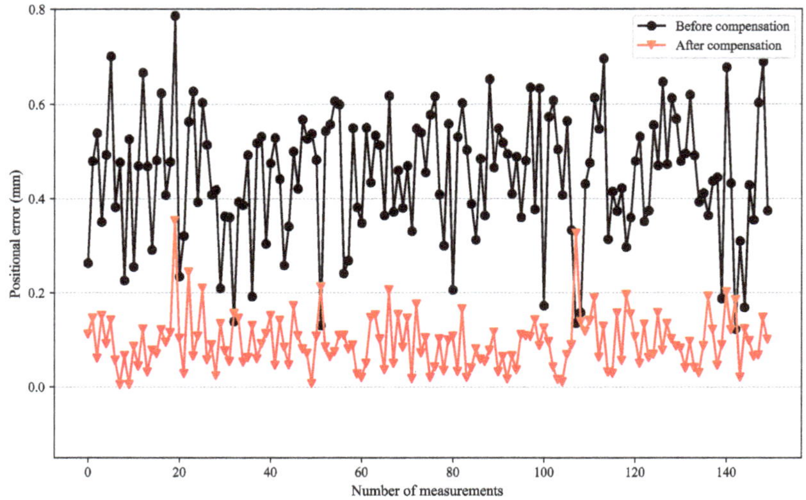

Figure 18. Absolute position errors of the robot end effector.

5. Conclusions

Based on deep belief networks using an off-line compensation method, a compensation algorithm for the absolute positioning accuracy of industrial robots is proposed. It predicts and compensates for the absolute positioning error of industrial robots based on the DBN and DE algorithm. The number of hidden layers, hidden-layer nodes, learning rate, momentum factors, RBM iterations, and DBN fine-tuning iterations are optimized. The position error model of industrial robots is established. Combined with the off-line feedback compensation method, the proposed method is verified experimentally using the KR6_R700 sixx_CR industrial robot and the AT901-B laser tracker.

After compensation, the absolute positioning error of the robot end is reduced by 82.14%, from 0.469 mm to 0.084 mm. The absolute positioning accuracy of the industrial robot is improved. This indicates the proposed approach is advantageous for performing more precise operation tasks. The results of this paper can be used to improve the absolute positioning accuracy of industrial robots, which is of great help in improving the motion accuracy and force control performance of robots.

Considerations of the off-line compensation method include an experimental environment free of vibration, the allowable operating temperature of the robot, and the higher accuracy of the laser tracker. Future work can further consider industrial robot load, motion speed, acceleration, ambient temperature, or other factors that affect the absolute positioning accuracy of the robot. Deep learning can be integrated into the robot's motion control system. The training model can be deployed in the control algorithm. Realizing the intelligent prediction and real-time compensation of robot errors is a direction of great research value.

Author Contributions: Conceptualization, Y.T. and H.L.; methodology, Y.T.; software, H.L. and S.C.; validation, J.L.; formal analysis, Q.Q.; investigation, W.X.; resources, H.L.; data curation, S.C.; writing—original draft preparation, H.L.; writing—review and editing, H.L.; visualization, S.C.; supervision, W.X.; project administration, Y.T. and W.X.; funding acquisition, Y.T. and W.X. All authors have read and agreed to the published version of the manuscript.

Funding: This research was funded by the Ministry of Industry and Information Technology of the People's Republic of China, National Key Research and Development Plan "Intelligent Robot" Project No. 2022YFB4700402. and No. 2019YFB1310100.

Data Availability Statement: All data have been included in the manuscript.

Acknowledgments: The authors would like to thank all the colleagues who contributed to this research.

Conflicts of Interest: The authors declare no conflict of interest.

References

1. Mubarak, M.F.; Petraite, M. Industry 4.0 Technologies, Digital Trust and Technological Orientation: What Matters in Open Innovation? *Technol. Forecast Soc. Chang.* **2020**, *161*, 120332. [CrossRef]
2. Papakostas, N.; Constantinescu, C.; Mourtzis, D. Novel Industry 4.0 Technologies and Applications. *Appl. Sci.* **2020**, *10*, 6498. [CrossRef]
3. Jaskó, S.; Skrop, A.; Holczinger, T.; Chován, T.; Abonyi, J. Development of Manufacturing Execution Systems in Accordance with Industry 4.0 Requirements: A Review of Standard- and Ontology-Based Methodologies and Tools. *Comput. Ind.* **2020**, *123*, 103300. [CrossRef]
4. Rosin, F.; Forget, P.; Lamouri, S.; Pellerin, R. Impacts of Industry 4.0 Technologies on Lean Principles. *Int. J. Prod. Res.* **2019**, *58*, 1644–1661. [CrossRef]
5. Moosavi, J.; Bakhshi, J.; Martek, I. The Application of Industry 4.0 Technologies in Pandemic Management: Literature Review and Case Study. *Healthc. Anal.* **2021**, *1*, 100008. [CrossRef]
6. Klerkx, L.; Rose, D. Dealing with the Game-Changing Technologies of Agriculture 4.0: How Do We Manage Diversity and Responsibility in Food System Transition Pathways? *Glob. Food Sec.* **2020**, *24*, 100347. [CrossRef]
7. Javaid, M.; Haleem, A.; Khan, I.H.; Suman, R. Understanding the Potential Applications of Artificial Intelligence in Agriculture Sector. *Adv. Agrochem.* **2023**, *2*, 15–30. [CrossRef]
8. Strong, R.; Wynn, J.T.; Lindner, J.R.; Palmer, K. Evaluating Brazilian Agriculturalists' IoT Smart Agriculture Adoption Barriers: Understanding Stakeholder Salience Prior to Launching an Innovation. *Sensors* **2022**, *22*, 6833. [CrossRef]
9. Ronaghi, M.; Ronaghi, M.H. Investigating the Impact of Economic, Political, and Social Factors on Augmented Reality Technology Acceptance in Agriculture (Livestock Farming) Sector in a Developing Country. *Technol. Soc.* **2021**, *67*, 101739. [CrossRef]
10. Osinga, S.A.; Paudel, D.; Mouzakitis, S.A.; Athanasiadis, I.N. Big Data in Agriculture: Between Opportunity and Solution. *Agric. Syst.* **2022**, *195*, 103298. [CrossRef]
11. Ahn, J.; Briers, G.; Baker, M.; Price, E.; Sohoulande Djebou, D.C.; Strong, R.; Piña, M.; Kibriya, S. Food Security and Agricultural Challenges in West-African Rural Communities: A Machine Learning Analysis. *Int. J. Food Prop.* **2022**, *25*, 827–844. [CrossRef]
12. Zheng, T.; Ardolino, M.; Bacchetti, A.; Perona, M. The Applications of Industry 4.0 Technologies in Manufacturing Context: A Systematic Literature Review. *Int. J. Prod. Res.* **2021**, *59*, 1922–1954. [CrossRef]
13. Frank, A.G.; Dalenogare, L.S.; Ayala, N.F. Industry 4.0 Technologies: Implementation Patterns in Manufacturing Companies. *Int. J. Prod. Econ.* **2019**, *210*, 15–26. [CrossRef]
14. Javaid, M.; Haleem, A.; Singh, R.P.; Suman, R. Substantial Capabilities of Robotics in Enhancing Industry 4.0 Implementation. *Cogn. Robot.* **2021**, *1*, 58–75. [CrossRef]
15. Zhang, T.; Yu, Y.; Yang, L.X.; Xiao, M.; Chen, S.Y. Robot Grinding System Trajectory Compensation Based on Co-Kriging Method and Constant-Force Control Based on Adaptive Iterative Algorithm. *Int. J. Precis. Eng. Manuf.* **2020**, *21*, 1637–1651. [CrossRef]
16. Wang, W.; Tian, W.; Liao, W.; Li, B. Pose Accuracy Compensation of Mobile Industry Robot with Binocular Vision Measurement and Deep Belief Network. *Optik* **2021**, *238*, 166716. [CrossRef]
17. Qi, J.; Chen, B.; Zhang, D. Compensation for Absolute Positioning Error of Industrial Robot Considering the Optimized Measurement Space. *Int. J. Adv. Robot Syst.* **2020**, *17*. [CrossRef]
18. Kong, L.B.; Yu, Y. Precision Measurement and Compensation of Kinematic Errors for Industrial Robots Using Artifact and Machine Learning. *Adv. Manuf.* **2022**, *10*, 397–410. [CrossRef]
19. Cao, C.T.; Do, V.P.; Lee, B.R. A Novel Indirect Calibration Approach for Robot Positioning Error Compensation Based on Neural Network and Hand-Eye Vision. *Appl. Sci.* **2019**, *9*, 1940. [CrossRef]
20. Chen, D.; Yuan, P.; Wang, T.; Cai, Y.; Xue, L. A Compensation Method for Enhancing Aviation Drilling Robot Accuracy Based on Co-Kriging. *Int. J. Precis. Eng. Manuf.* **2018**, *19*, 1133–1142. [CrossRef]
21. Chen, D.; Yuan, P.; Wang, T.; Ying, C.; Tang, H. A Compensation Method Based on Error Similarity and Error Correlation to Enhance the Position Accuracy of an Aviation Drilling Robot. *Meas. Sci. Technol.* **2018**, *29*, 085011. [CrossRef]
22. Shen, N.Y.; Guo, Z.M.; Li, J.; Tong, L.; Zhu, K. A Practical Method of Improving Hole Position Accuracy in the Robotic Drilling Process. *Int. J. Adv. Manuf. Technol.* **2018**, *96*, 2973–2987. [CrossRef]
23. Chen, D.; Wang, T.; Yuan, P.; Sun, N.; Tang, H. A Positional Error Compensation Method for Industrial Robots Combining Error Similarity and Radial Basis Function Neural Network. *Meas. Sci. Technol.* **2019**, *30*, 125010. [CrossRef]
24. Wang, L.; Tang, Z.; Zhang, P.; Liu, X.; Wang, D.; Li, X. Double Extended Sliding Mode Observer-Based Synchronous Estimation of Total Inertia and Load Torque for PMSM-Driven Spindle-Tool Systems. *IEEE Trans. Ind. Inf.* **2022**, *19*, 8496–8507. [CrossRef]
25. Fu, S.; Li, Y.; Zhang, M.; Hu, J.; Hua, F.; Tian, W. Robot Positioning Error Compensation Method Based on Deep Neural Network. *J. Phys. Conf. Ser.* **2020**, *1487*, 012045. [CrossRef]
26. LI, B.; TIAN, W.; ZHANG, C.; HUA, F.; CUI, G.; LI, Y. Positioning Error Compensation of an Industrial Robot Using Neural Networks and Experimental Study. *Chin. J. Aeronaut.* **2022**, *35*, 346–360. [CrossRef]

27. Wang, W.; Tian, W.; Liao, W.; Li, B.; Hu, J. Error Compensation of Industrial Robot Based on Deep Belief Network and Error Similarity. *Robot Comput. Integr. Manuf.* **2022**, *73*, 102220. [CrossRef]
28. Qi, J.; Chen, B.; Zhang, D. A Calibration Method for Enhancing Robot Accuracy Through Integration of Kinematic Model and Spatial Interpolation Algorithm. *J. Mech. Robot* **2021**, *13*, 061013. [CrossRef]
29. Adel, M.; Khader, M.M.; Algelany, S. High-Dimensional Chaotic Lorenz System: Numerical Treatment Using Changhee Polynomials of the Appell Type. *Fractal. Fract.* **2023**, *7*, 398. [CrossRef]
30. Adel, M.; Khader, M.M.; Assiri, T.A.; Kallel, W. Numerical Simulation for COVID-19 Model Using a Multidomain Spectral Relaxation Technique. *Symmetry* **2023**, *15*, 931. [CrossRef]
31. Khader, M.M.; Inc, M.; Adel, M.; Akinlar, M.A. Numerical Solutions to the Fractional-Order Wave Equation. *Int. J. Mod. Phys. C* **2023**, *34*, 2350067. [CrossRef]
32. Adel, M.; Srivastava, H.M.; Khader, M.M. Implementation of an Accurate Method for the Analysis and Simulation of Electrical R-L Circuits. *Math. Methods Appl. Sci.* **2023**, *46*, 8362–8371. [CrossRef]
33. Adel, M.; Khader, M.M.; Ahmad, H.; Assiri, T.A.; Adel, M.; Khader, M.M.; Ahmad, H.; Assiri, T.A. Approximate Analytical Solutions for the Blood Ethanol Concentration System and Predator-Prey Equations by Using Variational Iteration Method. *AIMS Math.* **2023**, *8*, 19083–19096. [CrossRef]
34. Ibrahim, Y.F.; Abd El-Bar, S.E.; Khader, M.M.; Adel, M. Studying and Simulating the Fractional COVID-19 Model Using an Efficient Spectral Collocation Approach. *Fractal. Fract.* **2023**, *7*, 307. [CrossRef]
35. Min, K.; Ni, F.; Chen, Z.; Liu, H.; Lee, C.-H. A Robot Positional Error Compensation Method Based on Improved Kriging Interpolation and Kronecker Products. *IEEE Trans. Ind. Electron.* **2023**, 1–10. [CrossRef]
36. Zhou, J.; Zheng, L.; Fan, W.; Zhang, X.; Cao, Y. Adaptive Hierarchical Positioning Error Compensation for Long-Term Service of Industrial Robots Based on Incremental Learning with Fixed-Length Memory Window and Incremental Model Reconstruction. *Robot Comput. Integr. Manuf.* **2023**, *84*, 102590. [CrossRef]
37. Li, R.; Ding, N.; Zhao, Y.; Liu, H. Real-Time Trajectory Position Error Compensation Technology of Industrial Robot. *Measurement* **2023**, *208*, 112418. [CrossRef]
38. Ma, S.; Deng, K.; Lu, Y.; Xu, X. Error Compensation Method of Industrial Robots Considering Non-Kinematic and Weak Rigid Base Errors. *Precis. Eng.* **2023**, *82*, 304–315. [CrossRef]
39. Hinton, G.E.; Osindero, S.; Teh, Y.W. A Fast Learning Algorithm for Deep Belief Nets. *Neural Comput.* **2006**, *18*, 1527–1554. [CrossRef]
40. Gao, S.; Xu, L.; Zhang, Y.; Pei, Z. Rolling Bearing Fault Diagnosis Based on SSA Optimized Self-Adaptive DBN. *ISA Trans.* **2022**, *128*, 485–502. [CrossRef]
41. Wang, Y.; Pan, Z.; Yuan, X.; Yang, C.; Gui, W. A Novel Deep Learning Based Fault Diagnosis Approach for Chemical Process with Extended Deep Belief Network. *ISA Trans.* **2020**, *96*, 457–467. [CrossRef]
42. Liu, J.; Wu, N.; Qiao, Y.; Li, Z. Short-Term Traffic Flow Forecasting Using Ensemble Approach Based on Deep Belief Networks. *IEEE Trans. Intell. Transp. Syst.* **2022**, *23*, 404–417. [CrossRef]
43. Storn, R.; Price, K. Differential Evolution—A Simple and Efficient Heuristic for Global Optimization over Continuous Spaces. *J. Glob. Optim.* **1997**, *11*, 341–359. [CrossRef]
44. Ahmad, M.F.; Isa, N.A.M.; Lim, W.H.; Ang, K.M. Differential Evolution: A Recent Review Based on State-of-the-Art Works. *Alex. Eng. J.* **2022**, *61*, 3831–3872. [CrossRef]
45. Bilal; Pant, M.; Zaheer, H.; Garcia-Hernandez, L.; Abraham, A. Differential Evolution: A Review of More than Two Decades of Research. *Eng. Appl. Artif. Intell.* **2020**, *90*, 103479. [CrossRef]
46. Deng, W.; Shang, S.; Cai, X.; Zhao, H.; Song, Y.; Xu, J. An Improved Differential Evolution Algorithm and Its Application in Optimization Problem. *Soft Comput.* **2021**, *25*, 5277–5298. [CrossRef]
47. Khaparde, A.R.; Alassery, F.; Kumar, A.; Alotaibi, Y.; Khalaf, O.I.; Pillai, S.; Alghamdi, S. Differential Evolution Algorithm with Hierarchical Fair Competition Model. *Intell. Autom. Soft Comput.* **2022**, *33*, 1045–1062. [CrossRef]
48. Fang, Z.; Roy, K.; Mares, J.; Sham, C.W.; Chen, B.; Lim, J.B.P. Deep Learning-Based Axial Capacity Prediction for Cold-Formed Steel Channel Sections Using Deep Belief Network. *Structures* **2021**, *33*, 2792–2802. [CrossRef]
49. Tong, Z.; Xu, P.; Denœux, T. An Evidential Classifier Based on Dempster-Shafer Theory and Deep Learning. *Neurocomputing* **2021**, *450*, 275–293. [CrossRef]
50. Du, Y.W.; Zhong, J.J. Generalized Combination Rule for Evidential Reasoning Approach and Dempster–Shafer Theory of Evidence. *Inf. Sci.* **2021**, *547*, 1201–1232. [CrossRef]
51. Deng, X.; Jiang, W.; Wang, Z. Zero-Sum Polymatrix Games with Link Uncertainty: A Dempster-Shafer Theory Solution. *Appl. Math. Comput.* **2019**, *340*, 101–112. [CrossRef]
52. Gudiyangada Nachappa, T.; Tavakkoli Piralilou, S.; Gholamnia, K.; Ghorbanzadeh, O.; Rahmati, O.; Blaschke, T. Flood Susceptibility Mapping with Machine Learning, Multi-Criteria Decision Analysis and Ensemble Using Dempster Shafer Theory. *J. Hydrol.* **2020**, *590*, 125275. [CrossRef]
53. Pan, Y.; Zhang, L.; Li, Z.W.; Ding, L. Improved Fuzzy Bayesian Network-Based Risk Analysis with Interval-Valued Fuzzy Sets and D–S Evidence Theory. *IEEE Trans. Fuzzy Syst.* **2020**, *28*, 2063–2077. [CrossRef]
54. Xiao, F. Generalization of Dempster–Shafer Theory: A Complex Mass Function. *Appl. Intell.* **2020**, *50*, 3266–3275. [CrossRef]

55. Xiao, F. A New Divergence Measure for Belief Functions in D–S Evidence Theory for Multisensor Data Fusion. *Inf. Sci.* **2020**, *514*, 462–483. [CrossRef]
56. Feng, R.; Xu, X.; Zhou, X.; Wan, J. A Trust Evaluation Algorithm for Wireless Sensor Networks Based on Node Behaviors and D-S Evidence Theory. *Sensors* **2011**, *11*, 1345–1360. [CrossRef]
57. Wang, H.; Deng, X.; Jiang, W.; Geng, J. A New Belief Divergence Measure for Dempster–Shafer Theory Based on Belief and Plausibility Function and Its Application in Multi-Source Data Fusion. *Eng. Appl. Artif. Intell.* **2021**, *97*, 104030. [CrossRef]
58. Slavkovic, N.; Zivanovic, S.; Kokotovic, B.; Dimic, Z.; Milutinovic, M. Simulation of Compensated Tool Path through Virtual Robot Machining Model. *J. Braz. Soc. Mech. Sci. Eng.* **2020**, *42*, 374. [CrossRef]

electronics

MDPI

Article

A Data-Driven Approach Using Enhanced Bayesian-LSTM Deep Neural Networks for Picks Wear State Recognition

Dong Song [1,2,*] and Yuanlong Zhao [3,*]

[1] China Coal Research Institute, Beijing 100013, China
[2] Shanxi Tiandi Coal Machinery Co., Ltd., Taiyuan 030006, China
[3] CCIC London Co., Ltd., London NW9 4AJ, UK
* Correspondence: tymkysd@126.com (D.S.); zhaoyuanlong@cciclondon.com (Y.Z.)

Abstract: Picks are key components for the mechanized excavation of coal by mining machinery, with their wear state directly influencing the efficiency of the mining equipment. In response to the difficulty of determining the overall wear state of picks during coal-mining production, a data-driven wear state identification model for picks has been constructed through the enhanced optimization of Long Short-Term Memory (LSTM) networks via Bayesian algorithms. Initially, a mechanical model of pick and coal-rock interaction is established through theoretical analysis, where the stress characteristic of the pick is analyzed, and the wear mechanism of the pick is preliminarily revealed. A method is proposed that categorizes the overall wear state of picks into three types based on the statistical relation of the actual wear amount and the limited wear amount. Subsequently, the vibration signals of the cutting drum from a bolter miner that contain the wear information of picks are decomposed and denoised using wavelet packet decomposition, with the standard deviation of wavelet packet coefficients from decomposed signal nodes selected as the feature signals. These feature signals are normalized and then used to construct a feature matrix representing the vibration signals. Finally, this constructed feature matrix and classification labels are fed into the Bayesian-LSTM network for training, thus resulting in the picks wear state identification model. To validate the effectiveness of the Bayesian-LSTM deep learning algorithm in identifying the overall picks wear state of mining machinery, vibration signals from the X, Y, and Z axes of the cutting drum from a bolter miner at the C coal mine in Shaanxi, China, are collected, effectively processed, and then input into deep LSTM and Back-Propagation (BP) neural networks respectively for comparison. The results showed that the Bayesian-LSTM network achieved a recognition accuracy of 98.33% for picks wear state, showing a clear advantage over LSTM, BP network models, thus providing important references for the identification of picks wear state based on deep learning algorithms. This method only requires the processing and analysis of the equipment parameters automatically collected from bolter miners or other mining equipment, offering the advantages of simplicity, low cost, and high accuracy, and providing a basis for a proper picks replacement strategy.

Keywords: data-driven approach; picks wear state recognition; wavelet packet decomposition; Bayesian-LSTM

Citation: Song, D.; Zhao, Y. A Data-Driven Approach Using Enhanced Bayesian-LSTM Deep Neural Networks for Picks Wear State Recognition. *Electronics* **2023**, *12*, 3593. https://doi.org/10.3390/electronics12173593

Academic Editor: Domenico Ursino

Received: 30 July 2023
Revised: 21 August 2023
Accepted: 22 August 2023
Published: 25 August 2023

1. Introduction

The advancement of technologies such as the Internet of Things (IoT), 5G, Big Data, Cloud Computing, and Artificial Intelligence (AI) has promoted the integration and innovation of a new generation of information technology and coal-mining machinery technology, providing specific technical approaches for the digital transformation and upgrade of coal-mining equipment. The picks are a key component of coal-mining equipment for mechanized coal extraction, and their wear state directly affects the efficiency of the mining equipment. In the process of cutting coal-rock, the picks crush and cut coal-rock under the action of strong thrust, suffering from severe impact and high stress, and experience intense friction with the coal wall. There is a strong non-linear coupling effect and friction wear behavior between the picks and the coal-rock, which can easily lead to the wear failure of the picks [1]. According to statistics, wear failure accounts for as much as 75–90% of all failure modes of picks [2]. In actual coal-mining production, due to factors such as coal-rock characteristics and different picks' installation angles, the wear degree of the picks at different positions of the mining equipment's cutting drum will inevitably vary during the cutting process. For picks that wear out quickly, if they are not replaced in time, this will lead to increased wear on the other picks, seriously affecting the cutting efficiency of mining equipment. Replacing picks immediately requires stopping the operation of mining equipment, which also impacts work efficiency [3,4]. Coal-mining enterprises currently rely mainly on manual experience to decide whether to replace picks, and to prevent work efficiency from being affected by multiple pick replacements, they can only adopt the strategy of replacing all picks of different wear levels at once, leading to significant economic waste [5]. Therefore, if the wear state of the mining machinery picks can be accurately identified, it will not only allow real-time understanding of the wear state of the picks on the cutting drum, ensuring the efficient operation of mining equipment, but could also help to propose a scientific picks replacement strategy, significantly reducing production costs for enterprises.

Many scholars have conducted extensive research on the wear mechanism of picks and the prediction of pick life. Dewangan [6] used an electron microscope and X-ray energy dispersive spectroscopy to scan and analyze the images before and after the wear of the pick, revealing the wear mechanism of the pick and the method of predicting wear volume. Zhang et al. [7] used PFC 3D software 5.0 to simulate the cutting process of the pick, conducted cutting experiments on different coated picks, calculated the mass loss before and after cutting, and then predicted and analyzed the life of the pick. Qin et al. [8] proposed a reliability model for the competitive failure of picks under random load impact by considering the effects of sustained impact, variable rate acceleration degradation, and hard failure threshold changes on pick wear. Tian et al. [9] proposed a degradation model based on the Gamma process to describe the wear and tear of picks on the tunneling machine, realizing the prediction of the remaining life of picks.

In recent years, with the development of sensor technology, some scholars have used machine-learning methods to study the identification of the picks wear state. By studying the features of vibration, acoustic emission, cutting force, power, and current signals during the cutting process of picks, they have obtained indirect indicators reflecting the wear of picks, thus achieving the identification of the picks wear state. Zhang et al. [10–12] built experimental devices, extracted triaxial vibration signals, infrared temperature signals, and current signals of picks with different wear degrees during the cutting process, constructed a multi-feature signal sample database for picks with different degrees of wear, and established a pick wear degree identification model based on the BP neural network. Jin et al. [13] used an acoustic emission signal acquisition device to collect signals from cutting four different proportions of coal-rock specimens, applied three-layer wavelet packet decomposition and reconstruction technology to process the signals, and used D-S evidence theory to intelligently identify the degree of picks' wear.

In summary, regarding the identification of the wear state of picks, existing research focuses on one hand on using statistical methods to calculate the wear volume of picks and predict their lifespan, and on the other hand on identifying the wear state of individual picks based on multi-source information fusion. However, due to the constraints of the underground application environment in coal mines, less attention is paid to the overall wear state evaluation of the picks of mining equipment in coal-mine production. Moreover, in the application of machine-learning methods, the commonly used methods in the existing research are shallow learning algorithms, including Support Vector Machines (SVM), Hidden Markov Models (HMM), and BP neural networks. Compared with deep learning models, traditional machine learning and shallow learning algorithms have obvious disadvantages in terms of data-processing capacity, non-linear processing capabilities, and convergence performance. In addition, the adaptive feature learning characteristic of deep learning methods effectively avoids the limitations of manual feature extraction, gets rid of the dependence on prior knowledge, and has a higher recognition accuracy and model generalization capabilities [14].

To further enhance the efficiency and accuracy of identifying the wear state of picks, this paper proposes a model for identifying the overall wear state of picks based on the Bayesian-LSTM deep learning neural network. The main contributions are as follows:

(1) We employed a theoretical analysis method to establish a mechanical model of pick and coal-rock interaction, analyzing the stress characteristics of the picks and revealing the wear mechanism of the picks. In light of the overall wear characteristics of the picks, we achieved the rapid classification of three types of picks wear state through the statistical relationship between the pick wear amount and the limited pick wear amount.
(2) We used the wavelet packet decomposition method to decompose and denoise the vibration signals from the cutting drum of a bolter miner, which contain extensive picks wear information. We then used the standard deviation of the decomposed signal node wavelet packet coefficients as the feature signals. These feature signals, after normalization, were used to construct a feature matrix representing the vibration signals.
(3) We utilized the Bayesian algorithm for its advantages in handling uncertain data, integrating it into the LSTM network to construct a Bayesian-LSTM network. By inputting the constructed feature matrix and classification labels into the Bayesian-LSTM model for training, the recognition results demonstrated a higher accuracy compared to both LSTM and BP neural networks.

The brief structure of this article is as follows: Section 1 introduces the background of the research content and the importance of the research. Section 2 shows the related work about this topic. Section 3 presents the interaction model of the pick and coal-rock during the cutting process, the overall wear state evaluation index, and the feature signals selection of picks wear based on wavelet packet decomposition. Section 4 introduces the Bayesian-LSTM network model. Section 5 presents the analysis of validation and comparison of results. Section 6 provides the conclusions of this article.

2. Related Work

Based on the above, in the research of picks wear state recognition, the current recognition methods are only conducted within a small sample range. When the data volume is too large, computational difficulties arise, which cannot meet the demand for handling massive data in the recognition of the overall wear state of picks [15–17]. Therefore, the utilization of deep learning for recognizing the overall wear state of picks presents a significant advantage. Currently, deep learning algorithms have begun to be used in areas like machine tool wear state recognition. For instance, Huang et al. [18] proposed a new method for tool wear prediction based on a deep convolutional neural network and multi-domain feature fusion, constructing a high-accuracy tool wear prediction model combining adaptive feature fusion and automatic continuous prediction. Furthermore, Ma et al. [19] used milling force signals to establish a tool wear prediction model based on convolutional bidirectional LSTM networks, achieving highly accurate prediction results. On the basis of deep learning models, some researchers have attempted to use optimization algorithms to address the reliance of recognition models on large data samples. Wu et al. [20] optimized LSTM networks using a particle swarm optimization (PSO) algorithm and applied an improved polynomial threshold function to denoise tool acceleration vibration signals, thus achieving tool wear quantity prediction and wear state classification. Due to the picks wear state information typically being a time series signal, certain researchers have employed a 1D convolutional neural network (CNN) for the feature classification of temporal signals in related fields. Abdeljaber et al. [21] presented a compact 1D CNN architecture that integrates feature extraction and classification modules, enabling automatic extraction of optimal image-sensitive features directly from raw acceleration signals, utilized for real-time vibration-induced damage monitoring and localization, with a demonstrated outstanding performance and an exceptionally high computational efficiency. Yuan et al. [22] introduced a 1D CNN model for rapid and accurate comprehensive damage assessment post-earthquake. Their results revealed that the prediction accuracy of the 1D CNN model is comparable to that of 2D CNN models, yet with an over 90% reduced computation time and an over 69% resource usage reduction. Abdoli et al. [23] introduced a 1D CNN-based approach for environmental sound classification that directly captures audio signal patterns through convolutional layers, achieving an average accuracy of 89% with fewer data than traditional feature-based methods.

Through comparative analysis of the use of deep learning methods for tool wear state recognition, it can be observed that most of the recognition models still employ traditional structures such as CNN and LSTM. Some choose to combine optimization algorithms like PSO and Genetic Algorithms (GA) to address convergence problems during weight training. When considering the selection of input parameters, the vast majority of studies still rely on the research experience of their predecessors, without considering the impact of different input parameter combinations on the output results [24–27]. Additionally, these network models yield fixed weight matrices after training, and these weight matrices are no longer updated. The model cannot allocate different weights based on the change in inputs, so its generalization ability when faced with different tasks can be significantly constrained [28]. The LSTM deep learning network, with its unique memory units and gate mechanisms, is adept at capturing dependencies in time series, offering a distinct advantage in processing temporal data [29–32]. However, the randomness introduced by environmental factors and parameter choices might compromise the accuracy of the recognition results [33]. In recent years, several researchers have incorporated Bayesian theory into LSTM deep learning networks to estimate weights and biases. This approach shifts the neural network parameter estimation from point estimation to probability distribution, enabling the network to evaluate the certainty or uncertainty of results. Consequently, this enriches the deep learning network's formidable data-fitting capability, further enhancing its learning precision. Li et al. [34] proposed a method that leverages Bayesian-LSTM to perform Stochastic Variational Inference (SVI) on process-based hydrological models. By constructing a residual model, they sought to refine the predictions of uncertainty in hydrological models. The results demonstrated that this method provided a highly reliable uncertainty interval. Compared to the Bayesian linear regression model, Bayesian-LSTM offered superior uncertainty estimation. Yang et al. [35] introduced a HiBayes-LSTM method containing an FIE component to capture past and future time dependencies. By collecting large-scale HTRO datasets, they extended the weights of the LSTM network to a probabilistic model, ensuring uncertainty in the HM direction of the head trajectory predictions. Experimental outcomes revealed that HiBayes-LSTM notably outperformed nine other methods in predicting ODIs' significance.

3. Preliminaries

This section analyzes the wear mechanism of picks by establishing a mechanical model of the pick and coal-rock mass, and proposes a classification method for the overall wear state of the picks of mining machinery. At the same time, a method for selecting the characteristic parameters of the picks wear state is provided.

3.1. Pick and Coal-Rock Interaction Model

Picks are the main tools for mining machinery to cut coal-rock. The tip of pick, that is, the top of the alloy head cone, is mainly used to wedge into the coal-rock mass. After the pick wedges into the coal-rock mass, it comes into contact with the coal-rock, and a pick and coal-rock interaction model is shown in Figure 1. The uncut coal-rock mass shows unevenness, some relatively hard and sharp coal-rock particles are pressed into the tip surface under the action of the normal load, and they perform a cutting action on the pick tip during the drum movement process. The long-term reciprocating cutting action causes the material on the pick surface to continuously peel off, thus intensifying the wear of the pick. Therefore, the friction force on the alloy head is the main cause of the wear of the pick tip.

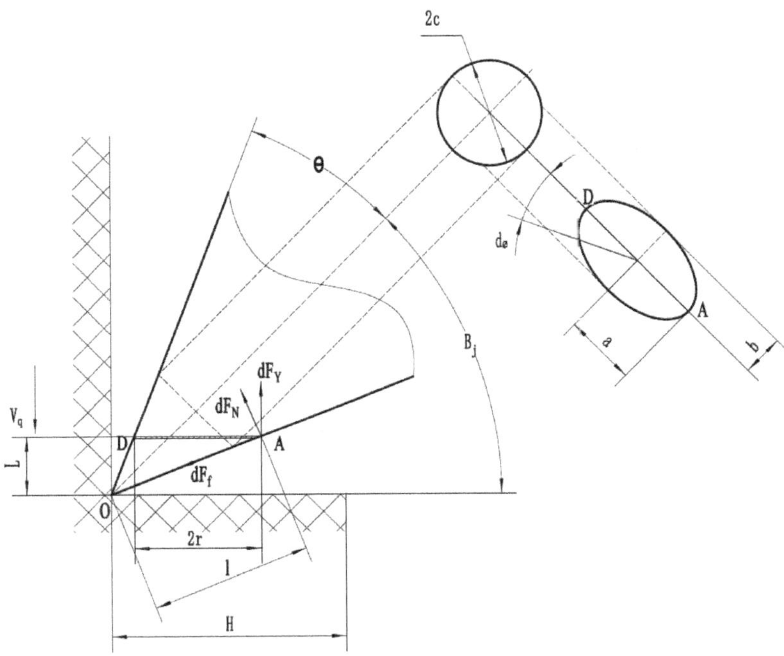

Figure 1. Interaction model of rotary pick and coal-rock mass.

According to the classic plane-cutting model of pick [36], it is assumed that the friction coefficient between the coal-rock mass and the pick is μ, and the relationship between the surface pressure stress q of the coal-rock mass and its compressive strength u is:

$$\mathrm{q} = u(\cos\theta - \mu\sin\theta) \tag{1}$$

where θ is the semi-cone angle of the pick tip.

The radius c of the circular hole of the pick tip in the coal-rock mass can be expressed as follows:

$$c = \frac{2t_q H}{u(\cos\theta - \mu\sin\theta)} \tag{2}$$

where H is the cutting thickness of the pick, and t_q is the tensile stress of the coal-rock mass.

As can be seen from Figure 1, when the pick on the drum is in a rotating cutting state, the cross-section AD perpendicular to the instantaneous cutting speed V_q direction on the pick body is elliptical. Take the differential element AD on the contact surface between the pick and the coal-rock mass for research, denoted as δA, according to the differential principle:

$$\delta A = r\delta\phi\delta l \tag{3}$$

where ϕ is the fracture angle, l is the length from the tip to a point on the pick body, and r is the radius of the cross-sectional circle.

According to the relative positional relationship, the semi-axes a and b of the ellipse can be expressed as follows:

$$a = \frac{c}{\cos B_j},\ b = c\left(1 - \tan B_j \tan\theta\right) \tag{4}$$

where B_j is the cutting angle of the cutting pick.

For ease of analysis, this ellipse is equivalently treated as a circle with a radius of a_0 to obtain:

$$a_0 = \frac{c\left[\cos\theta + \cos\left(\theta + B_j\right)\right]}{2\cos\theta\cos\beta_j} \tag{5}$$

From this, it can be determined that after considering the frictional force and cutting angle, the cutting force element on its conical surface when the pick is rotating and cutting is:

$$F_Y = dF_N \sin\theta + dF_f \cos\theta = \frac{(\sin\theta + \mu\cos\theta)}{(\cos\theta - \mu\sin\theta)} q\delta A = 2t_q \frac{H(\sin\theta + \mu\cos\theta)}{(\cos\theta - \mu\sin\theta)\sin\theta}\delta\varphi\delta r \tag{6}$$

where dF_f, dF_Y, and dF_N are the frictional force element, cutting force element, and normal pressure element on the pick surface, respectively.

After integrating Equation (6), the total horizontal force on the conical surface in interaction between the pick and the coal-rock mass is obtained:

$$\begin{aligned} F_Y = \int dF_Y &= 2t_q \frac{H(\sin\theta+\mu\cos\theta)}{(\cos\theta-\mu\sin\theta)\sin\theta} \int_0^{2\pi} d\varphi \int_0^{a_0} dr \\ &= 2\pi t_q c \frac{H(\sin\theta + \mu\cos\theta)}{(\cos\theta - \mu\sin\theta)\sin\theta}\left[\frac{\cos\theta + \cos\left(\theta + B_j\right)}{\cos\theta\cos B_j}\right] \end{aligned} \tag{7}$$

As can be seen from Equation (7), the cutting resistance of the pick in the rotating cutting condition is a quadratic function of its cutting thickness, it is directly proportional to the square of the tensile stress of the coal-rock mass and the ratio of the compressive strength, and it has a complex trigonometric function relationship with the cutting angle. The traction resistance of the pick is about (0.5–0.8) F_Y, and the lateral force is about (0.1–0.2) F_Y.

3.2. Overall Wear State Evaluation Index

As analyzed in the previous section, the interaction between the pick and the coal-rock mass is extremely complex, and its wear types mainly include abrasive wear, erosion wear, and fatigue wear, among which abrasive wear accounts for about 70~75% of the total wear volume. Abrasive wear refers to the phenomenon or process that causes surface material loss during the interaction between abrasives or hard micro-protrusions and the worn material surface. If the removed material is very small, it is usually referred to as micro-cutting. In the abrasive wear process of the pick, the abrasive can remove the pick material in one interaction. Therefore, the micro-cutting abrasive wear theoretical model can be used to study the macroscopic coal-rock cutting process of the pick.

When only considering the wear of the pick tip, the wear trend of the alloy head is shown in Figure 2. As can be seen, the wear of the alloy head is like a plane parallel to the coal-rock surface layer cutting the alloy head. The coal-rock surface continuously laminates the alloy head, while the height of the alloy head being cut gradually increases. When the layer cutting height changes from 1 mm to 3 mm, a significant change in its wear volume occurs, which can be represented by the cross-sectional area of the plane and the alloy head. The larger the wear area, the more wear the alloy head has.

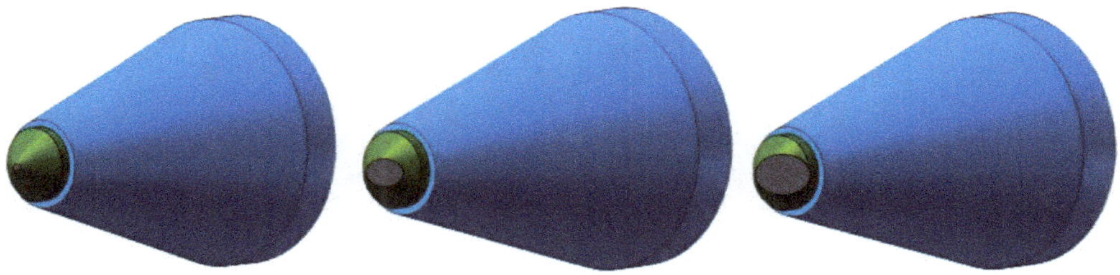

(a) (b) (c)

Figure 2. Trend of pick wear. (**a**) Slight wear, (**b**) Moderate wear, (**c**) Severe wear.

From the above analysis, it can be seen that for the wear of the pick tip, the contact area between the pick and the coal-rock is the main influencing factor. The larger the area of contact between the pick and the coal-rock, the more wear, that is, the more the area of contact between the pick tip and the coal-rock can reflect the wear amount of the pick. Therefore, without considering the self-rotation ability of the pick during the cutting process, the wear coefficient η of a single pick can be represented by the following equation:

$$\eta = \frac{S}{S_{\text{lim}}} \approx \frac{L}{L_{\text{lim}}} \tag{8}$$

where S is the contact area with the coal-rock, S_{lim} is the limited contact area with the coal-rock, L is the layer cutting thickness, and L_{lim} is the limited layer cutting thickness. Based on this, this article proposes to establish an overall wear state coefficient H based on the wear coefficient η of a single pick and use it to evaluate the overall wear situation of picks, thereby achieving a method of quickly obtaining the overall wear degree of picks during coal-mine production.

$$H = \sum_{i=1}^{N} S_i \eta_i / \sum_{i=1}^{N} S_i \tag{9}$$

where S_i is the total number of each type of pick and η_i is the single pick wear rate. According to the field pick replacement experience of engineering cases, before and after pick replacement, the overall picks wear state can be divided into three levels: slight wear, moderate wear, and severe wear. The range of the overall wear coefficient H corresponding to the determined various wear states is shown in Table 1.

Table 1. Classification of overall picks wear state.

Wear State	Overall Wear Coefficient H	Replacement Strategy
Slight wear	0~0.3	No action
Moderate wear	0.3~0.5	Check
Severe wear	0.5~1	Replace

3.3. Selection of Picks Wear Feature Signal

During the cutting process of mining machinery, the tip of the pick bears a high concentrated stress. Due to the small contact area of the pick tip, the picks violently rub against the coal-rock mass during the cutting process and generate vibration, accompanied by the propagation of vibration waves. Under certain cutting parameters, the vibration signals generated by the cutting drum with different global wear levels must be different. Therefore, this article chooses the vibration signal of the cutting drum to effectively identify the overall wear state of picks.

Wavelet packet analysis is a refined signal analysis method that can decompose the collected non-linear parameter signals into different scales, obtain the node features of the signals at each scale, and form a feature parameter group. The principle of extracting pick wear features with wavelet packets is as follows:

The expression of the wavelet packet function is

$$\mu_{j+1,k}^{n}(t) = 2^{\frac{j+1}{2}}\mu^{n}(2^{j+1}t - k) \tag{10}$$

where j is the scale parameter, n is the oscillation parameter, k is the translation parameter, and t is the time variable.

The wavelet packet function satisfies the double scale equation:

$$\begin{cases} \mu_{2n}(t) = \sqrt{2}\sum\limits_{k\in Z} h(k)\mu_{n}(2t-k) \\ \mu_{2n+1}(t) = \sqrt{2}\sum\limits_{k\in Z} g(k)\mu_{n}(2t-k) \end{cases} \tag{11}$$

In the formula, $h(k)$ is the coefficient of the low-pass filter, $g(k)$ is the coefficient of the high-pass filter, and $\{\mu_n(t)\}_{n\in Z}$ is the orthogonal wavelet packet.

The projection of the original parameter $x(t)$ signal on $\{\mu_n(t)\}_{n\in Z}$, that is, the wavelet packet coefficient is

$$d_j^k = \int_{-\infty}^{+\infty} x(t)\cdot\mu_{j+1}^{n}(t)dt \tag{12}$$

The algorithm of wavelet packet decomposition is

$$\begin{cases} d_j^{2n}(k) = \sum h(l-2k)d_{j+1}^{n}(l) \\ d_j^{2n+1}(k) = \sum g(l-2k)d_{j+1}^{n}(l) \end{cases} \tag{13}$$

This article relies on engineering examples to process the cutting vibration signal, compares the node wavelet packet coefficients of the signal with other signal feature parameters, and finally proposes to use the standard deviation of the vibration signal wavelet packet coefficients as a recognition indicator to identify the overall wear state of picks.

The feature vector definition for the overall wear state identification of picks is:

$$T(x(t),j,r) = \sqrt{\frac{1}{n}\sum_{k=1}^{n}[d_{j,r}(k) - \overline{d_{j,r}}]^2} \tag{14}$$

where $T(x(t),j,r)$ is the standard deviation of the wavelet packet coefficients of signal $x(t)$ at the node (j,r), $d_{j,r}(k)$ is the k-th wavelet packet coefficient of the signal $x(t)$ at the node (j,r), and $\overline{d_{j,r}}$ is the average of the wavelet packet coefficients of the signal $x(t)$ at the node (j,r).

4. Methods

The section illustrates the structure and characteristics of the LSTM neural network, proposes a Bayesian-LSTM neural network optimized by the Bayesian algorithm, and also provides the process of an overall picks wear state recognition model based on wavelet packet decomposition and Bayesian-LSTM.

4.1. LSTM Network

The LSTM deep learning network is a novel type of Recurrent Neural Network (RNN) with its inherent recursive traits. The LSTM model introduces three gates (input gate, forget gate, output gate) to control the historical information transmission between neural units, thereby avoiding the gradient explosion and gradient vanishing problems that may occur in the training process of traditional recursive neural networks [37]. It effectively handles the long-term dependency relationships present in long sequences. The vibration signal data used in this paper are a kind of time series data, which reflect the change trend in the wear condition of the picks over time. Therefore, this paper chooses to use the LSTM network for the state recognition of time series.

Within the architecture of an LSTM model, every unit holds a cell, which essentially acts as its memory store. The manner in which memory units in an LSTM are read and modified is controlled by three critical components: the input gate, the forget gate, and the output gate. Typically, sigmoid or tanh functions depict their operations. To illustrate, the operational process of an LSTM unit proceeds as such: at each interval, it absorbs two forms of external data-the current state and the preceding LSTM's hidden state. Additionally, an internal input, the state of the memory unit, is also fed to each gate. Following the receipt of these input data, the gates compute the data from diverse sources, and the outcomes determine their activation status. The input gate's input is manipulated via a nonlinear function, which then amalgamates with the memory unit state that the forget gate has handled, creating a novel memory unit state. Ultimately, the memory unit state, after being processed by a nonlinear function and dynamically managed by the output gate, becomes the LSTM unit's output. As a result, LSTM networks possess the capacity to retain long-term dependencies as they can selectively eliminate certain data, maintain beneficial information, and relay it to the subsequent step via the output gate. The basic structure of LSTM is shown in Figure 3.

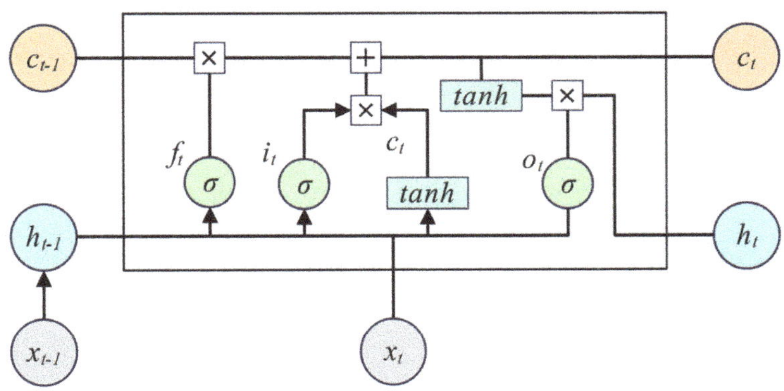

Figure 3. The basic structure of LSTM.

The data transfer within the LSTM neural unit follows these equations:

$$\text{Input gate}: \ \mathbf{i_t} = \sigma(\mathbf{W_{xi}x_t} + \mathbf{W_{hi}h_{t-1}} + \mathbf{W_{ci}c_{t-1}} + \mathbf{b_i}) \quad (15)$$

$$\text{Forget gate}: \ \mathbf{f_t} = \sigma(\mathbf{W_{xf}x_t} + \mathbf{W_{hf}h_{t-1}} + \mathbf{W_{cf}c_{t-1}} + \mathbf{b_f}) \quad (16)$$

$$\text{Output gate}: \ \mathbf{o_t} = \sigma(\mathbf{W_{xo}x_t} + \mathbf{W_{ho}h_{t-1}} + \mathbf{W_{co}c_t} + \mathbf{b_o}) \quad (17)$$

$$\text{Cell memory state}: \ \mathbf{c_t} = \mathbf{f}_t\mathbf{c_{t-1}} + \mathbf{i_t}\tanh(\mathbf{W_{xc}x_t} + \mathbf{W_{hc}h_{t-1}} + \mathbf{b_c}) \quad (18)$$

$$\text{Cell output}: \ \mathbf{h_t} = \mathbf{o_t}\tanh(\mathbf{c_t}) \quad (19)$$

In these equations, $\mathbf{W_{xc}}$, $\mathbf{W_{xi}}$, $\mathbf{W_{xf}}$, $\mathbf{W_{xo}}$ are weight matrices connected to the input signal $\mathbf{x_t}$; $\mathbf{W_{hc}}$, $\mathbf{W_{hi}}$, $\mathbf{W_{hf}}$, $\mathbf{W_{ho}}$ are weight matrices connected to the output signal $\mathbf{h_t}$ of the hidden layer; $\mathbf{W_{ci}}$, $\mathbf{W_{cf}}$, $\mathbf{W_{co}}$ are diagonal matrices connecting the output vectors of the neuron activation function and gate function; $\mathbf{b_i}$, $\mathbf{b_c}$, $\mathbf{b_f}$, $\mathbf{b_o}$ are bias vectors; and σ is the activation function, usually a tanh or sigmoid function.

4.2. Parameter Optimization Based on Bayesian Theory

The core idea of parameter optimization based on Bayesian theory is to treat LSTM as a Bayesian model, place prior distribution on the network weights and bias parameters of LSTM, and then use variational inference to infer the posterior distribution of the parameters given the data [38].

Bayesian theory considers $\boldsymbol{\theta}$ as a random variable that can be described by a probability distribution. According to Bayes' formula,

$$p(\boldsymbol{\theta} \mid \boldsymbol{y}_0) = \frac{p(\boldsymbol{y}_0 \mid \boldsymbol{\theta})p(\boldsymbol{\theta})}{p(\boldsymbol{y}_0)} \propto p(\boldsymbol{y}_0 \mid \boldsymbol{\theta})p(\boldsymbol{\theta}) \tag{20}$$

where $p(\boldsymbol{\theta} \mid \boldsymbol{y}_0)$ is the posterior distribution, $p(\boldsymbol{\theta})$ is the prior distribution, $p(\boldsymbol{y}_0)$ is the evidence or normalization constant, and its calculation formula is as follows:

$$p(\boldsymbol{y}_0) = \int p(\boldsymbol{y}_0 \mid \boldsymbol{\theta})p(\boldsymbol{\theta})d\boldsymbol{\theta} \tag{21}$$

Since the evidence is in integral form, it is mostly non-integrable except in some ideal situations. Therefore, the method of variational inference is often used, that is, a set of distributions are introduced to approximate the posterior distribution of parameters $p(\boldsymbol{\theta} \mid \boldsymbol{y}_0)$, denoted as $q(\boldsymbol{\theta} \mid \boldsymbol{\Lambda})$, where $\boldsymbol{\Lambda} = [\Lambda_1, \cdots \Lambda_N]$ is the variational parameter matrix corresponding to the model parameters $\boldsymbol{\theta} = [\theta_1, \cdots, \theta_N]$.

The difference between the original distribution and the variational distribution is generally measured by Kullback–Leibler (KL) divergence:

$$D_{KL}[q(\boldsymbol{\theta} \mid \boldsymbol{\Lambda}) \parallel p(\boldsymbol{\theta} \mid \boldsymbol{y}_0)] = -L(\boldsymbol{\Lambda}) + \lg p(\boldsymbol{y}_0) \tag{22}$$

In this equation, $L(\boldsymbol{\Lambda})$ is the Evidence Lower Bound (ELBO). It can be seen that the smaller the divergence, the greater the variational lower bound, indicating that the variational distribution is closer to the original distribution. The maximum value of the evidence lower bound is obtained to obtain the optimal distribution.

The idea of variational inference is to follow the gradient of variational parameters, express the gradient as an expected value, and use the Monte Carlo method to estimate this expectation. An unbiased gradient estimation is obtained by sampling from the variational distribution, which saves the analytical calculation of the variational lower bound. The objective function of variational inference is:

$$L(\boldsymbol{\Lambda}) = E_q[\lg p(\boldsymbol{y}_0, \boldsymbol{\theta}) - \lg(q(\boldsymbol{\theta} \mid \boldsymbol{\Lambda}))] \tag{23}$$

In this equation, E_q is the expectation about $(q(\boldsymbol{\theta} \mid \boldsymbol{\Lambda})$, and $p(\boldsymbol{y}_0, \boldsymbol{\theta})$ is the joint distribution of $\boldsymbol{y}_0$ and $\boldsymbol{\theta}$.

If Λ is the free parameter of $q(\boldsymbol{\theta} \mid \boldsymbol{\Lambda})$, the gradient of the lower bound of the distribution can be expressed as:

$$\nabla(L(\boldsymbol{\Lambda})) = E_{q(\Lambda)}[\nabla \ln q(\Lambda) \ln \frac{p(\boldsymbol{y}_0, \boldsymbol{\theta})}{q(\Lambda)}] \tag{24}$$

According to the Monte Carlo sampling method, the gradient of the variational lower bound is

$$\nabla(L(\boldsymbol{\Lambda})) = \frac{1}{N}\sum_{i=1} N[\nabla \ln q(\Lambda) \ln \frac{p(\boldsymbol{y}_0, \boldsymbol{\theta})}{q(\Lambda)}] \tag{25}$$

Therefore, for stochastic variational inference, the execution process of variational inference is

$$\Lambda_{t+1} = \Lambda_t + \rho_t \frac{1}{N}\sum_{i=1} N[\nabla \ln q(\Lambda) \ln \frac{p(\boldsymbol{y}_0, \boldsymbol{\theta})}{q(\Lambda)}] \tag{26}$$

where ρ_t is the learning rate. When the change in the free parameters Λ is less than a given tolerance, the calculation stops. Based on the inferred network weights and bias parameters from the posterior distribution, the network can continue to train according to the LSTM algorithm.

In summary, the proposed picks wear state recognition model is shown in Figure 4, and the process of the picks wear state recognition model based on wavelet packet decomposition and Bayesian optimization of LSTM is as follows:

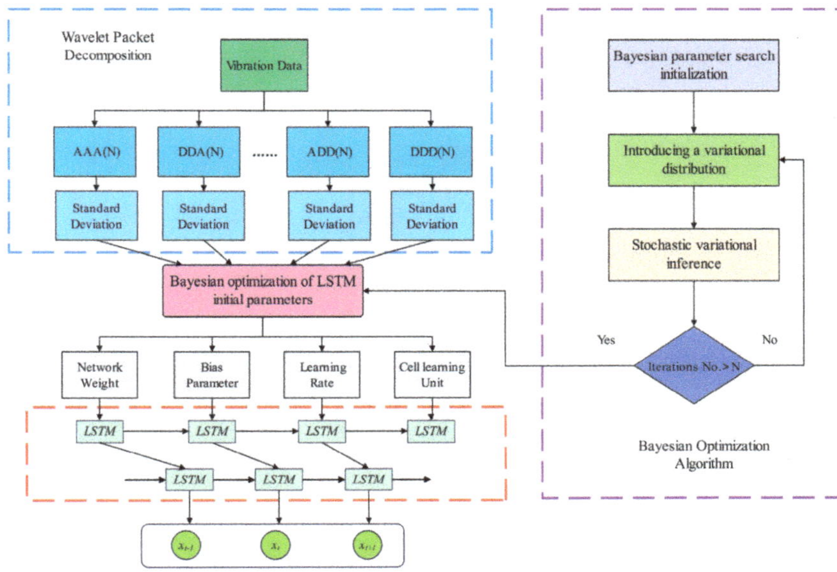

Figure 4. The framework of picks wear state recognition model.

(1) Use wavelet packet decomposition to decompose the original signal of cutting vibration, and choose the standard deviation of wavelet packet coefficients as the feature signal of the neural network;
(2) Establish the parameter seeking model of the LSTM network, and use Bayesian optimization theory to optimally seek parameters for the initial parameters of the LSTM network;
(3) Build and initialize the LSTM and fully connected layer network based on the parameter seeking result, and set the hyperparameters of the network;
(4) Train the network on the sample training set, and use the trained network to perform classification testing on the test samples.

5. Engineering Verification

To verify the effectiveness and efficiency of the proposed overall picks wear state recognition model, extensive engineering experiments on real datasets have been implemented against the classic methods under differently labeled rations.

5.1. Data Acquisition

In order to verify the effectiveness of the picks wear recognition method proposed in this article, the real parameters of the bolter miner of Shaanxi C mine in China are selected for verification. The roof of this mine is moderately stable, using a bolter miner for tunneling, the coal seam thickness is about 5 m, and the average daily advance is about 50 m. The properties of coal-rock mass in this mine are relatively stable, and the average daily number of pick replacements is found to be roughly equivalent through statistics, proving that the C mine is quite suitable for conducting picks wear state recognition experiments.

In the experiment, we continued to adopt the strategy of centrally changing the picks. By statistically measuring the wear volume of the picks before each shift, and calculating the overall wear coefficient H of the picks on drum cutting according to Formula (9), it was found that if the picks were not replaced for 2 days, the overall wear coefficient could reach 0.3. If extended to more than 3 days, the wear coefficient could reach 0.5. This preliminarily proved that delaying the replacement of picks can accelerate their wear. Therefore, the field collected the X, Y, and Z directional vibration acceleration of the cutting drum of the bolter miner immediately after replacing the picks, and 2 and 3 days later, respectively representing slight wear, moderate wear, and severe wear levels. The vibration sensor location is shown in Figure 5.

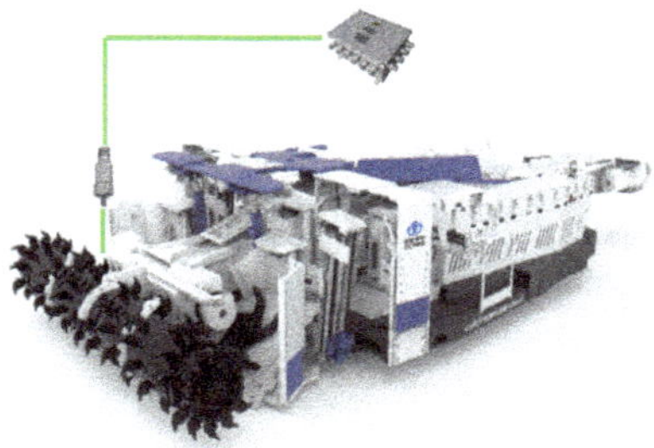

Figure 5. The vibration sensor location on bolter miner.

During the field tests, 2 s of X, Y, and Z directional vibration data were recorded every minute under each working condition, ensuring the data collection was in a stable state. In total, 100 sets of data were recorded under each working condition, with a total of two detection tests conducted, forming a total of 600 sets of characteristic data. The collected Y-directional raw vibration data are shown in Figure 6.

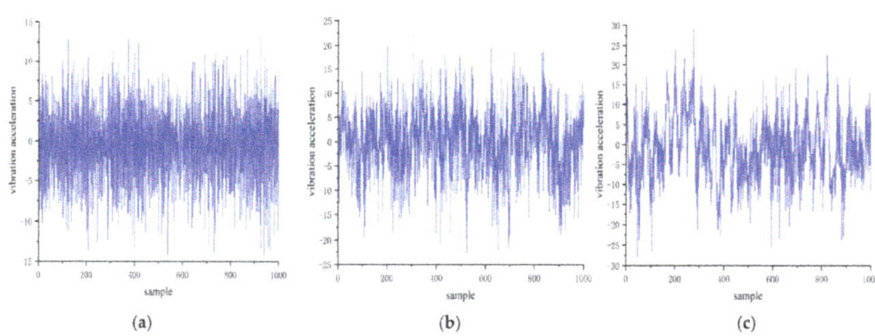

Figure 6. Y-directional raw vibration data. (**a**) Slight wear, (**b**) Moderate wear, (**c**) Severe wear.

5.2. Data Processing

As can be seen from Figure 6, the amplitude of the drum vibration signal is larger with an increased picks wear degree. In the process of collecting the vibration acceleration curve of the cutting drum, errors may occur due to factors such as the environment and noise, which cause inaccuracy in the signal. Merely utilizing time-domain analysis cannot adequately analyze the vibration signal. To more accurately identify the degree of wear of the cutting picks, this paper converts the acquired time-domain signal into a frequency-domain signal for further analysis to obtain more fitting evaluation parameters.

Wavelet packet analysis divides the signal into detailed hierarchical divisions to improve signal processing capabilities. This study chose to perform wavelet packet decomposition on the time-domain signals of the vibration acceleration of the cutting drum under three picks wear states, selecting DB wavelet basis. It was found that when n equals 8, the time-domain waveform of wavelet packet decomposition is the smoothest and the frequency characteristics are ideal; thus, this paper selected DB8 as the wavelet basis. After signal decomposition, each node is respectively recorded as (3, 0), (3, 1), (3, 2), (3, 3), (3, 4),

(3, 5), (3, 6), and (3, 7). Figure 7 shows the wavelet packet decomposition diagram of the Y-directional vibration signals of picks with moderate wear.

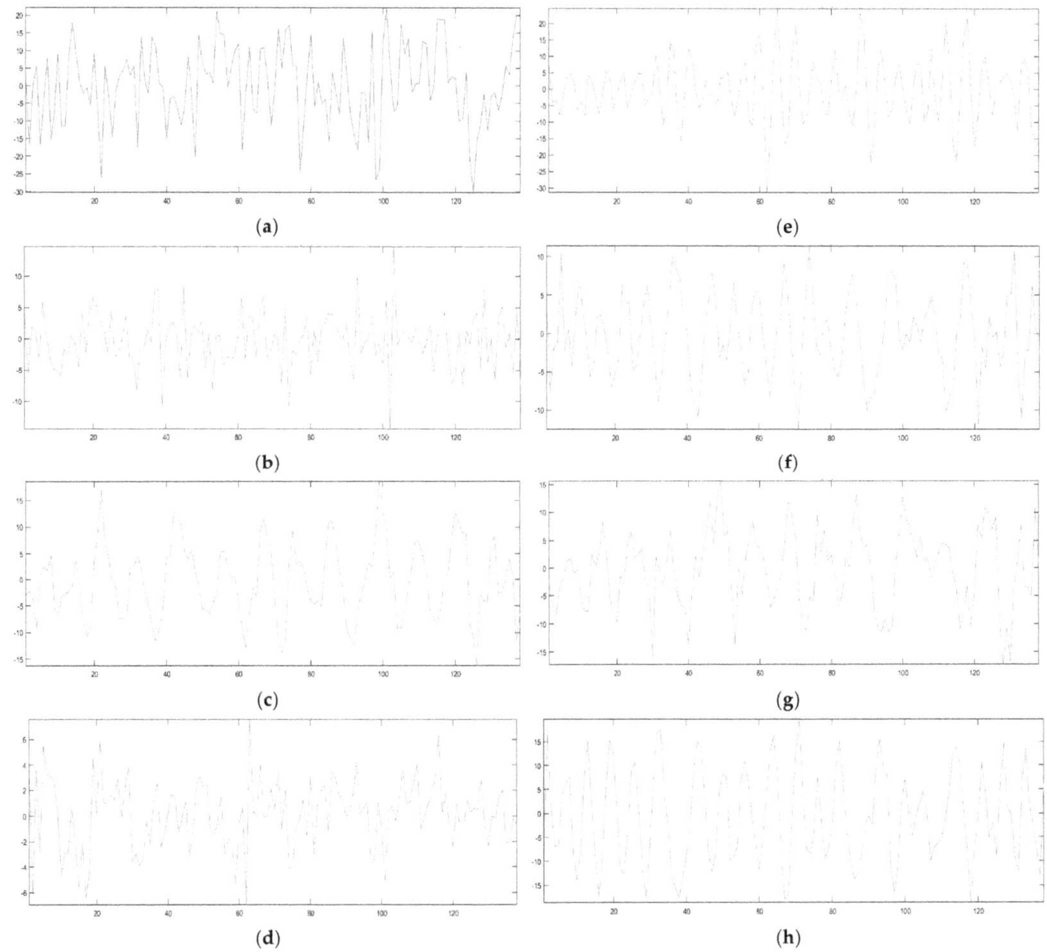

Figure 7. Wavelet packet decomposition diagram of Y-directional vibration signals for picks with moderate wear. (**a**) Coefficients of Packet (3, 0), (**b**) Coefficients of Packet (3, 1), (**c**) Coefficients of Packet (3, 2), (**d**) Coefficients of Packet (3, 3), (**e**) Coefficients of Packet (3, 4), (**f**) Coefficients of Packet (3, 5), (**g**) Coefficients of Packet (3, 6), (**h**) Coefficients of Packet (3, 7).

Based on the wavelet packet decomposition coefficients, the standard deviation of the decomposition wavelet packet coefficients can be obtained according to Formula (14), as shown in Table 2. The above data form a 25 × 600 feature matrix, where the first 24 columns are standard deviations of wavelet packet coefficients, and the last column is the classification label. The process involved randomly selecting 480 groups of data as the training set and 120 groups of data as the test set.

Table 2. Standard deviations of wavelet packet coefficients for each picks wear state.

Wear State	No.	Standard Deviations of Wavelet Packet Coefficients in X Direction			Standard Deviations of Wavelet Packet Coefficients in Y Direction			Standard Deviations of Wavelet Packet Coefficients in Z Direction		
		V1	...	V8	V1	...	V8	V1	...	V8
Slight Wear	1	3.9498	...	2.3341	4.7793	...	2.3901	5.1094	...	2.5387
	...	...	...	...	...	...	...	...	...	...
	200	2.9066	...	2.2601	3.4995	...	2.4891	4.6941	...	1.6829
Moderate Wear	1	4.5183	...	1.8889	5.5397	...	2.9457	3.9481	...	2.2765
	...	...	...	...	...	...	...	...	...	...
	200	3.8656	...	2.1931	5.2076	...	2.9488	4.2075	...	2.0691
Severe Wear	1	4.1303	...	2.6881	8.1387	...	3.5480	4.5001	...	2.6433
	...	...	...	...	...	...	...	...	...	...
	200	3.3412	...	2.8546	7.7809	...	3.8907	4.7600	...	3.8120

5.3. Overall Picks Wear State Recognition Model

Before importing the data, it is necessary to normalize the standard deviation data of the decomposed wavelet packet coefficients, as shown below:

$$x_t^* = \frac{x_t - x_{\min}}{x_{\max} - x_{\min}} \tag{27}$$

Bayesian-LSTM runs in a Python environment and is built based on the open-source machine learning library PyTorch and the probabilistic model Pyro. The hyperparameter settings are as shown in Table 3.

Table 3. Hyperparameter settings of the Bayesian-LSTM network.

Hyperparameter	Settings
Hidden layer	6
Learn rate	0.001
Epoch	1000
Sample Num	10

To verify the recognition effect of the Bayesian-LSTM network, a deep LSTM network is chosen for comparison analysis. In the deep LSTM network, the settings of hyperparameters such as the learning rate and the number of hidden layers are consistent. The training results of the two networks are shown in Figure 8.

As shown in Figure 8, with the increase in iterations, the Bayesian-LSTM network decreases very quickly. Compared to the standard LSTM network, it achieves a higher accuracy at a faster rate and the accuracy at each measurement point is higher than that of the standard LSTM. To quantify the accuracy of the two prediction models, a comparison of their accuracy rates is shown in Figure 9.

As can be seen from the above figures, given a certain set of hyperparameters, the classification accuracy of the Bayesian-LSTM model is 98.33%, while the LSTM model's classification accuracy is 89.16%. In the aforementioned conclusions, the recognition accuracy of the LSTM model is relatively low, which is because the weight parameters of the LSTM model are fixed and have not yet been optimized. If we use the Adam algorithm [39] to update and iterate the weights, and use Softmax as the classifier, the final set of hyperparameters for the optimized LSTM recognition model would be as given in Table 4 below.

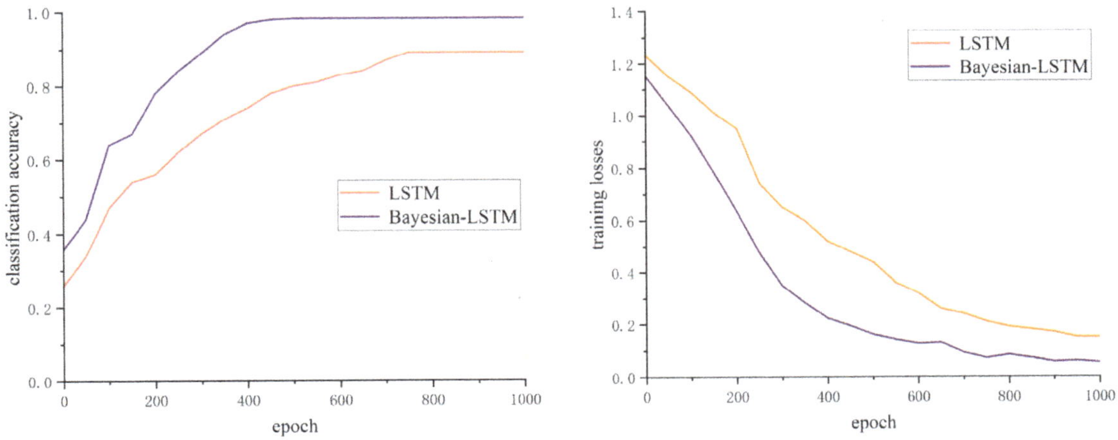

Figure 8. Comparison of the training results of the two networks.

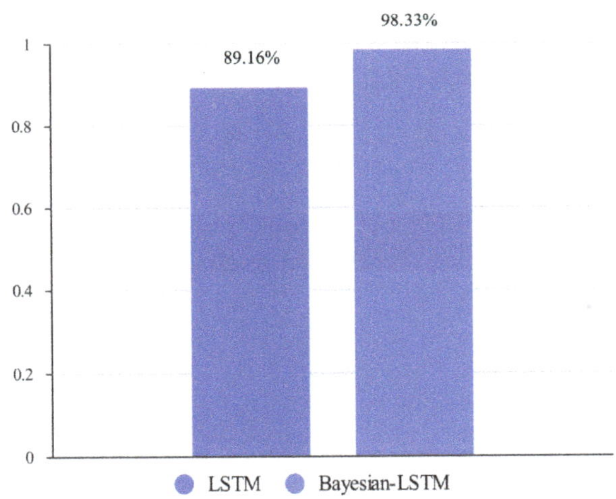

Figure 9. Comparison of accuracy rates for two networks.

The final confusion matrix of the LSTM network optimized by Adam is shown in Figure 10.

In order to further verify the accuracy and generalization ability of the Bayesian-LSTM deep learning network in the recognition of the picks wear state, the obtained results are compared with the classification results of the optimized LSTM and BP networks. The comparison results are as shown in Table 5 below.

Table 4. Hyperparameter settings of the optimized LSTM network.

Hyperparameter	Settings
Input size	24
Classification No.	3
Hidden layer	10
Learn rate	0.01
Epoch	500
Dropout	0.1
Optimizer	Adam

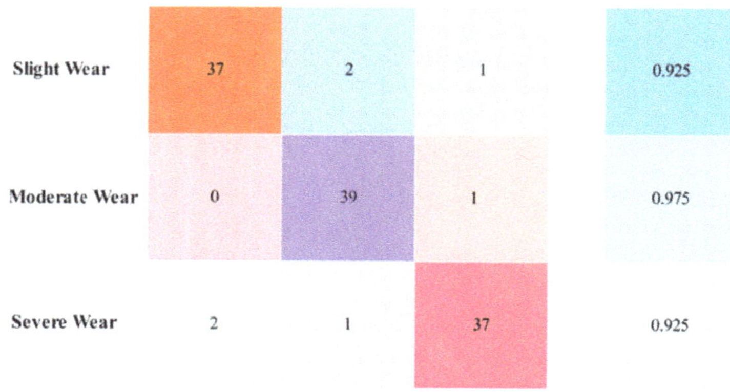

Figure 10. Confusion matrix with the optimized LSTM network.

Table 5. Comparison of the accuracy of overall picks wear recognition under different algorithms.

Network	Recognition Accuracy
BP	84.16%
Optimized LSTM	94.16%
Bayesian-LSTM	98.33%

From the table, it can be seen that the recognition accuracy of optimized LSTM and Bayesian-LSTM are higher than that of the BP network, proving that deep learning networks have a better accuracy when dealing with nonlinear data. However, in the macro view, the classification accuracy of the LSTM network on small-sample data is not ideal and has certain limitations. When the Bayesian theory is introduced, the Bayesian-LSTM model effectively reduces the model overfitting caused by sparse data and noise and provides an uncertainty quantification for prediction, effectively improving its recognition accuracy.

6. Conclusions

Accurate identification of the overall picks wear state is a core task in achieving intelligent upgrades of mining equipment. This study utilized theoretical analysis methods to research the mechanical model of the interaction between the pick and the coal-rock, preliminarily revealing the wear mechanism of the cutting picks. Based on this, we proposed a classification judgment method for three types of overall pick wear state. This study proposed an overall picks wear state recognition method based on Bayesian-LSTM. Using the vibration signals of the bolter miner's cutting drum as the basis for recognition, we used status labels and feature matrices to train the recognition model. The trained Bayesian-LSTM recognition model can effectively recognize the overall picks wear state. Compared to deep LSTM and BP, this method has a higher recognition accuracy.

In conclusion, this method only requires processing and analyzing equipment parameters automatically collected by mining machinery such as a bolter miner during its working process. It has the advantages of being easy to implement, low-cost, and highly accurate, providing a basis for the correct pick replacement strategy. However, there are still several challenging issues in theoretical and practical research, and recommended research works in the future as follows:

(1) It is necessary to research more feature parameters that can reflect the wear state of picks, such as current data on the cutting motor and pressure signals of the hydraulic cylinder.
(2) It is essential to study further efficient signal processing methods that can reduce the data disturbance caused by coal-mine scenes and to further improve the accuracy of picks wear state recognition.

Author Contributions: Conceptualization, D.S. and Y.Z.; writing-original draft preparation, D.S.; visualization, D.S. and Y.Z.; project administration, D.S.; funding acquisition, D.S. All authors have read and agreed to the published version of the manuscript.

Funding: This research was funded by the China National Key R&D Program (Grant No. 2020YFB1314000) and the Research Project Supported by Shanxi Scholarship Council of China (Grant No. 2022-186).

Data Availability Statement: The data used to support the findings of this study are available from the corresponding author upon request.

Acknowledgments: The partial study was completed at the National Engineering Laboratory for Coal Mining and Excavation Machinery Equipment, and the author would like to thank the laboratory for its assistance.

Conflicts of Interest: The authors declare no conflict of interest.

References

1. Dogruoz, C.; Bolukbasi, N.; Rostami, J.; Acar, C. An experimental study of cutting performances of worn picks. *Rock Mech. Rock Eng.* **2016**, *49*, 213–224. [CrossRef]
2. Holmberg, K.; Kivikytö-Reponen, P.; Härkisaari, P.; Valtonen, K.; Erdemir, A. Global energy consumption due to friction and wear in the mining industry. *Tribol. Int.* **2017**, *115*, 116–139. [CrossRef]
3. Liu, S.; Ji, H.; Liu, X.; Jiang, H. Experimental research on wear of conical pick interacting with coal-rock. *Eng. Fail. Anal.* **2017**, *74*, 172–187. [CrossRef]
4. Zhao, L.; He, J.; Hu, J.; Liu, W. Effect of pick arrangement on the load of shearer in the thin coal seam. *J. China Coal Soc.* **2011**, *36*, 1401–1406.
5. Krauze, K.; Mucha, K.; Wydro, T.; Pieczora, E. Functional and operational requirements to be fulfilled by conical picks regarding their wear rate and investment costs. *Energies* **2021**, *14*, 3696. [CrossRef]
6. Dewangan, S.; Chattopadhyaya, S. Characterization of wear mechanisms in distorted conical picks after coal cutting. *Rock Mech. Rock Eng.* **2016**, *49*, 225–242. [CrossRef]
7. Zhang, Q.; Fan, Q.; Gao, H.; Wu, Y.; Xu, F. A study on pick cutting properties with full-scale rotary cutting experiments and numerical simulations. *PLoS ONE* **2022**, *17*, e0266872. [CrossRef]
8. Qin, Y.; Zhang, X.; Zeng, J.; Shi, G.; Wu, B. Reliability analysis of mining machinery pick subject to competing failure processes with continuous shock and changing rate degradation. *IEEE Trans. Reliab.* **2022**, *72*, 795–807. [CrossRef]
9. Tian, Y.; Wei, X.; Hao, T.; Jiayao, Z. Study on wear degradation mechanism of roadheader pick. *Coal Sci. Technol.* **2019**, *47*, 129–134.
10. Zhang, Q.; Gu, J.; Liu, J.; Liu, Z.; Tian, Y. Pick wear condition identification based on wavelet packet and SOM neural network. *J. China Coal Soc.* **2018**, *43*, 2077–2083.
11. Zhang, Q.; Zhang, X.; Tian, Y.; Liu, Z. Research on recognition of pick cutting wear degree based on LVQ neural network. *Chin. J. Sens. Actuators* **2018**, *31*, 1721–1726.
12. Zhang, Q.; Yu, W.; Wang, C. Research on identification of pick wear degree of road header based on PNN neural network. *Coal Sci. Technol.* **2019**, *47*, 37–44.
13. Jin, L.; Cao, Y.; Qi, Y.; Yu, T.; Gu, J.; Zhang, Q. Identification of pick wear state based on acoustic emission and DS evidence theory. *Coal Sci. Technol.* **2020**, *48*, 120–128.
14. Lecun, Y.; Bengio, Y.; Hinton, G. Deep learning. *Nature* **2015**, *521*, 436–444. [CrossRef] [PubMed]
15. Su, H.; Qi, W.; Hu, Y.; Sandoval, J.; Zhang, L.; Schmirander, Y.; Chen, G.; Aliverti, A.; Knoll, A.; Ferrigno, G.; et al. Towards mode-free tool dynamic identification and calibration using multi-layer neural network. *Sensors* **2019**, *19*, 3636. [CrossRef]
16. Liu, X.; Jing, W.; Zhou, M.; Li, Y. Multi-scale feature fusion for coal-rock recognition based on completed local binary pattern and convolution neural network. *Entropy* **2019**, *21*, 622. [CrossRef]
17. Achmad, P.; Ryo, F.; Hideki, A. Image based identification of cutting tools in turning-milling machines. *J. Jpn. Soc. Precis. Eng.* **2019**, *85*, 159–166.
18. Huang, Z.; Zhu, J.; Lei, J.; Li, X.; Tian, F. Tool wear predicting based on multi-domain feature fusion by deep convolutional neural network in milling operations. *J. Intell. Manuf.* **2020**, *31*, 953–966. [CrossRef]
19. Ma, J.; Luo, D.; Liao, X.; Zhang, Z.; Huang, Y.; Lu, J. Tool wear mechanism and prediction in milling TC18 titanium alloy using deep learning. *Measurement* **2021**, *173*, 108554. [CrossRef]
20. Wu, F.; Nong, H.; Ma, C. Tool wear prediction method based on particle swarm optimization long and short time memory model. *J. Jilin Univ.* **2023**, *53*, 989–997.
21. Abdeljaber, O.; Avci, O.; Kiranyaz, S.; Gabbouj, M.; Inman, D.J. Real-time vibration-based structural damage detection using one-dimensional convolutional neural networks. *J. Sound Vib.* **2017**, *388*, 154–170. [CrossRef]
22. Yuan, X.; Tanksley, D.; Li, L.; Zhang, H.; Chen, G.; Wunsch, D. Faster post-earthquake damage assessment based on 1D convolutional neural networks. *Appl. Sci.* **2021**, *11*, 9844. [CrossRef]
23. Abdoli, S.; Cardinal, P.; Koerich, A. End-to-end environmental sound classification using a 1D convolutional neural network. *Expert Syst. Appl.* **2019**, *136*, 252–263. [CrossRef]

24. Zhu, Q.; Li, H.; Wang, Z.; Chen, J.F.; Wang, B.J.P.S.T. Short-term wind power forecasting based on LSTM. *Power Syst. Technol.* **2017**, *41*, 3797–3802.
25. Brili, N.; Ficko, M.; Klanènik, S. Automatic identification of tool wear based on thermography and a convolutional neural network during the turning process. *Sensors* **2021**, *21*, 1917. [CrossRef] [PubMed]
26. Casado-Vara, R.; Martin del Rey, A.; Pérez-Palau, D.; de-la-Fuente-Valentín, L.; Corchado, J.M. Web traffic time series forecasting using LSTM neural networks with distributed asy nchronous training. *Mathematics* **2021**, *9*, 421. [CrossRef]
27. Yang, T.; Chen, J.; Deng, H.; Lu, Y. UAV abnormal state detection model based on timestamp slice and multi-separable CNN. *Electronics* **2023**, *12*, 1299. [CrossRef]
28. Bie, F.; Du, T.; Lyu, F.; Pang, M.; Guo, Y. An integrated approach based on improved CEEMDAN and LSTM deep learning neural network for fault diagnosis of reciprocating pump. *IEEE Access* **2021**, *9*, 23301–23310. [CrossRef]
29. Marani, M.; Zeinali, M.; Songmene, V.; Mechefske, C.K. Tool wear prediction in high-speed turning of a steel alloy using long short-term memory modelling. *Measurement* **2021**, *177*, 109329. [CrossRef]
30. Najafi, M.; Jalali, S.M.E.; KhaloKakaie, R.; Forouhandeh, F. Prediction of cavity growth rate during underground coal gasification using multiple regression analysis. *Int. J. Coal Sci. Technol.* **2015**, *2*, 318–324. [CrossRef]
31. Gers, F.; Sshmidhuber, J.; Cummins, F. Learning to forget: Continual prediction with LSTM. *Neural Comput.* **2000**, *12*, 2451–2471. [CrossRef] [PubMed]
32. Schoot, R.; Depaoli, S.; King, R.; Kramer, B.; Märtens, K.; Tadesse, M.G.; Vannucci, M.; Gelman, A.; Veen, D.; Willemsen, J.; et al. Bayesian statistics and modelling. *Nat. Rev. Methods Primers* **2021**, *1*, 1. [CrossRef]
33. Song, Y.; Zhang, J.; Zhao, X.; Wang, J. An accelerator for semi-supervised classification with granulation selection. *Electronics* **2023**, *12*, 2239. [CrossRef]
34. Li, D.; Marshall, L.; Liang, Z.; Sharma, A.; Zhou, Y. Bayesian LSTM with stochastic variational inference for estimating model uncertainty in process-based hydrological models. *Water Resour. Res.* **2021**, *57*, e2021WR029772. [CrossRef]
35. Yang, L.; Xu, M.; Guo, Y.; Deng, X.; Gao, F.; Guan, Z. Hierarchical Bayesian LSTM for head trajectory prediction on omnidirectional images. *IEEE Trans. Pattern Anal. Mach. Intell.* **2021**, *44*, 7563–7580. [CrossRef] [PubMed]
36. Evans, I. A theory of the cutting force for point-attack picks. *Int. J. Rock Mech. Min. Sci.* **1984**, *2*, 67–71. [CrossRef]
37. Hocheiter, S.; Schmidhuber, J. Long short-term memory. *Neural Comput.* **1997**, *9*, 1735–1780. [CrossRef]
38. Wu, X.; Marshall, L.; Sharma, A. The influence of data transformations in simulating total suspended solids using Bayesian inference. *Environ. Model. Softw.* **2019**, *121*, 104493. [CrossRef]
39. Kingma, D.; Ba, J. Adam: A method for stochastic optimization. *arXiv* **2014**, arXiv:1412.6980.

electronics

Article

Improving Question Answering over Knowledge Graphs with a Chunked Learning Network

Zicheng Zuo [1], Zhenfang Zhu [1], Wenqing Wu [2], Wenling Wang [3], Jiangtao Qi [1] and Linghui Zhong [1,*]

1 School of Information Science and Electrical Engineering, Shandong Jiao Tong University, Jinan 250104, China; zuozicheng1997@163.com (Z.Z.)
2 School of Economic and Management, Nanjing University of Science and Technology, Nanjing 210094, China
3 Chinese Lexicography Research Center, Lu Dong University, Yantai 264025, China
* Correspondence: 205068@sdjtu.edu.cn

Abstract: The objective of knowledge graph question answering is to assist users in answering questions by utilizing the information stored within the graph. Users are not required to comprehend the underlying data structure. This is a difficult task because, on the one hand, correctly understanding the semantics of a problem is difficult for machines. On the other hand, the growing knowledge graph will inevitably lead to information retrieval errors. Specifically, the question-answering task has three difficulties: word abbreviation, object complement, and entity ambiguity. An object complement means that different entities share the same predicate, and entity ambiguity means that words have different meanings in different contexts. To solve these problems, we propose a novel method named the Chunked Learning Network. It uses different models according to different scenarios to obtain a vector representation of the topic entity and relation in the question. The answer entity representation that yields the closest fact triplet, according to a joint distance metric, is returned as the answer. For sentences with an object complement, we use dependency parsing to construct dependency relationships between words to obtain more accurate vector representations. Experiments demonstrate the effectiveness of our method.

Keywords: question answering; knowledge graph embedding; chunked learning network

Citation: Zuo, Z.; Zhu, Z.; Wu, W.; Wang, W.; Qi, J.; Zhong, L. Improving Question Answering over Knowledge Graphs with a Chunked Learning Network. *Electronics* **2023**, *12*, 3363. https://doi.org/10.3390/electronics12153363

Academic Editor: Davide Astolfi

Received: 19 July 2023
Revised: 4 August 2023
Accepted: 4 August 2023
Published: 6 August 2023

1. Introduction

Large-scale knowledge graphs like Freebase [1], DBPedia [2], Yago [3], and NELL [4] contain many facts from the real world, which makes question answering based on knowledge graphs (referred to as KGQA) a vital task. A complex data structure and a large number make it difficult for ordinary users to obtain a large amount of useful knowledge. KBQA (Knowledge Base Question Answering) and KGQA are both related to question-answering systems, but they focus on different types of knowledge repositories. In KBQA, a question is posed in a structured query (such as SPARQL [5]), and the goal is to retrieve the precise answer from the structured knowledge base. The system needs to understand the question, map it to the appropriate entities and relationships in the knowledge base, and retrieve the relevant information to provide an accurate answer. KGQA, on the other hand, extends beyond traditional knowledge bases and deals with more flexible and dynamic knowledge graphs. Knowledge graphs are also structured representations of information but are more expressive and allow for richer relationships and contextual information. They are often represented using semantic web technologies, such as the Resource Description Framework (RDF) or property graphs. Knowledge graphs can incorporate data from various sources and are more capable of representing complex and interconnected knowledge. In KGQA, the question-answering system needs to understand the question, navigate the knowledge graph, and perform more sophisticated reasoning to arrive at the correct answer. KGQA systems often employ natural language processing techniques, graph-based reasoning, and deep-learning methods to handle the complexities of the knowledge graph.

For example, consider the question, "Which actors have won an Oscar and also starred in a Christopher Nolan movie?" To answer, the KGQA system would need to reason through the knowledge graph, identifying entities related to actors, Oscar awards, and Christopher Nolan movies to find the correct answer.

The accurate understanding of question semantics and the effective filtering of interfering information are essential for successfully answering questions in KGQA. Currently, commonly used methods rely on semantic parsing [6–9] and information retrieval. The core concept behind semantic parsing is the conversion of natural language into a sequence of formal logical forms. Through the bottom-up analysis of logical forms, a logical form that can express the semantics of the entire problem is obtained, and the corresponding query sentence is used in the knowledge graph. This method is based on a relatively simple statistical method and has a greater dependence on data. Most importantly, it cannot map the relationship from natural language phrases to complex knowledge graphs. Secondly, supervised learning is needed when obtaining answers. And we need to train a classifier to score the generated logical form. To train such a powerful semantic parsing classifier, a great deal of training data is necessary. Whether it is Freebase [1] or WebQuestion [6], these two datasets have relatively few question and answer pairs. To address this issue, Zhang et al. [10] proposed a structural information constraint, which applies the structural information of the problem to path reasoning based on reinforcement. Zhen et al. [11] adopted a complementary approach, integrating a broader information retrieval model and a highly precise semantic parsing model, eliminating the need for manual template intervention.

The information retrieval method [12–14] is used to extract entities from the question and then search for the entities in the knowledge graph to obtain entity-centric subgraphs. Any node or edge in the subgraph can be a candidate answer. By observing the question and extracting information according to certain rules or templates, the feature vector of the question is obtained and a classifier is established. Then, the candidate answers are filtered by the feature vector of the input question to obtain the final answer [15]. However, KGQA needs to perform a multi-hop search to obtain the target entity when faced with missing inference chains. This makes the time and space complexity of the algorithm grow exponentially.

In addition, the same word can have different meanings in different contexts. We call this phenomenon entity ambiguity. For example, the meaning of an apple in Cook's hand and an apple in Newton's head are completely different. In daily life, people are used to using abbreviations instead of full names, such as Newton instead of Isaac Newton. This causes the algorithm to obtain a narrower entity search space. The diversity of predicates will produce a broader entity search space. When the same predicate connects different entities, its representation will be different, which requires the algorithm to be more robust.

We solve the above difficulties in two ways: (1) By embedding entities and relationships into the same vector space as the knowledge graph, we can naturally solve the problems caused by abbreviations, because similar entities can learn the same vector representation. And entities in different contexts will also obtain different vector representations. (2) Through the application of dependency parsing, a connection is established between entities and predicates. Following this, we incorporate the semantics of entities into the predicates, resulting in distinct weights being assigned to the relationships between various entities. We divide the question into two parts, the entity and the predicate, and then use different neural network methods to deal with these two parts, so our method is called the Chunked Learning Network (CLN).

This paper makes the following contributions:

- To address the distinctions in vector representation between entities and predicates, we employ separate modules for learning entities and predicates when tackling a question;
- By utilizing dependency parsing, we establish connections between entities and predicates, incorporating entity semantics into predicates to derive distinct weights for their relationships;

- The effectiveness of the CLN is demonstrated through experiments conducted on datasets containing both simple and complex questions.

This paper focuses on addressing the challenges in knowledge graph question answering (KGQA) by proposing an innovative solution. The introduction provides an overview of the current problems in KGQA and discusses existing methods, highlighting their limitations and unresolved issues. The related technologies section offers a comprehensive review of knowledge graphs, question-answering systems, and relevant methodologies. The proposed method section presents a novel approach, emphasizing head entity learning, relation learning, and their integration. The experimental evaluation section presents the experimental setup, dataset description, and performance analysis, demonstrating the superiority of the proposed method. This structured paper contributes to the advancement of KGQA by addressing challenges, introducing relevant technologies, proposing an innovative method, and validating its effectiveness through experiments.

2. Related Work

2.1. Question Answering over Knowledge Graphs

The Austrian linguist Edgar W. Schneider is credited with coining the term "knowledge graph" as early as 1972. In 2012, Google introduced their knowledge graph, which incorporates DBpedia, Freebase, and other sources. KGQA utilizes triples stored in the knowledge graph to answer natural language questions. Knowledge graphs usually represent knowledge in the form of triples. The general format of triples is (head entity, relation, tail entity), such as (Olympic Winter Games, Host city and the number of sessions, Beijing 24th), where "Olympic Winter Games" is the head entity, "Beijing 24th" is the tail entity, and "Host city and the number of sessions" is the relationship between the two entities. We use the lowercase letters h, r, and t to represent the head entity, relation, and tail entity, respectively, and (h, r, t) represents a triple in the knowledge graph. In previous work [16], transforming a multi-constraint question into a multi-constraint query graph was proposed. Since these multi-constraint rules require manual design and the rules are not scalable, this method does not perform well with large-scale knowledge graphs. Bordes A et al. [17] proposed a system that learns to answer questions using fewer multi-constraint rules to improve scalability. It uses a low-dimensional space to project the subgraph generated by the head entity for question answering. Then, it calculates the relevance score and determines the final answer by sorting. Likewise, so as not to be constrained by manual design rules, Bordes A et al. [18] developed a model that maps the questions to vector feature representations. A similarity function is learned during training to score questions and corresponding triples. The question is scored using all candidate triples at test time, and the highest-scoring entity is selected as the answer. But the vector representation of the question adopts a method similar to the bag-of-words model, which ignores the language order of the question (for example, the expressions of the two questions "who is George W. Bush's father?" and "Whose father is George W. Bush?" obtained by this method are the same, but the meanings of the two questions are obviously different). To focus on the order of words in the question, Dai Z et al. [19] use a Bidirectional Gate Recurrent Unit [20] (hereinafter referred to as Bi-GRU) to model the feature representation vector of the sentence and convert a simple single-fact QA question analysis into probabilistic questions. However, when the knowledge graph is incomplete, it is difficult to find the appropriate answer through probability. Based on the latest graph representation technology, Sun H et al. [21] described a method that extracts answers from subgraphs related to questions and linked texts, and they obtained good results. When the knowledge graph is incomplete, this method is effective, but external knowledge is not always obtainable. Recently, some works [22,23] used knowledge graph embedding to deal with question answering. With knowledge graph embedding, the potential semantic information can be retained, and the incompleteness of the knowledge graph can be handled. But the above methods model the problem and candidate relations separately without considering the word-level interactions between them, which may lead to local optimal results. Xie et al. [24] used a convolution-based topic entity extraction model to eliminate the noise problem in the process of extracting entities. Qiu et al. [25] proposed a global–local attention relationship detection model, using a local module to learn the features of word-level interactions and a global module to capture the nonlinear relationship between the question and the candidate relationship in the knowledge graph. Zhou et al. [26] proposed a deep fusion model based on knowledge graph embedding, which combines topic detection and predicate matching in a unified framework, where the model shares multiple parameters for joint training at the same time.

2.2. Knowledge Graph Embedding

TransE [26] is the most representative method of knowledge graph embedding that represents entities and relations as low-dimensional vectors and learns to translate entities through relations, aiming to capture semantic relationships between them. TransH [27] tries to solve the limitations of TransE in handling complex relationships, such as one-to-many, many-to-one, and many-to-many, by allowing an entity to have different representations under different relationships. However, TransH still assumes that the entities and relationships are in the same semantic space, which restricts the representation capabilities of TransH to some extent. TransR [28] treats an entity as a combination of multiple attributes, and different relationships focus on different attributes of the entity. TransR assumes that different relationships have different semantic spaces. For each triplet, the entity should be projected into the corresponding relationship space first, and then the translation from the head entity to the tail entity should be established. Previous research, like [23], has shown that TransE performs better on Freebase [1].

Another algorithm commonly used for embedding knowledge graphs is ComplEx [29]. ComplEx is an extension of the TransE algorithm that models relationships as complex-valued embeddings. It represents entities and relationships as complex-valued vectors, where each component consists of a real part and an imaginary part. By using complex-valued embeddings, ComplEx is able to capture rich semantic interactions and multi-relational patterns in the knowledge graph. The scoring function of ComplEx is based on the Hermitian dot product, which measures the plausibility of a triple (head entity, relationship, tail entity). ComplEx has been shown to be effective in capturing both one-to-one and many-to-many relationships. DistMult [30] is a simplified variant of ComplEx that models relationships as diagonal matrices. In DistMult, each relationship is represented by a diagonal matrix, where the diagonal elements capture the interaction between the head and tail entities. DistMult assumes that relationships are symmetric and does not consider complex interactions. The scoring function of DistMult is based on the dot product between the head entity and relationship embeddings, followed by element-wise multiplication and summing of the resulting vector. DistMult is computationally efficient and performs well on knowledge graph completion tasks. RotatE is an algorithm designed specifically for knowledge graphs with symmetric relationships, such as family relationships or sibling relationships. It represents relationships as rotations in the complex plane. RotatE [31] assumes that the head and tail entity embeddings can be rotated by a specific angle to represent the relationship between them. The scoring function of RotatE calculates the element-wise circular correlation between the embeddings, and the plausibility of a triple is determined based on the resulting score. By explicitly modeling rotational patterns, RotatE is effective in capturing symmetric relationships in the knowledge graph.

2.3. Syntactic Analysis

How can the machine be made to accurately understand the semantics of natural language? Manning CD et al. [32] propose a convenient dependency analysis method. The method can give the basic forms of words, mark the structures of sentences according to phrases and grammatical dependencies, and discover relationships between entities. Sun K et al. [33] combined the above method with Graph Convolutional Networks [34] (hereinafter referred to as GCNs) and used them in aspect-level sentiment analysis. Verberne et al. [35] added a reordering step to existing paragraph retrieval methods. When reordering, a ranking algorithm is used to calculate the question's score, and syntactic features are added to the question as weights. Arif et al. [36] used tree kernels (i.e., partial tree kernels (PTKs), subtree kernels (STKs), and subset tree kernels (SSTKs)) to consider the syntactic structure between them to solve the answer-reordering problem. Alberto et al. [37] calculated the similarity between trees based on the number of substructures shared between two syntactic trees and used this similarity to identify problems related to a new problem. To enhance downstream dependency analysis, a novel skeleton grammar has been proposed [38], which effectively represents the high-level structure of intricate problems. This lightweight formalization, along with a BERT-based parsing algorithm, contributes to the improvement of the analysis. For question-answering tasks, we make an improved dependency matrix better suited for concise and structured interrogatives by using it as the input of the GCN.

3. Chunked Learning Network

In this section, we first give an overall description of our model, introducing what the model contains and the role of each module. Section 3.2 introduces the architecture of the head-entity-learning model in detail, including the role of each component and the formulas involved. Section 3.3 introduces the architecture of the relation-learning model and the difference between this module and the head-entity-learning model. Section 3.4 introduces the structure of the pruning operation and the role of this module. Section 3.5 introduces the implementation process of the answer selection module.

3.1. CLN Overview

The CLN, as a deployable component, consists of four parts and trains the model through knowledge graph embedding. The main idea is shown in Figure 1. Consider the input question, "Which Olympic Winter Games was held in Beijing?" The pruning module identifies the entity "Olympics Winter Games" in the question, and the head-entity- and relation-learning modules learn their vector representations separately. Then, the learned vector representations are combined with the knowledge graph, and the answer selection module selects the closest fact triplet to return as the answer. When analyzing a question, our assumption is that it comprises a solitary entity. During data processing, for the convenience of subsequent processing, we designate consecutive entities as a single entity. The pruning operation marks the entity in the question to reduce the search space. Then, it learns the vector representation of the entity and relation in the question in the embedding space. Ultimately, a meticulously crafted joint distance metric is employed to locate the tail entity based on the $h + r \approx t$ equation, subsequently returning it as the answer.

3.2. Head-Entity-Learning Module

Given a question with length **L**, our goal is to restore the entity representation in the same space as the knowledge graph. The vector should represent the head entity of the question as closely as possible. We first map its **i** tokens to the sequence of the word-embedding vector $X = [x_1, \cdots, x_L]$. To take into account the sequential importance of words in the question, we use the Bidirectional Simple Recurrent Unit [39] (hereinafter referred to as Bi-SRU) to retain the global information of the question. Taking the forward direction as an example, Equations (1)–(4) show the details of calculating $\overrightarrow{h_i} \in \mathbf{R}^{d_h}$ where d_h is the dimension of the hidden-state output of the Bi-SRU.

$$f_i = \sigma\left(W_{x_f x_i} + W_{h_f \overrightarrow{h}_{i-1}} + b_f\right) \tag{1}$$

$$r_i = \sigma\left(W_{x_r x_i} + W_{h_r \overrightarrow{h}_{i-1}} + b_r\right) \tag{2}$$

$$s_i = f_i \odot s_{i-1} + (1 - f_i) \odot W_{h_s \overrightarrow{h}_{i-1}} \tag{3}$$

$$\overrightarrow{h_i} = r_i \odot g(s_i) + (1 - r_i) \odot x_i \tag{4}$$

where f_i, r_i, and s_i are the forget gate, reset gate, and internal state, respectively. σ and $\mathbf{g}()$ represent activation functions, and $\odot$ denotes the Hadamard product.

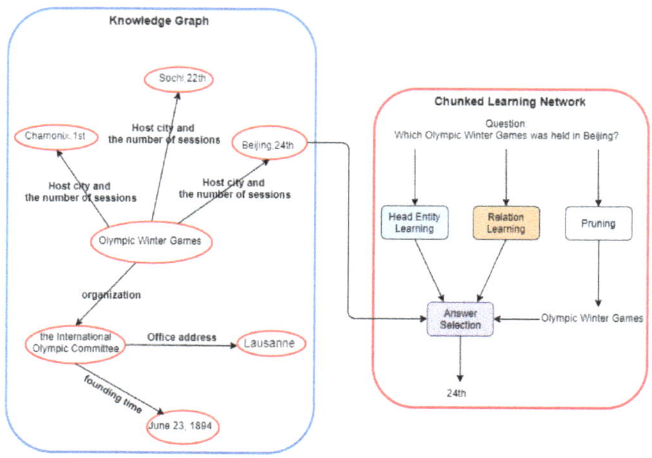

Figure 1. Overview of CLN. The **right** half shows the components of the model, and the **left** half is a simple schematic diagram of the knowledge graph.

By combining the hidden states from both the forward and backward directions, we obtain the concatenated representation, denoted by $h_i = \left[\overrightarrow{h_i}; \overleftarrow{h_i}\right]$. The Bi-SRU is complemented by a convolutional neural network (hereinafter referred to as CNN) module, which captures nearby contextual information that is in proximity to the entity. Equation (5) shows the calculation process of the j-th feature map of the l-th layer.

$$c_j^l = ReLU\left(\sum_{i=1}^{L} c_i^{l-1} * k_{ij}^l + b_j^l\right) \tag{5}$$

where c_i^{l-1} represents the i-th input of the $(l-1)$-th layer (when $l = 1, c_i^{l-1} = h_i$), the symbol $*$ represents the convolutional operation, k_{ij}^l represents the weight of convolution kernel j corresponding to the i-th input feature, and b_j^l is the bias of the convolution kernel. In the network described in this paper, we employ the rectified linear unit (ReLU) to compute feature maps.

After the convolution operation, we replace the pooling layer with an attention layer and apply its result to h_i. Equations (6) and (7) illustrate this process. The weight and bias of this layer are denoted by w and b, respectively. In this way, not only can the information of the entity be extracted, but the contextual information can also be preserved to a certain extent.

$$\alpha_i = \frac{exp(c_i)}{\sum_{i=1}^{L} exp(c_i)} \tag{6}$$

$$e_i = tanh\left(w_e^T \alpha_i h_i + b_e\right) \tag{7}$$

The result e_i is then used as the target vector of the i-th token, and Equation (8) represents using the average of the target vectors of all tokens as the predictive representation of the entity.

$$\widehat{e_h} = \frac{1}{L} \sum_{i=1}^{L} e_i^T \tag{8}$$

$\widehat{e_h}$ represents the learned entity vector representation. We independently train this module so that the vector representations of entities in sentences are as close as possible to the representations of entities in triples. The head-entity-learning module of the CLN is depicted in Figure 2.

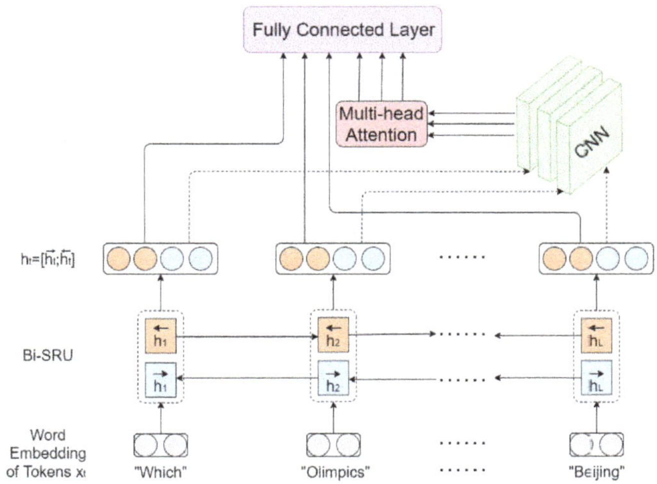

Figure 2. Proposed head-entity-learning module.

3.3. *Relation-Learning Module*

For relation-modeling problems, semantic parsing is the foundation of traditional methods of mapping relationships or for dictionary construction. Since the user's questions are not restricted, the predicates for the new questions may differ from the predicates in the training data. We therefore use the global information preserved in the knowledge graph embedding space to represent relational information in the question.

Similar to Equations (1)–(4), we obtain the vector representation $h_i \in \mathbf{R}^{d_{2h}}$ of the sentence through the Bi-SRU. Because the context of entities often contains relation features, and entities typically have similar features and attributes, the CNN extracts features from local regions by using convolutional kernels with shared weights. Conversely, as relationships between entities tend to be non-local and encompass the entire graph, the GCN can aggregate information from neighboring nodes and propagate this information to other nodes in the graph, thereby modeling complex relational patterns. To maximize the effect of dependency analysis, we construct the result into a dependency matrix as the adjacency matrix of the graph to spread node feature information.

The same predicate is associated with different entities, each making distinct contributions, so we capture the dependency between the predicate and the entity through syntactic analysis to obtain the adjacency matrix $\mathbf{A} \in \mathbf{R}^{L \times L}$. The adjacency matrix of each word and itself is set to self-loop; that is, the diagonal value of $\mathbf{A}$ is 1. These constraints between words are aggregated through a graph convolutional layer into a vector representation of predicates. With these constraints, different vector representations can be produced by the same predicate when connecting different entities. The calculation process of feature fusion is shown by Equations (9) and (10).

$$\widetilde{g_i^l} = \sum_{j=1}^{L} A_{ij} W^l g_j^{l-1} \tag{9}$$

$$\widetilde{g_i^l} = ReLU\left(\frac{\widetilde{g_i^l}}{d_i + 1} + b^l\right) \tag{10}$$

where $g_j^{l-1} \in R^{2d_h}$ is the representation of the j-th token obtained from the previous GCN layer (when $l = 1, g_j^{l-1} = x_j$), $g_i^l \in R^{2d_h}$ is the output of the current GCN layer, $d_i = \sum_{j=1}^{L} A_{ij}$ is the degree of the i-th token in the dependency tree, and W^l and b^l are the weight and bias matrices in the GCN layers, respectively.

Similar to Equations (6)–(8), we fuse the output of the GCN layer to h_i through the attention mechanism as the representation of a single token, and the mean of all representations is the vector representation of the relation. Equations (11)–(13) show the details of the calculation process.

$$\beta_i = \frac{exp\left(\sum_{i=1}^{L} g_i^l\right)}{\sum_{i=1}^{L} exp\left(\sum_{i=1}^{L} g_i^l\right)} \tag{11}$$

$$p_i = tanh\left(w_p^T \beta_i h_i + b_p\right) \tag{12}$$

$$\widehat{p_l} = \frac{1}{L} \sum_{i=1}^{L} p_i^T \tag{13}$$

Through independent training, we ensure that the relation representation is as close as possible to the representation of the relation in the triplet. Figure 3 shows the relation-learning module of the CLN. As shown in Equation (14), both head-entity-learning and relation-learning modules use MSEloss as the loss function during head entity training, where $\widehat{p_l} = \widehat{e_h}$, and $p_l = e_h$.

$$loss(\widehat{p_l}, p_l) = (\widehat{p_l} - p_l)^2 \tag{14}$$

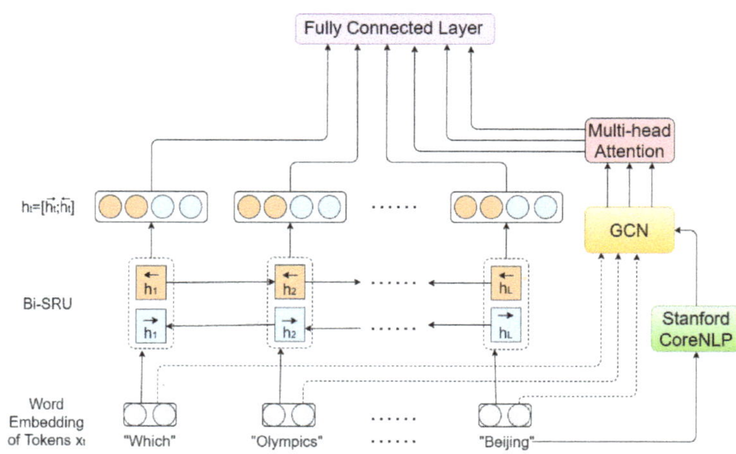

Figure 3. Proposed relation-learning module.

3.4. *Pruning Module*

In Section 3.1, we introduced the objective of this module, which is to recognize entities by identifying one or more consecutive words in the question. This enables us to designate the entire search space as a collection of multiple entities sharing similar or identical names. To streamline the module, we employ "Bi-LSTM (Bidirectional Long Short-Term Memory) + softmax" exclusively for named entity recognition. The question **q**, along with its associated entities, serves as the training data for the pruning model. Given that the topic entity in the question is contiguous, the model will identify the continuous words in the test set as either entities or as components of the correct topic entity. Hence, all words that are identical to the topic entity or contain the topic entity will be treated as candidate entities.

The entity and non-entity tokens (HED_{entity} and HED_{non}) obtained will be passed to the answer selection module.

3.5. Answer Selection Module

This module receives the output of the first three modules as input and finds the answer from the knowledge graph based on the principle of $f(e_h, p_l) \approx e_t$. The above process is achieved by using the joint distance metric proposed in [23]. A candidate fact is a fact triple whose head entity belongs to the candidate head entity. Consider **C** as the set comprising all candidate facts.

$$minimize_{h,l,t \in C} \|p_l - \hat{p}_l\|_2 + \beta_1 \|e_h - \hat{e_h}\|_2 + \beta_2 \|f(e_h, p_l) - \hat{e_t}\|_2 - \beta_3 sim\left[n(h), HED_{entity}\right] - \beta_4 sim[n(l), HED_{non}] \quad (15)$$

As shown in Equation (15), p_l and e_h are the relation and entity embeddings in the knowledge graph, respectively. The $sim[x, y]$ function measures the similarity between two strings, and $n(x)$ returns the name of an entity or predicate. β_1, β_2, β_3, and β_4 are predefined weights used to balance the contribution of each item.

4. Experiments and Analysis

In this section, we first verify the effectiveness of our model on public datasets and then analyze the reasons why our model can improve accuracy.

4.1. Datasets

The knowledge graphs and datasets used in the experiments can be downloaded through public channels.

FB5M [1]: The data in Freebase contain a lot of topics and types of knowledge, including information about humans, media, geographical locations, and so on. In our study, we utilized FB5M, which is among the more expansive subsets of Freebase.

SimpleQuestions [7]: This dataset comprises over 10,000 Freebase-related questions, with the issues within the dataset being summarized using facts and articles as references.

FB5M was employed as the knowledge graph in our study, and TransE was used for knowledge graph embedding to learn entity and relation representations. The performance of the model is measured by the accuracy of finding the ground truth.

4.2. Overall Results

Now, we will discuss the performance of the CLN. We take several representative KGQA methods as baseline models and compare the results. These works include the Bi-GRU rank model from Dai et al. [19], the Memory Network approach from Bordes et al. [7], the character-level CNN from Yin et al. [40], the character-level encoder–decoder from Golub et al. [41], the KG embedding method from Huang et al. [23], and the transformer-based question encoder from Li et al. [42]. Table 1 presents quantitative disparities in the performance of various methods on the dataset.

Table 1. Performance of different methods on SimpleQuestions.

Methods	Accuracy
Cfo [19]	0.626
MemNNs [7]	0.639
AMPCNN [40]	0.672
Character-level [41]	0.703
KEQA [23]	0.749
Te-biltm [42]	0.751
CLN (ours)	**0.753 (+11.4%)**

Note: Since the Freebase API is no longer available, thanks to Huang et al. [23] for re-evaluating the Freebase-API-based models of Cfo [19] and AMPCNN [40] for new results.

Accuracy is calculated by comparing the predicted entity–relationship pair with the ground truth, and the result is considered correct only when the entity–relationship pair given by the model conforms to the ground truth. From the results in Table 1, we can see that our model outperforms the previous methods. Compared with the accuracy when SimpleQuestions was released [7], the accuracy of the CLN was improved by 11.4%. This is due to a more complex neural network design and the use of different models for different subtasks.

4.3. Comparison of Baselines

To represent the baseline models more clearly, according to the results in Table 1, we introduce the differences between the CLN and the baseline models and then explain the reasons for the improved accuracy.

Cfo [19]: A Bi-GRU is used to rank candidate predicates. When the knowledge graph is incomplete, it is difficult to obtain the ranking through the established probability model. Our model uses knowledge map embedding, which can have a similar vector representation of words with similar meaning.

MemNNs [7]: This approach acquires entity and predicate representations from training questions and compares the entities and predicates within the new question with the previously acquired vectors. However, this method requires a large amount of data to train the classifier, and we use different models to train entities and predicates, which will improve the accuracy with the same amount of data.

AMPCNN [40]: This method uses a character-level CNN to match topic entities in fact candidates with entity descriptions in questions. However, the description of the relationship may not be limited to a part of the text, long-distance dependencies may exist, and the length is inconvenient to estimate. So, we use the GCN to capture long-distance dependencies for better accuracy.

Character-level [41]: The authors designed a character-level encoder–decoder frame, where each character corresponds to a one-hot vector. This makes the parameters of the model larger and uses more resources. We use word-level encoding to guaranteed accuracy without using pre-trained language models.

KEQA [23]: Using KG embedding, the model learns how to express the entity and predicate of the question. On this basis, we use different models to deal with entities and relations, making the model targeted.

Te-biltm [42]: The authors changed the encoder part of the transformer to Bi-LSTM, which can make the encoding of words obtain directional information for relation extraction in question answering. We are not limited to a single subtask but use different models to complete different subtasks, and the results from each module are fused to obtain the answer to the question, which is also the theme of this paper.

4.3.1. Statistical t-Test

We performed a statistical significance test using SPSS software to validate the results of our proposed method. Our objective was to assess the statistical significance of the accuracy achieved through our approach. The t-test was employed to generate a p-value, which measures the probability of the observed results being due to chance. A lower p-value indicates a higher likelihood of statistical significance. To evaluate our method, we compared the results obtained from the CLN-based model with those presented in Table 1. Notably, the accuracy achieved yielded a p-value of 0.036. These findings substantiate the statistical significance of the improvements attained by our proposed method.

4.3.2. Ablation Study

We removed different parts of our model and verified statistical significance, and the results obtained are shown in Table 2. Due to the effect of the CNN and multi-head attention, the accuracy of our head-entity-learning model is improved by 0.4%. In addition, the use of the Bi-SRU shortens the training time. As mentioned in the previous paragraph, the use of more complex models increases the accuracy of relation learning by 0.4%. The improvements of these two models have improved the final accuracy. We can see that there is a statistically significant improvement over the baseline when both CLN modules exist at the same time.

Table 2. Ablation experiment results.

Head-Entity-Learning Module, Accuracy	Relation-Learning Module, Accuracy	Total Accuracy
Bi-GRU + attention, 0.644	Bi-GRU + attention, 0.815	0.749
CLN_entity, 0.647	CLN_relation, 0.818	0.753
Bi-GRU + attention, 0.644	CLN_relation, 0.818	0.751
CLN_entity, 0.647	Bi-GRU + attention, 0.815	0.752

4.3.3. Qualitative Analysis

Different models obtain the embeddings of the same sentence at the same epoch, as shown in Figure 4. The scatter plot indicates that the enhanced model effectively discerns the distinctions between words, and points with similar values represent the predicate fusion with the representation of the entity. It is not difficult to find that after the "Bi-GRU + attention" sentence representation, the model needs a lot of data to learn weights. The improved model only needs a small number of samples to learn the vector representation of sentences better and faster.

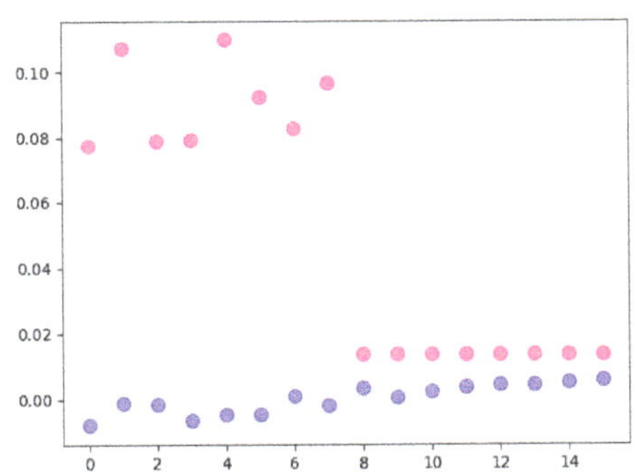

Figure 4. Result analysis of different modules. The horizontal axis refers to the words in the sentence, the vertical axis refers to the vector representation of the word, and the right half of the sentence represents the words that are padded to make the sentences the same length. The relationship-learning module's sentence representation with the CLN is represented by pink dots, while the "Bi GRU+attention" sentence representation is depicted by blue dots.

We delve into the joint impact of semantic parsing and the GCN using accuracy as an example. In Figure 5, we can see that in the initial stage of training, the accuracy of the relation-learning module rises rapidly, thanks to the combined effect of semantic parsing and the GCN. When the relationship between words in all sentences is constructed, the change in accuracy is relatively smooth. This trend of the change can also be seen through the change in the loss of the head-entity-learning module. But this method is only suitable for relational construction, so we only use this feature in relation-learning models.

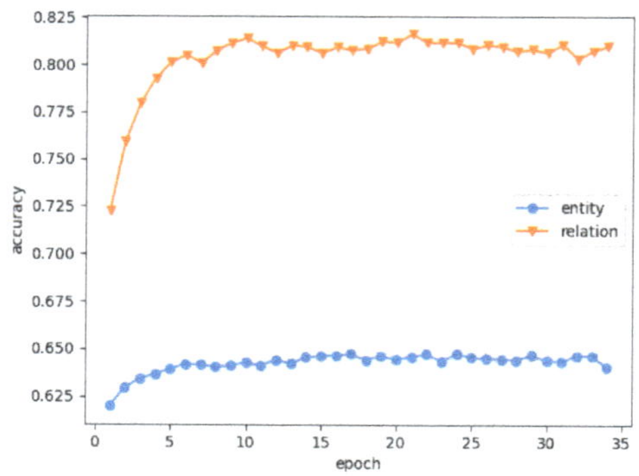

Figure 5. Variation in accuracy of head-entity-learning module and relation-learning module.

From the loss curves of the two modules in Figure 6, we can see that the loss of the head-entity-learning module decreases rapidly at the beginning of training and then tends to be flat, which indicates that the module has achieved good performance. At the same time, the relation-learning module loss drops rapidly and remains largely unchanged in the following periods, indicating that when the relational construction of words in all sentences is completed, other parts of the model can also support relational learning well.

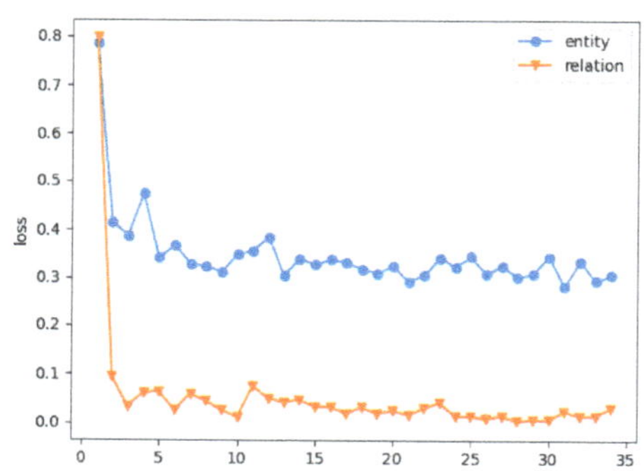

Figure 6. Variation in loss of head-entity-learning module and relation-learning module. In order to facilitate observation, we expanded the value of loss by 1000 times.

4.3.4. Error Analysis

In cases where the knowledge graph does not contain the required information to answer the question, the model cannot implement question answering. Most of the time, it is because the answer to the question is not unique. For example, consider the question, "Which actor was born in Warsaw?" Experiments show that the model can correctly learn the vector representation of "Warsaw" in the head-entity-learning module and "location.location.people_born_here" in the relation-learning module. But only one item, $\|f(e_h, p_l) - \hat{e}_t\|_2$, in the answer selection module involves tail entities, so it cannot represent a large set of answers.

4.3.5. Component Introduction Experiment

The purpose of this experiment is to assess the viability of our proposed model, which has a small parameter count and can be seamlessly integrated as a component into other models. Our objective is to verify the applicability of our method on different datasets. We conducted experiments using the WebQuestion dataset and introduced the proposed component into two existing models: EmbedKGQA [22] and TransferNet [43]. In EmbedKGQA, we incorporated the results of the relation-learning module into the inference module. For TransferNet, we introduced the outputs of the head-entity-learning module and the relation-learning module into step t using an attention mechanism. The results obtained are shown in Table 3.

Table 3. Hits@1 results on WebQuestionsSP.

Methods	WebQuestionsSP
EmbedKGQA	66.6
EmbedKGQA + CLN	67
TransferNet	71.4
TransferNet + CLN	71.6

The diverse architectures and parameter settings of different models can lead to variations in the performance of the introduced component within each model. A component that exhibits promising performance in one model may not achieve its optimal effectiveness when placed in another model. Furthermore, the design and functionality of other components within the model can also impact the performance of the introduced component. If there is a close interaction or dependency between the other components in the model and the specific component being introduced, placing that component in different models may yield different effects on its performance. It is worth noting that both our proposed model and EmbedKGQA leverage knowledge graph embeddings. Thanks to the shared utilization of knowledge graph embeddings, which enhances the models' ability to capture semantic relationships and facilitate reasoning capabilities, the introduced component exhibits enhanced effectiveness when integrated into our model and EmbedKGQA.

5. Conclusions

We propose a Chunked Learning Network for KGQA in this paper. The objective is to address the challenge of machines struggling to comprehend the semantic meaning of a question. The model incorporates the vector representation of entities and predicates into the question by utilizing the knowledge graph embedding. It employs distinct processing methods for different word types within the question. Words with similar meanings, such as word abbreviations, exhibit similar vector representations within the vector space. Additionally, the graph convolutional neural network assigns varying weights to capture the dependency relationship between words, thereby enhancing the contextual impact on each word. The experimental results demonstrate that our method enhances KGQA accuracy on datasets, and the proposed components indicate a promising direction for future research. However, our method currently falls short in entity recognition accuracy and faces challenges in coping with the expanding knowledge graph. To overcome this challenge, we plan to take into account the dynamic properties of the knowledge graph, as they are frequently updated in real-world scenarios.

Author Contributions: Writing—original draft, Z.Z. (Zicheng Zuo); Writing—review & editing, Z.Z. (Zhenfang Zhu), W.W. (Wenqing Wu), W.W. (Wenling Wang), J.Q. and L.Z. All authors have read and agreed to the published version of the manuscript.

Funding: This research received no external funding.

Data Availability Statement: The data that support the findings of this study is openly available at https://github.com/ZuoZicheng/CLN, accessed on 18 July 2023.

Conflicts of Interest: The authors declare no conflict of interest.

References

1. Bollacker, K.; Evans, C.; Paritosh, P.; Sturge, T.; Taylor, J. Freebase: A collaboratively created graph database for structuring human knowledge. In Proceedings of the 2008 ACM SIGMOD International Conference on Management of Data, Vancouver, BC, Canada, 9–12 June 2008; pp. 32–58.
2. Lehmann, J.; Isele, R.; Jakob, M.; Jentzsch, A.; Kontokostas, D.; Mendes, P.N.; Hellmann, S.; Morsey, M.; Van Kleef, P. ; Auer, S. Dbpedia—A large-scale, multilingual knowledge base extracted from wikipedia. *Semantic Web.* **2015**, *6*, 167–195. [CrossRef]
3. Fabian, M.; Gjergji, K.; Gerhard, W. Yago: A core of semantic knowledge unifying wordnet and wikipedia. In Proceedings of the 16th International World Wide Web Conference, Banff, AL, Canada, 8–12 May 2007; pp. 697–706.
4. Carlson, A.; Betteridge, J.; Kisiel, B.; Settles, B.; Hruschka, E.R.; Mitchell, T.M. Toward an architecture for never-ending language learning. In Proceedings of the 24th AAAI Conference on Artificial Intelligence, Atlanta, GA, USA, 11–15 July 2010; pp. 317–330.
5. Cyganiak, R. A relational algebra for SPARQL. *Digit. Media Syst. Lab. Lab. Bristol* **2005**, *35*, 9.
6. Berant, J.; Chou, A.; Frostig, R.; Liang, P. Semantic parsing on freebase from question-answer pairs. In Proceedings of the 2013 conference on empirical methods in natural language processing, Grand Hyatt Seattle, Seattle, WA, USA, 18–21 October 2013; pp. 1533–1544.
7. Bordes, A.; Usunier, N.; Chopra, S.; Weston, J. Large-scale simple question answering with memory networks. *arXiv* **2015**, arXiv:1506.02075.
8. Gomes, J., Jr.; de Mello, R.C.; Ströele, V.; de Souza, J.F. A hereditary attentive template-based approach for complex Knowledge Base Question Answering systems. *Expert Syst. Appl.* **2022**, *205*, 117725. [CrossRef]
9. Sui, Y.; Feng, S.; Zhang, H.; Cao, J.; Hu, L.; Zhu, N. Causality-aware Enhanced Model for Multi-hop Question Answering over Knowledge Graphs. *Knowl.-Based Syst.* **2022**, *250*, 108943. [CrossRef]
10. Zhang, J.; Zhang, L.; Hui, B.; Tian, L. Improving complex knowledge base question answering via structural information learning. *Knowl.-Based Syst.* **2022**, *242*, 108252. [CrossRef]
11. Zhen, S.; Yi, X.; Lin, Z.; Xiao, W.; Su, H.; Liu, Y. An integrated method of semantic parsing and information retrieval for knowledge base question answering. In Proceedings of the China Conference on Knowledge Graph and Semantic Computing, Online, 25 August 2021; pp. 44–51.
12. Kim, Y.; Bang, S.; Sohn, J.; Kim, H. Question answering method for infrastructure damage information retrieval from textual data using bidirectional encoder representations from transformers. *Autom. Constr.* **2022**, *134*, 104061. [CrossRef]
13. Alsubhi, K.; Jamal, A.; Alhothali, A. Deep learning-based approach for Arabic open domain question answering. *PeerJ Comput. Sci.* **2022**, *8*, e952. [CrossRef]
14. Kim, E.; Yoon, H.; Lee, J.; Kim, M. Accurate and prompt answering framework based on customer reviews and question-answer pairs. *Expert Syst. Appl.* **2022**, *203*, 117405. [CrossRef]
15. Yao, X.; Van Durme, B. Information extraction over structured data: Question answering with freebase. In Proceedings of the 52th Annual Meeting of the Association for Computational Linguistics, Baltimore, MD, USA, 23–25 June 2014; pp. 956–966.
16. Eberhard, D.; Voges, E. Digital single sideband detection for interferometric sensors. In Proceedings of the 26th International Conference on Computational Linguistics, Osaka, Japan, 11–16 December 2016; pp. 2503–2514.
17. Bordes, A.; Chopra, S.; Weston, J. Question answering with subgraph embeddings. *arXiv* **2014**, arXiv:1406.3676.
18. Bordes, A.; Weston, J.; Usunier, N. Open question answering with weakly supervised embedding models. In Proceedings of the Joint European Conference on Machine Learning and Knowledge Discovery in Databases, Nancy, France, 15–19 September 2014; pp. 165–180.
19. Dai, Z.; Li, L.; Xu, W. Cfo: Conditional focused neural question answering with large-scale knowledge bases. *arXiv* **2016**, arXiv:1506.02075.
20. Cho, K.; Van Merriënboer, B.; Gulcehre, C.; Bahdanau, D.; Bougares, F.; Schwenk, H.; Bengio, Y. Learning phrase representations using RNN encoder-decoder for statistical machine translation. *arXiv* **2014**, arXiv:1406.1078.
21. Sun, H.; Dhingra, B.; Zaheer, M.; Mazaitis, K.; Salakhutdinov, R.; Cohen, W.W. Open domain question answering using early fusion of knowledge bases and text. *arXiv* **2018**, arXiv:1506.02075.
22. Saxena, A.; Tripathi, A.; Talukdar, P. Improving multi-hop question answering over knowledge graphs using knowledge base embeddings. In Proceedings of the 58th Annual meeting Of the Association for Computational Linguistics, Online, 5–10 July 2020; pp. 4498–4507.
23. Huang, X.; Zhang, J.; Li, D.; Li, P. Knowledge graph embedding based question answering. In Proceedings of the 12th ACM International Conference on Web Search and Data Mining, Melbourne, VIC, Australia, 11–15 February 2019; pp. 105–113.
24. Xie, Z.; Zeng, Z.; Zhou, G.; Wang, W. Topic enhanced deep structured semantic models for knowledge base question answering. *Sci. China Inf. Sci.* **2017**, *60*, 1–15. [CrossRef]
25. Qiu, C.; Zhou, G.; Cai, Z.; Sogaard, A. A Global–Local Attentive Relation Detection Model for Knowledge-Based Question Answering. *IEEE Trans. Artif. Intell.* **2021**, *2*, 200–212. [CrossRef]
26. Zhou, G.; Xie, Z.; Yu, Z.; Huang, J.X. DFM: A parameter-shared deep fused model for knowledge base question answering. *Inf. Sci.* **2021**, *547*, 103–118. [CrossRef]
27. Bordes, A.; Usunier, N.; Garcia-Duran, A.; Weston, J.; Yakhnenko, O. Translating embeddings for modeling multi-relational data. *Adv. Neural Inf. Process. Syst.* **2013**, *26*.

28. Wang, Z.; Zhang, J.; Feng, J.; Chen, Z. Knowledge graph embedding by translating on hyperplanes. In Proceedings of the 28th AAAI Conference on Artificial Intelligence, Québec City, QC, Canada, 27–31 July 2014; Volume 28.
29. Trouillon, T.; Welbl, J.; Riedel, S.; Gaussier, É.; Bouchard, G. Complex embeddings for simple link prediction. In Proceedings of the International Conference on Machine Learning, New York City, NY, USA, 19–24 June 2016; pp. 2071–2080.
30. Yang, B.; Yih, W.-T.; He, X.; Gao, J.; Deng, L. Embedding entities and relations for learning and inference in knowledge bases. *arXiv* **2014**, arXiv:1412.6575.
31. Sun, Z.; Deng, Z.-H.; Nie, J.-Y.; Tang, J. Rotate: Knowledge graph embedding by relational rotation in complex space. *arXiv* **2019**, arXiv:1902.10197.
32. Lin, Y.; Liu, Z.; Sun, M.; Liu, Y.; Zhu, X. Learning entity and relation embeddings for knowledge graph completion. In Proceedings of the 29th AAAI Conference on Artificial Intelligence, Austin, TX, USA, 25–29 January 2015; Volume 29.
33. Manning, C.D.; Surdeanu, M.; Bauer, J.; Finkel, J.R.; Bethard, S.; McClosky, D. The Stanford CoreNLP natural language processing toolkit. In Proceedings of the 52th Annual Meeting of the Association for Computational Linguistics: System Demonstrations, Baltimore, MD, USA, 22–27 June 2014; pp. 55–60.
34. Kipf, T.N.; Welling, M. Semi-supervised classification with graph convolutional networks. *arXiv* **2016**, arXiv:1609.02907.
35. Sun, K.; Zhang, R.; Mensah, S.; Mao, Y.; Liu, X. Aspect-level sentiment analysis via convolution over dependency tree. In Proceedings of the 9th International Joint Conference on Natural Language Processing, Hong Kong, China, 3–7 November 2019; pp. 5679–5688.
36. Verberne, S.; Boves, L.W.j.; Oostdijk, N.H.J.; Coppen, P.A.J.M. Using syntactic information for improving why-question answering. In Proceedings of the 22nd International Conference on Computational Linguistics, Manchester, UK, 18–22 August 2008.
37. Arif, R.; Bashir, M. Question Answer Re-Ranking using Syntactic Relationship. In Proceedings of the 15th International Conference on Open Source Systems and Technologies, Online, 1–15 December 2021; pp. 1–6.
38. Sun, Y.; Li, P.; Cheng, G.; Qu, Y. Skeleton parsing for complex question answering over knowledge bases. *J. Web Semant.* **2022**, *72*, 100698. [CrossRef]
39. Lei, T.; Zhang, Y.; Wang, S.I.; Dai, H.; Artzi, Y. Simple recurrent units for highly parallelizable recurrence. *arXiv* **2017**, arXiv:1709.02755.
40. Yin, W.; Yu, M.; Xiang, B.; Zhou, B.; Schütze, H. Simple question answering by attentive convolutional neural network. *arXiv* **2016**, arXiv:1606.03391.
41. Golub, D.; He, X. Character-level question answering with attention. *arXiv* **2016**, arXiv:1604.00727.
42. Li, J.; Qu, K.; Li, K.; Chen, Z.; Fang, S.; Yan, J. Knowledge graph question answering based on TE-BiLTM and knowledge graph embedding. In Proceedings of the 5th International Conference on Innovation in Artificial Intelligence, Xiamen, China, 5–8 March 2021; pp. 164–169.
43. Shi, J.; Cao, S.; Hou, L.; Li, J.; Zhang, H. Transfernet: An effective and transparent framework for multi-hop question answering over relation graph. *arXiv* **2021**, arXiv:2104.07302.

electronics

Article

Centrifugal Navigation-Based Emotion Computation Framework of Bilingual Short Texts with Emoji Symbols

Tao Yang [1], Ziyu Liu [2,*], Yu Lu [3] and Jun Zhang [3]

[1] Education Information Technology Center, China West Normal University, Nanchong 637002, China; yangt@cwnu.edu.cn
[2] School of Electronic and Information Engineering, China West Normal University, Nanchong 637002, China
[3] School of Computer Science, China West Normal University, Nanchong 637002, China; luyu@stu.cwnu.edu.cn (Y.L.); zhangjun@stu.cwnu.edu.cn (J.Z.)
* Correspondence: liuziyu@stu.cwnu.edu.cn

Abstract: Heterogeneous corpora including Chinese, English, and emoji symbols are increasing on platforms. Previous sentiment analysis models are unable to calculate emotional scores of heterogeneous corpora. They also struggle to effectively fuse emotional tendencies of these corpora with the emotional fluctuation, generating low accuracy of tendency prediction and score calculation. For these problems, this paper proposes a Centrifugal Navigation-Based Emotional Computation framework (**CNEC**). **CNEC** adopts Emotional Orientation of Related Words (**EORW**) to calculate scores of unknown Chinese/English words and emoji symbols. In **EORW**, t neighbor words of the predicted sample from one element in the short text are selected from a sentiment dictionary according to spatial distance, and related words are extracted using the emotional dominance principle from the t neighbor words. Emotional scores of related words are fused to calculate scores of the predicted sample. Furthermore, **CNEC** utilizes Centrifugal Navigation-Based Emotional Fusion (**CNEF**) to achieve the emotional fusion of heterogeneous corpora. In **CNEF**, how the emotional fluctuation occurs is illustrated by the trigger angle of centrifugal motion in physical theory. In light of the corresponding relationship between the trigger angle and conditions of the emotional fluctuation, the fluctuation position is determined. Lastly, emotional fusion with emotional fluctuation is carried out by a **CNEF** function, which considers the fluctuation position as a significant position. Experiments demonstrate that the proposed **CNEC** effectively computes emotional scores for bilingual short texts with emojis on the Weibo dataset collected.

Keywords: bilingual short text with emoji; emotional fusion; emotional fluctuation; emotional computation

Citation: Yang, T.; Liu, Z.; Lu, Y.; Zhang, J. Centrifugal Navigation-Based Emotion Computation Framework of Bilingual Short Texts with Emoji Symbols. *Electronics* **2023**, *12*, 3332. https://doi.org/10.3390/electronics12153332

Academic Editor: Arkaitz Zubiaga

Received: 18 June 2023
Revised: 2 August 2023
Accepted: 2 August 2023
Published: 3 August 2023

1. Introduction

Sentiment analysis (SA), also referred to as opinion mining, is an academic field aimed at extracting users' views, attitudes, and emotions towards events of interest using specific rules and techniques applied to textual data. Sentiment analysis terminology was first mentioned in 2003, and emotional features related to specific topics from documents were extracted by Nasukawa and Yi [1]. The tasks of sentiment analysis are primarily classified into three levels: document-level, sentence-level, and aspect-level analysis. Additionally, the objective of sentiment analysis for customer reviews is a binary classification task that aims to determine the polarity of opinions and belongs to the sentence-level sentiment analysis category [2]. The task of multi-class text categorization [3] was presented, equal to the rating-inference problem. In addition, the focus of aspect-based sentiment analysis [4] is on the aspect words in the sentence, assigning polarity to all the aspects.

In the current situation of data expansion and AI agglomeration, SA as a part of AI is applied to the various walks of life. Sentiment analysis is utilized on movie reviews to perform fine-grained analysis to determine both the sentiment orientation and sentiment strength of the reviewer towards various aspects of a movie [4]. Besides, the identification and extraction of covert social networks represents a critical challenge within the realm of government security in the domain of e-government [5]. On the social platform, users' opinions reflect on hot spot news or public messages. Moreover, the frequency of emojis in posted messages is increasing. Especially in response to hot events, posts containing emojis emerge one after another [6]. Furthermore, regarding emoji usage on social platforms, sarcasm detection is also a concern of researchers [7]. The architecture is trained using two embeddings, namely word and emoji embeddings, and combines an LSTM with the loss function of SVM for sarcasm detection.

While sentiment analysis has been widely studied, most researchers have focused on analyzing a single corpus. This paper, however, presents a novel frame **CNEC** to perform the sentiment analysis of examining bilingual and emoji-containing text on social media platforms. **CNEC** employs the Emotional Orientation of Related Words (**EORW**) technique designed in previous work [8] to calculate emotional scores of Chinese phrases or English words. Additionally, the emoji is mapped to its true meaning in the form of a word by *LinkMap*. It is designed to associate text and emojis. Following that, emotional score of the emoji is calculated by **EORW**, driven by **CNEC**. In addition, **CNEC** utilizes the centrifugal motion framework in physics to describe emotional fluctuations. Then, **CNEC** uses **CNEF** to fuse emotional scores of different corpora to obtain the emotional score of the predicted text. Based on **Maximum Density Dominance**, emotional scores of texts free of emotional fluctuations can also be handled. The contribution to the field is unique at present, setting our approach apart from the existing literature.

2. Related Work

In response to traditional machine learning methods, Bayesian and support vector machines [9] are commonly used for sentiment analysis. In addition, a feature set selection method for social network sentiment analysis based on information gain, bigram, and object-oriented extraction methods was introduced. In order to solve the problem of performance degradation caused by aspect-based methods that cannot reasonably adapt the general vocabulary to the context of aspect-based datasets, Mohammad Erfan Mowlaei et al. [10] proposed two extended methods of dictionary generation methods for aspect-oriented problems—statistical methods and their previously proposed genetic algorithms, which fused the above vocabulary with prominent static words to classify the aspects in the comments. The ALGA algorithm was proposed [11] to address the task of polarity classification of Weibo emotions. This algorithm constructs an adaptive emotional vocabulary and seeks to identify the optimal emotional vocabulary for the task. An aspect-based hybrid approach to sentiment analysis that integrates domain vocabulary and rules was proposed [12] to analyze the entities of intelligent application reviews, extract important aspects from comments, achieve sentiment classification, and finally produce summary results, in order to understand the needs and expectations of their customers. A double feed-forward neural network [13] was used to pass output layer information to a two-layer neural network to optimize and process information for emotional classification.

A scholarly approach was developed to emphasize sarcasm, as outlined in the study [7]. This approach employed an architectural framework that integrated two types of embeddings: word embeddings and emoji embeddings. The framework leveraged the power of LSTM (Long Short-Term Memory) in conjunction with a loss function derived from SVM (Support Vector Machines). A deep learning framework [14] was proposed for analyzing product reviews on the YouTube social media platform, which could automatically collect, filter, and analyze reviews of a specific product from YouTube. Aminu Da'u et al. [15] proposed a recommendation system for the time-consuming problem and accuracy problem generated in the process of aspect-based opinion mining in user comments, which adopts a deep learning method based on aspect-weighted opinion mining. This method uses deep learning methods to extract aspects of products and underlying weighted user opinions from review text, and fuse them into extended collaborative filtering (CF) technology to improve the recommendation system. With the intention of catching up with the speed of streaming data generated on social media platforms and analyzing users' emotions on topics, Ajeet Ram Pathak et al. [16] proposed a theme-level sentiment analysis model based on deep learning. The proposed model uses the online latent semantic index of regularization constraints to extract topics at the sentence level, and then applies the topic-level attention mechanism to the LSTM network for sentiment analysis. An improved sentiment analysis method [17] was presented to classify sentence type using BiLSTM-CRF and CNN for different types of emotions. This method divided sentences into different types and then performed sentiment analysis on each type of sentence. Bin Liang et al. [18] proposed a SenticNet-based graph convolutional network to build graph neural networks by integrating sentiment knowledge in SenticNet to enhance the dependent graph of sentences. On this basis, the sentiment enhancement graph model considers the dependence between contextual words and aspect words and the emotional information between opinion words and aspects. Since the influence of contextual intersentence associations was considered, an aspect-level sentiment analysis model [19] was proposed with aspect-specific contextual position information, which could extract the influence of the contextual association of each sentence in the document on the aspect sentiment polarity of individual sentences. A two-way LSTM (ET-Bi-LSTM) [20] emotion analysis model was designed for expression text integration in order to accurately classify the emotion of microblog comments with emoticons in microblog social networks. A model for predicting sentiment polarity on social media, which incorporates an emoji-aware attention-based GRU network, was proposed [21]. Bi-LSTM-SNP [22] was designed to address the challenge of capturing contextual semantic correlation between aspect word and content words more effectively. Except for the backbone model of LSTM, a graph convolutional neural network model [23–26] was widely used on aspect-level sentiment analysis. Moreover, based on ensemble learning, the sentiment analysis task was accomplished by combining the Bi-LSTM and Graph Convolutional Neural Network (GCN) techniques [27]. In addition, there are some works using fuzzy theory to perform the sentiment analysis [28–30].

3. Centrifugal Navigation-Based Emotion Computation Framework of Bilingual Short Texts with Emoji Symbols

In Section 3.1, this paper analyzes problems of emotional bilingual texts with emojis and describes methods how to tackle the problems. Section 3.2 explains how to perform the emotional computation of unknown phrases and emoji symbols, adopting **EORW**. Section 3.3 depicts emotional fluctuations by the centrifugal motion and illustrates emotional computation of texts stuck in emotional fluctuations.

3.1. Problem Analysis and Basic Definition

Problem Analysis. For computing emotional score, the pipeline of the traditional model has five steps: (1) word segmentation; (2) word retrieval; (3) output words and their scores; (4) emotional fusion of all words; (5) output the emotion score of the text. There are two problems with the traditional model, as depicted in Figure 1.

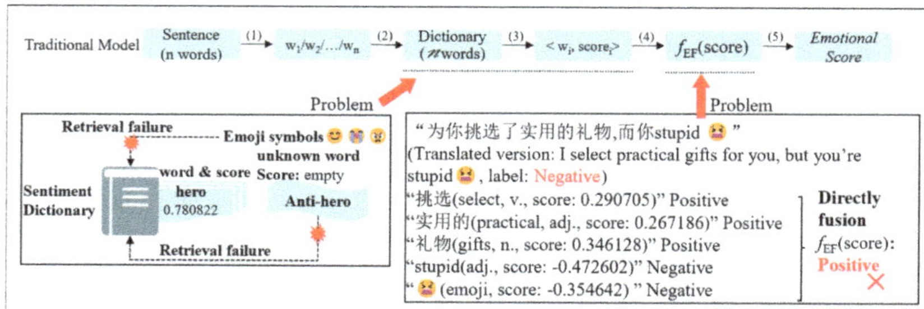

Figure 1. The problems addressed in this article. The English word "anti-hero" is not stored in the sentiment dictionary used by the experiment. Subsequently, the emotional score of "Anti-hero" can't be loaded by the sentiment dictionary. In other words, the score is empty. On the contrary, the English word "hero" is contained by the dictionary this paper uses. Afterwards, [score(hero) = 0.780822] is retrieved by the dictionary where scores are normalized in advance to between −1 and 1. In the mixed text "为你挑选了实用的礼物, 而你stupid 😫" (Translated version: I select practical gifts for you, but you're stupid 😫.) with the negative tendency, keywords are "挑选 (select, v., score: 0.290705)", "实用的 (practical, adj., score: 0.267186)", "礼物 (gifts, n., score: 0.346128)", "stupid (adj., score: −0.472602)" and "😫 (emoji, score: −0.354642)". If the emotional scores are directly summed, the result would be 0.076775, which does not match the label of this sentence. Besides, the emotional fluctuation on the position of "stupid" cannot be directly illustrated.

Problem 1. *The sentiment dictionary cannot encompass the emotional scores of all words that belong to Chinese or English. In addition, the direct calculation of emotional scores for emojis is not feasible due to the absence of a specialized sentiment dictionary for emojis.*

Problem 2. *Traditional emotional fusion adopts an average method [31] or maximum-value strategy [32], resulting in insufficiently accurate scores.*

From what has been presented, this paper aims to address three issues: firstly, emotional computation for unknown words and emojis; secondly, describing the process of sentiment fluctuation; and thirdly, sentiment calculation for texts containing sentiment fluctuations. As for these issues, firstly, **EORW** is employed in this article by **CNEC** to compute emotional scores of unknown words. After the transformation of texts through LinkMap designed by this paper, the emotional scores of emojis' textual forms are calculated using **EORW**. Then, inspired by the circular motion, the centrifugal motion process in circular motion is adopted by **CNEC** to portray emotional fluctuation. Finally, emotional scores of mixed texts stuck in emotional fluctuations are computed using emotional fusion by **CNEF**.

Basic Definition. For facilitating understanding and description, this subsection defines some parameters and functions.

Definition 1. *Emotional Dictionary Dic. Dic = {w_1, w_2, …, w_N}, where w_i represents the i-th word, and N = size of Dic.*

Definition 2. *Unknown Word uw. According to Function Rtr(·), the score of uw is empty, as depicted in Equation (1).*

$$Rtr(uw, Dic) = empty \tag{1}$$

Definition 3. *Word Vector V_i of w_i. V_i is calculated by a BERT-base model (The structure of the BERT-base model is illustrated in Figure 2), as shown in Equation (2).*

$$V_i = BERTembd(w_i) \tag{2}$$

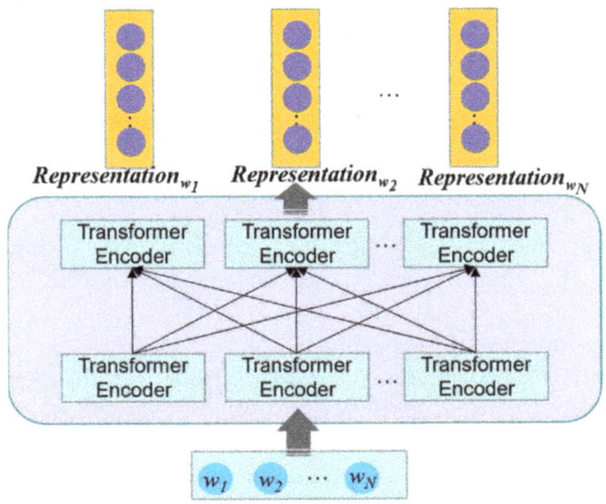

Figure 2. Embeddings of extraction.

Definition 4. *Word Similarity. Sim(w_i, w_j) is utilized to compute word similarity between w_i and w_j in space. The value domain of Sim(w_i, w_j) ∈ [−1, 1]. The closer the result is to 1, the greater the word similarity.*

$$Sim(w_i, w_j) = \frac{\sum_{i=1}^{D} V_i * V_j}{\sqrt{\sum_{i=1}^{D} (V_i)^2} * \sqrt{\sum_{i=1}^{D} (V_j)^2}} \tag{3}$$

where D represents the dimension of word vectors.

Definition 5. *Neighbor Words Set, **NW-Set**. **NW-Set** stores the first t words with word similarity close to 1. **NW-Set** = {w_1, w_2, ..., w_t}, t ∈ N+. Rank(·) function is to choose the t nearest words.*

$$NW - Set = \{\overbrace{Rank(Sim(w_i, w_j))}^{t}\} \tag{4}$$

Definition 6. *Mutually Exclusive Subset, **MES$_i$**. **MES$_1$** and **MES$_2$** have opposite emotional tendencies. For example, when the tendency of **MES$_1$** is positive, the tendency of **MES$_2$** is negative. Conversely, when the tendency of **MES$_1$** is negative, the tendency of **MES$_2$** is positive.*

$$\begin{cases} [MES_1 = \{w_1^1, \ldots, w_1^x\}, ten(w_1^i) = positive, i = 1, \ldots, x] \cap [MES_2 = \{w_2^1, \ldots, w_2^y\}, ten(w_2^i) = negative, i = 1, \ldots, y] \\ [MES_1 = \{w_1^1, \ldots, w_1^x\}, ten(w_1^i) = negative, i = 1, \ldots, x] \cap [MES_2 = \{w_2^1, \ldots, w_2^y\}, ten(w_2^i) = positive, i = 1, \ldots, y] \end{cases} \tag{5}$$

*where w_i^θ represents the θ-th word in the i-th **MES**. θ ∈ [1, x] or [1, y], and x, y < t. ten(·) denotes the tendency.*

Definition 7. *Emotional Dominance. The set containing the largest quantity of words is considered as the emotionally dominant set—**Related Words Set (RS)**.*

$$RS = Dom(MES_1, MES_2) = \begin{cases} MES_1(x > y) \\ MES_2(x < y) \end{cases} \tag{6}$$

*where x represents the quantity of **MES$_1$**, and y represents the quantity of **MES$_2$**.*

3.2. Emotional Computation of Unknown Words and Emojis

Definition 8. ***EORW****. Emotional Orientation of Related Words. According to Word Similarity, the* ***NW-Set*** *of uw can be computed. Then, based on Emotional Dominance, the Related Words Set* ***RS*** *is calculated. Ultimately, the emotional score of uw is fused by emotional scores of words in* ***RS****.*

3.2.1. Emotional Computation of Unknown Words

The process for computing the emotional score of unknown vocabulary *uw* involves mapping its word vector E_{uw} in space using a BERT-base model, as illustrated in Equation (2), where the form of E_{uw} is $E_{uw} = (x_1, x_2, \ldots, x_D)$, $x_i \in$ R and D represents the dimension of the word vector E_{uw}.

Step 1. The word embedding E_{dic} of the emotional dictionary *Dic* is also calculated by *BERTembd*(·), shown in Equation (2), where Dic_{w_n} represents words in Dic, and the form of E_{dic} is similar to that of E_{uw}.

Step 2. To get the *k*-group neighbor words set (**NW-Set**), the principle of spatial similarity is utilized to compute the similarity between *uw* and Dic_{w_n}.The principle of spatial similarity is denoted by Equation (3), where E_{uw_i} and E_{dic_i} here represent the components of vector E_{uw} and E_{dic}, respectively.

Step 3. The Dic_{w_n} are organized in descending order based on the result of Equation (3), with the aim of identifying the top *t* words of Dic_{w_n}. *t* words considered as neighbors of *uw* are put into **NW-Set**, as shown in Equation (4), and this operation is repeated for *t* times.

Step 4. The form of **NW-Set** is shown as NW-Set = $\{w_1, \ldots, w_t\}$. After getting the **NW-Set**, **EORW** is applied to identify w_i in **NW-Set** with similar emotional tendencies.

Step 5. If the emotional tendencies of the *t* words are consistent, the emotional score of the unknown word is the average of the *k* groups' emotional scores. Conversely, if the *k* words' emotional tendencies are inconsistent, **MES**$_i$ can be calculated by Equation (5).

Step 6. From **MES**s, *Dom*(**MES**$_1$, **MES**$_2$) can cope with the dominant set **RS**, as shown in Equation (6).

Step 7. The emotional scores of *w* in **RS** are fused to calculate the final score S_{uw} of *uw*. S_{uw} is represented as Equation (7).

$$S_{uw} = \begin{cases} \frac{\sum_{i=1}^{t} e_i}{t} (consistent) \\ \frac{\sum_{i=1}^{x} e_i}{x} (exclusive) \end{cases} \quad (7)$$

where e_i (i = 1, 2 ...t) represents a score in a group of *k* words with consistent emotional tendencies, while e_i (i = 1, 2 ...x) represents a score in a group of *x* scores in **RS**.

Figure 3 illustrates the comprehensive workflow of computing the emotional score of unknown words.

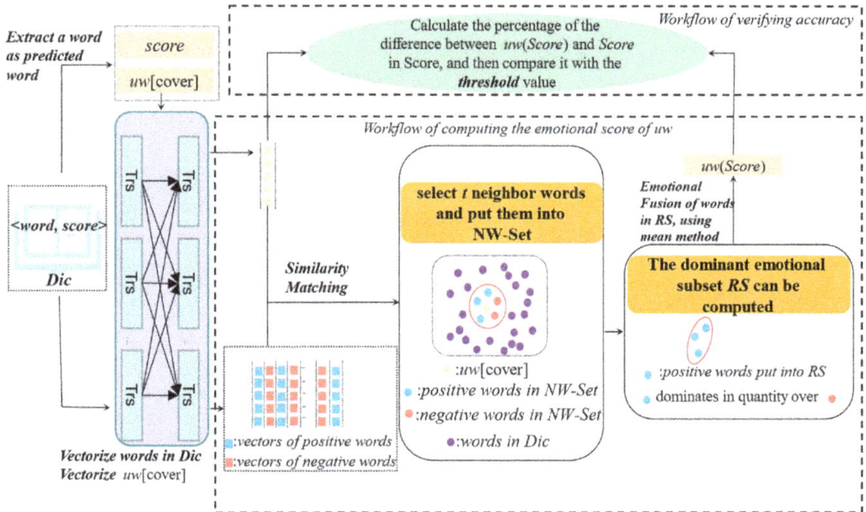

Figure 3. Emotional computation of unknown words. A *Word* and its *Score* are extracted as the experiment data *uw*[cover] and *score*. In light of Equation (2), the similarity value among the embeddings can be calculated. Through the **EORW** method, *uw*(*Score*) is computed.

3.2.2. Emotional Computation of Emojis

Given that the emotional score of emojis can't be directly calculated, it is necessary to rely on other methods. This paper adopts the **EORW** method to accomplish the goal of computing the emotional score of emojis. Based on **EORW**, an emoji is considered as a *uw*. Therefore, the meaning of emojis in textual form needs to be denoted. However, there are discrepancies between the Chinese shapes and English appearances of emojis emerging in textual data. In order to tackle this issue and make sentiment analysis easier, an emoji linking map *LinkMap* has been developed. Some emojis are shown in the following Table 1.

Table 1. Instances of emoji Linking Map *LinkMap*.

Emoji	Textual Meaning	English Shape in Textual Data
😀	[hah]	:grinning_face:
❤️	[love you]	:red_heart:
😵	[dizzy]	:dizzy_face:
😠	[angry]	:angry_face:
👏	[clap]	:clapping_hands:
😱	[shocked]	:face_screaming_in_fear:
👍	[like]	:thumbs_up:
😊	[smile]	:smiling_face_with_smiling_eyes:

In the textual data, the form of a Chinese emoji C_{emoji} is shown in Equation (8), and similarly the shape of an English emoji E_{emoji} is depicted in Equation (9).

$$C_{emoji} = [Ch\_w] \tag{8}$$

where Ch_w represents short Chinese words, illustrating the meaning of the emoji.

$$E_{emoji} =: w_i\_w_j : \tag{9}$$

where $w_i\_w_j$ illuminates that two words linked by underscore explain this emoji. w_i and w_j are English words.

Then, through *LinkMap*, *emofMeaning* can be denoted as Equation (10).

$$emofMeaning = LinkMap(emoji) \tag{10}$$

where *emofMeaning* represents the English phrase of the emoji meaning.

Due to the operation of *LinkMap*, the result reveals the form of emoji in textual representation *emofMeaning*[uw]. Following that, the **EORW** technique is utilized. The progress of **EORW** can be checked in Section 3.2.

As depicted in Figure 4, based on *LinkMap*, the emoji 'loudly crying face' is equal to [crying] and ':Loudly_Crying_Face:' and its meaning 'tear' computed. Subsequently, 'tear' is fed into the embedding layer, its vector calculated by a BERT-base in Equation (2). Then, through the **EORW** method, **RS**(crying) can be acquired. As a result, the emotional score of emoji 'loudly crying face' is calculated by the emotional fusion of w in **RS**(crying).

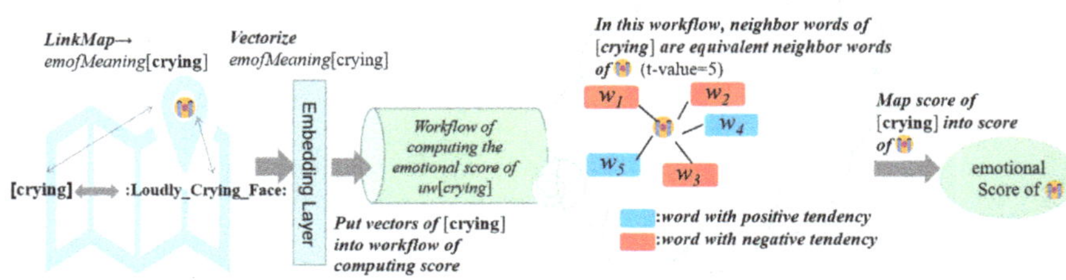

Figure 4. The pipeline of sentiment computation of the emoji named tear.

3.3. *Emotional Computation of Bilingual Short Texts with Emojis in Emotional Fluctuation*

This section describes how to compute the emotional score of the short text with emoji. According to Section 3.2.1, the scores of words that can't be retrieved from the emotional dictionary are calculated by the **EORW** method. Then emotional fluctuation is considered.

Step 1. The input sentence S is fed into the model, and its internal components are segmented into three distinct elements: the Chinese element C, the English element E, and the emoji element e. S is represented by the following equation, Equation (11).

$$S = [\overbrace{w_1, \ldots, w_i}^{C}, \overbrace{w_1, \ldots, w_k}^{E}, \overbrace{e}^{emoji}] \tag{11}$$

where w represents words in S, i words belong to C, k words belong to E, and e denotes the emoji.

Step 2. C can be represented by Equation (12), and E can be represented in the same way by Equation (13).

$$C = [w_1, \ldots, w_i] \tag{12}$$

$$E = [w_1, \ldots, w_k] \tag{13}$$

Step 3. This article supposes that nouns N, verbs V, adjectives Adj, and adverbs Adv have a significant impact on S. Therefore, these words are retained and extracted into a keyword's subset K. K has two forms in this paper: one is Kc and the other is Ke. Kc contains Chinese keywords, Ke contains English keywords, and Kc and Ke are respectively denoted as Equations (14) and (15).

$$K_C = [N_C, V_C, Adj_C, Adv_C] \tag{14}$$

$$K_e = [N_e, V_e, Adj_e, Adv_e] \tag{15}$$

where the subscripts C and e respectively represent C and E.

Step 4. *emofMeaning* is fed into the embedding layer, considered as English words using **EORW** to compute its emotional score.

Step 5. In C and E, after generating K, the emotional scores of words in K can be computed by the **EORW** technique, if these words don't belong to the emotional dictionary. Then, this paper checks whether emotion is fluctuating. If a fluctuation position exists, **CNEF** is designed to calculate the emotional score of the sentence. In other words, more attention is paid to the fluctuation position, considered as the most significant position. The tendencies of other elements are considered as other elements M.

Emotional Fluctuation and emotional fusion are introduced in the following sections.

Emotional Fluctuation. There are definitions to describe the emotional fluctuation and illustrate the categories of different emotional fluctuations.

Definition 9. *Emotional Fluctuation. A sentence $S = \{element_1, element_2, element_3\}$ is divided into three elements $element_1$, $element_2$, and $element_3$. Emotional fluctuation appears when the emotional tendencies of $element_1$, $element_2$, and $element_3$ are not consistent, such as ($element_1$, positive) ($element_2$, **negative**) $\leftarrow$ emotional fluctuation appears in S or ($element_1$, positive) ($element_2$, positive) ($element_3$, **negative**) $\leftarrow$ emotional fluctuation appears in S.*

Definition 10. *Emotional Fluctuation Position and Normal Position. The position of emotional inconsistency is defined as the Emotional Fluctuation Position. Other positions are considered as the Normal Position. For example, S = ($element_1$, positive, Normal Position) ($element_2$, positive, Normal Position) ($element_3$, **negative, Emotional Fluctuation Position**).*

The emotional fluctuations are classified into three categories by **CNEC**: (1) the second element is identified as the position of emotional fluctuation when the Chinese, English, and emoji demonstrate positive-negative-negative or negative-positive-positive emotional tendencies; (2) the last element is considered as the position of emotional fluctuation when the emotional tendencies of the Chinese, English, and emoji elements are negative-negative-positive or positive-positive-negative; (3) the emotions expressed in the text are generally reckoned to be consistent with the emotional tendencies of the beginning and end, when the emotional fluctuation occurs either at the beginning or end and the emotional tendencies of the beginning and end are consistent. These situations are shown in Figure 5.

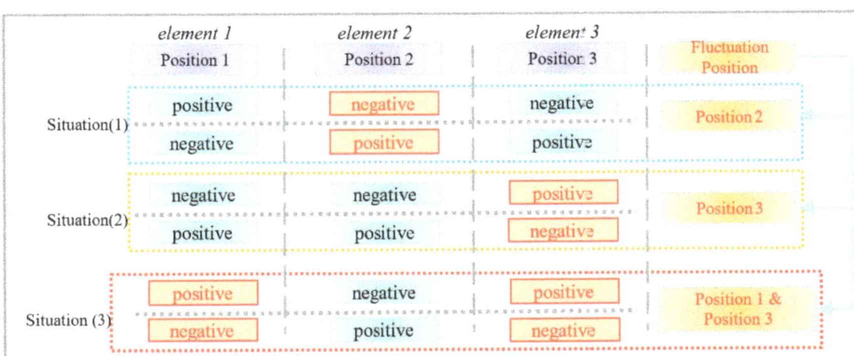

Figure 5. Fluctuation position of three elements.

Moreover, when confronted with fluctuation between two elements, a similar approach is taken as in the case of three elements. The primary adjustment is made to the element whose sentiment tendency has changed. In light of what has been previously presented, the fluctuation position Fl_P can be denoted as Equation (16). Equation (16) aligns with various situations depicted in Figure 5. Specifically, Case 1 corresponds to

M2, Case 2 corresponds to M3, and Case 3 corresponds to M1 and M3. Furthermore, in situations where only two modules are present, Case 4 of Equation (16) pertains to M2.

$$FI\_P = \begin{cases} M\,2\ (sgn(M\,1) > 0 \text{ and } sgn(M\,2) < 0 \text{ and } sgn(M\,3) < 0) \text{ or } (sgn(M\,1) < 0 \text{ and } sgn(M\,2) > 0 \text{ and } sgn(M\,3) > 0) \\ M\,3\ (sgn(M\,1) < 0 \text{ and } sgn(M\,2) < 0 \text{ and } sgn(M\,3) > 0) \text{ or } (sgn(M\,1) > 0 \text{ and } sgn(M\,2) > 0 \text{ and } sgn(M\,3) < 0) \\ M\,1\ \&\ M\,3\ (sgn(M\,1) > 0 \text{ and } sgn(M\,2) < 0 \text{ and } sgn(M\,3) > 0) \text{ or } (sgn(M\,1) < 0 \text{ and } sgn(M\,2) > 0 \text{ and } sgn(M\,3) < 0) \\ M\,2\ (sgn(M\,1) > 0 \text{ and } sgn(M\,2) < 0)) \text{ or } (sgn(M\,1) < 0 \text{ and } sgn(M\,2) > 0)) \end{cases} \tag{16}$$

where M_i represents the *i-th* element of three elements. Function $sgn(\cdot)$ stands for the sign of the emotional score.

Emotional Computation of Texts in Emotional Fluctuation. To elucidate the emotional fluctuation more effectively, a Centrifugal Navigation-Based Emotion Computation framework (**CNEC**) employs the centrifugal process of circular motion in physics to describe the phenomenon of emotional fluctuation. In the **CNEC** framework, m_f represents the emotional fluctuation element, which is analogous to an object m in circular motion. M illustrates other elements equal to the center of circular track, which is denoted as O in physics. In addition, R denotes the radius of the circle, which is determined by the distance between M and m_f. As M and m_f are not point particles in reality, their separation distance is considered to be the sum of their individual lengths. R can be represented by Equation (17).

$$R = length_x + length_y \tag{17}$$

where $length_x$ represents the length of m_f, and $length_y$ means the length of M.

When m_f undergoes uniform motion along a circular orbit, it illustrates that there is no emotional fluctuation between M and m_f. When there is emotional fluctuation between M and m_f, m_f, equivalent to the object m, moves in a centrifugal motion, departing from the circular orbit. The moment centrifugal motion occurs, the condition where the angle θ between the velocity direction of m_f and the line connecting m_f to the center O of the circle is greater than 90 degrees corresponds to Equation (16). In other words, the condition of emotional fluctuation is equal to the condition of centrifugal motion. As shown in Figure 6, an example of a bilingual text with an emoji that is in a fluctuation position illustrates the corresponding relationship between the workflow of emotional fluctuation and centrifugal motion in the **CNEC** framework.

S = "为你挑选了实用的礼物, 而你stupid 😫" (Translation: I select practical gifts for you, but you're stupid 😫), a sentence with an emoji, is put into the Model. Then, this study carries out splitting the sentence S, and puts the Chinese, English, and emoji into different elements through Equations (12) and (13). The format of the segment = $[C][E][C_{emoji}]$. Using word tokenization of BERT, C is split into individual words w_i stored in a set CW, i < N, where N represents the length of C. Similarly, k words w_k are stored in a set EW, k < M, where M represents the length of E. Then, a library Jieba is utilized to mark the part of speech of w_i in CW. As a result, N_c, V_c, Adj_c, and Adv_c are extracted from CW and saved into Kc. Moreover, the library NLTK is used to mark the part of speech of w_k in EW, so N_e, V_e, Adj_e, and Adv_e are extracted from EW into Ke. These conditions are shown in Equations (14) and (15). Later, Kc is fed into the embedding layer, the word vector set E_c fed into Equation (1) can be obtained, and the word vector set E_e obtained by Equation (1) in the same way.

On the basis of **EORW**, S_i^t can be calculated, which represents the emotional score of the i-th word with t neighbor words. Then the score S_{Kc} of Kc is fused by S_i^t, and the score S_{Ke} of Ke fused by S_i^t, as the following Equation (18).

$$S_{K_x} = \frac{\sum S_i^t}{\rho_x} \tag{18}$$

where ρ_c stands for the density of Kc that is equal to the length of Kc, S_{Kc} replaces S_{Kx}, and ρ_c replaces ρ_x. In addition, ρ_e stands for the density of Ke that is equal to the length of Ke, S_{Ke} replaces S_{Kx}, and ρ_e replaces ρ_e.

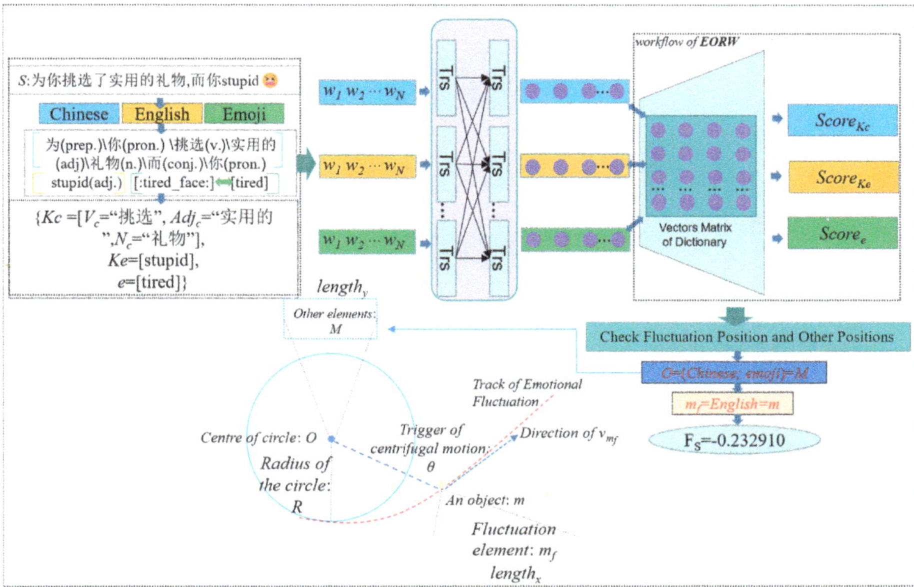

Figure 6. Centrifugal Navigation-Based Emotion Computation framework for emotional fluctuation.

When enters sequentially Equations (9) and (10), its *emofMeaning* can be computed. After that, *emofMeaning*[tired] is fed into embedding layer, obtaining $E_{emofMeaning[\text{tired}]}$. Then, emoji's score S_e can be calculated by **EORW**.

After collecting S_{Kc}, S_{Ke}, and S_e, the emotional fluctuation can be checked by Equation (16). While a fluctuation position exists, S_{Kc} represents the score of Kc corresponding to the score of M1, S_{Ke} represents the score of Ke corresponding to the score of M2, and S_e represents the score of e corresponding to the score of M3. Finally, the text score F_S can be calculated by Emotional Fusion, described in the next section.

As depicted in Table 2, each parameter of centrifugal motion corresponds to parameters in emotional fluctuation.

Table 2. Corresponding Parameters between emotional fluctuation and centrifugal motion.

Centrifugal Motion	Sentiment Computation
Centre of circle: O	**Other elements**: M
An object: m	**Fluctuation element**: m_f
Radius of the circle: R	**The sum of length of** M **and** m_f: $length_x + length_y$
Trigger of centrifugal motion: $\theta > 90$ degrees	**Trigger of emotion fluctuation**: Condition of Equation (16)

Centrifugal Navigation-Based Emotional Fusion. When fluctuation doesn't exist, it means that the tendencies among C, E, and e are consistent. Therefore, the final score F_S of a short text S is computed by **Maximum Density Dominance**, as shown in Equation (19).

$$F_S = Max(\frac{m_i}{\rho_i}),\ consistent\ \ tendency \tag{19}$$

where Function $Max(\cdot)$ determines the largest score of the three elements (m_1, m_2, m_3), and ρ_i represents the density of i-th element that is equal to the length of i-th element.

On the contrary, when one of emotional fluctuations emerges among the three elements, emotional fusion is utilized as shown in Equation (20).

$$F_S = \frac{sgn(m_f) * e^{score_{m_f}}}{\sum e^{score_M}},\ inconsistent\ \ tendency \tag{20}$$

where Function sgn(·) extracts emotional tendency.

Workflow of emotional computation about two elements. To facilitate understanding and examination of two elements, an instance of a monolingual corpus containing the emoji is presented below, as demonstrated in Figure 7. *S*, an English sentence with the emoji, is put into the Model. Then, this study carries out splitting the sentence *S*, and puts the English and emoji into different groups. The format of the segment = $[E][C_{emoji}]$. Using word tokenization, *E* is split into individual words w_i stored in a set *EW*, *i* < N, where N represents the length of *E*. Then, NLTK is utilized to mark the part of speech of w_i in *EW*. As a result, *N*, *V*, *Adj*, and *Adv* are extracted from *EW* and saved into *Ke*. Moreover, *C* is empty, so *Kc* belongs to the null set ∅. Later, *Ke* is fed into embedding layer, word vectors set E_e fed into Equation (2) can be obtained.

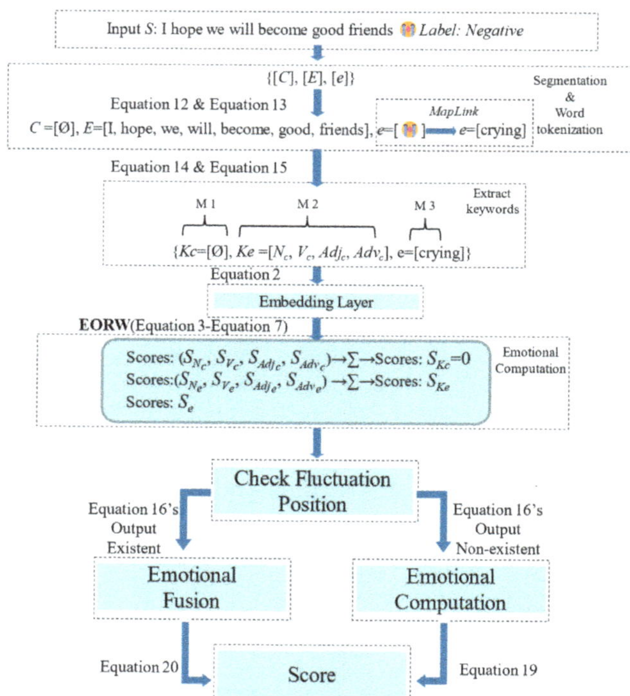

Figure 7. Emotional fluctuation of English text with an emoji.

On the basis of **EORW**, S_i^t can be calculated, which represents the emotional score of the *i*-th word with *t* neighbor words. Then the score S_{Kc} of *Kc* is fused by S_i^t as the following Equation (21).

$$S_{K_x} = \frac{\sum S_i^t}{\rho_x} \tag{21}$$

where ρ_e stands for the density of *Ke* that is equal to the length of *Ke*, S_{Ke} replaces S_{Kx}, and ρ_e replaces ρ_x.

When 😭 enters Equation (10), its *emofMeaning* can be computed. After that, *emofMeaning*[crying] is fed into embedding layer, obtaining $E_{emofMeaning[\text{crying}]}$. Then, the emoji's score S_e can be calculated by **EORW**.

After collecting S_{Kc} and S_e, the emotional fluctuation can be checked by Equation (16). While a fluctuation position exists, S_{Kc} represents the score of *Kc* corresponding to the score of M1, and S_e represents the score of *Ke* corresponding to the score of M2. Finally, text score F_S can be calculated by Equation (20).

Conversely, if the fluctuation doesn't exist, Equation (19) is adopted to compute the final score F_S.

4. Experiment

4.1. Dataset

To perform the computation of scores with expert knowledge, the sentiment dictionary with emotional scores is an essential tool. In this paper, the Boson dictionary including 114,767 words is utilized to retrieve emotional scores of words. In this dictionary, the format of data is [*word*][*score*]. The value range of scores is $(-7, 7)$. In order to facilitate the calculation, scores in the dictionary are normalized to the region of $[-1, 1]$.

We collect the short text dataset from Chinese platform WeiboWe imitate the dataset from github.com/SophonPlus/ChineseNlpCorpus to annotate our short text dataset, and extract the data that satisfies the task of this paper. Long texts over 50 words are deleted from the merged dataset, leaving about 903 texts with emoji. Data text has any combination of Chinese [C], English [E], and emojis [e], and the format of the texts = {[C, E, e] | [C, E] | [C, e] | [E, e]}, as shown in Table 3.

Table 3. Instances of Dataset.

Label	Text
Negative	为你挑选了实用的礼物, 而你stupid 😫
Negative	I hope we will become good friends 😭
Positive	#积极向上#日子还要继续, 开心点! 😃
Positive	@养熊猫的李桑Love you! 😊

4.2. Experiment and Result

In this section, we perform two groups of experiments, containing the selection of ***t*-value of EORW** and **Emotional Computation**. Regarding the selection of *t*-value experiment, multiple iterations of experiments are conducted to determine the optimal value of *t*. In response to the experiment of **Emotional Computation**, the emotional scores of the sentences in the dataset are computed, and these scores are utilized to assess whether the tendencies align with the tendencies associated with the labels.

In the first experiment, two samples about words and the emoji about emotional computation are shown as follows.

As shown in Table 4, angry and fire are taken as examples (six decimal places are retained and $t = 5$). When *uw* is angry, **NW-Set**(angry) = {irate, enraged, indignant, incensed, annoyed}. Then, **NW-Set**(angry) is fed into Equation (4). However, $\mathbf{MES}_i$(angry) = {Ø} on account of the fact that the emotional scores of w_i in **NW-Set**(angry) are consistent. As a result, S_{uw}(angry) can be calculated by the first case of Equation (6). Conversely, the emotional tendencies of neighbors of fire as *uw* are inconsistent. **NW-Set**(fire) = {firing, alarm, destroyed, fumes, fired}. It is noteworthy that the emotional tendency of "fumes" in **NW-Set** is different from others. Therefore, $\mathbf{MES}_1$(fire) = {firing, alarm, destroyed, fired}, and $\mathbf{MES}_2$(fire) = {fumes}. Based on Equation (5), **RS**(fire) = $\mathbf{MES}_1$(fire). At last, S_{uw}(fire) can be computed by the second case of Equation (6).

Table 4. Results of emotional computation about words.

Word	Neighbor	Emotion (Ei) and Score i		Score
angry	irate	E1	−0.726027	−0.479452
	enraged	E2	−0.397260	
	indignant	E3	−0.424658	
	incensed	E4	−0.479452	
	annoyed	E5	−0.369863	
fire	firing	E1	−0.315068	−0.452055
	alarm	E2	−0.315068	
	destroyed	E3	−0.534247	
	fumes	E4	0.041096 (discarded)	
	fired	E5	−0.643836	

In Table 5, neighbor means *t* neighbor words of emofMeaning "crying" from the emoji 😭, Emotion represents the emotional score of neighbors of emofMeaning "crying", and ***S*** stands for the final score of emofMeaning "crying" that is equal to the emoji 😭.

Table 5. Emotional computation of emoji 'loudly crying face' (retaining six decimal places).

Emoji	emofMeaning	Neighbor	Emotion	S
😭	crying	sobbing	−0.369863	−0.388128
		cried	−0.369863	
		screaming	−0.369863	
		weeping	−0.452055	
		cries	−0.39726	
		cry	−0.506849	
		moaning	−0.041096	
		screamed	−0.287671	
		sob	−0.69863	

t-value of EORW. To accomplish the goal of computing the emotional score of *uw* and verify the feasibility of **EORW**, this paper covers up the emotional score of *w* in *Dic* to compare the emotional score calculated by **EORW** with the score stored in *Dic*.

For facilitating the test, the threshold is set to 30%. When the threshold *T* is lower than 30%, the result is computed correctly. *T* is denoted as Equation (22).

$$T = \frac{|label - uw[score]|}{label} \times 100\% \tag{22}$$

In this paper, the accuracy of a group is determined by the ratio of correctly computed words to the total number of words. In this condition, the symbol n_{ccw} is used to denote the number of words computed correctly, while N represents the total number of words in a given experimental group. The accuracy *Acc* is denoted as shown in Equation (23).

$$\begin{aligned} n_{ccw} &= len(\{w_{ccw} | T < 30\%\}) \\ Acc &= \tfrac{n_{ccw}}{N} \end{aligned} \tag{23}$$

In addition, this paper chooses a moderate *t*-value. Five hundred words *uw*[cover] are generated from the dictionary *Dic*. Ten groups of experiments are conducted with different *t* values, where *t* belongs to the set = [3, 5, 7, 9, 11, 13, 15].

As shown in Figure 8, when 300 words are randomly generated, the overall accuracy is between 74.3% and 85%. When *t* is 9, the upper bound of accuracy is about 82.7% and the lower bound of accuracy is about 79.3%. When the *t*-value is equal to 9, accuracy becomes stable. When 500 words are randomly generated, the overall accuracy is between 74.2% and 84%. When *t* is 9, the upper bound of accuracy is about 83.2% and the lower bound of accuracy is about 78%. From what has been presented, the upper and lower bounds of accuracy are closer than other *t* when *t* is 9, which means that **EORW** is the most stable in this condition.

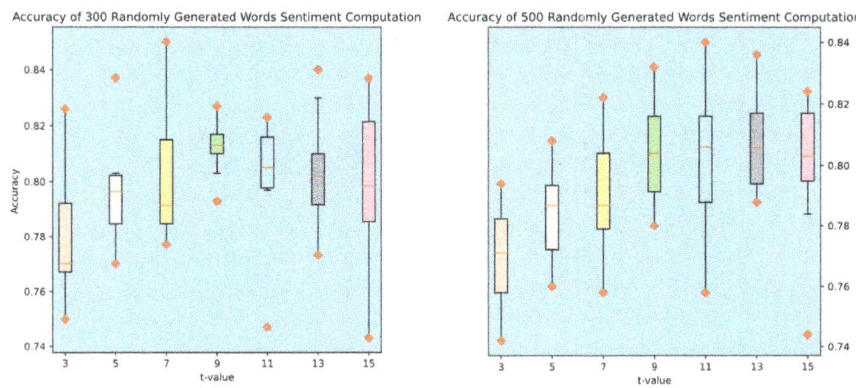

Figure 8. The accuracies of emotional computation of 300 and 500 randomly generated words from the emotional dictionary. Different *t*-values are adopted. The various box plots with distinct colors illustrate different t-values aligned with their respective positions.

Proportion of emoji usage. This paper presents a statistical analysis of the proportion of emoji usage in the collected dataset, as depicted in Figure 9. In the collected data, the utilization rate of 'loudly crying face' emoji has the highest usage frequency, accounting for 21% of the total. In contrast, it is obvious from Figure 9 that the 'kiss' emoji has a very low usage frequency, representing only 1.13% of the total. Additionally, other infrequently used emojis have been grouped into the "others" category, which accounts for approximately 15.18% of the total emoji usage.

😭 in the bilingual sentence *S* is transformed to *emofMeaning*[crying]. The processing results of emotional scores of [crying] calculated are depicted in Table 5. As shown in Table 6, the emoji 😫 in the bilingual sentence *S* can be similarly computed.

Result of emotional computation. After eliminating disordered data, the remaining dataset consists of 903 texts where 601 data labels are positive and 302 data labels are negative. In addition, some marks with bad influence on the experiment are stripped from the dataset. Then, the dataset is fed into the computation model. As a result, the accuracy of the emotional computation reaches about 98.67%.

In the bilingual sentence *S*, *Kc* and *Ke* respectively harbor three keywords and one keyword. Each phrase has nine neighbors that have their own scores. The presence of a score above 0 indicates a positive tendency, while a score below 0 signifies a negative tendency. Through **EORW**, the dominant emotional set also named as **RS** can be computed. As a result, the emotional scores of the keywords can be calculated as shown in Table 7. As shown in Figure 10, the bilingual sentence *S* contains four keywords (Chinese phrases: 挑选 (translation: select); 实用的 (translation: practical); 礼物 (translation: gifts). English word: stupid).

As depicted in Figure 11, the English sentence *S* has four keywords (hope, become, good, friends) extracted from *Ke*. According to **EORW**, **RS** can be computed. As a result, the emotional scores of the keywords can be calculated as shown in Table 8.

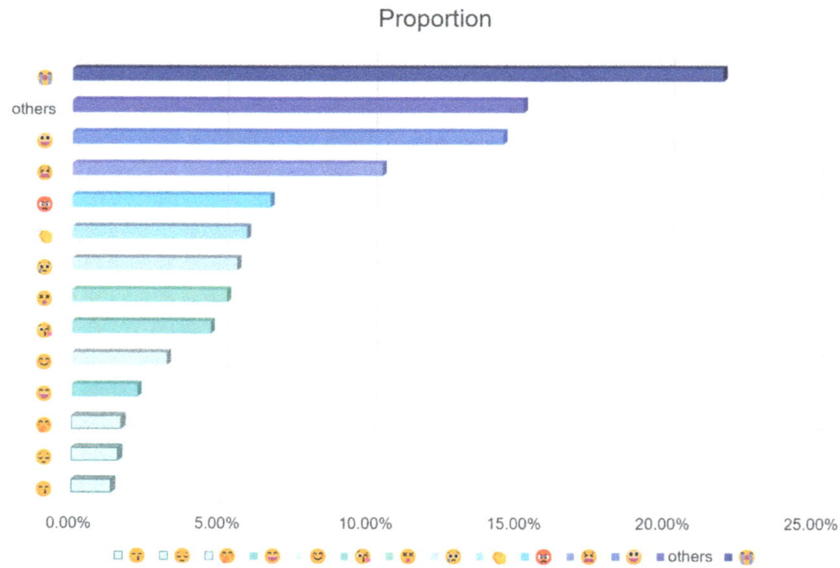

Figure 9. The proportion of emojis used in the dataset collected in this paper.

Table 6. The processing of emojis in the bilingual sentence S (retaining six decimal places).

Emoji	emofMeaning	Neighbor	Emotion	S
😫	tired	fatigued	−0.315068	−0.354642
		weary	−0.232877	
		bored	−0.232877	
		frustrated	−0.589041	
		irritated	−0.479452	
		annoyed	−0.369863	
		impatient	−0.260274	
		exhausted	−0.342466	
		jaded	−0.369863	

Table 7. Emotional scores of keywords in bilingual sentence S (retaining six decimal places).

Keyword	Score
select (Original Chinese: 挑选)	0.290705
Practical (Original Chinese: 实用的)	0.267186
Gifts (Original Chinese: 礼物)	0.346129
stupid	−0.472603

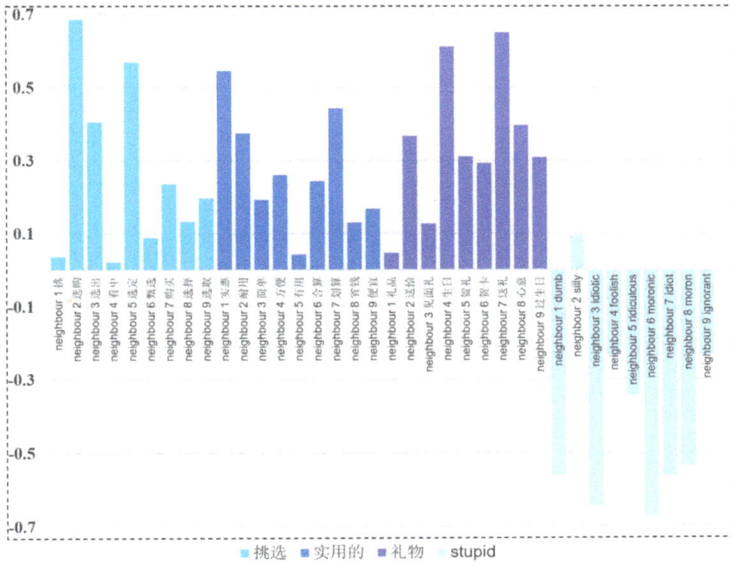

Figure 10. Emotional scores of neighbors of w_i in *Kc* and neighbors of w_k in *Ke*. This figure depicts that the bilingual sentence *S* has four keywords extracted respectively from *Kc* and *Ke*. In the Chinese element, The neighbors in NW-Set of select (Original Chinese: 挑选): {挑 (choose), 选购 (optional), 看中 (fancy), 选定 (pick out), 甄选 (pick), 购买 (purchase), 选择 (option), 选取 (select and extract)}; The neighbors in NW-Set of Practical (Original Chinese: 实用的): {实惠 (affordable), 耐用 (durable), 简单 (easy), 方便 (convenient), 有用 (useful), 合算 (be a bargain), 划算 (favorable), 省钱 (economical), 便宜 (cheap)}; The neighbors in NW-Set of Gifts (Original Chinese: 礼物): {礼品 (souvenirs), 送给 (give sb), 见面礼 (a gift such as is usually given to sb. on first meeting him), 生日 (birthday), 贺礼 (congratulatory gift), 贺卡 (congratulation card), 送礼 (give a present), 心意 (compliments or gifts), 过生日 (celebrate a birthday)}.

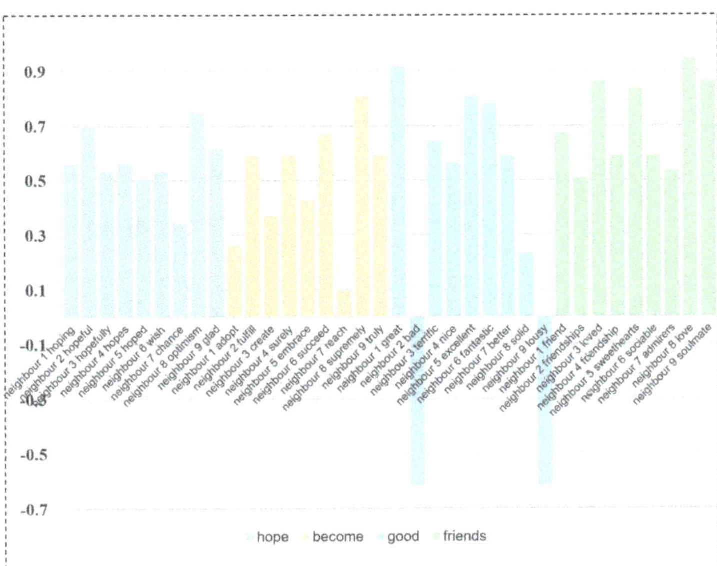

Figure 11. Emotional scores of neighbors of w_i in *Ke*. This figure depicts that the English sentence *S* has four keywords extracted from *Ke*.

Table 8. Emotional scores of keywords in the English sentence S (retaining six decimal places).

Keyword	Score
hope	0.567732
become	0.488584
good	0.647750
friends	0.710807

According to Equation (22), the emotional score of the Chinese and English corpora can be respectively computed. Through Equation (16), fluctuation can be checked. Finally, in light of Equation (21), the result of the emotional fusion is computed. The emotional processing of the bilingual sentence and English sentence are noted in Table 9.

Table 9. Results of emotional scores of different elements (retaining six decimal places).

Sentence Type	Label	S_{M1}	S_{M2}	S_{M3}	Fl_P	F_S
Bilingual	negative	0.301340	−0.472603	−0.354642	M2	−0.232910
English	negative	0.603718	−0.388129	Empty	M2	−0.270548

As demonstrated in Table 10, **CNEC** outperforms the average strategy by 1.11%. In addition, **CNEC** improves by 21.04% compared with the maximum-value method. In response to emotional fusion, **CNEC** has more competitive advantages. Furthermore, the experimental outcomes are compared by utilizing several prominent deep learning models that are frequently employed in the field, as shown in Table 11. The BERT model is built upon the BERT-base model, with a learning rate of 2×10^{-5}, a batch size of 32, and a total of 5 training epochs. RoBERTa-base is adopted in this paper, with a learning rate of 2×10^{-5}, a batch size of 16, and a total of 5 training epochs. In $\text{BERT}_{\text{prompting}}$, BERT-base is also the backbone of this model, with a learning rate of 2×10^{-5}, a batch size of 8, and a total of 5 training epochs. According to prompt learning, the $\text{BERT}_{\text{prompting}}$'s template is designed as "How {} it was.". For all methods, a cross-entropy loss is used for the loss function during training for the sentiment analysis, as shown in Equation (24).

$$Loss = \frac{1}{N}\sum_i -[y_i \cdot \log(p_i) + (1 - y_i) \cdot \log(1 - p_i)] \quad (24)$$

where N denotes the number of samples, y_i represents the true label of the i-th sample, and p_i represents its probability.

Table 10. Comparable accuracy of different emotional fusion methods.

Average	Maximum-Value	CNEC
0.9756	0.7763	**0.9867**

Table 11. Comparison of results on dataset with deep learning models.

BERT	RoBERTa	$\text{BERT}_{\text{Promping}}$ (Template: How {} it was)	CNEC
0.8106	0.9700	0.8671	**0.9867**

5. Discussion

This paper pays attention to the computation of emotion scores aiming at bilingual short texts with emoji symbols. According to traditional methods or some deep learning elements, they cannot tackle the problem of computing scores of bilingual data incorporating emoji symbols. Moreover, mainstream approaches of deep learning just compute the score of a single corpus based on their own rules. This study employs three deep learning models to conduct a comparative analysis with the suggested approach. The BERT model's pre-training data primarily consists of a vast English corpus, which poses limitations when dealing with texts that contain multiple languages and emojis. Nonetheless, by incorporating a template into BERT, its expressive capabilities are enhanced by approximately 5%. Conversely, the RoBERTa model, which excludes the NSP task and incorporates a larger training corpus, exhibits superior performance when handling extremely short sentences. However, these approaches excel primarily in classification tasks and do not effectively utilize existing knowledge that contains emotional scores for emotion scoring. On the contrast, the **CNEC** framework relies on the professional knowledge of the emotional dictionary and utilizes the **EORW** method to improve the defects of the emotional dictionary on bilingual data. Furthermore, **CNEC** drives **EORW** to compute emotional scores of emoji symbols. In addition, **CNEC** can illustrate emotional fluctuation. In addition, based on **CNEF**, the emotional fusion is utilized to compute the emotional score of bilingual short texts with emoji. The reason for utilizing centrifugal motion in this paper to illustrate the emotional fluctuation is due to the consistency of emotions within each part of a short text. If the emotion of a part fluctuates, it is akin to an object in circular motion suddenly lacking the centripetal force necessary to maintain its inertia, and thus moving centrifugally. The reason why the accuracy of the experiment can reach 98.67% is that these emotional tendencies of short texts are more explicit than those of long texts. Additionally, this dataset from Weibo doesn't contain implicit emotion such as emojis with positive tendency used to express negative tendency. As a result, emotional scores of emojis or phrases are stationary for every sentence where they exist. Furthermore, the quantity of datasets with a bilingual corpus with emojis is limited, and most data involve a single language with emojis. Besides, as emotion dictionaries are constructed based on specific knowledge categories, selecting different emotion dictionaries may result in situations where a phrase has different emotional scores, which in turn imposes specific constraints on the emotional score range of the phrase.

In the future, how to tackle different implicit meanings of phrases and emojis is to be considered as the most significant point. In addition, how to classify emotions at a fine-grained level is also a matter of great importance in future work. It is also worth studying the issue of harmonizing the grading differences among different dictionaries for various fields as much as possible.

Author Contributions: Conceptualization, T.Y. and Z.L.; methodology, T.Y. and Z.L.; software, T.Y. and Z.L.; validation, Z.L. and Y.L.; formal analysis, T.Y.; investigation, J.Z.; resources, Z.L.; data curation, T.Y. and Z.L.; writing—original draft preparation, Z.L.; writing—review and editing, T.Y.; supervision, T.Y. All authors have read and agreed to the published version of the manuscript.

Funding: This work was supported by the Sichuan Science and Technology Program under Grant No. 2022YFG0322, China Scholarship Council Program (Nos. 202001010001 and 202101010003) and the Innovation Team Funds of China West Normal University (No. KCXTD2022-3).

Data Availability Statement: Not applicable.

Conflicts of Interest: The authors declare no conflict of interest.

References

1. Nasukawa, T.; Yi, J. Sentiment analysis: Capturing favorability using natural language processing. In Proceedings of the 2nd International Conference on Knowledge Capture, Sanibel Island, FL, USA, 23–25 October 2003; pp. 70–77.
2. Hu, M.; Liu, B. Mining and summarizing customer reviews. In Proceedings of the Tenth ACM SIGKDD International Conference on Knowledge Discovery and Data Mining, Seattle, DC, USA, 22–25 August 2004; pp. 168–177.
3. Pang, B.; Lee, L. Seeing stars: Exploiting class relationships for sentiment categorization with respect to rating scales. *arXiv* **2005**, arXiv:0506075.

4. Thet, T.T.; Na, J.C.; Khoo, C.S. Aspect-based sentiment analysis of movie reviews on discussion boards. *J. Inf. Sci.* **2010**, *36*, 823–848. [CrossRef]
5. Alguliyev, R.M.; Aliguliyev, R.M.; Niftaliyeva, G.Y. Extracting social networks from e-government by sentiment analysis of users' comments. *Electron. Gov. Int. J.* **2019**, *15*, 91–106. [CrossRef]
6. Li, M.; Ch'ng, E.; Chong, A.Y.L.; See, S. Multi-class Twitter sentiment classification with emojis. *Ind. Manag. Data Syst.* **2018**, *118*, 582. [CrossRef]
7. Jain, D.K.; Kumar, A.; Sangwan, S.R. TANA: The amalgam neural architecture for sarcasm detection in indian indigenous language combining LSTM and SVM with word-emoji embeddings. *Pattern Recognit. Lett.* **2022**, *160*, 11–18. [CrossRef]
8. Yang, T.; Liu, Z.; Chen, Q.; Ma, X.; Deng, H. Emotional Computation of Unfamiliar Word based on Emotional Orientation of Related Words. In Proceedings of the 2nd International Conference on Artificial Intelligence, Big Data and Algorithms, Nanjing, China, 17–19 June 2022.
9. Le, B.; Nguyen, H. Twitter sentiment analysis using machine learning techniques. In Proceedings of the 3rd International Conference on Computer Science, Applied Mathematics and Applications-ICCSAMA, Hanoi, Vietnam, 19–20 December 2015; pp. 279–289.
10. Mowlaei, M.E.; Abadeh, M.S.; Keshavarz, H. Aspect-based sentiment analysis using adaptive aspect-based lexicons. *Expert Syst. Appl.* **2020**, *148*, 113234. [CrossRef]
11. Keshavarz, H.; Abadeh, M.S. ALGA: Adaptive lexicon learning using genetic algorithm for sentiment analysis of microblogs. *Knowl.-Based Syst.* **2017**, *122*, 1–16. [CrossRef]
12. Alqaryouti, O.; Siyam, N.; Abdel Monem, A.; Shaalan, K. Aspect-based sentiment analysis using smart government review data. *Appl. Comput. Inform.* **2020**. [CrossRef]
13. Revathy, G.; Alghamdi, S.A.; Alahmari, S.M.; Yonbawi, S.R.; Kumar, A.; Haq, M.A. Sentiment analysis using machine learning: Progress in the machine intelligence for data science. *Sustain. Energy Technol. Assess.* **2022**, *53*, 102557. [CrossRef]
14. Mai, L.; Le, B. Joint sentence and aspect-level sentiment analysis of product comments. *Ann. Oper. Res.* **2021**, *300*, 493–513. [CrossRef]
15. Da'u, A.; Salim, N.; Rabiu, I.; Osman, A. Weighted aspect-based opinion mining using deep learning for recommender system. *Expert Syst. Appl.* **2020**, *140*, 112871.
16. Pathak, A.R.; Pandey, M.; Rautaray, S. Topic-level sentiment analysis of social media data using deep learning. *Appl. Soft Comput.* **2021**, *108*, 107440. [CrossRef]
17. Chen, T.; Xu, R.; He, Y.; Wang, X. Improving sentiment analysis via sentence type classification using BiLSTM-CRF and CNN. *Expert Syst. Appl.* **2017**, *72*, 221–230. [CrossRef]
18. Liang, B.; Su, H.; Gui, L.; Cambria, E.; Xu, R. Aspect-based sentiment analysis via affective knowledge enhanced graph convolutional networks. *Knowl.-Based Syst.* **2022**, *235*, 107643. [CrossRef]
19. Huang, B.; Guo, R.; Zhu, Y.; Fang, Z.; Zeng, G.; Liu, J.; Shi, Z. Aspect-level sentiment analysis with aspect-specific context position information. *Knowl.-Based Syst.* **2022**, *243*, 108473. [CrossRef]
20. Li, X.; Zhang, J.; Du, Y.; Zhu, J.; Fan, Y.; Chen, X. A novel deep learning-based sentiment analysis method enhanced with Emojis in microblog social networks. *Enterp. Inf. Syst.* **2023**, *17*, 2037160. [CrossRef]
21. Li, D.; Rzepka, R.; Ptaszynski, M.; Araki, K. Emoji-Aware Attention-based Bi-directional GRU Network Model for Chinese Sentiment Analysis. In Proceedings of the LaCATODA/BtG@ IJCAI, Macao, China, 24–29 August 2019; pp. 11–18.
22. Huang, Y.; Liu, Q.; Peng, H.; Wang, J.; Yang, Q.; Orellana-Martín, D. Sentiment classification using bidirectional LSTM-SNP model and attention mechanism. *Expert Syst. Appl.* **2023**, *221*, 119730. [CrossRef]
23. Gu, T.; Zhao, H.; He, Z.; Li, M.; Ying, D. Integrating external knowledge into aspect-based sentiment analysis using graph neural network. *Knowl.-Based Syst.* **2023**, *259*, 110025. [CrossRef]
24. Zhao, M.; Yang, J.; Zhang, J.; Wang, S. Aggregated graph convolutional networks for aspect-based sentiment classification. *Inf. Sci.* **2022**, *600*, 73–93. [CrossRef]
25. Zhou, T.; Law, K.M. Semantic Relatedness Enhanced Graph Network for aspect category sentiment analysis. *Expert Syst. Appl.* **2022**, *195*, 116560. [CrossRef]
26. Xu, L.; Pang, X.; Wu, J.; Cai, M.; Peng, J. Learn from structural scope: Improving aspect-level sentiment analysis with hybrid graph convolutional networks. *Neurocomputing* **2023**, *518*, 373–383. [CrossRef]
27. Xu, Y.; Yao, E.; Liu, C.; Liu, Q.; Xu, M. A novel ensemble model with two-stage learning for joint dialog act recognition and sentiment classification. *Pattern Recognit. Lett.* **2023**, *165*, 77–83. [CrossRef]
28. Zhang, Y.; Liu, Y.; Li, Q.; Tiwari, P.; Wang, B.; Li, Y.; Pandey, H.M.; Zhang, P.; Song, D. CFN: A Complex-Valued Fuzzy Network for Sarcasm Detection in Conversations. *IEEE Trans. Fuzzy Syst.* **2021**, *29*, 3696–3710. [CrossRef]
29. Qiu, J.; Ji, W.; Lam, H.-K. A New Design of Fuzzy Affine Model-Based Output Feedback Control for Discrete-Time Nonlinear Systems. *IEEE Trans. Fuzzy Syst.* **2023**, *31*, 1434–1444. [CrossRef]
30. Khalifa, T.R.; El-Nagar, A.M.; El-Brawany, M.A.; El-Araby, E.A.G.; El-Bardini, M. A Novel Hammerstein Model for Nonlinear Networked Systems Based on an Interval Type-2 Fuzzy Takagi–Sugeno–Kang System. *IEEE Trans. Fuzzy Syst.* **2020**, *29*, 275–285. [CrossRef]

31. Turney, P.D. Thumbs up or thumbs down? Semantic orientation applied to unsupervised classification of reviews. *arXiv* **2002**, arXiv:0212032.
32. Wang, H.; Liu, B.; Li, C.; Yang, Y.; Li, T. Learning with noisy labels for sentence-level sentiment classification. *arXiv* **2019**, arXiv:1909.00124.

electronics

Article

An Enhancement Method in Few-Shot Scenarios for Intrusion Detection in Smart Home Environments

Yajun Chen [1], Junxiang Wang [1,*], Tao Yang [2], Qinru Li [3] and Nahian Alom Nijhum [4]

1 School of Electronic Information Engineering, China West Normal University, Nanchong 637001, China; scnccyj@cwnu.edu.cn
2 Education and Information Technology Center, China West Normal University, Nanchong 637001, China; yangt@cwnu.edu.cn
3 School of Computer Science, China West Normal University, Nanchong 637001, China; qrli@cwnu.edu.cn
4 School of Software Engineering, China West Normal University, Nanchong 637001, China; 31nahian@gmail.com
* Correspondence: wangjx@stu.cwnu.edu.cn

Abstract: Different devices in the smart home environment are subject to different levels of attack. Devices with lower attack frequencies confront difficulties in collecting attack data, which restricts the ability to train intrusion detection models. Therefore, this paper presents a novel method called EM-FEDE (enhancement method based on feature enhancement and data enhancement) to generate adequate training data for expanding few-shot datasets. Training intrusion detection models with an expanded dataset can enhance detection performance. Firstly, the EM-FEDE method adaptively extends the features by analyzing the historical intrusion detection records of smart homes, achieving format alignment of device data. Secondly, the EM-FEDE method performs data cleaning operations to reduce noise and redundancy and uses a random sampling mechanism to ensure the diversity of the few-shot data obtained by sampling. Finally, the processed sampling data is used as the input to the CWGAN, and the loss between the generated and real data is calculated using the Wasserstein distance. Based on this loss, the CWGAN is adjusted. Finally, the generator outputs effectively generated data. According to the experimental findings, the accuracy of J48, Random Forest, Bagging, PART, KStar, KNN, MLP, and CNN has been enhanced by 21.9%, 6.2%, 19.4%, 9.2%, 6.3%, 7%, 3.4%, and 5.9%, respectively, when compared to the original dataset, along with the optimal generation sample ratio of each algorithm. The experimental findings demonstrate the effectiveness of the EM-FEDE approach in completing sparse data.

Keywords: data enhancement; few-shot data; smart home; generative adversarial networks; intrusion detection

Citation: Chen, Y.; Wang, J.; Yang, T.; Li, Q.; Nijhum, N.A. An Enhancement Method in Few-Shot Scenarios for Intrusion Detection in Smart Home Environments. *Electronics* **2023**, *12*, 3304. https://doi.org/10.3390/electronics12153304

Academic Editor: Domenico Ursino

Received: 21 June 2023
Revised: 19 July 2023
Accepted: 30 July 2023
Published: 31 July 2023

1. Introduction

With the development of Internet of Things (IoT) technology, the application of smart home scenarios is becoming increasingly widespread [1]. The number of smart home devices connected to home networks rapidly increases, leading to a surge in network scale and data traffic. These factors exacerbate security threats such as network attacks and privacy breaches, presenting new security challenges [2]. Additionally, different types of smart home devices differ in various aspects, posing unique security challenges for each device. For example, smart locks may face security issues like password cracking and fingerprint recognition attacks, while smart appliances may encounter concerns related to electrical safety and control signal security.

There is currently a limited amount of research focusing on smart home security, with most studies primarily concentrating on hardware design and passive defense measures. For instance, the authors of [3] propose a method for privacy risk assessment and risk control measures to address privacy security concerns. The authors of [4] describe a provable security authentication scheme to ensure the security of the smart home environment. These studies, which are passive defense mechanisms, can improve the security of smart homes to some extent, but they do not fully address all the security issues.

Intrusion detection, as a typical representative of active defense, is one of the critical technologies for safeguarding the security of smart home systems [5]. It overcomes the limitations of traditional network security techniques in terms of real-time responsiveness and dynamic adaptability. Monitoring and identifying abnormal behavior in network traffic enables timely detection and prevention of malicious attacks. Therefore, designing an efficient intrusion detection model is of paramount importance in ensuring the security of smart home systems. Traditional machine learning-based intrusion detection algorithms are relatively straightforward to train, widely adopted, and demonstrate high efficiency and reliability in practical applications [6,7]. On the other hand, intrusion detection algorithms based on deep learning exhibit superior detection performance, but their exceptional performance relies heavily on a significant amount of training data [8–10].

Currently, there is no comprehensive framework for research on smart home security, and it still faces several challenges [11,12]. Due to the varying attack frequencies of different devices in smart homes, there is an imbalance in collecting network traffic data, with some devices having a deficient proportion of attack data compared to normal data. The insufficient quantity of data makes it difficult to effectively train intrusion detection models, resulting in a decline in their performance [13,14]. Therefore, this paper proposes an enhancement method, EM-FEDE, applied to smart home intrusion detection in few-shot scenarios. Firstly, the EM-FEDE method analyzes the historical intrusion detection records of smart homes to determine whether there are features indicative of device types and data types in the captured data and then adaptively extends the features to achieve format alignment of device data. Secondly, the EM-FEDE method performs data cleaning operations to reduce noise and redundancy by removing duplicate entries and normalizing the data. Furthermore, the method adjusts the random sampling mechanism to ensure the diversity of the few-shot data obtained through sampling. Finally, the processed sampling data is used as input for the CWGAN, a variant of GAN that improves data generation through modifications in the loss function and optimization algorithms. The Wasserstein distance, which measures the dissimilarity between two probability distributions, is employed by CWGAN to calculate the loss between the generated data (fake data) and the real data. Based on this loss, the CWGAN is adjusted, and the generator of the CWGAN outputs effectively generates data. The main contributions of this paper are as follows:

- This paper proposes a feature enhancement module to improve the data quality in the dataset by analyzing historical intrusion detection records of smart homes, adaptively extending feature columns for the smart home devices dataset, and performing data cleaning on the dataset;
- This paper proposes a data enhancement module to generate valid data to populate the dataset using conditional Wasserstein GAN to realize the operation of data enhancement for few-shot data;
- The effectiveness of the EM-FEDE method is evaluated using a typical smart home device dataset, N-BaIoT. The performance of the original dataset and the expanded dataset using the EM-FEDE method on each intrusion detection model is compared to conclude that the classifier's performance is higher for the expanded dataset than the original dataset;
- The experiments demonstrate that expanding the dataset using the EM-FEDE method is crucial and effective in improving the performance of attack detection. This work successfully addresses the problem of few-shot data affecting the performance of intrusion detection models.

2. Related Works

2.1. Intrusion Detection Methods for Smart Homes

Intrusion detection methods for smart homes have gained significant attention in recent years as a popular research direction in the field of smart homes, and many scholars have conducted relevant research [15–17]. Many methods utilize sensors and network communication functions within smart home devices to detect intrusions by monitoring user behavior, device status, and other relevant data.

In 2021, the authors of [18] proposed an intrusion detection system that uses bidirectional LSTM recursive behavior to save the learned information and uses CNN to perfectly extract data features to detect anomalies in smart home networks. In 2021, the authors of [19] proposed a two-layer feature processing method for massive data and a three-layer hybrid architecture composed of binary classifiers in smart home environments to detect malicious attack environments effectively. In 2022, the authors of [20] proposed an intelligent two-tier intrusion detection system for the IoT. Using the feature selection module combined with machine learning, both flow-based and packet-based, it can minimize the time cost without affecting the detection accuracy. In 2023, the authors of [21] proposed an effective and time-saving intrusion detection system using an ML-based integrated algorithm design model. This model has high accuracy, better time efficiency, and a lower false alarm rate. In 2023, the authors of [14] proposed a transformer-based NIDS method for the Internet of Things. This method utilizes a self-attention mechanism to learn the context embedding of input network features, reducing the negative impact of heterogeneous features.

Even though numerous scholars have obtained commensurate outcomes pertaining to the issue of smart home security, such research endeavors were executed with ample data and did not consider the predicament of limited samples attributable to the shortage of data emanating from various devices in smart homes. As a result, it is difficult for intrusion detection models to assimilate the data feature, and the suggested models of the research endeavors above are unsuitable for situations involving few-shot data.

2.2. GAN-Based Data Enhancement Methods

In machine learning and deep learning, the size of the dataset is a critical factor affecting the performance of the model. However, obtaining large-scale labeled datasets will require a large workforce and resources. Researchers have been exploring data enhancement techniques to expand the dataset and improve model performance. Among these techniques, data enhancement methods using Generative Adversarial Networks (GAN) [22] proposed by Ian J. Goodfellow et al. in 2014 have gained significant attention. GAN-based data enhancement methods have shown promising results in enhancing the performance of intrusion detection models by generating generated data that can be used to supplement the limited labeled dataset.

GAN consists of a discriminator network and a generator network. The goal of the discriminant network is to accurately determine whether a sample is from real or fake data. The purpose of the generator network is to generate samples whose sources cannot be distinguished by the discriminant network. In GAN, the Generator uses random noise Z as input data, and its output is fake sample data $G(z)$. The discriminator receives real sample data x and fake sample data $G(z)$ as inputs and obtains loss by determining whether the data is real or fake by using the backpropagation algorithm to update the GAN parameters based on the loss function.

In recent years, more and more research has applied GAN for data enhancement to improve the performance and robustness of machine learning models. In 2021, the authors of [23] proposed using the ACGAN model to solve the problem of the imbalanced distribution of 1D intrusion detection sample data, which improved the average detection accuracy of some classification models. In 2021, the authors of [24] proposed an improved DCGAN model with higher stability and sample balance to achieve higher classification accuracy for a few samples. In 2022, the authors of [25] proposed a new generation of methods that use a class of classification models to determine the authenticity of facial images. This method improves cross-domain detection efficiency while maintaining source-domain accuracy. In 2023, the authors of [26] proposed an attention-self-supervised learning-aided classifier generative adversarial network algorithm to expand the samples to improve the defect recognition ability of small sample data sets. In 2023, the authors of [27] proposed a generative model for generating virtual marker samples by combining supervised variational automatic encoders with Wasserstein GAN with a gradient penalty. This model can significantly improve the prediction accuracy of soft sensor models for small-sample problems.

Although scholars have made many achievements using GAN for data enhancement, their applications are mainly carried out on images. In network security, there is still a lack of research on data enhancement using GAN. In addition, the implementation of GANs for data augmentation in the field of smart home intrusion detection has not been fully explored, thereby limiting their potential to solve problems in this field.

3. EM-FEDE Method

3.1. Problem Analysis

Figure 1 shows a typical smart home environment. A diverse array of smart home devices is linked to a gateway, which in turn is connected to the Internet via routers, and the data collected by these devices is subsequently sent to terminals for user analysis.

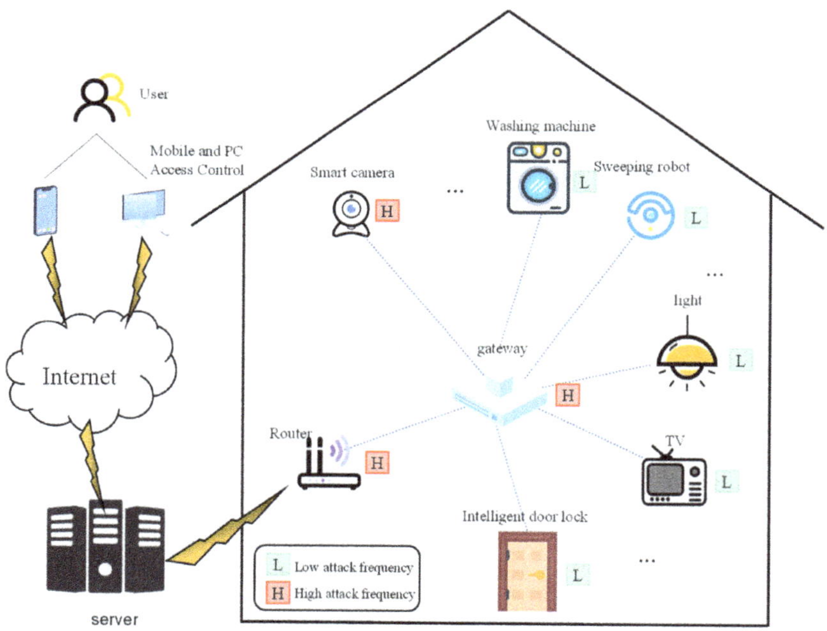

Figure 1. Smart home topology diagram.

Smart home devices differ in functionality and operational characteristics, exhibiting distinct working hours and data throughput. Attackers take various factors into account, such as device usage frequency and attack complexity, when exploiting vulnerabilities in different types of devices. This leads to the devices being subjected to varying frequencies of network attacks. The N-BaIoT dataset [28] of typical IoT devices shows the variance in data throughput and traffic collected among different devices. As depicted in Figure 2, a smart doorbell device (device1) and a smart camera device (device2) exhibit different data throughput, with device1 having a lower data throughput. Consequently, the amount of data collected by device1 is significantly less than that collected by device2 (338,599 vs.

1,075,936). Moreover, the number of data points generated by different attack behaviors also varies based on the attack frequencies of the devices. For instance, Figure 3 shows that attack1 (a UDP attack by the Gafgyt botnet) and attack2 (a UDP attack by the Mirai botnet) both utilize vulnerabilities to carry out DDoS attacks. However, attack2 is more effective and straightforward, resulting in a higher frequency of occurrence. Therefore, attack1 has far fewer data samples (255,111 vs. 1,229,999).

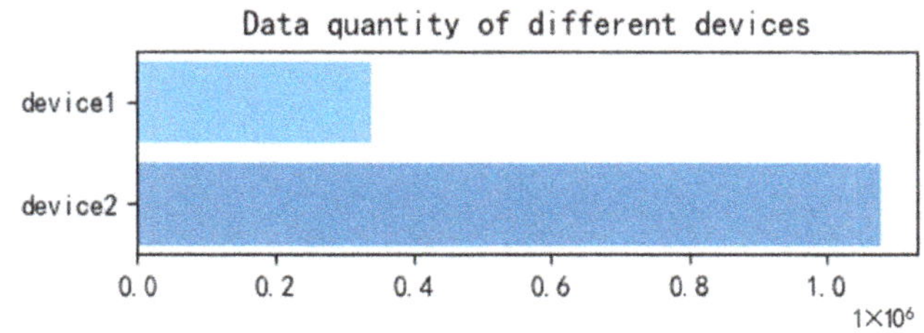

Figure 2. Number of data points from different devices.

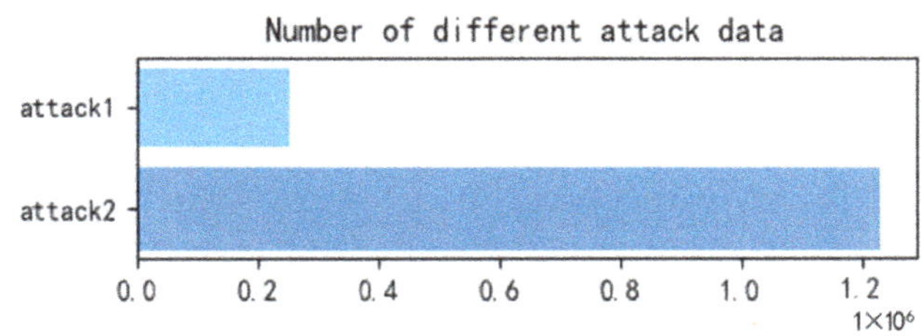

Figure 3. Number of different attack data.

Through the utilization of authentic datasets, information retrieval, and prior knowledge, the present study presents an account of the operational and safety conditions of various commonplace smart home devices in Table 1. The tabulation highlights that distinctive devices within the smart home setting exhibit assorted data throughput and attack frequencies. Additionally, diverse categories of smart devices are susceptible to differing attack behaviors, which results in a dissimilar amount of attack-related data. This situation leads to marked discrepancies in the data collected between various devices and attack types. For example, in the case of smart light bulbs, detecting and identifying attacks on these devices effectively is challenging due to the limited amount of attack data available. This scarcity of data is a result of the relatively low number of attacks that have been observed on this particular type of device. On the other hand, for smart door lock devices that experience a high frequency of attacks, more attack data is typically collected. However, there may still be instances of infrequent attack behaviors of a specific type (such as DDoS attacks commonly observed on smart cameras). These infrequent attack behaviors generate a small amount of attack data, which can be considered a sample size. As the tally of interconnected smart home devices continues to increase, these disparities become more prominent. Accordingly, during the process of flow data collection, specific devices are often unable to generate sufficient attack data, which impairs the efficacy of intrusion detection models. This limitation ultimately has a bearing on the overall security and stability of the smart home environment. Therefore, addressing the challenge of few-shot data resulting from a shortage of attack-related data is a critical research direction in the field of smart home device network security.

Table 1. Smart device working status table.

Devices	Working Hours (h)	Data Throughput	Frequency of Attack
Router	24	Larger	Higher
Gateway	20	Larger	Higher
Light	14	Smaller	Lower
TV	8	Larger	Lower
Intelligent door lock	3	Smaller	Lower
Floor sweeper	2	Smaller	Lower
Washing machine	2	Smaller	Lower
Smart camera	24	Larger	Higher

We propose the EM-FEDE method, as depicted in Figure 4. The method consists of three modules: the feature enhancement module, the data enhancement module, and the intrusion detection module. The feature enhancement module is assigned the task of processing the raw data by optimizing and filtering it. The data enhancement module focuses on generating samples, thereby expanding the dataset by adding fake samples. The intrusion detection module is responsible for identifying attacks and is trained using the expanded dataset, resulting in an improved ability of the model to classify and recognize various types of attacks. The EM-FEDE method employs symbols and their meanings, as listed in Table 2.

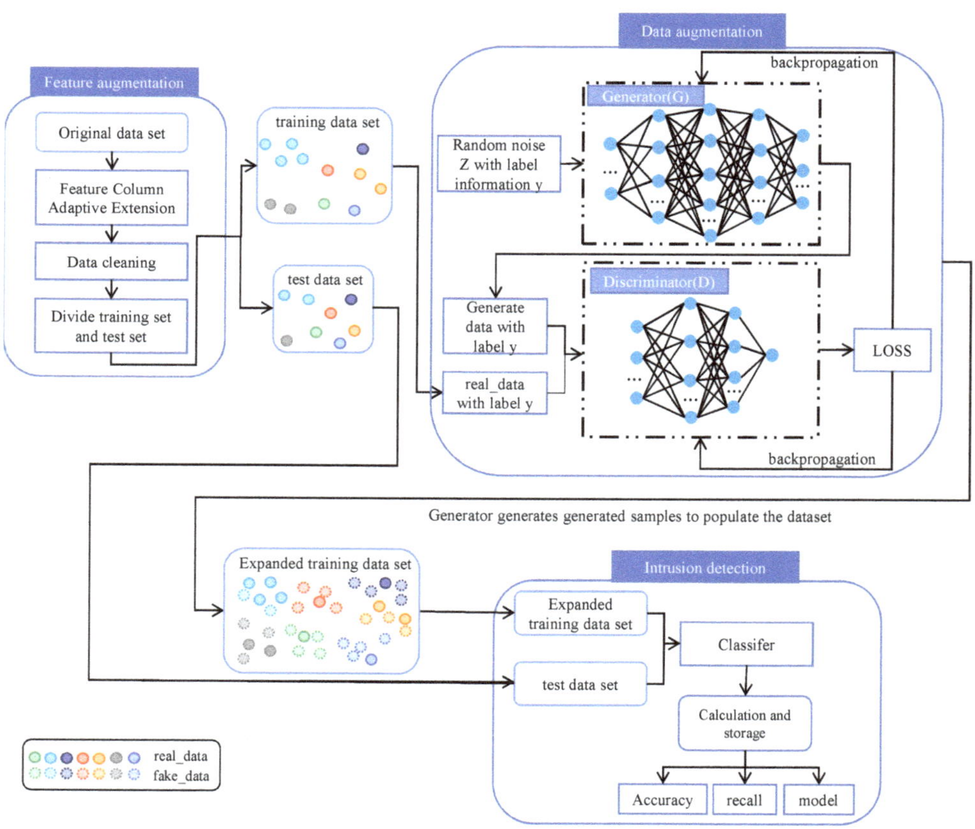

Figure 4. EM-FEDE method.

Table 2. Symbols used in the EM-FEDE method.

Symbols	Description
R	Historical intrusion detection records.
$SearchF(x)$	Used to determine the existence of the device class and the data class in x. It returns 1 if features are present and 0 otherwise.
Insert()	Insert operation.
$Class\_Label(x)$	Obtain the corresponding class from the information in x.
a_i	The device class feature column.
bi	The data class feature column.
$LabelEncoding(x)$	The function is used for mapping during the process of numericalization in x.
$FE\_Duplicate(x)$	The function is used for removing duplicate data in x.
$FE\_Normalization(x)$	The function is used for normalizing the data in x.
L	1-Lipschitz function.
P_{real}	Real data distribution.
P_z	Data distribution of input noise.
G(z)	Fake sample data generated by the generator.
$D(x)$	The probability that the discriminator determines that x belongs to the real data.
Z	Noise vector of the a priori noise distribution P_z.
$\prod(P_{real}, P_g)$	Joint probability distribution of real data and generated data.
$Fake\_data$	Generated data with label y_fake.

3.2. Feature Enhancement

This section focuses on the specific implementation of the EM-FEDE method in terms of feature enhancement.

In Section 3.1, it was discussed that various smart home devices exhibit differing data throughput and attack frequencies. Once the traffic data from these devices is captured, it is typically stored in a pcap file format. The format of the pcap file is shown in Figure 5. While the pcap file contains information such as timestamps, source addresses, and destination addresses, it lacks the ability to indicate device and data classes, resulting in the inability to label traffic. As a result, intrusion detection models that utilize supervised learning methods cannot directly utilize this data for training purposes. This challenge is also present in the N-BaIoT standard dataset, which includes eleven types of data collected from nine types of IoT devices (including one type of normal data and ten types of attack data). The dataset fails to provide feature columns that indicate the device class of each sample and distinguish the data class. To address the challenge of being unable to use raw data for training intrusion detection models and to improve data quality, this study proposes the EM-FEDE method's feature enhancement module. This module achieves feature enhancement through R analysis, feature-adaptive expansion, and data cleaning, thereby optimizing the data and indicating missing class features.

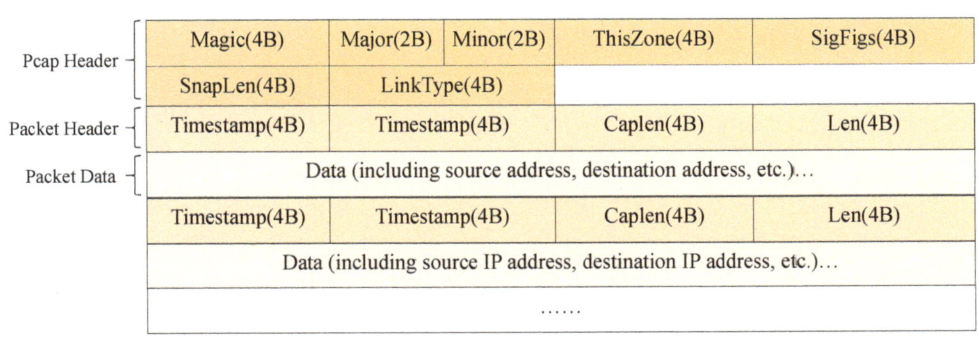

Figure 5. Common pcap file format.

To address the inability to directly use the raw data for training intrusion detection models and improve data quality, this study proposes the feature enhancement module of the EM-FEDE method. It achieves feature enhancement through R analysis, feature-adaptive expansion, and data cleaning. This process optimizes the data and enables the identification of missing class features.

The following is the specific process of feature enhancement:

Step 1. The LF is used to indicate the device class, and the data class in R is determined by Equation (1). If $LF = 1$, go to Step 3; if $LF = 0$, go to Step 2.

$$LF = SearchF(R); \tag{1}$$

Step 2. If direct prior knowledge (E) is available regarding the class of device and data, the device class feature (ai) and data class feature (bi) can be added to R through E. In the absence of such knowledge, the captured traffic data is analyzed to gather relevant information. As different attacks take place at different timestamps and distinct source IP addresses represent unique device characteristics, the timestamp and source IP address are treated as prior knowledge E. The device class feature (ai) and data class feature (bi) are then added to R, utilizing E. The specific equations pertaining to this process are illustrated in (2) and (3).

$$[ai, bi] = Class\_Label(E), \tag{2}$$

$$R = [R \cup Insert(ai) \cup Insert(bi)]; \tag{3}$$

Step 3. Numerical, de-duplication, and normalization of R by Equations (4)–(6).

$$R = LabelEncoding(R), \tag{4}$$

$$R = FE\_Duplicate(R), \tag{5}$$

$$R = FE\_Normalization(R), \tag{6}$$

In this step, we utilized Equation (4) to carry out numerical operations to convert non-numerical data in R into numerical data for the purpose of training the model. To tackle the problem of duplicate data, we applied Equation (5) to eliminate redundant data and minimize its impact on the results during data analysis. Additionally, we normalized the data using Equation (6) to ensure that all feature data was of the same magnitude and reduce the influence of noise on the results;

Step 4. Divide the training set and the test set, and output.

Figure 6 illustrates the process of feature enhancement using the N-BaIoT dataset as an example. The dataset contains $data1$ = {138.9020131, 72.11292822, ..., 0}, where 138.9020131 represents the value of *MI_dir_L5_weight*, 72.11292822 represents the value of *MI_dir_L5_mean*, 0 represents the value of *HpHp_L0.01_pcc*. There are 115 dimensional features in $data1$, and the specific steps are as follows:

Step 1. There is no common feature used to indicate the device class and attack class in N-BaIoT, $LF = 0$, so jump to Step 2;

Step 2. The N-BaIoT dataset contains prior knowledge E that enables us to determine the device class and data class of the dataset. Based on Equations (2) and (3), we added feature columns "*device*" and "*Label*". *Data1* corresponds to mirai_attacks syn attacks on the Ecobee_Thermostat device. As a result, we obtain $data1'$ = {138.9020131, 72.11292822, ..., 0,"Ecobee_Thermostat","mirai_attacks syn"};

Step 3. The obtained $data1'$ contains non-numerical data, so it is numericalized by Equation (3) to obtain $data1'$ = {138.9020131, 72.11292822, ..., 0, 2, 8}, where 2 represents the value of *device*, 8 represents the value of *Label*. Then the $data1'$ is de-duplicated and normalized by Equations (4) and (5), and finally the $data'''$ = {0.3972691, 0.0116122, ..., 0.380514, 2, 8} after feature enhancement is obtained;

Step 4. Output the training set and test set.

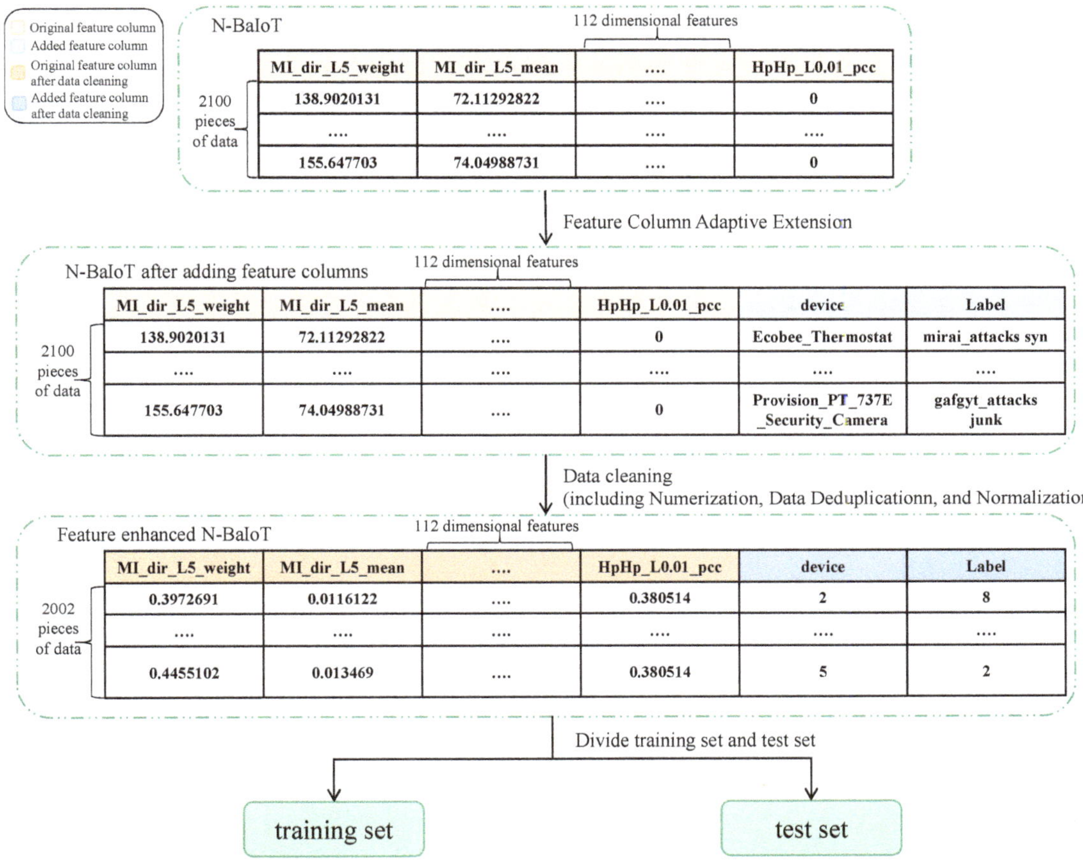

Figure 6. Dataset N-BaIoT feature enhancement process.

Feature enhancement is beneficial for reducing the superfluous information present in the data by processing and transforming the original features. This leads to the normalization of data, enhances its quality and usability, and provides a more dependable foundation for subsequent data enhancement and intrusion detection.

3.3. Data Enhancement

This section presents a detailed account of the practical implementation of data enhancement utilizing the EM-FEDE method. The data enhancement framework is shown in Figure 7.

Figure 7 is composed of three primary components. The first component is the input section, where the original data is enhanced by feature enhancement and utilized as input for the subsequent model training. The second component constitutes the CWGAN section. It contains two key components: the generator and the discriminator. The generator generates a variety of fake samples, while the discriminator is responsible for distinguishing between real and fake samples. The third component is the output section, where the generator produces diverse and authentic fake samples following iterative training of the CWGAN. These samples are then integrated into the original dataset, resulting in the enhanced dataset as the final output.

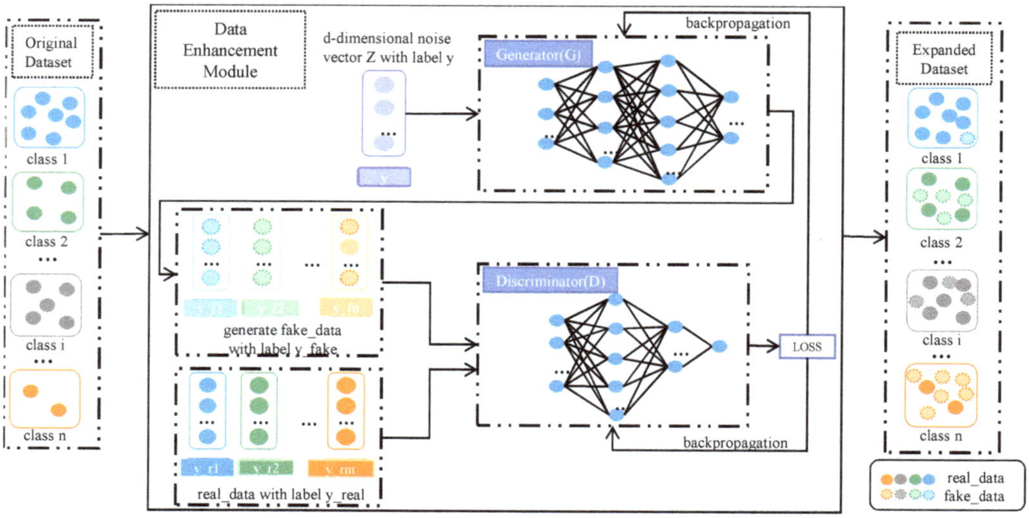

Figure 7. Data enhancement framework.

To quantify the disparity between the real data distribution and the fake data distribution, the EM-FEDE method employs the Wasserstein distance, which is expressed as Equation (7):

$$Wasserstein(P_{real}, P_z) = inf_{\gamma \sim \prod(P_{real}, P_z)} E_{(x,y)\sim\gamma}[||x - y||], \quad (7)$$

That is, for any joint probability distribution γ there exist edge probability distributions P_{real} and P_g. Two sample points, x and y, can be sampled from the edge distribution, and the value of the Wasserstein distance is a lower boundary on the expectation of the x and y distances.

The Wasserstein distance is a metric that quantifies the dissimilarity between two probability distributions. It is smaller when the distributions are more similar. Even if the two distributions have no overlap, the Wasserstein distance can still be computed, unlike the Jensen–Shannon divergence, which cannot handle this case. This property has been leveraged by the EM-FEDE method, which incorporates the Wasserstein distance into the loss function of the CWGAN. As a result, the neural network structure is improved, and the objective function is represented by an equation:

$$L_{CWGAN} = E_{x\sim P_{real}}[D(x|y)] - E_{z\sim P_z}[D(G(z|y))], \quad (8)$$

where $D \in L$, x is the sample from the real data distribution, P_{real}, and y is the conditional variable, i.e., the class characteristics of the data.

The following are the main steps of the data enhancement process:

Step 1. The training set in R, after undergoing the feature enhancement process, is utilized as the training data for the CWGAN. The generator and discriminator, both of which employ multilayer perceptron models, are defined as two neural network models. Equation (8) is employed to determine the objective function of the EM-FEDE method;

Step 2. Training the discriminator. The process of training the discriminator is illustrated in Figure 8. It involves inputting a set of randomly generated *fake_data* samples and *real_data* samples of sizes n and m, respectively, into the discriminator. The loss values of both sets of data are computed using Equation (9) and subsequently used to update the discriminator's parameters:

$$Loss_{Discriminator} = maxL_{CWGAN} = min(-L_{CWGAN}), \\ = E_{z\sim P_z}[D(G(z|y))] - E_{x\sim P_{real}}[D(x|y)] ; \quad (9)$$

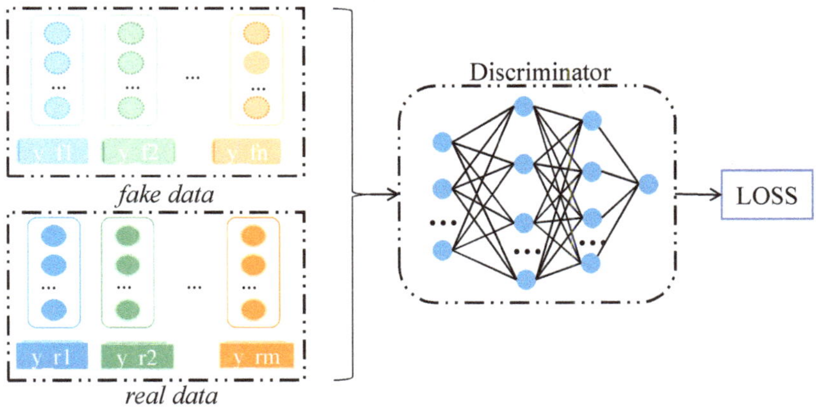

Figure 8. Discriminator process.

Step 3. Training the generator. The process of training the generator is illustrated in Figure 9. The generator is trained by generating a d-dimensional noise vector Z with label y as input, producing a set of *fake_data* samples of size n. These *fake_data* samples, along with the *real_data* samples, are then input into the discriminator. The loss value for this set of *fake_data* is computed using Equation (10), and the generator's parameters are updated accordingly.

$$Loss_{Generator} = min_{CWGAN} = -E_{z\sim P_z}[D(G(z|y))]; \quad (10)$$

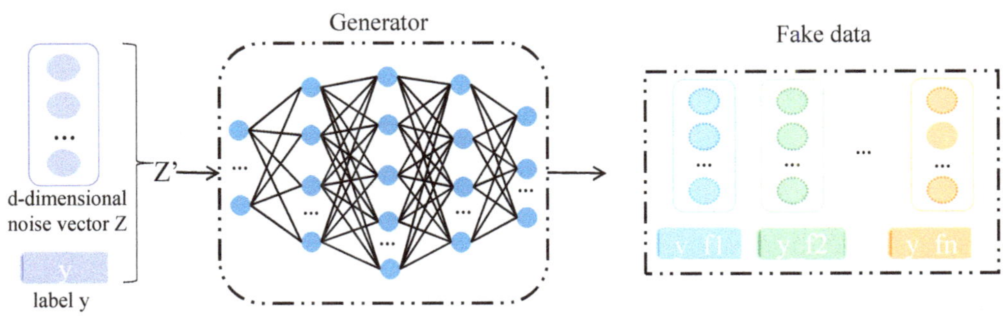

Figure 9. Generator process.

Step 4. The training process iterates Steps 2 and 3 repeatedly until the predetermined number of iterations or loss of convergence is reached. Conversely, by generating a new set of *fake_data* and returning $R = [R \cup fake\_data]$.

Following the process of data enhancement, the imbalanced original dataset is enriched with fake data, effectively ensuring a more even distribution of data across all classes within the dataset.

In the EM-FEDE method, the computational cost of feature enhancement is negligible, so its computational complexity depends mainly on the CWGAN part of the data enhancement module. For the EM-FEDE method, the gradients of the generator and discriminator need to be computed and updated. In each epoch, O($|g_\omega| + |g_\theta|$) floating-point operations are required (where g_ω is the gradient of the generator and g_θ is the gradient of the discriminator), and thus its overall complexity is O($|R| \cdot (|g_\omega| + |g_\theta|) \cdot Ne$)(where $|R|$ is the training dataset and Ne is the total number of training times). Algorithm 1 gives the detailed algorithmic flow of the EM-FEDE method.

Algorithm 1: EM-FEDE

Input: $\alpha = 0.0005$, the learning rate; $n = 50$, the batch size; $c = 0.01$, the clipping parameter; ω_0, initial discriminator parameters; θ_0, initial generator parameters; $Ne = 1000$, the training cycles.
Output: Expanded R
Process:
1. Calculate LF by Equation (1)
2. If $LF = 0$
3. Add feature columns that are helpful for classification to R through Equations (2)–(4)
4. Numerization, de-duplication, and normalization by Equations (5)–(7)
5. Divide the processed R into training sets and test sets
6. End if
7. While θ has not converged or $epoch < Ne$ do
8. $epoch$++
9. Sample of m noise samples$\{z_1, \ldots, z_n\} \sim P_Z$ a batch of prior data
10. Sample of m examples$\{(x_1,y_1), \ldots, (x_n,y_n)\} \sim P_{real}$ a batch from the real data
11. Update the discriminator D by ascending its stochastic gradient (g_ω)
12. $g_\omega = \nabla\omega\left[\frac{1}{m}\sum_{i=1}^{m} f_\omega(x_i|y_i) - \frac{1}{m}\sum_{i=1}^{m} f_\omega(g_\theta(z_i|y_i))\right]$
13. $\omega = \omega + \alpha * RMSProp(\omega, g_\omega)$
14. $\omega = clip(\omega, -c, c)$
15. Sample of m noise samples$\{z_1, \ldots, z_m\} \sim P_Z$ a batch of prior data.
16. Update the generator G by ascending its stochastic gradient (g_θ)
17. $g_\theta = -\nabla\theta\frac{1}{m}\sum_{i=1}^{m} f_\omega(g_\theta(z_i|y_i))$
18. $\theta = \theta - \alpha * RMSProp(\theta, g_\theta)$
19. End while
20. Generate sample data for each class through the generator to populate R
21. Train the expanded R on different classifiers to obtain various evaluation indicators

4. Results

4.1. N-BaIoT Dataset Description

The N-BaIoT dataset, released in 2018, consists of network traffic samples extracted from nine real IoT devices, featuring normal traffic from these devices and five varieties of attack traffic from the gafgyt and mirai botnet families. Figures 10 and 11 illustrate the differences in the data distribution across different traffic types and devices.

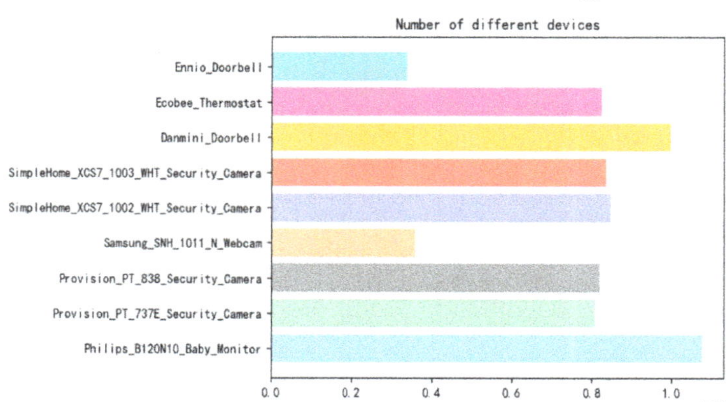

Figure 10. Distribution of the number of different devices.

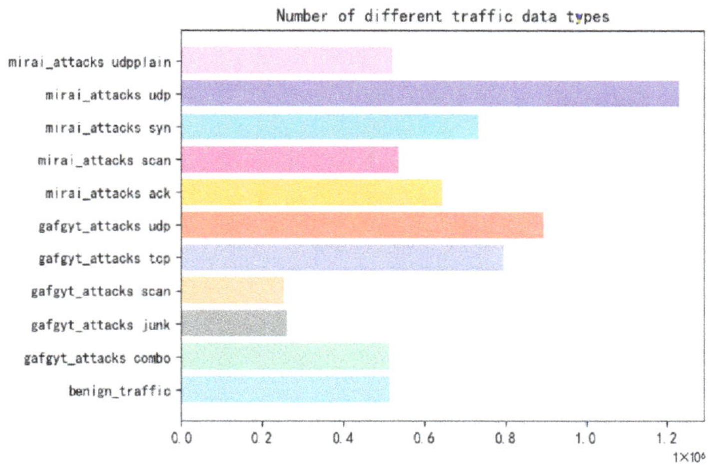

Figure 11. Distribution of the number of different traffic types.

The N-BaIoT dataset comprises extracted functions derived from raw IoT network traffic information. Upon receipt of each packet, a synopsis of the protocol and the host's behavior is computed with respect to the transmission of each packet. The contextual information of the data packet is then represented by a set of statistical features that are generated whenever a data packet arrives. Specifically, the arrival of each data packet leads to the extraction of 23 statistical features from five distinct time windows, namely, 100 ms, 500 ms, 1.5 s, 10 s, and 1 min. These five 23-dimensional vectors are subsequently concatenated into a single 115-dimensional vector.

The N-BaIoT dataset has been obtained in a real-world IoT setting, thus ensuring a high level of authenticity and representativeness. It serves as a standardized dataset that can be used by researchers to evaluate and enhance the efficacy of intrusion detection systems for IoT devices.

4.2. Data Preprocessing

The normalization method used in Equation (6) is Min–Max normalization. The specific formula for normalization is shown below:

$$x_i = \frac{x_i - x_{min}}{x_{max} - x_{min}}, \tag{11}$$

where x_i is the current feature, x_{min} is the minimum eigenvalue in the same dimension, and x_{max} is the maximum eigenvalue in the same dimension.

To replicate the scarcity of data in smart home devices in the real world and to guarantee that the dataset gathered from sampling includes samples from all categories, this research employs stratified sampling. This method ensures that each sample has an equal opportunity to be selected while maintaining the randomness of the samples. Ultimately, 2860 data samples were randomly chosen from the dataset as representative examples. The training and test data were then divided in a 7:3 ratio, and the sample distribution of the training and test sets can be found in Table 3.

Table 3. Sample distribution of training and test sets.

Traffic Type Name	Number of Training Sets	Number of Test Sets
benign_traffic	1054	325
gafgyt_attacks combo	136	60
gafgyt_attacks junk	122	75

Table 3. *Cont.*

Traffic Type Name	Number of Training Sets	Number of Test Sets
gafgyt_attacks scan	124	78
gafgyt_attacks tcp	97	56
gafgyt_attacks udp	91	51
mirai_attacks ack	76	42
mirai_attacks scan	83	40
mirai_attacks syn	76	42
mirai_attacks udp	73	49
mirai_attacks udpplain	70	40

4.3. Experimental Environment

In order to verify the feasibility of the model in this paper, experiments were conducted in the experimental environment shown in Table 4.

Table 4. Experimental environment configuration.

Category	Parameters
CPU	Intel(R) Xeon(R) CPU E5-2650 v4 @ 2.20 GHz
RAM	64 GB
Programming Tools	Jupyter Notebook
Programming Languages	Python3.8
Deep Learning Framework	Pytorch1.8
Machine Learning Platform	Weka3.9
Data Processing Library	Numpy, pandas, etc.

4.4. Network Structure

The structural details of the CWGAN model designed in this article are presented in Table 5. The structural details of the MLP and CNN classifiers are shown in Table 6.

Table 5. Model parameters of the generator and discriminator.

G/D	Structure	Size
Generator	Input layer	50
	Hidden layer 1(Tanh())	128
	Hidden layer 2(Tanh())	256
	Hidden layer 3(Tanh())	128
	Output layer(Tanh())	116
Discriminator	Input layer	116
	Hidden layer 1(Tanh())	128
	Hidden layer 2(Tanh())	128
	Output layer	1

Table 6. Model parameters of the classifier.

Classifier	Structure	Size
MLP	Input layer	116
	Hidden layer 1(Tanh())	128
	Hidden layer 2(Tanh())	128
	Output layer	11
CNN	Input layer	116
	Conv1D(Relu())	32
	Pooling layer	32

Table 6. *Cont.*

Classifier	Structure	Size
CNN	Conv1D(Relu())	32
	Pooling layer	32
	Flatten	224
	Dense	50
	Dense	11

4.5. Results and Analysis

This study aimed to assess the effectiveness of the EM-FEDE method in enhancing the intrusion detection classifier. To accomplish this, machine learning and deep learning classification algorithms were employed to evaluate the dataset. The evaluation metrics used in this study, namely *Accuracy, Precision, Recall, and F1 Score*, are widely accepted in the field. The formulas for calculating each metric are provided below:

$$Accuracy = \frac{TN + TP}{TN + FP + FN + TP}, \tag{12}$$

$$Precision = \frac{TP}{FP + TP}, \tag{13}$$

$$Recall = \frac{TP}{FN + TP}, \tag{14}$$

$$F1\ Score = 2 * \frac{Prescision\ *\ Recall}{Prescision + Recall}, \tag{15}$$

where *TP* indicates the number of true positives; *TN* indicates the number of true negatives in the sample; *FN* indicates the number of false negatives; and *FP* indicates the number of false positives.

To test the effectiveness of the fake samples, the intrusion detection classifier was trained on both the original training set and the training set enhanced by the EM-FEDE method. Subsequently, the enhancement effect of the EM-FEDE method was evaluated by assessing the comprehensive classification performance of the intrusion detection classifier using the test set. Multiple sets of data were generated for experiments, each with different ratios of fake samples, as documented in Table 7.

Table 7. Sample size at different generated sample ratios.

Dataset (Generated Sample Ratios)	Number of Fake Samples	Number of Samples after Expansion
x (Original sample size)	0	2002
$2x$	2004	4006 (2002 + 2004)
$3x$	4006	6008 (2002 + 4006)
$4x$	6118	8120 (2002 + 6118)
$5x$	8010	10,012 (2002 + 8010)
$6x$	10,012	12,014 (2002 + 10,012)
$7x$	12,014	14,016 (2002 + 12,014)
$8x$	14,016	16,018 (2002 + 14,016)
$9x$	16,018	18,020 (2002 + 16,018)
$10x$	18,009	20,011 (2002 + 18,009)

The distributions of the original data training set and the expanded training set are shown in Figures 12 and 13, respectively, using a generated sample ratio of $5x$ as an example.

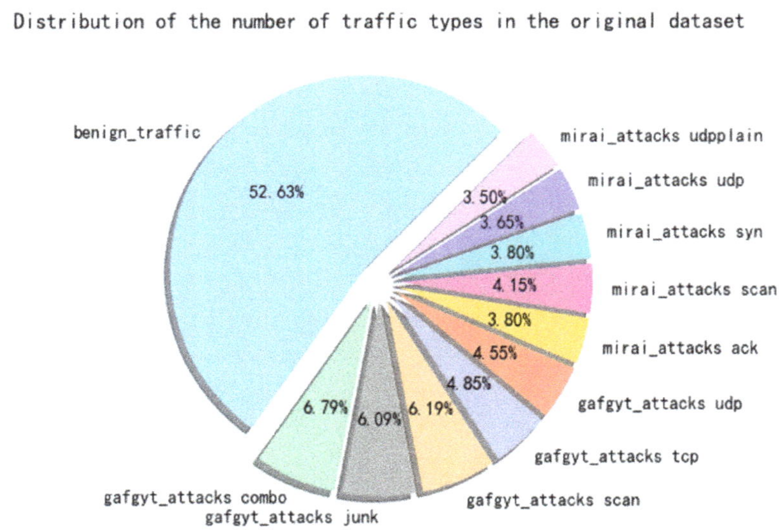

Figure 12. Data sample distribution of the original training set.

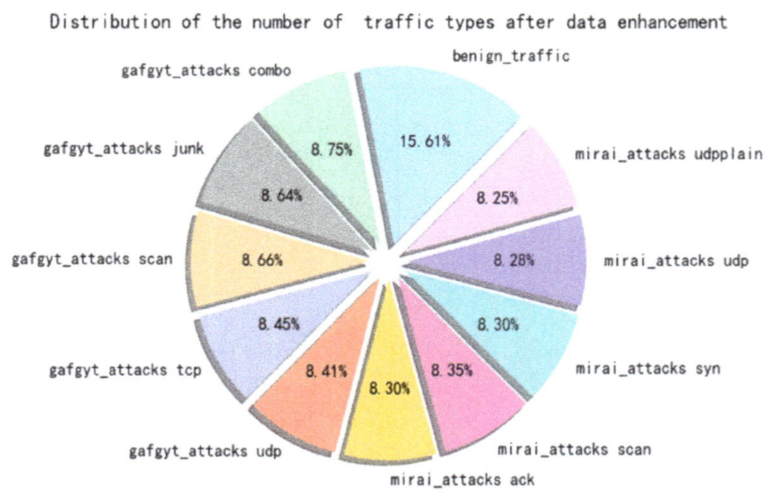

Figure 13. Sample distribution of training set data with a generated sample ratio of $5x$.

The study conducted a comparison of multi-classification results between the original dataset and the generated sample ratio of $5x$, as presented in Table 8. We evaluate the EM-FEDE method using various machine learning algorithms, where J48 is a decision tree algorithm, Random Forest is the Random Forest algorithm, Bagging is an integrated learning algorithm, PART is an algorithm that extracts rules in a dataset using incomplete decision trees, KStar is an instance-based classification algorithm, KNN is the K Nearest Neighbors algorithm, MLP is a Multi-Layer Perceptron Machine, and CNN is a Convolutional Neural Network. The results demonstrated that J48, Random Forest, Bagging, PART, KStar, KNN, MLP, and CNN showed an accuracy improvement of 16.4%, 4.9%, 10.7%, 9.2%, 4.9%, 4.4%, 3.1%, and 5.7%, respectively.

Table 8. Comparison of multi-classification results between the original dataset of size *x* and the mixed dataset with a generated sample ratio of 5*x* (the precision and F1 Score of some algorithms are unknown (?), which is due to the presence of Nan values in the calculation of precision, i.e., a denominator of 0). This scenario can occur when the algorithm fails to classify any sample into a particular class or when it wrongly classifies all samples in that class.

Dataset	Algorithm	Accuracy	Precision	Recall	F1 Score
N-BaIoT	J48	0.624	?	0.624	?
	Random Forest	0.755	?	0.756	?
	Bagging	0.655	?	0.655	?
	PART	0.673	?	0.673	?
	KStar	0.789	0.699	0.701	0.699
	KNN	0.768	0.773	0.768	0.771
	MLP	0.811	0.711	0.706	0.708
	CNN	0.712	0.726	0.673	0.698
N-BaIoT after EM-FEDE method processing	J48	**0.788**	**0.788**	**0.788**	**0.788**
	Random Forest	**0.804**	?	**0.804**	?
	Bagging	**0.762**	?	**0.762**	?
	PART	**0.765**	**0.795**	**0.765**	**0.779**
	KStar	**0.838**	**0.831**	**0.838**	**0.834**
	KNN	**0.812**	**0.796**	**0.812**	**0.803**
	MLP	**0.842**	**0.731**	0.678	0.703
	CNN	**0.769**	**0.828**	**0.736**	**0.779**

The evaluation of the experiments was carried out using various classification algorithms, including KNN, KStar, Bagging, PART, J48, Random Forest, MLP, and CNN. The evaluated results for datasets enhanced with different generated sample ratios are shown in Figure 14.

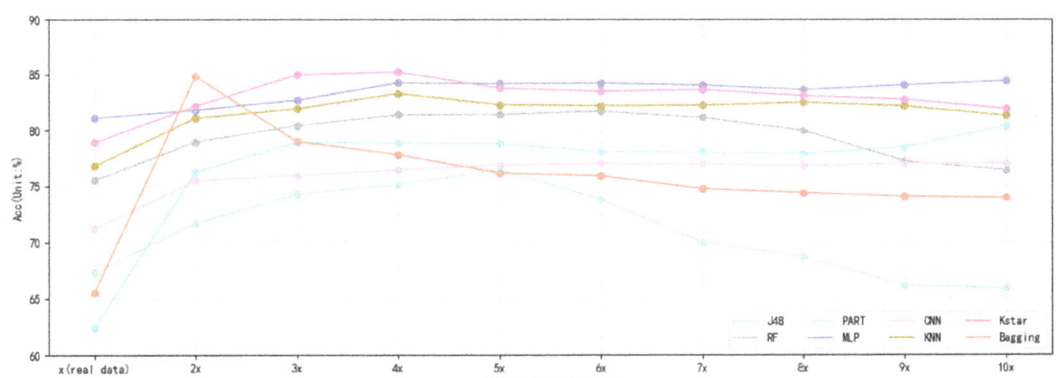

Figure 14. Experimental results.

As shown in Figure 14, the utilization of the EM-FEDE method has improved the accuracy of various classification algorithms. This improvement was observed when expanding the dataset compared to the original dataset. Additionally, the optimal sample ratio for achieving the best performance varies across different classification algorithms. With an increase in the number of generated samples, the accuracy of each classification algorithm gradually increases. The accuracy of J48 has increased from 62.39% (*x*) to 80.43% (10*x*), RF has increased from 75.55% (*x*) to 81.73% (6*x*), PART has increased from 67.28% (*x*) to 76.49% (5*x*), MLP has increased from 81.09% (*x*) to 84.45% (10*x*), CNN has increased from 71.18% (*x*) to 77.05% (10*x*), KNN has increased from 76.8% (*x*) to 83.31% (4*x*), KStar has increased from 78.9% (x) to 85.24% ($4x$), and Bagging has increased from 65.5% (x) to 84.86% ($2x$).

However, when the generated sample ratio becomes too large, the accuracy of some classification algorithms slightly decreases compared to a smaller generated sample ratio. The accuracy of RF decreases from 81.73% ($6x$) to 76.44% ($10x$), PART decreases from 76.49% ($5x$) to 65.91% ($10x$), KNN decreases from 83.31% ($4x$) to 81.32% ($10x$), KStar decreases from 85.24% ($4x$) to 81.94% ($10x$), and Bagging decreases from 84.86% ($2x$) to 73.98% ($10x$).

The accuracy of several classification algorithms such as RF, PART, KNN, KStar, and Bagging initially improves as the number of generated samples increases until they reach their optimal generated sample ratio, after which the accuracy decreases. This trend occurs due to the presence of fake data, which can negatively affect the quality of the data. The generator model aims to approximate the distribution of real data as closely as possible, but if the quantity of fake data becomes too large, the generator model can experience mode collapse. This phenomenon indicates that the fake data becomes excessively similar, and increasing the data further no longer improves the classifier's performance. Instead, it can lead to a decrease in classification accuracy due to noise in the fake data.

In contrast, J48, MLP, and CNN exhibit a gradual increase in accuracy. J48, a machine learning classifier based on feature partitioning, is typically sensitive to diversity and complexity. MLP and CNN, as deep learning classifiers, possess stronger representational and generalization capabilities. An increase in fake data leads to an increase in the training data for classifiers. This increase provides more opportunities for the classifiers to learn from different data distributions and features, leading to more complex and deeper levels of feature representation. Consequently, the classifiers' accuracy improves.

The variability in the best generated sample ratios is evident across different algorithms, as illustrated in Figure 14. Table 9 presents the accuracy of said ratios, in contrast to the original dataset, for various algorithms. The accuracy of J48, Random Forest, Bagging, PART, KStar, KNN, MLP, and CNN improved by 21.9%, 6.2%, 19.4%, 9.2%, 6.3%, 7%, 3.4%, and 5.9%, respectively. It is worth noting that the extended dataset demonstrated an overall higher accuracy in comparison to the original dataset when scaled to the best generated sample ratio of each algorithm.

Table 9. The accuracy of multi-classification is compared between the original data set and the mixed data set with the optimal generation sample ratio of each algorithm.

Algorithm	Optimal Generation Sample Ratio nx ($1 \leq n \leq 10$)	Accuracy of the Original Dataset	Accuracy of the Mixed Dataset with the Optimal Generation Sample Ratio	The Percentage of Growth
J48	$10x$	0.624	0.843	21.9%
Random Forest	$6x$	0.755	0.817	6.2%
Bagging	$2x$	0.655	0.849	19.4%
PART	$5x$	0.673	0.765	9.2%
KStar	$4x$	0.789	0.852	6.3%
KNN	$4x$	0.768	0.833	7%
MLP	$10x$	0.811	0.845	3.4%
CNN	$10x$	0.712	0.771	5.9%

SMOTE [29] is an oversampling method that generates new samples to expand the dataset based on the relationship between samples, and CGAN [30] is an extension of GAN for conditional sample generation. This part of the experiment examined the impact of different generated sample ratios on accuracy in J48 and Bagging for mixed datasets created using SMOTE, CGAN, and the proposed method. Additionally, we compared it with the same number of datasets containing only real data to prove the effectiveness of the proposed method in this paper. The experimental results are shown in Figures 15 and 16.

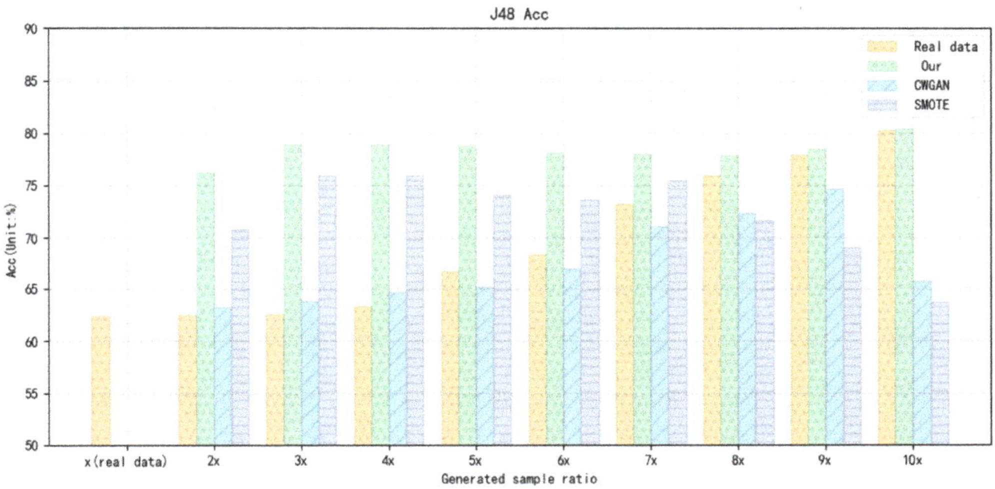

Figure 15. Accuracy of J48 for real and expanded datasets.

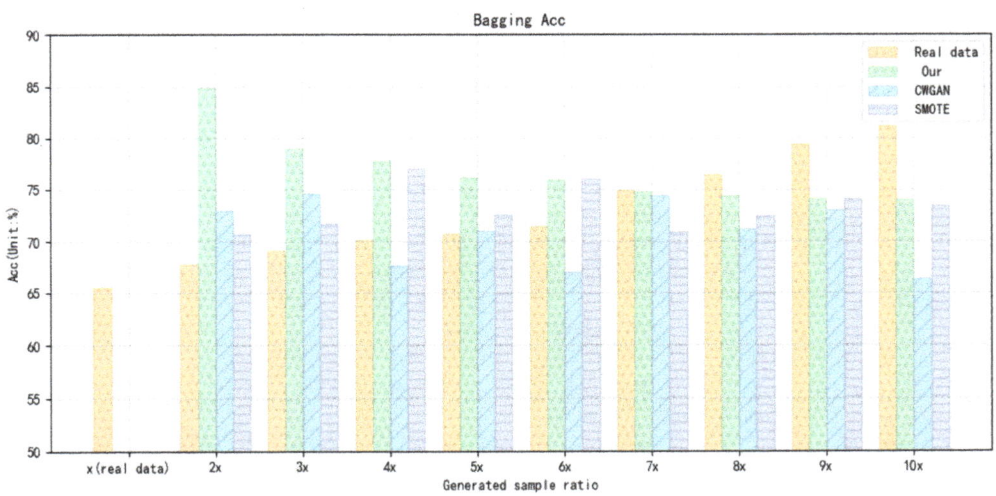

Figure 16. Accuracy of Bagging for real and expanded datasets.

Based on the results presented in Figure 15, in our method, it is evident that the accuracy of the enhanced dataset, which includes a combination of fake data and real data at the generated sample ratios of nx ($n = 2, 3, \ldots, 10$), is superior to that of an equivalent number of instances from the original dataset. At lower generated sample ratios, the mixed dataset exhibits significantly improved accuracy on J48 in comparison to an equivalent number of instances from the real dataset. As the generated sample ratio increases, the accuracy of the mixed dataset on J48 exhibits fluctuation, albeit within a small range, and ultimately reaches a plateau. Although the mixed dataset continues to outperform the real dataset in terms of accuracy on J48, its advantage diminishes as the generated sample ratio becomes larger.

Regarding the CWGAN, at lower generated sample ratios nx ($n = 2, 3, 4$), the accuracy of the mixed dataset in J48 slightly improves compared to the same number of instances of the real dataset. However, for generated sample ratios of nx ($n = 4, \ldots, 10$), the accuracy of the mixed dataset at J48 is lower than that of the equivalent number of real datasets, and the performance of the real dataset is significantly better than that of the mixed dataset as the generated sample ratio increases.

Regarding the SMOTE, at the generated sample ratio of nx ($n = 2, \ldots, 7$), the accuracy of the mixed dataset in J48 is significantly higher than that of the real dataset with the same number of samples. At the generated sample ratio of nx ($n = 8, 9, 10$), the accuracy of the hybrid dataset starts to decrease and is lower than the equivalent number of real datasets.

Based on the experimental results, we can conclude that the J48 algorithm has more capacity for learning the supplementary feature information that is provided by the expanded dataset. This attribute of the algorithm contributes to an improved understanding of the dataset's traits and patterns, thereby leading to an enhancement of the classifier's performance. In addition to this, the introduction of a small quantity of artificial data has been observed to have a beneficial effect on the model's ability to generalize, and it can also serve to mitigate the effects of overfitting and noisy data. However, it should be noted that there is a threshold beyond which the quantity of artificially generated data becomes sufficient, and further increments of such data do not yield any improvement in the accuracy of the intrusion detection model.

The results in Figure 16 show that the accuracy of the mixed dataset with generated sample ratio nx ($n = 2, \ldots, 6$) on the Bagging algorithm is better than that of the corresponding number of real datasets in the method of this paper. However, for the generated sample ratio nx ($n = 7, \ldots, 10$), the accuracy of the mixed dataset is lower than that of the corresponding number of real datasets. The experimental results reveal that the optimal generated sample rate for the Bagging algorithm using the method in this paper is $2x$. Moreover, the accuracy of Bagging decreases and stabilizes as the generated sample rate increases.

Regarding the CWGAN, the accuracy of the mixed dataset is higher than the same number of instances of the real dataset for the generation sample rate nx ($n = 2, 3, 5$). However, for the generating sample ratio of nx ($n = 4, 6, \ldots, 10$), the accuracy of the mixed dataset is lower than the accuracy of the same number of real datasets. The results indicate that the best generated sample ratio for the Bagging algorithm using the CWGAN is $3x$.

Regarding the SMOTE, the accuracy of the mixed dataset is higher for the generation sample ratio nx ($n = 2, \ldots, 6$) compared to the same number of instances of the real dataset. For the generation sample ratio nx ($n = 7, \ldots, 10$), the accuracy of the mixed dataset is lower than the accuracy of the same number of instances of the real dataset. From the experimental results, it can be concluded that the optimal generation sample ratio for Bagging on SMOTE is $4x$.

When the generated sample ratio nx ($n = 7, \ldots, 10$) is too large, the accuracy of both the methods in this paper, CWGAN and SMOTE on Bagging, is lower than the equivalent number of real datasets. Despite the decrease in accuracy, the accuracy of this paper's method and SMOTE is still higher than that of the original dataset x. By comparing this paper's method, CWGAN, and SMOTE, it can be concluded that this paper's method exhibited better performance.

Based on our experimental results, we can conclude that utilizing fake data for data enhancement can significantly enhance the accuracy of the classifier, particularly when the expansion multiplier is small. However, the introduction of fake data may result in noise, and its proportion increases with the expansion multiplier. This difference between real and fake data can make it challenging to provide sufficient useful feature information, which can, in turn, impede the ability of the model to learn the data features. Ultimately, this can lead to a reduction in the accuracy of the classifier.

The SMOTE algorithm analyzes the minority class samples and manually synthesizes new samples to add to the dataset based on the minority class samples. This technique of generating new samples through oversampling helps prevent overfitting. However, it may generate the same number of new samples for each minority class sample, resulting in increased overlap between classes and the creation of samples that do not offer useful information. The CGAN method improves the data generation process by incorporating additional information to guide the model. However, the training process of CGAN is not very stable, and the quality of the generated data can vary. In contrast, the EM-FEDE method proposed in this paper uses the CWGAN approach to generate data with greater diversity. It also provides more informative samples and is more stable during training, resulting in higher-quality generated data compared to CGAN. To summarize, the effectiveness of the EM-FEDE method has been demonstrated, making it suitable for training datasets for intrusion detection models. However, it is crucial to consider that the optimal generated sample ratio may differ based on the particular algorithm and model in use. To attain the highest level of accuracy and performance for a given intrusion detection algorithm or model, it is essential to undertake a meticulous evaluation and selection of the most fitting generated sample ratio. This selection and evaluation process is crucial to guaranteeing optimal outcomes.

5. Discussion

The present article discusses the issue of few-shot data on smart home devices and the challenges this poses for intrusion detection models. Specifically, the study highlights how the security dataset collected from traffic information often lacks data, which limits the performance of intrusion detection models. To address this issue, the article proposes a method called EM-FEDE, which enhances the dataset and effectively mitigates the impact of few-shot data on intrusion detection performance, improving security in smart home environments. The study evaluates the performance of datasets enhanced with different generated sample ratios and analyzes the effect of using enhanced datasets for intrusion detection model training. Furthermore, the article examines the influence of different generated sample ratios on classification performance for specific classification algorithms. The results indicate that the optimal generated sample ratio may vary depending on the algorithm and model used. Based on the obtained results, it can be concluded that the proposed method shows promising performance in solving few-shot data. In addition to intrusion detection, it can be applied to different domains, such as sentiment analysis tasks where the samples of various sentiment categories are highly imbalanced and underwater target recognition tasks where the samples are too small to train an effective model.

In this paper, the specific details regarding the optimal expansion multiplier and the ratio of generated data to real data for various classification algorithms are not extensively explored. Thus, future studies will focus on optimizing the intrusion detection model by selecting more suitable classification algorithms to enhance detection accuracy. Additionally, further research will be conducted to determine the appropriate enhancement factors and ratios between generated and real data during the data enhancement process.

Author Contributions: Conceptualization, T.Y. and J.W.; methodology, Y.C. and J.W.; software, Y.C., T.Y. and J.W.; validation, J.W., T.Y. and Y.C.; formal analysis, J.W.; investigation, J.W.; resources, Y.C.; data curation, J.W.; writing—original draft preparation, J.W. and T.Y.; writing—review and editing, J.W., Q.L. and N.A.N.; supervision, T.Y. and J.W. All authors have read and agreed to the published version of the manuscript.

Funding: This work was supported by the Sichuan Science and Technology Program under Grant No. 2022YFG0322, China Scholarship Council Program (Nos. 202001010001 and 202101010003), the Innovation Team Funds of China West Normal University (No. KCXTD2022-3), the Nanchong Federation of Social Science Associations Program under Grant No. NC22C280, and China West Normal Universi-ty 2022 University-level College Student Innovation and Entrepreneurship Training Program Project under Grant No. CXCY2022285.

Data Availability Statement: Data are unavailable due to privacy.

Acknowledgments: Thanks to everyone who contributed to this work.

Conflicts of Interest: The authors declare no conflict of interest.

References

1. Cvitić, I.; Peraković, D.; Periša, M.; Jevremović, A.; Shalaginov, A. An Overview of Smart Home IoT Trends and related Cybersecurity Challenges. *Mob. Netw. Appl.* **2022**. [CrossRef]
2. Hammi, B.; Zeadally, S.; Khatoun, R.; Nebhen, J. Survey on smart homes: Vulnerabilities, risks, and countermeasures. *Comput. Secur.* **2022**, *117*, 102677. [CrossRef]
3. Wang, Y.; Zhang, R.; Zhang, X.; Zhang, Y. Privacy Risk Assessment of Smart Home System Based on a STPA–FMEA Method. *Sensors* **2023**, *23*, 4664. [CrossRef] [PubMed]
4. Wu, T.Y.; Meng, Q.; Chen, Y.C.; Kumari, S.; Chen, C.M. Toward a Secure Smart-Home IoT Access Control Scheme Based on Home Registration Approach. *Mathematics* **2023**, *11*, 2123. [CrossRef]
5. Li, Y.; Zuo, Y.; Song, H.; Lv, Z. Deep learning in security of internet of things. *IEEE Internet Things J.* **2021**, *9*, 22133–22146. [CrossRef]
6. Chkirbene, Z.; Erbad, A.; Hamila, R.; Gouissem, A.; Mohamed, A.; Guizani, M.; Hamdi, M. A weighted machine learning-based attacks classification to alleviating class imbalance. *IEEE Syst. J.* **2020**, *15*, 4780–4791. [CrossRef]
7. Zivkovic, M.; Tair, M.; Venkatachalam, K.; Bacanin, N.; Hubálovský, Š.; Trojovský, P. Novel hybrid firefly algorithm: An application to enhance XGBoost tuning for intrusion detection classification. *PeerJ Comput. Sci.* **2022**, *8*, e956.
8. Li, X.K.; Chen, W.; Zhang, Q.; Wu, L. Building auto-encoder intrusion detection system based on random forest feature selection. *Comput. Secur.* **2020**, *95*, 101851.
9. Wang, Z.; Liu, Y.; He, D.; Chan, S. Intrusion detection methods based on integrated deep learning model. *Comput. Secur.* **2021**, *103*, 102177. [CrossRef]
10. Tsimenidis, S.; Lagkas, T.; Rantos, K. Deep learning in IoT intrusion detection. *J. Netw. Syst. Manag.* **2022**, *30*, 8. [CrossRef]
11. Heartfield, R.; Loukas, G.; Budimir, S.; Bezemskij, A.; Fontaine, J.R.; Filippoupolitis, A.; Roesch, E. A taxonomy of cyber-physical threats and impact in the smart home. *Comput. Secur.* **2018**, *78*, 398–428. [CrossRef]
12. Touqeer, H.; Zaman, S.; Amin, R.; Hussain, M.; Al-Turjman, F.; Bilal, M. Smart home security: Challenges, issues and solutions at different IoT layers. *J. Supercomput.* **2021**, *77*, 14053–14089. [CrossRef]
13. Cao, X.; Luo, Q.; Wu, P. Filter-GAN: Imbalanced Malicious Traffic Classification Based on Generative Adversarial Networks with Filter. *Mathematics* **2022**, *10*, 3482. [CrossRef]
14. Wang, M.; Yang, N.; Weng, N. Securing a Smart Home with a Transformer-Based IoT Intrusion Detection System. *Electronics* **2023**, *12*, 2100. [CrossRef]
15. Guebli, W.; Belkhir, A. Inconsistency detection-based LOD in smart homes. *Int. J. Semant. Web Inf. Syst. IJSWIS* **2021**, *17*, 56–75. [CrossRef]
16. Madhu, S.; Padunnavalappil, S.; Saajlal, P.P.; Vasudevan, V.A.; Mathew, J. Powering up an IoT-enabled smart home: A solar powered smart inverter for sustainable development. *Int. J. Softw. Sci. Comput. Intell. IJSSCI* **2022**, *14*, 1–21. [CrossRef]
17. Tiwari, A.; Garg, R. Adaptive Ontology-Based IoT Resource Provisioning in Computing Systems. *Int. J. Semant. Web Inf. Syst. IJSWIS* **2022**, *18*, 1–18. [CrossRef]
18. Elsayed, N.; Zaghloul, Z.S.; Azumah, S.W.; Li, C. Intrusion detection system in smart home network using bidirectional lstm and convolutional neural networks hybrid model. In Proceedings of the 2021 IEEE International Midwest Symposium on Circuits and Systems (MWSCAS), Lansing, MI, USA, 9–11 August 2021; IEEE: Piscataway, NJ, USA, 2021; pp. 55–58.
19. Shi, L.; Wu, L.; Guan, Z. Three-layer hybrid intrusion detection model for smart home malicious attacks. *Comput. Electr. Eng.* **2021**, *96*, 107536. [CrossRef]
20. Alani, M.M.; Awad, A.I. An Intelligent Two-Layer Intrusion Detection System for the Internet of Things. *IEEE Trans. Ind. Inform.* **2022**, *19*, 683–692. [CrossRef]
21. Rani, D.; Gill, N.S.; Gulia, P.; Arena, F.; Pau, G. Design of an Intrusion Detection Model for IoT-Enabled Smart Home. *IEEE Access* **2023**, *11*, 52509–52526. [CrossRef]
22. Goodfellow, I.; Pouget-Abadie, J.; Mirza, M.; Xu, B.; Warde-Farley, D.; Ozair, S.; Courville, A.; Bengio, Y. Generative adversarial nets. *Adv. Neural Inf. Process. Syst.* **2014**, *27*, 2672–2680.
23. Fu, W.; Qian, L.; Zhu, X. GAN-based intrusion detection data enhancement. In Proceedings of the 2021 33rd Chinese Control and Decision Conference (CCDC), Kunming, China, 22–24 May 2021; IEEE: Piscataway, NJ, USA, 2021; pp. 2739–2744.
24. Zhang, L.; Duan, L.; Hong, X.; Liu, X.; Zhang, X. Imbalanced data enhancement method based on improved DCGAN and its application. *J. Intell. Fuzzy Syst.* **2021**, *41*, 3485–3498. [CrossRef]
25. Li, S.; Dutta, V.; He, X.; Matsumaru, T. Deep Learning Based One-Class Detection System for Fake Faces Generated by GAN Network. *Sensors* **2022**, *22*, 7767. [CrossRef] [PubMed]
26. Yang, W.; Xiao, Y.; Shen, H.; Wang, Z. An effective data enhancement method of deep learning for small weld data defect identification. *Measurement* **2023**, *206*, 112245. [CrossRef]
27. Jin, H.; Huang, S.; Wang, B.; Chen, X.; Yang, B.; Qian, B. Soft sensor modeling for small data scenarios based on data enhancement and selective ensemble. *Chem. Eng. Sci.* **2023**, *279*, 118958. [CrossRef]
28. Meidan, Y.; Bohadana, M.; Mathov, Y.; Mirsky, Y.; Shabtai, A.; Breitenbacher, D.; Elovici, Y. N-BaIoT—Network-Based Detection of IoT Botnet Attacks Using Deep Autoencoders. *IEEE Pervasive Comput.* **2019**, *17*, 12–22. [CrossRef]

29. Chawla, N.V.; Bowyer, K.W.; Hall, L.O.; Kegelmeyer, W.P. SMOTE: Synthetic minority over-sampling technique. *J. Artif. Intell. Res.* **2002**, *16*, 321–357. [CrossRef]
30. Mirza, M.; Osindero, S. Conditional generative adversarial nets. *arXiv* **2014**, arXiv:1411.1784.

electronics

Article

A Network Clustering Algorithm for Protein Complex Detection Fused with Power-Law Distribution Characteristic

Jie Wang [1,*], Ying Jia [1], Arun Kumar Sangaiah [2,3,*] and Yunsheng Song [4]

1 School of Information, Shanxi University of Finance and Economics, Taiyuan 030006, China; 13007526396@163.com
2 International Graduate Institute of Artificial Intelligence, National Yunlin University of Science and Technology, Douliou 64002, Taiwan
3 Department of Electrical and Computer Engineering, Lebanese American University, Byblos 1102-2801, Lebanon
4 School of Information Science and Engineering, Shandong Agricultural University, Taian 271018, China; songys@sdau.edu.cn
* Correspondence: 20191031@sxufe.edu.cn (J.W.); aksangaiah@ieee.org (A.K.S.); Tel.: +86-351-7666-126 (J.W.)

Abstract: Network clustering for mining protein complexes from protein–protein interaction (PPI) networks has emerged as a prominent research area in data mining and bioinformatics. Accurately identifying complexes plays a crucial role in comprehending cellular organization and functionality. Network characteristics are often useful in enhancing the performance of protein complex detection methods. Many protein complex detection algorithms have been proposed, primarily focusing on local micro-topological structure metrics while overlooking the potential power-law distribution characteristic of community sizes at the macro global level. The effective use of this distribution characteristic information may be beneficial for mining protein complexes. This paper proposes a network clustering algorithm for protein complex detection fused with power-law distribution characteristic. The clustering algorithm constructs a cluster generation model based on scale-free power-law distribution to generate a cluster with a dense center and relatively sparse periphery. Following the cluster generation model, a candidate cluster is obtained. From a global perspective, the number distribution of clusters of varying sizes is taken into account. If the candidate cluster aligns with the constraints defined by the power-law distribution function of community sizes, it is designated as the final cluster; otherwise, it is discarded. To assess the prediction performance of the proposed algorithm, the gold standard complex sets CYC2008 and MIPS are employed as benchmarks. The algorithm is compared to DPClus, IPCA, SEGC, Core, SR-MCL, and ELF-DPC in terms of F-measure and Accuracy on several widely used protein–protein interaction networks. The experimental results show that the algorithm can effectively detect protein complexes and is superior to other comparative algorithms. This study further enriches the connection between analyzing complex network topology features and mining network function modules, thereby significantly contributing to the improvement of protein complex detection performance.

Keywords: data mining; network clustering; protein complex detection; power-law distribution; topological characteristics

Citation: Wang, J.; Jia, Y.; Sangaiah, A.K.; Song, Y. A Network Clustering Algorithm for Protein Complex Detection Fused with Power-Law Distribution Characteristic. *Electronics* **2023**, *12*, 3007. https://doi.org/10.3390/electronics12143007

Academic Editor: Ping-Feng Pai

Received: 17 June 2023
Revised: 6 July 2023
Accepted: 6 July 2023
Published: 8 July 2023

1. Introduction

Cells rely on the interaction of multiple proteins for life activities. A protein complex, formed through interactions, consists of molecules with similar functions. Detecting protein complexes in protein–protein interaction (PPI) networks facilitates the exploration of the relationships between network structures and function modules. Moreover, it plays a crucial role in annotating the proteins with unknown functions and gaining insights into the organization and functionality of cells [1].

Researchers have proposed many experimental methods to identify the interactions between proteins, including yeast two-hybrid (Y2H) [2,3] and tandem affinity purification (TAP) [4]. These methods have generated a vast amount of protein–protein interaction (PPI) data, which serve as valuable support for the application of data mining techniques in protein complex detection.

A PPI dataset is usually abstracted as an undirected network, wherein proteins are nodes and the interactions between proteins are edges. A PPI network contains different protein function modules [5]. Generally, a protein complex is a biological functional module [6] comprising two or more proteins that share the same function. Proteins in the same protein complex exhibit strong connections, whereas the proteins belonging to different complexes have weaker connections. Detecting protein complexes from PPI networks aims to discover sets of proteins with dense connections. This process can be viewed as a network clustering task, wherein clusters are determined based on topological features, where the connection strength within a cluster is greater than that between clusters [7,8]. This process yields disjoint or overlapping clusters as its outcome [9].

Various network clustering algorithms for identifying protein complexes have been developed. In general, these algorithms include graph partition algorithms, density-based local search algorithms, and algorithms based on graph embedding [10–12].

The clustering algorithm based on graph partition divides nodes into clusters according to an objective function, aiming to identify an optimal partitioned network. It maximizes the similarity between nodes within each cluster while minimizing the similarity between different clusters. One well-known algorithm in this category is the Markov algorithm (MCL) [13,14]. MCL begins by constructing the initial flow matrix based on a PPI network and then simulates random flow through the network using the concept of random walk to partition the entire network into sub-graphs with high connectivity probability. The collection of nodes within each sub-graph represents a protein complex. However, MCL does not handle overlapping clusters. To address this limitation, the soft regularized MCL (SR-MCL) algorithm was developed, which enables the identification of overlapping clusters.

The density-based local search clustering algorithm focuses on identifying dense sub-graphs based on the characteristic of connection density. Among the various network clustering methods, one approach aims to find *k*-closely connected sub-network modules, such as the Closely Connected Percolation Method (CPM) [15]. CPM initially identifies closely connected subnets within the network and subsequently identifies *k*-closely connected subnet modules based on these initial subnets. A few approaches are also known as the seed expansion method. They select a node as a seed and expand around the seed to a cluster according to certain rules. One example of the seed expansion method is the density peak clustering (DPClus) algorithm [16]. DPClus introduces the concept of "cluster periphery" in protein interaction networks. It assigns edge weights based on common neighbor counts between interacting proteins, while node weights are determined by the sum of their adjacent edges' weights. The peripheral value of a node within a cluster is determined as the ratio of its adjacent nodes to the total number of nodes in the cluster. The algorithm starts by selecting the highest-weighted node as the seed for the initial cluster. Edge weights are influenced by common neighbor counts, and node weights reflect the density of immediate neighbors. If nodes satisfy both the custom threshold for local density and the threshold for cluster peripheral value, DPClus iteratively adds the nodes to obtain the final cluster. To account for the minimum diameter and average node distance characteristics of protein complexes, the improved DPClus algorithm (IPCA) [17] enhances DPClus through the integration of sub-graph diameters and interaction probabilities, which provide insights into the density of the network. Other methods in this category include SEGC [18], Core [19], etc.

Network clustering algorithms based on graphs embed map network nodes onto a lower-dimensional vector space by encoding their properties [20]. This mapping preserves the topological characteristics of the nodes as much as possible. Subsequently, a network clustering was performed in this transformed vector space [21,22]. One example of such an algorithm is the ensemble learning framework for density peak clustering (ELF-DPC) [23]. ELF-DPC first maps the PPI network to the vector space and constructs a weighted network to identify core edges. By integrating structural modularity and trained voting regression models, the algorithm creates an ensemble learning model. ELF-DPC then expands the core edges into clusters based on this learning model.

The PPI network, as a type of complex network, exhibits intricate network topology characteristics [24–26]. The fundamental features used to describe the network topology are primarily derived into three levels. Firstly, micro-topological structure metrics focus on individual nodes or edges, including measures such as node degree and centrality [27,28]. Secondly, meso-topological metrics analyze groups of nodes, such as community structure [29], modules, and motifs. Lastly, macro-topological metrics consider the entire network, encompassing aspects such as degree distribution and community size distribution. Developing a network clustering algorithm that incorporates these network features can enhance the accuracy of community detection [30]. At present, seed expansion methods can effectively utilize network features. However, existing algorithms mainly consider local micro-topological structure features [31] and ignore the potential distribution characteristics of community size at a macro-global level. The distribution of community sizes in the PPI network exhibits a certain correlation with power-law distribution [32].

In this paper, we present a novel network clustering approach that incorporates the characteristics of power-law distribution to identify protein complexes. Our proposed algorithm, named GCAPL, encompasses two main stages: cluster generation and cluster determination. During the cluster generation stage, the GCAPL algorithm incorporates node degree and clustering coefficient to assign weights to nodes. The unclustered nodes with the highest weight were selected as seeds. Following that, a cluster generation model leveraging the scale-free power-law distribution was given to discovery clusters with dense centers and sparse peripheries. Through an iterative process, candidate nodes were added to the seeds to form candidate clusters using the cluster generation model. In the cluster determination stage, we constructed a power-law distribution function about the distribution of cluster sizes and the cluster total number. The function acts as a criterion to regulate the presence of clusters of various sizes. By applying the power-law distribution function, we can assess whether a candidate cluster qualifies as a final cluster.

This paper makes several significant contributions: (1) Integrating multiple available basic micro-topological structural information into the k-order neighborhood of a node for seed selection; (2) Constructing a cluster generation model considering scale-free power-law distribution to obtain inherent organization information of functional modules; (3) Giving a cluster determination model based on macro-topological structure characteristic of the number distribution of clusters of different sizes to constrain final clusters; (4) Verifying the proposed network clustering algorithm fused with topological structural information could effectively mine functional modules by the experiment results on the real datasets.

The other sections of our paper are as follows. Section 2 introduces preliminary concepts and symbols. Section 3 presents a network clustering algorithm fused with power-law distribution characteristics. Section 4 reports the relevant experiments to verify the effectiveness of the network clustering algorithm. Section 5 provides conclusions.

2. Preliminary

A PPI network is represented by an undirected network G = (V, E), with V as the set of proteins (nodes) and E as the set of interactions (edges) between proteins. *Dia(G)* represents the diameter of the network G, which corresponds to the maximum value in the shortest path between any two nodes in the network G. The k-adjacent nodes set of a given node v_i is denoted as $NE_k(v_i)$, and it is defined by

$$NE_k(v_i) = \begin{cases} NE(v_i) & if\ k = 1 \\ NE_{k-1}(v_i) \cup \{v_j \in V | distance(v_i, v_j) = k\} & if\ k > 1 \end{cases} \tag{1}$$

where $distance(v_i, v_j)$ represents the length of the distance between nodes v_i and v_j.

The clustering coefficient of v_i [33] is

$$CCE(v_i) = \frac{2|ES(H(v_i))|}{|NE(v_i)|(|NE(v_i)| - 1)} \tag{2}$$

where $H(v_i)$ represents sub-graph created by the directly adjacent node set $NE(v_i)$, and $ES(H(v_i)) = \{(v_j, v_l)|v_j, v_l \in NE(v_i), (v_j, v_l) \in E\}$. A network's clustering coefficient $CCE(v_i)$ is calculated as the average value of the clustering coefficients of all nodes in the node set V, i.e., $\overline{CCE(G)} = \sum_{i=1}^{|V|} CCE(v_i)$. In order to facilitate readers' reading of this paper, some main symbols and their corresponding meanings are listed in Table 1.

Table 1. Main symbols and their corresponding meanings.

Symbols	Meaning
$G = (V, E)$	Network G is composed of a collection of nodes V and set of edges E.
v_i	Node i in a certain node set.
(v_i, v_j)	The edge between nodes i and j.
$distance(v_i, v_j)$	The shortest path distance between nodes i and j.
NE_k	Set of k-neighbors.
$ES(M)$	The set of edges within sub-graph M.
CCE	The clustering coefficient for a node
ND	The degree of a node in the network.
$w(.)$	The weight for a node or an edge
X_{size}	Set of cluster sizes
Y_{num}	Set of cluster numbers
$CT(u, M)$	The tightness measure of node u with respect to sub-graph M.
$CS(v)$	Node set generated by the selected seed v.
$Dia(G)$	The diameter of a network G
PC	Final cluster set
λ	Rate of change

3. Methods

GCAPL algorithm consists of two stages: cluster generation and cluster determination. In the first stage, the algorithm calculates the weights of nodes and edges by incorporating micro-topological structure metrics. A seed is the node that has the highest weight among the unclustered nodes. The seed is expanded by a cluster generation model considering a scale-free power-law distribution to a candidate cluster. In the second stage, we established the cluster determination model with a power-law distribution of the cluster numbers with different cluster sizes. This cluster decision model was utilized to determine the final clusters. Figure 1 shows the algorithm flow chat.

3.1. Cluster Generation

In the cluster generation stage, the GCAPL algorithm initially selects seeds based on node weights and subsequently expands these seeds using the cluster generation model to obtain candidate clusters.

To identify a suitable seed node, a node with a higher weighted degree may be a good seed node in network community mining. A node with a higher weighted degree may serve as a useful seed node in network community mining. The weighted degree of a node v_i is calculated based on its directly adjacent edges and the weights associated with these edges, and was defined as:

$$w(v_i) = \sum_{v_j \in NE(v_i)} w(v_i, v_j) \tag{3}$$

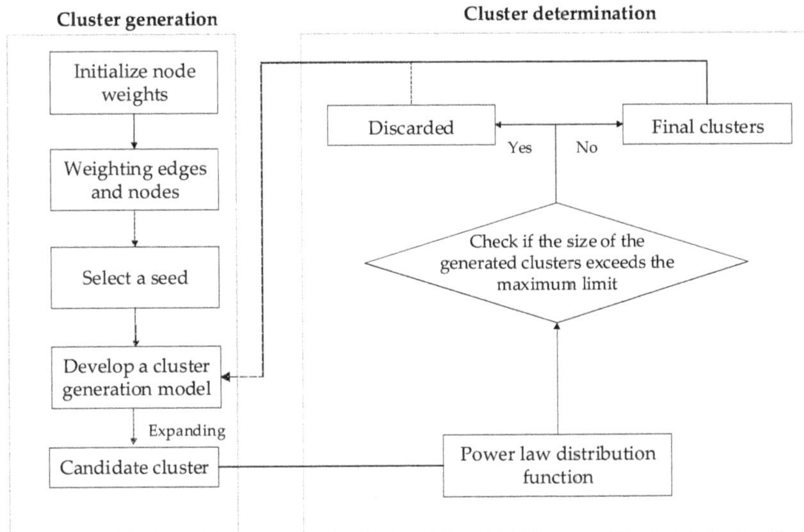

Figure 1. Algorithm flow chart.

For an edge (v_i, v_j), the endpoints of the edge and the common adjacent nodes between these endpoints are tightly surrounding this edge. We can obtain the edge weight of (v_i, v_j) according to the importance of these nodes in topological characteristics. The micro-topological structure metrics, such as clustering coefficient and node degree, are employed to capture the topological characteristics and assign weights to nodes. For a dense submodule in a network, nodes with high clustering coefficients and low node degrees may serve as important central nodes. The topological characteristics of a node v_i is expressed by the ratio of its clustering coefficient $CCE(v_i)$ to its node degree $ND(v_i)$, i.e., $CCE(v_i)/ND(v_i)$. More comprehensively, the global information of a network is introduced. A network *G*'s clustering coefficient *FC(G)* is defined as the average value of the clustering coefficients of all nodes in the node set *V*, i.e., $FG(G) = \overline{CCE(G)}$. Similarly, the *G*'s node degree *FD(G)* is the average of all node degrees in the network, i.e., $FD(G) = \overline{ND(G)}$. In the network G, the connection strength of a node v_i is related to $\frac{CCE(v_i)}{\overline{CCE(G)}} \times \frac{\overline{ND(G)}}{ND(v_i)}$. Therefore, the weight of the edge (v_i, v_j) can be defined as follows:

$$w(v_i, v_j) = \frac{CCE(v_i)}{\overline{CCE(G)}} \times \frac{\overline{ND(G)}}{ND(v_i)} + \frac{CCE(v_j)}{\overline{CCE(G)}} \times \frac{\overline{ND(G)}}{ND(v_j)} + \sum_{u \in NE(v_i) \cap NE(v_j)} \frac{CCE(v_u)}{\overline{CCE(G)}} \times \frac{\overline{ND(G)}}{ND(v_u)} \quad (4)$$

Furthermore, Equation (4) from the previous section only considers the information of the node's direct neighbors. To highlight the importance of an edge within a large network module, the edge weight in its *t* neighborhood can be defined as follows:

$$\begin{aligned} w^t(v_i, v_j) = \; & w^{t-1}(v_i) \times \frac{CCE(v_i)}{\overline{CCE(G)}} \times \frac{\overline{ND(G)}}{ND(v_i)} + \\ & w^{t-1}(v_j) \times \frac{CCE(v_j)}{\overline{CCE(G)}} \times \frac{\overline{ND(G)}}{ND(v_j)} + \\ & \sum_{u \in NE(v_i) \cap NE(v_j)} w^{t-1}(u) \times \frac{CCE(v_u)}{\overline{CCE(G)}} \times \frac{\overline{ND(G)}}{ND(v_u)} \end{aligned} \quad (5)$$

Here, t is a predefined parameter that determines the extent of the neighborhood. After the t-th iteration, the node weight can be defined as:

$$w^t(v_i) = \sum_{v_j \in NE(v_i)} w^t(v_i, v_j) \tag{6}$$

Initially, the node weights are set to $w^0(v_i) = 1$ for all nodes, indicating that the initial importance of all nodes is the same.

Once the node weight calculation is completed, the next step is to select a seed node v from the node set V whose node weight is highest. Following that, the seed node is used to establish the cluster generation model, which allows for the expansion of the seed into a candidate cluster.

The cluster generation model aims to expand seed nodes into candidate clusters based on connection strength. The obtained seed node v serves as the initial cluster *CS(v)*, and candidate nodes from the neighborhood *NE(C(v))* are considered for addition based on the compactness of *CS(v)* and the connection strength between *CS(v)* and a candidate node u to expand the initial cluster *CS(v)*. The compactness g of the cluster *CS(v)* quantifies the connection density within the cluster and is defined as $g(u, CS(v)) = |NE(u) \cap V(CS(v))| / |V(CS(v))|$, where $V(CS(v))$ represents a set of nodes that make up *C(v)*, and $|NE(u)|$ denotes the node u's direct neighbor nodes. The connection strength h of a candidate node u reflects the peripheral edges of the cluster and is defined as $h(u, CS(v)) = |NE(u) \cap V(CS(v))| / |NE(u)|$. The cluster generation model requires a variable function to combine the compactness of the cluster and the peripheral edges of the cluster, so that as the cluster size increases, the contribution of the cluster's compactness to the cluster generation gradually decreases while the contribution of the cluster's peripheral connections to the cluster generation gradually increases. A suitable choice for this function is the scale-free power-law distribution function, which is a monotonic function. It serves as a foundation for constructing the variable function that effectively fuses the above two kinds of connection information. A power-law distribution function is $y = cx^{-k}$ and let $c = 1/\lambda$, $k = ND(v)$, $x = V(CS(v)) - 1$, then we can define the variable function as:

$$\beta(CS(v)) = \frac{1}{\lambda \times \sqrt[ND(v)]{V(CS(v)) - 1} + 1} \tag{7}$$

where λ is a parameter to control the change of $\beta(CS(v))$. Then, define the cluster generation model as:

$$\begin{aligned} CT(u, CS(v)) = \beta(CS(v)) g(u, CS(v)) + \\ (1 - \beta(CS(v))) h(u, CS(v)) \end{aligned} \tag{8}$$

When $\beta(CS(v))$ is set to 1, CT tends to prioritize the formation of dense clusters. On the other hand, when $\beta(CS(v))$ is set to 0, nodes with lower degrees are more likely to be added to *CS(v)*. The $\beta(CS(v))$ enables the cluster generation model to find both dense clusters and clusters with dense cores and sparse peripheries, providing flexibility in capturing different types of cluster structures. For each candidate node u and threshold $\mu \in [0, 1]$, if $CT(u, CS(v)) > \mu$ and $Dia([CS(v) \cup \{u\}]) \leq \delta$ (δ is a user-defined threshold), then the node u is added to the cluster *CS(v)*. This process is repeated for each node in *NE(CS(v))*, resulting in the initial formation of a candidate cluster *CS(v)*.

3.2. Cluster Determination

In complex networks, the distribution of community size exhibits heterogeneity. Smaller communities tend to be more abundant in number, while larger communities are relatively scarce. This inverse relationship between size and number also holds in PPI networks, where the sizes and numbers of protein complexes are inversely proportional. It is assumed that the number of complexes follows a power-law distribution that is defined as follows:

$$y = cx^{-k} \tag{9}$$

where x and y represent positive random variables.

Let the size of a protein complex be X_{size}. The corresponding number of the complexes under this size is given by a cluster determination model:

$$Y_{num} = cX_{size}{}^{-k} \tag{10}$$

where c and k are positive parameters.

The cluster determination model aims to effectively regulate the number of clusters, considering their varying sizes, from a global perspective. To accomplish this, we defined two sequences: $X_{size} = \{x^1_{size}, x^2_{size}, \ldots, x^n_{size}\}$ is a predefined sequence with uniform values representing the cluster sizes, and $Y_{num} = \{y^1_{num}, y^2_{num}, \ldots, y^n_{num}\}$ is a sequence obtained through the cluster generation model representing the corresponding cluster numbers.

Let the cluster *CS(v)*'s size be denoted as $|V(CS(v))|$ and $error_{size}$ be a parameter that refers to the allowable difference or deviation in the size of a cluster. Following that, we can find a value x^i_{size} with $|V(CS(v))| \in [x^i_{size} - error_{size}, x^i_{size} + error_{size}]$ in X_{size}, assuming that y'_{num} clusters of size x^i_{size} have been generated at the current stage. We calculated the maximum number of clusters y^i_{num} corresponding to a given cluster size x^i_{size} according to the power-law distribution function. If $y'_{num} \leq y^i_{num}$, the candidate cluster *CS(v)* is considered a final cluster. Otherwise, *CS(v)* is discarded.

The two stages of cluster generation and cluster determination are repeated alternately until all nodes have been clustered.

3.3. Complexity Analysis

The GCAPL algorithm utilizes linked lists to construct a graph. First, it calculates the weights of all nodes using Formula (6). Following that, it selects the node with the highest weight as the seed and treats it as the initial cluster. Subsequently, following the cluster generation model, neighbor nodes of the initial cluster are incrementally added to create candidate clusters. Finally, the algorithm determines the final clusters by the cluster determination model. The specific process of the GCAPL algorithm is shown in Algorithm 1.

Algorithm 1: GCAPL Algorithm.

Input: Network $G = (V, E)$, Parameters *iter*, λ, μ for cluster generation, Parameters c, k, $error_{size}$ for Cluster determination
output: Set of final clusters *PC*
Initialize $PC = \varnothing$, and the unclustered nodes set, $UV = V$;
Compute edge and node weights by utilizing information within the t-neighborhood;
Determine the cluster size set $X_{size} = \{x^1_{size}, x^2_{size}, \ldots, x^n_{size}\}$;
Calculate the upper limit of the number of clusters $Y_{num} = \{y^1_{num}, y^2_{num}, \ldots, y^n_{num}\}$, corresponding to the cluster size X_{size} using Equation (9);
while $UV \neq \varnothing$, do
    Select a node v with the largest weight in *UV* as a seed, and the initial cluster is *CS(v)*;
    Iteratively select the node set *AN* among the neighbor nodes of *CS(v)*, such that each node u in *AN* satisfies $CT(u, H) > \mu$ and $Dia([CS(v) \cup \{u_i\}]) \leq \delta$;
    $CS(v) = CS(v) \cup AN$;
    Compute the cluster *CS(v)*'s size as $|V(CS(v))|$, and compute the number of generated clusters with size $|V(CS(v))|$ as y'_{num};
    Find x^i_{size} in X_{size}, and $|V(CS(v))| \in \left[x^i_{size} - error_{size}, x^i_{size} + error_{size}\right]$
    Compute the number of generated clusters of size $|V(CS(v))|$ as y'_{num}
    if $y'_{num} \leq y^i_{num}$ **then**
        $PC = PC \cup \{CS(v)\}$, $UV = UV - CS(v)$
return *PC*

The time cost of the GCAPL algorithm lies in two parts: cluster generation and cluster determination.

Assuming a network G has n nodes and m edges. In the cluster generation stage, the node weighting process revealed a time cost of $O(k \times \overline{ND} \times n) = O(k \times m)$. The time cost of seed selection based on node weights is $O(n \times \log n)$. The expansion of seeds into clusters also has a time cost of $O(n \times \log n)$. Therefore, $O(|PC| \times n \times \log n)$ is the total time complexity of the cluster generation phase.

In the cluster determination phase, the worst-case scenario is when each candidate cluster size needs to be compared with each element in the sequence X_{size}. As a result, this phase revealed a time cost of $O(n \times |X_{size}|)$. Therefore, algorithm GCAPL's overall time complexity is $O(|PC| \times n \times \log n)$, considering both the cluster generation and cluster determination phases.

4. Experiments and Results

4.1. Datasets

The protein interaction networks used in the experiments are presented in Table 2. These datasets were processed to remove self-intersections and duplicate interactions.

Table 2. Datasets of protein interaction networks.

	Gavin02 [34]	Gavin06 [35]	K-Extend [36]	BioGRID [37]
Proteins	1352	1430	3672	4187
Interactions	3210	6531	14,317	20,454

The gold standard complex datasets CYC2008 [38] and MIPS [39] were utilized for parameter analysis and evaluation of the clustering results.

4.2. Evaluation Metrics

The evaluation of the effectiveness of the GCAPL algorithm was performed using the F-measure and Accuracy metrics as evaluation criteria.

The F-measure [40] provides a balanced measure of precision and recall. It serves as a quantitative metric of the agreement between a predicted complex set and a benchmark complex set, capturing the level of similarity between them. Precision measures the agreement between the generated clusters and known complexes, while recall quantifies the agreement between the known complexes and the generated clusters.

Given the generated cluster as $PC = \{PC_1, PC_2, \ldots, PC_p\}$ and the gold standard complex as $TC = \{TC_1, TC_2, \ldots, TC_l\}$, the affinity score within the neighborhood $NA(PC_i, TC_j)$ is employed for quantifying the similarity between the generated cluster PC_i and the standard complex TC_j, and $NA(PC_i, TC_j) = |PC_i \cap TC_j|^2 / |PC_i| \times |TC_j|$, $i \in \{1, 2, \ldots, p\}$, $j \in \{1, 2, \ldots, l\}$. A higher $NA(PC_i, TC_j)$ value indicates a stronger resemblance between PC_i and TC_j. Assuming a threshold of $\mu = 0.2$ [40,41], if $NA(PC_i, TC_j) \geq \mu$, PC_i and TC_j can be considered as matched. Let M_C represent the set of correct predictions, where each generated cluster exhibits some correspondence with at least one known protein complex in the set TC, and $M_C = \{PC_i | PC_i \in PC \wedge \exists j (TC_j \in TC \wedge NA(PC_i, TC_j) \geq \mu)\}$. Additionally, let M_{CO} be the set of known complexes, where each complex matches at least one complex in the generated cluster set PC, and $M_{CO} = \{TC_j | TC_j \in TC \wedge \exists i (PC_i \in PC \wedge NA(PC_i, TC_j) \geq \mu)\}$.

Precision is quantitatively calculated as the ratio of the number of correctly predicted instances to the total number of predicted instances, i.e., $Precision = |M_C| / |PC|$. Recall is defined as $Recall = |M_{CO}| / |TC|$. F-measure is quantitatively calculated as

$$F - measure = 2 \times Precision \times Recall / (Precision + Recall) \tag{11}$$

Accuracy, as another evaluation metric, is computed as the geometric mean of the positive predictive value (PPV) and sensitivity (Sn). PPV represents the proportion of correctly identified positive instances among the predicted instances, while Sn measures the proportion of correctly identified positive instances among all actual positive instances.

Suppose T is a $p \times l$ matrix, in which the i-th row of T represents the i-th prediction cluster PC_i and the j-th column represents the j-th annotation complex TC_j. T_{ij} denotes the count of shared proteins between the predicted complex PC_i and the known complex TC_j and quantifies the degree of overlap or similarity between these two complexes. PPV is characterized by

$$PPV = \frac{\sum_{i=1}^{p}\sum_{j=1}^{l}\left(T_{ij} \times \max_{j=1}^{l}\left(\frac{T_{ij}}{\sum_{j=1}^{l} T_{ij}}\right)\right)}{\sum_{i=1}^{p}\sum_{j=1}^{l} T_{ij}} \tag{12}$$

Sn is defined as

$$Sn = \frac{\sum_{j=1}^{l}\left(|TC_j| \times max_{i=1}^{p}\left(\frac{T_{ij}}{|TC_j|}\right)\right)}{\sum_{j=1}^{l}|TC_j|} \tag{13}$$

Accuracy [39] is then calculated as

$$Accuracy = \sqrt{PPV \times Sn} \tag{14}$$

4.3. Parametric Analysis

GCAPL encompasses several predefined parameters, including $c, k \in [2,3]$, $error_{size}$, $iter$, $\lambda \in [0,1]$, and $\mu \in [0,1]$. The coefficients c and k correspond to the coefficients and exponents of the power-law distribution function, respectively. The $error_{size}$ is a cluster size error. The $iter$ refers to the count of repetitive steps. The λ stands for an adaptive parameter. The μ is defined as the compactness threshold. The BioGrid dataset serves as a standard protein interaction network dataset, wherein all interactions are derived from reliable and precise low-throughput theoretical interactions. Consequently, on this dataset, the parameter optimization aims to maximize the value of $F - measure + Accuracy$, prompting a thorough parameter analysis to identify the optimal parameter value.

The analysis of parameters c, k, and $error_{size}$ was performed to investigate the impact of these parameters on the algorithm. The coefficient c and the exponent k were utilized to generate the sequences X_{size} and Y_{num} based on the power-law distribution function. Meanwhile, the parameter $error_{size}$ was employed to regulate the error tolerance in cluster size. Initially, the analysis focuses on varying c and k while keeping the parameter $error_{size}$ constant. Subsequently, the investigation shifts to studying the influence of the parameter $error_{size}$ while maintaining c and k at constant values.

We first fixed $error_{size} = 6$, and experiments were conducted on the BioGrid PPI network to investigate the impact of the parameters c and k. The values of c ranged from 100 to 250, while k varied from 2.0 to 3.0. These experiments aimed to assess how the changes in c and k influenced the results and outcomes of the study. When the values of $c = 200$ and $k = 2.2$ are set, the $F - measure + Accuracy$ metric attains a higher value. Next, we first fixed $c = 200$, $k = 2.2$ and $F - measure + Accuracy$ is maximized at $error_{size} = 6$. We set $c = 200$, $k = 2.2$, $error_{size} = 6$. In Figure 2a, the impact of parameters c and k on $F - measure + Accuracy$ is illustrated, with $error_{size} = 6$. The relationship between the parameter $error_{size}$ and $F - measure + Accuracy$ are depicted in Figure 2b, with $c = 200$ and $k = 2.2$.

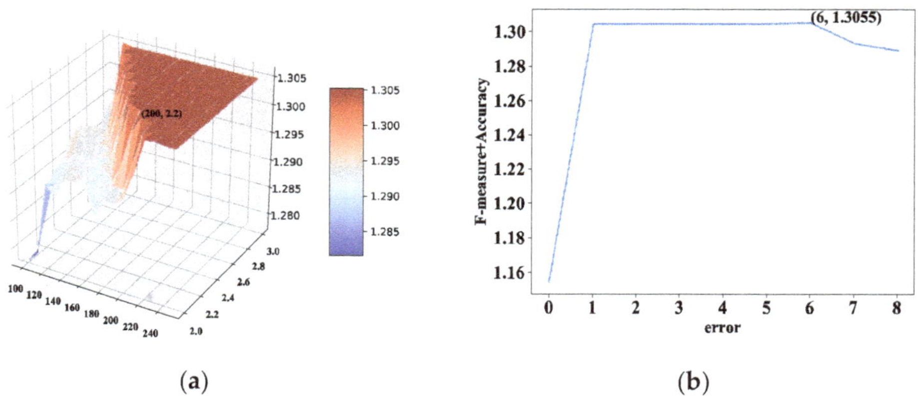

Figure 2. Performance impact analysis of parameters on BioGRID dataset: (**a**) analyze c and k; (**b**) analyze $error_{size}$.

Next, we kept the values of $c = 200$, $k = 2.2$, and $error_{size} = 6$ fixed, and analyzed the real-valued discrete parameters: the number of iterations $iter$, the adjustment parameter $\lambda \in [0, 1]$ of the change rate, and the tightness threshold $\mu \in [0, 1]$. Considering the interdependence among these parameters, an orthogonal matrix was employed to identify the optimal parameter combination with a high likelihood. During the experimental design phase, each parameter variable was treated as an independent factor. Feasible values corresponding to these factors are assigned as distinct levels. The complete set of parameter combinations represents the experimental space. An orthogonal array L36 (63 × 37) is employed, which comprises 36 parameter combinations. There are parameters, $iter \in \{1, 2, 3, 4, 5, 6\}$, $\lambda \in \{0.1, 0.2, 0.3, 0.4, 0.5, 0.6\}$, and $\mu \in \{0.1, 0.2, 0.3, 0.4, 0.5, 0.6\}$, that we exclusively consider the initial three columns of the orthogonal array to facilitate the analysis. Among the 36 parameter combinations, the one with the highest $F - measure + Accuracy$ is selected as the optimal configuration. Through the experiments, the parameters are set to $iter = 2$, $\lambda = 0.1$, and $\mu = 0.4$.

4.4. Power-Law Distribution Analysis

This subsection examines the power-law distribution of network clustering results, taking the BioGRID dataset as an example. The clustering result of this dataset was utilized to explore the relationship between the cluster size and number.

Assume that the cluster size is represented by x and the corresponding number of clusters is denoted by y. According to Equation (9), we have $y = cx^{-k}$. By performing logarithm operations on both sides of the equation, it represents that

$$\ln y = \ln c - k \ln x \tag{15}$$

It was observed that $\ln y$ and $\ln x$ exhibit a linear relationship. Thus, the analysis of the power-law distribution of x and y was transformed into a linear relationship analysis of $\ln x$ and $\ln y$.

In the clustering result of the BioGRID dataset, we took the logarithm of the cluster size x and the corresponding cluster number y, resulting in transformed variables $x' = \ln x$ and $y' = \ln y$. To explore whether there is a linear relationship between x' and y', a linear fitting method was performed on x' and y'. The results of the linear fitting analysis conducted on x' and y' is shown in Figure 3, providing valuable insights into the nature of their relationship.

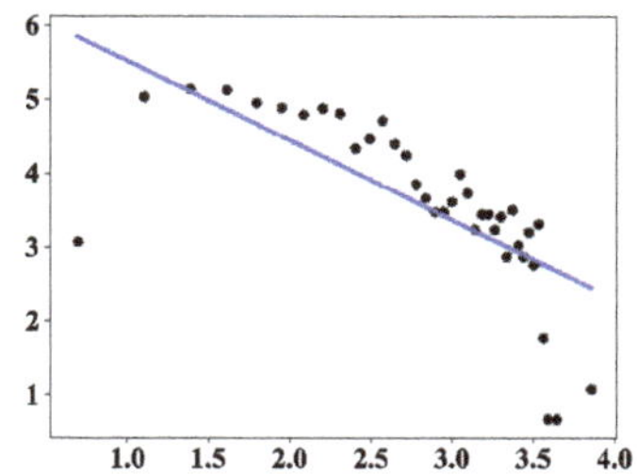

Figure 3. Fitting curve of x' and y'.

Table 3 presents the calculated p-value and R^2 for the linear fitting analysis conducted on x' and y'. A small p-value indicates a strong fit of the clustering result, demonstrating good fitting effectiveness. Similarly, a large value of R^2 suggests a favorable fit. In Table 3, the obtained p-value is 9.9×10^{-7}, and the value of R^2 is 0.5. Thus, the sizes of clusters generated by the proposed algorithm in the PPI network follow a power-law distribution, along with the corresponding numbers of these clusters.

Table 3. Fitting effect of x' and y'.

Criteria	Value
p-value	$9.90771462 \times 10^{-7}$
R^2	0.5001443526421876

4.5. Comparative Experiment

To assess the algorithm's performance, we compared the GCAPL algorithm with several other algorithms, namely DPClus, IPCA, SEGC, Core, SR-MCL, and ELF-DPC.

Figure 4a–d present the experimental results on the Gavin02, Gavin06, K-extend, and BioGRID datasets, using CYC2008 as the standard set. The results demonstrate that, compared to other algorithms, the GCAPL algorithm achieves comparable or higher F-measure and Accuracy values. The GCAPL algorithm performs well in terms of $F - measure + Accuracy$. Compared with other algorithms, the *F-measure* + *Accuracy* of GCAPL exhibits an average improvement of 13.12%, 6.97%, 14.43%, and 14.39% on Gavin02, Gavin06, K-extend, and BioGRID. In addition, the SEGC algorithm demonstrates lower F-measure and Accuracy performance compared to GCAPL on the Gavin02, K-extend, and BioGRID datasets. On the Gavin06 dataset, the DPClus algorithm performs better than other algorithms, except for the GCAPL algorithm. The GCAPL algorithm has a similar framework to the two algorithms mentioned above, and incorporating macro-topological information contributes to improving complex detection performance. By considering both the micro-topological structure of a network and the macro-topological structure feature of the power-law distribution, the GCAPL algorithm effectively detects protein complexes.

Figure 5a–d illustrate the evaluation results of the DPClus, IPCA, SEGC, Core, SR-MCL, ELF-DPC, and GCAPL algorithms on the Gavin02, Gavin06, K-extend, and BioGRID datasets, respectively, using MIPS as the standard set. The GCAPL algorithm consistently exhibits superior values of F-measure and Accuracy across the four different PPI datasets compared to compared algorithms. Compared with other algorithms, the *F-measure* + *Accuracy* of GCAPL exhibits an average increase of 9.90%, 7.01%, 14.34%, and 13.63% on Gavin02, Gavin06, K-extend, and BioGRID. This indicates that the GCAPL algorithm performs well in terms of its ability to detect protein complexes.

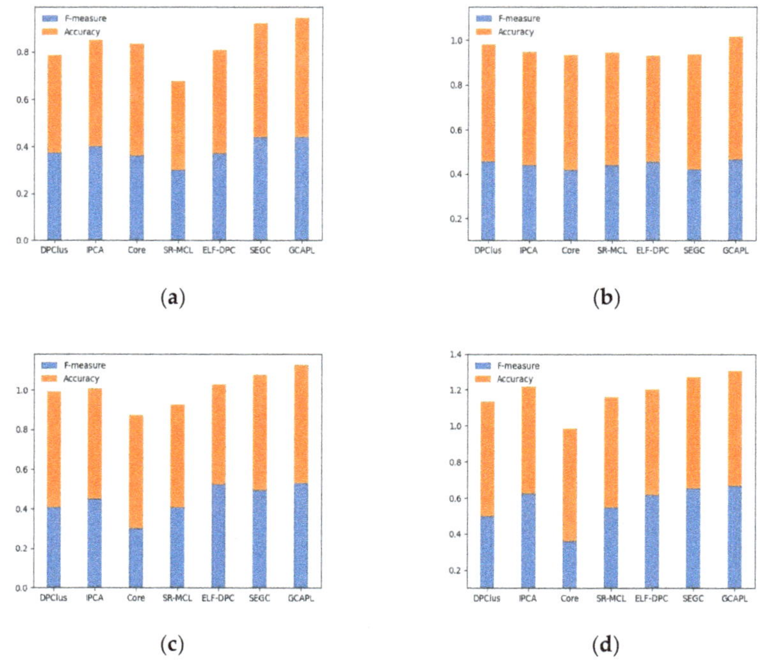

Figure 4. CYC2008 as benchmarks. Evaluation results by different algorithms on (**a**) Gavin02; (**b**) Gavin06; (**c**) K-extend; (**d**) BioGRID.

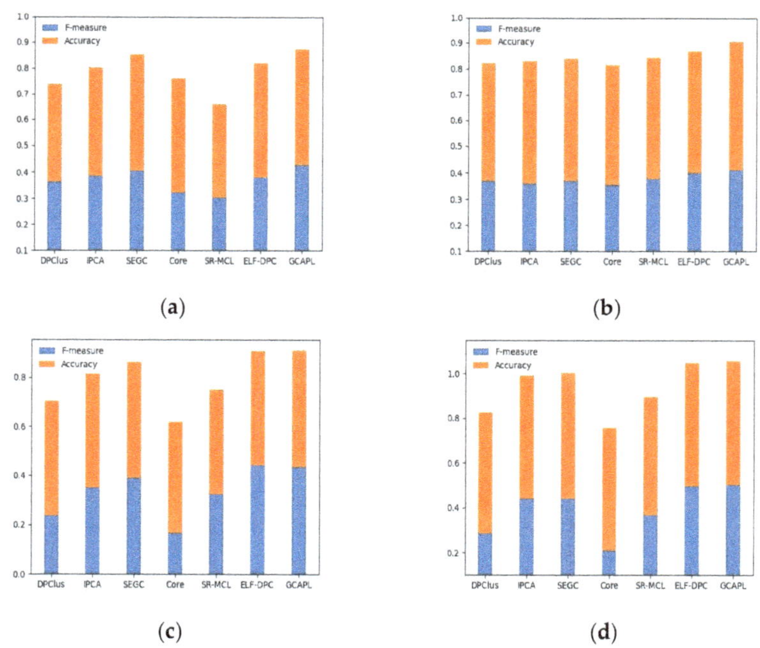

Figure 5. MIPS as benchmarks: Evaluation results by different algorithms on (**a**) Gavin02; (**b**) Gavin06; (**c**) K-extend; (**d**) BioGRID.

In summary, the GCAPL algorithm has good performance in detecting protein complexes. The GCAPL algorithm uses not only micro-topological structure metrics but also the macro-topological structure characteristic of the power-law distribution about clusters, and it can obtain better results in complex detection. The GCAPL algorithm further explores the relationship between network topological characteristics and functional modules in PPI networks, which is of great significance for improving the accuracy of protein complex detection.

4.6. Examples of Predicted Complexes

In this subsection, four predicted protein complexes with different sizes detected by the GCAPL algorithm are exhibited, and their corresponding network topology structures are shown in Figure 6. The predicted complex in Figure 6a is a fully interconnected network. Figure 6b shows a cluster that has a dense sub-graph with a relatively sparse periphery. Figure 6c,d show two clusters that are dense sub-graphs. Table 4 presents the Gene Ontology annotations of these predicted protein complexes in three aspects of biological processes, molecular functions, and cell components with corresponding significance *p*-values. The obtained *p*-values are notably small, indicating that these clusters have significant biological significance. The effectiveness of the GCAPL algorithm is demonstrated in its ability to identify protein complexes with multiple network structures.

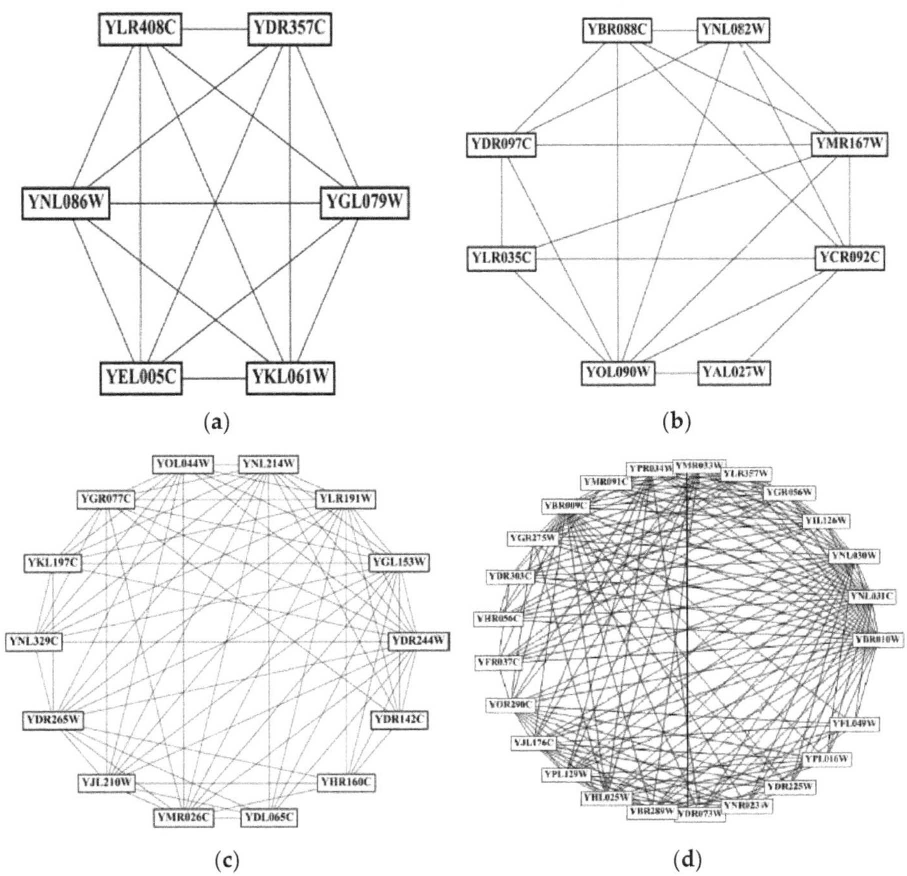

Figure 6. Examples of predicted protein complexes: (**a**) cluster a; (**b**) cluster b; (**c**) cluster c; (**d**) cluster d.

Table 4. Gene ontology annotations of the four predicted protein complexes.

ID	Processes		Functions		Components	
	Gene Ontology Term	***p*-Value**	**Gene Ontology Term**	***p*-Value**	**Gene Ontology Term**	***p*-Value**
a	endosome organization (GO:0007032)	4.04×10^{-15}	molecular function (GO:0003674)	0.00194	BLOC complex (GO:0031082)	1.06×10^{-19}
b	DNA repair (GO:0006281)	1.37×10^{-9}	DNA binding (GO:0003677)	9.64×10^{-8}	nucleus (GO:0005634)	0.00506
c	protein targeting to peroxisome (GO:0006625)	1.14×10^{-35}	Binding (GO:0005488)	0.00206	microbody part (GO:0044438)	3.31×10^{-29}
d	DNA-templated transcription (GO:0006351)	2.39×10^{-22}	DNA binding (GO:0003677)	6.11×10^{-7}	nuclear chromosome part (GO:0044454)	1.59×10^{-36}

5. Conclusions

Detecting protein complexes is of great significance for understanding biological mechanisms. This paper proposes a network clustering algorithm fused with power-law distribution for protein complex detection. The algorithm begins by calculating node weights, taking into account micro-topological structure metrics. Subsequently, the algorithm selects the non-clustered nodes with the higher weights as seeds and forms initial clusters around the seeds. Next, the algorithm greedily adds candidate nodes into the initial clusters based on the characteristics of scale-free power-law distribution to generate candidate clusters. A power-law distribution function, based on the macro-topological structure feature of power-law distribution about cluster size and number, is established to guide the cluster generation process. The power-law distribution function is employed to determine whether a candidate cluster qualifies as a final cluster. Compared with other algorithms, the *F-measure* + *Accuracy* of GCAPL improves by an average of 12.23% and 10.97% on the CYC2008 and MIPS benchmarks, respectively. The experimental analysis reveals that the proposed algorithm exhibits distinct advantages over other approaches.

The GCAPL algorithm mainly considers the biological network whose community size conforms to the power-law distribution characteristics. The algorithm does not take into account other distribution characteristics of the community size and fully considers the preferential attachment. The above information may further improve the performance of our algorithm to detect protein complexes. In addition, in real PPI networks, the connections between nodes are subject to constant changes, leading to variations in network topological structures. To mine functional modules in dynamic PPI networks, our future work will also focus on constructing dynamic networks and developing dynamic protein complex identification methods.

Author Contributions: Conceptualizing the algorithm, designing the method and revising the draft, J.W.; implementation of the computer code and writing the original draft, Y.J.; revising the manuscript, A.K.S.; visualizing and curating data, Y.S. All authors have read and agreed to the published version of the manuscript.

Funding: This paper was funded by the National Natural Science Foundation of China (No. 62006145); the Scientific and Technological Innovation Programs of Higher Education Institutions in Shanxi, China (No. 2020L0245); the Youth Science Foundation of Shanxi University of Finance and Economics, China (No. QN-202016); and Shandong Provincial Natural Science Foundation, China (No. ZR2020MF146).

Data Availability Statement: The datasets used in this study are publicly available and downloaded from the BioGRID database (https://downloads.thebiogrid.org/BioGRID, accessed on 1 March 2023), MIPS database (http://mips.gsf.de, accessed on 8 September 2019), and CYC2008 complexes database (http://wodaklab.org/cyc2008/, accessed on 12 April 2023).

Acknowledgments: This study received support from the Teaching and Research Department of Computer Science and Technology, Shanxi University of Finance and Economics, and all authors would like to express their gratitude for this.

Conflicts of Interest: The authors declare no conflict of interest.

References

1. Wu, L.; Huang, S.; Wu, F.; Jiang, Q.; Yao, S.; Jin, X. Protein Subnuclear Localization Based on Radius-SMOTE and Kernel Linear Discriminant Analysis Combined with Random Forest. *Electronics* **2020**, *9*, 1566. [CrossRef]
2. Ito, T.; Chiba, T.; Ozawa, R.; Yoshida, M.; Hattori, M.; Sakaki, Y. A comprehensive two-hybrid analysis to explore the yeast protein interactome. *Proc. Natl. Acad. Sci. USA* **2001**, *98*, 4569–4574. [CrossRef] [PubMed]
3. Causier, B.; Davies, B. Analysing protein-protein interactions with the yeast two-hybrid system. *Plant Mol. Biol.* **2002**, *50*, 855–870. [CrossRef] [PubMed]
4. Puig, O.; Caspary, F.; Rigaut, G.; Rutz, B.; Bouveret, E.; Bragado-Nilsson, E.; Wilm, M.; Séraphin, B. The tandem affinity purification (TAP) method: A general procedure of protein complex purification. *Methods* **2001**, *24*, 218–229. [CrossRef] [PubMed]
5. Rahiminejad, S.; Maurya, M.R.; Subramaniam, S. Topological and functional comparison of community detection algorithms in biological networks. *BMC Bioinform.* **2019**, *20*, 212. [CrossRef]
6. Spirin, V.; Mirny, L.A. Protein complexes and functional modules in molecular networks. *Proc. Natl. Acad. Sci. USA* **2003**, *100*, 12123–12128. [CrossRef]
7. Bai, L.; Cheng, X.; Liang, J.; Guo, Y. Fast graph clustering with a new description model for community detection. *Inf. Sci.* **2017**, *388–389*, 37–47. [CrossRef]
8. Saxena, A.; Prasad, M.; Gupta, A.; Bharill, N.; Patel, O.P.; Tiwari, A.; Er, M.J.; Ding, W.; Lin, C.-T. A review of clustering techniques and developments. *Neurocomputing* **2017**, *267*, 664–681. [CrossRef]
9. Emmons, S.; Kobourov, S.; Gallant, M.; Börner, K. Analysis of network clustering algorithms and cluster quality metrics at scale. *PLoS ONE* **2016**, *11*, e0159161. [CrossRef]
10. Bhowmick, S.S.; Seah, B.S. Clustering and summarizing protein-protein interaction networks: A survey. *IEEE Trans. Knowl. Data Eng.* **2016**, *28*, 638–658. [CrossRef]
11. Pan, Y.; Guan, J.; Yao, H.; Shi, Y.; Zhou, Y. Computational methods for protein complex prediction: A survey. *J. Front. Comput. Sci. Technol.* **2022**, *16*, 1–20.
12. Manipur, I.; Giordano, M.; Piccirillo, M.; Parashuraman, S.; Maddalena, L. Community Detection in Protein-Protein Interaction Networks and Applications. *IEEE/ACM Trans. Comput. Biol. Bioinform.* **2021**, *20*, 217–237. [CrossRef]
13. Liu, G.; Wong, L.; Chua, H.N. Complex discovery from weighted PPI networks. *Bioinformatics* **2009**, *25*, 1891–1897. [CrossRef]
14. Bader, G.D.; Hogue, C.W.V. An automated method for finding molecular complexes in large protein interaction networks. *BMC Bioinform.* **2003**, *4*, 2. [CrossRef]
15. Palla, G.; Derényi, I.; Farkas, I.; Vicsek, T. Uncovering the overlapping community structure of complex networks in nature and society. *Nature* **2005**, *435*, 814–818. [CrossRef] [PubMed]
16. Amin, A.U.; Shinbo, Y.; Mihara, K.; Kurokawa, K.; Kanaya, S. Development and implementation of an algorithm for detection of protein complexes in large interaction networks. *BMC Bioinform.* **2006**, *7*, 207. [CrossRef]
17. Li, M.; Chen, J.-E.; Wang, J.-X.; Hu, B.; Chen, G. Modifying the DPClus algorithm for identifying protein complexes based on new topological structures. *BMC Bioinform.* **2008**, *9*, 398. [CrossRef] [PubMed]
18. Wang, J.; Zheng, W.; Qian, Y.; Liang, J. A seed expansion graph clustering method for protein complexes detection in protein interaction networks. *Molecules* **2017**, *22*, 2179. [CrossRef]
19. Leung, H.C.; Xiang, Q.; Yiu, S.M.; Chin, F.Y. Predicting protein complexes from PPI data: A core-attachment approach. *J. Comput. Biol.* **2009**, *16*, 133–144. [CrossRef] [PubMed]
20. Yue, L.; Jun, X.; Sihang, Z.; Siwei, W.; Xifeng, G.; Xihong, Y.; Ke, L.; Wenxuan, T.; Wang, L.X. A survey of deep graph clustering: Taxonomy, challenge, and application. *arXiv* **2022**, arXiv:2211.12875.
21. Sun, H.; He, F.; Huang, J.; Sun, Y.; Li, Y.; Wang, C.; He, L.; Sun, Z.; Jia, X. Network embedding for community detection in attributed networks. *ACM Trans. Knowl. Discov. Data* **2020**, *14*, 1–25. [CrossRef]
22. Kumar, S.; Panda, B.S.; Aggarwal, D. Community detection in complex networks using network embedding and gravitational search algorithm. *J. Intell. Inf. Syst.* **2021**, *57*, 51–72. [CrossRef]
23. Wang, R.; Ma, H.; Wang, C. An ensemble learning framework for detecting protein complexes from PPI networks. *Front. Genet.* **2022**, *13*, 839949. [CrossRef]
24. Liu, X.; Yang, Z.; Zhou, Z.; Sun, Y.; Lin, H.; Wang, J.; Xu, B. The impact of protein interaction networks' characteristics on computational complex de-tection methods. *J. Theor. Biol.* **2018**, *439*, 141–151. [CrossRef]

25. Cherifi, H.; Palla, G.; Szymanski, B.K.; Lu, X. On community structure in complex networks: Challenges and opportunities. *Appl. Netw. Sci.* **2019**, *4*, 117. [CrossRef]
26. Huang, Z.; Zhong, X.; Wang, Q.; Gong, M.; Ma, X. Detecting community in attributed networks by dynamically exploring node attributes and topological structure. *Knowl.-Based Syst.* **2020**, *196*, 105760. [CrossRef]
27. Ghalmane, Z.; Cherifi, C.; Cherifi, H.; El Hassouni, M. Centrality in complex networks with overlapping community structure. *Sci. Rep.* **2019**, *9*, 10133. [CrossRef]
28. Rajeh, S.; Savonnet, M.; Leclercq, E.; Cherifi, H. Characterizing the interactions between classical and community-aware centrality measures in complex networks. *Sci. Rep.* **2021**, *11*, 10088. [CrossRef] [PubMed]
29. Girvan, M.; Newman, M.E.J. Community structure in social and biological networks. *Proc. Natl. Acad. Sci. USA* **2002**, *99*, 7821–7826. [CrossRef]
30. Sangaiah, A.K.; Rezaei, S.; Javadpour, A.; Zhang, W. Explainable AI in big data intelligence of community detection for digitalization e-healthcare services. *Appl. Soft Comput.* **2023**, *136*, 110119. [CrossRef]
31. Ma, J.; Fan, J. Local optimization for clique-based overlapping community detection in complex networks. *IEEE Access* **2019**, *8*, 5091–5103. [CrossRef]
32. Kustudic, M.; Xue, B.; Zhong, H.; Tan, L.; Niu, B. Identifying Communication Topologies on Twitter. *Electronics* **2021**, *10*, 2151. [CrossRef]
33. Watts, D.J.; Strogatz, S.H. Collective dynamics of 'small-world' networks. *Nature* **1998**, *393*, 440–442. [CrossRef]
34. Gavin, A.C.; Bösche, M.; Krause, R.; Grandi, P.; Marzioch, M.; Bauer, A.; Schultz, J.; Rick, J.M.; Michon, A.M.; Cruciat, C.M.; et al. Functional organization of the yeast proteome by systematic analysis of protein complexes. *Nature* **2002**, *415*, 141–147. [CrossRef] [PubMed]
35. Gavin, A.-C.; Aloy, P.; Grandi, P.; Krause, R.; Boesche, M.; Marzioch, M.; Rau, C.; Jensen, L.J.; Bastuck, S.; Dümpelfeld, B.; et al. Proteome survey reveals modularity of the yeast cell machinery. *Nature* **2006**, *440*, 631–636. [CrossRef]
36. Krogan, N.J.; Cagney, G.; Yu, H.; Zhong, G.; Guo, X.; Ignatchenko, A.; Li, J.; Pu, S.; Datta, N.; Tikuisis, A.P.; et al. Global landscape of protein complexes in the yeast Saccharomyces cerevisiae. *Nature* **2006**, *440*, 637–643. [CrossRef]
37. Stark, C.; Breitkreutz, B.J.; Reguly, T.; Boucher, L.; Breitkreutz, A.; Tyers, M. BioGRID: A general repository for interaction datasets. *Nucleic Acids Res.* **2006**, *34* (Suppl. S1), D535–D539. [CrossRef]
38. Pu, S.; Wong, J.; Turner, B.; Cho, E.; Wodak, S.J. Up-to-date catalogues of yeast protein complexes. *Nucleic Acids Res.* **2009**, *37*, 825–831. [CrossRef] [PubMed]
39. Brohée, S.; Van Helden, J. Evaluation of clustering algorithms for protein-protein interaction networks. *BMC Bioinform.* **2006**, *7*, 488. [CrossRef] [PubMed]
40. Li, X.; Wu, M.; Kwoh, C.-K.; Ng, S.-K. Computational approaches for detecting protein complexes from protein interaction networks: A survey. *BMC Genom.* **2010**, *11*, S3. [CrossRef] [PubMed]
41. Ma, X.; Gao, L. Predicting protein complexes in protein interaction networks using a core-attachment algorithm based on graph communicability. *Inf. Sci.* **2012**, *189*, 233–254. [CrossRef]

electronics

Article

Graph Convolution Network over Dependency Structure Improve Knowledge Base Question Answering

Chenggong Zhang [1,2,*], Daren Zha [2], Lei Wang [2], Nan Mu [2], Chengwei Yang [3], Bin Wang [4] and Fuyong Xu [4,*]

1 Institute of School of Cyber Security, University of Chinese Academy of Sciences, Beijing 100043, China
2 Institute of Information Engineering, Chinese Academy of Sciences, Beijing 100864, China; zhadaren@iie.ac.cn (D.Z.); wanglei@iie.ac.cn (L.W.); munan@iie.ac.cn (N.M.)
3 School of Management Science and Engineering, Shandong University of Finance and Economics, Jinan 250014, China; yangchengwei2006@163.com
4 School of Information Science and Engineering, Shandong Normal University, Jinan 250358, China; wang_bean_068@163.com
* Correspondence: zcg870108@163.com (C.Z.); fyxu0908@outlook.com (F.X.)

Abstract: Knowledge base question answering (KBQA) can be divided into two types according to the type of complexity: questions with constraints and questions with multiple hops of relationships. Previous work on knowledge base question answering have mostly focused on entities and relations. In a multihop question, it is insufficient to focus solely on topic entities and their relations since the relation between words also contains some important information. In addition, because the question contains constraints or multiple relationships, the information is difficult to capture, or the constraints are missed. In this paper, we applied a dependency structure to questions that capture relation information (e.g., constraint) between the words in question through a graph convolution network. The captured relation information is integrated into the question for re-encoding, and the information is used to generate and rank query graphs. Compared with existing sequence models and query graph generation models, our approach achieves a 0.8–3% improvement on two benchmark datasets.

Keywords: dependency structure; graph convolution network; question answering

Citation: Zhang, C.; Zha, D.; Wang, L.; Mu, N.; Yang, C.; Wang, B.; Xu, F. Graph Convolution Network over Dependency Structure Improve Knowledge Base Question Answering. *Electronics* **2023**, *12*, 2675. https://doi.org/10.3390/electronics12122675

Academic Editor: Ping-Feng Pai

Received: 20 April 2023
Revised: 1 June 2023
Accepted: 2 June 2023
Published: 14 June 2023

1. Introduction

The rapid development of information technology has created a need to accurately extract information from large-scale data, making question answering (QA) systems an important area of research. In the 1960s, QA systems primarily relied on expert systems, which involved numerous rules or templates. As technology advanced, QA systems shifted towards information-retrieval-based approaches. Retrieval-based QA systems rely on keyword matching and information extraction to analyze surface-level meaning and to extract answers from relevant documents. However, these systems can only provide answers to predefined questions.

To overcome this limitation, large-scale commercial engines have been developed. Community-based QA systems, which are built upon keyword matching retrieval, utilize historical questions from users and recommend answers to new questions. In recent years, the growth of the World Wide Web has led to the accumulation of vast amounts of high-quality data. This has paved the way for the emergence of extensive knowledge bases (KBs) that contain structured data. Natural language questions can be mapped to structured queries on these knowledge bases. KBQA (knowledge base question answering) aims to correctly understand the semantics of user questions and to use fact retrieval, matching, and reasoning techniques within the knowledge base to find answers.

In summary, as information technology continues to advance, QA systems have evolved from rule-based expert systems to retrieval-based approaches. With the availability of large-scale knowledge bases, KBQA systems have emerged to effectively understand user questions and to provide accurate answers through fact retrieval and reasoning within the knowledge base. The main process of KBQA is shown in Figure 1.

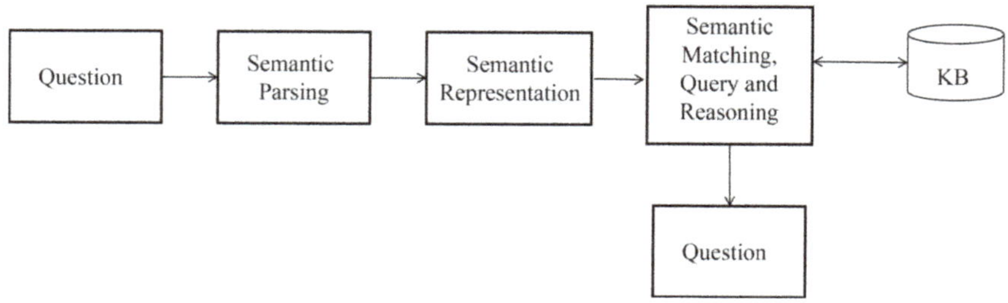

Figure 1. Main process of KBQA.

A knowledge base (KB) stores many complex structured information sets commonly represented by triples (entity, entity, and the relations between them). The task of knowledge base question answering (KBQA) is to answer the users' natural language questions using a knowledge base. For example, as shown in Figure 2, the triple starring (Jackie Chan, New Fist of Fury), release date (New Fist of Fury, 8 July 1976), and directed by (New Fist of Fury, Lo wei) can be used to answer the question "Who was the director of Jackie Chan's first starring film?".

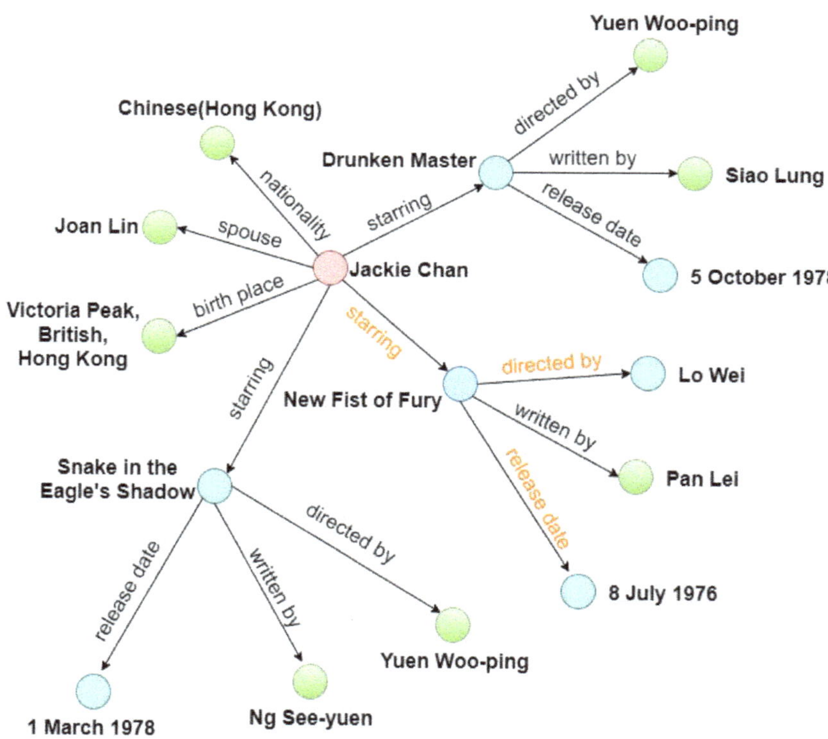

Figure 2. Regarding the triples involved in the question "Who was the director of Jackie Chan's first starring film?" in the knowledge graph; bold letters represent entities, pink circles represent topic entities, blue circles represent traversed entities, green circles represent irrelevant entities, and orange letters represent critical paths.

Previous work [1–3] on KBQA mainly focused on external resources, pattern matching, or the construction of handcrafted features [4,5] to address simple questions. These methods need labeled logical supervision. However, these methods have difficulty dealing with complex questions containing constraints, e.g., "the first" in the question "Who is the first president of the United States".

To address constraints in natural language questions, staged query graph generation methods [6–8] have been proposed. These methods first identify the single-hop relation path and then add constraints to the relation path from a query graph. The reply is obtainable by executing the query graph in the knowledge base. However, in reality, there are questions of not only single relations but also multihop relations, such as "Who is the wife of the founder of Facebook?" There are two hops between the answer and "Facebook", namely, "wife" and "founder". To answer this type of question, the longer relation path has to be considered, which will increase the search space exponentially. The beam search method was introduced by References [9,10] to reduce the search space by considering the best matching relation to reduce the number of multihop relation paths. Lan et al. [11] proposed modifying the staged query graph generation method to deal with longer relation paths and large search spaces. However, allowing a longer relationship path causes constraints to be ignored or connected with the wrong entity, resulting in errors in the prediction of the intermediate relationships. If the prediction of the intermediate relationship is wrong, the subsequent prediction will also be wrong. In query graph generation's operation, it is therefore particularly significant to analyze the relation between words.

A dependency tree can help the model capture the long-distance relationship between words. Models that use the dependency parses [12,13] have been demonstrated to be very effective in relationship extraction, since they capture long-distance semantic relations. Multihop questions generally contain constraints and multiple relations. For example, for the query "What posts did John Adams hold before he was president?", the constraint is "before", and the answer is related to "John Adams" via two hops, namely, "president" and "job". To solve this situation, the relations between words need to be focused on to reach the correct answers. We use the dependency analysis of the input question to assist the model in selecting relations. An efficient graph convolution operation [14] was used to encode the input question's dependency structure to extract the entity-centered representation.

In this paper, to focus on the relationship between words and the constraints in a question due to a long relation path, we propose a dependency structure for a question based on a graph convolution network (GCN), which encodes the dependency formation above the input query with efficient graph convolution actions to improve the attention paid to the constraints in a question and then guides the actions of the query graph generation and final ranking. This study makes three research contributions:

- For underutilization of the relationships between words in the question, we propose a question answering method on a knowledge base by applying GCNs, which permits it to efficiently pool information above arbitrary dependency formations and to produce a more effective sequence vector representation.
- For the problem of an incorrect relation selection in the process of query graph generation, we analyze the dependency structure to establish the relation between words and use the structure to obtain a more effective representation to further affect the ranking and action selection of the query graph.
- On the WebQuestionsSP (WQSP) and ComplexQuestions (CQ) datasets, our method performs well, and it is more effective in ranking query graphs.

The remainder of this paper is organized as follows. Related work about KBQA is introduced in Section 2. Section 3 describes the proposed methods in this paper. Section 4 introduces the experiments and shows the results in this paper. Section 5 concludes this paper and provides suggestions about KBQA.

2. Related Work

The current approaches that are proposed to deal with the KBQA task can be approximately classified into two categories: semantic parsing (SP) and embedding-based approaches [15,16]. These systems [17,18] are effective and provide an in-depth explanation of the query, but they need reinforcement learning or expensive data annotations. However, most SP-based approaches rely on aspects or handcrafted rules that limit their scalability and transferability.

Recently, embedding-based methods [19,20] for KBQA have become increasingly popular. Unlike SP-based methods, embedding-based approaches first allocate competitors from the KG, depicting these competitors as distributed representations, and then choose and rank these representations. Some embedding-based models directly predict solutions [21,22], while others concentrate on separating relation trails and require further procedures to obtain an answer [7,23]. Our method follows the same procedure as embedding-based models and regards query graph generation as a multistep relation path extraction process. References [9,10,24] proposed considering better relations. Lan et al. (2020) [11] proposed modifying a query graph generation process from longer relations. However, the current method is defective in its action accuracy for query graph generation. Extending the relationship path and allowing for longer relationship paths means increasing intermediate relationships, and the information in the question may be omitted. Therefore, capturing the relationship between words is particularly important in the process of forming query graphs because it affects whether the information in the question is fully utilized.

Our work also uses a dependency structure to help model the captured relations between words. A dependency tree can help the relation extraction model capture the long-distance relations between words. One common approach [12,13] is exploiting structure features on parsed tree below the lowest common ancestor (LCA).

Our method is based on the existing query graph generation process method. We add a dependency structure to the query to obtain the relation between the words and to further improve the attention paid to the constraints in a question. Compared with previous methods, we introduce the dependency structure of the question and analyze it through a graph convolution network to focus more attention on the constraints. In summary, to obtain a more effective representation, a graph convolution network is used, which allows for efficiently pooling information from an arbitrary dependency structure to achieve an effective action and to increase the accuracy of the intermediary relation selection in the query graph generation process.

3. Method

3.1. Query Graph Generation

Formally, our method followed Lan et al. (2020) [11], which is an extension of the existing staged query graph generation method. We use beam search to iteratively generate candidate query graphs. The grounded entity represents the existing entity in the knowledge base. The existential variable and lambda variable are ungrounded entities, where the lambda variable represents the answer. Finally, the aggregate function is used to perform function operations on specific entities, which usually captures some numerical features.

We assume that a set of query graphs is generated after the $k-th$ iteration, denoted as G_k. At the $k+1$ iteration, we apply, extend, connect, and aggregate (the details are shown in Figure 3) actions to grow G_k by one more edge and nodes. The extended action is used to extend the core relation path by finding the relation. The action of the connection is to find other grounded entities in the question and to connect them to the existing nodes. We denote G'_{k+1} as the resulting query graph. After each iteration, a large number of query graphs with applied actions will be generated. We use graph convolutional networks (explained in Section 3.2) to select query graphs that use the correct action, which will affect their scores.

Then, we describe how the query graph is generated. At every iteration, the actions {*extend, connect, aggregate*} will apply to query graph candidates. As shown in Figure 3, we show how the three actions act on the query graph (in fact, there is no sequence for the three actions) for the question "Who was the director of Jackie Chan's first starring film?". First, in query graph (a), starting from a grounding entity "Jackie Chan", a core relation path is found to connect entities and answers. If there are no redundant constraint words and other relations, the answer is x. However, because the question contains other relations, query graph (b) applied an extended action to extend the core relation path. The query graph (c) applies a connection action to find other grounded entities in the question and connects them to the existing nodes. The query graph (d) applies an aggregate action to add constraint nodes to the grounded entity or existential variable.

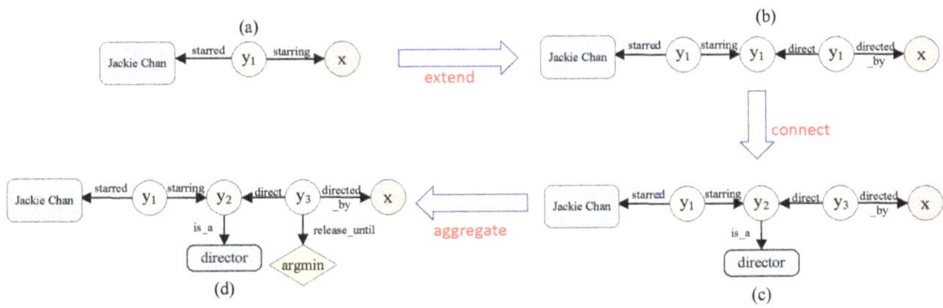

Figure 3. A possible sequence of the graph generation for "Who was director of Jackie Chan's first starring film?" Note that (**b–d**) are the results of the extend, connect and aggregate actions, respectively.

In practice, the order of each action is not fixed, so several potential query graphs will be generated. It is very important to select the correct action sequence and to determine the correct query graph. This will affect the correctness of the final result query graph because query graph candidates may contain intermediate relations and incorrect entities. Following our intuition mentioned in the first subsection, to enhance the generation of the query graph and to improve the accuracy of the intermediate relations, we employ the dependency structure of the input question.

3.2. Dependency Structure of a Question Based on a GCN

The dependency structure helps models capture the relations between words. First, we represent the input question as a dependency structure. An example is shown in Figure 4 (here, we set it as an undirected graph). We can see that the "film" is related to "starring", "first", "Chan", etc. The words of each neighboring node are related. The process is shown in Figure 5. First, we convert query graph g into a sequence of tokens g_t. We represent an input question $Q = \{q_i\}_{i=1}^{\{Q\}}$ as a sequence of word embeddings q_i. Then, we use BERT (language model) [25] to encode the concat of the question and query graph as h_q, which is the sequence of the hidden states.

GCN [26,27] is an adaptation of the convolutional neural network for encoding graphs. Given a graph with n nodes. We employ the convolution action to obtain the dependency trees. In a GCN with an l-layer, we represent the input vector of the i-th node of the l-th layer as $h_i^{(l-1)}$, and the output vector is expressed as h_i^l. In addition, a normalization operation is performed before the data are transferred into the nonlinear layer, and self-circulation is added to each node in the graph. The convolution action can be formulated as follows:

$$h_i^{(l)} = pool(\sigma(\sum_{j=1}^{n} \tilde{A}W^{(l)}h_j^{(l-1)}/d_i + b^{(l)})) \quad (1)$$

This operation is superimposed onto Layer l to obtain a deep GCN network, where we set $h_1^{(0)}, \ldots, h_n^{(0)}$ to be the input word vector obtained by BERT and $h_1^{(L)}, \ldots, h_n^{(L)}$ as the output word representations. All operations can be efficiently applied through matrix multiplication, making the method suitable for batch computing and running on a GPU. Thus far, we have obtained the question representation containing the relation between words, which is used to affect the selection of the relations in the ranking of the query graph. In addition, the representation also captures the edge information needed by the selection relation.

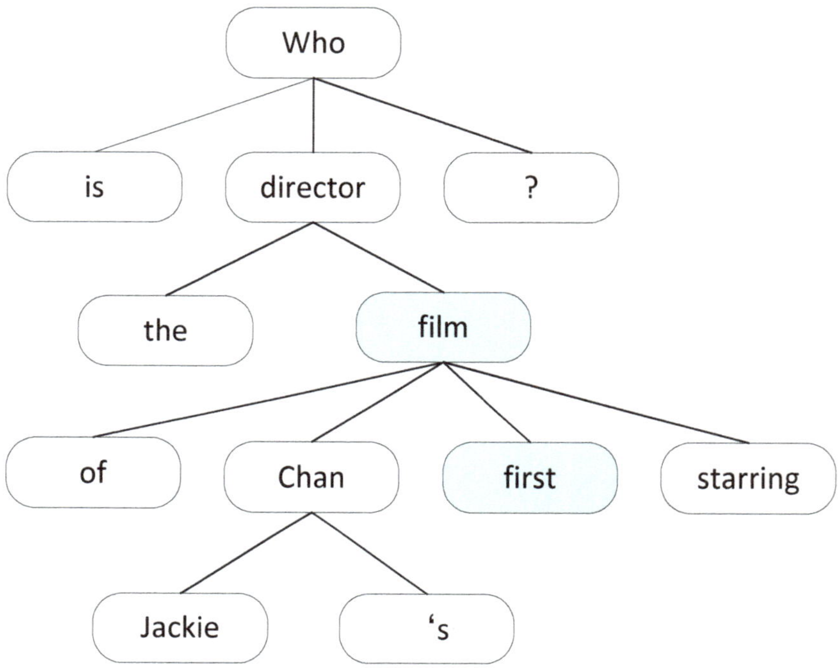

Figure 4. The dependency structure of the question "Who was the director of Jackie Chan's first starring film?" We treat the dependency graph as undirected.

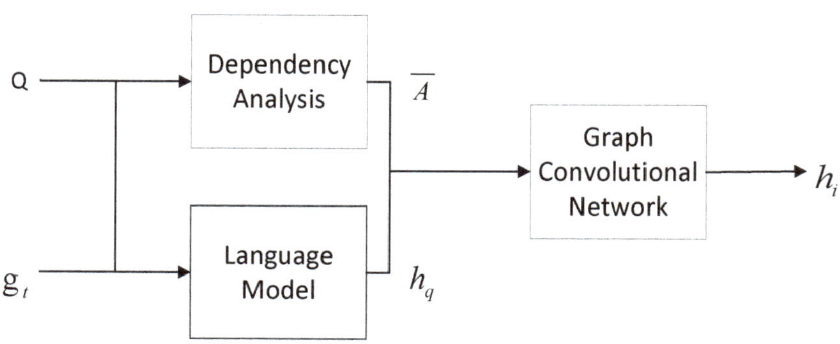

Figure 5. Overview of the dependency structure of a question based on a GCN.

3.3. Query Graph Ranking

After each query graph extension, we need to rank the candidate query graphs $g \in G_t^{'}$, which follows the sequence of operations taken by the construction g. For example, the query graph (a) in Figure 3 is expressed as (Jackie Chan, starred, starring). Before the ranking, we integrate the information extracted from the graph convolution network into the question vector:

$$v_x = h_q + h^{(l)} \tag{2}$$

$$v_q = MLP(v_x) \tag{3}$$

where h_q is the question vector, $h^{(l)}$ is the output vector from the GCN, and MLP(·) denotes an MLP layer. Then, we derive a vector v_g for each graph and put it into FFN. Finally, we calculate the probability with softmax.

3.4. Learning

Without any correct query graph, we use question–answer pairs to train our model. Inspried by Das et al. (2018) [28], we use a RL (reinforcement learning) algorithm to obtain $p_\theta(g \| v_q)$ so that the query graph can fit the problem better. θ is the learnable parameters. As our focus is not on the model's optimization approach, but on a novel graph's application based method to KBQA, the procedure of model learning and RL exploration is not described in detail.

4. Experiments

4.1. Datasets and Settings

WebQuestionsSP(WQSP) [8] WebQuestionsSP includes 5810 train samples. WQSP annotates SPARQL query statements for each answer and removes some questions with ambiguities, unclear intentions or no clear answer. WebQuestionsSP was created for the task of question answering over structured data, specifically targeting Freebase, a large knowledge base. Each sample in WebQuestionsSP is associated with a SPARQL query statement that retrieves the answer from Freebase. To ensure the quality and clarity of the dataset, certain questions that had ambiguities, unclear intentions, or no clear answer were removed during the annotation process. This helps to maintain a reliable and focused dataset for training and evaluating question answering models. The statistic of WQSP is shown as Table 1.

Table 1. The QA pair distributions of WebQuestionsSP (WQSP) ComplexQuestions (CQ) dataset.

	WQSP	CQ
Total QA pairs	4737	2100
Training set QA pairs	3098	1300
Test set QA pairs	1639	800

ComplexQuestions(CQ) [6] is used to increase the complexity of the question. On the basis of webquestions, complex questions introduce the constraint types, explicit or implicit time constraints, multi-entity constraints, and aggregate class constraints (the sum of the maximum value) and provide the logical form of a query.

We need to first discover the entities in the query and then link them to the corresponding entities in the KB. We use the existing tools to learn the linking model by training questions and their answers. For superlative linking and temporal expressions, we apply a superlative word list and regular expressions simply. We use a pretrained BERT vector to initialize the word embedding with a size of 768. We set the dropout ratio and the size of the hidden layer to 0.1 and 768, respectively.

4.2. Experimental Results and Comparison

Better results are achieved by our method on the two datasets, as shown in Table 2. The performance of our method on WQSP achieves an F1 of 74.8. Our method outperforms previous state-of-the-art methods significantly on the CQ by achieving an F1 of 44.2. It is important to note that our method is more effective when handling complicated KBs and questions. We compared our method with those of References [6–8], which have staged query graph generation methods that cannot handle the complex questions. Reference [29] focused on multihop relations; however, without the limitation of the method, the search space grows exponentially. Chen et al. (2019) [9] used a beam search to face the multihop questions, but it did not effectively handle the issue of constraints. We compared our method with that of Bhutani et al. (2019) [30], which constructs complex query patterns using a set of simple queries. We also compared our method with that of Ansari et al. (2019) [31], which generates query graphs token by token. The most important thing is to compare our method with Lan et al. (2020) [11], which allows longer relation paths to modify the graph generation method and uses the beam search to reduce the search space; however, in query graph generation, Lan et al. (2020) [11] is not optimal in the selection of the relations in each iteration because a longer relation path means more relation choices. Although effective for multihop questions, these methods sometimes ignore the constraints for the questions with constraints. Consequently, we apply GCN to effectively fuse the information on the dependency structure and to encode the dependency structure, which is helpful for relation selection. Additionally, the constraints in the question are easier to capture through an analysis of the dependency structure. Our method not only focuses on reducing the search space but also increases the relation accuracy selection in the query graph generation process, which affects the query graph ranking. Table 2 shows that our method not only works well on complex questions but also works well on the WQSP, which proves the robustness of our method.

Table 2. Results on different QA datasets.

Method	Dataset	
	WQSP (F1)	**CQ (F1)**
[8]	69.0	-
[6]	-	40.9
[7]	-	42.8
[29]	67.9	-
[9]	68.5	35.3
[30]	60.3	-
[31]	72.6	-
[11]	74.0	43.3
Our	74.8	44.2

4.3. Qualitative Analysis

The questions containing constraints are extracted from the CQ (approximately 25%) and WQSP (approximately 10%) test datasets to verify the effectiveness of our method for questions containing constraints. Table 3 shows the performance of the questions with constraints on the test dataset of CQ and WQSP. In Lan's [11] method, the relationship between words is not captured, and the constraints in some problems are omitted, leading to a lower accuracy on questions with constraints. Compared with Lan's [11] method, our method captures the relationship between words and has high sensitivity to the constraints in the question, so the accuracy of a question with constraints is higher. We also discuss the validity of the dependency structure of questions based on the GCN. By comparing the generated query graph, our method is proved to be effective.

Table 3. Performance of question with constraints on the test dataset of CQ and WQSP.

Method	CQ	WQSP
Lan et al. (2020) [11]	0.715	0.640
Our method	0.730	0.670

To summarize, our method not only affects the selection of relations in the graph generation process but also affects the ranking of the final query graph and even successfully captures some constraints that are difficult to capture. Therefore, our method is proven to be effective. Our method successfully affects the query graph generation process by convoluting the dependency structure of the question. In addition, the results show that our system performs stably and works well on not only multi-constraint questions but also on simple questions.

4.4. Error Analysis

We sampled 100 error cases randomly and obtained the following two types of errors. First, due to the query graph generation strategy, it is difficult to generate a query graph for some questions without predicate relations in the knowledge graph, which are approximately 63% of the questions. Second, the wrong query graph is generated due to the wrong entity or expression link, which is approximately 32% of the query graphs. For example, for the question "What guitar does Corey Taylor play?", there is no obvious constraint word in the question that leads to the wrong query graph.

5. Conclusions

In this paper, we proposed a graph convolution operation on a dependency structure of the question to obtain relation information between words and then integrated the relation information into the question vector to generate and rank the query graph. Our proposed methods have a dual objective of reducing the search space and improving the accuracy of relation selection during the query graph generation process. This, in turn, has a direct impact on the ranking of query graphs. Through experimentation, the results have demonstrates the effectiveness of our approach in addressing both complex questions and the WQSP dataset, thereby highlighting the robustness of our method. Notably, our method has shown a significant improvement over previous baseline methods.

Our methods also have its own weaknesses. One such weakness may be in the handling of certain types of questions or datasets that require specialized treatment or have unique characteristics. Additionally, there may be limitations in terms of scalability and efficiency when dealing with extremely large-scale datasets or in scenarios with real-time constraints. These weaknesses provide opportunities for future research and improvement. In future work, we plan to explore additional enhancements. One aspect we will focus on is pruning dependency structures to eliminate unnecessary information, which can help streamline the processing and improve efficiency. Furthermore, we aim to increase the accuracy of answer prediction, ensuring more precise and reliable responses. By continuously refining and expanding our approach, we anticipate further advancements in the field of question answering systems.

Author Contributions: Conceptualization, C.Z.; methodology, C.Z.; software, D.Z.; validation, D.Z., L.W. and C.Z.; formal analysis, N.M.; investigation, C.Z.; resources, C.Z.; writing—original draft preparation, C.Y., C.Z.; writing—review and editing, C.Z., B.W., F.X. All authors have read and agreed to the published version of the manuscript.

Funding: This work was supported in part by the National Social Science Foundation under Award 19BYY076; in part by the Key R & D project of Shandong Province 2019 JZZY010129; in part by the Shandong Natural Science Foundation under Award ZR2021MF064, Award ZR2021MF064, and Award ZR2021QG041; and in part by the Shandong Provincial Social Science Planning Project under Award 19BJCJ51, Award 18CXWJ01, and Award 18BJYJ04. This project is also supported by Major Science and Technology Demonstration Projects: Intelligent Perception Technology in

Complex Dynamic Scenes and IT Application Demonstration in Emergency Management and Social Governance, No. 2021SFGC0102).

Data Availability Statement: The data presented in this study are openly available in [6,8].

Conflicts of Interest: The authors declare no conflict of interest.

References

1. Bordes, A.; Usunier, N.; Chopra, S.; Weston, J. Large-scale simple question answering with memory networks. *arXiv* **2015**, arXiv:1506.02075. [CrossRef].
2. Cai, Q.; Alexander, Y. Large-scale semantic parsing via schema matching and lexicon extension. In Proceedings of the Annual Meeting of the Association for Computational Linguistics, Sofia, Bulgaria, 4–9 August 2013; pp. 423–433.
3. Krishnamurthy, J.; Mitchel, T.M. Weakly supervised training of semantic parsers. In Proceedings of the Conference on Empirical Methods in Natural Language Processing, Jeju Island, Republic of Korea, 12–14 July 2012; pp. 754–765.
4. Abujabal, A.; Yahya, M.; Riedewald, M.; Weikum, G. Automated template generation for question answering over knowledge graphs. In Proceedings of the 26th International Conference on World Wide Web, Perth, Australia, 3–7 April 2017; pp. 1191–1200. [CrossRef]
5. Hu, S.; Zou, L.; Yu, J.X.; Wang, H.; Zhao, D. Answering natural language questions by subgraph matching over knowledge graphs. *IEEE Trans. Knowl. Data Eng.* **2017**, *30*, 824–837. [CrossRef]
6. Bao, J.; Duan, N.; Yan, Z.; Zhou, M.; Zhao, T. Constraint-based question answering with knowledge graph. In Proceedings of the COLING, Osaka, Japan, 11–16 December 2016; pp. 2503–2514.
7. Luo, K.; Lin, F.; Luo, X.; Zhu, K.Q. Knowledge base question answering via encoding of complex query graphs. In Proceedings of the Conference on Empirical Methods in Natural Language Processing, Brussels, Belgium, 31 October–4 November 2018; pp. 2185–2194. [CrossRef]
8. Yih, W.-T.; Chang, M.-W.; He, X.; Gao, J. Semantic Parsing via Staged Query Graph Generation: Question Answering with Knowledge Base. In Proceedings of the Annual Meeting of the Association for Computational Linguistics, Beijing, China, 26–31 July 2015.
9. Chen, Z.-Y.; Chang, C.-H.; Chen, Y.-P.; Nayak, J.; Ku, L.-W. UHop: An unrestricted-hop relation extraction framework for knowledge-based question answering. *arXiv* **2019**, arXiv:1904.01246. [CrossRef].
10. Lan, Y.; Wang, S.; Jiang, J. Multi-hop knowledge base question answering with an iterative sequence matching model. In Proceedings of the IEEE International Conference on Data Mining (ICDM), Beijing, China, 8–11 November 2019; pp. 359–368. [CrossRef]
11. Lan, Y.; Jiang, J. Query graph generation for answering multi-hop complex questions from knowledge bases. In Proceedings of the Annual Meeting of the Association for Computational Linguistics, Online, 5–10 July 2020. [CrossRef]
12. Miwa, M.; Bansal, M. End-to-End Relation Extraction using LSTMs on Sequences and Tree Structures. In Proceedings of the Annual Meeting of the Association for Computational Linguistics, Berlin, Germany, 7–12 August 2016. [CrossRef]
13. Xu, K.; Feng, Y.; Huang, S.; Zhao, D. Semantic Relation Classification via Convolutional Neural Networks with Simple Negative Sampling. *Comput. Sci.* **2015**, *71*, 941–949. [CrossRef]
14. Youcef, D.; Gautam, S.; Wei, L.J.C. Fast and accurate convolution neural network for detecting manufacturing data. *IEEE Trans. Ind. Inform.* **2020**, *17*, 2947–2955. [CrossRef]
15. Peng, H.; Chang, M.; Yih, W.T. Maximum margin reward networks for learning from explicit and implicit supervision. In Proceedings of the Conference on Empirical Methods in Natural Language Processing, Copenhagen, Denmark, 7–11 September 2017; pp. 2368–2378. [CrossRef]
16. Sorokin, D.; Gurevych, I. Modeling semantics with gated graph neural networks for knowledge base question answering. In Proceedings of the 27th International Conference on Computational Linguistics, Association for Computational Linguistics, Santa Fe, NM, USA, 20–26 August 2018; pp. 3306–3317. [CrossRef]
17. Iyyer, M.; Yih, W.-T.; Chang, M.-W. Search-based neural structured learning for sequential question answering. In Proceedings of the Annual Meeting of the Association for Computational Linguistics, Vancouver, BC, Canada, 30 July–4 August 2017; pp. 1821–1831. [CrossRef]
18. Krishnamurthy, J.; Dasigi, P.; Gardner, M. Neural semantic parsing with type constraints for semi-structured tables. In Proceedings of the Conference on Empirical Methods in Natural Language Processing, Copenhagen, Denmark, 7–11 September 2017; pp. 1516–1526. [CrossRef]
19. Moqurrab, S.A.; Ayub, U.; Anjum, A.; Asghar, S.; Srivastava, G. An accurate deep learning model for clinical entity recognition from clinical notes. *IEEE J. Biomed. Health Inform.* **2021**, *25*, 3804–3811. [CrossRef] [PubMed]
20. Wang, F.; Wu, W.; Li, Z.; Zhou, M. Named entity disambiguation for questions in community question answering. *Knowl.-Based Syst.* **2017**, *126*, 68–77. [CrossRef]
21. Bast, H.; Haussmann, E. More accurate question answering on freebase. In Proceedings of the 24th ACM International Conference on Information and Knowledge Management, Melbourne, Australia, 18–23 October 2015; pp. 1431–1440. [CrossRef]
22. Chakraborty, N.; Lukovnikov, D.; Maheshwari, G.; Trivedi, P.; Lehmann, J.; Fischer, A. Introduction to neural network based approaches for question answering over knowledge graphs. *arXiv* **2019**, arXiv:1907.09361. [CrossRef].

23. Chen, H.-C.; Chen, Z.-Y.; Huang, S.-Y.; Ku, L.-W.; Chiu, Y.-S.; Yang, W.-J. Relation extraction in knowledge base question answering: From general-domain to the catering industry. In Proceedings of the International Conference on HCI in Business, Government, and Organizations, Las Vegas, NV, USA, 15 July 2018; pp. 26–41. [CrossRef]
24. Yang, Z.; Garg, H.; Li, J.; Srivastava, G.; Cao, Z. Investigation of multiple heterogeneous relationships using a q-rung orthopair fuzzy multi-criteria decision algorithm. *Neural Comput. Appl.* **2021**, 33, 10771–10786. [CrossRef]
25. Devlin, J.; Chang, M.W.; Lee, K.; Toutanova, K. Bert: Pre-training of deep bidirectional transformers for language understanding. *arXiv* **2019**, arXiv:1810.04805. [CrossRef].
26. Kipf, T.N.; Welling, M. Semi-supervised classification with graph convolutional networks. *arXiv* **2016**, arXiv:1609.02907. [CrossRef].
27. Marcheggiani, D.; Ivan, T. Encoding Sentences with Graph Convolutional Networks for Semantic Role Labeling. In Proceedings of the 2017 Conference on Empirical Methods in Natural Language Processing, Copenhagen, Denmark, 7–11 September 2017; pp. 1506–1515. [CrossRef]
28. Das, R.; Dhuliawala, S.; Zaheer, M.; Vilnis, L.; Durugkar, I.; Krishnamurthy, A.; Smola, A.; McCallum, A. Go for a walk and arrive at the answer: Reasoning over paths in knowledge bases using reinforcement learning. *arXiv* **2018**, arXiv:1711.05851.
29. Lan, Y.; Wang, S.; Jiang, J. Knowledge base question answering with topic units. In Proceedings of the International Joint Conference on Artificial Intelligence, Macao, China, 10–16 August 2019; pp. 5046–5052. [CrossRef]
30. Bhutani, N.; Suhara, Y.; Tan, W.-C.; Halevy, A.Y.; Jagadis, H.V. Open Information Extraction from Question-Answer Pairs. In Proceedings of the 17th Annual Conference of the North American Chapter of the Association for Computational Linguistics: Human Language Technologies (NAACL-HLT 2019), Minneapolis, MN, USA, 3–5 June 2019; pp. 2294–2305.
31. Ahmed, G.A.; Saha, A.; Kumar, V.; Bhambhani, M.; Sankaranarayanan, K.; Chakrabarti, S. Neural Program Induction for KBQA Without Gold Programs or Query Annotations. In Proceedings of the International Joint Conference on Artificial Intelligence, Macao, China, 10–16 August 2019; pp. 4890–4896. [CrossRef]

electronics

MDPI

Article

A Collaborative Multi-Granularity Architecture for Multi-Source IoT Sensor Data in Air Quality Evaluations

Wantong Li, Chao Zhang *, Yifan Cui and Jiale Shi

School of Computer and Information Technology, Shanxi University, Taiyuan 030006, China; wantongli2023@163.com (W.L.); cuiyifan0808@163.com (Y.C.); shijiale868@163.com (J.S.)
* Correspondence: czhang@sxu.edu.cn

Abstract: Air pollution (AP) is a significant environmental issue that poses a potential threat to human health. Its adverse effects on human health are diverse, ranging from sensory discomfort to acute physiological reactions. As such, air quality evaluation (AQE) serves as a crucial process that involves the collection of samples from the environment and their analysis to measure AP levels. With the proliferation of Internet of Things (IoT) devices and sensors, real-time and continuous measurement of air pollutants in urban environments has become possible. However, the data obtained from multiple sources of IoT sensors can be uncertain and inaccurate, posing challenges in effectively utilizing and fusing this data. Meanwhile, differences in opinions among decision-makers regarding AQE can affect the outcome of the final decision. To tackle these challenges, this paper systematically investigates a novel multi-attribute group decision-making (MAGDM) approach based on hesitant trapezoidal fuzzy (HTrF) information and discusses its application to AQE. First, by combining HTrF sets (HTrFSs) with multi-granulation rough sets (MGRSs), a new rough set model, named HTrF MGRSs, on a two-universe model is proposed. Second, the definition and property of the presented model are studied. Third, a decision-making approach based on the background of AQE is constructed via utilizing decision-making index sets (DMISs). Lastly, the validity and feasibility of the constructed approach are demonstrated via a case study conducted in the AQE setting using experimental and comparative analyses. The outcomes of the experiment demonstrate that the presented architecture owns the ability to handle multi-source IoT sensor data (MSIoTSD), providing a sensible conclusion for AQE. In summary, the MAGDM method presented in this article is a promising scheme for solving decision-making problems, where HTrFSs possess excellent information description capabilities and can adequately describe indecision and uncertainty information. Meanwhile, MGRSs serve as an outstanding information fusion tool that can improve the quality and level of decision-making. DMISs are better able to analyze and evaluate information and reduce the impact of disagreement on decision outcomes. The proposed architecture, therefore, provides a viable solution for MSIoTSD facing uncertainty or hesitancy in the AQE environment.

Keywords: granular computing; multi-granulation rough set; hesitant trapezoidal fuzzy set; air quality evaluation

Citation: Li, W.; Zhang, C.; Cui, Y.; Shi, J. A Collaborative Multi-Granularity Architecture for Multi-Source IoT Sensor Data in Air Quality Evaluations. *Electronics* **2023**, *12*, 2380. https://doi.org/10.3390/electronics12112380

Academic Editor: Franco Cicirelli

Received: 17 April 2023
Revised: 13 May 2023
Accepted: 23 May 2023
Published: 24 May 2023

1. Introduction

AP [1,2] is a matter of paramount concern to both the environment and public health, brought about by the contamination of air by chemical, physical, or biological agents. This deleterious phenomenon is known to have far-reaching implications in the agricultural industry [3], as it has been demonstrated to cause acid rain, reduced crop production, and inferior soil fertility. Notably, AP is a leading contributor to the global climate change crisis, resulting in more severe weather patterns across the globe [4]. Recent studies have provided compelling evidence to suggest that exposure to AP is linked to several negative health outcomes, including developmental delays in children [5], increased risk of mental illnesses such as depression [6], and poor reproductive health in females [7]. In this light,

AP is currently one of the most significant risk factors affecting global health. According to a survey by the European Environment Agency (EEA) in 2020, 96% of city residents in the European Union were exposed to higher than recommended levels of fine particulate matter, according to the World Health Organization (WHO) [8], resulting in 238,000 premature deaths. Furthermore, WHO has conducted extensive research on the effects of AP and found that environmental and household AP cause approximately 6.7 million deaths per year, with 2.4 billion people exposed to hazardous levels of household AP.

AQE is an indispensable tool for comprehending the state of air quality and forecasting its future trends. The objective of AQE is to mitigate the deleterious impact of AP and foster a healthy atmospheric environment, and, thus, it has become a prominent research topic in recent years [9,10]. Numerous scholars have conducted research in various directions via using different approaches in the context of AQE. For instance, Oprea [11] utilized an expert system to carry out research on knowledge modeling for better analysis of AP in city regions. Wang et al. [12] proposed a deep convolutional neural network method for predicting AP. Gu et al. [13] suggested a new fuzzy multiple linear regression model to forecast the air quality index [14–16]. However, AQE, using neural networks, necessitates an adequate amount of training examples to ensure sufficient training of the model, and AQE based on expert systems requires frequent manual maintenance and manipulation of the AQE knowledge base, which hinder the accuracy guarantee. With the widespread adoption of IoT technologies [17–20], sensor networks have become increasingly popular for collecting air quality data from multiple sources. IoT sensors [21] are capable of collecting and transmitting data in real time, providing a dynamic understanding of AP patterns by continuous and high-resolution measurements of air quality parameters. The use of MSIoTSD allows for a more comprehensive and accurate assessment of air quality. However, the accuracy and reliability of MSIoTSD [22] can be affected by various factors, leading to uncertain data. Moreover, effectively utilizing and fusing MSIoTSD presents a challenge. In contrast, AQE, using fuzzy methods, not only overcomes the limitations of the aforementioned approaches, but can also effectively deal with multi-source uncertain data. Furthermore, AQE is influenced by several factors, including different locations, attributes, and times, which can be established as a typical MAGDM problem.

This paper primarily examines and resolves the AQE issue from three perspectives. First, we investigate a fuzzy approach applied to AQE in the context of HTrFSs during the information description process. Second, we use MGRSs to fuse multiple sources of AQE data during the information fusion stage process. Finally, we employ DMISs to diminish the impact of inconsistent opinions of individual decision-makers within a decision group on the decision outcome during the information analysis process. Based on the analysis above, we recall the components of HTrFSs and MGRSs below.

1.1. A Brief Review of HTrFSs

Zadeh [23] proposed the theory of fuzzy sets to describe various fuzzy concepts of reality in 1965. However, classical fuzzy sets have their own limitations when dealing with multiple sources of uncertain information. Therefore, scholars have explored generalized fuzzy sets in depth [24–27]. Hesitant fuzzy sets (HFSs), which represent generalized fuzzy sets describing hesitant information, were proposed by Torra [28]. HFSs permit the membership degree of elements belonging to a set to consist of multiple possible values. Since the creation of HFSs, many experts have researched HFSs from various perspectives, and a series of achievements have been obtained. For instance, Divsalar et al. [29] presented a novel TODIM approach using the Choquet integral in a probabilistic hesitant fuzzy environment. Krishankumar et al. [30] proposed a novel decision framework with completely unknown weight information in the context of interval-valued probabilistic HFSs. Ahmad et al. [31] proposed an innovative resolution of multi-objective optimization issues by applying hesitant fuzzy aggregation operators. Fuzzy data serve as a type of data that are imprecise or have uncertain sources. Trapezoidal fuzzy numbers (TrFNs) have more advantages in describing fuzzy data compared with simple real-valued numbers, as they contain a specific interval with a full membership rank. Therefore, Ye [32] introduced the concept of HTrFSs, which takes advantage of the unique benefits of TrFNs and HFSs. The distinctive advantages of HTrFSs in dealing with uncertain information have prompted scholars to conduct a substantial number of theoretical and practical explorations [33,34].

1.2. A Brief Review of MGRSs

Granular computing (GrC) has emerged as a structured solution model for addressing large-scale complicated problems by simulating human thinking and replacing exact solutions with feasible satisfactory approximations that meet the needs of actual problems. As a novel concept and computing paradigm in artificial intelligence, GrC has revolutionized the traditional understanding of computing and holds great value in tackling complex problems [35,36]. Zadeh introduced the concept of fuzzy information granulation in 1979 [37], and, after years of research, officially presented the concept of GrC in 1997 [38]. With continuous research and development efforts, GrC has been widely utilized and refined [39–41].

The rough set theory is a prominent GrC model that was first proposed by Pawlak [42] in 1982. However, due to the overly stringent requirements for equivalence relations in classical rough set models, scholars have extended rough sets from different aspects [43–45]. In terms of relations, Qian et al. introduced optimistic and pessimistic styles of MGRSs [46] to describe issues through multiple binary relations, thereby enhancing the ability of multi-source information systems to handle uncertain information. In recent years, many generalized MGRSs have been developed to cater to the diverse needs of users [47–50]. In terms of universes, decision-makers can express decision information more accurately on two universes [51] than on a single universe. Sun and Ma [52] further proposed the theory of MGRSs on two-universe. The MGRSs on two-universe can effectively describe complex real-world information. For instance, in AQE, it is essential to consider the relationship between locations and air pollutants. This relationship consists of two distinct types of objects that belong to different universes. Furthermore, MGRSs on two-universe offer both optimistic and pessimistic information fusion strategies, which are valuable for risk-seeking and risk-averse decision-making, respectively. This allows the integration of diverse opinions from various experts, leading to a consensus through the integration of multiple binary relations. Thus, MGRSs on two-universe serve as an excellent information fusion strategy. Therefore, the research on the MGRSs theory of two universes and its application has made significant progress [53–55]. Moreover, the probabilistic soft logic has also shown remarkable performance in handling uncertain information and integrating multiple sources of information. In recent years, researchers have applied this approach to a wide range of fields. For instance, Gu et al. [56] proposed a novel approach for extracting temporal information about complex medicine by integrating the probabilistic soft logic and textual feature feedback. Alshukaili et al. [57] presented a technique for structuring linked data search results by leveraging the probabilistic soft logic. Fakhraei et al. [58] developed a network-based method for predicting drug-targeted interactions using the probabilistic soft logic.

1.3. Study Motivations

This paper explores an MAGDM approach based on HTrF MGRSs and its applications in AQE. In the following, we introduce some of the main study motivations.

1. As AQE plays a crucial role in measuring air quality to reduce air pollution, there is a pressing need to explore further methods in AQE. Consequently, we intend to propose a new collaborative multi-granularity architecture to AQE.
2. HTrFSs demonstrate superior capabilities in handling hesitant and uncertain data, while MGRSs exhibit excellent performance in multi-source information fusion. Thus, we intend to synergistically combine HTrFSs and MGRSs to present a novel model.

3. Considering that the opinions of different experts within a decision-making group may differ significantly, it is imperative to utilize DMISs to mitigate the impact of disagreement on the outcome of decisions.

1.4. Contributions of This Article

By combining the above research motivations, this article presents the following innovative ideas.

1. An HTrF MGRS two-universe model is proposed, and some properties and definitions are discussed.
2. A novel MAGDM method is constructed by utilizing HTrF, MGRSs, and DMISs, and applying them to the AQE.

This paper is structured as below. In Section 2, we review the fundamental concept of HTrFSs and MGRSs on two-universe. In Section 3, we develop the concept of HTrF MGRSs and introduce the related properties. In Section 4, we present an MAGDM method based on HTrF MGRSs. Then, we give an application of the presented approach to AQE and analyze it in comparison with other approaches in Section 5. In Section 6, we summarize the article and discuss several options for research in the future.

Furthermore, we add a table to this paper to provide readers with easy-to-find explanations of any abbreviations used in the text, as shown in Abbreviations section.

2. Basic Knowledge

For a better understanding, this section introduces the fundamental concepts of HTrFSs and MGRSs.

2.1. HTrFSs

HTrFSs have shown flexibility in handling hesitant, inaccurate information. Before introducing the notion of HTrFSs, we first present the TrFN.

Definition 1 ([59]). *A fuzzy number $\tilde{a} = (a, b, c, d)$ is called a TrFN when its membership function is denoted as:*

$$\mu_{\tilde{a}}(x) = \begin{cases} 0, & (x < a \text{ or } x > d) \\ (x-a)/(b-a), & (a \le x < b) \\ 1, & (b \le x \le c) \\ (x-d)/(c-d), & (c < x \le d) \end{cases}, \tag{1}$$

where $0 \le a \le b \le c \le d \le 1$, a and the closed interval $[b, c]$ and d stand for the lower, mode, upper limits of $\tilde{a}$, respectively.

Afterward, we review the fundamental understanding of HTrFSs.

Definition 2 ([32]). *Suppose U is a universe. An HTrFS on U is expressed as:*

$$E = \{\langle x, h_E(x)\rangle | x \in U\}, \tag{2}$$

where $h_E(x) : U \to Trap[0,1]$ represents the possible degree of memberships of x in E, and $Trap[0,1]$ is the set containing all trapezoidal values in $[0,1]$. Moreover, $h_E(x)$ is named as an HTrF element, and the set of all HTrFSs on U is expressed as $HTrF(U)$.

As the rules for the operations of HTrFSs support decision-making processes to efficiently analyze data, we present the laws for the operations of HTrFSs below.

Definition 3 ([32]). *Suppose U is a universe. $\forall E_1, E_2 \in HTrF(U)$, then:*

1. *The complement of* E_1*, expressed as* E_1^c*, is given by* $\forall x \in U$. $h_{E_1^c}(x) = \sim h_{E_1}(x) = \left\{(1-a_{E_1}^f, 1-b_{E_1}^f, 1-c_{E_1}^f, 1-d_{E_1}^f) \mid f = 1,2,\ldots l\right\}$.
2. *The intersection of* E_1 *and* E_2*, expressed as* $E_1 \cap E_2$*, is given by* $\forall x \in U$, $h_{E_1 \cap E_2}(x) = h_{E_1}(x) \wedge h_{E_2}(x) = \left\{(a_{E_1}^f \wedge a_{E_2}^f, b_{E_1}^f \wedge b_{E_2}^f, c_{E_1}^f \wedge c_{E_2}^f, d_{E_1}^f \wedge d_{E_2}^f) \mid f = 1,2,\ldots l\right\}$.
3. *The union of* E_1 *and* E_2*, expressed as* $E_1 \cup E_2$*, is given by* $\forall x \in U$, $h_{E_1 \cup E_2}(x) = h_{E_1}(x) \vee h_{E_2}(x) = \left\{(a_{E_1}^f \vee a_{E_2}^f, b_{E_1}^f \vee b_{E_2}^f, c_{E_1}^f \vee c_{E_2}^f, d_{E_1}^f \vee d_{E_2}^f) \mid f = 1,2,\ldots l\right\}$.

The utilization of score functions represents a pivotal approach for the selection of the optimal alternative in HTrF MAGDM problems; hence, we discuss the following notion of HTrF score functions.

Definition 4 ([32]). *For an HTrF element,* $h_E(x)$, $S(h_E(x)) = \frac{1}{4\#(h_E(x))}\sum_{\tilde{a} \in h_E(x)} \tilde{a}$, *where* $\#(h_E(x))$ *is the number of TrFNs in* $h_E(x)$*. For two HTrF elements,* $h_E(x)$ *and* $h_F(x)$*, if* $S(h_E(x)) \geq S(h_F(x))$*, then* $h_E(x) \geq h_F(x)$.

To compare two HTrFSs, it is necessary to propose a new definition; hence, we present the notion of HTrF subsets below.

Definition 5. *Suppose U is a universe.* $\forall E, F \in HTrF(U)$*, if* $h_E(x) \preceq h_F(x)$ *is true for every* $x \in U$*, such that* $h_E(x) \preceq h_F(x) \Leftrightarrow a_E^f \leq a_F^f, b_E^f \leq b_F^f, c_E^f \leq c_F^f, d_E^f \leq d_F^f$ $f = 1, 2, \ldots l$*, then E is mentioned as an HTrF subset of F, expressed as* $E \subseteq F$*. It is evident that* $\subseteq$ *is anti-symmetric, reflexive, and transitive on* $HTrF(U)$.

2.2. MGRSs on Two-Universe

MGRSs on two-universe have outstanding performance in multi-source information fusion, and we recall the fundamental definitions of MGRSs on two-universe below.

Definition 6 ([51]). *Suppose* U, V *are two universes, and R is a binary compatibility relation family over* $U \times V$*, in respect to a family of binary mapping* $F_k : U \to 2^V$ $u \mapsto \{v \in V \mid (u,v) \in R_k\}$, $R_k \in R$, $k = 1,2,\ldots,n$*. Then, the MG approximation space on two-universe is expressed as* (U, V, R).

Definition 7 ([52]). *Suppose* F_1 *and* F_2 *are two binary mappings over* $U \times V$. $\forall Y \subseteq V$*, the pessimistic and optimistic lower and upper MG approximations in the matter of* (U, V, R) *are expressed as:*

$$\underline{apr}^P_{F_1+F_2}(Y) = \{x \in U \mid F_1(x) \subseteq Y \wedge F_2(x) \subseteq Y\}; \tag{3}$$

$$\overline{apr}^P_{F_1+F_2}(Y) = \underline{apr}^P_{F_1+F_2}(Y^c)^c; \tag{4}$$

$$\underline{apr}^O_{F_1+F_2}(Y) = \{x \in U \mid F_1(x) \subseteq Y \vee F_2(x) \subseteq Y\}; \tag{5}$$

$$\overline{apr}^O_{F_1+F_2}(Y) = \underline{apr}^O_{F_1+F_2}(Y^c)^c, \tag{6}$$

The pair $(\underline{apr}^P_{F_1+F_2}(Y), \overline{apr}^P_{F_1+F_2}(Y))$ and $(\underline{apr}^O_{F_1+F_2}(Y), \overline{apr}^O_{F_1+F_2}(Y))$ are referred to as a pessimistic MGRS on two-universe and an optimistic MGRS on two-universe, respectively.

This section offers a comprehensive overview of the theoretical foundations underlying TrFNs, HTrFSs, HTrF score functions and subsets, as well as MGRSs on two-universe. First, we present the definition of TrFNs and proceed to discuss the fundamental definition and operation rules of HTrFSs. Subsequently, we introduce the definition of HTrF score functions and HTrF subsets. Moreover, we provide a brief introduction to the basic concepts of optimistic and pessimistic MGRSs on two-universe.

3. HTrF MGRSs on Two-Universe

This section systematically discusses the notion of HTrF MGRSs on two-universe. First, a definition of HTrF relations (HTrFRs) on two-universe is given.

Definition 8. *An HTrFR R on $U \times V$ is given by:*

$$R = \{\langle (x,y), h_R(x,y)\rangle | (x,y) \in U \times V\}, \tag{7}$$

where $h_R(x,y) : U \times V \to Trap[0,1]$ represents the possible degree of memberships of $(x,y) \in U \times V$. For convenience, the set of all HTrFRs on $U \times V$is expressed as $HTrFR(U \times V)$.

Then, we extend the HTrFRs on two-universe to the context of MGRSs.

3.1. Optimistic HTrF MGRSs on Two-Universe

Definition 9. *Suppose U, V are two universes, $R_k \in HTrFR(U \times V)(k = 1,2,\ldots,n)$ is an HTrFR over $U \times V$. Moreover, an HTrF MG approximation space on two-universe is expressed as (U, V, R_k). $\forall E \in HTrF(V)$, the definitions of optimistic HTrF MG lower and upper approximations on two-universe of E are given below:*

$$\underline{\sum_{k=1}^{n} R_k}^{O}(E) = \left\{ \left\langle x, h_{\underline{\sum_{k=1}^{n} R_k}^{O}(E)}(x) \right\rangle | x \in U \right\}, \tag{8}$$

$$\overline{\sum_{k=1}^{n} R_k}^{O}(E) = \left\{ \left\langle x, h_{\overline{\sum_{k=1}^{n} R_k}^{O}(E)}(x) \right\rangle | x \in U \right\}, \tag{9}$$

where $h_{\underline{\sum_{k=1}^{n} R_k}^{O}(E)}(x) = \vee_{k=1}^{n} \wedge_{y \in V} \{h_{R_k^c}(x,y) \vee h_E(y)\}$; $h_{\overline{\sum_{k=1}^{n} R_k}^{O}(E)}(x) = \wedge_{k=1}^{n} \vee_{y \in V} \{h_{R_k}(x,y) \wedge h_E(y)\}$.

The pair $\left(\underline{\sum_{k=1}^{n} R_k}^{O}(E), \overline{\sum_{k=1}^{n} R_k}^{O}(E) \right)$ indicates an optimistic HTrF MGRS on two-universe of E in the matter of (U, V, R_k).

Theorem 1. *Suppose U, V are two universes, $R_k \in HTrFR(U \times V)(k = 1,2, \ldots, n)$ is an HTrFR over $U \times V$. $\forall E, F \in HTrF(V)$, the optimistic HTrF MG lower and upper approximations on two-universe meet these properties:*

(1) $\underline{\sum_{k=1}^{n} R_k}^{O}(E^c) = \left(\overline{\sum_{k=1}^{n} R_k}^{O}(E)\right)^c$, $\overline{\sum_{k=1}^{n} R_k}^{O}(E^c) = \left(\underline{\sum_{k=1}^{n} R_k}^{O}(E)\right)^c$;

(2) $E \subseteq F \Rightarrow \underline{\sum_{k=1}^{n} R_k}^{O}(E) \subseteq \underline{\sum_{k=1}^{n} R_k}^{O}(F)$, $E \subseteq F \Rightarrow \overline{\sum_{k=1}^{n} R_k}^{O}(E) \subseteq \overline{\sum_{k=1}^{n} R_k}^{O}(F)$;

(3) $\underline{\sum_{k=1}^{n} R_k}^{O}(E \cap F) = \underline{\sum_{k=1}^{n} R_k}^{O}(E) \cap \underline{\sum_{k=1}^{n} R_k}^{O}(F)$, $\overline{\sum_{k=1}^{n} R_k}^{O}(E \cup F) = \overline{\sum_{k=1}^{n} R_k}^{O}(E) \cup \overline{\sum_{k=1}^{n} R_k}^{O}(F)$;

(4) $\underline{\sum_{k=1}^{n} R_k}^{O}(E \cup F) \supseteq \underline{\sum_{k=1}^{n} R_k}^{O}(E) \cup \underline{\sum_{k=1}^{n} R_k}^{O}(F)$, $\overline{\sum_{k=1}^{n} R_k}^{O}(E \cap F) \subseteq \overline{\sum_{k=1}^{n} R_k}^{O}(E) \cap \overline{\sum_{k=1}^{n} R_k}^{O}(F)$.

Proof.

(1) $\forall x \in U$, we have $\underline{\sum_{k=1}^{n} R_k}^{O}(E^c) = \{\langle x, \vee_{k=1}^{n} \wedge_{y\in V} \{h_{R_k^c}(x,y) \vee h_{E^c}(y)\}\rangle | x \in U\}$ $= \{\langle x, \vee_{k=1}^{n} \wedge_{y\in V} \{(\sim h_{R_k}(x,y)) \vee (\sim h_E(y))\}\rangle | x \in U\} = \{\langle x, \vee_{k=1}^{n} \wedge_{y\in V} \{\sim (h_{R_k}(x,y) \wedge h_E(y))\}\rangle | x \in U\} = \{\langle x, \sim (\wedge_{k=1}^{n} \vee_{y\in V} \{(h_{R_k}(x,y) \wedge h_E(y))\})\rangle | x \in U\} = (\overline{\sum_{k=1}^{n} R_k}^{O}(E))^c$.

$\overline{\sum_{k=1}^{n} R_k}^{O}(E^c) = (\underline{\sum_{k=1}^{n} R_k}^{O}(E))^c$ is similarly obtained.

(2) Because of $E \subseteq F$, depending on Definition 5, we have $h_E(y) \preceq h_F(y) \Leftrightarrow a_E^f \le a_F^f, b_E^f \le b_F^f, c_E^f \le c_F^f, d_E^f \le d_F^f$, so $\vee_{k=1}^{n} \wedge_{y\in V} \left\{a_{R_k^c}^f \vee a_E^f\right\} \le \vee_{k=1}^{n} \wedge_{y\in V} \left\{a_{R_k^c}^f \vee a_F^f\right\}$ and $\vee_{k=1}^{n} \wedge_{y\in V} \left\{b_{R_k^c}^f \vee b_E^f\right\} \le \vee_{k=1}^{n} \wedge_{y\in V} \left\{b_{R_k^c}^f \vee b_F^f\right\}$ and $\vee_{k=1}^{n} \wedge_{y\in V} \left\{c_{R_k^c}^f \vee c_E^f\right\} \le \vee_{k=1}^{n} \wedge_{y\in V} \left\{c_{R_k^c}^f \vee c_F^f\right\}$ and $\vee_{k=1}^{n} \wedge_{y\in V} \left\{d_{R_k^c}^f \vee d_E^f\right\} \le \vee_{k=1}^{n} \wedge_{y\in V} \left\{d_{R_k^c}^f \vee d_F^f\right\}$. Therefore, we have $E \subseteq F \Rightarrow \underline{\sum_{k=1}^{n} R_k}^{O}(E) \subseteq \underline{\sum_{k=1}^{n} R_k}^{O}(F)$.

$E \subseteq F \Rightarrow \overline{\sum_{k=1}^{n} R_k}^{O}(E) \subseteq \overline{\sum_{k=1}^{n} R_k}^{O}(F)$ is similarly obtained.

(3) $\forall x \in U$, we have $\underline{\sum_{k=1}^{n} R_k}^{O}(E \cap F) = \{\langle x, \vee_{k=1}^{n} \wedge_{y\in V} \{h_{R_k^c}(x,y) \vee h_{E\cap F}(y)\}\rangle | x \in U\} = \{\langle x, \vee_{k=1}^{n} \wedge_{y\in V} \{h_{R_k^c}(x,y) \vee (h_E(y) \wedge h_F(y))\}\rangle | x \in U\} = \{\langle x, (\vee_{k=1}^{n} \wedge_{y\in V} \{a_{R_k^c}^f \vee (a_E^f \wedge a_F^f)\}, \vee_{k=1}^{n} \wedge_{y\in V} \{b_{R_k^c}^f \vee (b_E^f \wedge b_F^f)\}, \vee_{k=1}^{n} \wedge_{y\in V} \{c_{R_k^c}^f \vee (c_E^f \wedge c_F^f)\}, \vee_{k=1}^{n} \wedge_{y\in V} \{d_{R_k^c}^f \vee (d_E^f \wedge d_F^f)\})\rangle | x \in U\} = \{\langle x, (\vee_{k=1}^{n} \wedge_{y\in V} \{(a_{R_k^c}^f \vee a_E^f) \wedge (a_{R_k^c}^f \vee a_F^f)\}, \vee_{k=1}^{n} \wedge_{y\in V} \{(b_{R_k^c}^f \vee b_E^f) \wedge (b_{R_k^c}^f \vee b_F^f)\}, \vee_{k=1}^{n} \wedge_{y\in V} \{(c_{R_k^c}^f \vee c_E^f) \wedge (c_{R_k^c}^f \vee c_F^f)\}, \vee_{k=1}^{n} \wedge_{y\in V} \{(d_{R_k^c}^f \vee d_E^f) \wedge (d_{R_k^c}^f \vee d_F^f)\})\rangle | x \in U\} = \{\langle x, h_{\underline{\sum_{k=1}^{n} R_k}^{O}(E)}(x)\rangle | x \in U\} \wedge \{\langle x, h_{\underline{\sum_{k=1}^{n} R_k}^{O}(F)}(x)\rangle | x \in U\} = \underline{\sum_{k=1}^{n} R_k}^{O}(E) \cap \underline{\sum_{k=1}^{n} R_k}^{O}(F)$ similarly, $\overline{\sum_{k=1}^{n} R_k}^{O}(E \cup F) = \overline{\sum_{k=1}^{n} R_k}^{O}(E) \cup \overline{\sum_{k=1}^{n} R_k}^{O}(F)$ is obtained.

(4) Based on the above findings, it is easily obtained that $\underline{\sum_{k=1}^{n} R_k}^{O}(E \cup F) \supseteq \underline{\sum_{k=1}^{n} R_k}^{O}(E) \cup \underline{\sum_{k=1}^{n} R_k}^{O}(F)$ and $\overline{\sum_{k=1}^{n} R_k}^{O}(E \cap F) \subseteq \overline{\sum_{k=1}^{n} R_k}^{O}(E) \cap \overline{\sum_{k=1}^{n} R_k}^{O}(F)$.

□

In Theorem 1, (1) states the complement of optimistic HTrF MGRSs on two-universe; (2) states the monotone of optimistic HTrF MGRSs on two-universe in the matter of various HTrF targets; (3) states the multiplication of optimistic HTrF MGRSs on two-universe; (4) states the addition of optimistic HTrF MGRSs on two-universe.

Theorem 2. *Suppose U, V are two universes, $R_k, R_k' \in HTrFR(U \times V)(k = 1, 2, \ldots, n)$ are two HTrFRs on $U \times V$. If $R_k \subseteq R_k'$, $\forall E \in HTrF(V)$, the properties below are satisfied:*

(1) $\underline{\sum_{k=1}^{n} R_k'}^{O}(E) \subseteq \underline{\sum_{k=1}^{n} R_k}^{O}(E), \forall E \in HTrF(V)$;

(2) $\overline{\sum_{k=1}^{n} R_k{}'}^{O}(E) \supseteq \overline{\sum_{k=1}^{n} R_k}^{O}(E), \forall E \in HTrF(V)$.

Proof. Because of $R_k \subseteq R_k{}'$, depending on Definition 5, we have $a^f_{R_k{}^c} \geq a^f_{R_k{}'^c}$, and $b^f_{R_k{}^c} \geq b^f_{R_k{}'^c}$, and $c^f_{R_k{}^c} \geq c^f_{R_k{}'^c}$, and $d^f_{R_k{}^c} \geq d^f_{R_k{}'^c} \forall (x,y) \in (U \times V)$.

Thus, it can be seen that $\underline{\sum_{k=1}^{n} R_k}^{O}(E) = \{\langle x, \vee_{k=1}^{n} \wedge_{y\in V} \{h_{R_k{}^c}(x,y) \vee h_E(y)\}\rangle | x \in U\} = \{\langle x, (\vee_{k=1}^{n} \wedge_{y\in V} \{a^f_{R_k{}^c} \vee a^f_E\}, \vee_{k=1}^{n} \wedge_{y\in V} \{b^f_{R_k{}^c} \vee b^f_E\}, \vee_{k=1}^{n} \wedge_{y\in V} \{c^f_{R_k{}^c} \vee c^f_E\}, \vee_{k=1}^{n} \wedge_{y\in V} \{d^f_{R_k{}^c} \vee d^f_E\})\rangle | x \in U\} \geq \{\langle x, (\vee_{k=1}^{n} \wedge_{y\in V} \{a^f_{R'^c} \vee a^f_E\}, \vee_{k=1}^{n} \wedge_{y\in V} \{b^f_{R'^c} \vee b^f_E\}, \vee_{k=1}^{n} \wedge_{y\in V} \{c^f_{R'^c} \vee c^f_E\}, \vee_{k=1}^{n} \wedge_{y\in V} \{d^f_{R'^c} \vee d^f_E\})\rangle | x \in U\} = \underline{\sum_{k=1}^{n} R_k{}'}^{O}(E)$. Therefore, we have $\underline{\sum_{k=1}^{n} R_k{}'}^{O}(E) \subseteq \underline{\sum_{k=1}^{n} R_k}^{O}(E)$.

Similarly, $\overline{\sum_{k=1}^{n} R_k{}'}^{O}(E) \supseteq \overline{\sum_{k=1}^{n} R_k}^{O}(E)$ is obtained. □

Theorem 2 states that the optimistic HTrF MG lower and upper approximations on two-universe exhibit monotonicity in the matter of the monotonic forms of multiple HTrFRs.

3.2. Pessimistic HTrF MGRSs on Two-Universe

Definition 10. *Suppose U, V are two universes. $R_k \in HTrFR(U \times V)(k = 1,2,\ldots,n)$ is an HTrFR over $U \times V$. Moreover, an HTrF MG approximation space on two-universe is expressed as (U,V,R_k). $\forall E \in HTrF(V)$, the definitions of pessimistic HTrF MG lower and upper approximations on two-universe of E are given below:*

$$\underline{\sum_{k=1}^{n} R_k}^{P}(E) = \left\{ \left\langle x, h_{\underline{\sum_{k=1}^{n} R_k}^{P}(E)}(x) \right\rangle | x \in U \right\}; \tag{10}$$

$$\overline{\sum_{k=1}^{n} R_k}^{P}(E) = \left\{ \left\langle x, h_{\overline{\sum_{k=1}^{n} R_k}^{P}(E)}(x) \right\rangle | x \in U \right\}, \tag{11}$$

where $h_{\underline{\sum_{k=1}^{n} R_k}^{P}(E)}(x) = \wedge_{k=1}^{n} \wedge_{y\in V} \{h_{R_k{}^c}(x,y) \vee h_E(y)\}$; $h_{\overline{\sum_{k=1}^{n} R_k}^{P}(E)}(x) = \vee_{k=1}^{n} \vee_{y\in V} \{h_{R_k}(x,y) \wedge h_E(y)\}$.

The pair $\left(\underline{\sum_{k=1}^{n} R_k}^{P}(E), \overline{\sum_{k=1}^{n} R_k}^{P}(E)\right)$ indicates a pessimistic HTrF MGRS on two-universe of E in the matter of (U,V,R_k).

Theorem 3. *Suppose U, V are two universes. $R_k \in HTrFR(U \times V)(k = 1,2,\ldots,n)$ is an HTrFR over $U \times V$. $\forall E, F \in HTrF(V)$, the pessimistic HTrF MG lower and upper approximations on two-universe meet these properties:*

(1) $\underline{\sum_{k=1}^{n} R_k}^{P}(E^c) = (\overline{\sum_{k=1}^{n} R_k}^{P}(E))^c$, $\overline{\sum_{k=1}^{n} R_k}^{P}(E^c) = (\underline{\sum_{k=1}^{n} R_k}^{P}(E))^c$;

(2) $E \subseteq F \Rightarrow \underline{\sum_{k=1}^{n} R_k}^{P}(E) \subseteq \underline{\sum_{k=1}^{n} R_k}^{P}(F)$, $E \subseteq F \Rightarrow \overline{\sum_{k=1}^{n} R_k}^{P}(E) \subseteq \overline{\sum_{k=1}^{n} R_k}^{P}(F)$;

(3) $\underline{\sum_{k=1}^{n} R_k}^{P}(E \cap F) = \underline{\sum_{k=1}^{n} R_k}^{P}(E) \cap \underline{\sum_{k=1}^{n} R_k}^{P}(F)$, $\overline{\sum_{k=1}^{n} R_k}^{P}(E \cup F) = \overline{\sum_{k=1}^{n} R_k}^{P}(E) \cup \overline{\sum_{k=1}^{n} R_k}^{P}(F)$;

(4) $\underline{\sum_{k=1}^{n} R_k}^{P}(E \cup F) \supseteq \underline{\sum_{k=1}^{n} R_k}^{P}(E) \cup \underline{\sum_{k=1}^{n} R_k}^{P}(F)$, $\overline{\sum_{k=1}^{n} R_k}^{P}(E \cap F) \subseteq \overline{\sum_{k=1}^{n} R_k}^{P}(E) \cap \overline{\sum_{k=1}^{n} R_k}^{P}(F)$.

In Theorem 3, (1) states the complement of pessimistic HTrF MGRSs on two-universe; (2) states the monotone of pessimistic HTrF MGRSs on two-universe in the matter of various HTrF targets; (3) states the multiplication of pessimistic HTrF MGRSs on two-universe; (4) states the addition of pessimistic HTrF MGRSs on two-universe.

Theorem 4. *Suppose U, V are two universes. $R_k, R_k' \in HTrFR(U \times V)(k = 1, 2, \ldots, n)$ are two HTrFRs on $U \times V$. If $R_k \subseteq R_k'$, $\forall E \in HTrF(V)$, the properties below are satisfied:*

(1) $\underline{\sum_{k=1}^{n} R_k'}^{P}(E) \subseteq \underline{\sum_{k=1}^{n} R_k}^{P}(E), \forall E \in HTrF(V)$;

(2) $\overline{\sum_{k=1}^{n} R_k'}^{P}(E) \supseteq \overline{\sum_{k=1}^{n} R_k}^{P}(E), \forall E \in HTrF(V)$;

Theorem 4 states the pessimistic HTrF MG lower and upper approximations on two-universe exhibit monotonicity in the matter of the monotonic forms of multiple HTrFRs.

3.3. Relationships between Optimistic and Pessimistic HTrF MGRSs on Two-Universe

Theorem 5. *Suppose U, V are two universes. $R_k \in HTrFR(U \times V)(k = 1, 2, \ldots, n)$ is an HTrFR over $U \times V$. $\forall E \in HTrF(V)$, the optimistic and pessimistic HTrF MG lower and upper approximations on two-universe meet these properties:*

(1) $\underline{\sum_{k=1}^{n} R_k}^{P}(E) \subseteq \underline{\sum_{k=1}^{n} R_k}^{O}(E)$;

(2) $\overline{\sum_{k=1}^{n} R_k}^{P}(E) \supseteq \overline{\sum_{k=1}^{n} R_k}^{O}(E)$.

Proof. $\forall x \in U$, $\underline{\sum_{k=1}^{n} R_k}^{O}(E) = \{\langle x, \vee_{k=1}^{n} \wedge_{y \in V} \{h_{R_k^c}(x,y) \vee h_E(y)\}\rangle | x \in U\} \geq \{\langle x, \wedge_{k=1}^{n} \wedge_{y \in V} \{h_{R_k^c}(x,y) \vee h_E(y)\}\rangle | x \in U\} = \underline{\sum_{k=1}^{n} R_k}^{P}(E)$. $\overline{\sum_{k=1}^{n} R_k}^{P}(E) \supseteq \overline{\sum_{k=1}^{n} R_k}^{O}(E)$ is similarly obtained. □

Theorem 5 states that the optimistic HTrF MG lower approximation includes the pessimistic HTrF MG lower approximation, and the pessimistic HTrF MG upper approximation includes the optimistic HTrF MG upper approximation.

Remark 1. *This section presents a novel model, named HTrF MGRSs, on two-universe. The model combines the advantages of HTrFSs with MGRSs, which serves as a powerful tool to effectively deal with the AQE issue. The HTrFSs integrate HFSs with TrFNs, which offer a robust and flexible way of representing uncertain and imprecise AQE MSIoTSD. Compared to other fuzzy numbers, TrFNs demonstrate higher stability and are less susceptible to minor parameter variations. Furthermore, the various shapes of their membership functions enable them to capture fuzzy concepts in a flexible manner, reflecting real-world scenarios more accurately. Meanwhile, HFSs enable the expression of expert knowledge fully, as they allow the assignment of multiple membership values to an object, thus effectively representing the uncertainty and fuzziness in human reasoning. In the AQE process, determining the optimal solution requires the evaluation results provided by different experts. However, these experts may have distinct viewpoints on AQE. MGRSs on two-universe are* *distinguished by their remarkable information fusion capabilities, which enable the integration of distinct evaluation results from numerous experts via the provision of pessimistic and optimistic strategies, ultimately leading to a consensus and agreement. In summary, the proposed HTrF MGRSs model on two-universe has the potential to improve AQE decision ability and provide sound conclusions for AQE.*

This section presents an HTrF MGRSs on two-universe model. Initially, we provide a precise definition of HTrFRs. Building upon this definition, we subsequently introduce two types of HTrF MGRSs, specifically optimistic and pessimistic HTrF MGRSs. In order to establish the theoretical foundation for the proposed model, we proceed to analyze and prove the fundamental properties of both optimistic and pessimistic HTrF MGRSs on two-universe.

4. The AQE Approach

The present section illustrates a novel MAGDM method to solve the AQE issue, utilizing the presented model based on HTrF MGRSs on two-universe. The key steps of our MAGDM method are as follows.

4.1. Application Model

We consider $U = \{x_1, x_2, \ldots, x_p\}$ as a set of geographical locations and $V = \{y_1, y_2, \ldots, y_q\}$ as a set of air quality attributes. Suppose $R_k \in HTrFR(U \times V)$ $(k = 1, 2, \ldots, n)$ are n HTrFRs on $U \times V$, which represents the HTrF AQE information provided by n experts. Then, we suppose $E \in HTrF(U \times V)$ is the air quality testing sample. Thus, an HTrF decision information system (U, V, R_k, E) in terms of AQE is obtained.

Subsequently, we propose an MAGDM method using HTrF MGRSs on two-universe. First, based on Definitions 9 and 10, we compute the optimistic and pessimistic HTrF MG lower and upper approximations on two-universe of E, respectively. Then, we attain the sets $\underline{\sum_{k=1}^{n} R_k}^{O}(E)$, $\overline{\sum_{k=1}^{n} R_k}^{O}(E)$, $\underline{\sum_{k=1}^{n} R_k}^{P}(E)$, and $\overline{\sum_{k=1}^{n} R_k}^{P}(E)$. Depending on the rules of operations in [32]: $h_E(x) \oplus h_F(x) = \bigcup_{\tilde{a}_E \in h_E(x), \tilde{a}_F \in h_F(x)} \{(a_E + a_F, b_E + b_F,\ c_E + c_F, d_E + d_F)\}$, we get the sets $\underline{\sum_{k=1}^{n} R_k}^{O}(E) \oplus \overline{\sum_{k=1}^{n} R_k}^{O}(E)$ and $\underline{\sum_{k=1}^{n} R_k}^{P}(E) \oplus \overline{\sum_{k=1}^{n} R_k}^{P}(E)$. Furthermore, according to the decision strategy proposed by Sun et al. [60], we propose the decision rules for AQE based on HTrF MGRSs over two universes. Initially, we indicated that:

$$T_1 = \left\{ l \,\middle|\, \max_{x_l \in U} \left\{ \underline{\sum_{k=1}^{n} R_k}^{O}(E)(x_l) \oplus \overline{\sum_{k=1}^{n} R_k}^{O}(E)(x_l) \right\} \right\} \tag{12}$$

$$T_2 = \left\{ j \,\middle|\, \max_{x_j \in U} \left\{ \underline{\sum_{k=1}^{n} R_k}^{P}(E)(x_j) \oplus \overline{\sum_{k=1}^{n} R_k}^{P}(E)(x_j) \right\} \right\} \tag{13}$$

$$T_3 = \left\{ i \,\middle|\, \max_{x_i \in U} \left\{ \left(\underline{\sum_{k=1}^{n} R_k}^{O}(E)(x_i) \oplus \overline{\sum_{k=1}^{n} R_k}^{O}(E)(x_i) \right) \oplus \left(\underline{\sum_{k=1}^{n} R_k}^{P}(E)(x_i) \oplus \overline{\sum_{k=1}^{n} R_k}^{P}(E)(x_i) \right) \right\} \right\} \tag{14}$$

where T_1, T_2, T_3 denote the DMISs that consist of subscripts of the biggest HTrF element in the corresponding HTrFSs $\underline{\sum_{k=1}^{n} R_k}^{O}(E) \oplus \overline{\sum_{k=1}^{n} R_k}^{O}(E)$, $\underline{\sum_{k=1}^{n} R_k}^{P}(E) \oplus \overline{\sum_{k=1}^{n} R_k}^{P}(E)$ and $\left(\underline{\sum_{k=1}^{n} R_k}^{O}(E) \oplus \overline{\sum_{k=1}^{n} R_k}^{O}(E) \right) \oplus \left(\underline{\sum_{k=1}^{n} R_k}^{P}(E) \oplus \overline{\sum_{k=1}^{n} R_k}^{P}(E) \right)$, respectively. In accordance with Definition 4, the computation of the values of the score function for the HTrF elements in the corresponding HTrFSs mentioned above is feasible. Subsequently, we can easily obtain the T_1, T_2, and T_3 index sets. Next, we will discuss the practical implications of the three DMISs described above. Optimistic MGRSs are founded on the principle of "seeking common ground while preserving differences", i.e., retaining both the same and inconsistent parts of the opinions given by different experts, which can be regarded as a relatively risky risk-seeking approach to information fusion; whereas, pessimistic MGRSs are founded on the principle of "seeking common ground while excluding differences", i.e., retaining the same parts of the opinions given by different experts and removing different opinions and claims, which can be regarded as a relatively conservative risk-averse approach to information fusion. Thus, T_1 is the optimistic evaluation result, T_2 is the pessimistic evaluation result, and T_3 is the weighted evaluation result of T_1 and T_2, with a weighted value of 0.5. According to the definitions above, the decision rules are given by:

1. In case $T_1 \cap T_2 \cap T_3 \neq \varnothing$, that $x_l(l \in T_1 \cap T_2 \cap T_3)$ is the optimal location.
2. In case $T_1 \cap T_2 \cap T_3 = \varnothing$, but also $T_1 \cap T_2 \neq \varnothing$, that $x_l(l \in T_1 \cap T_2)$ is the optimal location.
3. In case $T_1 \cap T_2 \cap T_3 = \varnothing$, but also $T_1 \cap T_2 = \varnothing$, that $x_l(l \in T_3)$ is the optimal location.

4.2. The Algorithm Based on HTrF MGRSs on Two-Universe for AQE

In the following, we summarize the specific steps of the proposed method, and Algorithm 1 is further listed in terms of the specific steps.

Input: An HTrF decision information system (U, V, R_k, E).

Output: The optimal location.

Step1: Calculate the optimistic and pessimistic HTrF MG lower and upper approximations on two-universe of E, respectively.

Step2: Calculate optimistic and pessimistic HTrF MGRSs on two-universe of E, respectively.

Step3: Calculate $\left(\underline{\sum_{k=1}^{n} R_k}^{O}(E) \oplus \overline{\sum_{k=1}^{n} R_k}^{O}(E)\right) \oplus \left(\underline{\sum_{k=1}^{n} R_k}^{P}(E) \oplus \overline{\sum_{k=1}^{n} R_k}^{P}(E)\right)$.

Step4: Calculation of three types of DMISs.

Step5: Obtain the optimal location according to the decision rules.

Algorithm 1 The algorithm based on HTrF MGRSs over two universes for AQE.

Require: An HTrF decision information system (U, V, R_k, E).
Ensure: The optimal location.
for $i = 1$ to p, $j = 1$ to n, $t = 1$ to q **do**
Compute $\underline{\sum_{k=1}^{n} R_k}^{O}(E)$, $\overline{\sum_{k=1}^{n} R_k}^{O}(E)$, $\underline{\sum_{k=1}^{n} R_k}^{P}(E)$, and $\overline{\sum_{k=1}^{n} R_k}^{P}(E)$, respectively.
end for
for $t = 1$ to p **do**
Compute $\underline{\sum_{k=1}^{n} R_k}^{O}(E) \oplus \overline{\sum_{k=1}^{n} R_k}^{O}(E)$ and $\underline{\sum_{k=1}^{n} R_k}^{P}(E) \oplus \overline{\sum_{k=1}^{n} R_k}^{P}(E)$, respectively.
end for
for $t = 1$ to p **do**
Compute $\left(\underline{\sum_{k=1}^{n} R_k}^{O}(E) \oplus \overline{\sum_{k=1}^{n} R_k}^{O}(E)\right) \oplus \left(\underline{\sum_{k=1}^{n} R_k}^{P}(E) \oplus \overline{\sum_{k=1}^{n} R_k}^{P}(E)\right)$.
end for
for $t = 1$ to p **do**
Calculate T_1, T_2 and T_3.
end for
Calculate $T_1 \cap T_2 \cap T_3$, $T_1 \cap T_2$, and determine the optimal location.

Remark 2. *In the above steps, we set the number of locations as p, the number of attributes as q, and the number of experts as n. The first step has a complexity of $O(pnq)$. For the subsequent steps, i.e., steps 2 to 4, the complexity is represented as $O(p)$. Then, the complexity of the last step is denoted as $O(1)$. Consequently, the overall complexity of the proposed algorithm is represented as $O(pnq)$.*

In this section, we introduce a novel MAGDM method based on HTrF MGRSs on two-universe. We begin by introducing the HTrF decision information system. Subsequently, we describe the specific steps of the proposed MAGDM method in detail. Then, we apply the proposed method to AQE and propose a specific algorithm for this domain. Additionally, we conduct a complexity analysis of the proposed algorithm to assess its computational efficiency.

5. Case Analysis

The present section showcases the viability of the proposed MAGDM approach within the realm of AQE by means of a practical case study. Additionally, a comprehensive series of comparative and experimental analyses are executed to validate the efficacy of the presented approach.

5.1. Case Study in the Background of AQE

This study utilizes the AQE data of 31 provincial capital cities in China (https://www.aqistudy.cn/historydata/, accessed on 16 April 2023). Specifically, we employ six air pollutants, namely $PM_{2.5}$, PM_{10}, SO_2, CO, NO_2, and O_3, to determine the level of air quality, and consider the data from February, March, June, September, October, and December in the years 2018, 2019, and 2020 as decision-makers. Then, we define the set of 31 cities as $U = \{x_1, x_2, \ldots, x_{31}\}$, while the set of attributes are defined as $V = \{y_1, y_2, y_3, y_4, y_5, y_6\}$, where y_1 represents $PM_{2.5}$, y_2 represents PM_{10}, y_3 represents SO_2, y_4 represents CO, y_5 represents NO_2, and y_6 represents O_3. Next, we assume $R = \{R_1, R_2, R_3\}$ as the evaluation information, and let the values of the air quality test sample E be the average values on V. Furthermore, we process the obtained data to convert it into fuzzy data, which is done by applying the formula: $\mu_{ij} = \left(b_{ij} - \min_i b_{ij}\right) / \left(\max_i b_{ij} - \min_i b_{ij}\right)$. Here, b_{ij} denotes the raw value of air pollutant y_j for city x_i, while μ_{ij} represents the corresponding fuzzy data. Subsequently, we convert the fuzzy data to HTrF data. In particular, we denote μ_{ij}^2, μ_{ij}^3, μ_{ij}^6, μ_{ij}^9, μ_{ij}^{10}, and μ_{ij}^{12} as the corresponding fuzzy data for February, March, June, September, October, and December, respectively. Next, we can obtain the corresponding HTrF element represented as $d_{ij} = \left\{(\mu_{ij}^3, \mu_{ij}^6, \mu_{ij}^9, \mu_{ij}^{12}), (\mu_{ij}^2, \mu_{ij}^6, \mu_{ij}^{10}, \mu_{ij}^{12})\right\}$. By following this process, we obtain the HTrF decision information system.

Then, we will follow the steps of the proposed algorithm to compute. First, we calculate $\underline{\sum_{k=1}^{n} R_k}^{O}(E)$, $\overline{\sum_{k=1}^{n} R_k}^{O}(E)$, $\underline{\sum_{k=1}^{n} R_k}^{P}(E)$, and $\overline{\sum_{k=1}^{n} R_k}^{P}(E)$, respectively. Next, we further get $\underline{\sum_{k=1}^{n} R_k}^{O}(E) \oplus \overline{\sum_{k=1}^{n} R_k}^{O}(E)$ and $\underline{\sum_{k=1}^{n} R_k}^{P}(E) \oplus \overline{\sum_{k=1}^{n} R_k}^{P}(E)$. Then, we calculate the values of the score function for the HTrF elements in the corresponding sets $\underline{\sum_{k=1}^{n} R_k}^{O}(E) \oplus \overline{\sum_{k=1}^{n} R_k}^{O}(E)$, $\underline{\sum_{k=1}^{n} R_k}^{P}(E) \oplus \overline{\sum_{k=1}^{n} R_k}^{P}(E)$, and $\left(\underline{\sum_{k=1}^{n} R_k}^{O}(E) \oplus \overline{\sum_{k=1}^{n} R_k}^{O}(E)\right) \oplus \left(\underline{\sum_{k=1}^{n} R_k}^{P}(E) \oplus \overline{\sum_{k=1}^{n} R_k}^{P}(E)\right)$, respectively. Thus, it is easy to get $T_1 \cap T_2 \cap T_3 = \{12\} \neq \varnothing$, which implies that x_{12} is the optimal location and x_{12} is Haikou City.

5.2. Comparative Analysis

To establish the efficacy of the presented approach, the present section demonstrates that the MAGDM approach based on HTrF MGRSs is efficient by comparing different similar methods through comparative experiments. Set $\left(\underline{\sum_{k=1}^{n} R_k}^{O}(E) \oplus \overline{\sum_{k=1}^{n} R_k}^{O}(E)\right) \oplus \left(\underline{\sum_{k=1}^{n} R_k}^{P}(E) \oplus \overline{\sum_{k=1}^{n} R_k}^{P}(E)\right)$ is a compromise that combines the optimistic and pessimistic scenarios, and its ranking results have the advantage of being comprehensive. Thus, we use the ranking results of the above set for comparative analysis.

5.2.1. Comparative Analysis with Classic HTrF MAGDM Approaches

First, we conduct a comparison of the proposed approach with several classical HTrF MAGDM methods, including HTrF averaging (HTrFA) operators, HTrF geometric (HTrFG) operators, HTrF Einstein averaging (HTrFEA) operators, HTrF Einstein geometric (HTrFEG) operators, the HTrF VIKOR method, the HTrF TOPSIS method, and the improved HTrF TOPSIS method. The HTrF VIKOR method ranks alternatives by maximizing group utility and minimizing individual regret values. The HTrF TOPSIS method sorts alternatives according to their distance from the ideal solution. Despite its effectiveness, the optimal solution selected by the HTrF TOPSIS method may not always be the nearest to the positive ideal solution and the most distant from the negative ideal solution simultaneously. Thus, the improved HTrF TOPSIS approach utilizes an improved approach for calculating the relative closeness coefficient based on the HTrF TOPSIS approach. Our comparative findings are shown in Figure 1.

It is apparent from Figure 1 that there is minimal disparity between the ranking conclusions derived from various methodologies. The approach presented in this study aligns with the optimal schemes adopted by HTrFG operators, HTrFEG operators, the HTrF VIKOR method, the HTrF TOPSIS method, and the improved HTrF TOPSIS method. While the optimal scheme selected by HTrFA operators and HTrFEA operators does not correspond with the methodology proposed in this paper, the overall trend remains consistent. This observation provides compelling evidence for the validity of the presented methodology.

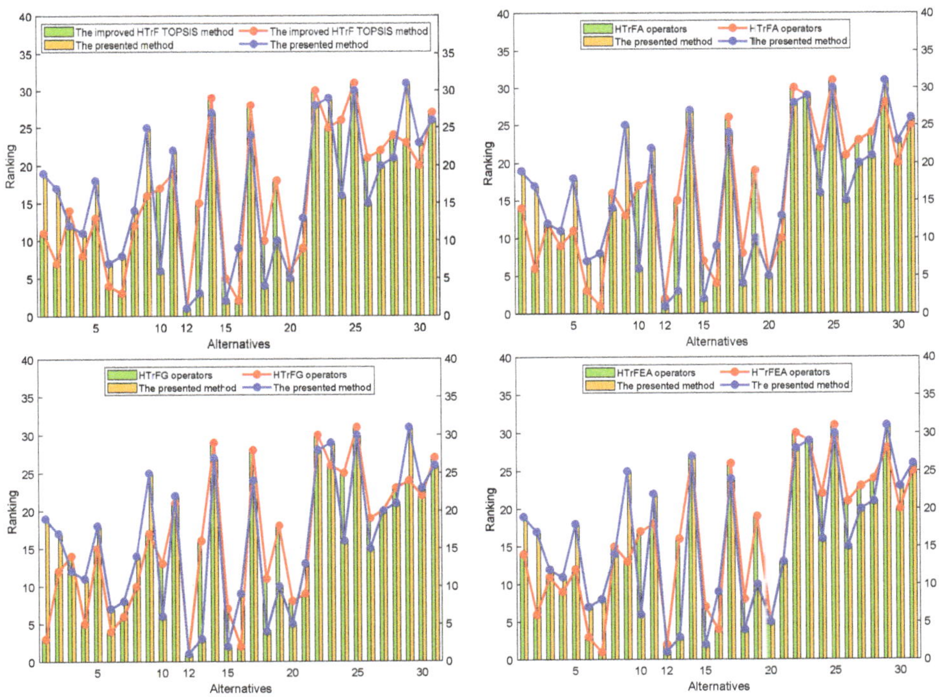

Figure 1. *Cont.*

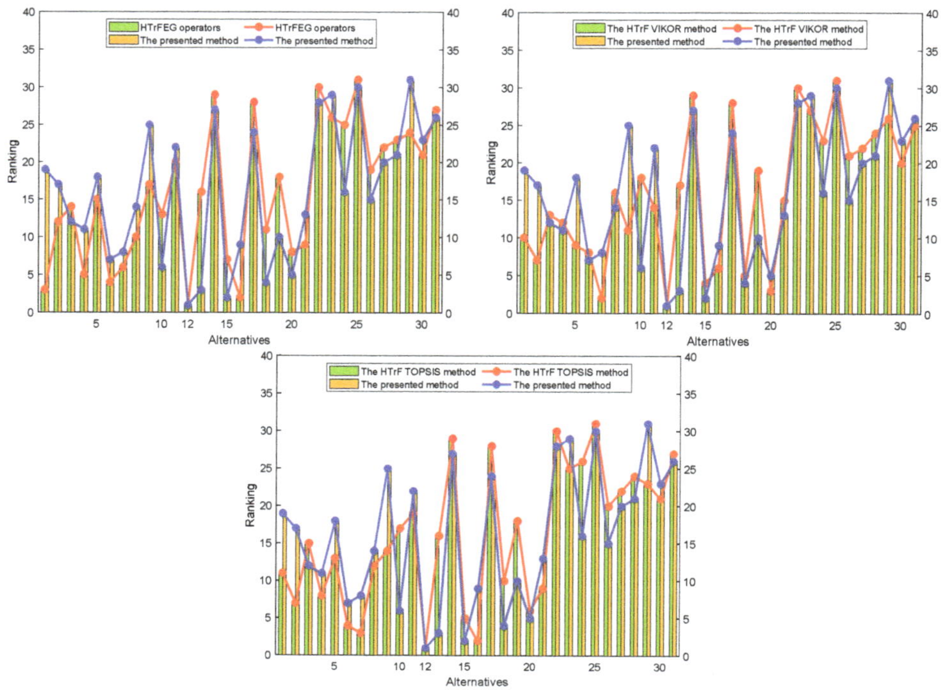

Figure 1. Comparison with classical HTrF MAGDM methods.

5.2.2. Comparative Analysis with the HTrF MABAC Method

This study next employs the Multi-Attributive Border Approximation area Comparison (MABAC) [61] approach to conduct a comparative analysis. The fundamental assumption of this method is to define the distance between the alternatives and the boundary approximation zone. In the MABAC method, each alternative is evaluated and ranked based on the difference in the specified distances. Notably, the MABAC method is distinguished by its mathematical simplicity and the stability of its evaluation results. It also takes into consideration the potential value of gains and losses and produces comprehensive results. Therefore, it is essential to compare the presented approach with the MABAC approach to validate the efficacy of the proposed approach. The conclusions of the comparison between this presented method and the HTrF MABAC method are shown in Figure 2.

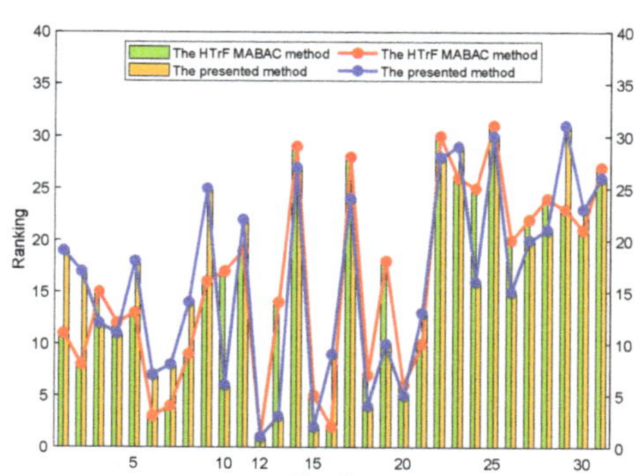

Figure 2. Comparison with the HTrF MABAC method.

From Figure 2, it is incontrovertible that the presented methodology and the HTrF MABAC approach demonstrate a congruous overall trend, and, more notably, select the same optimal scheme. This observation serves as further evidence of the efficacy and soundness of the proposed methodology.

5.3. Experimental Analysis

The Spearman correlation coefficient is widely utilized in statistical analysis as a non-parametric measure for evaluating the correlation between two variables. It employs a monotonic function to gauge the correlation strength between two statistical variables. In this article, we apply the Spearman correlation coefficient to assess the relevance of the method presented in this article in comparison to other similar methods. The number of ranking positions for the presented approach and other approaches are denoted as $Ind(X_l)(l = 1, 2, \ldots, n)$ and $Ind(x_l)$, respectively. Hence, the Spearman correlation coefficient can be conveniently calculated as $\rho = 1 - \frac{6\sum(Ind(x_l) - Ind(X_l))^2}{n(n^2-1)}$. The results of this analysis are shown in Table 1 and Figure 3.

Table 1. Spearman correlation coefficient between the presented approach and similar approaches.

Different Methods	Spearman Correlation Coefficient
The HTrF MABAC method	0.8121
HTrFA operators	0.8048
HTrFEA operators	0.7989
The improved HTrF TOPSIS method	0.7915
HTrFG operators	0.7903
HTrFEG operators	0.7883
The HTrF TOPSIS method	0.7806
The HTrF VIKOR method	0.7641

According to the above experimental analysis, the correlation between the presented MAGDM approach and other comparable approaches is relatively strong, which validate the validity and stability of the proposed MAGDM approach.

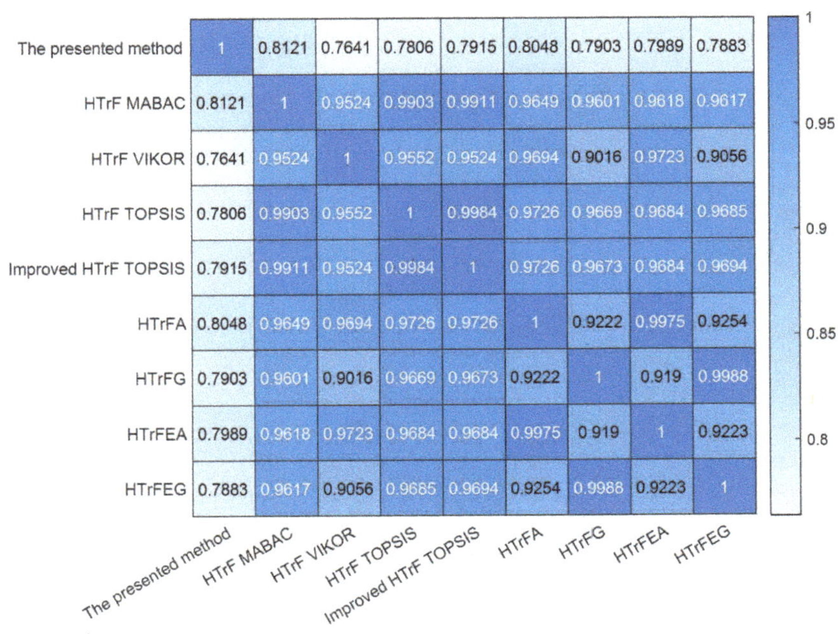

Figure 3. Spearman correlation coefficient between different methods.

Remark 3. *It is important to note that the Equation* $\rho = 1 - \frac{6\sum(Ind(x_l) - Ind(X_l))^2}{n(n^2-1)}$ *can only be applied when all n ranks are unique integers, with* $Ind(x_l) - Ind(X_l)$ *representing the difference between the two ranks of each observation and n indicating the total number of observations. This condition must be satisfied in order for the equation to be valid and accurate in the calculation of the Spearman correlation coefficient.*

Furthermore, we summarize the advantages of the presented method, as shown in Table 2.

Table 2. The advantage of different approaches.

Diverse Approaches	Ranking	Decision-Making Risks	Group Decisions-Making	Uncertain Information	Reduction of Divergence
HTrF MABAC	√	×	√	√	×
HTrFA	√	×	√	√	×
HTrFEA	√	×	√	√	×
Improved HTrF TOPSIS	√	×	√	√	×
HTrFG	√	×	√	√	×
HTrFEG	√	×	√	√	×
HTrF TOPSIS	√	×	√	√	×
HTrF VIKOR	√	×	√	√	×
The presented method	√	√	√	√	√

5.4. Discussion

Deep learning algorithms, such as Convolutional Neural Networks, have been successful in various applications. However, for MAGDM problems, they may not always be the optimal solution. It should be noted that Convolutional Neural Networks and their variations require the data to be divided into training and testing sets. While the conventional division practice is usually to allocate 80% of the data to training and 20% to testing, discrepancies in the ratio of training to testing data allocation, as well as the stochasticity of the division process, may lead to dissimilar outcomes.

Regarding the data used in this study, the dataset included weather information for 367 cities in China from December 2013 onwards. However, for the purpose of demonstration, we selected some data from 31 provincial capital cities from 2018 to 2020 as our sample. We took great care to ensure that the selected sample represents the characteristics and distribution of the entire dataset. Nevertheless, future studies could use larger datasets or incorporate additional attributes to improve the accuracy and generalizability of the proposed method.

The experimental results outlined above demonstrate that the decision-making method based on HTrF MGRSs on two-universe represents a comprehensive utilization of the strengths of HTrFSs and MGRSs. First, HTrFSs offer significant advantages over other fuzzy sets by allowing for a more precise representation of fuzzy or imprecise information through TrFNs. Moreover, HTrFSs combine the advantages of HFSs to enable decision-makers to express their hesitations or uncertainties during the decision-making process, thus enabling them to consider all possible scenarios and make more informed decisions. Then, the MGRSs on two-universe approach serves as an excellent information fusion strategy that integrates the perspectives of different experts to arrive at a final conclusion. Furthermore, we leverage DMISs to mitigate the impact of disagreements among the experts within the expert group on the evaluation outcomes. By incorporating DMISs, the presented MAGDM method offers a multifaceted evaluation scheme to experts, allowing them to attain more sensible and precise evaluation results. In summary, the MAGDM method presented in this article substantially reduces the uncertainty involved in decision-making and enhances its accuracy and reliability. By combining the advantages of HTrFSs, MGRSs, and DMISs, the proposed approach provides a viable option for assessment and decision-making in situations of uncertainty and fuzziness. The proposed approach demonstrates the potential for solving decision problems in various domains.

Regarding the AQE in different cities, the experimental results indicate that the air quality in Haikou is relatively good, whereas the air quality in Xining and Taiyuan is relatively poor. First, the successful experience of Haikou city demonstrates that economic development and environmental protection are not mutually exclusive. Therefore, the government should actively strengthen ecological construction and protection to improve air quality. Second, with sustained government control, recent data reveals that the overall air quality in Xining has improved, suggesting the critical role of governance in improving air quality. For cities such as Taiyuan, where coal is the main source of energy and coal burning and industrial pollution are the main sources of pollution, the government should actively promote the transformation of the energy structure, reduce dependence on coal, promote clean energy, control the emissions from industrial pollution sources, and promote other measures to reduce the emission of atmospheric pollutants. In summary, the government should formulate and implement relevant policies and measures to improve urban air quality and enhance residents' quality of life.

This section presents a comprehensive case study that demonstrates the validity and feasibility of the proposed MAGDM method within the domain of AQE. The evaluation employs comparative and experimental analysis to showcase the effectiveness of the proposed approach. We begin by providing a detailed description of the experimental procedure. Subsequently, we conduct a comparative analysis, where we compare and contrast the proposed MAGDM method with several classical HTrF MAGDM methods and the HTrF MABAC method. Moreover, we compute the Spearman correlation coefficient and plot a graph that compares the proposed method with other similar methods. The advantages of the proposed method are also presented in tabular format. Finally, a comprehensive discussion and analysis is presented, which includes a discussion of the limitations of deep learning methods, a detailed analysis of the datasets used in this paper, the potential of the proposed method, and the implications of this paper's research for government work.

6. Conclusions

AQE plays a crucial role in creating and maintaining a clean atmospheric environment. In this article, we introduce a novel MAGDM method to AQE. First, we propose an HTrF MGRS on two-universe model by combining the advantages of HTrFSs in information representation and MGRSs in information fusion. Then, we investigate the fundamental definitions and properties of optimistic and pessimistic HTrF MGRSs on two-universe. Afterward, we present a general approach to the AQE decision problem. Finally, we conduct several numerical analyses, using AQE-related datasets, to showcase the feasibility, effectiveness, and stability of the presented MAGDM approach.

While the proposed architecture presents a promising solution for AQE, there are still several challenging issues in theoretical and practical research. We recommend the exploration of the following research directions in the future:

1. Realistic decision-making scenarios are diverse; hence, it is essential to extend the application of the presented MAGDM approach to other real-world contexts, such as water quality testing, forest fire prediction, disease diagnosis, etc.
2. Further exploration of property reduction methods and uncertainty measures for HTrF MGRSs on two-universe has important implications for the application of the presented MAGDM method to other uncertain and complicated decision scenarios.
3. Large-scale MAGDM can leverage the complementary knowledge structures of large groups of people to enhance the precision and objectivity of decision-making. As such, it is imperative to explore large-scale MAGDM to tackle intricate practical situations.

Author Contributions: Conceptualization, C.Z.; software, W.L.; formal analysis, W.L., Y.C. and J.S.; investigation, J.S.; writing—original draft preparation, W.L.; writing—review and editing, C.Z.; visualization, Y.C. All authors have read and agreed to the published version of the manuscript.

Funding: This research was partially funded by the 20th Undergraduate Innovation and Entrepreneurship Training Program of Shanxi University (No. X2022020043), the Special Fund for Science and Technology Innovation Teams of Shanxi (No. 202204051001015).

Data Availability Statement: The dataset utilized in this research is available from https://www.aqistudy.cn/historydata/ (accessed on 16 April 2023).

Conflicts of Interest: The authors declare no conflict of interest.

Abbreviations

AP	Air pollution	AQE	Air quality evaluation
MAGDM	Multi-attribute group decision-making	HTrF	Hesitant trapezoidal fuzzy
HTrFSs	Hesitant trapezoidal fuzzy sets	HTrFRs	Hesitant trapezoidal fuzzy relations
MGRSs	Multi-granulation rough sets	EEA	European environment agency
HFSs	Hesitant fuzzy sets	GrC	Granular computing
WHO	World health organization	TrFNs	Trapezoidal fuzzy numbers
HTrFA	Hesitant trapezoidal fuzzy averaging	HTrFG	Hesitant trapezoidal fuzzy geometric
HTrFEA	Hesitant trapezoidal fuzzy Einstein averaging	HTrFEG	Hesitant trapezoidal fuzzy Einstein geometric
DMISs	Decision-making index sets	IoT	Internet of Things
MSIoTSD	Multi-source Internet of Things sensor data	MABAC	Multi-attributive border approximation area comparison

References

1. de Santos, U.P.; Arbex, M.A.; Braga, A.L.F.; Mizutani, R.F.; Cançado, J.E.D.; Terra-Filho, M.; Chatkin, J.M. Environmental air pollution: Respiratory effects. *J. Bras. Pneumol.* **2021**, *47*, e20200267. [CrossRef] [PubMed]
2. González-Martín, J.; Kraakman, N.J.R.; Pérez, C.; Lebrero, R.; Muñoz, R. A state–of–the-art review on indoor air pollution and strategies for indoor air pollution control. *Chemosphere* **2021**, *262*, 128376. [CrossRef]
3. Wei, W.; Wang, Z. Impact of industrial air pollution on agricultural production. *Atmosphere* **2021**, *12*, 639. [CrossRef]
4. Michetti, M.; Gualtieri, M.; Anav, A.; Adani, A.; Benassi, B.; Dalmastri, C.; D'Elia, I.; Piersanti, A.; Sannino, G.; Zanini, G.; et al. Climate change and air pollution: Translating their interplay into present and future mortality risk for Rome and Milan municipalities. *Sci. Total Environ.* **2022**, *830*, 154680. [CrossRef] [PubMed]
5. Caleyachetty, R.; Lufumpa, N.; Kumar, N.; Mohammed, N.I.; Bekele, H.; Kurmi, O.; Wells, J.; Manaseki-Holland, S. Exposure to household air pollution from solid cookfuels and childhood stunting: A population-based, cross-sectional study of half a million children in low- and middle-income countries. *Int. Health* **2022**, *14*, 639–647. [CrossRef] [PubMed]
6. Latham, R.M.; Kieling, C.; Arseneault, L.; Rocha, T.B.M.; Beddows, A.; Beevers, S.D.; Danese, A.; de Oliveira, K.; Kohrt, B.A.; Moffitt, T.E.; et al. Childhood exposure to ambient air pollution and predicting individual risk of depression onset in UK adolescents. *J. Psychiatr. Res.* **2021**, *138*, 60–67. [CrossRef]
7. Ahmed, M.; Shuai, C.; Abbas, K.; Rehman, F.U.; Khoso, W.M. Investigating health impacts of household air pollution on woman's pregnancy and sterilization: Empirical evidence from Pakistan, India, and Bangladesh. *Energy* **2022**, *247*, 123562. [CrossRef]
8. Goshua, A.; Akdis, C.A.; Nadeau, K.C. World Health Organization global air quality guideline recommendations: Executive summary. *Allergy* **2022**, *77*, 1955–1960. [CrossRef]
9. Huang, W.; Li, T.; Liu, J.; Xie, P.; Du, S.; Teng, F. An overview of air quality analysis by big data techniques: Monitoring, forecasting, and traceability. *Inf. Fusion* **2021**, *75*, 28–40. [CrossRef]
10. Zhu, J.; Chen, L.; Liao, H. Multi-pollutant air pollution and associated health risks in China from 2014 to 2020. *Atmos. Environ.* **2022**, *268*, 118829. [CrossRef]
11. Oprea, M. A case study of knowledge modelling in an air pollution control decision support system. *AI Commun.* **2005**, *18*, 293–303.
12. Wang, W.; Mao, W.; Tong, X.; Xu, G. A novel recursive model based on a convolutional long short-term memory neural network for air pollution prediction. *Remote Sens.* **2021**, *13*, 1284. [CrossRef]
13. Gu, Y.; Zhao, Y.; Zhou, J.; Li, H.; Wang, Y. A fuzzy multiple linear regression model based on meteorological factors for air quality index forecast. *J. Intell. Fuzzy Syst.* **2021**, *40*, 10523–10547. [CrossRef]

14. Ma, J.; Ma, X.; Yang, C.; Xie, L.; Zhang, W.; Li, X. An air pollutant forecast correction model based on ensemble learning algorithm. *Electronics* **2023**, *12*, 1463. [CrossRef]
15. Gu, Y.; Li, B.; Meng, Q. Hybrid interpretable predictive machine learning model for air pollution prediction. *Neurocomputing* **2022**, *468*, 123–136. [CrossRef]
16. Tao, Y.; Wu, Y.; Zhou, J.; Wu, M.; Wang, S.; Zhang, L.; Xu, C. How to realize the effect of air pollution control? A hybrid decision framework under the fuzzy environment. *J. Clean. Prod.* **2021**, *305*, 127093. [CrossRef]
17. Martín-Baos, J.Á.; Rodriguez-Benitez, L.; García-Ródenas, R.; Liu, J. IoT based monitoring of air quality and traffic using regression analysis. *Appl. Soft Comput.* **2022**, *115*, 108282. [CrossRef]
18. Sangaiah, A.K.; Rostami, A.S.; Hosseinabadi, A.A.R.; Shareh, M.B.; Javadpour, A.; Bargh, S.H.; Hassan, M.M. Energy-Aware geographic routing for Real-Time workforce monitoring in industrial informatics. *IEEE Internet Things J.* **2021**, *8*, 9753–9762. [CrossRef]
19. Sangaiah, A.K.; Javadpour, A.; Ja'fari, F.; Pinto, P.; Zhang, W.; Balasubramanian, S. A hybrid heuristics artificial intelligence feature selection for intrusion detection classifiers in cloud of things. *Clust. Comput.* **2023**, *26*, 599–612. [CrossRef]
20. Lin, M.; Huang, C.; Xu, Z.; Chen, R. Evaluating IoT platforms using integrated probabilistic linguistic MCDM method. *IEEE Internet Things J.* **2020**, *7*, 11195–11208. [CrossRef]
21. Schilt, U.; Barahona, B.; Buck, R.; Meyer, P.; Kappani, P.; Möckli, Y.; Meyer, M.; Schuetz, P. Low-Cost sensor node for air quality monitoring: Field tests and validation of particulate matter measurements. *Sensors* **2023**, *23*, 794. [CrossRef] [PubMed]
22. Dmytryk, N.; Leivadeas, A. A generic preprocessing architecture for Multi-Modal IoT sensor data in artificial general intelligence. *Electronics* **2022**, *11*, 3816. [CrossRef]
23. Zadeh, L.A. Fuzzy sets. *Inf. Control* **1965**, *8*, 338–353. [CrossRef]
24. Li, W.; Zhai, S.; Xu, W.; Pedrycz, W.; Qian, Y.; Ding, W.; Zhan, T. Feature selection approach based on improved fuzzy C-Means with principle of refined justifiable granularity. *IEEE Trans. Fuzzy Syst.* **2022**. [CrossRef]
25. Li, W.; Zhou, H.; Xu, W.; Wang, X.Z.; Pedrycz, W. Interval Dominance-Based feature selection for Interval-Valued ordered data. *IEEE Trans. Neural Netw. Learn. Syst.* **2022**. [CrossRef]
26. Lin, M.; Wang, H.; Xu, Z. TODIM-based multi-criteria decision-making method with hesitant fuzzy linguistic term sets. *Artif. Intell. Rev.* **2020**, *53*, 3647–3671. [CrossRef]
27. Lin, M.; Li, X.; Chen, R.; Fujita, H.; Lin, J. Picture fuzzy interactional partitioned Heronian mean aggregation operators: An application to MADM process. *Artif. Intell. Rev.* **2022**, *55*, 1171–1208. [CrossRef]
28. Torra, V. Hesitant fuzzy sets. *Int. J. Intell. Syst.* **2010**, *25*, 529–539. [CrossRef]
29. Divsalar, M.; Ahmadi, M.; Ebrahimi, E.; Ishizaka, A. A probabilistic hesitant fuzzy Choquet integral-based TODIM method for multi-attribute group decision-making. *Expert Syst. Appl.* **2022**, *191*, 116266. [CrossRef]
30. Krishankumar, R.; Ravichandran, K.S.; Gandomi, A.H.; Kar, S. Interval-valued probabilistic hesitant fuzzy set-based framework for group decision-making with unknown weight information. *Neural Comput. Appl.* **2021**, *33*, 2445–2457. [CrossRef]
31. Ahmad, F.; Adhami, A.Y.; John, B.; Reza, A. A novel approach for the solution of multiobjective optimization problem using hesitant fuzzy aggregation operator. *RAIRO-Oper. Res.* **2022**, *56*, 275–292. [CrossRef]
32. Ye, J. Multicriteria decision-making method using expected values in trapezoidal hesitant fuzzy setting. *J. Converg. Inf. Technol.* **2013**, *8*, 135–143.
33. Deli, I. Bonferroni mean operators of generalized trapezoidal hesitant fuzzy numbers and their application to decision-making problems. *Soft Comput.* **2021**, *25*, 4925–4949. [CrossRef]
34. Deli, I.; Karaaslan, F. Generalized trapezoidal hesitant fuzzy numbers and their applications to multi criteria decision-making problems. *Soft Comput.* **2021**, *25*, 1017–1032. [CrossRef]
35. Zhang, C.; Ding, J.; Zhang, J.; Sangaiah, A.K.; Li, D. Fuzzy intelligence learning based on bounded rationality in IoMT systems: A case study in parkinson's disease. *IEEE Trans. Comput. Soc. Syst.* **2022**. [CrossRef]
36. Zhang, C.; Li, X.; Sangaiah, A.K.; Li, W.; Wang, B.; Cao, F.; Shangguan, X. Collaborative fuzzy linguistic learning to Low-Resource and robust decision system based on bounded rationality. *ACM Trans. Asian Low-Resour. Lang. Inf. Process.* 2023. [CrossRef]
37. Zadeh, L.A. Fuzzy sets and information granularity. *Adv. Fuzzy Set Theory Appl.* **1979**, *11*, 3–18.
38. Zadeh, L.A. Toward a theory of fuzzy information granulation and its centrality in human reasoning and fuzzy logic. *Fuzzy Sets Syst.* **1997**, *90*, 111–127. [CrossRef]
39. Zhan, T. Granular-based state estimation for nonlinear fractional control systems and its circuit cognitive application. *Int. J. Cogn. Comput. Eng.* **2023**, *4*, 1–5. [CrossRef]
40. Zhang, C.; Li, D.; Liang, J. Interval-valued hesitant fuzzy multi-granularity three-way decisions in consensus processes with applications to multi-attribute group decision making. *Inf. Sci.* **2020**, *511*, 192–211. [CrossRef]
41. Ren, X.; Li, D.; Zhai, Y. Research on mixed decision implications based on formal concept analysis. *Int. J. Cogn. Comput. Eng.* **2023**, *4*, 71–77. [CrossRef]
42. Pawlak, Z. Rough sets. *Int. J. Comput. Inf. Sci.* **1982**, *11*, 341–356. [CrossRef]
43. Bai, J.; Sun, B.; Chu, X.; Wang, T.; Li, H.; Huang, Q. Neighborhood rough set-based multi-attribute prediction approach and its application of gout patients. *Appl. Soft Comput.* **2022**, *114*, 108127. [CrossRef]
44. Abdullah, S.; Al-Shomrani, M.M.; Liu, P.; Ahmad, S. A new approach to three-way decisions making based on fractional fuzzy decision-theoretical rough set. *Int. J. Intell. Syst.* **2022**, *37*, 2428–2457. [CrossRef]

45. Bai, H.; Li, D.; Ge, Y.; Wang, J.; Cao, F. Spatial rough set-based geographical detectors for nominal target variables. *Inf. Sci.* **2022**, *586*, 525–539. [CrossRef]
46. Qian, Y.; Liang, J.; Yao, Y.; Dang, C. MGRS: A multi-granulation rough set. *Inf. Sci.* **2010**, *180*, 949–970. [CrossRef]
47. Bai, W.; Zhang, C.; Zhai, Y.; Sangaiah, A.K. Incomplete intuitionistic fuzzy behavioral group decision-making based on multi-granulation probabilistic rough sets and MULTIMOORA for water quality inspection. *J. Intell. Fuzzy Syst.* **2023**, *44*, 4537–4556. [CrossRef]
48. Zhang, C.; Ding, J.; Zhan, J.; Li, D. Incomplete three-way multi-attribute group decision making based on adjustable multigranulation Pythagorean fuzzy probabilistic rough sets. *Int. J. Approx. Reason.* **2022**, *147*, 40–59. [CrossRef]
49. Zhang, C.; Bai, W.; Li, D.; Zhan, J. Multiple attribute group decision making based on multigranulation probabilistic models, MULTIMOORA and TPOP in incomplete q-rung orthopair fuzzy information systems. *Int. J. Approx. Reason.* **2022**, *143*, 102–120. [CrossRef]
50. Li, W.; Xu, W.; Zhang, X.; Zhang, J. Updating approximations with dynamic objects based on local multigranulation rough sets in ordered information systems. *Artif. Intell. Rev.* **2022**, *55*, 1821–1855. [CrossRef]
51. Pei, D.; Xu, Z.B. Rough set models on two universes. *Int. J. Gen. Syst.* **2004**, *33*, 569–581. [CrossRef]
52. Sun, B.; Ma, W. Multigranulation rough set theory over two universes. *J. Intell. Fuzzy Syst.* **2015**, *28*, 1251–1269. [CrossRef]
53. Sun, B.; Zhou, X.; Lin, N. Diversified binary relation-based fuzzy multigranulation rough set over two universes and application to multiple attribute group decision making. *Inf. Fusion* **2020**, *55*, 91–104. [CrossRef]
54. Yang, D.; Cai, M.; Li, Q.; Xu, F. Multigranulation fuzzy probabilistic rough set model on two universes. *Int. J. Approx. Reason.* **2022**, *145*, 18–35. [CrossRef]
55. Zhang, C.; Li, D.; Liang, J. Multi-granularity three-way decisions with adjustable hesitant fuzzy linguistic multigranulation decision-theoretic rough sets over two universes. *Inf. Sci.* **2020**, *507*, 665–683. [CrossRef]
56. Gu, J.; Wang, D.; Hu, D.; Gao, F.; Xu, F. Temporal extraction of complex medicine by combining probabilistic soft logic and textual feature feedback. *Appl. Sci.* **2023**, *13*, 3348. [CrossRef]
57. Alshukaili, D.; Fernandes, A.A.A.; Paton, N.W. Structuring linked data search results using probabilistic soft logic. In Proceedings of the 15th International Semantic Web Conference (ISWC 2016), Kobe, Japan, 17–21 October 2016; pp. 3–19.
58. Fakhraei, S.; Huang, B.; Raschid, L.; Getoor, L. Network-Based Drug-Target interaction prediction with probabilistic soft logic. *IEEE/ACM Trans. Comput. Biol. Bioinform.* **2014**, *11*, 775–787. [CrossRef]
59. Dubois, D.; Prade, H. Ranking fuzzy numbers in the setting of possibility theory. *Inf. Sci.* **1983**, *30*, 183–224. [CrossRef]
60. Sun, B.; Ma, W.; Zhao, H. A fuzzy rough set approach to emergency material demand prediction over two universes. *Appl. Math. Model.* **2013**, *37*, 7062–7070. [CrossRef]
61. Pamučar, D.; Ćirović, G. The selection of transport and handling resources in logistics centers using Multi-Attributive Border Approximation area Comparison (MABAC). *Expert Syst. Appl.* **2015**, *42*, 3016–3028. [CrossRef]

electronics

Article

A Variable Structure Multiple-Model Estimation Algorithm Aided by Center Scaling

Qiang Wang, Guowei Li, Weitong Jin, Shurui Zhang * and Weixing Sheng

School of Electronic and Optical Engineering, Nanjing University of Science and Technology, Nanjing 210094, China
* Correspondence: shuruizhang@njust.edu.cn

Abstract: The accuracy for target tracking using a conventional interacting multiple-model algorithm (IMM) is limited. In this paper, a new variable structure of interacting multiple-model (VSIMM) algorithm aided by center scaling (VSIMM-CS) is proposed to solve this problem. The novel VSIMM-CS has two main steps. Firstly, we estimate the approximate location of the true model. This is aided by the expected-mode augmentation algorithm (EMA), and a new method—namely, the expected model optimization method—is proposed to further enhance the accuracy of EMA. Secondly, we change the original model set to ensure the current true model as the symmetry center of the current model set, and the model set is scaled down by a certain percentage. Considering the symmetry and linearity of the system, the errors produced by symmetrical models can be well offset. Furthermore, narrowing the distance between the true model and the default model is another effective method to reduce the error. The second step is based on two theories: symmetric model set optimization method and proportional reduction optimization method. All proposed theories aim to minimize errors as much as possible, and simulation results highlight the correctness and effectiveness of the proposed methods.

Keywords: variable structure of interacting multiple-model; symmetric model set optimization method; proportional reduction optimization method; expected model optimization method

Citation: Wang, Q.; Li, G.; Jin, W.; Zhang, S.; Sheng, W. A Variable Structure Multiple-Model Estimation Algorithm Aided by Center Scaling. *Electronics* **2023**, *12*, 2257. https://doi.org/10.3390/electronics12102257

Academic Editor: Dimitris Apostolou

Received: 27 March 2023
Revised: 11 May 2023
Accepted: 12 May 2023
Published: 16 May 2023

1. Introduction

Multiple-model (MM) is an advanced method to solve many problems, especially the target tracking problem [1,2]. Compared with traditional algorithms combined with radar systems [3,4], MM's power of MM comes from the teamwork of multi parallel estimators [5], not only a single estimator. The MM approach has been presented for more than fifty years, and it was first proposed in [6,7]. It has a mature framework system [8,9], and the parallel structure using Bayesian filters proves its great performance. Usually, the model set designed in advance or generated in real-time is used to cover the possible true models. Then, the system dynamics can be described as hybrid systems [10,11] with discrete modes and continuous states. The model set used during the target tracking process has a big influence on estimation results [12], and a better model set often leads to more precise tracking results. The overall estimation is the combination of all estimations from the parallel-running Bayesian filters [13,14]. In recent decades, the MM methods have achieved rapid development [15]. The MM has been used widely because of its completeness of method and easiness of implementation. In addition, it goes through three stages altogether [16]: static MM (SMM), interacting MM (IMM) and variable-structure interacting of MM (VSIMM).

Compared with SMM, the biggest advantage of IMM is considering the jumps between models [6,7], and this drawback was fixed by Blom and Bar-Shalom [17]. Many advanced methods of IMM have been proven that improving tracking accuracy and not increasing computation burden can be conducted at the same time [18–20]. A reweighted interacting multiple model algorithm [21], which is a recursive implementation of a maximum a posteriori (MAP) state sequence estimator, is a competitive alternative to the popular IMM algorithm and GPB methods. Considering the non-Gaussian white noise, the interacting multiple model based on maximum correntropy Kalman filter (IMM-MCKF) [22] is presented to combine interacting multiple model with MCKF to deal with the impulsive noise [23]. Furthermore, to overcome the small kernel bandwidth, changing the kernel is also presented in it. Not only that, some emerging technologies, such as neural network [24] combined with MM also achieves rapid development. For example, the multiple model tracking algorithm of the multiple processing switch is a reliable method to improve the tracking precision [25]. However, the inherent defect, fixed model structure, of IMM limits its development to a large extent. In many real-world scenarios, the true model space is not discrete [26], but continues and uncountable. It is hard to design a model set to cover all possible models. Therefore, it seems natural to design a model set including huge models to fit the true model space perfectly. However, too many models leads to competition between models, even worse than few models [5]. In addition, the surging computation burden is another severe problem. Loosely speaking, the major objective of MM is to find a method using as few models as possible to achieve better performance of target tracking.

FIMM and SMM both use fixed model set at all time, without considering the properties of true model space. VSIMM [27–29] is a successful method to consider. On the one hand, it uses a limited number of models to reduce computation burden. On the other hand, the model set generated on the current is closer than the original model set. Generally, the VSIMM adapts the continuous mode space. It seems that tracking precision and acceptable computation burden are both solved perfectly. The key problem of VSIMM is how to design a highly cost-effective model set adaptive (MSA) mechanism. The model-group switching algorithm (MGS) has been used widely in a large class of problems with hybrid (continuous and discrete) uncertainties [30]. Compared with FIMM, the main advantage of MGS is reducing computation significantly, but it has limited performance improvements. The expected-mode augmentation covers a large continuous mode space by a relatively small number of models at a given accuracy level [31]. Though the expected model is closer than any other model, the results may even deteriorate because it is not precise enough. Similarly, the model set is augmented by a variable model intended to best match the unknown true model, and the algorithm is named the equivalent-model augmentation algorithm (EqMA) [32]. Compared with IMM and EMA, the EqMA has stronger timeliness. The likely-model set algorithm is also an adaptive method for VSIMM, and its cost-effectiveness is more valid than many other methods [33]. However, those three algorithms mentioned in [33] are more difficult to implement. A method using hypersphere-symmetric model-subset and axis-symmetric model-subset is presented as the fundamental model-subset for multiple models estimation with fixed structure, variable structure, and moving bank [34]. Different kinds of VSIMM algorithms provide reasonable methods to achieve model sets, and MSA is still an open topic to study.

With the gradual upgrading of radar detection systems, the tracking of maneuvering targets requires better accurate results. VSIMM-CS not only meets the requirements well in terms of tracking accuracy but also hardly adds any computation. In this paper, a variable structure of multiple-model algorithm aided by center scaling (VSIMM-CS) is proposed for providing a rational method to generate model sets in real-time. Considering the properties of the linear system [35] and the effect of distance between the model set and the true model on final error, we provide a symmetric model set optimization method and proportional reduction optimization method. For the Kalman filter [36], it is a linear system, and if the true model is at the geometric center of the model set, any two symmetric models produce opposite errors. Therefore, if the model set has an even number of models, the overall error converges to 0. From another point of view, the error of a single filter is related to the distance of the model from the actual model; thus, it is reasonable to scale down the model set by a certain percentage. Those two theories are based on the current true model. Therefore, to find a model that is closer to the true model, the expected model optimization method is proposed. The modified model is closer to the real model than the original model.

For many existing methods, such as FIMM, VSIMM-CS has excellent cost performance ratio without more computation in the design of the initial model set. Compared with many VSIMM algorithms, just like LMS proposed in [33], VSIMM has better implementability with the same computation. In general, VSIMM-CS shows high precision and universality, and it is also easy to implement.

The remaining parts of the paper are organized as follows: Section 2 introduces the processes of the IMM and VSIMM. Section 3 provides three optimization methods: symmetric model set optimization method, proportional reduction optimization method, and expected model optimization method to prove the feasibility. Section 4 presents the process of VSIMM-CS. Finally, Section 5 provides the conclusion.

2. Multiple-Model Algorithm

In this section, the processing of VSIMM and FIMM are briefly introduced.

2.1. The Process of FIMM

If the model set $M = \{m^{(1)}, m^{(2)}, ..., m^{(N)}\}$ is determined in advance, the model probability transition matrix Π is also determined. The transition probability from i-th model ($m^{(i)}$) to j-th model ($m^{(j)}$) is π_{ij}. If a multiple-model system has M models, each of them can be denoted as

$$\begin{aligned} x_k^{(i)} &= F_{k-1}^{(i)} x_{k-1}^{(i)} + G_{k-1}^{(i)} (a_{k-1} + w_{k-1}^{(i)}) \\ z_k^{(i)} &= H_k^{(i)} x_k^{(i)} + v_k^{(i)} \\ i &= 1, ..., N \end{aligned} \tag{1}$$

where $x = (x, \dot{x}, y, \dot{y})'$ is the target state; $a = (a_x, a_y)'$ is the acceleration; the process noise is $w_k \sim N[0, Q]$; the measurement value is z and its random measure error is $v \sim N[0, R]$; F represents the state transition matrix, G represents the acceleration input matrix, and H is the observation matrix.

Assume that the best target estimation, the state estimation covariance matrix and the model probability of $m^{(i)}$ are $\hat{x}_{k|k}^{(i)}$, $p_{k|k}^{(i)}$ and $u_k^{(i)}$ at the time k, respectively. Then, the overall state estimation and state estimation covariance are

$$\hat{x}_{k|k} = \sum_i x_{k|k}^{(i)} u_k^{(i)} \tag{2}$$

$$p_{k|k} = \sum_i u_k^{(i)} [p_{k|k}^{(i)} + (\hat{x}_{k|k} - \hat{x}_{k|k}^{(i)})(\hat{x}_{k|k} - \hat{x}_{k|k}^{(i)})'] \tag{3}$$

2.2. The Process of VSIMM

The biggest difference between FIMM and VSIMM is whether the model set changes through the target tracking process. For a maneuvering target, its mode space of acceleration S is very large and even uncountable. In most cases, the real model does not fall on the model set. Thus, using a limited model set M to approach S is unreasonable. Simply increasing the number of models does not improve the results, and it has a high probability to degrade the performance, even worse than very few models. The significant advantages of VSIMM are high precision, less computation, and strong adaptability.

If we obtain the current model set M_k through a specific method, and $M_k = \{m_k^{(j)}, j = 1, ..., n_k\}$. The model set M_k at any time is included in the total model set M. Then, the model transition probability is $\pi_{ij} = p\{m_k^{(j)} | m_{k-1}^{(i)}\} = p\{m_k^{(j)} | m_{k-1}^{(i)}, s_k \in M\}$. The overall state estimation and state estimation covariance based on $\hat{x}_{k|k}^{(i)|M_k}$, $p_{k|k}^{(i)|M_k}$ and $u_k^{(i)|M_k}$ are, respectively,

$$\hat{x}_{k|k} = \sum_{m^{(i)} \in M_k} \hat{x}_{k|k}^{(i)|M_k} u_k^{(i)|M_k} \tag{4}$$

$$p_{k|k} = \sum_{m^{(i)} \in M_k} [p_{k|k}^{(i)|M_k} + (\hat{x}_{k|k}^{(i)|M_k} - \hat{x}_{k|k}) (\hat{x}_{k|k}^{(i)|M_k} - \hat{x}_{k|k})'] u_{k|k}^{(i)|M_k} \tag{5}$$

3. Model Optimization Method

In this section, we introduce three methods to optimize the model sets, including symmetric model set optimization method, proportional reduction optimization method and expected model optimization method.

3.1. Symmetric Model Set Optimization Method

A linear system can be described as shown in Figure 1. In addition, X is input and output is $AX + B$. If input is $-X$, the output is $-Ax + B$. It is clear that the output of the sum of $-X$ and X is $2B$. If AX is the error produced by the system, two opposite inputs eliminate errors well.

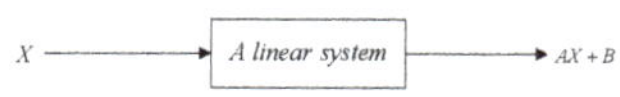

Figure 1. A linear system.

It is assumed that two model sets $M^{(1)}$ and $M^{(2)}$, where $M^{(1)} = \{m_1^{(1)}, m_2^{(1)}, ..., m_N^{(1)}\}$ and $M^{(2)} = \{m_1^{(2)}, m_2^{(2)}, ..., m_N^{(2)}\}$. The current motion mode is s_k. $M^{(1)}$ and $M^{(2)}$ are centrosymmetric and axially symmetric, respectively. However, the centers of symmetries are different. $M^{(1)}$'s center of symmetry is (0,0) and $M^{(2)}$ is s_k. Obviously, the connection between $M^{(1)}$ and $M^{(2)}$, as shown in Figure 2, is

$$m_i^{(1)} + s_k = m_i^{(2)}, i = 1, ..., N \tag{6}$$

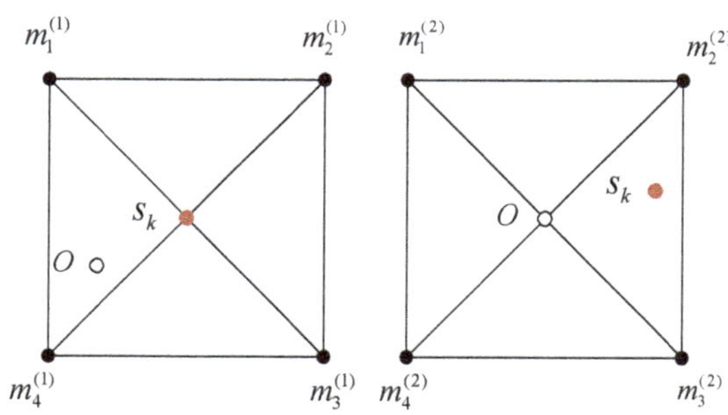

Figure 2. Example of $M^{(1)}$ and $M^{(2)}$.

For $M^{(1)}$, the distance between s_k and $m_i^{(1)}$ is different, and the magnitude of the distance is

$$|m_{q^1}^{(1)} - s_k| \leq |m_{q^2}^{(1)} - s_k| \leq \cdots \leq |m_{q^N}^{(1)} - s_k| \tag{7}$$

For IMM algorithm, the connection between the model probability $u^{(i)}$ and the distance $|m^{(i)} - s_k|$ is

$$|m^{(i)} - s_k| \propto \frac{1}{u^{(i)}} \tag{8}$$

Then, the following relation can be determined as

$$u_{q^1}^{(1)} \leq u_{q^2}^{(1)} \leq ... \leq u_{q^N}^{(1)} \tag{9}$$

For each model $m_i^{(1)}$, they have different errors $\hat{x}_{k|k}^{(i)|M^{(1)}} - s_k = \delta_i^{(1)}$; thus, the overall estimation error is

$$ERROR^{(1)} = \sum_i u_i^{(1)} \delta_i^{(1)} \tag{10}$$

Clearly, according to (8), the overall estimation error has been reduced to a certain extent. However, since the system is linear and it is asymmetrical with respect to s_k, the error of each $m_i^{(1)}$ is not eliminated well.

Since $M^{(2)}$ holds symmetric properties, the relationships can be obtained as

$$\begin{cases} m^{(2)}_{p_1^{(1)}} - s_k = \ldots = m^{(2)}_{p_1^{(2/N_1)}} - s_k \\ = s_k - m^{(2)}_{p_1^{(2/N_1+1)}} \ldots = s_k - m^{(2)}_{p_1^{(N_1)}} = \delta^{(2)}_{p_1} \\ \vdots \\ m^{(2)}_{p_i^{(1)}} - s_k = \ldots = m^{(2)}_{p_i^{(2/N_i)}} - s_k \\ = s_k - m^{(2)}_{p_i^{(2/N_i+1)}} \ldots = s_k - m^{(2)}_{p_i^{(N_i)}} = \delta^{(2)}_{p_i} \\ \vdots \\ m^{(2)}_{p_n^{(1)}} - s_k = \ldots = m^{(2)}_{p_n^{(2/N_n)}} - s_k \\ = s_k - m^{(2)}_{p_n^{(2/N_n+1)}} \ldots = s_k - m^{(2)}_{p_n^{(N_n)}} = \delta^{(2)}_{p_n} \end{cases} \tag{11}$$

where $N = \sum_i N_i$, and N_i is even number. In this linear system, $m^{(2)}_{p_j^{(i)}}$ and $m^{(2)}_{p_j^{(i+N_j/2)}}$ produce two opposite errors $\varepsilon^{(2)}_{p_j}$ and $-\varepsilon^{(2)}_{p_j}$, and

$$\begin{aligned} \varepsilon^{(2)}_{p_j} &= x_{k|k}^{(p_j)|M^{(2)}} - s_k \\ -\varepsilon^{(2)}_{p_j} &= x_{k|k}^{(p_{j+N_j/2})|M^{(2)}} - s_k \end{aligned} \tag{12}$$

Theoretically, if $M^{(2)}$ is strictly symmetric with respecting to s_k, the overall errors are equal to 0, without considering the system noise.

3.2. Proportional Reduction Optimization Method

Suppose two model sets $M^{(1)} = \{m_1^{(1)}, m_2^{(1)}, ..., m_N^{(1)}\}$ and $M^{(2)} = \{m_1^{(2)}, m_2^{(2)}, ..., m_N^{(2)}\}$, and the current model is s_k. The relationship between $M^{(1)}$ and $M^{(2)}$ is

$$\frac{m_i^{(1)} - s_k}{m_i^{(2)} - s_k} = \alpha > 1 \tag{13}$$

Then, the model sets obeying the equation above can be called relative position invariant model sets. The most important feature is that the position of the model sets relative of s_k does not change, as shown in Figure 3.

It is obvious that $M^{(2)}$ is more likely to have better performance than $M^{(1)}$. However, the precondition is that the topology formed by $M^{(1)}$ includes all possible real models, as shown in Figure 4. If the topology is tangent to the true mode space S, this situation can be called the critical point, namely, $\alpha = \alpha_0$.

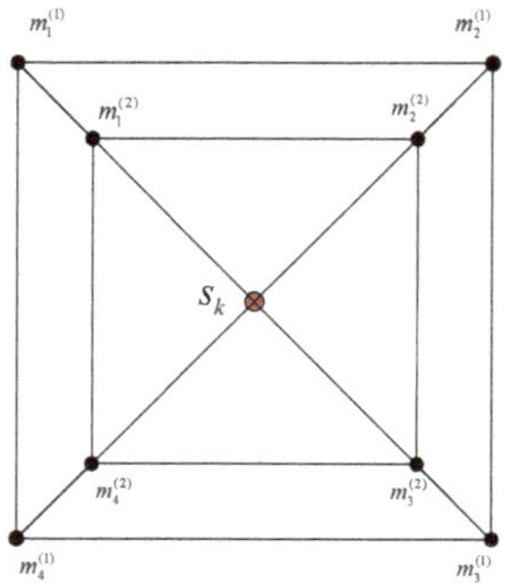

Figure 3. Topology structure of $M^{(1)}$ and $M^{(2)}$.

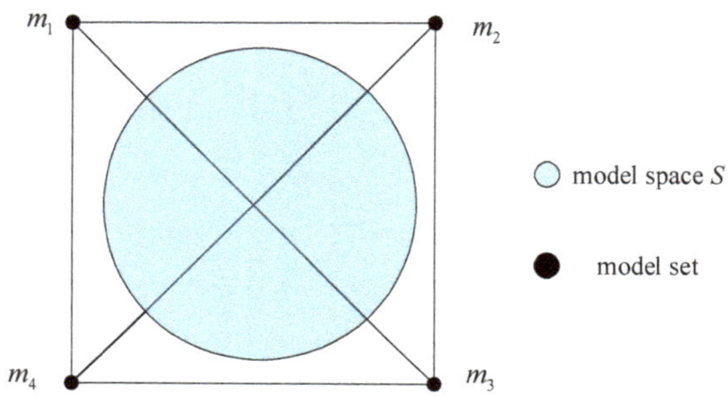

Figure 4. Topology of a model set including all model space S.

Thus, if some noise could be tolerated, the following relationships are determined as

$$\begin{cases} u_1^{(1)} = u_1^{(2)} \\ u_2^{(1)} = u_2^{(2)} \\ \vdots \\ u_N^{(1)} = u_N^{(2)} \end{cases} \tag{14}$$

For IMM, the distance between m_i and s_k directly affects the final error

$$|m_i - s_k| \propto |\hat{x}_{k|k}^{(i)} - s_k| \tag{15}$$

From (13), if the value of α is suitable, it is easy to deduce as

$$|\hat{x}_{k|k}^{(i)|M^{(2)}} - s_k| < |\hat{x}_{k|k}^{(i)|M^{(1)}} - s_k| \tag{16}$$

The connection between overall estimation error of $M^{(1)}$ and $M^{(2)}$ is

$$|ERROR^{(2)}| < |ERROR^{(1)}| \tag{17}$$

Generally, if the model set is scaled down, its performance improves. However, the scale should not be too small, or even exceed the critical point; otherwise, it may make the covariance matrix irreversible.

3.3. Expected Model Optimization Method

If the theories referenced in Sections 3.1 and 3.2 is feasible, the common condition is: the current model s_k is known or a model is found to approach s_k as close as possible. Then, an effective method to approach the current model s_k, called expected-mode augmentation, is given in [31]. The current expected model m_e is the weighted sum of all the models at the current time

$$m_e = \sum_i u_i m_i \tag{18}$$

and it is closer than any model to s_k, namely, $|m_e - s_k < \min\{|m_1 - s_k|, |m_2 - s_k|, \ldots, |m_N - s_k|\}$. However, the value of $|m_e - s_k|$ is not small enough, as shown in Figure 5. Expected model m_e may appear in anywhere in expected model space. If $m_e + \gamma = s_k$, where γ is the error, it is necessary to find a method to reduce γ to a certain extent.

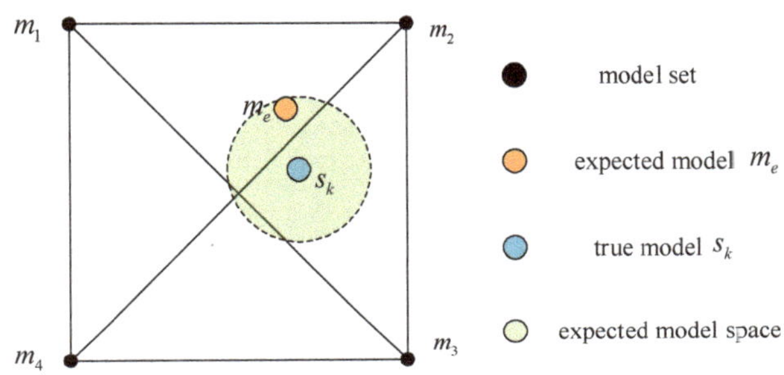

Figure 5. Example of the mentioned hypothesis.

If m_e is scaled a little bit, let m_e becomes λm_e, where λ is scaling factor. Thus, (18) is rewritten as

$$s_k - \lambda m_e = \gamma - (\lambda - 1)m_e \tag{19}$$

and the conditions are

$$\begin{cases} \lambda > 1, \gamma < 0 \\ 0 < \lambda < 1, \gamma > 0 \end{cases} \tag{20}$$

Obviously, the error γ reduces to $\gamma - (\lambda - 1)m_e$. If λ is chosen reasonably, the error becomes small enough and λm_e becomes close enough to s_k, namely, $\frac{\lambda m_e - s_k}{\lambda m_e + s_k} \approx 0$. What is worth noting is increasing or decreasing λ blindly contributes to serious mistakes, including even worse performance.

4. Variable Structure of Interacting Multiple-Model Algorithm Aided by Center Scaling

In this section, we introduce a new VSIMM algorithm: VSIMM-CS. The Sections 3.1–3.3 provide three model optimization theories. The main idea of VSIMM-CS is finding the λm_e to approach the current real model s_k as close as possible, and moving the original model set M and scaling it to obtain the current model set $M^{(k)}$. The current model set $M^{(k)}$ could achieve better performance than the original model set M. The main function of the original model set M is locating the expected model m_e.

The VSIMM algorithm has a clear framework of inputs and outputs. The process of Section 2.2 can be denoted as

$$\begin{aligned} &VSIMM[M^{(k-1)}, M^{(k)}] : \\ &\{\hat{x}_{k|k}^{i|M^{(k)}}, p_{k|k}^{i|M^{(k)}}, u_{k|k}^{i|M^{(k)}}\} \\ &= VSIMM(\hat{x}_{k-1|k-1}^{i|M^{(k-1)}}, p_{k-1|k-1}^{i|M^{(k-1)}}, u_{k-1|k-1}^{i|M^{(k-1)}}) \end{aligned} \tag{21}$$

Obviously, if $M^{(k)} = M^{(k-1)}$, the processing becomes to FIMM. Thus, the FIMM is a special case of VSIMM.

$$\begin{aligned} &FIMM[M]: \\ &\{\hat{x}^{i|M}_{k|k}, p^{i|M}_{k|k}, u^{i|M}_{k|k}\} \\ &= FIMM(\hat{x}^{i|M}_{k-1|k-1}, p^{i|M}_{k-1|k-1}, u^{i|M}_{k-1|k-1}) \end{aligned} \tag{22}$$

The novel VSIMM-CS always has an original model set M, which is taken part in the whole processing of target tracking, and $M^{(k)}$ is always generated by M. $M^{(k)}$ and $M^{(k-1)}$ are probably completely different. The steps of VSIMM-CS are shown in Algorithm 1.

Algorithm 1 VSIMM-CS Process

S1: Increase the time counter k by 1.
S2: Run the $FIMM[M]$ cycle, and obtain the outputs: $\hat{x}^i_{k+1|k+1}$, $u^i_{k+1|k+1}$ and $p^i_{k+1|k+1}$, based on the model set M.
S3: Obtain the expected model m_e by using EMA algorithm, which is the weighted summation of all models.

$$m_e = \sum_i u^i_{k+1|k+1} m_i$$

S4: Select the suitable values of α and λ to generate the current model set $M^{(k+1)}$. $M^{(k+1)} = \alpha\{m_i + \lambda m_e, i = 1, 2, ..., N\}$
S5: Run the $VSIMM[M^{(k)}, M^{(k+1)}]$ cycle to obtain the final results.

$$\hat{x}_{k+1|k+1} = \sum_{m^{(i)} \in M^{(k+1)}} \hat{x}^{(i)|M^{(k+1)}}_{k+1|k+1} u^{(i)|M^{(k+1)}}_{k}$$

$$\begin{aligned} p_{k+1|k+1} = \sum_{m^{(i)} \in M^{(k+1)})} & [p^{(i)|M^{(k+1)}}_{k+1|k+1} + (\hat{x}^{(i)|M^{(k+1)}}_{k+1|k+1} - \hat{x}_{k+1|k+1}) \\ & (\hat{x}^{(i)|M_{(k+1)}}_{k+1|k+1} - \hat{x}_{k+1|k+1})'] u^{(i)|M^{(k+1)}}_{k+1|k+1} \end{aligned}$$

S6: Go to **S1**.

The algorithm complexity is $T(n) = 2n^2 + 16n$, and n equals to the number of models.

The most important part of the process is model set generation. The values of λ and α have a huge impact on the performance of the model set. Therefore, it is unreasonable to choose λ and α blindly. The biggest advantage of VSIMM-CS is that the model set is generated in real-time without increasing huge computational complexity. In addition, if the original model set is designed rational and the number of models is few, the proposed VSIMM-CS can achieve rewarding results and keep the calculation volume within a reasonable range at the same time. The rational combination of λ and α achieve great performance in terms of precision.

5. Simulation Results

In this section, a reasonable simulation process is given: firstly, the target state, the measurement equations and their specific parameters are presented. Secondly, the original model set is given, including the probability transition matrix. Thirdly, the target motion state and performance criterion are presented. Finally, different simulation results are analyzed.

The target state and the measurement equations are, respectively, given as

$$x_{k+1} = F^{(j)}_k x_k + G^{(j)}_k a^{(j)}_k + w^{(j)}_k \tag{23}$$

$$z_k = H_k^{(j)} x_k + v_k^{(j)} \tag{24}$$

where $w_k^{(j)} \sim N[0, 0.01]$; $v_k^{(j)} \sim N[0, 1250]$; $a_k^{(i)}$ is target acceleration input; $F = \begin{bmatrix} 1 & T & 0 & 0 \\ 0 & 1 & T & 0 \\ 0 & 0 & 1 & T \\ 0 & 0 & 0 & 1 \end{bmatrix}$ and $G = \begin{bmatrix} 0.5 & 1 & 0 & 0 \\ 0 & 0 & 0.5 & 1 \end{bmatrix}'$. T is time interval.

The original model set M included 4 models as

$$\left\{ \begin{array}{ll} m_1 = [-10, 10]' & m_2 = [10, 10]' \\ m_4 = [-10, -10]' & m_3 = [10, -10]' \end{array} \right\} \tag{25}$$

and its topology structure is shown in Figure 6

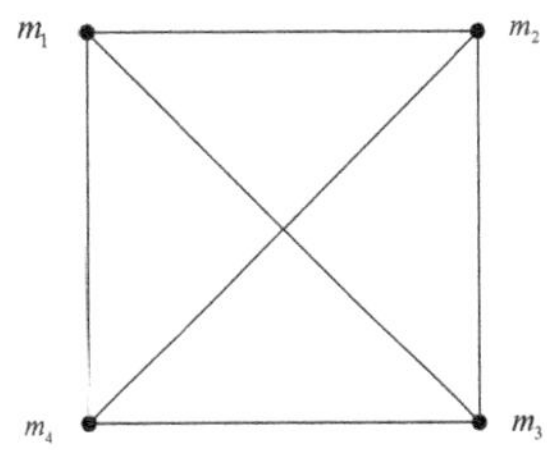

Figure 6. Topology structure of M.

Its probability transition matrix Π is $\begin{bmatrix} 0.85 & 0.05 & 0.05 & 0.05 \\ 0.05 & 0.85 & 0.05 & 0.05 \\ 0.05 & 0.05 & 0.85 & 0.05 \\ 0.05 & 0.05 & 0.05 & 0.85 \end{bmatrix}$. The Table 1 illustrates an ensemble of maneuver trajectories for compare the existing and proposed algorithm.

Table 1. Parameters of deterministic scenarios.

Time k (s)	a_x (m/s^2)	a_y (m/s^2)
1–50	0	0
50–100	5	5
100–150	3	−7
150–200	7	−2
200–250	4	1
250–300	−4	−2
300–350	0	0

If $\bar{e}$ represents the average error of the whole processing, it can be denoted as

$$\bar{e} = \sum_{k=k_1}^{k_N} \sqrt{\left((\hat{x}_k - x_k)' \begin{bmatrix} 1 & 0 & 0 & 0 \\ 0 & 0 & 0 & 0 \\ 0 & 0 & 1 & 0 \\ 0 & 0 & 0 & 0 \end{bmatrix} \right) * \left(\begin{bmatrix} 1 & 0 & 0 & 0 \\ 0 & 0 & 0 & 0 \\ 0 & 0 & 1 & 0 \\ 0 & 0 & 0 & 0 \end{bmatrix} (\hat{x}_k - x_k) \right)} / (k_N - k_1) \tag{26}$$

where $\hat{x}_k$ is the best estimation at time k, x_k is the target state; and $N = 300$. Therefore, different λ and α produce different $\bar{e}$, as shown in Table 2. The unit of $\bar{e}$ is meters, and we just consider the position error.

Table 2. The average error $\bar{e}$ of the tracking process.

λ \ α	0.5	0.6	0.7	0.8	0.9
0.8	16.72	13.00	10.45	8.59	7.19
1.9	13.29	9.89	7.55	5.84	4.56
3	10.16	6.96	4.76	3.15	1.98
4.1	7.21	4.16	2.09	0.98	1.52
5.2	4.32	1.53	1.86	3.14	4.09

Different combinations of λ and α present different performances. Assume λ = 0.8, 1.9, 3, 4.1, 5.2 and $\alpha = 0.5, 0.6, 0.7, 0.8, 0.9$. Their root mean square errors (RMSE) are shown in Figure 7 with 50 times Montecarlo experiment.

In most cases, the VSIMM-CS shows great performance on target tracking, and its errors are far lower than FIMM using 4 models. In some cases, the error is close to 0, and the proposed VSIMM-CS achieves a perfect performance. Thus, sacrificing a certain amount of computation for high accuracy is an acceptable approach.

When $\lambda = 0.8$, as shown in Figure 7a,b, most values of α do not present better performance than IMM-4. Especially, when $\alpha = 0.5$, it becomes the worst one and it produces twice as many errors as IMM-4. As α is increasing, the errors reduce gradually. Until $\alpha = 0.8$, the error curve is as same as IMM-4. When $\alpha = 0.9$, it becomes better than IMM-4. Obviously, in the situation of $\lambda = 0.8$, VSIMM-CS does not show its high precision. When $\lambda = 1.9$ in Figure 7c,d, compared with $\lambda = 0.8$, the tracking precision has been improved. Noting that when $\alpha = 0.7$ or 0.8, the errors are smaller than IMM-4. However, when $\alpha = 0.5$, it is still the worst one among them. It is surprising that, except $\alpha = 0.5$ when $\lambda = 3$, other values of α have greater performance than IMM-4. The relationship of these five algorithms has not changed. $\alpha = 0.9$ is still the best one and $\alpha = 0.5$ is still the worst one. Only when $\alpha = 0.5$, the performance is worse than IMM-4. While the minimum position error is about two meters, it is not close enough to 0. Clearly, up to now, λ does not achieve the most suitable value. When $\lambda = 4.1$, all VSIMM-CS algorithms have better performance than IMM-4, even $\alpha = 0.5$, $\alpha = 0.7$, and $\alpha = 0.9$ have similar tracking precision, and $\alpha = 0.9$ is not the best one anymore. When $\alpha = 0.7$, the position error is smaller than that in any situation as shown in Figure 7. When the true model does not equal to $(0,0)'$, the position error is also very close to 0. It seems that the best combination is $\lambda = 4.1$ and $\alpha = 0.8$ in this system. In this case, the conclusion, referenced in Section 3, has been fully demonstrated. As shown in Figure 7i,j, VSIMM-CS still shows its great performance. All the error curves are less than half of IMM-4's, and $\alpha = 0.6$ has the highest accuracy. The simulation result of $\alpha = 0.7$ is very close to $\alpha = 0.6$. However, the best one shown in Figure 7g,h does not maintain its advantage in Figure 7i,j.

For $\alpha = 0.5$, as λ changes, its precision does not become better, except $\lambda < 5.2$. In these simulation results, it is always the worst one among VSIMM-CS algorithms. The main reason for this is that the value of α is too small. Similarly, $\alpha = 0.6$ still does not meet the requirements. Compared with $\alpha = 0.5$, its performance becomes better. It is not difficult to find that VSIMM-CS takes twice as much computation as IMM-4. However, in most cases, the promotion of $\alpha = 0.6$ is very limited, even it is worse than IMM-4 when $\lambda = 0.9$ or 1.9. Only when $\lambda = 5.2$ does it have the best performance, which is shown in Figure 7i,j. When $\alpha = 0.7$, the precision becomes acceptable. Though it is not the best one in all situations, the performance is greatly promoted. As λ increases, the tracking precision of the proposed algorithm becomes better. When $\alpha = 0.8$ and $\lambda \leq 3$, it only performs worse than $\alpha = 0.9$. Especially when $\lambda = 4.1$, its performance becomes the best one. However, if the value of λ is too large, referenced as Figure 7ij, its advantage disappears. When $\alpha = 0.9$, its performance is always the best one in Figure 7a–f. Similarly, when $\lambda = 5.2$, so much large value of λ would deteriorate the performance.

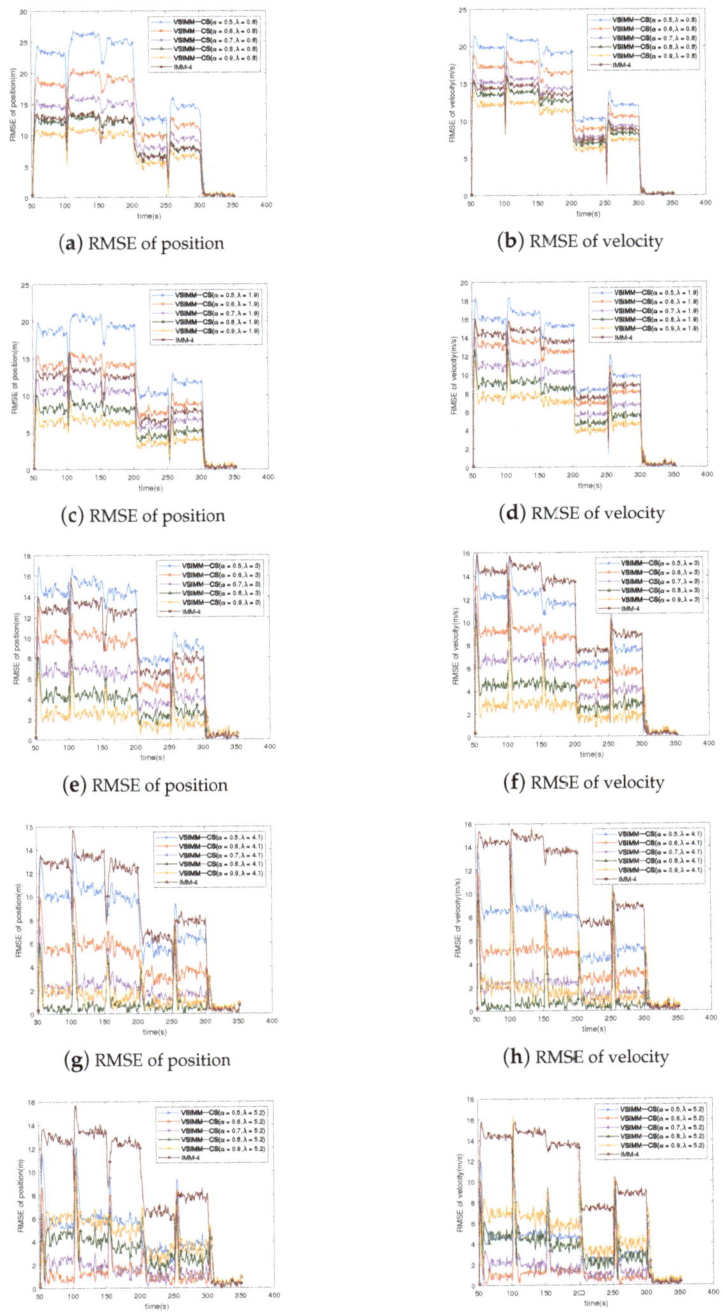

(**a**) RMSE of position (**b**) RMSE of velocity

(**c**) RMSE of position (**d**) RMSE of velocity

(**e**) RMSE of position (**f**) RMSE of velocity

(**g**) RMSE of position (**h**) RMSE of velocity

(**i**) RMSE of position (**j**) RMSE of velocity

Figure 7. Root mean square error of VSIMM-CS.

A good demonstration of the average error $\bar{e}$ of the unworkable portfolio is shown in Table 2. If $\alpha < 0.7$, the critical point would not bee meet the condition at certain times.

When $\alpha = 0.7, 0.8, 0.9$, respectively, there is a high probability that the proposed algorithm achieves better performance.

In general, VSIMM-CS certainly achieves a big performance boost compared with FIMM, if λ and α are selected rationally. Different systems tend to have different suitable combinations of λ and α. Smaller α or larger λ do not promote the performance. According to the current system, the reasonable λ and α always reflect satisfactory results.

6. Conclusions

In this paper, a new variable structure interacting multiple-model algorithm, VSIMM-CS, is proposed. Its model set is generated in real-time, and the generated model set is based on the original model set. Considering the error properties of a linear system and the symmetry of model set structure, the two theories called proportional reduction optimization method and symmetric model set optimization method are presented. The main purpose of the two theories is to reduce errors. Without considering the effect of noise, VSIMM-CS eliminates errors perfectly. To better locate the real model, the expected model optimization method is proposed. The excepted model generated by this method is closer than any other model. Simulation results show different combinations of α and λ have different performances. In most cases, VSIMM-CS achieved better tracking results. It is acceptable to sacrifice a certain amount of computation for high accuracy. A huge performance boost can be obtained by the precise selection of α and λ. When the performance achieves in the optimal situation, α and λ are 0.8 and 4.1, respectively. In different simulation conditions, the results may be different. Many factors may influence the values of α and λ, such as noise, original model set, and true mode space, etc., and our following research aims to focus on these factors. However, the unreasonable selection of α and λ leads to worse results. Simulation results also highlight the rationality and feasibility of this novel approach.

Author Contributions: Writing—original draft preparation, Q.W.; investigation, Q.W. and G.L.; writing—review and editing, G.L., W.J. and S.Z.; project administration, S.Z. and G.L.; supervision, W.S.; funding acquisition, S.Z. All authors have read and agreed to the published version of the manuscript.

Funding: This work was supported in part by the National Natural Science Foundation of China under Grants 62001227, 61971224 and 62001232.

Data Availability Statement: Not applicable.

Acknowledgments: The authors thank the reviewers for their great help on the article during its review progress.

Conflicts of Interest: The authors declare no conflict of interest.

References

1. Gorji, A.A.; Tharmarasa, R.; Kirubarajan, T. Performance measures for multiple target tracking problems. In Proceedings of the 14th International Conference on Information Fusion, Chicago, IL, USA, 5–8 July 2011; pp. 1–8.
2. Poore, A.B.; Gadaleta, S. Some assignment problems arising from multiple target tracking. *Math. Comput. Model.* **2006**, *43*, 1074–1091. [CrossRef]
3. Huang, X.; Tsoi, J.K.; Patel, N. mmWave Radar Sensors Fusion for Indoor Object Detection and Tracking. *Electronics* **2022**, *11*, 2209. [CrossRef]
4. Wei, Y.; Hong, T.; Kadoch, M. Improved Kalman filter variants for UAV tracking with radar motion models. *Electronics* **2020**, *9*, 768. [CrossRef]
5. Li, X.R.; Bar-Shalom, Y. Multiple-model estimation with variable structure. *IEEE Trans. Autom. Control* **1996**, *41*, 478–493.
6. Magill, D. Optimal adaptive estimation of sampled stochastic processes. *IEEE Trans. Autom. Control* **1965**, *10*, 434–439. [CrossRef]
7. Lainiotis, D. Optimal adaptive estimation: Structure and parameter adaption. *IEEE Trans. Autom. Control* **1971**, *16*, 160–170. [CrossRef]
8. Tudoroiu, N.; Khorasani, K. Satellite fault diagnosis using a bank of interacting Kalman filters. *IEEE Trans. Aerosp. Electron. Syst.* **2007**, *43*, 1334–1350. [CrossRef]
9. Kirubarajan, T.; Bar-Shalom, Y.; Pattipati, K.R.; Kadar, I. Ground target tracking with variable structure IMM estimator. *IEEE Trans. Aerosp. Electron. Syst.* **2000**, *36*, 26–46. [CrossRef]

10. Grossman, R.L.; Nerode, A.; Ravn, A.P.; Rischel, H. *Hybrid Systems*; Springer: Berlin/Heidelberg, Germany, 1993; Volume 736.
11. Branicky, M.S. Introduction to hybrid systems. In *Handbook of Networked and Embedded Control Systems*; Birkhäuser: Basel, Switzerland, 2005; pp. 91–116.
12. Li, X.R. Multiple-model estimation with variable structure. II. Model-set adaptation. *IEEE Trans. Autom. Control* **2000**, *45*, 2047–2060.
13. Labbe, R. Kalman and bayesian filters in python. *Chap* **2014**, *7*, 4.
14. Zhang, G.; Lian, F.; Gao, X.; Kong, Y.; Chen, G.; Dai, S. An Efficient Estimation Method for Dynamic Systems in the Presence of Inaccurate Noise Statistics. *Electronics* **2022**, *11*, 3548. [CrossRef]
15. Rong Li, X.; Jilkov, V. Survey of maneuvering target tracking. Part V. Multiple-model methods. *IEEE Trans. Aerosp. Electron. Syst.* **2005**, *41*, 1255–1321. [CrossRef]
16. Bar-Shalom, Y. *Multitarget-Multisensor Tracking: Applications and Advances*; Artech House, Inc.: Norwood, MA, USA, 2000; Volume iii.
17. Blom, H.A.; Bar-Shalom, Y. The interacting multiple model algorithm for systems with Markovian switching coefficients. *IEEE Trans. Autom. Control* **1988**, *33*, 780–783. [CrossRef]
18. Ma, Y.; Zhao, S.; Huang, B. Multiple-Model State Estimation Based on Variational Bayesian Inference. *IEEE Trans. Autom. Control* **2019**, *64*, 1679–1685. [CrossRef]
19. Wang, G.; Wang, X.; Zhang, Y. Variational Bayesian IMM-filter for JMSs with unknown noise covariances. *IEEE Trans. Aerosp. Electron. Syst.* **2019**, *56*, 1652–1661. [CrossRef]
20. Li, H.; Yan, L.; Xia, Y. Distributed robust Kalman filtering for Markov jump systems with measurement loss of unknown probabilities. *IEEE Trans. Cybern.* **2021**, *52*, 10151–10162. [CrossRef]
21. Johnston, L.; Krishnamurthy, V. An improvement to the interacting multiple model (IMM) algorithm. *IEEE Trans. Signal Process.* **2001**, *49*, 2909–2923. [CrossRef]
22. Fan, X.; Wang, G.; Han, J.; Wang, Y. Interacting Multiple Model Based on Maximum Correntropy Kalman Filter. *IEEE Trans. Circuits Syst. II Express Briefs* **2021**, *68*, 3017–3021. [CrossRef]
23. Davis, R.R.; Clavier, O. Impulsive noise: A brief review. *Hear. Res.* **2017**, *349*, 34–36. [CrossRef]
24. Nie, X. Multiple model tracking algorithms based on neural network and multiple process noise soft switching. *J. Syst. Eng. Electron.* **2009**, *20*, 1227–1232.
25. Mazor, E.; Averbuch, A.; Bar-Shalom, Y.; Dayan, J. Interacting multiple model methods in target tracking: A survey. *IEEE Trans. Aerosp. Electron. Syst.* **1998**, *34*, 103–123. [CrossRef]
26. Gao, W.; Wang, Y.; Homaifa, A. Discrete-time variable structure control systems. *IEEE Trans. Ind. Electron.* **1995**, *42*, 117–122.
27. Li, X.R.; Bar-Shakm, Y. Mode-set adaptation in multiple-model estimators for hybrid systems. In Proceedings of the 1992 American Control Conference, Chicago, IL, USA, 24–26 June 1992; pp. 1794–1799.
28. Pannetier, B.; Benameur, K.; Nimier, V.; Rombaut, M. VS-IMM using road map information for a ground target tracking. In Proceedings of the 2005 7th International Conference on Information Fusion, Philadelphia, PA, USA, 25–28 July 2005; Volume 1, 8p.
29. Xu, L.; Li, X.R. Multiple model estimation by hybrid grid. In Proceedings of the 2010 American Control Conference, Baltimore, MD, USA, 30 June–2 July 2010; pp. 142–147.
30. Li, X.R.; Zwi, X.; Zwang, Y. Multiple-model estimation with variable structure. III. Model-group switching algorithm. *IEEE Trans. Aerosp. Electron. Syst.* **1999**, *35*, 225–241.
31. Li, X.R.; Jilkov, V.P.; Ru, J. Multiple-model estimation with variable structure-part VI: expected-mode augmentation. *IEEE Trans. Aerosp. Electron. Syst.* **2005**, *41*, 853–867.
32. Lan, J.; Li, X.R. Equivalent-Model Augmentation for Variable-Structure Multiple-Model Estimation. *IEEE Trans. Aerosp. Electron. Syst.* **2013**, *49*, 2615–2630. [CrossRef]
33. Li, X.R.; Zhang, Y. Multiple-model estimation with variable structure. V. Likely-model set algorithm. *IEEE Trans. Aerosp. Electron. Syst.* **2000**, *36*, 448–466.
34. Sun, F.; Xu, E.; Ma, H. Design and comparison of minimal symmetric model-subset for maneuvering target tracking. *J. Syst. Eng. Electron.* **2010**, *21*, 268–272. [CrossRef]
35. Callier, F.M.; Desoer, C.A. *Linear System Theory*; Springer Science & Business Media: Berlin, Germany, 2012.
36. Kalman, R.E. A new approach to linear filtering and prediction problems. *J. Basic Eng.* **1960**, *82*, 35–45. [CrossRef]

electronics

MDPI

Article

An Accelerator for Semi-Supervised Classification with Granulation Selection

Yunsheng Song [1,2], Jing Zhang [1,*], Xinyue Zhao [1] and Jie Wang [3]

1 School of Information Science and Engineering, Shandong Agricultural University, Tai'an 271018, China; sys_sd@126.com (Y.S.); zhaoxinyue1006@163.com (X.Z.)
2 Key Laboratory of Huang-Huai-Hai Smart Agricultural Technology, Ministry of Agriculture and Rural Affairs, Shandong Agricultural University, Tai'an 271018, China
3 School of Information, Shanxi University of Finance and Economics, Taiyuan 030006, China; 20191031@sxufe.edu.cn
* Correspondence: zhangjing_12138@163.com

Abstract: Semi-supervised classification is one of the core methods to deal with incomplete tag information without manual intervention, which has been widely used in various real problems for its excellent performance. However, the existing algorithms need to store all the unlabeled instances and repeatedly use them in the process of iteration. Thus, the large population size may result in slow execution speed and large memory requirements. Many efforts have been devoted to solving this problem, but mainly focused on supervised classification. Now, we propose an approach to decrease the size of the unlabeled instance set for semi-supervised classification algorithms. In this algorithm, we first divide the unlabeled instance set into several subsets with the information granulation mechanism, then sort the divided subsets according to the contribution to the classifier. Following this order, the subsets that take great classification performance are saved. The proposed algorithm is compared with the state-of-the-art algorithms on 12 real datasets, and experiment results show it could get a similar prediction ability but have the lowest instance storage ratio.

Keywords: semi-supervised classification; co-training method; instance selection; granular computing; information granulation

Citation: Song, Y.; Zhang, J.; Zhao, X.; Wang, J. An Accelerator for Semi-Supervised Classification with Granulation Selection. *Electronics* **2023**, *12*, 2239. https://doi.org/10.3390/electronics12102239

Academic Editors: Chao Zhang, Wentao Li, Huiyan Zhang and Tao Zhan

Received: 16 April 2023
Revised: 11 May 2023
Accepted: 12 May 2023
Published: 15 May 2023

1. Introduction

Co-training is a semi-supervised learning technique in which two classifiers are trained on separate, complementary views of the same data, with the idea that the two views contain different but complementary information [1–5]. In the context of co-training, a view refers to a different representation of the same data. For example, if the data is a set of documents, one view could be the text of the documents, while the other view could be the meta-data associated with the documents, such as the author or the date of publication. The basic idea behind co-training is that the two classifiers learn from each other and the labeled data so that they become more accurate over time. In each iteration of co-training, the classifiers make predictions on the unlabeled data, and the most confident predictions are used to label more data. This newly labeled data is then used to retrain both classifiers, and the process repeats. Research on co-training has shown that it can be an effective technique for semi-supervised learning, especially in domains where the two views of the data are indeed complementary.

It is also worth mentioning that co-training has been applied to a variety of domains, including natural language processing (NLP), computer vision, and bioinformatics. In the field of NLP, co-training has been used to improve the performance of sentiment analysis, text classification, and topic modeling [6]. In computer vision, co-training has been used for image classification and object detection [7]. In bioinformatics, co-training has been used for protein function prediction and gene expression analysis [8]. One of the strengths of co-training is its ability to handle large and complex datasets, where traditional supervised learning methods may struggle. For instance, in NLP, co-training has been shown to be effective when dealing with imbalanced datasets, where the number of positive instances is much smaller than the number of negative instances. In such scenarios, co-training can effectively leverage the information contained in the unlabeled data to improve the performance of the classifier. Another area of application for co-training is in data privacy, where it is often the case that only a limited amount of labeled data is available for training machine learning models. In these scenarios, co-training can effectively leverage the information contained in the unlabeled data to improve the performance of the classifier, without compromising privacy or security [9].

In recent years, several variations and extensions of co-training have been proposed to address its limitations and improve its performance. For instance, some researchers have proposed using multiple views of the data rather than just two to capture more information and make the semi-supervised learning process more robust [10]. Another line of research has focused on developing new co-training algorithms that are able to handle noisy or conflicting views of the data [11]. These algorithms aim to identify and discard unreliable predictions made by one of the classifiers so that the other classifier can make better predictions in the absence of high-quality supervision. Additionally, there has been a growing interest in using deep learning models for co-training. For instance, one approach is to use generative models, such as Generative Adversarial Networks (GANs), to generate synthetic samples that can be used to augment the labeled data [12]. By using these synthetic samples in co-training, it is possible to effectively increase the size of the labeled data, leading to improved performance. Meanwhile, co-training can handle high-dimensional, complex data representations with deep learning models. For instance, some researchers have proposed using deep neural networks as the classifiers in co-training and have shown that this can lead to improved performance in various applications, including image classification, sentiment analysis, and document classification [13]. Overall, the field of co-training and semi-supervised learning is rapidly evolving, and there is a wealth of ongoing research aimed at improving the performance and robustness of these algorithms. As such, it is an exciting and promising area of study for anyone interested in machine learning and data science.

Although co-training plays an important role in the semi-supervised classification task, large-scale data poses a huge challenge to the efficiency of its modeling [14]. Existing co-training-based semi-supervised classification algorithms usually need to traverse all unlabeled samples multiple times to find high-confidence elements or valuable classification information, but large-scale unlabeled instances make it difficult to achieve efficient modeling. Some literature proposes using different subsets of unlabeled samples after division to improve the efficiency of the algorithm; it does not consider the differences in the contribution of different unlabeled samples to the algorithm. However, it takes a great challenge for traditional semi-supervised classification algorithms based on co-training to handle large-scale data in terms of compatibility, effectiveness, and timeliness. Instance selection as an important data reduction method can solve the large-scale classification problem by reducing the labeled instances depending on enough label information to achieve the aim [15,16]. Therefore, traditional instance selection methods cannot be applied to the semi-supervised classification problem because there exists a small number of labeled instances with little labeled information. Moreover, each instance is seen as a basic processing unit to judge whether it is selected or not [17]. It is difficult to follow this approach to dealing with large-scale unlabeled instances, and there is a need to solve this problem from a new perspective.

Granular computing is a methodology for processing and analyzing complex data by partitioning it into smaller, more manageable pieces [18–22]. These smaller pieces, or granules, can then be further analyzed and processed to provide insights into the original data. The goal of granular computing is to simplify complex problems by reducing their complexity to more manageable pieces. This approach has been applied to a variety of problems in machine learning, including clustering, classification, and feature selection. Meanwhile, granular computing and co-training are both techniques that can be used to improve classification accuracy. Granular computing can be used to reduce the complexity of the data by partitioning it into smaller, more manageable pieces [23]. Once the data has been partitioned into granules, co-training can be used to train multiple models on each granule. This approach can be particularly effective in semi-supervised learning applications where labeled data is limited. Nevertheless, the contribution of each kind of information granularity with a large difference to the classifier has not received sufficient attention, so its efficiency has dropped dramatically, and information could be redundant [24,25].

For the problem based on the above analysis, this paper has proposed an effective instance selection for a co-training-based semi-supervised classification task using the granulation mechanism, which deals with the large data using the information particles as the basic processing unit rather than each instance and considers the different contribution of granularity to the classifier. The contribution of this paper is as follows:

- Proposing a progressive instance selection mechanism to reduce unlabeled instances by the significant variation in classification accuracy.
- Giving a novel unlabeled information granulation mechanism based on the extent to which the unlabeled instance improves the performance of the classifier, and it avoids the influence of human subjective factors.
- Adaptive determining in which unlabeled information granulation is ultimately saved according to its contribution to the classification performance.
- Verifying the proposed method could largely reduce the unlabeled data size and keep the original classification performance by the experiment result on the real datasets.

The rest of this paper is listed as follows. Section 2 introduces related work about co-training-based semi-supervised classification algorithms. Section 3 analyzes the effect of unlabeled instances on the classifier and has proposed an effective instance selection for co-training-based semi-supervised classification. Section 4 verifies the effectiveness of the proposed method. Section 5 concludes this paper.

2. Related Work

A co-training-based semi-supervised classification algorithm needs to cooperate with different classifiers from multiple perspectives at the same time to realize the utilization of unlabeled data, and it has become the focus of research with its higher effectiveness [3,26]. According to the different learning strategies, the existing co-training algorithms are mainly divided into two categories: the ones based on the sample set augmentation and the ones based on regularization.

Co-training algorithms based on sample set augmentation, which use classifiers from different perspectives to select high-confidence unlabeled samples and corresponding prediction labels from the unlabeled sample set, alternately assign the newly added samples to different classifiers for retraining and finally repeat the above process until the prediction results converge. In such algorithms, how to efficiently select labeled samples with high confidence is the bottleneck that restricts the efficiency of the algorithm. Paper [27] divides the sample space into a set of equivalence classes and uses cross-validation to determine how to label unlabeled samples. Paper [28] uses voting to select unlabeled samples with high confidence; In order to improve the robustness of the collaborative training algorithm, and papers [29,30] use filtering to screen the newly added unlabeled samples instead of using them all [31].

Co-training algorithms based on regularization use the information provided by different perspective classifiers as the regularization term of the learning object, and transform the semi-supervised multi-view learning problem into an optimization problem [32,33]. In order to improve the training efficiency of such algorithms, Sun et al. [34] propose a sequential training method that uses the union of different unlabeled sample subsets and labeled sample set L on the basis of dividing the unlabeled sample set into ten subsets of equal size, the union of the first unlabeled sample subset and set L is first used for modeling, and then the next unlabeled sample set, some elements of the utilized unlabeled sample set participate in the modeling, and the union modeling of set L. Finally, repeat the previous step until all unlabeled sample subsets are utilized. Existing difference-based semi-supervised classification algorithms need to traverse all unlabeled samples multiple times to find high-confidence elements or valuable classification information, but the massive scale of unlabeled data makes it difficult to achieve efficient modeling. Although some literature proposes to use different subsets of unlabeled samples after division to improve the efficiency of the algorithm, it does not consider the differences in the contribution of different unlabeled samples to the algorithm.

In conclusion, the existing large-scale co-training-based supervised classification algorithms mainly improve the training efficiency from the view of optimization design. However, the time complexion is difficult to reduce and obtain a greater improvement for the large problem, and it still suffers from the large training burden of using the whole of the unlabeled instances to participate in the training process.

3. Main Content

For the given training set T, which is the union of the labeled instance set $L = \{x_1, x_2, \cdots, x_l\}$ and the unlabeled instance set $U = \{x_{l+1}, x_{l+2}, \cdots, x_{l+u}\}$, where x_i denotes the training instance, l and u are the number of labeled instances and unlabeled instances, and $i = 1, 2, \cdots, l+u$. Semi-supervised classification algorithms simultaneously use the labeled instance set L, and unlabeled instance set U to train a classifier $f(x)$ with good performance. A co-training-based semi-supervised classification algorithm uses the idea of compatible complementarity of multiple views to learn the final classifier. It assumes that the data has multiple sufficient and conditional independence views, and the classifier trained on one view can offer supplemental information to the classifiers on the other view. The supplemental information is achieved by selecting the most trusted unlabeled instances and pseudo-labels. Nevertheless, several iterations are required, and each iteration must scan the whole of the unlabeled instances set to the most trusted instances. The large-scale unlabeled instance carries difficulty in efficiently learning the final classifier.

Instance selection, as one of the most important data preprocessing technologies to reduce dataset size, is widely used for classification problems, as is the fact that the contribution of training instances with the different locations in the space to learn a classifier varies greatly. Numerous studies have shown that the training instances can be divided into critical instances and non-critical instances, where critical instances mainly define the class boundary and separate the instances of the same label from the ones from other labels [16]. Meanwhile, the number of critical instances is significantly smaller than that of non-critical instances in most real-world datasets. Therefore, the process requires an effective way to reduce the training set to a relatively small subset by selecting critical instances and preserving the original data information. Compared with the performance on the original training set, the efficiency of training the classifier on the reduced set can be significantly improved on the reduced subset.

Traditional instance selection methods for supervised classification tasks start with each instance as the most basic processing unit, critical instances are selected by the contribution of each labeled instance to the classifier. The contribution of an instance to the classifier is usually measured by its location in the input space and the difference with its nearest neighbors on the label. Although the instance selection is very efficient for supervised classification tasks, it is difficult to apply directly to semi-supervised classification tasks because of its limitations. Different from supervised classification, there exists a large number of unlabeled instances and few labeled instances for the semi-supervised classification tasks. Only labeled instances take labeled information to the learner, and this information is vital to learn a classifier with good performance, so it cannot reduce labeled instances. Otherwise, the generation ability of the obtained classifier could significantly decrease. On other hand, traditional instance selection needs the labeled information of each instance to execute data reduction, while this condition is not met for the unlabeled instances. Moreover, the way of treating each instance as the basic process unit is undesirable for large-scale problems because it is very time-consuming.

To overcome this difficulty, we have proposed a novel instance selection with the granulation mechanism. This proposed method consists of two key processes: unlabeled information granulation and granulation selection.

3.1. Unlabeled Information Granulation

Unlabeled instance selection aims to reduce the unlabeled instance set U, and the original unlabeled information remains relatively unchanged. Unlabeled information is expressed by the contribution of unlabeled instances to learn the classifier for semi-supervised learning, and it is closely related to the feature of semi-supervised classification algorithms. For co-training-based semi-supervised classification algorithms, the contribution of the unlabeled instance to learn the final classifier mainly depends on the determination of predictive label consistency of the classifiers trained with different views, as well as its location in the input space. The unlabeled instances located in different regions in the input space have different contributions to the classifier [35]. Figure 1 shows a 2-dimension binary classification problem to learn a linear classifier, where two labeled instances from different classes are represented by blue circle and yellow circle. The unlabeled instances nearby the decision boundary have much more of a contribution than the ones far from the decision boundary.

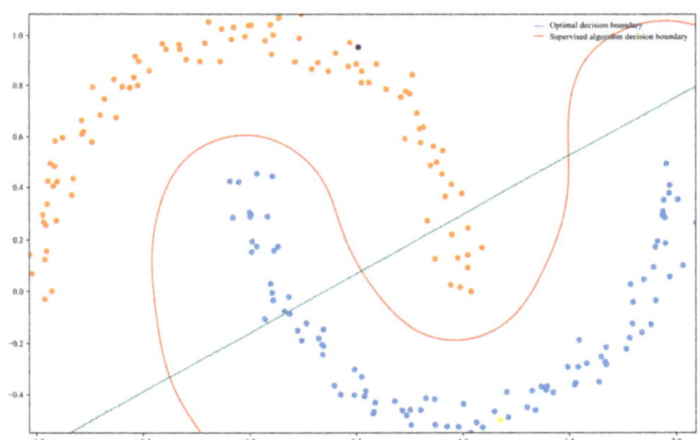

Figure 1. An example of a binary semi-supervised classification problem.

This difference in the contribution of unlabeled instances to the classifier yields the possibility of executing an instance selection. Compared with the abundant labeled information of the labeled instances, unlabeled instances bring a limited classification contribution to the learner. Due to the presence of a large number of unlabeled instances with limited classification information, it is difficult to select critical unlabeled instances with their contribution one by one. Furthermore, semi-supervised classification should not reduce unlabeled instances one by one from an execution efficiency perspective. This process is a disaster, especially for classification algorithms with high time complexity. Therefore, we adopt the idea of granular computing to divide the unlabeled instance set U into m disjoint subsets U_i, $U = \bigcup \sum_{i=1}^{m} U_i$. All the instances in the same subset U_i are considered as a basic information granularity to participate in the learning process. In this way, it can greatly improve learning efficiency by only processing a small number of units. Meanwhile, the contribution of a subset U_i is easy to obtain compared with the single unlabeled instance.

Data partition, as one of the important data granulation techniques, plays a crucial role in granular computing. There are three key factors to performing data partition for the co-training-based semi-supervised classification tasks.

- Divided unlabeled instance subsets have the unbalanced information for the final classifier to obtain a relatively small number of aim subsets to achieve data reduction.
- Number of divided subsets is determined by the characteristics of unlabeled instances and the contribution to the classifier rather than the subjective prior determination.
- Data partition should use the contribution of the unlabeled instances to the semi-supervised classifier and close together with the distinguishing feature of co-training.

Therefore, we utilize a similar framework as a Tri-Training [36] method to perform data partition. Each initial decision tree classifier $f_r(x)$ is independently trained on the different set B_r sampled from the labeled set L using Bootstrap sampling method, where r = 1, 2, 3. Owning to the feature of Bootstrap sampling, the sample set B_r has a large difference, as well as the classifier $f_r(x)$ on it. Then it iteratively retrains each classifier with the enlarged labeled set L, which is created by introducing several confident unlabeled instances and their pseudo-label until none of the classifiers changes. The confident unlabeled instance and its pseudo-label obtained by each classifier are provided by the remaining two classifiers. Specifically, if the two classifiers have the same prediction for the same unlabeled instances, these instances are considered to have high labeling confidence and are added to the labeled training set of the third classifier. In this way, we can estimate the frequency $fre(x_i)$ at which each unlabeled instance x_i is selected as a confident element during this process. Finally, the unlabeled set U is divided into several subsets according to the condition that all the unlabeled instances x_i in the same subset have the same frequency. The pseudocode of the proposed method is presented in Algorithm 1.

A decision tree (DT) is chosen as the basic classifier for the Tri-training algorithm for its unique advantages in Algorithm 1, that is, learning features, high efficiency, and instability. Both the conditional probability distribution information for the class and the local geometry information in the input space are used to learn the classifier of DT, and this kind of information is very comprehensive. Furthermore, the time complexion of DT is approximately linear of time complexion to efficiently process large-scale data. Finally, the instability of DT is sensitive to data change for its instability, this is constructive to instance selection [37].

The measurement $fre(x_i)$ is the frequency at which each unlabeled instance x_i is selected as the confident instance for three classifiers in the whole training process, and it reflects each unlabeled instance x_i to learn the final classifier. The large value of frequency $fre(x_i)$ means the unlabeled instance x_i is always chosen and has a large contribution to the final classifier. A different value of $fre(x_i)$ also indicates different degrees of effect on classification performance. Different from previous methods to evaluate the contribution with a real number, the measurement metrics takes a limited integer value. It is constructive to divide the unlabeled set U into several subsets according to the possible value of the measurement $fre(x_i)$. Moreover, the number $n = \max_{x_i \in U} fre(x_i) - \min_{x_i \in U} fre(x_i)$ of discrete values of the measurement $fre(x_i)$ is not subjectively predetermined; it depends on the effect of unlabeled instances on the classifier.

Algorithm 1 Unlabeled information granulation algorithm

Input : Training set $T = L \bigcup U$, where the labeled set $L = \{x_1, x_2, \cdots, x_l\}$ and the unlabeled instance set $U = \{x_{l+1}, x_{l+2}, \cdots, x_{l+u}\}$, decision-tree classification algorithm.

Output: The divided unlabeled instance subsets U_h, $U = \bigcup_{h=1}^{n} U_h$.

Initialization : $fre(x_i) = 0$ for $x_i \in U$, $L_1 = L_2 = L_3 = L$, $Update = 1$;
foreach $r \in \{1,2,3\}$ **do**
  Train the decision tree classifier $f_r(x)$ on the set B_r sampled from the set L using Bootstrap sampling method;
end
while *Update= 1* **do**
  foreach $r \in \{1,2,3\}$ **do**
    $Update_r = 0$;
    foreach $x_i \in U$ **do**
      if $f_j(x_i) = f_k(x_i)$ $(j, k \neq r)$ **then**
        $fre(x_i) = fre(x_i) + 1$, $L_r = L_r \bigcup \{(x_i, f_j(x_i))\}$, $Update_r = 1$;
      end
    end
  end
  foreach $r \in \{1,2,3\}$ **do**
    if $Update_r = 1$ **then**
      Re-train the decision tree classifier $f_r(x)$ on the new set L_r, $Update = 1$;
    end
  end
end
$n = Maxf - Minf + 1$, where $Maxf = \max_{x_i \in U} fre(x_i)$ and $Minf = \min_{x_i \in U} fre(x_i)$;
foreach $h \in \{1, 2, \cdots, n\}$ **do**
  $U_h = \{x_i \in U : fre(x_i) = h - 1 + Minf\}$;
end
Return *The divided unlabeled instance subset* U_h;

3.2. Granulation Selection

After the unlabeled data granulation with data partition, divided subsets of unlabeled instances must be finally chosen to train the semi-supervised classifier. It is undesirable to randomly select several divided subsets as the final result for different contributions to the classifier. According to the same value of $f(x_i)$ of the instances $x_i \in U_h$, the order of contribution from smallest to largest is $U_1, U_2, \cdots, U_m$, where m is the number of the divided unlabeled instance subsets. To keep the full information of the unlabeled set U with the smaller number of divided unlabeled instance subsets as much as possible, we adopt the way of one subset by one subset in reverse order. In this way, it is constructive to search the smaller number of subsets with much more auxiliary classification information to the learner in the following process. Moreover, this method relies solely on the value $f(x_i)$ over the divided subset to select results without any limitation to the classifier.

Let $Acc(U_g)$ be the measurement of classification performance for the classifier trained on the set U_h and the labeled set L, the change on the classification performance $\Delta(U_{g-1}) = Acc(U_g \bigcup U_{g-1}) - Acc(U_{g-1})$ between U_g and U_{g-1}, $g = 2, 3, \cdots, m$. $Acc(U_g)$ evaluates the effect of the set U_h to the semi-supervised learner classification performance, and $\Delta(U_g - 1)$ indicates the effect of adding set U_{g-1} to set U_g. If the value of $\Delta(U_{g-1})$ is small relative to $Acc(U_g)$, then merging set U_{g-1} with set U_g is difficult to train a strong semi-supervised classifier. Therefore, the following condition (1) is set to judge whether it merges set U_{g-1} with set U_g or not.

$$\frac{Acc(U_g \bigcup U_{g-1}) - Acc(U_{g-1})}{Acc(U_g)} \geq \alpha, \tag{1}$$

where $\alpha \in (0,1)$, $g = 2,3,4,\cdots,m$. Many papers suggest that the critical value $\alpha = 0.05$ to obtain a significant change in performance [38]. Above all, the pseudocode of the granulation selection is presented in Algorithm 2.

In Algorithm 2, the early stopping condition is used to prevent performing too many iterations. The classifier trained on set $L \bigcup U_g \bigcup U_{g-1}$ may improve the classification performance of the one on the set $L \bigcup U_g$ because it adds more unlabeled sample information from the set U_{g-1}. Moreover, the unsupervised information of the set U_g that is constructive to improve the classification performance could be more than the set U_{g-1}, where $g = 2,3,\cdots,m$. Therefore, the subset U_j is difficult to satisfy condition (1) if the previous subset U_i cannot meet, where $1 \leq j < i \leq m$. In this way, this selection process can be terminated early and obtain a lower number of divided unlabeled instance subsets.

Algorithm 2 Granulation selection algorithm

Input : Training set $T = L \bigcup U$, where the labeled set $L = \{x_1, x_2, \cdots, x_l\}$ and the unlabeled instance set $U = \{x_{l+1}, x_{l+2}, \cdots, x_{l+u}\}$, decision-tree classification algorithm, and semi-supervised classification algorithm f, critical value $\alpha = 0.05$.

Output: The reduced set U_s of the set U.

Run Algorithm 1 with decision tree to get m divided subsets U_h, $U = \bigcup_{h=1}^{m} U_h$;
Train semi-supervised classifier with the set U_m to obtain Acc^*, $U_s = U_m$;
foreach $g \in \{m-1,\cdots,2\}$ **do**
  $\tilde{U}_g = U_s \bigcup U_g$;
  Train semi-supervised classifier with the set $\tilde{U}_g$ to obtain $Acc(\tilde{U}_g)$;
  if $\frac{Acc(\tilde{U}_g) - Acc^*}{Acc^*} \geq \alpha$ **then**
    $U_s = \tilde{U}_g$;
    $Acc^* = Acc(\tilde{U}_g)$;
  else
    Get out of the loop
  end
end
Return *The set* U_s;

3.3. Complexity Analysis

Complexity analysis is very important for evaluating the classifier, and it starts with two main steps of the proposed method. The first step includes the frequency in which an unlabeled instance is selected as a trust element and the frequency discretization, where the former mainly depends on the time complexion of the basic classifier and the latter is linear with the number of unlabeled instances. A decision tree is selected as the basic classifier in this method of the approximately linear time complexity $O(dml\log(l))$ with the size l of labeled instances and m features. Meanwhile, the efficiency of a decision tree is very high to predict the label for the unlabeled instances with the time complexity $O(d)$, where d is the depth of the tree. Thus, it can efficiently process big data, which has massive unlabeled data, to offer the pseudo-labels of the linear time complexion that is linearly related to the size of the data. Therefore, the time complexity of the first step is $O(dml\log(l) + du)$. The time complexity of the second step relies on the complexity of the adopted semi-supervised algorithm and the number of iterations n. In each iteration, the semi-supervised classifier is trained using the subset of unlabeled set U rather than all the instances, and its time complexity is proportional to the size of the labeled instance subset. Meanwhile, the early stopping condition is constructive in reducing the number of iterations. In conclusion, the time complexity of the proposed method is approximately linear with the number and the dimension of labeled instances and unlabeled instances, and it is proportional to the time complexity of the adopted semi-supervised classifier that is used to get the classification accuracy.

4. Experiments

To verify the effectiveness and efficiency of the proposed algorithm for real problems, extensive experiments on real datasets have been implemented against the typical method under differently labeled rations.

4.1. Experiment Setup

Twelve large datasets of different types are randomly selected to evaluate the performance of the algorithms from the KEEL-dataset repository [31] and LIBSVM-dataset repository [39], where each data has larger than 4000 instances. The basic information of the selected datasets is listed in Table 1.

Table 1. Summary of twelve datasets.

Dataset	Size	Features	Classes
combined	98,528	101	3
connect-4	67,557	126	3
covtype	581,012	54	7
letter	20,000	16	26
optdigits	5620	65	10
pendigits	10,992	17	10
phoneme	5404	5	2
ring	7400	21	2
seismic	98,528	51	3
texture	5500	41	11
usps	9298	257	10
winequality	4898	11	7

Further, a typical image dataset of five generic categories called NORB [40] is selected to test the performance of the proposed method for high-dimensional datasets. The following Figure 2 shows examples of the training image and test image of the dataset.

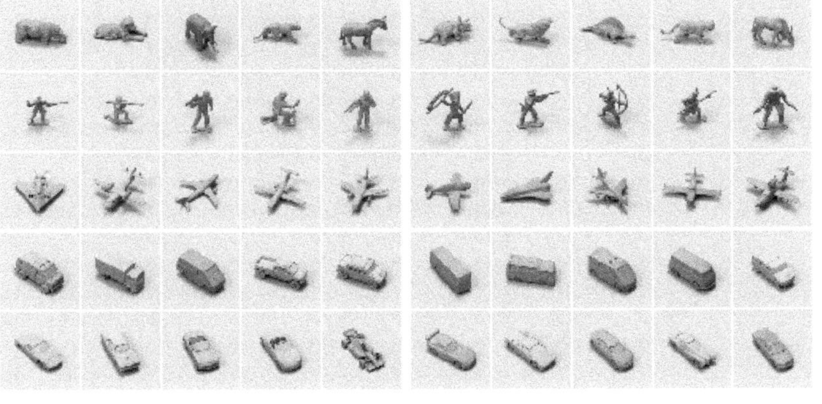

Figure 2. The examples of NORB dataset.

For each dataset, about 3/4 of the data is selected as the training set and the rest as the test set, where each training set is the union of the labeled subset L and the unlabeled subset U. To effectively verify the generalization performance of the proposed algorithm for real data, the proportion of labeled instances (PL) to the total training instances has different values. According to the suggestion of the paper [36], the value of PL includes 20%, 40%, 60%, and 80%.

The most commonly used method for evaluating semi-supervised classification algorithms is the performance measurement of algorithms on the datasets. Classification accuracy (Acc) and Cohen's kappa (Kappa) [41] on the test set are used to measure the generation ability of algorithms, and executing time (ET) in seconds of learning the classifier is applied to estimate training efficiency. Besides the above two performance indicators, the number of the selected unlabeled instances to learn the semi-supervised classifier is also an important measure of the performance of instance selection. Therefore, the proportion of the selected unlabeled instances (PS) to the total unlabeled instances is adopted to eliminate the impact of dataset size.

Tri-training (Tri) [36] is selected as the representative co-training semi-supervised classification algorithm for its good performance, and the improved Tri-training algorithm trained on the result based on the proposed instance selection is denoted as ISTri. To verify whether there is a significant difference in the performance between Tri and ISTri on different data, the Wilcoxon signed rank test [42] is selected for its weaker data distribution assumptions and good statistical performance on the real datasets, where the null hypothesis is that the proposed algorithm performance is significantly different from each of the other algorithms on the same multiple datasets. The p-value of the test is computed to judge whether the null hypothesis is rejected or not under the given significant level α. If the p-value is larger than α, then the null hypothesis should be accepted. Otherwise, the null hypothesis is rejected, and there exists a significant difference between the proposed algorithm and another algorithm. The significant level $\alpha = 0.05$. All the experiment is executed in Python 3.10 on Windows 10 on a PC of Intel(R) Xeon(R) Silver 4280 CPU (2.10 GHz) and 160 GB RAM.

4.2. Experimental Analysis

This section concretely compares the performance of the proposed algorithm with the Tri algorithm from the perspective of classification performance, execution time, and proportion of the selected unlabeled instances on twelve medium-dimensional datasets, as well as a typical high-dimensional dataset. Furthermore, the effect of the proportion of the selected unlabeled instances on the algorithm's performance is also studied.

4.2.1. Classification Performance

Classification accuracy and Cohen's Kappa are two common measurements to evaluate the classification performance of the classifier, where the latter is an important complement to the former for the class-imbalance problem.

Table 2 lists the classification accuracy on the selected data sets, where the mean and median of classification accuracy on all the datasets are at the back of this table. Meanwhile, the p-value of the Wilcoxon signed rank test between the two algorithms is also listed in the last row of Table 2. Figure 3 intuitively shows the comparison of classification accuracy of the two algorithms on different datasets under different label proportions.

The following comparative analysis is done from a single perspective and a holistic perspective. It can be found that the ISTri algorithm obtains very similar classification accuracy with the Tri algorithm on each dataset under the same labeled rate PL = 0.2, 0.4, 0.6, and 0.8 from Figure 2. To compare the classification accuracy of different algorithms from a holistic perspective, descriptive statistical analysis is made. The means of the classification accuracy of the ISTri algorithm on all the datasets under different PLs are 0.851, 0.873, 0.889, and 0.894, and the ones of the Tri algorithm are 0.857, 0.874, 0.886, and 0.893. This numeric result also corroborates the absolute difference between the two algorithms on the mean of classification accuracy under the same PL value is very small. Meanwhile, the medians of the classification accuracy of the ISTri algorithm on all the datasets under different PLs are 0.890, 0.920, 0.933, and 0.947, and the ones of the Tri algorithm are 0.896, 0.919, 0.930, and 0.942. Therefore, the absolute difference between the two algorithms on the median of classification under the same PL value is also very small.

Table 2. Classification accuracy of two algorithms on the selected datasets.

Data	PL = 0.2		PL = 0.4		PL = 0.6		PL = 0.8	
	Tri	**ISTri**	**Tri**	**ISTri**	**Tri**	**ISTri**	**Tri**	**ISTri**
combined	0.815	0.821	0.823	0.824	0.831	0.830	0.834	0.837
connect-4	0.792	0.801	0.823	0.822	0.829	0.837	0.845	0.848
covtype	0.896	0.894	0.923	0.924	0.938	0.939	0.947	0.948
letter	0.895	0.886	0.932	0.927	0.951	0.947	0.960	0.959
optdigits	0.967	0.951	0.977	0.975	0.984	0.984	0.980	0.974
pendigits	0.978	0.976	0.988	0.986	0.990	0.990	0.992	0.990
phoneme	0.859	0.845	0.871	0.877	0.898	0.903	0.900	0.905
ring	0.917	0.905	0.915	0.917	0.922	0.928	0.938	0.945
seismic	0.729	0.731	0.731	0.734	0.741	0.744	0.744	0.746
texture	0.934	0.918	0.966	0.961	0.971	0.973	0.972	0.973
usps	0.930	0.921	0.951	0.947	0.949	0.949	0.959	0.963
winequality-white	0.567	0.565	0.585	0.583	0.624	0.638	0.641	0.642
Mean	0.857	0.851	0.874	0.873	0.886	0.889	0.893	0.894
Median	0.896	0.890	0.919	0.920	0.930	0.933	0.942	0.947
p-value	0.064		0.519		0.058		0.129	

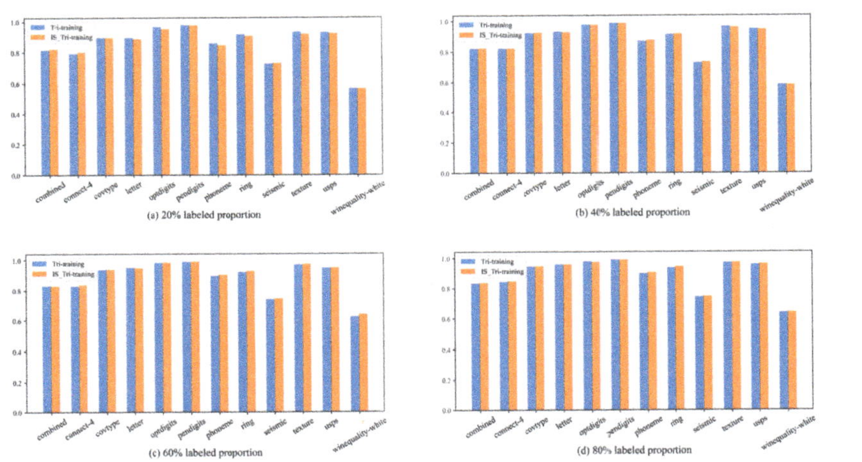

Figure 3. The comparison of classification accuracy between two algorithms on the selected datasets.

Finally, the Wilcoxon signed rank test between two algorithms classification accuracy is made to avoid the effect of the subjective judgment. p-values of this test under different PL are 0.060, 0.720, 0.375, and 0.206; these values are all larger than the given significant level of 0.05. Thus, there exists no significant difference in the classification accuracy between two algorithms.

Besides classification accuracy, Cohen's kappa is also used to evaluate the classification performance of the learner, which can solve the problem that accuracy does not compensate for hits that can be attributed to mere chance. Similar to the result of Table 2, Table 3 lists Kappa of two algorithms under different labeled proportions, as well as the descriptive statistics and p-value of the Wilcoxon signed-rank test. Figure 4 shows the comparison of the Kappa of the two algorithms.

Figure 4 shows the ISTri algorithm also obtains quite a similar Kappa as the Tri algorithm on each dataset under the same labeled rate PL = 0.2, 0.4, 0.6, and 0.8. A descriptive statistical analysis of kappa is made to compare the classification accuracy of different algorithms from a holistic perspective. The means of the Kappa of the ISTri algorithm on all the datasets under different PL are 0.739, 0.777, 0.806, and 0.814, and the ones of the Tri algorithm are 0.744, 0.775, 0.798, and 0.810. The absolute difference between the two algorithms on the mean of Kappa under the same PL value is very small from this numeric result. Meanwhile, the medians of the classification accuracy of the ISTri algorithm on all the datasets under different PL are 0.818, 0.855, 0.878, and 0.904, and the ones of the Tri algorithm are 0.833, 0.852, 0.872, and 0.895. Thus, the absolute difference between the two algorithms on the median of classification under the same PL value is also very small. Therefore, there exists a small difference between these two algorithms about Kappa from the above descriptive statistical results.

To make a more objective comparative evaluation, Wilcoxon signed rank test between two algorithms classification accuracy is made. p-values of this test under different PLs are 0.168, 0.519, 0.028, and 0.041, where the first two values are both larger than the given significant level of 0.05 and the latter two are smaller than 0.05. So, there exists no significant difference in the classification accuracy between two algorithms under PL = 0.2 and 0.4, while it exists no significant difference in the classification accuracy between two algorithms under PL = 0.6 and 0.8. Combing with the medians of Kappa on all the datasets, the ISTri algorithm achieves a better performance than the Tri algorithm under PL = 0.6 and 0.8.

The reason for the fact that the ISTri algorithm gets no significant difference classification in accuracy with the Tri algorithm corroborates the effectiveness and availability of the proposed unlabeled instance selection. It chooses the unlabeled instance subset by selecting the frequently confident ones identified by two other classifiers, where these selected unlabeled instances take much more ancillary information to the classifier than others. In other words, the proposed instance selection method obtains enough classification information as all the unlabeled instances so that the ISTri algorithm gets a similar classification performance as Tri algorithm.

Table 3. Kappa of two algorithms on the selected datasets.

Data	PL = 0.2		PL = 0.4		PL = 0.6		PL = 0.8	
	Tri	ISTri	Tri	ISTri	Tri	ISTri	Tri	ISTri
combined	0.708	0.717	0.720	0.722	0.733	0.740	0.737	0.743
connect-4	0.371	0.415	0.478	0.486	0.519	0.550	0.555	0.568
covtype	0.831	0.827	0.875	0.876	0.900	0.902	0.915	0.916
letter	0.891	0.881	0.929	0.924	0.949	0.944	0.958	0.958
optdigits	0.963	0.945	0.975	0.972	0.982	0.982	0.978	0.972
pendigits	0.976	0.973	0.987	0.985	0.989	0.989	0.991	0.989
phoneme	0.649	0.613	0.694	0.705	0.752	0.765	0.758	0.771
ring	0.834	0.810	0.830	0.835	0.844	0.855	0.876	0.891
seismic	0.566	0.570	0.571	0.577	0.588	0.591	0.590	0.594
texture	0.926	0.908	0.962	0.956	0.967	0.970	0.968	0.970
usps	0.922	0.912	0.945	0.940	0.943	0.943	0.954	0.959
winequality-white	0.294	0.294	0.339	0.344	0.408	0.437	0.439	0.443
Mean	0.744	0.739	0.775	0.777	0.798	0.806	0.810	0.814
Median	0.833	0.818	0.852	0.855	0.872	0.878	0.895	0.904
p-value	0.168		0.519		0.028		0.041	

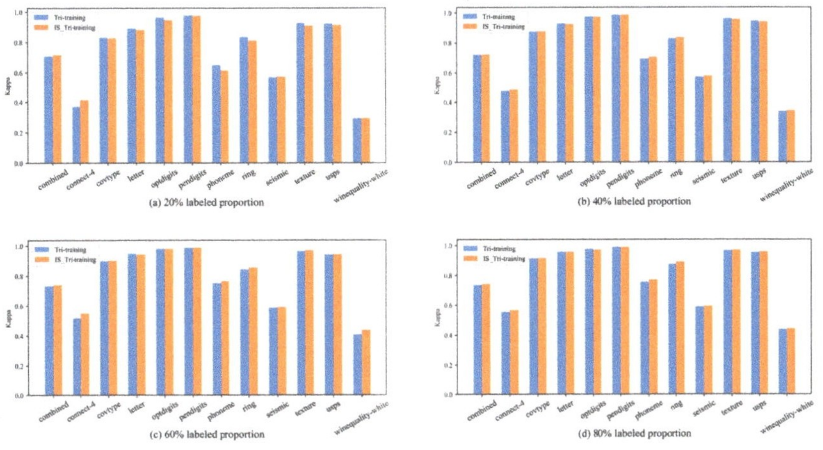

Figure 4. The comparison of Kappa between two algorithms on the selected datasets.

4.2.2. Selection Rate

The proportion of the selected instances to the original instances is an indicator to evaluate the reduction performance of instance selection. In our proposed method, the selected unlabeled instances and labeled instances reconstitute the training set that is used to efficiently learn the classifier. Therefore, the reduction performance of unlabeled instance selection affects the training efficiency of the classifier. Table 4 lists the PS of the proposed method under different labeled proportions.

It can be found that the value of PS on each dataset is significantly smaller than one under different PLs from Table 4. This result indicates the proposed method does obviously reduce the unlabeled instances, and the reformed training set with the selected unlabeled instances and labeled instances is smaller than the original set. Own to the different characteristics of the dataset, the values of PS on different datasets have significant differences. Especially for the winequality–white dataset, the proposed method finally saves a storage rate of up to 25% unlabeled instances. To carefully evaluate the selection proportion of the proposed algorithm from a global perspective, descriptive statistics are also computed on all the datasets. The means of PS on all the datasets under different PL are 0.539, 0.598, 0.618, and 0.648, and the medians are 0.568, 0.627, 0.661, and 0.693. The proposed instance selection method can reduce at least 30% of unlabeled instances from the average state.

There exist two reasons for the higher reduction ratio. Firstly, all the unlabeled instances have different levels of contributions to learning the classifier, and the number of unlabeled instances with large contributions is less than the ones with small contributions. On the other hand, our method aims to select the unlabeled instances with high confidence which have high contributions to the classifier. Thus, this experiment result has confirmed the effectiveness of the proposed algorithm.

Table 4. PS of ISTri algorithm on the selected dataset.

Data	PL = 0.2	PL = 0.4	PL = 0.6	PL = 0.8
combined	0.486	0.522	0.491	0.517
connect-4	0.585	0.641	0.629	0.672
covtype	0.670	0.728	0.761	0.808
letter	0.513	0.603	0.671	0.696
optdigits	0.531	0.603	0.651	0.718
pendigits	0.733	0.808	0.847	0.857
phoneme	0.590	0.653	0.682	0.693
ring	0.552	0.614	0.618	0.668
seismic	0.395	0.425	0.406	0.433
texture	0.686	0.743	0.745	0.779
usps	0.591	0.645	0.708	0.693
winequality–white	0.131	0.191	0.212	0.241
Mean	0.539	0.598	0.618	0.648
Median	0.568	0.627	0.661	0.693

4.2.3. Training Efficiency

Training efficiency is also an important indicator to evaluate the performance of the classification algorithms, where the execution time (in seconds) on the selected data is the common metric to measure training efficiency. Table 5 lists the execution time of two methods under different PLs on the selected datasets, and the simple statistical result is also listed in the bottom two rows of this table.

Table 5. ET of two algorithms on the selected datasets.

Data	PL = 0.2		PL = 0.4		PL = 0.6		PL = 0.8	
	Tri	ISTri	Tri	ISTri	Tri	ISTri	Tri	ISTri
combined	643.075	322.514	588.245	388.624	1328.337	795.159	1365.614	943.540
connect-4	11.757	5.905	14.787	8.466	18.317	10.916	21.232	13.493
covtype	768.997	395.715	1050.317	605.054	1336.295	815.328	1645.704	1036.018
letter	19.530	8.666	24.332	13.058	28.442	16.382	32.392	19.873
optdigits	6.763	3.204	7.974	4.502	9.265	5.615	10.407	6.582
pendigits	10.211	5.318	12.743	7.457	14.877	9.088	17.419	10.739
phoneme	5.076	2.556	6.293	3.506	7.417	4.386	8.888	5.325
ring	17.100	7.655	22.455	12.564	28.235	16.736	34.692	21.797
seismic	395.678	149.781	548.484	287.580	571.574	367.503	1038.199	511.116
texture	9.762	5.089	12.413	7.262	15.022	9.156	17.441	11.058
usps	45.562	24.265	59.183	38.108	73.571	50.979	89.867	66.263
winequality–white	6.142	2.468	7.805	3.846	9.200	5.121	10.698	6.351
Mean	161.638	77.761	196.253	115.002	286.713	175.531	357.713	221.013
Median	14.428	6.780	18.621	10.515	23.276	13.649	26.812	16.683

Table 5 shows that the ISTri algorithm has much less execution time on each dataset under the same value of PL. Meanwhile, the means of ET of the ISTri algorithm on all the datasets under different PL are 77.761, 115.005, 175.531, and 221.013, while the ones of the Tri algorithm are 161.638, 196.253, 286.73, and 221.013. Additionally, the medians of ET of the ISTri algorithm on all the datasets under different PL are 6.780, 10.515, 13.649, and 16.683, while the ones of the Tri algorithm are 14.428, 18.621, 23.276, and 26.812. This descriptive statistical result also corroborates the ISTri algorithm being able to obtain much less execution time than the Tri algorithm. The execution time of algorithms is affected by the dataset size, and its value is positively correlated with the amount of data.

To effectively compare the training efficiency of the algorithm, a speedup ratio named SR = ET(Tri)/ET(ISTri) is defined, where ET(Tri) and ET(ISTri) are the execution time of the Tri algorithm and IStri algorithm on the same dataset. This new relative indicator can eliminate the effect of data volume on the algorithm performance, and it evaluates the difference between the two algorithms' performance from a relative perspective. Table 6 lists the SR between two algorithms under different labeled proportions.

Table 6. SR between two algorithms on the selected datasets.

Data	PL = 0.2	PL = 0.4	PL = 0.6	PL = 0.8
combined	1.994	1.764	1.671	1.447
connect-4	1.991	1.747	1.678	1.574
covtype	1.943	1.736	1.639	1.588
letter	2.254	1.863	1.736	1.630
optdigits	2.111	1.771	1.650	1.581
pendigits	1.920	1.709	1.637	1.622
phoneme	1.986	1.795	1.691	1.669
ring	2.234	1.787	1.687	1.592
seismic	2.642	1.907	1.555	2.031
texture	1.918	1.709	1.641	1.577
usps	1.878	1.553	1.443	1.356
winequality-white	2.489	2.029	1.797	1.684
Mean	2.113	1.760	1.652	1.613
Median	1.993	1.759	1.660	1.590

The result of Table 6 shows that the value of SR on each data is significantly greater than one on each dataset under different PLs, and it confirms that the proposed algorithm obtains higher training efficiency than the original algorithm. Especially, the ISTri algorithm obtains a training efficiency of more than two times higher than the Tri algorithm on dataset letter, optdigits, ring, seismic, and winequality–white under PL = 0.2. ISTri algorithm also obtains nearly 1.5 times higher training efficiency than the Tri algorithm on most datasets when PL = 0.2, 0.4, and 0.6. Simple statistical result lists that the means of SR on all the dataset under different PLs are 2.113, 1.760, 1.652, and 1.613, and the medians are 1.993, 1.759, 1.660, and 1.590. Therefore, the ISTri algorithm achieves a training efficiency of more than 0.5 times higher than the original algorithm from a global perspective.

The reason for the higher training efficiency of the ISTri algorithm is that it uses the reduced unlabeled instance subset rather than the original unlabeled instance set to learn the classifier. As we all know, the training time of the classifier is negatively correlated with the training set size. The larger the training set, the longer the training time. For the semi-supervised classification tasks, unlabeled instances make up a large proportion of the training set. Moreover, the proposed instance selection method can effectively and efficiently compress unlabeled instances while retaining most of the information valid for the classifier, and this result can be verified by the low proportion of the selected unlabeled instances.

4.2.4. High-Dimensional Problem

The proposed method has obtained a good performance on twelve medium-dimensional datasets in the previous experiment. In this section, a high-dimensional representative image classification dataset called NORB is selected, in which each image is converted to a 2047-dimensional vector by the package SciPy. Table 7 lists classification accuracy, Kappa, execution time, and selection ratio under different labeled ratios.

Table 7. The performance of two algorithms on NORB.

PL	Acc		Kappa		ET		PS-ISTri
	Tri	ISTri	Tri	ISTri	Tri	ISTri	
0.2	0.980	0.971	0.974	0.968	481.778	271.731	0.326
0.4	0.987	0.982	0.984	0.979	621.516	514.607	0.424
0.6	0.994	0.992	0.992	0.990	795.907	686.806	0.499
0.8	0.996	0.995	0.994	0.993	936.799	836.467	0.514

As the result in Table 7 shows, the ISTri algorithm efficiently and effectively processes the high-dimensional problems and achieves comparable results to the Tri algorithm. The absolute difference in Acc between two algorithms under different values of PL are 0.009, 0.005, 0.002, and 0.001, as well as the one on Kappa, are 0.006, 0.005, 0.002, and 0.001. In the worst case, the largest difference between Acc and Kappa is 0.009 and 0.006, and this difference is very small relative to the overall performance of the algorithm. Therefore, there exists a negligible difference between Acc and Kappa under different values of PL. The execution time of the ISTri algorithm is much less than the Tri algorithm under the same PL. The ratio SR between them is 1.773, 1.208, 1.159, and 1.120; all the values are larger than one. Therefore, the ISTri algorithm obtains higher training efficiency than the Tri algorithm. The last column of Table 7 also lists the unlabeled selection proportion; the value of PS is 0.326, 0.424, 0.499, and 0.514; the values are significantly smaller than one. To sum up, the proposed instance selection method greatly reduces the size of unlabeled instances while it can preserve the classification information to learn the classifier. Meanwhile, the execution time of the IStri algorithm is much less than the Tri algorithm under each value of PL, and the ratio SR between them is also larger than 1. Moreover, the selection ratio of unlabeled instances under different PLs is 0.326, 0.424, 0.499, and 0.514, which shows that the ISTri algorithm uses fewer unlabeled instances to constitute the training set. This result demonstrates that the ISTri algorithm has a higher training efficiency than the Tri algorithm.

4.2.5. Effect of Labeled Proportion

The proposition of the labeled instances to all the training instances plays an important role in the performance of the learner for the semi-supervised classification tasks. Therefore, we study the effect of PL on the classifier from three metrics: classification performance, selection rate, and training efficiency. Figures 5–7 separately show the effect of PL on three metrics. Moreover, the Friedman test is used to compare whether there exists a significant difference in each metric under the different values of PL or not, where the null hypothesis is that there does not exist a significant difference against the alternative that there is a significant difference.

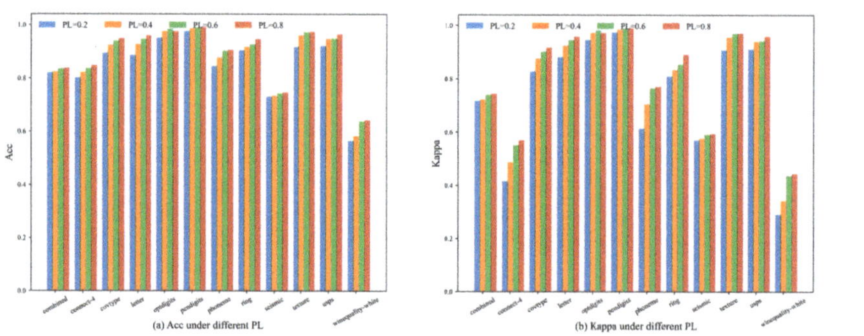

Figure 5. Classification performance of ISTri algorithm under different PL.

Figure 5 shows the change in classification performance of the proposed method under different PLs, where the left fig describes classification accuracy and the right fig for Kappa. There exists a noticeable difference in the value of Acc on almost all datasets except data combine, pendgitits, and seismic from Figure 5a. The value of Kappa also has a significant change on each dataset under different PL, especially for dataset connext-4, phoneme, winequality–white from Figure 5b. Meanwhile, the p-values of the Friedman test on Acc and Kappa are 4.02×10^{-7} and 5.49×10^{-7}, both smaller than the given significant level of 0.05. So, PL affects the classification performance of the proposed method. Moreover, Figure 5 also shows the value of Acc is positively correlated to PL on these datasets, i.e., its value significantly increases by the increasing PL, as well as this similar result for Kappa. The descriptive statistics of Acc and Kappa over all the datasets under different values of PLs also verify this result from Tables 3 and 4. The labeled instances take much more valuable label information that is critical to learn the classifier than unlabeled instances, so PS plays a key role in the classification performance of the classifier for semi-supervised classification problems. This fact explains why the classification performance of the ISTri algorithm is positively correlated with PS. Nevertheless, the ISTri algorithm still obtains no significant difference from the Tri algorithm.

Figure 6 shows the change of the metric PS under different PLs. The value of PS fluctuates greatly on each dataset, and this result is also proved by the numeric results in Table 4. The *p*-value of the Friedman test on PS is 1.38×10^{-6}. smaller than the given significant level of 0.05. Therefore, there exists a significant difference in PS under different values of PL. Similar to the performance of Acc and Kappa under different PLs, the value of PS is also positively correlated with PL. The unlabeled instances selection of the ISTri algorithm mainly depends on the agreement on the pseudo-labels offered by the classifiers on the labeled instance subsets, where the parameter PL controls the number of labeled instances. The classification ability of multi-view classifiers trained on the labeled instance subsets increasingly improves as the enlarging value of PL so that the likelihood that predictive labels for each unlabeled instance are the same could increase obviously. In this way, the final selection of unlabeled data increases significantly.

The change in speedup ratio (SR) under different PLs is shown in Figure 7, where the baseline SR = 1 is also plotted on it. It can be found that all the value of SR on each dataset under different values of PL is larger than one. The metric SR has significantly different values on each dataset under different PLs, and it can be validated by the result of Table 6. The *p*-value of the Friedman test on SR is 3.73×10^{-7}, smaller than the given significant level of 0.05. Therefore, there exists a significant difference in SR under different values of PL. Meanwhile, SR is negatively correlated with PL on each dataset from Figure 7. SR evaluates the ratio of the execution time between the ISTri algorithm and the Tri algorithm, and the main difference between them is the number of unlabeled instances that are used to learn the classifier. The selected number of unlabeled instances continues to increase with the increasing value of PL for the ISTri algorithm, and it also induces its execution time to get longer. This result explains the reason that SR is negatively correlated with PL.

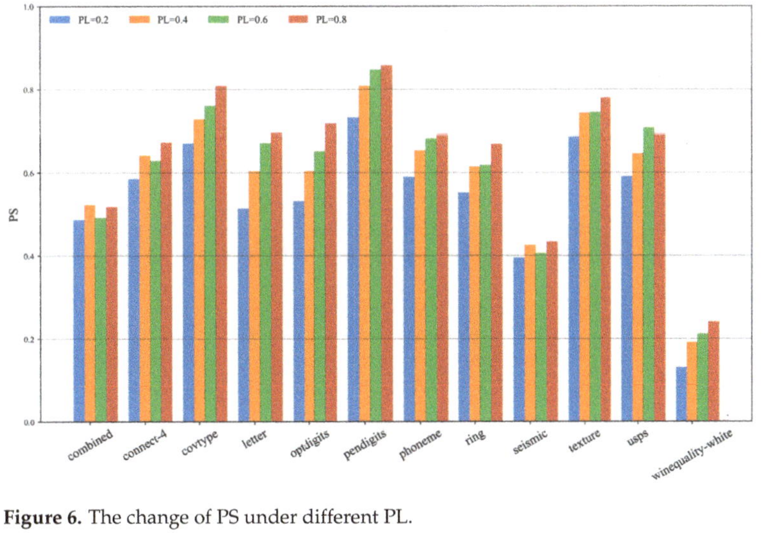

Figure 6. The change of PS under different PL.

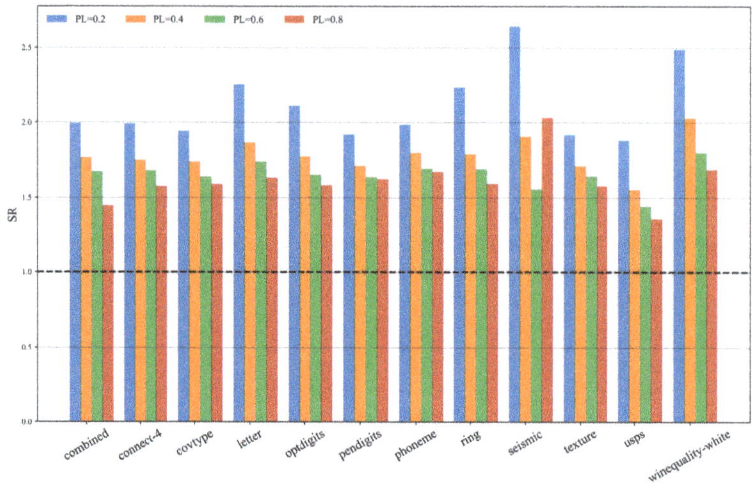

Figure 7. The change of SR under different PL.

5. Conclusions

For the problem of massive unlabeled instances bringing a great challenge to efficiently train co-training-based semi-supervised classification algorithms, this paper has developed an unlabeled instance selection algorithm based on the granulation mechanism. Different from the previous approaches from the view of algorithm optimization, it takes advantage of data reduction to avoid the difficulty of using domain knowledge to improve the efficiency of algorithms. The proposed method treats the unlabeled instances with the same frequency at which the trust instance is selected as the basic information granulation rather than each unlabeled instance; it is constructive to significantly improve execution efficiency. The selection of each unlabeled instance subset into the training set depends on its contribution to the current classification performance; this operation is guaranteed to have strong adaptability for different datasets and algorithms. The advantage of the proposed method is verified by the experiment results on the medium-dimensional and high-dimensional datasets. Especially it has a comparable classification performance with the typical algorithm, while it has high execution efficiency and fewer unlabeled instances within the training set. The proposed method can be widely used for driverless car obstacle recognition, mobile phone face recognition, temperature monitoring in greenhouses, and other large-scale application scenarios. Finally, this paper provides a potentially effective solution to improve the training efficiency of other kinds of semi-supervised classification algorithms. Future research work will explore the application of proposed algorithms in practical systems such as text classification, image classification, and pattern recognition.

Author Contributions: Writing the original draft and data preparation, Y.S.; writing the review and editing, J.Z.; oversight and leadership responsibility for the research activity planning and execution, X.Z.; implementation of the computer code and supporting algorithms, J.W. All authors have read and agreed to the published version of the manuscript.

Funding: National Natural Science Foundation of China: 62006145; Shandong Provincial Natural Science Foundation, China: ZR2020MF146.

Data Availability Statement: The selected datasets in this paper are public, and they can be freely downloaded at LIBSVM-dataset repository (https://www.csie.ntu.edu.tw/~cjlin/libsvmtools/datasets/, 12 April 2023), KEEL-dataset repository (https://sci2s.ugr.es/keel/datasets.php, 12 April 2023) and NORB (https://cs.nyu.edu/~yann/research/norb/, 12 April 2023).

Acknowledgments: This paper was completed by Key Laboratory of Huang-Huai-Hai Smart Agricultural Technology of Ministry of Agriculture and Rural Affairs, Shandong Agricultural University. We thank the school for its support and help.

Conflicts of Interest: This paper represents the opinions of the authors and does not mean to represent the position or opinions of the Shandong Agricultural University.

References

1. Blum, A.; Mitchell, T. Combining labeled and unlabeled data with co-training. In Proceedings of the Eleventh Annual Conference on Computational Learning Theory, Madison, MI, USA, 24–26 July 1998; pp. 92–100.
2. Prasetio, B.H.; Tamura, H.; Tanno, K. Semi-supervised deep time-delay embedded clustering for stress speech analysis. *Electronics* **2019**, *8*, 1263. [CrossRef]
3. Ning, X.; Cai, W.; Zhang, L.; Yu, L. A review of research on co-training. *Concurr. Comput. Pract. Exp.* **2021**, *21*, e6276. [CrossRef]
4. Ng, K.W.; Furqan, M.S.; Gao, Y.; Ngiam, K.Y.; Khoo, E.T. HoloVein—Mixed-reality venipuncture aid via convolutional neural networks and semi-supervised learning. *Electronics* **2023**, *12*, 292. [CrossRef]
5. Li, L.; Zhang, W.; Zhang, X.; Emam, M.; Jing, W. Semi-supervised remote sensing image semantic segmentation method based on deep learning. *Electronics* **2023**, *12*, 348. [CrossRef]
6. Lang, H.; Agrawal, M.N.; Kim, Y.; Sontag, D. Co-training improves prompt-based learning for large language models. In Proceedings of the 39th International Conference on Machine Learning, Baltimore, MD, USA, 17–23 July 2022; pp. 11985–12003.
7. Fan, J.; Gao, B.; Jin, H.; Jiang, L. Ucc: Uncertainty guided cross-head co-training for semi-supervised semantic segmentation. In Proceedings of the IEEE/CVF Conference on Computer Vision and Pattern Recognition, New Orleans, LA, USA, 18–24 June 2022; pp. 9947–9956.
8. Sheikh Hassani, M.; Green, J.R. Multi-view Co-training for microRNA prediction. *Sci. Rep.* **2019**, *9*, 10931. [CrossRef]
9. Wang, H.; Shen, H.; Li, F.; Wu, Y.; Li, M.; Shi, Z.; Deng, F. Novel PV power hybrid prediction model based on FL Co-Training method. *Electronics* **2023**, *12*, 730. [CrossRef]
10. Sun, S.; Jin, F. Robust co-training. *Int. J. Pattern Recognit. Artif. Intell.* **2011**, *25*, 1113–1126. [CrossRef]
11. Dong, Y.; Jiang, L.; Li, C. Improving data and model quality in crowdsourcing using co-training-based noise correction. *Inf. Sci.* **2022**, *583*, 174–188. [CrossRef]
12. Cui, K.; Huang, J.; Luo, Z.; Zhang, G.; Zhan, F.; Lu, S. GenCo: Generative co-training for generative adversarial networks with limited data. In Proceedings of the AAAI Conference on Artificial Intelligence, Vancouver, BC, Canada, 22 February–1 March 2022; Volume 36, pp. 499–507.
13. Han, T.; Xie, W.; Zisserman, A. Self-supervised co-training for video representation learning. *Adv. Neural Inf. Process. Syst.* **2020**, *33*, 5679–5690.
14. Li, B.; Wang, J.; Yang, Z.; Yi, J.; Nie, F. Fast semi-supervised self-training algorithm based on data editing. *Inf. Sci.* **2023**, *626*, 293–314. [CrossRef]
15. Li, Y.; Maguire, L. Selecting critical patterns based on local geometrical and statistical information. *IEEE Trans. Pattern Anal. Mach. Intell.* **2010**, *33*, 1189–1201.
16. Garcia, S.; Derrac, J.; Cano, J.; Herrera, F. Prototype selection for nearest neighbor classification: Taxonomy and empirical study. *IEEE Trans. Pattern Anal. Mach. Intell.* **2012**, *34*, 417–435. [CrossRef] [PubMed]
17. Li, Y.; Liang, D. Safe semi-supervised learning: a brief introduction. *Front. Comput. Sci.* **2019**, *13*, 669–676. [CrossRef]
18. Liang, J.; Qian, Y.; Li, D.; Qinghua, H. Theory and method of granular computing for big data mining. *Sci. China Inf. Sci.* **2015**, *45*, 188–198.
19. Yao, Y. Three-way granular computing, rough sets, and formal concept analysis. *Int. J. Approx. Reason.* **2020**, *116*, 106–125. [CrossRef]
20. Zhang, Z.; Gao, J.; Gao, Y.; Yu, W. Two-sided matching decision making with multi-granular hesitant fuzzy linguistic term sets and incomplete criteria weight information. *Expert Syst. Appl.* **2021**, *168*, 114311. [CrossRef]
21. Chu, X.; Sun, B.; Chu, X.; Wu, J.; Han, K.; Zhang, Y.; Huang, Q. Multi-granularity dominance rough concept attribute reduction over hybrid information systems and its application in clinical decision-making. *Inf. Sci.* **2022**, *597*, 274–299. [CrossRef]
22. Sangaiah, A.K.; Javadpour, A.; Ja'fari, F.; Pinto, P.; Zhang, W.; Balasubramanian, S. A hybrid heuristics artificial intelligence feature selection for intrusion detection classifiers in cloud of things. *Clust. Comput.* **2023**, *26*, 599–612. [CrossRef]
23. Song, Y.; Zhang, J.; Zhang, C. A survey of large-scale graph-based semi-supervised classification algorithms. *Int. J. Cogn. Comput. Eng.* **2015**, *45*, 1355–1369. [CrossRef]
24. Zheng, W.; Qian, F.; Zhao, S.; Zhang, Y. M-GWNN: Multi-granularity graph wavelet neural networks for semi-supervised node classification. *Neurocomputing* **2021**, *453*, 524–537. [CrossRef]
25. Zhu, P.; Zhang, W.; Wang, Y.; Hu, Q. Multi-granularity inter-class correlation based contrastive learning for open set recognition. *Int. J. Softw. Inf.* **2022**, *12*, 157–175. [CrossRef]
26. Zhao, J.; Xie, X.; Xu, X.; Sun, S. Multi-view learning overview: Recent progress and new challenges. *Inf. Fusion* **2017**, *38*, 43–54. [CrossRef]

27. Zhou, Y.; Goldman, S. Democratic co-learning. In Proceedings of the 16th IEEE International Conference on Tools with Artificial Intelligence, Boca Raton, FL, USA, 15–17 November 2004; pp. 594–602.
28. Li, M.; Zhou, Z. Improve computer-aided diagnosis with machine learning techniques using undiagnosed samples. *IEEE Trans. Syst. Man Cybern.-Part A Syst. Hum.* **2007**, *37*, 1088–1098. [CrossRef]
29. Xu, X.; Li, W.; Xu, D.; Tsang, I.W. Co-labeling for multi-view weakly labeled learning. *IEEE Trans. Pattern Anal. Mach. Intell.* **2015**, *38*, 1113–1125. [CrossRef]
30. Ma, F.; Meng, D.; Xie, Q.; Li, Z.; Dong, X. Self-paced co-training. In Proceedings of the 34th International Conference on Machine Learning, Sydney, Australia, 6–11 August 2017; pp. 2275–2284.
31. Derrac, J.; Garcia, S.; Sanchez, L.; Herrera, F. Keel data-mining software tool: Data set repository, integration of algorithms and experimental analysis framework. *J. Mult. Valued Log. Soft Comput.* **2015**, *17*, 255–287.
32. Ye, H.; Zhan, D.; Miao, Y.; Jiang, Y.; Zhou, Z. Rank consistency based multi-view learning: A privacy-preserving approach. In Proceedings of the 24th ACM International on Conference on Information and Knowledge Management, Melbourne, Australia, 19–23 October 2015; pp. 991–1000.
33. Tang, J.; Tian, Y.; Zhang, P.; Liu, X. Multiview privileged support vector machines. *IEEE Trans. Neural Netw. Learn. Syst.* **2017**, *29*, 3463–3477. [PubMed]
34. Sun, S.; Shawe-Taylor, J. Sparse semi-supervised learning using conjugate functions. *J. Mach. Learn. Res.* **2010**, *11*, 2423–2455.
35. Van Engelen, J.E.; Hoos, H.H. A survey on semi-supervised learning. *Mach. Learn.* **2020**, *109*, 373–440. [CrossRef]
36. Zhou, Z.; Li, M. Tri-training: Exploiting unlabeled data using three classifiers. *IEEE Trans. Knowl. Data Eng.* **2005**, *17*, 1529–1541. [CrossRef]
37. Breiman, L. Heuristics of instability and stabilization in model selection. *Ann. Stat.* **1996**, *24*, 2350–2383. [CrossRef]
38. Song, Y.; Liang, J.; Lu, J.; Zhao, X. An efficient instance selection algorithm for k nearest neighbor regression. *Neurocomputing* **2017**, *251*, 26–34. [CrossRef]
39. Chang, C.C.; Lin, C.J. LIBSVM: A library for support vector machines. *Acm Trans. Intell. Syst. Technol.* **2011**, *2*, 1–27. [CrossRef]
40. LeCun, Y.; Huang, F.J.; Bottou, L. Learning methods for generic object recognition with invariance to pose and lighting. In Proceedings of the 2004 IEEE Computer Society Conference on Computer Vision and Pattern Recognition, Washington, DC, USA, 27 June–2 July 2004; Volume 2.
41. Ben-David, A. A lot of randomness is hiding in accuracy. *Eng. Appl. Artif. Intell.* **2007**, *20*, 875–885. [CrossRef]
42. Demšar, J. Statistical comparisons of classifiers over multiple data sets. *J. Mach. Learn. Res.* **2006**, *7*, 1–30.

electronics MDPI

Article

Flight Delay Prediction Model Based on Lightweight Network ECA-MobileNetV3

Jingyi Qu *, Bo Chen, Chang Liu and Jinfeng Wang

Tianjin Key Laboratory of Advanced Signal Processing, Civil Aviation University of China, Tianjin 300300, China
* Correspondence: jyqu@cauc.edu.cn

Abstract: In exploring the flight delay problem, traditional deep learning algorithms suffer from low accuracy and extreme computational complexity; therefore, the deep flight delay prediction algorithm is difficult to directly deploy to the mobile terminal. In this paper, a flight delay prediction model based on the lightweight network ECA-MobileNetV3 algorithm is proposed. The algorithm first preprocesses the data with real flight information and weather information. Then, in order to increase the accuracy of the model without increasing the computational complexity too much, feature extraction is performed using the lightweight ECA-MobileNetV3 algorithm with the addition of the Efficient Channel Attention mechanism. Finally, the flight delay classification prediction level is output via a Softmax classifier. In the experiments of single airport and airport cluster datasets, the optimal accuracy of the ECA-MobileNetV3 algorithm is 98.97% and 96.81%, the number of parameters is 0.33 million and 0.55 million, and the computational volume is 32.80 million and 60.44 million, respectively, which are better than the performance of the MobileNetV3 algorithm under the same conditions. The improved model can achieve a better balance between accuracy and computational complexity, which is more conducive mobility.

Keywords: delay prediction model; lightweight neural network; lightweight attention mechanism

Citation: Qu, J.; Chen, B.; Liu, C.; Wang, J. Flight Delay Prediction Model Based on Lightweight Network ECA-MobileNetV3. *Electronics* **2023**, *12*, 1434. https://doi.org/10.3390/electronics12061434

Academic Editor: Baris Aksanli

Received: 15 February 2023
Revised: 14 March 2023
Accepted: 16 March 2023
Published: 17 March 2023

1. Introduction

In recent years, China's air traffic industry has grown rapidly with the implementation of the 13th Five-Year Plan for Civil Aviation [1]. However, the number of flights continues to grow, but the normal rate of flights is becoming lower and lower. During this period, the Civil Aviation Administration carried out total control of flight slots and adjusted flight structure, and the problem of flight delays was alleviated. According to a report from the Civil Aviation Work Conference 2022 held by the Civil Aviation Administration of China [2], since 2020, due to the impact of the epidemic, the number of flights has significantly decreased abnormally, so flight delays during the epidemic are not considered. In addition, China will overtake the United States as the largest air transport organization in 2029, according to research from the International Air Transport Association (IATA) [3]. With the COVID-19 epidemic under effective control, the volume of air traffic will also increase rapidly. Therefore, the speed of air traffic recovery and the projections of international reports firmly reflect the urgent traffic demand of China's air traffic industry. Serious flight delays are likely to trigger "mass incidents of air passengers" [4–6], thus endangering the public safety of the airport and the personal safety of the passengers. Understanding flight delays in advance has become a pressing issue for civil aviation. To this end, a large number of studies have been carried out by domestic and foreign scholars in related fields.

The traditional flight delay prediction methods mainly include statistical inference, simulation and modeling, and machine learning methods [7]. Xu et al. [8] proposed a permutation and incremental permutation SVM algorithm considering the demand of flight volume and real-time refreshment of flight data and validated it on manual data. Finally, the accuracy of flight delay prediction can reach more than 80%. Similarly,

Luo 's team [9] and Luo 's team [10] also gradually considered using support vector machines or improved support vectors to analyze flight delays. In view of the irregular dynamic distribution attributes of flight data, Cheng et al. [11] proposed a classification prediction model of flight delay based on C4.5 decision tree to avoid the impact of flight distribution changes on the algorithm model, which is a certain improvement compared with the traditional Bayesian algorithm. Nigam and Govinda [12] analyzed the flight data and meteorological data of several airports in the United States and used the logistic regression algorithm in the machine learning algorithm to predict the flight departure delay. Khanmohammadi et al. [13] proposed a flight delay prediction model based on an improved Artificial Neural Network (ANN), and they used multiple linear N-1 coding to preprocess complex airport data models. Wu et al. [14] proposed a flight delay spread prediction model based on CBAM-CondenseNet, which enhances the transmission of deep information in the network structure by adopting a channel and spatial attention mechanism to increase the prediction accuracy. When using deep learning to predict flight delay, these scholars chose a relatively deep learning network, which requires a lot of computing time and resources. They can only choose to deploy the flight delay algorithm to the PC terminal. However, for the deployment requirements of mobile terminals, there is no trade-off between the accuracy and computational complexity of the prediction algorithm.

Recently, experts from home and abroad have carried out in-depth research and innovation in lightweight materials. In the beginning, experts used knowledge distillation, model pruning and other methods to try out algorithms. The former first trains a Net-T network and then uses network distillation to obtain a smaller Net-S network, thus achieving the effect of a simplified model. The latter simplified the model through channel pruning and other operations on the trained model [15–19]. Lightweight convolutional neural networks are an emerging branch of deep learning algorithms. This type of network applies lightweight operations to the algorithm itself and continuously innovates the algorithm from within so that it can maximize accuracy and continuously meet the computational power requirements of mobile devices. For example, the peerless team [20] proposed the ShuffleNet series algorithm, which uses the Channel Shuffle and Channel split operations [21] to speed up network and feature reuse, and the lightweight neural network Efficientnet algorithms presented by the Google team [22]. This algorithm comprehensively considers the input data size, network depth, and width, and proposes a model compound scaling method to control model computing power and ensure accuracy. Iandola et al. [23] proposed a lightweight SqueezeNet algorithm that proposes a Fire module structure to design the network structure. In this structure, to extract features and lessen model computation, single-layer and double-layer convolutions were used. The Google team [24] proposed MobileNet series algorithms, which are extremely influential in lightweight neural networks, which use deep separable convolution and SE attention mechanism for feature extraction and combine with structures, such as inverted residual error, which considerably improve the accuracy and computational performance of the model [25,26]. Excellent results have been achieved in face recognition, image classification, target detection, etc. [27–29].

To sum up, in view of the problems of low prediction accuracy and high computational complexity of the existing flight delay prediction algorithms, which are not conducive to deployment on mobile devices and other devices, this paper proposes an improved lightweight ECA-MobilenetV3 algorithm, which replaces the SE model with a lightweight ECA (Efficient Channel Attention) module, effectively reducing the computational complexity of the model without losing accuracy; it lays a foundation for the application of the model in mobile devices. The experiment uses real domestic meteorological data and flight data for analysis and verification.

The organizational structure of this paper is as follows: Section 1 introduces the background and significance of the paper, as well as the research status at home and abroad. Section 2 proposes and introduces the ECA-MobileNetV3 network model. Section 3 introduces the building process of a flight delay prediction model in detail. Section 4 shows the analysis of the experimental results and the application of the model. Section 5 summarizes the work of this paper and describes the future work.

2. Design of the ECA-MobileNetV3 Network Model

2.1. The Overall Structure of the Network

The network structure of the MobileNetV3 algorithm is shown in Figure 1a. Inheriting the three advantages of deep separable convolution, inverted residue structure, and linear bottleneck structure of MobileNetV2 network, the algorithm adds the SE attention mechanism in each inverted residue model [30]. In this paper, considering the non-lightweight nature of the SE attention mechanism, we propose an improved lightweight ECA-MobileNetV3 network, which replaces the SE module with the lightweight ECA attention mechanism [31], and the improved algorithm structure is shown in Figure 1b. The ECA-MobileNetV3 algorithm uses 1-dimensional convolution and cross-channel interaction methods to obtain channel importance, which effectively reduces the computational complexity of the model while ensuring the accuracy of the model.

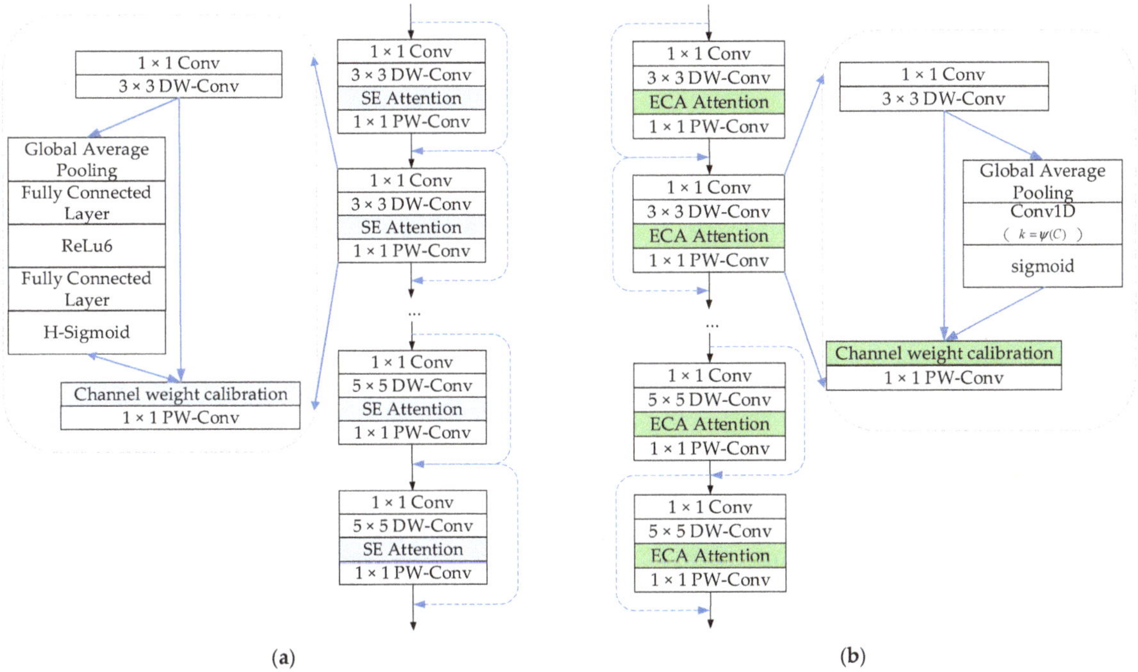

Figure 1. Comparison of the backbone network structure of the two algorithms before and after the improvement. (**a**) Backbone network structure of the MobileNetV3. (**b**) Backbone network structure of the ECA-MobileNetV3.

The ECA-MobileNetV3 algorithm uses a deep convolution kernel of a different size for the inverted residual structure. As can be seen from the structural configuration table for the ECA-MobileNetV3 algorithm listed in Table 1, the size of the deep convolution kernel in the inverted residual module 1, module 2, and module 3 is $[3 \times 3]$, while the size of the deep convolution kernel in the remaining inverted residual module is $[5 \times 5]$. Width Multiplier is the hyper parameter in the MobileNetV3 network; by adjusting its size, one can change the channel number of the output matrix in each layer of the whole network, so as to quickly change the model size; the number of output feature matrix channels is $N_K \in (16, 16, 24, 24, 40, 40, 40, 48, 48, 96, 96, 96, 88, 1280, 5)$. The α denotes the channel factor, which is the hyper parameter in the MobileNetV3 network. By adjusting its size, the number of channels in the output matrix in each layer of the network can be changed, so that the model size can be changed quickly.

Table 1. Configuration table of the flight delay prediction model based on the ECA-MobileNetV3 algorithm.

Network Layer		Output Matrix Size $D_F \times D_F$	Convolutional Kernel Size $D_K \times D_K$	Number of Output Matrix Channels N_K
Input		$8 \times 8/8 \times 9$	-	1
Traditional convolutional layer		$8 \times 8/8 \times 9$	$[3 \times 3]$	$\alpha \times 16$
Inverted residuals module 1	Traditional convolutional layer Deep convolutional layer ECA module Convolutional layer by point	$8 \times 8/8 \times 9$	$[1 \times 1]$ $[3 \times 3]$ - $[1 \times 1]$	$\alpha \times 16$
Inverted residuals module 2	Traditional convolutional layer Deep convolutional layer ECA module Convolutional layer by point	$8 \times 8/8 \times 9$	$[1 \times 1]$ $[3 \times 3]$ - $[1 \times 1]$	$\alpha \times 24$
		……		
Inverted residuals module 11	Traditional convolutional layer Deep convolutional layer ECA module Convolutional layer by point	$8 \times 8/8 \times 9$	$[1 \times 1]$ $[5 \times 5]$ - $[1 \times 1]$	$\alpha \times 96$
Traditional convolutional layer		$8 \times 8/8 \times 9$	$[1 \times 1]$	$\alpha \times 88$
GAP (Global Average Pooling)		$8 \times 8/8 \times 9$	-	-
Fully Connected Layer		1×1	-	$\alpha \times 1280$
Dropout layer		1×1	-	-
Fully Connected Layer		1×1	-	5

2.2. Lightweight ECA Module

There is an SE module in the MobileNetV3 algorithm, in which the feature matrix is first dimensioned down and then dimensioned up to obtain the weight channel importance. However, the dimensionality reducing operation between the two completely connected layers is not conducive to the weight learning of the channel and will lose certain feature information. Moreover, the model's calculation load will increase a little with the fully connected layer. Therefore, this paper considers using the lightweight attention mechanism ECA module to replace the SE attention mechanism module in MobileNetV3. The ECA module also functions as a channel attention mechanism, completing the acquisition of channel weights through an adaptive one-dimensional convolution and a cross-channel interaction technique without dimensionality reduction. The model can effectively reduce the computational complexity while maintaining the property.

Figure 2 depicts the ECA module's general structure. Assuming that the feature matrix before the input to the ECA attention mechanism is $X \in R(H \times W \times C)$, through global average pooling, the features first reduce the width and height of the feature matrix. The model then enters the adaptive one-dimensional convolution calculation to complete the acquisition of feature weights, as shown in Formula (1), in which adaptation refers to the adaptive selection of k adjacent channels in the process of obtaining channel weights, as shown in Formula (2).

$$W_c = \sigma(C1D_k(Z_c)) \tag{1}$$

$$k = \psi(C) = \left| \frac{\log_2(C)}{\gamma} + \frac{b}{\gamma} \right|_{odd} \tag{2}$$

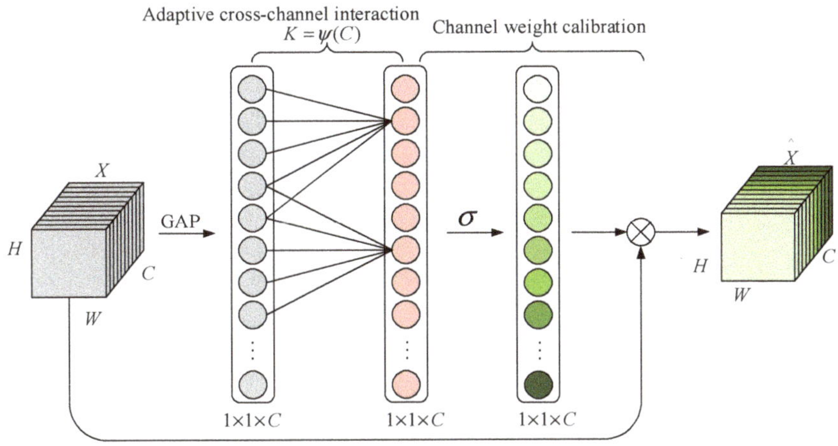

Figure 2. Structural diagram of the ECA attention mechanism.

In Formula (1), W_c represents the channel-acquired weights, $\sigma(\bullet)$ the Sigmoid activation function, $C1D_k(\bullet)$ the adaptive one-dimensional convolution, and Z_c the feature matrix after global average pooling. In Formula (2), k represents the number of local cross-channel interactions, which is the size of the one-dimensional convolutional kernel, C represents the number of channels in the feature matrix, and γ and b represent constants. The experiment is set to 2 and 1, respectively, according to the requirements of the original paper.

2.3. Network Training of Forward Propagation

An inverted residue module inside ECA-MobileNetV3 consists of $[1 \times 1]$ traditional convolution, $[3 \times 3]/[5 \times 5]$ deep convolution, ECA attention mechanism, and $[1 \times 1]$ point-by-point convolution. Each layer is convolved with a BN regularization layer and an activation function layer. The ECA-MobileNetV3 algorithm has 11 reverse residual modules. In the first three reverse residual modules, the ReLU function is chosen as the activation function for the first conventional convolutional layer and the second depth layer, as shown in Formula (3), and the H-Swish function as the activation function, as shown in Formula (4); all successive convolutional layers use the linear function as an activation function and Sigmoid function for channel weight, as shown in Formula (5).

$$\mathrm{Re}LU(x) = \max(x, 0) \tag{3}$$

$$H\text{-}Swish(x) = x\frac{\mathrm{Re}LU6(x+3)}{6} \tag{4}$$

$$Sigmoid(x) = \frac{1}{1+e^{-x}} \tag{5}$$

Through the above description, we can obtain the feature matrix in the calculation process, the convolution layer can be expressed as the following Formulas (6) and (7), which can derive a residual module after three convolution operations, and they can be represented by Formulas (8)–(10).

$$z_j^l = \sum_k W_{jk}^l \otimes y_k^{l-1} + b_j^l \tag{6}$$

$$y_j^l = \sigma(BN(\sum_k W_{jk}^l \otimes y_k^{l-1} + b_j^l)) \tag{7}$$

$$y_j^1 = \sigma(BN(\sum_k W_{jk}^1 \otimes X_k + b_j^1)) \tag{8}$$

$$y_j^2 = \sigma(BN(\sum_k W_{jk}^2 \otimes y_k^1 + b_j^2)) \tag{9}$$

$$y_j^3 = line(BN(\sum_k W_{jk}^3 \otimes y_k^2 + b_j^3)) \tag{10}$$

where W_{jk}^l represents the weight of the k-th feature to the j-th feature in the layer $l-1$, b_j^l represents the bias of the j-th feature in the layer l, z_j^l represents the output value before the k-th feature in layer l passes the activation function, $\sigma(\bullet)$ represents the activation function, and y_k^{l-1} represents the mapping value of the k-th feature in the layer $l-1$ after the activation function.

In addition, the feature matrix enters the ECA module after entering the deconvolutional module and passing through traditional convolutional layers and deep convolutional layers. The ECA module lies between deep convolution and pointwise convolution. As a complete calculation unit for acquiring channel weights, its forward propagation process is shown in Formula (11):

$$y_j = y_k^{conv} \otimes sigmoid(C1D(GAP(y_k^{conv}))) \tag{11}$$

where y_k^{conv} represents the feature matrix after the deep convolution operation, and the second half of the formula represents the feature weights acquired through the ECA module.

2.4. Network Training of Back Propagation

After the forward propagation of the ECA-MobileNetV3 algorithm is completed to obtain the predicted value of the model, the loss function between it and the true value is computed. Then, the chain derivative rule is used to obtain the chain derivative of the error term of the training samples, and the weight parameters and the bias are updated continuously until the network model converges. Chain derivative rule can also be called BP (back propagation) [32]. The error term δ_j^l between layers $l+1$ and l is calculated according to the rule, as shown in Formula (12). The chain analysis results of weight and bias within a residual module are shown in Formulas (13) and (14):

$$\delta_j^l = \frac{\partial J}{\partial z_j^l} = \frac{\partial J}{\partial y_j^l}\sigma(z_j^l) = BN(\sum_k W_{jk}^{l+1} \otimes \delta_k^{l+1} + b_j^l) \otimes \sigma'(z_j^l) \tag{12}$$

$$\frac{\partial J}{\partial W_{jk}^1} = (\delta_j^4 \otimes W_j^4 + \delta_j^3 \otimes W_j^3 + \delta_j^2 \otimes W_j^2) \otimes y_j \tag{13}$$

$$\frac{\partial J}{\partial b_{jk}^1} = (\delta_j^4 \otimes W_j^4 + \delta_j^3 \otimes b_j^3 + \delta_j^2 \otimes b_j^2) \tag{14}$$

where J represents the loss function, δ_j^l represents the error value of the j-th eigenvalue in layer l, W_{jk}^{l+1} represents the weight of neurons from k-th to j-th feature in layer l, and $\otimes$ represents the multiplication between matrices.δ_j^2, δ_j^3, and δ_j^4, respectively, represent the error terms between traditional convolutional layer, deep convolutional layer, ECA module, and point-by-point convolutional layer. According to Formulas (13) and (14), the weight and bias can be updated from back to forward, respectively.

3. Flight Delay Prediction Model Based on ECA-MobileNetV3

The overall structure of the flight delay prediction model based on ECA-MobileNetV3 is shown in Figure 3. The flight delay prediction model is divided into three parts [33]: data processing, feature extraction of the delay prediction model, and classification prediction of the model.

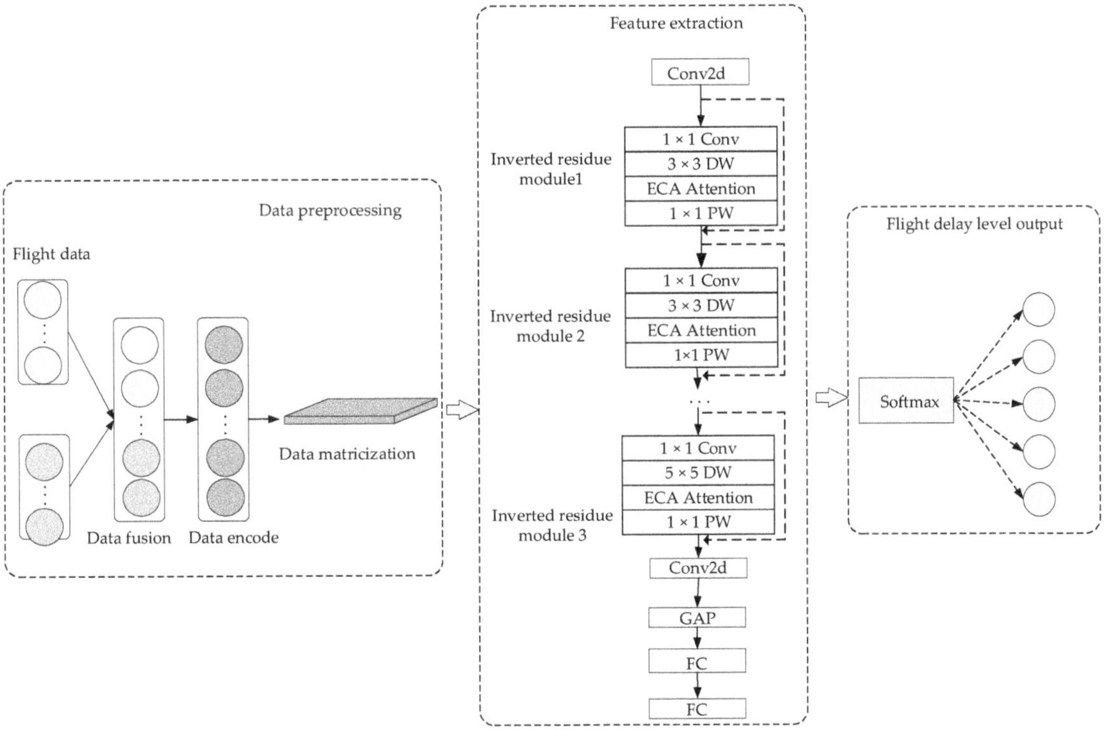

Figure 3. Flight delay prediction model based on the ECA-MobileNetV3 algorithm.

3.1. Data Preprocessing

The dataset used in this paper is mainly the flight dataset integrated with meteorological information and independently built. The data acquisition mainly comes from two sources: the flight dataset from March 2018 to March 2019 provided by the East China Air Traffic Administration and the meteorological dataset observed by the Automatic Weather Observation System (AWOS) [34], and the flight dataset from September 2019 to October 2020 provided by the North China Air Traffic Administration and the corresponding meteorological dataset. The flight dataset integrated with meteorological information contains multiple characteristic variables, including flight number, departure airport, destination airport, departure time, arrival time, delay time, flight status, etc. It also includes meteorological data, such as temperature, humidity, wind speed, precipitation, etc. According to the different sources of data acquisition, the dataset is divided into the Shanghai Hongqiao Airport dataset provided by the East China Air Traffic Control Bureau and the Beijing–Tianjin–Hebei Airport Cluster dataset provided by the North China Air Traffic Control Bureau. The Shanghai Hongqiao Airport dataset contains 301,594 sample data, and the Beijing–Tianjin–Hebei Airport Cluster dataset contains 1,048,576 sample data. The dataset also contains missing values and duplicate values, the flight data provided by the air traffic control bureau is manually recorded, and the meteorological data are collected by the airport's sensor equipment. There will be errors and omissions in the manually recorded data, and some data will be missing and incomplete due to sensor failure. During data integration processing, the same data may also be recorded repeatedly, so the dataset needs to be cleaned and processed to ensure data quality.

A series of preprocessing operations is performed on the dataset before feeding into the lightweight convolutional neural network algorithm. Figure 4 shows a data preprocessing flowchart. The whole process can be divided into: data cleaning, data fusion, data encoding, and matrix quadrature. For the dataset of Shanghai Hongqiao Airport, the flights of Shanghai Hongqiao Airport should be extracted from the original dataset according to the planned departure airport and planned arrival airport according to the four-character code of civil aviation airport "ZSSS" (Shanghai Hongqiao International Airport). Similarly, for the Beijing–Tianjin–Hebei airport cluster dataset, flights from major airports in Beijing, Tianjin, and Shijiazhuang were, respectively, extracted according to the four-character code of civil aviation airport "ZBAA" (Beijing Capital International Airport), "ZBAD" (Beijing Daxing International Airport), "ZBTJ" (Tianjin Binhai International Airport), and "ZBSJ" (Shijiazhuang Zhengding International Airport), and then the subsequent data pretreatment work was carried out.

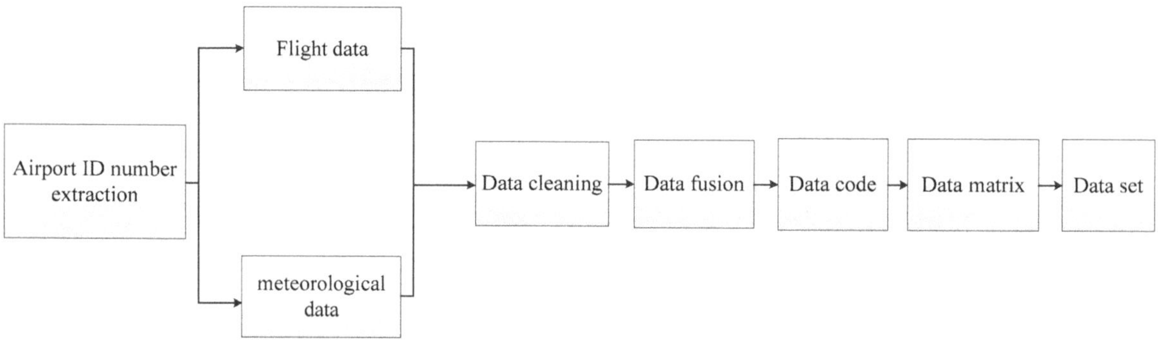

Figure 4. Data processing flowchart.

The first step in data preprocessing is data cleaning: attribute columns with many nulls in the dataset, duplicate data and attribute deletion, and other operations. The second step is data fusion operation: set the time attribute in the meteorological data as the association primary key I, set the planned start time and planned landing time in the flight data as the association primary key II according to the airport ID, and then conduct the association fusion between the primary key I and the primary key II. In order to enhance the data, the 10 min meteorological information is fused in this paper to enlarge the feature information of the fused data. The third step is the data encoding operation: Considering that the categorical data in the dataset contain low-base data and high-base data, as well as the numerical attributes of the data, the mixed encoding methods of Min–Max coding [35] and CatBoost coding [36] are adopted in this paper to encode the dataset, so as to ensure that the data remain in the same dimensional range before input into the algorithm. There is also no dimensional explosion. The fourth step is the data matrix operation: since the MobileNetV3 algorithm belongs to the convolutional neural network, its input data is required to be in the form of a matrix, so the dataset in this paper needs to be converted from the form of vector to the form of matrix before input into the algorithm, so as to meet the input requirements.

3.2. Feature Extraction of the Delay Prediction Model

The processed dataset needs to be transformed into tensor form and fed into the model for feature extraction and training. The specific feature extraction process is as follows: After the ECA-MobileNetV3 network model accepts the input characteristic matrix, the characteristic matrix first passes through the first standard convolution layer, which converts the input characteristic matrix into a set of characteristic graphs and then activates it immediately following a nonlinear activation function. Next, these feature maps will pass through multiple inverse residual modules composed of a standard convolution layer, deep convolution layer, ECA attention mechanism module, and point-by-point convolution layer and activation function. These convolution layers with deep separability can effectively reduce the model parameters and calculation amount, and multiple inverse residual models can extract features at different levels. During this period, the ECA-MobileNetV3 network uses an ECA module and feature fusion technology to fuse feature maps at different levels to improve the expression ability of feature maps. Then, the feature map passes through a global average pooling layer. The global average pooling layer can reduce the dimension of the feature map into a vector. Finally, this vector maps the feature vector to the target category through a fully connected layer classifier to complete the classification task.

3.3. Classification Prediction

The pre-processed dataset needs to be transformed into tensor form and fed into the model for training. According to the relevant definition of flight delay in Normal Flight Management Regulations [37] issued by the Civil Aviation Administration of China in 2017, on this basis, this paper subdivides the delay situation into five different time periods. In this paper, the five levels of flight delay are taken as the labels of the dataset, and the flight arrival delay time is taken as the flight delay time T. It defines the difference between the actual arrival time and the planned arrival time.

According to the classification of flight delay levels given in Table 2 and the sample number of each delay level in the two datasets, when T is less than 15 min, it is considered as delay-free; that is, the flight delay level is 0 and the label is 0. When T is between 15 and 60 min, it is considered to be slightly delayed; that is, the flight delay level is 1 and the label is 1. When T is between 60 and 120 min, it is considered moderate delay; that is, the flight delay level is 2 and the label is 2. When T is between 120 and 240 min, it is considered to be highly delayed; that is, the flight delay level is 3 and the label is 3. When T is above 240 min, it is considered a severe delay, that is, a flight delay level of 4 with a label of 4.

Table 2. Flight delay level classification.

Flight Delay Grade	Flight Delay Time T (minute)	Hongqiao Airport	Beijing–Tianjin–Hebei Airport Cluster
0 (No delay)	$T \leq 15$	242,873	898,033
1 (Mild delay)	$15 < T \leq 60$	34,388	91,362
2 (Moderate delay)	$60 < T \leq 120$	14,904	32,053
3 (Highly delayed)	$120 < T \leq 240$	7379	16,932
4 (Heavy delay)	$T > 240$	2049	10,195

The flight delay prediction algorithm then uses the Softmax classifier to determine the flight delay level. Softmax function is a commonly used activation function, which is often used for the final output of multi-classification problems. The original Softmax classifier formula is shown in (15), where x_i represents the i-th sample, q represents the number of categories, and j represents the number of categories. The Softmax function can map a q-dimension vector to a q-dimension probability distribution, where the value of each element represents the probability size of the category. Therefore, the classifier can compute a probability value for each delay level, and the highest value is used as each datum's final result. The Softmax classifier formula is shown in (16):

$$softmax(x_i) = \frac{e^{x_i}}{\sum_{j=1}^{q} e^{x_i}} \quad (15)$$

$$h_\theta(x) = \begin{bmatrix} p(y^{(i)} = 1)x^{(i)};\theta \\ p(y^{(i)} = 2)x^{(i)};\theta \\ p(y^{(i)} = 3)x^{(i)};\theta \\ \cdots \\ p(y^{(i)} = q)x^{(i)};\theta \end{bmatrix} = \frac{1}{\sum_{j=1}^{q} e^{\theta_j^T x(i)}} \begin{bmatrix} e^{\theta_1^T x(i)} \\ e^{\theta_2^T x(i)} \\ e^{\theta_3^T x(i)} \\ \cdots \\ e^{\theta_q^T x(i)} \end{bmatrix} \quad (16)$$

Among them, $h_{\theta}(x)$ is the final output of the flight delay prediction model, θ is the optimal parameter obtained by the model, i represents the serial number of data quantity, and q represents the classification number of flight delay level.

4. Interpretation

4.1. Experimental Environment and Model Parameter Configuration

The computer used in this paper was set up as follows under the described experimental setting: the processor was an Intel Xeon E5-1620 with a CPU frequency of 3.60 GHz; memory 16.004 GB; the OS is Ubuntu16.04. The graphics accelerator GeForce GTX TITAN Xp; the deep learning development framework is Tensorflow 2.3.0. The sample size of the Shanghai Hongqiao Airport dataset used in the experiment is 301,089, the feature attribute quantity is 64, and the size after matrix is 8×8; the sample size of Beijing–Tianjin–Hebei airport cluster dataset is 1,650,797, the feature attribute quantity is 72, and the size after matrix is 8×9. The specific experimental parameter configurations used to train the model are shown in Table 3 below.

Table 3. Table of configuration of experimental parameters.

Parameter Name	Hongqiao Airport Parameters Take the Value	Beijing–Tianjin–Hebei Airport Group Parameter Value
Iteration number	300	150
Train_test_split	9:1	9:1
Loss function	Cross entropy	Cross entropy
Optimizer	Adam	Adam
Learning rate	0.001	0.000001
Dropout	0.2	0.2
Training batch volume	256	128
Test batch volume	256	128
Width Multiplier α	0.50/0.75/1.00	0.50/0.75/1.00

4.2. Evaluation Index of the Model

Loss value and accuracy rate are evaluation metrics that characterize how well a deep learning algorithm fits. The loss value is mainly used to measure the difference between the predicted result of the model and the actual value and can be calculated from the loss function, which is negatively correlated with accuracy, with higher accuracy leading to smaller loss values. The percentage of samples that produced accurate predictions compared to all samples is known as the accuracy rate. The formula is shown in (17), where C represents the predicted correct sample.

$$Accuracy = \frac{\sum C}{N} \tag{17}$$

Computational complexity can describe the hardware consumption at runtime. The higher the complexity, the more memory is occupied and the higher the processing time required. It is mainly divided into spatial complexity and time complexity: spatial complexity is expressed in terms of the number of parameters. The number of parameters of single-layer convolutional layer and single-layer fully connected layer in the algorithm can be approximated as Formulas (18) and (19). The time complexity is expressed in computational quantities, which might be understood as the quantity of FLOPs (Floating Point Operations). The computation amount of single-layer convolutional layer and single-layer fully connected layer can be approximated as Formulas (20) and (21).

$$P_C = D_K \times D_K \times C_F \times N_K \tag{18}$$

$$P_Q = D_F \times D_F \times C_F \times N_K \tag{19}$$

$$F_C = D_F \times D_F \times C_F \times N_K \times D_K \times D_K \tag{20}$$

$$F_Q = D_F \times D_F \times C_F \times N_K \times 1 \times 1 \quad (21)$$

In Formulas (18) and (19), P_C and P_Q are the number of parameters of single-layer convolutional layer and single-layer fully connected layer, respectively, D_K is the convolutional kernel size in the current layer, C_F is the number of input feature channels of the current layer, N_K is the number of output feature channels of the current layer, and D_F is the input feature size of the current layer. In Formulas (20) and (21), F_C and F_Q are the calculated amount of single-layer convolutional layer and single-layer full connection layer, respectively. Thus, 1 represents the output feature size of the full connection layer, and the other parameters have the same meaning in the parameter number formula.

4.3. Loss Values and Accuracy Rates

The validation will be performed on the Shanghai Hongqiao Airport dataset and the Beijing–Tianjin–Hebei Airport dataset.

Based on the Shanghai Hongqiao Airport dataset, the accuracy and the magnitude of loss values in the MobileNetV3 algorithm and ECA-MobileNetV3 algorithm with different channel factors are given in Table 4. According to Table 4, from the longitudinal analysis, the accuracy of MobileNetV3 and ECA-MobileNetV3 algorithms gradually increases and the loss value gradually decreases as the channel factor becomes larger, and the accuracy of the MobileNetV3 algorithm reaches the highest at 98.87% when the channel factor is 1.00. The ECA-MobileNetV3 algorithm achieves the highest accuracy of 98.97% at a channel factor of 0.75. From a cross-sectional perspective, the accuracy of the ECA-MobileNetV3 algorithm with the addition of the ECA attention mechanism module is higher than that of the original MobileNetV3 algorithm for the same number of channel factors, and it can be seen that the improved algorithm does not lose accuracy on a single-airport dataset such as the Shanghai Hongqiao Airport dataset.

Table 4. Comparison table of accuracy and loss values for different-width multipliers on Shanghai Hongqiao Airport dataset.

Width Multiplier	MobileNetV3		ECA-MobileNetV3	
	Accuracy	Loss Value	Accuracy	Loss Value
0.50	98.00%	0.0716	98.41%	0.0675
0.75	98.53%	0.0553	98.97%	0.0445
1.00	98.87%	0.0419	98.90%	0.0449

Based on the dataset of Shanghai Hongqiao Airport, the accuracy and loss curves of the MobileNetV3 algorithm and ECA-MobileNetV3 algorithm under different channel factors are, respectively, presented in Figures 5 and 6. According to the trend of the curves, at different channel factors, the accuracy rate gently increases while the loss value gently decreases. The loss values and accuracies of MobileNetV3 and ECA-MobileNetV3 tend to stabilize when the number of training rounds is around 300. From the experimental results, the MobileNetV3 algorithm has a loss value of about 0.0419 when the channel factor is 1.00. The highest accuracy was 98.87%. When the channel factor is 0.75, the lowest loss value of the ECA-MobileNetV3 algorithm is about 0.0449, and the highest accuracy is 98.90%. Compared with the MobileNetV3 algorithm, the accuracy of the ECA-MobileNetV3 algorithm with attention mechanism is slightly improved and the loss value is slightly increased.

Based on the Beijing–Tianjin–Hebei airport cluster dataset, according to Table 5, from the longitudinal analysis, as the channel factor becomes larger, the accuracy of the two algorithms gradually increases and the loss value gradually decreases. Further, the accuracy rates of the MobileNetV3 algorithm and ECA-MobileNetV3 algorithm reach the highest when the channel factor is 1.00, and the accuracy rate of the MobileNetV3 algorithm reaches 96.60%; the accuracy rate of the ECA-MobileNetV3 algorithm reaches 96.81%. From a cross-sectional perspective, the accuracy of the ECA-MobileNetV3 algorithm is slightly lower than that of the MobileNetV3 algorithm at channel factor numbers of 0.50 and 0.75, and the accuracy of the improved algorithm is 0.18% lower than that before the improvement at a channel factor of 0.50. At a channel factor of 1.00, the accuracy of the ECA-MobileNetV3 algorithm is slightly higher than that of the MobileNetV3 algorithm, and the accuracy of the improved algorithm is 0.21% higher than that before the improvement. Therefore, on the whole, the improved ECA-MobileNetV3 algorithm has a minor loss in accuracy and still has some advantages in a multi-airport-associated cluster dataset such as the Beijing–Tianjin–Hebei airport cluster dataset.

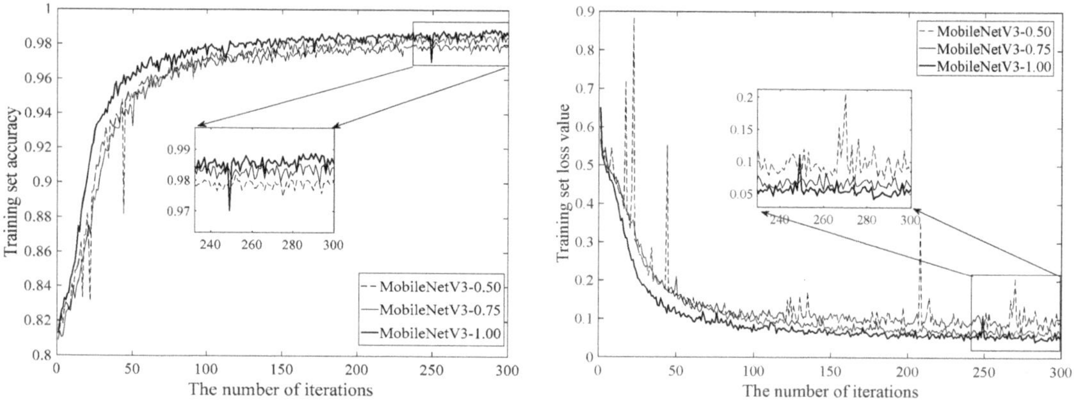

(a) (b)

Figure 5. Comparison of loss values and accuracy for different-width multipliers based on the MobileNetV3 algorithm on Shanghai Hongqiao Airport dataset. (**a**) Accuracy comparison of different-width multipliers. (**b**) Loss value comparison of different-width multipliers.

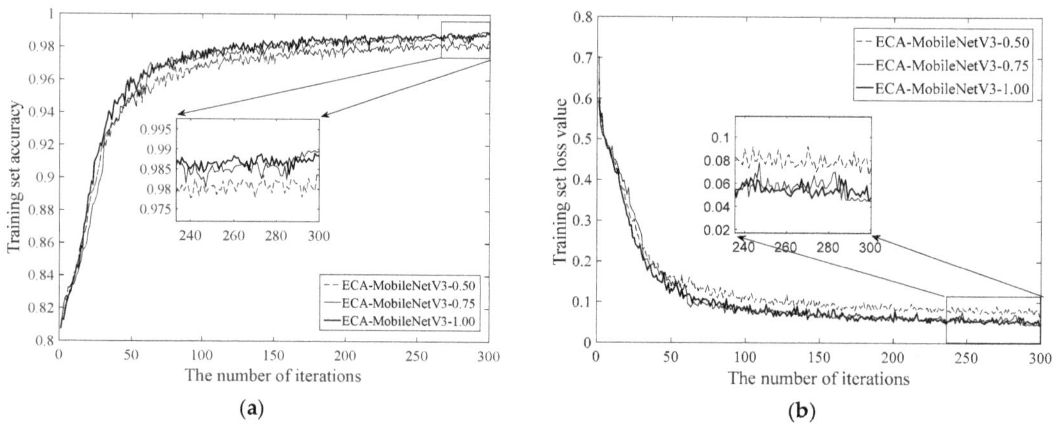

(a) (b)

Figure 6. Comparison of loss values and accuracy for different-width multipliers based on the ECA-MobileNetV3 algorithm on Shanghai Hongqiao Airport dataset. (**a**) Accuracy comparison of different-width multipliers. (**b**) Loss value comparison of different-width multipliers.

Based on the Beijing–Tianjin–Hebei airport cluster dataset, the accuracy and loss curves of the MobileNetV3 algorithm and ECA-MobileNetV3 algorithm under different channel factors are, respectively, presented in Figures 7 and 8. According to the trend of the curves, at different channel factors, the accuracy rate gently increases while the loss value gently decreases. The loss values and accuracies of MobileNetV3 and ECA-MobileNetV3 tend to stabilize when the number of training rounds is around 150. From the experimental results, the MobileNetV3 algorithm has a loss value of about 0.0819 when the channel factor is 1.00. The highest accuracy was 96.60%. When the channel factor is 1.00, the lowest loss value of the ECA-MobileNetV3 algorithm is about 0.0813, and the highest accuracy is 96.81%. Compared with the MobileNetV3 algorithm, the accuracy of the ECA-MobileNetV3 algorithm with an attention mechanism is slightly improved, while the loss value is slightly decreased.

Table 5. Comparison table of accuracy and loss values for different width multipliers on Beijing–Tianjin–Hebei airport cluster dataset.

Width Multiplier	MobileNetV3		ECA-MobileNetV3	
	Accuracy	Loss Value	Accuracy	Loss Value
0.50	96.40%	0.0932	96.22%	0.1049
0.75	96.56%	0.0871	96.55%	0.0878
1.00	96.60%	0.0819	96.81%	0.0813

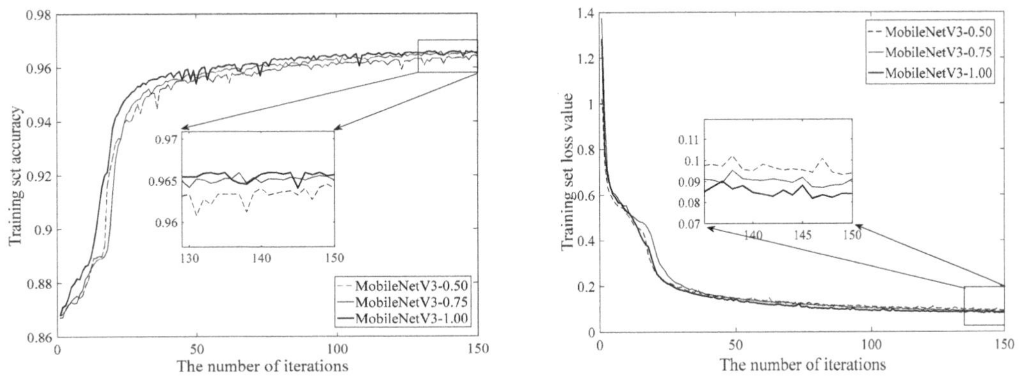

(a) (b)

Figure 7. Comparison of loss values and accuracy for different-width multipliers based on the MobileNetV3 algorithm on Beijing–Tianjin–Hebei airport cluster dataset. (**a**) Accuracy comparison of different-width multipliers. (**b**) Loss value comparison of different-width multipliers.

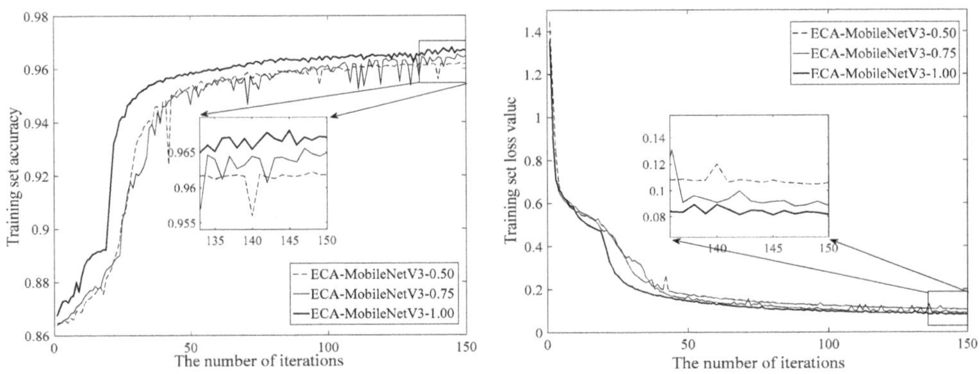

(a) (b)

Figure 8. Comparison of loss values and accuracy for different-width multipliers based on the ECA-MobileNetV3 algorithm on Beijing–Tianjin–Hebei airport cluster dataset. (**a**) Accuracy comparison of different-width multipliers. (**b**) Loss value comparison of different-width multipliers.

4.4. Complex Calculation of the Model

Validation will be performed on the Shanghai Hongqiao Airport dataset and the Beijing–Tianjin–Hebei Airport dataset.

Based on the Shanghai Hongqiao Airport dataset, Table 6 displays the suggested algorithm's accuracy and computational complexity before and after the enhancement. Vertically, as the channel factor increases, the MobileNetV3 algorithm's accuracy rises together with the model's complexity. The model complexity and accuracy of the ECA-MobileNetV3 algorithm increase with the channel factor. However, for channel factors of 0.75 and 1.0, the accuracy rate does not improve significantly due to the complexity of the model but gradually stabilizes. From a horizontal perspective, under the same channel factor, the ECA-MobileNetV3 model can efficiently minimize the number of parameters and computation without sacrificing accuracy.

Table 6. Algorithmic complexity comparison for different-width multipliers on Shanghai Hongqiao Airport dataset.

Width Multiplier	MobileNetV3			ECA-MobileNetV3		
	Params(M)	FLOPs(M)	Accuracy	Params(M)	FLOPs(M)	Accuracy
0.50	0.29	16.43	98.00%	0.17	16.21	98.41%
0.75	0.60	33.31	98.53%	0.33	32.80	98.97%
1.00	1.01	54.66	98.87%	0.55	53.76	98.90%

Based on the Beijing–Tianjin–Hebei airport cluster dataset, the computational complexity and accuracy of the proposed algorithm before and after the improvement are shown in Table 7. From the longitudinal point of view, as the channel factor increases, the complexity of the MobileNetV3 model increases and the accuracy improves. However, with channel factors of 0.75 and 1.00, the accuracy increases slightly and remains essentially stable. As the channel factor increases, the complexity of the ECA-MobileNetV3 model increases and the accuracy improves. When the channel factors are 0.75 and 1.0, the accuracy is significantly improved. From a horizontal perspective, the computational complexity of the ECA-MobileNetV3 model is effectively reduced with little loss in accuracy for the same channel factor.

Table 7. Algorithmic complexity comparison for different-width multipliers on Beijing–Tianjin–Hebei airport cluster dataset.

Width Multiplier	MobileNetV3			ECA-MobileNetV3		
	Params(M)	FLOPs(M)	Accuracy	Params(M)	FLOPs(M)	Accuracy
0.50	0.29	18.44	96.40%	0.17	18.22	96.22%
0.75	0.60	37.39	96.56%	0.33	36.88	96.55%
1.00	1.01	61.35	96.60%	0.55	60.44	96.81%

It is clear from the experimental findings on the aforementioned two datasets that the computational cost and precision of the proposed model are not only linear. We can find algorithms that better balance the accuracy and computational complexity of the model, which is also the direction of efforts in lightweight neural networks. By computing conditions on different mobile devices, it is possible to match flight delay prediction models of different sizes to maximize the model utilization.

4.5. Comparison of Different Network Models

Compared with traditional deep learning algorithms, the modified ECA-MobileNetV3 achieves better performance in terms of computational complexity and model accuracy when dealing with real domestic flight datasets with weather information fusion. In this regard, this paper verifies the single airport and airport group datasets, compares the ECA-MobileNetV3_1.00 model with the traditional ResNet [38], DenseNet [39] algorithm, and MobileNetV2 algorithm under the same channel factor and analyzes it from the following three aspects. ResNet and DenseNet are models that have been trained and widely used in large-scale datasets and have achieved good results in many computer vision and natural language processing tasks. Therefore, they are very representative models and can be used as benchmarks for other models. MobileNetV2, as the leader in the lightweight model, has been widely used in many mobile device applications. MobileNetV2 is the predecessor of MobileNetV3, which can verify whether the improvement in ECA-MobileNetV3 is effective. Taking MobileNetV2 as a comparative test can also provide reference and inspiration for more lightweight model design. The results are shown in Table 8.

Table 8. Comparison of the evaluation indicators for the different models.

Model Name	Shanghai Hongqiao Airport			Beijing-Tianjin-Heibei Airport Group		
	Params(M)	FLOPs(M)	Accuracy	Params(M)	FLOPs(M)	Accuracy
ResNet18	11.18	1429.40	95.56%	11.18	1608.07	94.33%
DenseNet121	7.04	498.36	94.94%	7.04	580.54	93.76%
MobileNetV2_1.00	2.26	70.75	99.06%	2.26	88.40	95.99%
MobileNetV3_1.00	1.01	54.66	98.87%	1.01	61.35	96.60%
ECA-MobileNetV3_1.00	0.55	53.76	98.90%	0.55	60.44	96.81%

In the Hongqiao Airport dataset, the accuracy of the ECA-MobileNetV3_1.00 algorithm increases by 3.34% and 3.96% and the number of citations decreases by 10.63 million and 6.49 million, respectively, compared to the other two traditional networks. The calculated amounts were reduced by 1375.64 million and 444.6 million, respectively. In the single airport dataset, it can be seen that the enhanced ECA-MobileNetV3 model outperforms in terms of accuracy and computational complexity. In addition, when compared to the MobileNetV2 method at the same channel factor, the accuracy of the enhanced algorithm is decreased by 0.16%, and the number of parameters and computational cost are decreased, respectively, by 171 million and 16.99 million. As compared to the MobileNetV2 algorithm at the same level, it can be seen that the upgraded algorithm achieves a better balance between model complexity and accuracy.

In the Beijing–Tianjin–Hebei airport cluster dataset, the accuracy of the ECA-MobileNetV3_1.00 model increases by 2.48% and 3.05% and the reference number decreases by 10.63 million and 6.49 million, respectively, compared to the other two traditional networks. The calculated amounts were reduced by 1547.63 million and 520.10 million, respectively. On the airport cluster dataset, it can be shown that the revised model performs quite well in the three evaluation measures mentioned above. Additionally, the ECA-MobileNetV3 algorithm's accuracy is increased by 0.82% when compared to MobileNetV2 at the same channel factor, while the algorithm's computational cost and parameter count are decreased by 1.71 million and 27.96 million, respectively. As is evident, the improved model also achieves a better balance of the above three metrics compared to the MobileNetV2 algorithm of the same level.

4.6. Application of the Model

At present, the flight delay prediction Web visualization system based on the flight delay prediction model of ECA-MobileNetV3 has been put into use in the air traffic control bureau. The system uses the flight delay prediction model studied in this paper to predict flight delay and then displays the predicted delay results on the web page through the Web visualization technology and can carry out statistical analysis on the historical delay information, so as to explore the deeper laws of delay generation, for example, to see in which time periods of the day and in which months of the year flight delays mainly occur. This application mainly takes advantage of the high prediction accuracy of the flight delay prediction model studied in this paper. The subsequent application direction will focus on the advantages of light weight. The lightweight model has the characteristics of fast prediction speed, less demand for computing resources, higher real-time performance, and portability. Therefore, this model can be deployed on some low-power devices, such as mobile devices and sensors. This can quickly process data input and quickly update forecast results and provide real-time information for airlines and the base to help them plan and manage flight missions.

5. Conclusions

This paper studies the lightweight neural network MobileNetV3 algorithm and the improved ECA-MobileNetV3 algorithm. By using the Shanghai Hongqiao Airport dataset and the Beijing–Tianjin–Hebei Airport Cluster dataset, for example, analysis and practical application of the model, the following conclusions are drawn: The algorithm proposed in this paper can effectively reduce the computational complexity in the model without loss of accuracy or with a small loss of accuracy by replacing the SE module in the original MobileNetV3 algorithm with a lightweight ECA attention mechanism module. Compared with the ResNet algorithm, DenseNet algorithm, and MobileNetV2 algorithm under the same channel factor, the improved ECA-MobileNetV3 algorithm has more advantages in computational complexity and accuracy. The flight delay prediction model based on ECA-the MobileNetV3 algorithm has the advantage of light weight compared with the flight delay prediction model that has been deployed now. The lightweight flight delay model can bring faster execution speed, fewer computing resources, higher real-time performance, and higher flexibility and portability, which can greatly lay the foundation for subsequent deployment on mobile terminals and other platform devices, and for airlines, the airport and passengers provide better service and better experience. However, there is still a lot of room for improvement in the process of research in this paper. On the one hand, the number of flight samples with different delay levels is quite different, which will affect the accuracy of flight prediction. It is necessary to consider the impact of sample imbalance on model training. On the other hand, the problem of flight delay is time-varying, and the prediction model needs to be updated at any time. The next step is to explore how to achieve a real-time update of the model and improve the practicability of the model.

Author Contributions: Methodology, J.Q.; validation, B.C. and C.L.; investigation, J.Q.; writing—original draft preparation, B.C., C.L. and J.W.; writing—review and editing, B.C. All authors have read and agreed to the published version of the manuscript.

Funding: This research was funded by the Tianjin Municipal Education Commission Scientific Research Program, grant number 2022ZD006, and the Fundamental Research Funds for the Central Universities, grant number 3122019185.

Data Availability Statement: Not applicable.

Conflicts of Interest: The authors declare no conflict of interest.

References

1. Five Major Development Tasks Clearly Clarified by the 13th Five-Year Plan for Civil Aviation. Available online: http://caacnews.com.cn/1/1/201612/t20161222_1207217.html (accessed on 22 December 2016).
2. Report on the 2022 National Civil Aviation Work Conference. Available online: http://www.caac.gov.cn/XWZX/MHYW/202201/t20220110_210827.html (accessed on 10 January 2022).
3. IATA: The Global Air Passenger Volume Will Reach 8.2 Billion Person-Times in 2007. Available online: https://www.ccaonline.cn/news/top/461390.html (accessed on 30 October 2018).
4. Zhang, Y. Study on Group Time Countermeasures caused by Abnormal Flights-Take CEA as An Example. Master's Thesis, East China University of Political Science and Law, Shanghai, China, 2018.
5. Zhang, M. Research on Optimization of Countermeasures for Handling Mass Incidents of Passengers caused by Flight Delays. Master's Thesis, Handong University of Finance and Economics, Jinan, China, 2016.
6. Li, X.; Liu, G.C.; Yan, M.C.; Zhang, W. Economic losses of airlines and passengers caused by flight delays. *Syst. Eng.* **2007**, *25*, 20–23.
7. Liu, B.; Ye, B.J.; Tian, Y. Overview of flight delay prediction methods. *Aviat. Comput. Technol.* **2019**, *49*, 124–128.

8. Xu, T.; Ding, J.L.; Gu, B.; Wang, J.D. Airport flight delay warning based on the incremental arrangement of support vector machines. *Aviat. J.* **2009**, *30*, 1256–1263.
9. Luo, Q.; Zhang, Y.H.; Cheng, H.; Li, C. Model of hub airport flight delay based on aviation information network. *Syst. Eng. Theory Pract.* **2014**, *34*, 143–150.
10. Luo, Z.S.; Chen, Z.J.; Tang, J.H.; Zhu, Y.W. A flight delay prediction study using SVM regression. *Transp. Syst. Eng. Inf.* **2015**, *15*, 143–149, 172.
11. Cheng, H.; Li, Y.M.; Luo, Q.; Li, C. Study on Prediction of arrival flight delay based on C4.5 decision tree method. *Syst. Eng. Theory Pract.* **2014**, *34*, 239–247.
12. Nigam, R.; Govinda, K. Cloud Based Flight Delay Prediction using logistic Regression. In Proceedings of the 2017 International Conference on Intelligent Sustainable Systems (ICISS), Palladam, India, 7–8 December 2017.
13. Khanmohammadi, S.; Tutun, S.; Kucuk, Y. A new multilevel input layer artificial neural network for predicting flight delays at JFK airport. *Procedia Comput. Sci.* **2016**, *95*, 237–244. [CrossRef]
14. Wu, R.B.; Zhao, Y.Q.; Qu, J.Y. Flight delay spread prediction model based on CBAM-CondenseNet. *J. Electron. Inf. Technol.* **2021**, *43*, 187–195.
15. Hition, G.; Vinyals, O.; Dean, J. Distilling the knowledge in a neural network. *Comput. Sci.* **2015**, *14*, 38–39.
16. Vongkulbhisal, J.; Vinayavekhin, P.; Visentini-scarzanella, M. Unifying Heterogeneous Classifiers with Heterogeneous Classifiers with Distillation. In Proceedings of the IEEE/CVF Conference on Computer Vision and Pattern Recognition (CVPR), Long Beach, CA, USA, 15–20 June 2019.
17. Zhuang, l.; Li, J.G.; Shen, Z.; Zhang, C.M. Learning Efficient Convolutional Networks Through Network Slimming. In Proceedings of the IEEE International Conference on Computer Vision (ICCV), Venice, Italy, 22–29 October 2017.
18. He, Y.; Zhang, X.; Sun, J. Channel Pruning for Accelerating Very Deep Neural Networks. In Proceedings of the IEEE International Conference on Computer Vision (ICCV), Venice, Italy, 22–29 October 2017.
19. Luo, J.H.; Wu, J.X.; Lin, W.Y. Thinet: A Filter Level Pruning Method for Deep Neural Network Compression. In Proceedings of the IEEE International Conference on Computer Vision (ICCV), Venice, Italy, 22–29 October 2017.
20. Zhang, X.; Zhou, X.; Lin, M. ShuffleNet: An Extremely Efficient Convolutional Neural Network for Mobile Devices. In Proceedings of the IEEE Conference on Computer Vision and Pattern Recognition (CVPR), Salt Lake City, UT, USA, 18–23 June 2018.
21. Ma, N.N.; Zhang, X.Y.; Zheng, H.T. ShuffleNetV2: Practical Guidelines for Efficient CNN Architecture Design. In Proceedings of the European Conference on Computer Vision (ECCV), Munich, Germany, 8–14 September 2018.
22. Tan, M.; Le, Q.E. EfficientNet: Rethinking Model Scaling for Convolutional Neural Networks. *arXiv* **2019**, arXiv:1905.11946.
23. Iandola, F.N.; Han, S.; Moskewicz, M.W. SqueezeNet: AlexNet-level Accuracy with 50x Fewer Parameters and <0.5 MB Model Size. *arXiv* **2016**, arXiv:1602.07360.
24. Howard, A.G.; Zhu, M.; Chen, B. MobileNets: Efficient Convolutional Neural Networks for Mobile Vision Applications. *arXiv* **2017**, arXiv:1704.04861.
25. Sandler, M.; Howard, A.; Zhu, M. MobileNetV2: Inverted Residuals and Linearbottlenecks. In Proceedings of the IEEE Conference on Computer Vision and Pattern Recognition (CVPR), Salt Lake City, UT, USA, 18–23 June 2018.
26. Howard, A.; Sandler, M.; Chu, G. Searching for Mobilenetv3. In Proceedings of the IEEE/CVF International Conference on Computer Vision (ICCV), Seoul, Republic of Korea, 27 October–3 November 2019.
27. Cai, Q.J.; Peng, C.; Shi, X.W. Based on the MobieNetV2 lightweight face recognition algorithm. *Comput. Appl.* **2020**, *40*, 65–68.
28. Qi, Y.K. Lightweight algorithm for pavement obstacle detection based on MobileNet and YOLOv3. *Comput. Syst. Appl.* **2022**, *31*, 176–184.
29. Hu, J.L.; Shi, Y.P.; Xie, S.Y.; Chen, P. Improved MobileNet face recognition system based on Jetson Nano. *Sens. Microsyst.* **2021**, *40*, 102–105.
30. Hu, J.; Shen, L.; Albanie, S.; Sun, G.; Wu, E. Squeeze-and-Excitation Networks. In Proceedings of the IEEE Conference on Computer Vision and Pattern Recognition (CVPR), Salt Lake City, UT, USA, 18–23 June 2018.
31. Wang, Q.; Wu, B.; Zhu, P. ECA-Net: Efficient Channel Attention for Deep Convolutional Neural Networks. In Proceedings of the IEEE/CVF Conference on Computer Vision and Pattern Recognition (CVPR), Seattle, WA, USA, 14–19 June 2020.
32. Rumelhart, D.E.; Hinton, G.E.; Williams, R.J. Learning representations by back-propagating errors. *Nature* **1986**, *323*, 533–536. [CrossRef]
33. Qu, J.; Zhao, T.; Ye, M. Flight delay prediction using deep convolutional neural network based on fusion of meteorological data. *Neural Process. Lett.* **2020**, *52*, 1461–1484. [CrossRef]
34. Quality Controlled Local Climatological Data. Available online: https://www.ncdc.noaa.gov/orders/qclcd/ (accessed on 13 February 2019).
35. Cao, L. Research on Flight Delay Prediction and Visualization method based on CliqueNet. Master's Thesis, Civil Aviation University of China, Tianjin, China, 2020.
36. Prokhorenkova, L.; Gusev, G.; Vorobev, A. CatBoost: Unbiased Boosting with Categorical Features. *Adv. Neural Inf. Process. Syst.* **2018**, *31*, 6638–6648.
37. Flight Normal Management Regulations. Available online: https://xxgk.mot.gov.cn/2020/jigou/fgs/202006/t20200623_3307796.html (accessed on 24 March 2016).

38. He, K.; Zhang, X.; Ren, S. Deep Residual Learning for Image Recognition. In Proceedings of the IEEE Conference on Computer Vision and Pattern Recognition (CVPR), Las Vegas, NV, USA, 26 June–1 July 2016.
39. Huang, G.; Liu, Z.; Weinberger, K.Q. Densely Connected Convolutional Networks. In Proceedings of the IEEE Conference on Computer Vision and Pattern Recognition (CVPR), Honolulu, HI, USA, 21–26 July 2017.

electronics

Article

Machine Learning-Based Prediction of Orphan Genes and Analysis of Different Hybrid Features of Monocot and Eudicot Plants

Qijuan Gao [1], Xiaodan Zhang [2], Hanwei Yan [3] and Xiu Jin [3,*]

1 School of Computer Science, Hefei Normal University, Hefei 230001, China
2 Anhui Province Key Laboratory of Smart Agricultural Technology and Equipment, Anhui Agricultural University, Hefei 230036, China
3 Key Laboratory of Crop Biology of Anhui Province, Anhui Agricultural University, Hefei 230036, China
* Correspondence: jinxiu123@ahau.edu.cn

Abstract: Orphan genes (OGs) may evolve from noncoding sequences or be derived from older coding material. Some shares of OGs are present in all sequenced genomes, participating in the biochemical and physiological pathways of many species, while many of them may be associated with the response to environmental stresses and species-specific traits or regulatory patterns. However, identifying OGs is a laborious and time-consuming task. This paper presents an automated predictor, XGBoost-A2OGs (identification of OGs for angiosperm based on XGBoost), used to identify OGs for seven angiosperm species based on hybrid features and XGBoost. The precision and accuracy of the proposed model based on fivefold cross-validation and independent testing reached 0.90 and 0.91, respectively, outperforming other classifiers in cross-species validation via other models, namely, Random Forest, AdaBoost, GBDT, and SVM. Furthermore, by analyzing and subdividing the hybrid features into five sets, it was proven that different hybrid feature sets influenced the prediction performance of OGs involving eudicot and monocot groups. Finally, testing of small-scale empirical datasets of each species separately based on optimal hybrid features revealed that the proposed model performed better for eudicot groups than for monocot groups.

Keywords: orphan genes (OGs); hybrid features; machine learning; angiosperm

Citation: Gao, Q.; Zhang, X.; Yan, H.; Jin, X. Machine Learning-Based Prediction of Orphan Genes and Analysis of Different Hybrid Features of Monocot and Eudicot Plants. *Electronics* **2023**, *12*, 1433. https://doi.org/10.3390/electronics12061433

Academic Editors: Chao Zhang, Wentao Li, Huiyan Zhang and Tao Zhan

Received: 16 February 2023
Revised: 11 March 2023
Accepted: 12 March 2023
Published: 17 March 2023

1. Introduction

Monocotyledonous and eudicotyledonous plants (monocots and eudicots) have morphological differences in the number and arrangement of their embryonic leaves. These are typically parallel in monocots and reticulate in eudicots; besides, monocots have a sheathing leaf base encircling the stem. Monocots diverged from their eudicot relatives in angiosperm evolution derived from the whole genome duplication (WGD), which contributed to increased diversification, environmental adaptation, and genomic novelty [1]. In the evolutionary process, orphan genes (OGs) can arise in a lineage and are prevalently expressed in many organisms [2]. In particular, taxonomically restricted OGs are widely distributed in angiosperm species, including eudicot and monocot groups, such as *Arabidopsis thaliana, Populus trichocarpa, Citrus sinensis, Triticum aestivum, Oryza sativa, cowpea, Camellia sinensis,* and *Saccharum spontaneum* [3–10]. Numerous studies of OGs have identified general trends in the sequence features of OGs across different species, including gene length, GC content, and introns, which are also vital for environmental adaptation, including biotic and abiotic stress [11–13]. Specifically, the OG Qua-Quine Starch (QQS) in *Arabidopsis thaliana* is known to regulate the ratio of protein and starch carbon. Being transferred and expressed in other species, QQS has been reported to change the metabolic process by regulating the allocation of carbon and nitrogen in proteins and carbohydrates and affecting the compounds in seeds and leaves, consequently improving crop yields [14].

Previous studies also revealed that OGs play a vital role in response to drought stress in *cowpea* and *Fusarium resistance* in *Triticum aestivum* [6,7].

OGs have usually been applied through BLAST (Basic Local Alignment Search Tool) sequence alignment, involving genome and transcriptome sequences for all analysis processes, including BLASTP, BLASTN, TBLASTX, and so on [15]. However, this method is time-consuming and requires considerable server-driven resources to identify OGs. Alternatively, OGs can be distinguished from nonorphan genes (NOGs), e.g., protein-coding genes, by more significant differences in gene length, exon number, GC content, and expression level [11]. Their analysis and further classification can be facilitated via machine learning-based methods, which have already been successfully applied to classifying biological datasets and solving various discrimination problems. Thus, such ensemble learning methods as Gradient Boosting Decision Tree (GBDT), Random Forest, and Adaboost have been used for biological prediction based on genome datasets. In particular, Zhu et al. used GBDT to classify tissue and cell types in cancer samples using a gene expression dataset, which performed similarly to other machine learning methods [16]. In contrast, the Extreme Gradient Boost (XGBoost) method adopted by Chen and Guestrin [17] outperformed numerous machine learning methods and found wide applications in data mining, regression, and classification domains. In addition, Gao et al. have used an effective model named SMOTE-ENN-XGBoost to predict the OGs of *A. thaliana* [18]. However, to the best of the authors' knowledge, it has yet to be carried out in the bioinformatic field of predicting OGs of different types of plant species.

In this study, OGs were measured by taking into account sequence features, which share some characteristics of other angiosperm species (shorter sequence length, fewer exon numbers, and lower GC content), while having fewer transcript support and lower expression than NOGs [12]. Then, these protein features were extracted, and the XGBoost-A2OG model was constructed and applied to the prediction of OGs in angiosperm species.

2. Related Works

Recently, machine learning methods have received considerable interest in the identification of OGs fields, which are an important source of genetics and contribute to evolutionary innovations. These methods include the Decision tree (DT) [19], Neural network (NN) [19], Convolutional Neural Network (CNN) with transformer [20], and ensemble learning method [20]. Besides, many researchers have been conducted to compare different machine learning algorithms or combined with other methods to accelerate the performance of identification of OGs.

Gao et al. proposed a novel ensemble method to predict the OGs of A. thaliana in bioinformatics studies. Then another deep learning method, CNN with transformer technique was successfully applied to identifying OGs in moso bamboo which used a convolutional neural network in combination with a transform neural network in protein sequences [19]. Their proposed approach provides better performance in a specific species.

In addition, decision trees and neural networks were employed to improve the accurate discovery of OGs by Casola et al. relying on basic sequence features obtained from DNA and protein sequences in three angiosperm families. The experimental results showed that both DT and NN classifiers achieve high levels of accuracy and recall in identifying OGs.

Recently, many studies have confirmed that OGs generated de novo in a species may be more prevalent than gene duplication and be one of the main ways of orphan generation [21–25]. Some researchers have found that in the newly evolved OGs in Arabidopsis, protein length is usually shorter, mainly due to the evolution of the orphan gene having fewer exons in the process, while in some species, the exon length is significantly shorter [26,27].

However, these researchers haven't focused on different families of angiosperm plants. To find a general method to identify a large number of plants of OGs based on a rapid accumulation of genomic data, we have analyzed some features regarding the genome and protein sequences that may affect the results in the classification process.

3. Materials and Methods

3.1. The Framework of the XGBoost-A2OG Model

The workflow used in this study and depicted in Figure 1 comprised the following five parts: data selection, data pre-processing, data modelling, model development, and model interpretation.

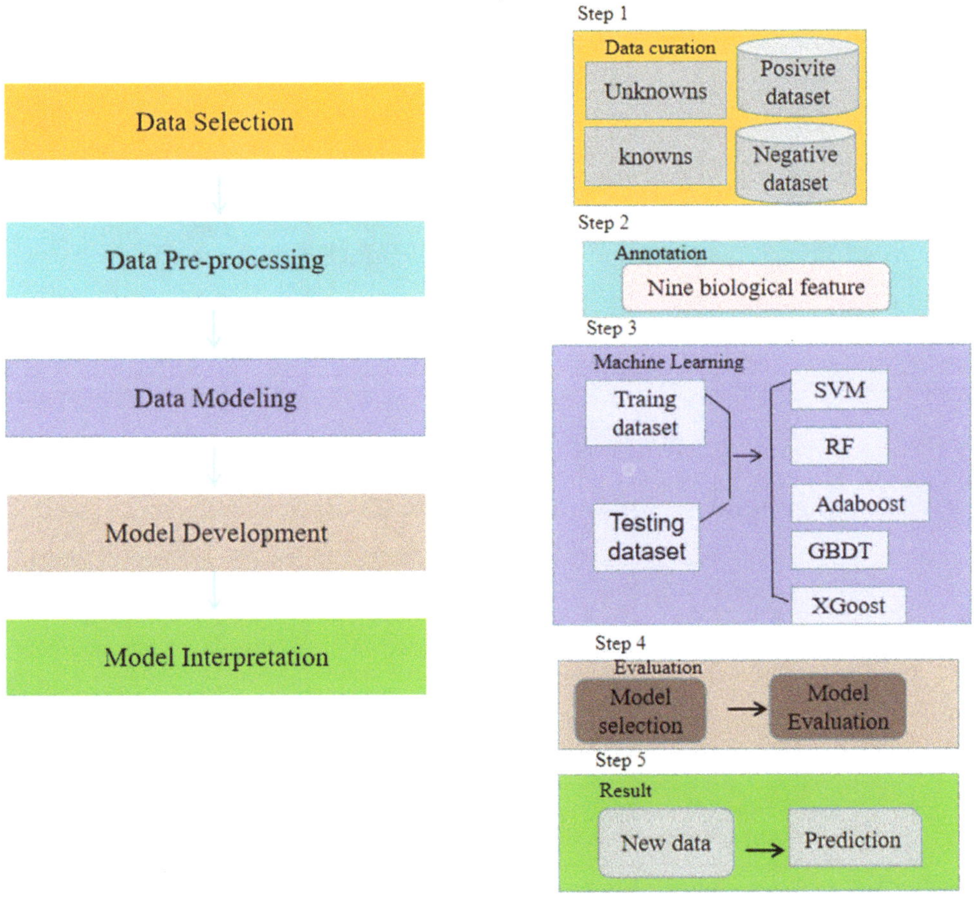

Figure 1. Workflow of the model framework.

3.2. Data Collection

This study collected protein sequences and gene annotation datasets for 136 plant species from Phytozome [28]. Non-redundant protein sequences (NR) were obtained from the NCBI database [29] and Ensemble Plants [30]. Next, BLASTp was used to identify OG based on a previous study [28] to search for homologs of all 401,834 gene annotations in seven plants (*Arabidopsis thaliana, Populus trichocarpa, Citrus sinensis, Camellia sinensis Sorghum bicolour, Oryza sativa, Zea mays*) (Figure 2) of the other 94 species released in Phytozome V12.1 with an E-value cutoff of 1×10^{-3}. Noteworthy is that the E-value or expectation value is a more inclusive value than probability, defining the number of times the query sequence is expected to match with the database sequences by random chance.

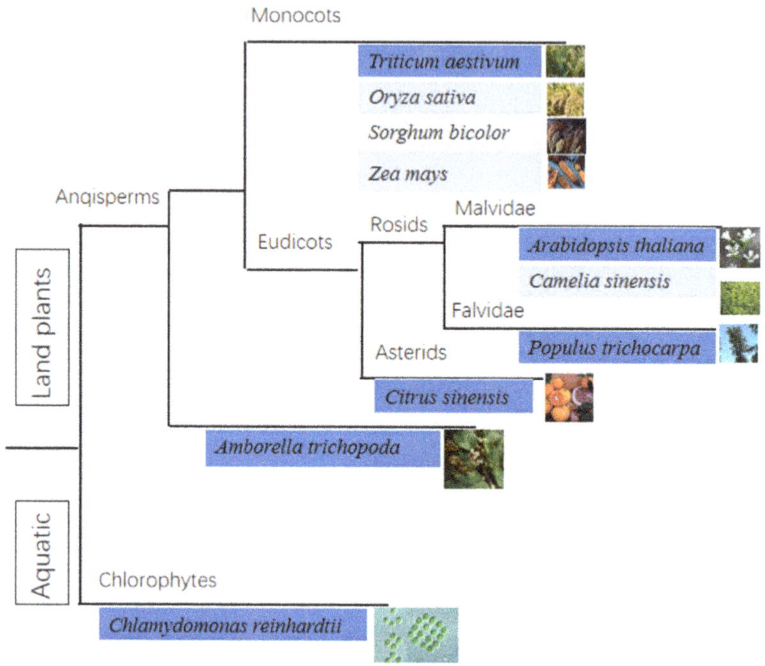

Figure 2. The phylogenetic tree of Monocotyledonous and eudicotyledonous plants.

The obtained 9022 OGs and 392,812 NOGs were identified with label 1 and label 0, respectively, to thoroughly train the ensemble learning model. All of them were combined to form the five plant species' OG datasets. Then, we extracted the characteristics of gene structure, cDNA sequence, and protein-coding genes of all five species from Phytozome and Ensemble plants, forming databases containing high annotation of plant genomes.

3.3. Ensemble Algorithm

XGBoost (Extreme Gradient Boost) is an ensemble learning technique for regression and classification problems based on the boosting algorithm [17]. The motivation is to classify data using the best hyperplane representing the most extensive separation between two classes. Unlike the traditional integrated decision tree algorithm, XGBoost adds a regular term to the loss function, which can control the complexity of the model and prevent the model from overfitting. The objective function is given to be optimized by the following formula:

(1) Taylor's formula to approximate the original goal.

$$obj(\theta) = \sum_{i=1}^{n} l(y_i, y_i,) \tag{1}$$

(2) Taylor expansion:

$$obj^{(t)} = \sum_{i=1}^{n} l(y_i, \hat{y}_i^{\,t-1}) + g_i f_t(x_i) + \frac{1}{2} h_i f_t^2(x_i)] + \Omega(f_t) + \text{constant} \cdots \tag{2}$$

(3) Among them, g_i, h_i are expressed as:

$$\left\{ \begin{array}{l} g_i = \partial_{\hat{y}^{(t-1)}} l(y_i, \hat{y}^{(t-1)}), \\ h_i = \partial 2_{\hat{y}^{(t-1)}} l(y_i, \hat{y}^{(t-1)}). \end{array} \right\} \tag{3}$$

(4) The formula of decision tree complexity calculation:

$$\Omega(f) = \gamma T + \frac{1}{2}\lambda\sum_{j=1}^{T} w_j^2 \tag{4}$$

(5) T is the number of leaf nodes, and w is the leaf node score. Substituting (2)–(4) into (1) the objective function:

$$obj^{(t)} = \sum_{j=1}^{T} [G_j w_j + \frac{1}{2}(H_j + \lambda) w_j^2] + \gamma T \tag{5}$$

(6) Among, $I_{ij} = \{i|q(x_i = j\}$, which represents the sample set belonging to the j-th leaf node.

$$G_j = \sum\nolimits_{i \in I_j} g_i, H_j = \sum\nolimits_{i \in I_j} h_i \tag{6}$$

(7) To minimize the objective function, set the derivative being 0 and find the optimal prediction score of each leaf node:

$$w_j^* = -\frac{G_j}{H_j + \lambda} \tag{7}$$

(8) Substitute the objective function again to get its minimum value:

$$obj^* = -\frac{1}{2}\sum_{j=1}^{T} \frac{G_j^2}{H_j + \lambda} + \gamma T \tag{8}$$

(9) Use *obj** to find the tree with the best structure and add it to the model and apply the greedy algorithm to find the optimal tree structure. Each time when trying to add a split to an existing leaf, the *Gain* is calculated as follows:

$$Gain(\Phi) = \frac{1}{2}[\frac{(\sum\limits_{i \subseteq I_L} g_i)^2}{\sum\limits_{i \subseteq I_L} h_i + \lambda} + \frac{(\sum\limits_{i \subseteq I_R} g_i)^2}{\sum\limits_{i \subseteq I_R} h_i + \lambda} - \frac{(\sum\limits_{i \subseteq I} g_i)^2}{\sum\limits_{i \subseteq I} h_i + \lambda}] - \gamma \tag{9}$$

When the XGBoost model was used in the experiment, the following parameters were adjusted to make the model perform its best performance. For example, one of the most critical parameters in this and other tree-based ensemble algorithms, such as GBDT, Random Forest (RF), and AdaBoost, is "learning_rate", which dramatically affects the model performance. Another parameter is "n_estimators", which is the number of iterations in training: too small or too large parameters will lead to underfitting or overfitting, respectively. The third critical parameter is "max_depth", which is the maximum depth of the tree. Its higher values make the tree model more complex and improve its fitting ability, but at the same time, it increases the risk of overfitting.

In contrast to XGBoost, the GBDT is a radial basis function kernel that adopts an automatic gamma value (which is the inner product coefficient in the polynomial) and soft margin parameter C = 1, which controls the trade-off between the slack variable penalty and the margin size. Random Forest (RF) is based on trees and is characterized by the square root of the number of features. In AdaBoost, the most critical parameters are "base_estimator", "n_estimators", and "learning_rate".

3.4. Data Preparation and Feature Selection Settings

Data pre-processing is the base step before mining data, including cleanup, integration, and transformation, as well as data discretization, missing value, and outlier processing. The first pre-processing stage focuses on detecting incomplete, accurate, consistent, and corrupt data and then modifying or deleting these false data with some techniques. Different datasets have various characteristics in actual research, so there are different ways to predict the data.

In this paper, we divide into two parts feature selection, one is the filter-based feature selection. This algorithm adopts some principles involving information, consistency, dependency, and distance for measuring the feature characteristics, which are generalized for various classifiers based on the independent features of the machine learning algorithm [31]. For example, a variation filter is to remove the features with small difference value and retain the features with large variance value, because the variance of each feature determines the different degree of the feature in a sample. When a feature in the data set is exposed to Bernoulli distribution (binary classification), it can be used the formula as follow:

$$\sigma = \mathrm{p}(1 - \mathrm{p}) \tag{10}$$

The classic Chi-square(Chi2) filter method is a statistical test for computing the correlation from two types of categorical data. Considering the inconsistency between the observed value and the expected value of the sampling frequency, such as the independent variable equal to i and the dependent variable equal to j, the statistic is constructed, Chi2 tests use the following formula to calculate the test statistic:

$$\kappa^2 = \frac{(A - E)^2}{E} \tag{11}$$

The other part is manual feature selection. In this section, we set three main experiments to evaluate the classifiers to validate the performance of classifying the OGs from various feature datasets with the proposed model.

Firstly, two sets of experiments were organized based on nine gene pair feature datasets involving GC, GC%, protein length, molecular weight (*Mw*(Da)), isoelectric point value (*pI*), exon number, average exon length, intron number, average intron length, gene length, and the output value as an assessment criterion, namely, AGI, for detecting the conditional relatedness between a pair of genes. For model training, the datasets were divided into two sections containing training and testing parts, and the target labels of AGI values were marked as 1 s and 0 s for the two types of gene pairs. The total datasets were divided into training, validating and testing processes using 5-fold cross-validation. The training dataset was used to develop the aforementioned statistical criteria for selected models. The testing dataset was applied to assess the performance of these models with the default parameters without tuning.

Secondly, to explore the importance of genomic and cDNA sequence features after selecting the optimal models, we used a feature selection method by removing one feature from "set_all" of features each time with no redundancy, such as set1 of feature data with no protein length, set2 with no protein of *Mw*(Da), set3 with no protein of *pI*, set4 with no exon number, and set5 with no GC%.

Finally, to validate this model for predicting the OGs of each plant species with specific feature sets, we selected seven testing datasets matched with seven plants (*Arabidopsis thaliana, Populus trichocarpa, Sorghum bicolour, Oryza sativa, Zea mays, Citrus sinensis*, and *Camellia sinensis*).

3.5. Validation Strategies and Evaluation Metrics

The confusion matrix is a matrix table (shown in Table 1) that is used to judge the validation of classification. The results of the prediction model are analyzed using four basic indicators: true positive (TP), true negative (TN), false positive (FP), and false negative (FN).

Table 1. Binary confusion matrix.

	Real Positive	Real Negative
Predict positive	TP	FP
Predict negative	FN	TN

We performed an initial statistical analysis to evaluate the prediction performance for binary classes and grasp the critical features. As the performance measures, stratified five-fold cross-validation was used for obtaining classification accuracy; however, accuracy was found to be an inappropriate evaluation metric for class-imbalanced datasets. Alternatively, precision, recall, F1-Score, and AUC (area under the ROC curve) parameters were used to evaluate the proposed method's feasibility, as in [32]. The AUC is the value of the area under the ROC curve (receiver operating characteristic) that reflects the probability of identifying correct and wrong results according to different thresholds, which is generally between 0.5–1. The quantized index value can better compare the performance of the classifiers: a high-performance classifier AUC value is close to 1properly reflected the test performance.

(i) Accuracy rate (accuracy rate of positive samples):

$$\text{Accuracy} = \frac{\text{TP} + \text{TN}}{\text{TP} + \text{TN} + \text{FP} + \text{FN}} \tag{12}$$

(ii) Recall rate (accuracy rate of positive samples):

$$\text{Recall} = \frac{\text{TP}}{\text{TP} + \text{FN}} \tag{13}$$

(iii) Precision (precision rate of positive samples):

$$\text{Precision} = \frac{\text{TP}}{\text{TP} + \text{FP}} \tag{14}$$

(iv) F1-score value:

$$\text{F1} = \frac{2\text{PR}}{\text{P} + \text{R}} \tag{15}$$

4. Results and Discussion

4.1. The Compared Features between OGs and NOGs in Different Species

After the above seven species of gene datasets were introduced, the next step was to arrange their variable features for constructing a prediction model. The annotations of all protein sequences, CD sequences, and gene characteristics included *GC*, *GC%*, gene length, *Mw* (Da), *pI*, average exon length, average intron length, and so on. Seven species were compared in regard to OGs and NOGs. As seen in Figure 3b–d, OGs had lower values of gene length, *Mw* (Da), and *GC%* than NOGs, with the opposite result on *pI* values (Figure 3a).

Another critical step was data selection. This paper selected nine features: (1) GC, (2) GC%, (3) protein length, (4) molecular weight (Mw (Da)), (5) isoelectric point value (pI), (6) exon number, (7) average exon length, (8) intron number, and (9) average intron length, which were denoted as 1 to 9, respectively, and recorded as V1–V9. The classes of orphan genes and nonorphan genes were recorded as 1 and 0, respectively. Since the datasets contained various types of features, and attribute units were inconsistent in dimensions, it was necessary to use the pre-processing method to standardize the data.

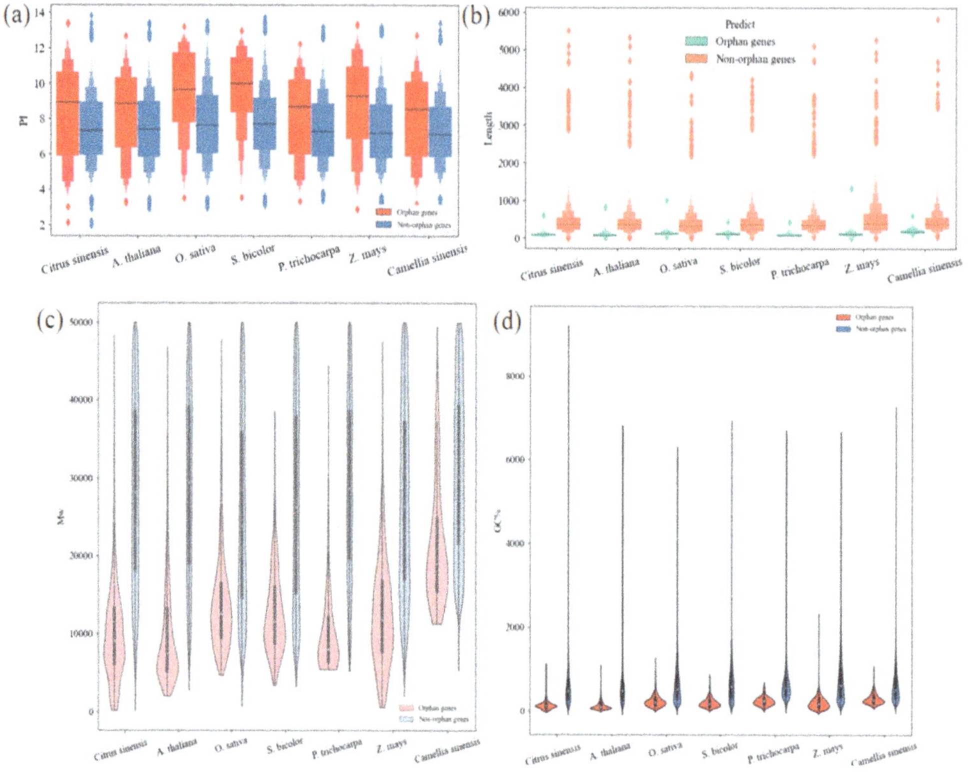

Figure 3. Comparison of various features of OGs and NOGs in seven species: (**a**) *pI*, (**b**) gene length, (**c**) *Mw*, (**d**) *GC%*.

4.2. Comparison with Other Methods for Predicting Cross-Species OGs

This study constructed a novel hybrid classification XGBoost-A2OG model for classifying the angiosperm species of OG distributions. We tested the dataset of *Arabidopsis thaliana*, *Populus trichocarpa*, *Sorghum bicolour*, *Oryza sativa*, *Zea mays*, *Citrus sinensis*, and *Camellia sinensis*, obtaining 6322 OGs from the public release of these species' protein sequences through the BLAST sequence alignment. To predict the OGs, XGBoost-A2OGs were trained using the following parameters: 200 estimators and a learning rate of 0.02 with a maximum depth of six. To optimize the parameters, the optimized XGBoost-A2OG models were trained by 5-fold nested cross-validation. In addition, we compared XGBoost-A2OGs with SVM and tree-based ensemble algorithms (GBDT, RF, and AdaBoost).

The results on the accuracy, precision, recall, and F1-score of the five models are listed in Table 2. Compared to the above four reference methods, the proposed XGBoost-A2OG model achieved competitive performance in recall and F1-score, outperforming them in precision. Thus, it more precisely distinguished normal OGs from NOGs, exhibiting the best classification effect on the AG datasets.

Table 2. Performance measure indices of the five models based on the same parameters of the training and test datasets.

Index	SVM	RF	GBDT	AdaBoost	XGBoost
Accuracy	0.88	0.85	0.88	0.88	0.91
Precision	0.86	0.79	0.87	0.87	0.90
Recall	0.91	0.97	0.89	0.88	0.91
F1-Score	0.88	0.87	0.88	0.88	0.91

Moreover, according to the area under the curve (AUC) value shown in Figure 4, the ROC and precision-recall (P-R) curves of the XGBoost model completely wrapped those of the other four methods (AdaBoost, GBDT, RF, and SVM), outperforming them by classification efficiency.

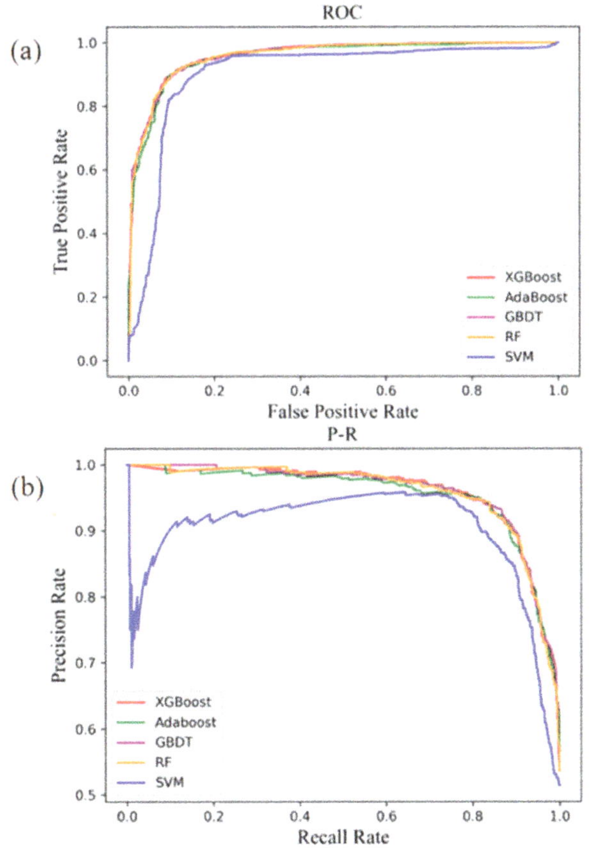

Figure 4. The ROC (**a**) and precision-recall (P-R) (**b**) curves of XGBoost, Adaboost, GBDT, RF, and SVM methods.

4.3. *Predicting OGs with Different Feature Sets in Eudicot and Monocot Species via XGBoost-A2OGs*

Some features might become noise, deteriorating the robustness and stability of the constructed model. Moreover, contribution rates of various features differ, the highest ones being the most lucrative for OGs' prediction. Therefore, this work presents two filter-based selection methods to remove irrelevant and redundant features in terms of both training processes. In particular, we selected two types of delegated species from the eudicot subclass (*P. trichocarpa* and *Camellia sinensis*) and monocot subclass (*O. sativa* and *S. bicolour*) applied with filter-based selection methods. Then the filtered feature are the same containing the GC, protein length, Mw (Da), and pI. Thus, the classification results on these selection methods with four species separately by variation and Chi2 method based on the XGBoost-A2OGs model are listed in Table 3.

Table 3. Performance measure indices of eudicot and monocot species for the training and testing datasets by filter method based on the same parameters.

Type	Species	Filter Method	Precision	Accuracy	AUC
Eudicots	*P. trichocarpa*	variation	0.92	0.93	0.94
Eudicots	*P. trichocarpa*	Chi2	0.9	0.92	0.94
Eudicots	*Camellia sinensis*	variation	0.82	0.69	0.85
Eudicots	*Camellia sinensis*	Chi2	0.82	0.69	0.85
Monocots	*O. sativa*	variation	0.78	0.83	0.9
Monocots	*O. sativa*	Chi2	0.78	0.83	0.9
Monocots	*S. bicolor*	variation	0.81	0.87	0.94
Monocots	*S. bicolor*	Chi2	0.81	0.87	0.94

Filter algorithms can scale for multiple dimensional datasets. However, the features selected by the filter method ignore the interaction among features, and individual scores in a filter-based method are assigned to each feature without considering its significance in combination with other shared features. Therefore, we further proposed an artificial group for feature selection to explore the contribution of each feature for different types of angiosperm. First of all, we also selected a eudicot subclass (*P. trichocarpa* and *Camellia sinensis*) and applied to them five sets of feature selection methods to identify the one with the optimal performance. The classification results on five sets of feature selection methods with two species separately based on XGBoost-A2OGs are listed in Table 4, where the Set3 of Camellia sinensis featured the lowest precision, accuracy, and AUC values (0.80, 0.69, and 0.85). Meanwhile, the Set5 of *P. trichocarpa* combined the highest respective values (precision of 0.9, accuracy of 0.92, and AUC = 0.94).

Table 4. Performance measure indices of eudicot species for the training and testing datasets by feature sets based on the same parameters.

Type	Species	Feature	Precision	Accuracy	AUC
Eudicots	*P. trichocarpa*	Set_all	0.9	0.9	0.92
Eudicots	*P. trichocarpa*	Set1	0.89	0.87	0.89
Eudicots	*P. trichocarpa*	Set2	0.9	0.9	0.91
Eudicots	*P. trichocarpa*	Set3	0.88	0.9	0.92
Eudicots	*P. trichocarpa*	Set4	0.9	0.9	0.94
Eudicots	*P. trichocarpa*	Set5	0.9	0.92	0.94
Eudicots	*Camellia sinensis*	Set_all	0.89	0.74	0.85
Eudicots	*Camellia sinensis*	Set1	0.83	0.68	0.84
Eudicots	*Camellia sinensis*	Set2	0.83	0.69	0.82
Eudicots	*Camellia sinensis*	Set3	0.80	0.69	0.85
Eudicots	*Camellia sinensis*	Set4	0.94	0.76	0.87
Eudicots	*Camellia sinensis*	Set5	0.89	0.74	0.88

As it was mentioned earlier, monocots have branched off from eudicots via whole genome duplication (WGD) [33]. Systematic identification of orphan genes in eudicots revealed that the optimal precision of *P. trichocarpa* and *Camellia sinensis* orphan genes were nearly 0.9 shown in Table 4. Five sets of feature selection methods were also applied to reveal the optimal feature selection performance with XGBoost-A2Ogs for the monocot group containing *O. sativa* and *S. bicolour*. The results are listed in Table 5, indicating that the Set5 feature selection in the monocot group yielded higher precision, accuracy, and AUC values than those obtained via the Set_all feature selection. The respective values of *S. bicolour* in Set5 (0.82, 0.87, and 0.94) exceeded those in Set_all (0.65, 0.73, and 0.6) by about 26, 19, and 57%, respectively.

Table 5. Performance measure indices of monocot species for the training and testing datasets by feature sets based on the same parameters.

Type	Species	Feature	Precision	Accuracy	AUC
Monocots	*O. sativa*	Set_all	0.78	0.83	0.9
Monocots	*O. sativa*	Set1	0.76	0.81	0.9
Monocots	*O. sativa*	Set2	0.76	0.81	0.88
Monocots	*O. sativa*	Set3	0.76	0.81	0.93
Monocots	*O. sativa*	Set4	0.76	0.81	0.9
Monocots	*O. sativa*	Set5	0.79	0.83	0.9
Monocots	*S. bicolor*	Set_all	0.65	0.73	0.6
Monocots	*S. bicolor*	Set1	0.65	0.73	0.6
Monocots	*S. bicolor*	Set2	0.65	0.73	0.62
Monocots	*S. bicolor*	Set3	0.65	0.73	0.62
Monocots	*S. bicolor*	Set4	0.65	0.73	0.6
Monocots	*S. bicolor*	Set5	0.82	0.87	0.94

Additionally, we further explored and compared these combined feature sets of four selected plant species, containing the eudicot and monocot species of evolutionary lineages. The results, plotted in Figure 5, strongly indicate that the featured protein of *pI*, which plays a vital role in determining molecular biochemical function, is essential for predicting OGs in eudicot genomes and further clarifying their biochemical function in eudicots via proteomic studies.

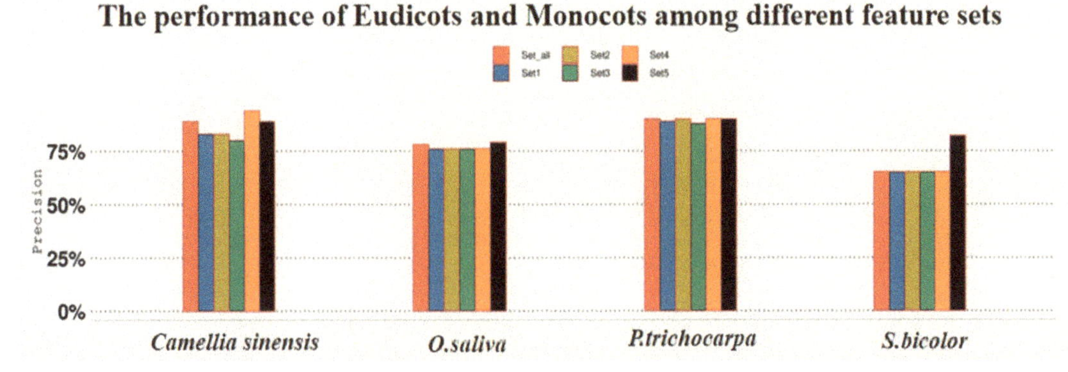

Figure 5. The performance precision for eudicots and monocots via different selected feature sets in angiosperm specie.

We also observed that GC content was more likely to impact prediction performance, as real OGs in monocot groups evolved from eudicots, such as *O. sativa* and *S. bicolour*. However, GC content is one of the critical compositional features of the genome and varies significantly among different genomes and regions within a genome [34,35].

Finally, to further validate the performance of the XGBoost-A2OG model for eudicot and monocot groups, we tested the model on the dataset of *Arabidopsis thaliana*, *Populus trichocarpa*, *Sorghum bicolour*, *Oryza sativa*, *Zea mays*, *Citrus sinensis*, and *Camellia sinensis* with feature set5 separately. The results are shown in Figure 6.

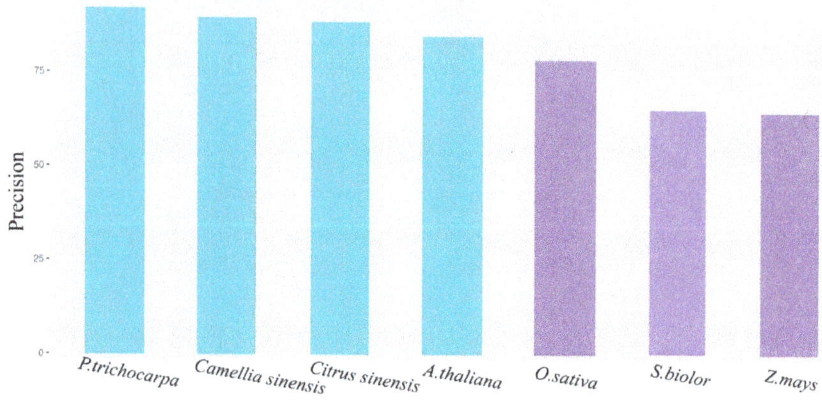

Figure 6. The testing performance in predicting OGs of various angiosperm species.

The precision of predicting OGs for different angiosperm species was not the same, indicating a higher reliability of XGBoost-A2OGs in identifying OGs of eudicot species (*P. trichocarpa*, *Camellia sinensis*, *Citrus sinensis*, and *A. thaliana*) than that of monocot species (*O. sativa*, *S. bicolour*, and *Z. mays*).

With a range of evolutionary processes, OGs can be derived in a lineage and provide lineage-specific adaptations. As mentioned above, there is some evidence that the sequence characteristics of orphan genes are common in two groups of angiosperm: eudicot and monocot species. However, some of them play different roles in identifying OGs based on the XGBoost-A2OG model due to differences in their evolution and origins. However, there is a lack of evidence on the mechanism of origin for the divergence of essential features of OGs between monocots and eudicots due to the rapid evolution of orphan genes.

5. Conclusions

Based on the background of enlarged genome sequences in angiosperm plants, this study proposed an XGBoost-A2OGs model to identify orphan genes (OGs) via the ensemble learning approach applied to several genome and cDNA features in angiosperm species, some of which have a consistent distribution. Cross-species models were trained on datasets of seven angiosperm species, performing better than SVM and other ensemble models (Adaboost, GBDT, and Random Forest). The proposed XGBoost-A2OGs method adopted makes multiple feature sets that have been proven helpful in OG identification and used feature selection to select the optimal feature subset. Thus, plant OGs exhibited discrepant results on combined features in eudicots (*P. trichocarpa* and *Camellia sinensis*) and monocots (*O. sativa* and *S. bicolour*) but still shared some features. Finally, the proposed method further established species-specific models with the optimal features on seven plants' datasets, which performed better on eudicot groups than on monocot ones.

In summary, XGBoost-A2OGs is a helpful method for identifying OGs from genome features. The feature importance of monocot and eudicot orphans was analyzed, providing a theoretical basis for the inheritance and variation of orphan genes in the process of evolution. In future work, with the rapid development of next-generation sequencing technologies, an ensemble learning approach with comparative genomics can be imported to obtain information on different types of angiosperm plants. Alternative deep learning algorithms, such as Transformer and LSTM, can also be applied to improve the potential performance. The follow-up study envisages incorporating some other essential features, such as gene expression, into the proposed model, which may significantly improve the efficiency of predicting OGs in angiosperm plants.

Author Contributions: Conceptualization, X.J.; methodology, Q.G.; software, X.Z.; H.Y.; writing—original draft preparation, Q.G.; writing—review and editing, Q.G. All authors have read and agreed to the published version of the manuscript.

Funding: This research was funded by commercial research fund named "High-throughput sequencing and metagenomic approaches for the study of functional health components of tea leaves" and grant number as 20223401002858.

Data Availability Statement: Not applicable.

Conflicts of Interest: The authors declare no conflict of interest.

References

1. Bomblies, K.; Madlung, A. Polyploidy in the Arabidopsis genus. *Chromosome Res. Int. J. Mol. Supramol. Evol. Asp. Chromosome Biol.* **2014**, *22*, 117–134. [CrossRef] [PubMed]
2. Wilson, G.A.; Bertrand, N.; Patel, Y.; Hughes, J.B.; Feil, E.J.; Field, D. Orphans as taxonomically restricted and ecologically important genes. *Microbiology* **2005**, *151*, 2499–2501. [CrossRef] [PubMed]
3. Donoghue, M.T.A.; Keshavaiah, C.; Swamidatta, S.H.; Spillane, C. Evolutionary origins of *Brassicaceae* specific genes in *Arabidopsis thaliana*. *BMC Evol. Biol.* **2011**, *11*, 47. [CrossRef] [PubMed]
4. Lin, W.L.; Cai, B.; Cheng, Z.M. Identification and characterization of lineage-specific genes in *Populus trichocarpa*. *Plant Cell Tissue Organ Cult.* **2013**, *116*, 217–225. [CrossRef]
5. Xu, Y.; Wu, G.; Hao, B.; Chen, L.; Deng, X.; Xu, Q. Identification, characterization and expression analysis of lineage-specific genes within sweet orange (*Citrus sinensis*). *BMC Genom.* **2015**, *16*, 995. [CrossRef]
6. Perochon, A.; Kahla, A.; Vranić, M.; Jia, J.; Malla, K.B.; Craze, M.; Wallington, E.; Doohan, F.M. A wheat NAC interacts with an orphan protein and enhances resistance to Fusarium head blight disease. *Plant Biotechnol. J.* **2019**, *17*, 1892–1904. [CrossRef]
7. Li, G.; Wu, X.; Hu, Y.; Muñoz-Amatriaín, M.; Luo, J.; Zhou, W.; Wang, B.; Wang, Y.; Wu, X.; Huang, L.; et al. OGs are involved in drought adaptations and ecoclimatic-oriented selections in domesticated cowpea. *J. Exp. Bot.* **2019**, *70*, 3101–3110. [CrossRef] [PubMed]
8. Shen, S.; Peng, M.; Fang, H.; Wang, Z.; Zhou, S.; Jing, X.; Zhang, M.; Yang, C.; Guo, H.; Li, Y.; et al. An *Oryza* specific hydroxycinnamoyl tyramine gene cluster contributes to enhanced disease resistance. *Sci. Bull.* **2021**, *66*, 2369–2380. [CrossRef] [PubMed]
9. Zhao, Z.; Ma, D. Genome-wide identification, characterization and function analysis of lineage-specific genes in the tea plant *Camellia sinensis*. *Front. Genet.* **2021**, *12*, 770570. [CrossRef]
10. Cardoso-Silva, C.B.; Aono, A.H.; Mancini, M.C.; Sforca, D.A.; da Silva, C.C.; Pinto, L.R.; de Souza, A.P. Taxonomically restricted genes are associated with responses to biotic and abiotic stresses in *Sugarcane* (*Saccharum* spp.). *bioRxiv* **2022**. [CrossRef]
11. Ma, S.W.; Yuan, Y.; Tao, Y.; Jia, H.Y.; Ma, Z.Q. Identification characterization and expression analysis of lineage-specific genes within *Triticeae*. *Genomics* **2020**, *112*, 1343–1350. [CrossRef] [PubMed]
12. Arendsee, Z.W.; Li, L.; Wurtele, E.S. Coming of age: OGs in plants. *Trends Plant Sci.* **2014**, *19*, 698–708. [CrossRef] [PubMed]
13. Jiang, M.; Li, X.; Dong, X.; Zu, Y.; Zhan, Z.; Iiao, Z.; Lang, H. Research advances and prospects of OGs in plants. *Front. Plant Sci.* **2022**, *13*, 947129. [CrossRef] [PubMed]
14. O'Conner, S.; Neudorf, A.; Zheng, W.; Qi, M.; Zhao, X.; Du, C.; Nettleton, D.; Li, L. From Arabidopsis to crops: The Arabidopsis QQS orphan gene modulates nitrogen allocation across species. In *Engineering Nitrogen Utilization in Crop Plants*; Springer: Cham, Switzerland, 2018; pp. 95–117.
15. Altschul, S.F.; Gish, W.; Miller, W.; Myers, E.W.; Lipman, D.J. Basic local alignment search tool. *J. Mol. Biol.* **1990**, *215*, 403–410. [CrossRef] [PubMed]
16. Zhu, S.L.; Dong, J.; Zhang, C.; Huang, Y.B.; Pan, W. Application of machine learning in the diagnosis of gastric cancer based on noninvasive characteristics. *PLoS ONE* **2020**, *15*, e0244869. [CrossRef]
17. Chen, T.; Guestrin, C. XGBoost: A scalable tree boosting system. In Proceedings of the 22nd ACM SIGKDD International Conference on Knowledge Discovery and Data Mining, San Francisco, CA, USA, 13–17 August 2016.
18. Gao, Q.; Jin, X.; Xia, E.; Wu, X.; Gu, L.; Yan, H.; Xia, Y.; Li, S. Identification of orphan genes in unbalanced datasets based on ensemble learning. *Front. Genet.* **2020**, *11*, 820. [CrossRef]
19. Casola, C.; Owoyemi, A.; Pepper, A.E.; Ioerger, T.R. Accurate identification of de novo genes in plant genomes using machine learning algorithms. *bioRxiv* **2022**. [CrossRef]
20. Zhang, X.; Xuan, J.; Yao, C.; Gao, Q.; Wang, L.; Jin, X.; Li, S. A deep learning approach for orphan gene identification in moso bamboo (*Phyllostachys edulis*) based on the CNN+ Transformer model. *BMC Bioinform.* **2022**, *23*, 162. [CrossRef]
21. Carvunis, A.R.; Rolland, T.; Wapinski, I.; Calderwood, M.A.; Yildirim, M.A.; Simonis, N.; Charloteaux, B.; Hidalgo, C.A.; Barbette, J.; Santhanam, B.; et al. Proto-genes and de novo gene birth. *Nature* **2012**, *487*, 370–374. [CrossRef]
22. Prabh, N.; Rödelsperger, C. De novo, divergence, and mixed origin contribute to the emergence of orphan genes in Pristionchus Nematodes. *G3* **2019**, *9*, 2277–2286. [CrossRef]
23. Schlötterer, C. Genes from scratch-the evolutionary fate of de novo genes. *Trends Genet.* **2015**, *31*, 215–219. [CrossRef]

24. Zhang, W.Y.; Gao, Y.X.; Long, M.Y.; Shen, B.R. Origination and evolution of orphan genes and de novo genes in the genome of Caenorhabditis elegans. *Sci. China Life Sci.* **2019**, *62*, 579–593. [CrossRef]
25. Singh, U.; Wurtele, E.S. How new genes are born. *Elife* **2020**, *9*, e55136. [CrossRef]
26. Albà, M.M.; Castresana, J. On homology searches by protein blast and the characterization of the age of genes. *BMC Evol. Biol.* **2007**, *7*, 53. [CrossRef]
27. Domazet-Lošo, T.; Brajković, J.; Tautz, D. A phylostrati graphy approach to uncover the genomic history of major adaptations in metazoan lineages. *Trends Genet.* **2007**, *23*, 533–539. [CrossRef]
28. Goodstein, D.M.; Shu, S.; Howson, R.; Neupane, R.; Hayes, R.D.; Fazo, J. Phytozome: A comparative platform for green plant genomics. *Nucleic Acids Res.* **2012**, *40*, D1178–D1186. [CrossRef]
29. Wheeler, D.L.; Barrett, T.; Benson, D.A.; Bryant, S.H.; Canese, K.; Church, D.M.; Yaschenko, E. Database resources of the National Center for Biotechnology Information. *Nucleic Acids Res.* **2005**, *33*, D39–D45. [CrossRef]
30. Bolser, D.; Staines, D.M.; Pritchard, E.; Kersey, P. Ensembl plants: Integrating tools for visualizing, mining, and analyzing plant genomics data. In *Plant Bioinformatics*; Humana Press: New York, NY, USA, 2016; pp. 115–140.
31. Halim, Z. An ensemble filter-based heuristic approach for cancerous gene expression classification. *Knowl.-Based Syst.* **2021**, *234*, 107560.
32. Ispandi, R.; Wahono, S. Application of genetic algorithms to optimize parameters in support vector machine to increase direct marketing predictions. *J. Intell. Syst.* **2015**, *1*, 115–119.
33. Chaw, S.M.; Chang, C.C.; Chen, H.L.; Li, W.H. Dating the monocot dicot divergence and the origin of core eudicots using whole chloroplast genomes. *J. Mol. Evol.* **2004**, *58*, 424–441.
34. Bowman, M.J.; Pulman, J.A.; Liu, T.L.; Childs, K.L. A modified GC-specific MAKER gene annotation method reveals improved and novel gene predictions of high and low GC content in Oryza sativa. *BMC Bioinform.* **2017**, *18*, 522. [CrossRef]
35. Singh, R.; Ming, R.; Yu, Q. Comparative analysis of GC content variations in plant genomes. *Trop. Plant Biol.* **2016**, *9*, 136–149. [CrossRef]

electronics

Article

UAV Abnormal State Detection Model Based on Timestamp Slice and Multi-Separable CNN

Tao Yang, Jiangchuan Chen *, Hongli Deng and Yu Lu

School of Computer Science, China West Normal University, Nanchong 637002, China
* Correspondence: chenjiangchuan@stu.cwnu.edu.cn

Abstract: With the rapid development of UAVs (Unmanned Aerial Vehicles), abnormal state detection has become a critical technology to ensure the flight safety of UAVs. The position and orientation system (POS) data, etc., used to evaluate UAV flight status are from different sensors. The traditional abnormal state detection model ignores the difference of POS data in the frequency domain during feature learning, which leads to the loss of key feature information and limits the further improvement of detection performance. To deal with this and improve UAV flight safety, this paper presents a method for detecting the abnormal state of a UAV based on a timestamp slice and multi-separable convolutional neural network (TS-MSCNN). Firstly, TS-MSCNN divides the POS data reasonably in the time domain by setting a set of specific timestamps and then extracts and fuses the key features to avoid the loss of feature information. Secondly, TS-MSCNN converts these feature data into grayscale images by data reconstruction. Lastly, TS-MSCNN utilizes a multi-separable convolution neural network (MSCNN) to learn key features more effectively. The binary and multi-classification experiments conducted on the real flight data, Air Lab Fault and Anomaly (ALFA), demonstrate that the TS-MSCNN outperforms traditional machine learning (ML) and the latest deep learning methods in terms of accuracy.

Keywords: unmanned aerial vehicle; anomaly detection; ALFA; CNN

Citation: Yang, T.; Chen, J.; Deng, H.; Lu, Y. UAV Abnormal State Detection Model Based on Timestamp Slice and Multi-Separable CNN. *Electronics* **2023**, *12*, 1299. https://doi.org/10.3390/electronics12061299

Academic Editor: Ping-Feng Pai

Received: 2 February 2023
Revised: 1 March 2023
Accepted: 7 March 2023
Published: 8 March 2023

1. Introduction

With the development of unmanned aerial vehicles (UAVs), their applications in civilian and military fields have expanded, including agriculture [1], transportation [2], and fire protection [3]. However, as UAVs play an increasingly important role, their flight safety problems have become more prominent [4]. Network attacks can lead to UAV failures, and physical component failures such as elevators and rudders can also affect UAV flight safety. For example, in June 2020, a US Air Force MQ-9 "Death" UAV crashed in Africa, causing a loss of USD 11.29 million [5]. In February 2022, a DJI civilian UAV crashed out of control, resulting in a personal economic loss of up to 16,300 RMB [6]. According to the Civil Aviation Administration of China, the number of registered UAVs in China alone has reached 8.3 million [7]. Therefore, it is necessary to establish a UAV safety detection model to ensure the safety and reliability of UAV flights. Improving the flight safety of UAVs has become a major research topic in the field of UAVs. Currently, a common method to ensure UAV flight safety is to monitor UAV flight data for anomalies [8]. Abnormal flight data indicates that the UAV may have hardware failure or misoperation, and timely identification of the cause of the failure can effectively prevent UAV flight accidents. Figure 1 shows the main components of a typical UAV anomaly detection system.

UAV flight data is mainly extracted from attitude estimation data of different UAV sensors [9,10], which include the POS data and the system status (SS) data. These data enable the detection of UAV flight status. The POS data consists of a triple of values in the x, y, and z directions, while the SS data contains only a single value. Additionally, these data are closely related to UAV guidance, navigation, and control (GNC) [11,12]. The early

UAV anomaly detection method was based on flight data rules; however, the rule-based anomaly detection method has a low detection performance [13]. To better ensure the flight safety of UAVs, ML and deep learning methods have been introduced into the research field of UAV safety. The development of these methods has opened up new ideas for the research of UAV anomaly detection. However, the traditional anomaly detection method ignored the difference between POS data and SS data used to evaluate the flight status of UAVs in the frequency domain, resulting in the loss of some key feature information in-flight data. This limitation restricts the performance of UAV anomaly detection models. To address these problems, this paper proposes a method of extracting frequency domain information by setting timestamp slices and proposes a UAV anomaly detection model based on a multi-separable convolution neural network fusion method. It should be noted that this paper takes the time of UAV failure as the dividing point and does not consider the recovery process.

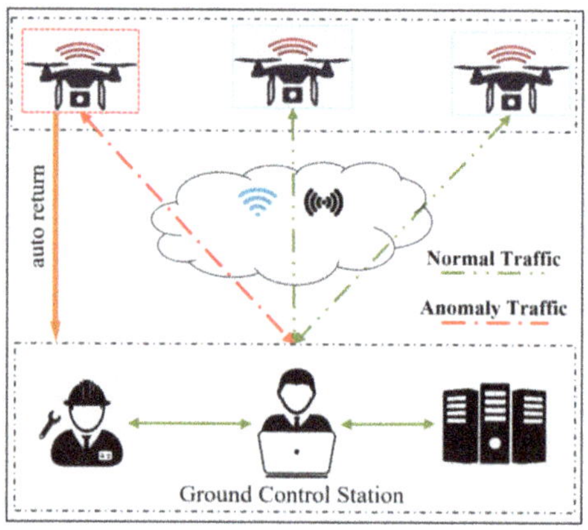

Figure 1. Main components in the UAV anomaly detection system.

In the next part of this paper, Section 2 describes the related research. Section 3 introduces the processing method of the ALFA dataset [14] and proposes the TS-MSCNN anomaly detection model. Section 4 carries out experiments from various angles and analyzes the experimental results of binary and multi-class classification. The final section provides a summary and conclusion of this paper.

2. Related Works

This section provides a review of research related to UAV anomaly detection, covering rule-based algorithms and those based on ML and deep learning methods.

Regarding rule-based algorithms, Chen et al. [15] investigated the impact of attackers' behavior on the effectiveness of malware detection technology and proposed a specification-based intrusion detection system that showed effective detection with high probability and low false positives. Mitchell et al. [16] considered seven threat models and proposed a specification-based intrusion detection system with specific adaptability and low runtime resource consumption. Sedjelmaci et al. [17] studied four attacks—false information propagation, GPS deception, jamming, and black hole and gray hole attacks—and designed and implemented a new intrusion detection scheme with an efficient and lightweight response, which showed high detection rates, low false alarm rates, and low communication overhead. This scheme was also able to detect attacks well in situations involving many UAVs and attackers.

In terms of the UAV anomaly detection model based on traditional ML methods, Liu et al. [18] proposed a real-time UAV anomaly detection method based on the KNN algorithm for the UAV flight sensor data stream in 2015, which has high efficiency and high accuracy. In 2016, Senouci et al. [19] focused on the two main problems of intrusion detection and attacker pop-up in the UAV-assisted network. The Bayesian game model was used to balance the intrusion detection rate and intrusion detection resource consumption. This method achieved a high detection rate and a low false positive rate. In 2019, Keifour et al. [20] released an initial version of the ALFA dataset [13] and proposed a real-time UAV anomaly detection model using the least squares method. This method does not need to assume a specific aircraft model and can detect multiple types of faults and anomalies. In 2021, Shrestha et al. [21] simulated a 5G network and UAV environment through the CSE-CIC-IDS-2018 network dataset, established a model for intrusion detection based on the ML algorithm, and also implemented the model based on ML into ground or satellite gateways. This research proves that the ML algorithm can be used to classify benign or malicious packets in UAV networks to enhance security.

However, some outliers can be difficult to detect using traditional machine learning (ML) techniques [22]. To address this challenge, deep learning (DL) methods have been increasingly used to improve the detection accuracy of UAV anomalies, especially when processing high-dimensional UAV flight data. In 2021, Park et al. [23] proposed a UAV anomaly detection model using a stacking autoencoder to address the limitations of the current rule-based model. This model mainly judges the normal and abnormal conditions of data through the loss of data reconstruction. The experimental results on different UAV data demonstrate the effectiveness of the proposed model. In 2022, Abu et al. [24] proposed UAV intrusion detection models in homogeneous and heterogeneous UAV network environments based on a convolutional neural network (CNN) using three types of UAV WIFI data records. The final experimental results demonstrate the effectiveness of the proposed model. Dudukcu et al. [25] utilized power consumption data and simple moving average data of the UAV battery sensor as the multivariate input of the time-domain convolution network to identify the anomaly of the instantaneous power consumption of the UAV battery. The simulation results show that the time-domain convolutional network can achieve good results in instantaneous power consumption prediction and anomaly detection when combining simple moving average data and UAV sensor data. In addition, some studies have explored the use of probability models, time series data, and data dimensions for anomaly detection, achieving effective results [26–28], which have important implications for this study.

All of the previously mentioned methods have been successful in detecting anomalies, but they have not taken into account the differences between the POS data and SS data used to evaluate UAV flight status in the frequency domain. This has resulted in the loss of some key feature information in the flight data, which limits the improvement of anomaly detection model performance. The differences in the frequency domain can be seen in two aspects: first, the feature information amount of the POS data and the SS data in the frequency domain is inconsistent in the same time domain; second, the data structure is different. The feature of POS data in the frequency domain is triple, while SS data is a single value. When the amount of feature information is inconsistent, a feature vector with variable length is generated, which leads to the loss of key feature information in the model training process. Additionally, the difference in data structure causes POS data and SS data to lose some key information due to the confusion of feature information during the anomaly detection model's feature extraction process.

To address the issues mentioned above, this paper proposes several solutions. Firstly, a specific timestamp size is set, and the frequency domain information of UAV data is divided and extracted to fuse key feature information, addressing the problem of inconsistency between POS data and SS data in the frequency domain. Secondly, POS and SS data are reconstructed into grayscale images. Lastly, the MSCNN is utilized to learn and fuse the key features of POS and SS data, overcoming the problem of key feature information loss caused by the structural differences between POS data and SS data. The following sections will provide a detailed description of these solutions.

3. TS-MSCNN Model Design

Taking into account the analysis presented above, this section proposes a TS-MSCNN anomaly detection model, which consists of two main components: a time stamp slice-based frequency domain information processing method for extracting and fusing key features of UAV flight data, and an MSCNN-based anomaly detection method for learning and fusing flight data features. The processing flow of the TS-MSCNN model is illustrated in Figure 2. The subsequent section will provide a detailed description of the model design.

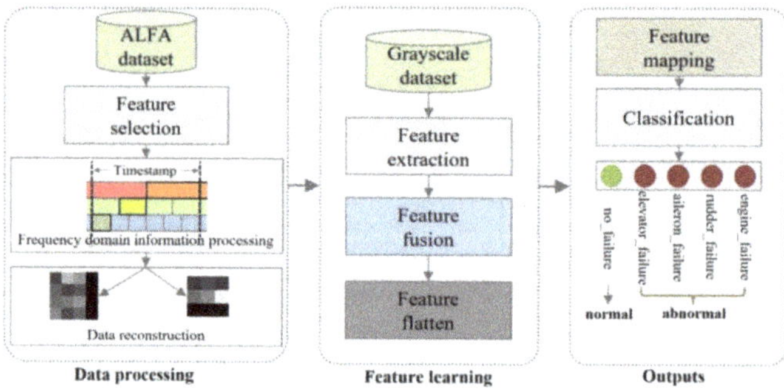

Figure 2. The block diagram of the proposed TS-MSCNN.

3.1. *UAV Flight Data Processing Methods*

3.1.1. Analysis of ALFA Dataset

The ALFA dataset comprises the original flight log of a fixed-wing UAV that operated in a real flight environment and can be roughly classified into five categories: no failure, engine failure, rudder failure, elevator failure, and aileron failure. The UAV was flown at Pittsburgh Airport in the United States. The dataset includes two types of data: SS data with only one numerical dimension and POS data with three numerical dimensions. The POS data contains latitude, longitude, elevation, heading angle (Phi), pitch angle (Omega), and roll angle (Kappa) data obtained during the UAV flight, which are mainly represented by different values in the X, Y, and Z directions. The original UAV flight log contains a multitude of features, which are not conducive to model training. Therefore, this paper uses the feature selection method in [23] to obtain the key features of UAV flight data shown in Table 1.

Table 1. Features selected from the ALFA.

Category	Feature Name	Description
POS Data	Magnetic Field (x, y, z)	The value of the magnetic field at axis x, y and z
	Linear Acceleration (x, y, z)	The linear acceleration at axis x, y and z
	Angular Velocity (x, y, z)	An angular velocity at axis x, y and z
	Velocity (x, y, z)	Measured velocity of axis x, y and z
System status Data	Fluid Pressure	The value of the pressure using fluid pressure sensors
	Temperature	The temperature of the battery
	Altitude Error	The error value of current altitude
	Airspeed Error	The error value of current airspeed
	Tracking Error (x)	The tracking error at x axis
	WP Distance	The distance between ideal location and current location

3.1.2. Frequency Domain Information Extraction and Fusion Method Based on Timestamp Slices

The frequency domain information of the original UAV data in the same time domain is different, so the fixed length feature vector cannot be formed, which leads to the loss of key feature information in the model training process. Suppose that at time t, by observing the temperature information of the UAV battery, $f_{temperature}$ can be expressed as a binary, that is, $f_{temperature} = \{temp_1, temp_2\}$. At different times, the value of the $f_{temperature}$ binary is different. According to the above representation, other flight data information from UAV, such as fluid pressure and magnetic field value, can be expressed as corresponding characteristic tuples, namely $f_{pressure} = \{pre_1, pre_2, pre_3, pre_4\}$, $f_{magnetic} = \{mag_1, mag_2, mag_3, mag_4, mag_5, mag_6\}$. These feature tuples are marked with inconsistent frequency domain feature information at the same time (as shown in Figure 3a). During the calculation process, features with more frequency domain information will cover other feature information values, leading to the loss of key information. Therefore, this paper will process based on the following methods.

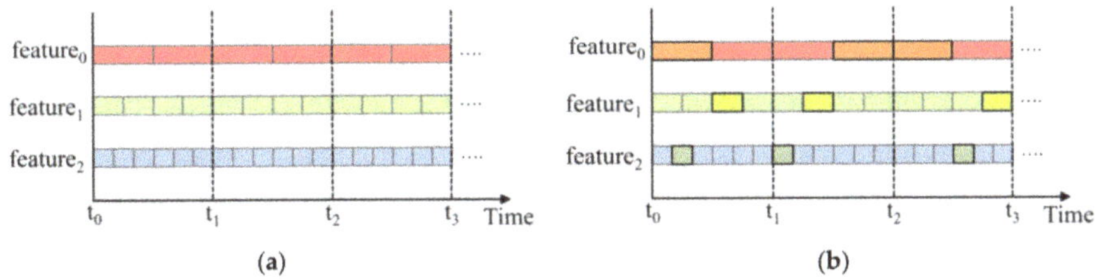

Figure 3. (**a**) Distribution of various features. (**b**) Extraction of the features in the timestamp.

Step 1: Feature information extraction in the frequency domain.

$$select(feature) = \{vij \mid when\ t = tk\ and\ (index\ (vij\ \&\ tk) \neq index\ (vij\ \&\ tk-1))\} \quad (1)$$

where v_{ij} represents the characteristic value, i represents the characteristic number, j represents the characteristic value number, *index()* represents the index of the characteristic value in the frequency domain, and t_k represents the time.

Step 2: Frequency domain information fusion.

$$v = \{select(feature_0) \cup select(feature_1) \ldots \cup select(feature_n) \mid when\ t = t_k\} \quad (2)$$

where n represents the characteristic number and t_k represents the time.

Figure 3b illustrates the results of information extraction and fusion at different time points. It shows that the same feature has different index positions in different timestamps, which preserves the differences between features in different time domains and enables the maximum amount of information to be obtained. In real UAV log data, POS data and other values change significantly, and there are more characteristic data in the same timestamp than in the SS data. Therefore, this paper extracts and fuses flight log data based on Equations (1) and (2), using the time span of the feature with the least amount of data as the time stamp unit. This approach ensures the difference between different features, as well as the consistency of frequency domain information of different features in the time domain, and the frequency domain difference of the same feature in different time domains.

3.1.3. Unbalanced Data Processing

Based on the idea presented in Section 1, this paper performed information extraction and fusion on the ALFA UAV flight dataset, and the results are shown in Figure 4a. The dataset had a serious data imbalance, with the largest percentage of abnormal data for engines being 58% of the entire dataset, the minimum percentage of abnormal data for elevators being 4%, and only 12% of the data being normal. This imbalance can lead to learning deviations in the anomaly detection model, causing the model to learn the features of the data with a high proportion while only learning a few features from the data with a low proportion. Therefore, this paper balanced the data distribution using the down-sampling method, and the resulting balanced data distribution is shown in Figure 4b.

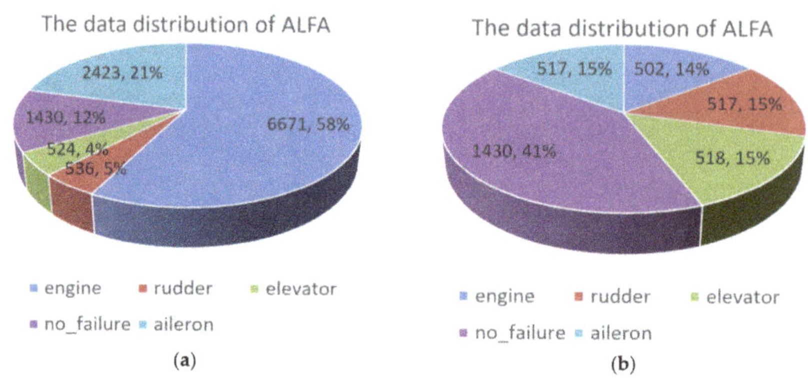

Figure 4. (**a**) The data distribution of ALFA. (**b**) The balanced ALFA.

3.1.4. Validation of Flight Data

To demonstrate the effectiveness of the obtained UAV flight data, this paper reproduces normal flight and flight with elevator failure using a UAV flight simulator. The configuration of the main parameters is shown in Table 2, and the flight path is illustrated in Figure 5. During the flight, when the elevator fails at a specific time, the UAV cannot complete the ascent and descent, so it can only maintain the same flight altitude. The trouble-free UAV completes the difficult flight activities by lifting and lowering. This paper simulates the flight trajectory of UAVs using the obtained data, and the trajectory has a noticeable difference in 2D and 3D space, thus demonstrating the difference of abnormal data of different UAVs and the effectiveness of the proposed UAV data processing method in this paper.

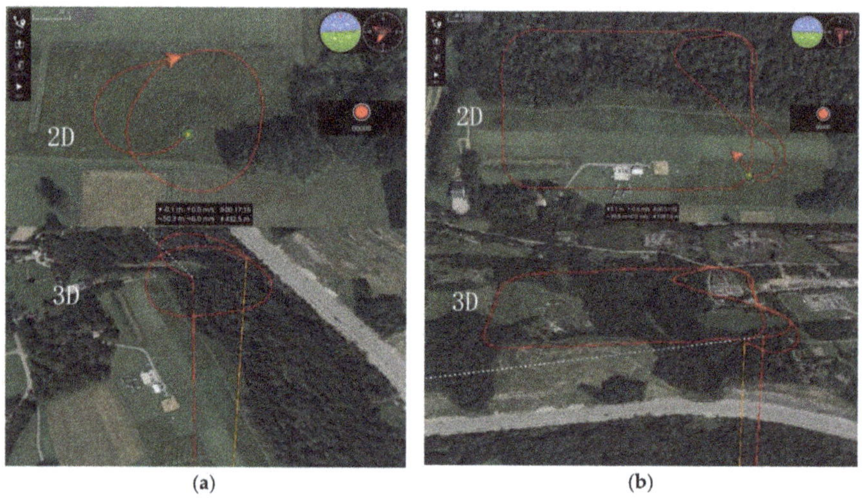

Figure 5. (**a**) The normal flight data. (**b**) The flight data of elevator failure.

Table 2. Parameter configuration of the simulator.

SITL Parameter	Default	Description
SIM_RC_FAIL	0.000000	Force RC failure
SIM_ACCEL_FAIL	0.000000	Force IMU ACC failure
SIM_ENGINE_MUL	1.000000	-
SIM_MAG1_DEVID	97,539.000000	1st Compass (0 to remove)
SIM_SPEEDUP	1.000000	Allows running sim SPEEDUP times faster
SIM_WIND_TURB	0.000000	Not implemented
SIM_GYR_FAIL_MSK	0.000000	Bitmask for setting a Gyro 1, 2, and/or 3 failure

3.2. Design of Anomaly Detection Model

3.2.1. Separable Convolutional

The separable convolution technique offers several advantages, including fewer parameters and lower computational cost, while also exhibiting high expressiveness in the field of texture image recognition [29]. Its primary structure consists of a channel convolution kernel that has the same size as the input image and a 1 × 1 convolution kernel used to fuse the channel convolution information, as shown in Figure 6a. The structure of the separable convolutional neural network (SCNN) is shown in Figure 6b. Compared to traditional convolutional neural networks, separable convolution networks require fewer parameters and consume less computational resources while maintaining classification accuracy, as illustrated in Figure 7.

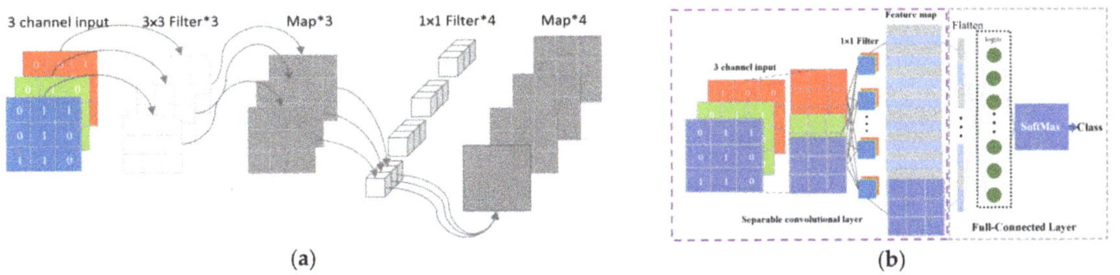

Figure 6. (**a**) The separable convolutions. (**b**) The separable convolutional neural network.

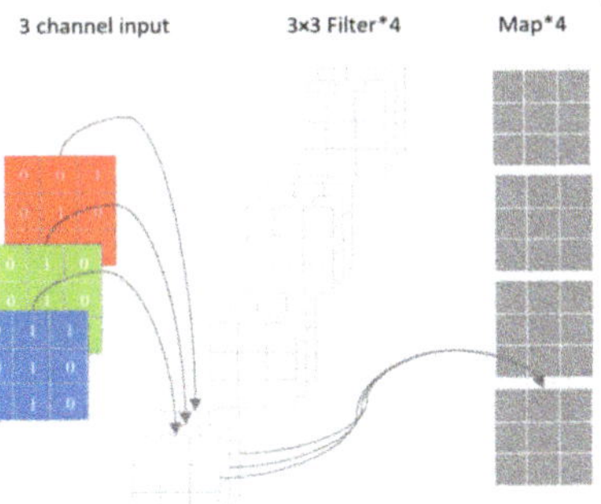

Figure 7. Traditional CNN convolution layers.

Set the input as M channels, the image size as $D_{f\_in} \times D_{f\_in}$, the convolution kernel as $N \times (M \times D_k \times D_k)$, and the output feature map as N channels and size $D_{f\_out} \times D_{f\_out}$. So, the parameters of the separable convolution are $D_k \times D_k \times M + M \times N$; the parameter quantity of the conventional convolution is $D_k \times D_k \times M \times N$. The calculation consumption of the separable convolution is $M \times D_k \times D_k \times D_{f\_out} \times D_{f\_out} + 1 \times 1 \times N \times D_{f\_out} \times D_{f\_out}$; the calculation consumption of the conventional convolution is $M \times D_k \times D_k \times D_{f\_out} \times D_{f\_out} \times N$.

The comparison of parameter quantity and computational consumption between separable convolution and conventional convolution is presented in Figures 8 and 9. It is evident that as the number of channels and convolution layers increases, the parameter quantity and computational consumption of the conventional convolution layer are much higher than those of the separable convolution layer, and the increase of the conventional convolution is exponential. This illustrates that the separable convolution layer can save more parameters and computational consumption than the conventional convolution layer and has a faster calculation speed. Consequently, this paper will devise an efficient model based on separable convolution.

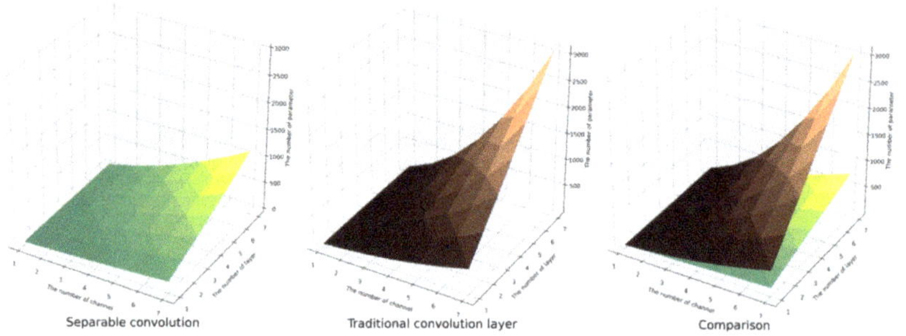

Figure 8. The influence of model structure on the number of parameters.

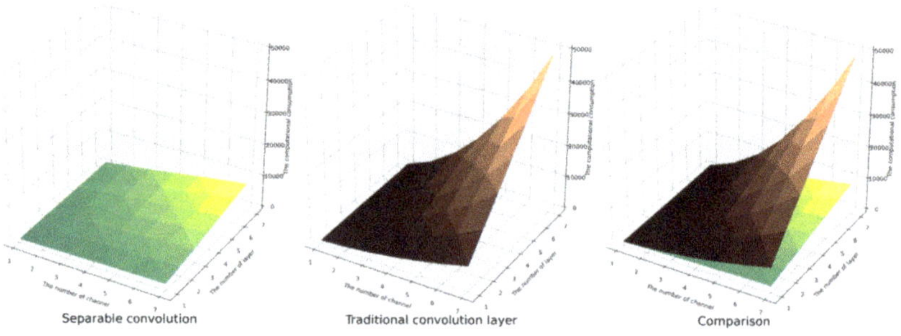

Figure 9. The influence of model structure on computational overhead.

3.2.2. Feature Extraction and Fusion Layer

Based on the analysis above, this section presents the design of the feature extraction and fusion layer (FEF) for POS data and SS data in UAV flight data using separable convolution, as illustrated in Figure 10. FEF mainly consists of multi-layer parallel separable convolutions and a feature fusion layer, and the number of separable convolution layers varies for each data image. The main methods of feature extraction and fusion calculations are as follows:

$$f = \sum_{k=1}^{p} max\left({w_{c_{2k}}}^{T} \max\left(\sum_{l=1}^{c_{1k}} {w_l}^{T} x_{i,j} + b_{c_{1k}}, 0 \right) + b_{c_{2k}}, 0 \right) \quad (3)$$

where (i, j) is the pixel index in the feature map, $x_{i,j}$ is the input slice centered on the position (i, j), c is the channel index in the feature map, and p is the separable convolutional parallel number.

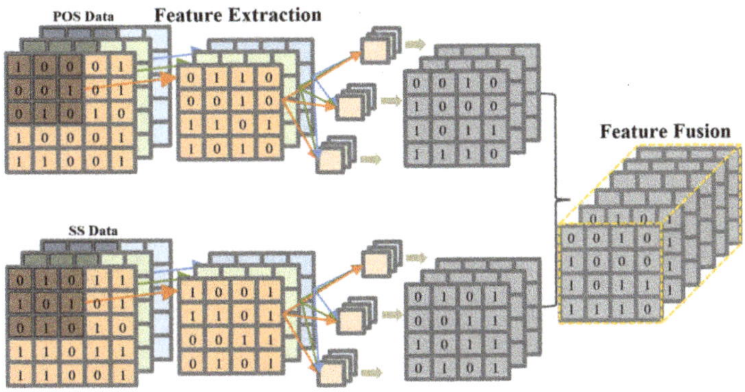

Figure 10. The feature extraction and fusion layer.

The FEF layer is designed to extract features from the grayscale image corresponding to the POS and SS data, and then fuse the two extracted features. The main fusion method involves concatenating the two feature maps. For instance, if there are 3 feature maps from the convolution layer for each of the POS and SS data, the resulting feature map size after fusion will be 6.

3.2.3. Feature Mapping and Classification Layer

Based on the fusion feature map of the FEF layer, this paper requires an effective feature mapping to the sample classification space. Therefore, this paper designed a Feature Mapping and Classification (FMC) layer, as illustrated in Figure 11a. The FMC layer is composed of three layers, namely the Flatten layer, the Fully Connected layer, and the Output layer. The Flatten layer maps the obtained feature map to a one-dimensional space. The Fully Connected layer acts as a classifier by fusing local information of features. The Output layer mainly uses the softmax function to map the calculated values of neurons to a probability space with a sum of 1. The working mode of the flattened layer is shown in Figure 11b. The classification calculation equation is as follows:

$$class = \max\left(\frac{e^{\max(w^T f' + b, 0)_i}}{\sum_{j=1}^{k} e^{\max(w^T f' + b, 0)_j}} \middle| i = 1, \ldots, k \right) \tag{4}$$

where f' represents one-dimensional characteristic data and k represents the number of sample categories.

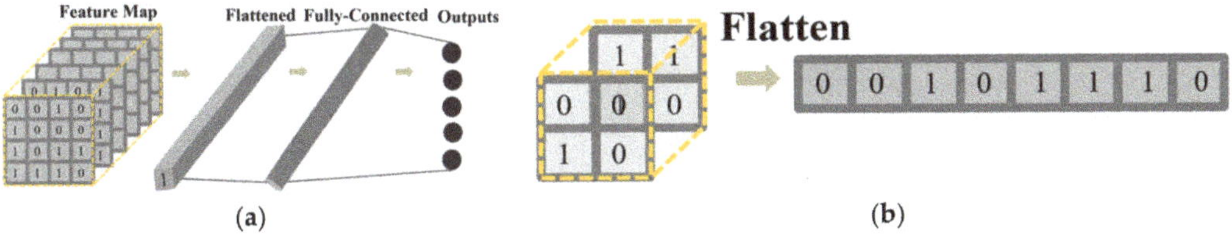

Figure 11. (**a**) Feature mapping and classification layer. (**b**) The way the feature flattens out.

3.2.4. TS-MSCNN Model Design

The complete design of the TS-MSCNN model is illustrated in Figure 12. During the training process, the model is validated using the verification set to ensure the accuracy of the training process. The loss rate threshold is set as the termination condition for the model training. Finally, the trained model is used to detect the test set and output the evaluation metrics. The process of the TS-MSCNN model, from training to anomaly detection, involves three main stages: forward propagation, backward propagation, and model testing, which can be broken down into the following six steps.

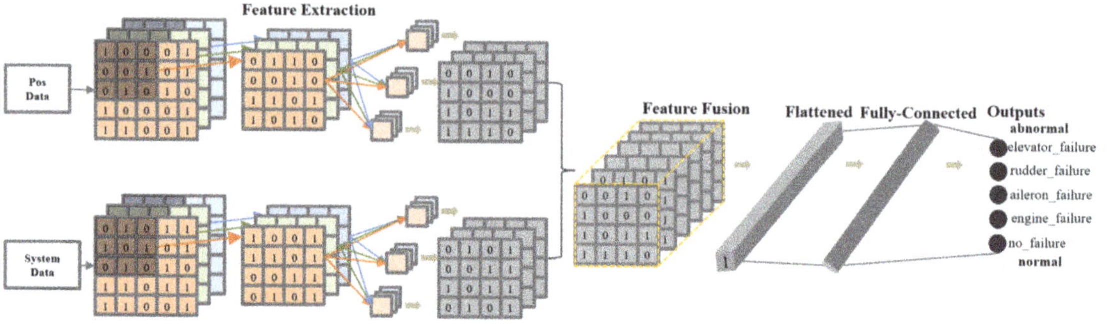

Figure 12. The structure of TS-MSCNN model.

Step 1: Feature data extraction and fusion. Set the timestamp slice, extract and fuse the UAV frequency domain information through Equations (1) and (2), and obtain the fixed length UAV flight data feature vector.
Step 2: Data to image. The POS and SS data of UAV are transformed into two-dimensional grayscale images by data reconstruction to adapt to model input.
Step 3: Feature extraction and fusion. The grayscale image features of UAV POS data and SS data are extracted and fused using the FEF layer pass-through Equation (3).
Step 4: Feature mapping and classification. The feature map from the FEF layer is flattened into one-dimensional data, and then the one-dimensional feature data is mapped to the sample category space using Equation (4) to achieve classification.
Step 5: Backpropagation and parameter updating. After classification, the cross-entropy loss function is first used to calculate the loss between the predicted and actual values. The cross-entropy loss function is given as Equation (5) (where $p(s_i)$ and $q(s_i)$, respectively, represent the real and predicted distributions of sample i, and H represents the final loss value. Backpropagation is then carried out according to the loss value. The Adam optimizer is adopted for the backward propagation to update the weight and bias of each layer.):

$$\mathrm{H}(p,q) = -\sum_{i=1}^{k} p(c_i)log(q(c_i)) \quad (5)$$

Step 6: Model testing. Input test data into the model to test the effect of the model.

4. Experiment

This study employs the PyTorch [30] deep learning library to train the TS-MSCNN and conventional CNN models. The experiments were conducted on an HP-Z480 workstation equipped with an Intel Xeon ® CPU and 64 GB of RAM. In this section, we will first introduce the evaluation metrics of the model and then demonstrate the performance of the TS-MSCNN model in binary and multi-classification tasks. We compare our model with conventional machine learning algorithms, conventional CNNs, and other relevant research results to verify its effectiveness. It should be noted that to adapt the convolutional structure for feature extraction, we convert the UAV flight data into a two-dimensional grayscale image using a data reconstruction method. Figure 13 displays the data reconstruction method and UAV image data, where the 'ALL' chart shows the image data used for the single model structure. The detailed experimental process will be discussed in the next section.

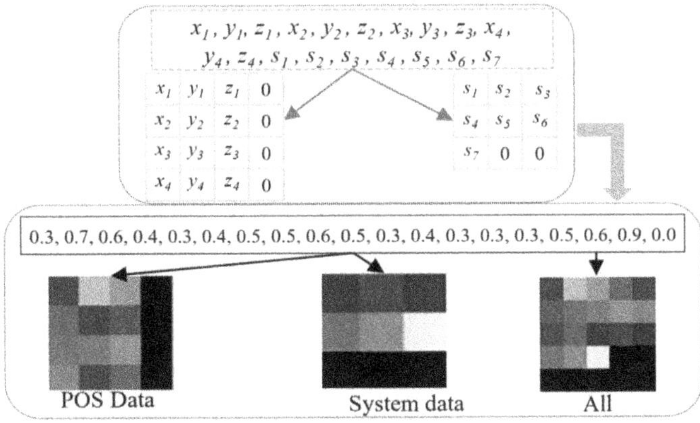

Figure 13. Two-dimensional UAV flight data.

4.1. Evaluation Metrics

The main performance metric used in this paper is accuracy, followed by Recall, F1-score, and Precision. TPs (true positive) refers to the number of abnormal records identified as abnormal. True negative (TNs) is the number of normal records that are considered normal. False positives (FPs) are the number of normal records identified as abnormal. False negatives (FNs) are the number of abnormal records identified as normal. The performance metrics used in this paper are defined as follows.

Accuracy: the percentage of the number of correctly classified records to the total number of records, as shown in Equation (6).

$$Accuracy = (TP + TN)/(TP + TN + FP + FN) \tag{6}$$

Recall: Measure how many positive examples in the sample are correctly predicted, that is, the proportion of all positive examples correctly predicted, as shown in Equation (7).

$$Recall = TP/(TP + FN) \tag{7}$$

Precision: It is used to measure how many samples with positive prediction are real positive samples, that is, the proportion of real positive samples in the results with positive prediction, as shown in Equation (8).

$$Precision = TP/(TP + FP) \tag{8}$$

F1-score: The F1-score measures the harmonic mean of the precision and recall, which serves as a derived effectiveness measurement, as shown in Equation (9).

$$F_1 = \frac{2 \times Precision \times Rcall}{Precision + Rcall} \tag{9}$$

4.2. Single SCNN Model for Binary Classification

In previous research, UAV flight data has been imaged. In this section, traditional CNN and SCNN models of a single model will be trained based on UAV image data. To better train the model, this paper sets the learning rate to 0.001 and the termination loss rate of model training to 0.001. Divide the processed ALFA dataset into a training set, test set and verification set according to the ratio of 6:3:1, and classify the data set into abnormal and normal. In addition, the number of convolution layers in various models is both designed as 3. Table 3 shows the experimental result of the CNN model and SCNN model and it also shows that separable convolution ensures the validity of the model while optimizing the model parameters and computing consumption.

Table 3. The accuracy of the single model.

Model	Accuracy
CNN	95.40%
SCNN	96.35%

Next, this paper will use conventional ML methods to detect binary anomalies based on UAV flight data. Among them, the main algorithms used are ZeroR, OneR, Naive-Bayes [31], KNN [32], J48 [33], RandomForest [34], RandomTree [35], and Adaboost [36]. Figure 14a shows the comparison between traditional ML algorithms and CNN and SCNN models. Additionally, the SCNN model is the best, with 96.35%. Obviously, the CNN model has great potential for detecting UAV anomalies, and it can accurately learn features from data. At the same time, the SCNN model based on separable convolution has higher accuracy.

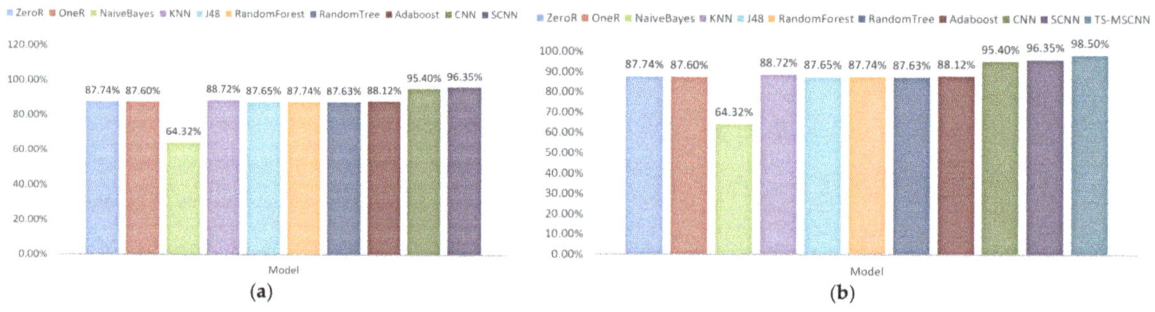

Figure 14. (**a**) Performance of the single model. (**b**) Performance of the TS-MSCNN and other models.

4.3. Multi-SCNN Fusion Model for Binary Classification

To enhance the accuracy of the UAV binary anomaly detection model, this paper proposes a TS-MSCNN model that leverages the characteristics of UAV flight data. Table 4 presents the performance of CNN, SCNN, and TS-MSCNN models in terms of binary classification. The TS-MSCNN model outperforms CNN and SCNN in all metrics. Furthermore, Figure 14b compares the TS-MSCNN model with other models, showing that the TS-MSCNN model achieves superior accuracy to other comparison algorithms, with the highest accuracy rate of 98.50%. The results demonstrate that the TS-MSCNN model effectively extracts and fuses features from UAV flight data and accurately detects anomalies.

Table 4. The detailed performance of CNN, SCNN, and TS-MSCNN.

Model	Accuracy	Class	Recall	Precision	F1-Score
CNN	95.40%	No_failure	99.50%	95.41%	97.41%
		failure	67.56%	95.27%	79.06%
SCNN	96.35%	No_failure	98.35%	96.53%	97.43%
		failure	76.06%	87.18%	81.24%
TS-MSCNN	98.50%	No_failure	99.24%	98.98%	99.11%
		failure	93.06%	94.76%	93.91%

4.4. Single SCNN Model for Multiclass Classification

The objective of UAV anomaly detection is to identify UAV faults and prevent potential losses. This paper conducts a multi-class anomaly detection experiment using the ALFA dataset, which includes multiple classes of objects. The dataset contains four types of abnormal flight data and one type of normal flight data. In this section, we implement a multi-classification experiment using a single-model SCNN and present the specific experimental results in Table 5. The results show that, in the case of multi-classification, the SCNN not only optimizes the convolution structure parameters and computational consumption but also ensures the effectiveness of the model and accurately detects anomalies across multiple classes.

Table 5. The accuracy of the single model.

Model	Accuracy
CNN	93.10%
SCNN	94.68%

Furthermore, this paper also employs traditional ML methods, consistent with those used above, to detect anomalies. Figure 15a presents the experimental results. Among them, the SCNN model achieved the best performance, with 94.68%. These results indicate that the SCNN model has advantages over traditional ML methods in processing high-dimensional UAV data. Moreover, the OneR algorithm obtains the lowest accuracy rate, as it only uses a specific feature in the training data as the classification basis.

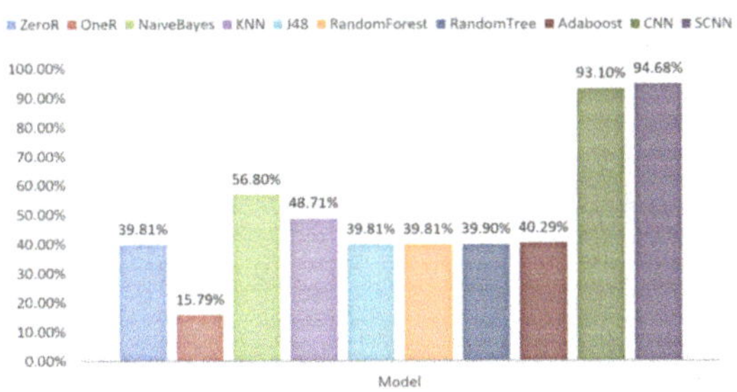

(**a**)

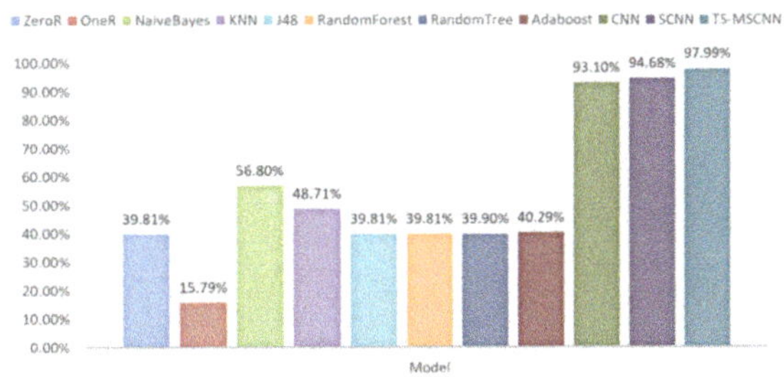

(**b**)

Figure 15. (**a**) Performance of the single model. (**b**) Performance of the TS-MSCNN and other models.

4.5. Multi-SCNN Fusion Model for Multiclass Classification

In the case of multi-classification, it has been shown that the single-structure anomaly detection model has limitations. To address this issue, this paper proposes using the feature fusion method described above to enhance the accuracy of the convolution-based anomaly detection model. The training and test sets used are consistent with those described above. Table 6 presents the detailed performance of the CNN, SCNN, and TS-MSCNN models in multi-classification. The TS-MSCNN model outperforms the CNN and SCNN models in all metrics. Furthermore, Figure 15b shows a comparison between the TS-MSCNN model and other models, where the TS-MSCNN model performs better than other comparison algorithms with the highest accuracy rate being 97.99%.

In addition, this paper compares the anomaly detection results of multi-classification and binary classification, as shown in Figure 16. It can be inferred that due to the more detailed classification of anomaly types, there are significant differences among the data types, which increases the challenge of model classification and leads to better experimental results in binary classification than in multi-classification. For the TS-MSCNN model, the results of the binary classification experiment are only 0.51 higher than those of the multi-classification experiment, which further verifies the effectiveness of the proposed TS-MSCNN model and demonstrates that it can accurately extract UAV flight data characteristics in both multi-classification and binary classification scenarios.

Table 6. The detailed performance of CNN, SCNN, and TS-MSCNN.

Model	Accuracy	Class	Recall	Precision	F_1-Score
CNN	93.10%	aileron_failure	94.33%	93.42%	93.87%
		elevator_failure	77.11%	90.14%	83.12%
		engine_failure	98.01%	96.19%	97.09%
		no_failure	91.50%	93.59%	92.53%
		rudder_failure	84.07%	88.95%	86.44%
SCNN	94.68%	aileron_failure	95.44%	93.50%	94.46%
		elevator_failure	75.90%	86.90%	81.03%
		engine_failure	97.91%	96.57%	97.24%
		no_failure_failure	91.28%	92.10%	91.69%
		rudder_failure	82.42%	90.91%	86.46%
TS-MSCNN	**97.99%**	aileron_failure	99.72%	96.39%	98.03%
		elevator_failure	90.36%	94.94%	92.59%
		engine_failure	98.98%	99.08%	99.03%
		no_failure	96.20%	97.07%	96.63%
		rudder_failure	91.76%	97.66%	94.62%

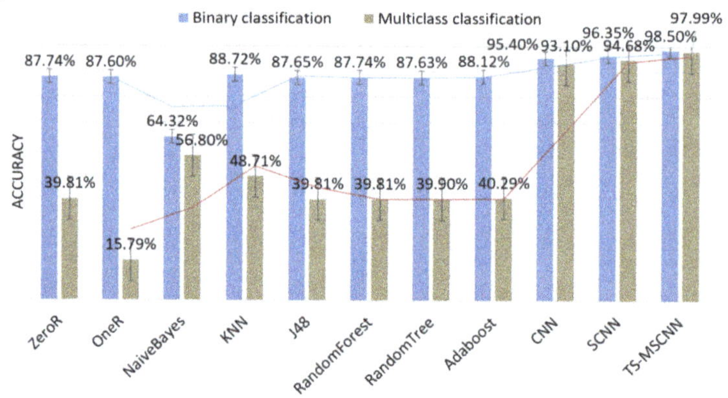

Figure 16. Comparison between the binary classification and the multiclass classification.

The research in [20] and [23] are similar to the research conducted in this paper. In order to compare the experimental results, Table 7 is presented. It is important to note that while [23] evaluates the area under the curve (AUC) of the receiver operating characteristic curve (ROC), this section supplements the AUC results for multiple classifications. The authors of [20] utilized a reduced version of the ALFA dataset, whereas [23] employed the same full version of the ALFA dataset as used in this paper. The experimental model proposed in this paper outperforms the other comparison algorithms. Overall, the experimental results show that the TS-MSCNN model proposed in this paper has achieved the desired purpose and is ready to be used for UAV flight anomaly detection.

Table 7. The accuracies of the TS-MSCNN and the other latest algorithm in multiclass classification.

Model	AUC				ACC
	Aileron_Failure	Elevator_Failure	Engine_Failure	Rudder_Failure	
TS-MSCNN	99.75%	98.35%	99.77%	98.14%	97.99%
Autoencoder [23]	75.09%	80.76%	76.46%	93.21%	/
Recursive Least Squares [20]	/	/	/	/	88.23%

5. Conclusions

UAV flight anomaly detection is a common safety measure to ensure the safety of UAV flights by identifying abnormal UAV flight data. However, the conventional anomaly detection model neglects the difference in POS data used to evaluate UAV flight status in the frequency domain, resulting in the loss of some crucial feature information that limits the improvement of the UAV anomaly detection model's accuracy. Therefore, without considering the recoverable operation of UAV, this paper proposes a TS-MSCNN anomaly detection model based on timestamp slice and the MSCNN. Firstly, by setting a specific timestamp size, this paper extracts and fuses the frequency domain key features of POS data and SS data in the UAV flight log time domain. Then, the POS data and SS data are transformed into two-dimensional grayscale images to serve as the input data of the TS-SCNN model through data reconstruction. Finally, the TS-SCNN model accurately learns and fuses UAV grayscale image data features. The final experimental results demonstrate that the TS-SCNN model outperforms the comparative algorithm in the experimental results of binary classification and multi-classification, which validates the effectiveness of the TS-SCNN model proposed in this paper.

The deep learning model used in anomaly detection has a high time complexity, and UAVs typically have limited resources. Therefore, in future research, the authors of this paper will investigate a lightweight UAV anomaly detection model, taking into account both the timeliness of the anomaly detection model and the computational resources required by the model. The goal is to develop an anomaly detection model that can meet the resource constraints of UAV-embedded systems.

Author Contributions: Conceptualization, J.C. and T.Y.; methodology, J.C., T.Y. and H.D.; writing—original draft, J.C. and Y.L; validation, J.C., T.Y., H.D. and Y.L.; writing—review and editing, J.C., T.Y., H.D. and Y.L.; data curation, T.Y., H.D. and Y.L.; supervision, Y.L.; project administration, J.C and Y.L. All authors have read and agreed to the published version of the manuscript.

Funding: This work was supported by the Sichuan Science and Technology Program under Grant No. 2022YFG0322, China Scholarship Council Program (Nos. 202001010001 and 202101010003), Sichuan Science and Technology Program under Grant No. 2020JDRC0075, the Innovation Team Funds of China West Normal University (No. KCXTD2022-3), the Nanchong Federation of Social Science Associations Program under Grant No. NC22C280, and the China West Normal University 2022 University-level College Student Innovation and Entrepreneurship Training Program Project under Grant No. CXCY2022285.

Data Availability Statement: Not applicable.

Acknowledgments: This paper was completed by the Key Laboratory of the School of Computer Science, China West Normal University. We thank the school for its support and help.

Conflicts of Interest: The authors declare that they have no conflicts of interest to report regarding this present study.

References

1. Kulbacki, M.; Segen, J.; Knieć, W.; Klempous, R.; Kluwak, K.; Nikodem, J.; Kulbacka, J.; Serester, A. Survey of drones for agriculture automation from planting to harvest. In Proceedings of the 2018 IEEE 22nd International Conference on Intelligent Engineering Systems (INES), Las Palmas de Gran Canaria, Spain, 21–23 June 2018; pp. 000353–000358.
2. Puri, A. A survey of unmanned aerial vehicles (UAV) for traffic surveillance. *Dep. Comput. Sci. Eng. Univ. S. Fla.* **2005**, 1–29.
3. Innocente, M.S.; Grasso, P. Self-organising swarms of firefighting drones: Harnessing the power of collective intelligence in decentralised multi-robot systems. *J. Comput. Sci.* **2019**, *34*, 80–101. [CrossRef]
4. Choudhary, G.; Sharma, V.; You, I.; Yim, K.; Chen, R.; Cho, J.H. Intrusion detection systems for networked unmanned aerial vehicles: A survey. In Proceedings of the 2018 14th International Wireless Communications & Mobile Computing Conference (IWCMC), Limassol, Cyprus, 25–29 June 2018; pp. 560–565.
5. Available online: www.popularmechanics.com (accessed on 15 December 2022).
6. Jimu News. Available online: http://www.ctdsb.net/ (accessed on 10 December 2022).
7. Civil Aviation Administration of China. Available online: www.caac.gov.cn (accessed on 20 December 2022).

8. Puranik, T.G.; Mavris, D.N. Identifying instantaneous anomalies in general aviation operations. In Proceedings of the 17th AIAA Aviation Technology, Integration, and Operations Conference, Atlanta, GA, USA, 25–29 June 2017; p. 3779.
9. Hamel, T.; Mahony, R. Attitude estimation on SO [3] based on direct inertial measurements. In Proceedings of the 2006 IEEE International Conference on Robotics and Automation, 2006. ICRA 2006, Orlando, FL, USA, 15–19 May 2006; pp. 2170–2175.
10. Garraffa, G.; Sferlazza, A.; D'Ippolito, F.; Alonge, F. Localization Based on Parallel Robots Kinematics as an Alternative to Trilateration. *IEEE Trans. Ind. Electron.* **2021**, *69*, 999–1010. [CrossRef]
11. Kendoul, F. Survey of advances in guidance, navigation, and control of unmanned rotorcraft systems. *J. Field Robot.* **2012**, *29*, 315–378. [CrossRef]
12. Alonge, F.; D'Ippolito, F.; Fagiolini, A.; Garraffa, G.; Sferlazza, A. Trajectory robust control of autonomous quadcopters based on model decoupling and disturbance estimation. *Int. J. Adv. Robot. Syst.* **2021**, *18*, 1729881421996974. [CrossRef]
13. Koubâa, A.; Allouch, A.; Alajlan, M.; Javed, Y.; Belghith, A.; Khalgui, M. Micro air vehicle link (mavlink) in a nutshell: A survey. *IEEE Access* **2019**, *7*, 87658–87680. [CrossRef]
14. Keipour, A.; Mousaei, M.; Scherer, S. Alfa: A dataset for uav fault and anomaly detection. *Int. J. Robot. Res.* **2021**, *40*, 515–520. [CrossRef]
15. Mitchell, R.; Chen, I.R. Specification based intrusion detection for unmanned aircraft systems. In Proceedings of the First ACM MobiHoc Workshop on Airborne Networks and Communications, Hilton Head, SC, USA, 11 June 2012; pp. 31–36.
16. Mitchell, R.; Chen, R. Adaptive intrusion detection of malicious unmanned air vehicles using behavior rule specifications. *IEEE Trans. Syst. Man Cybern. Syst.* **2013**, *44*, 593–604. [CrossRef]
17. Sedjelmaci, H.; Senouci, S.M.; Ansari, N. A hierarchical detection and response system to enhance security against lethal cyber-attacks in UAV networks. *IEEE Trans. Syst. Man Cybern. Syst.* **2017**, *48*, 1594–1606. [CrossRef]
18. Liu, Y.; Ding, W. A KNNS based anomaly detection method applied for UAV flight data stream. In Proceedings of the 2015 Prognostics and System Health Management Conference (PHM), Beijing, China, 21–23 October 2015; pp. 1–8.
19. Sedjelmaci, H.; Senouci, S.M.; Ansari, N. Intrusion detection and ejection framework against lethal attacks in UAV-aided networks: A Bayesian game-theoretic methodology. *IEEE Trans. Intell. Transp. Syst.* **2016**, *18*, 1143–1153. [CrossRef]
20. Keipour, A.; Mousaei, M.; Scherer, S. Automatic real-time anomaly detection for autonomous aerial vehicles. In Proceedings of the 2019 International Conference on Robotics and Automation (ICRA), Montreal, QC, Canada, 20–24 May 2019; pp. 5679–5685.
21. Shrestha, R.; Omidkar, A.; Roudi, S.A.; Abbas, R.; Kim, S. Machine-learning-enabled intrusion detection system for cellular connected UAV networks. *Electronics* **2021**, *10*, 1549. [CrossRef]
22. Chowdhury MM, U.; Hammond, F.; Konowicz, G.; Xin, C.; Wu, H.; Li, J. A few-shot deep learning approach for improved intrusion detection. In Proceedings of the 2017 IEEE 8th Annual Ubiquitous Computing, Electronics and Mobile Communication Conference (UEMCON), New York, NY, USA, 19–21 October 2017; pp. 456–462.
23. Park, K.H.; Park, E.; Kim, H.K. Unsupervised fault detection on unmanned aerial vehicles: Encoding and thresholding approach. *Sensors* **2021**, *21*, 2208. [CrossRef] [PubMed]
24. Abu Al-Haija, Q.; Al Badawi, A. High-performance intrusion detection system for networked UAVs via deep learning. *Neural Comput. Appl.* **2022**, *34*, 10885–10900. [CrossRef]
25. Dudukcu, H.V.; Taskiran, M.; Kahraman, N. Unmanned Aerial Vehicles (UAVs) Battery Power Anomaly Detection Using Temporal Convolutional Network with Simple Moving Average Algorithm. In Proceedings of the 2022 International Conference on INnovations in Intelligent SysTems and Applications (INISTA), Biarritz, France, 8–12 August 2022; pp. 1–5.
26. Zhang, C.; Li, D.; Liang, J.; Wang, B. MAGDM-oriented dual hesitant fuzzy multigranulation probabilistic models based on MULTIMOORA. *Int. J. Mach. Learn. Cybern.* **2021**, *12*, 1219–1241. [CrossRef]
27. Xie, H.; Hao, C.; Li, J.; Li, M.; Luo, P.; Zhu, J. Anomaly Detection for Time Series Data Based on Multi-granularity Neighbor Residual Network. *Int. J. Cogn. Comput. Eng.* **2022**, *3*, 180–187. [CrossRef]
28. Khan, W.; Haroon, M. An unsupervised deep learning ensemble model for anomaly detection in static attributed social networks. *Int. J. Cogn. Comput. Eng.* **2022**, *3*, 153–160. [CrossRef]
29. Sifre, L.; Mallat, S. Rigid-motion scattering for texture classification. *arXiv* **2014**, arXiv:1403.1687.
30. Pytorch. Available online: https://pytorch.org/ (accessed on 1 December 2022).
31. GJohn, P.L. Estimating continuous distributions in Bayesian classifiers. In Proceedings of the Eleventh Conference on Uncertainty in Artificial Intelligence, Montreal, QC, Canada, 18–20 August 1995; pp. 338–345.
32. Peterson, L.E. K-nearest neighbor. *Scholarpedia* **2009**, *4*, 1883. [CrossRef]
33. Quinlan, J.R. *C4. 5: Programs for Machine Learning*; Elsevier: Amsterdam, The Netherlands, 2014.
34. Breiman, L. Random forests. *Mach. Learn.* **2001**, *45*, 5–32. [CrossRef]
35. Aldous, D. The continuum random tree. II. An overview. *Stoch. Anal.* **1991**, *167*, 23–70.
36. Schapire, R.E. Explaining adaboost. In *Empirical Inference: Festschrift in Honor of Vladimir N. Vapnik*; Springer: Berlin/Heidelberg, Germany, 2013; pp. 37–52. [CrossRef]

electronics

MDPI

Article

A Context Awareness Hierarchical Attention Network for Next POI Recommendation in IoT Environment

Xuebo Liu, Jingjing Guo * and Peng Qiao *

Shanxi Information Industry Technology Research Institute Co., Ltd., Taiyuan 030012, China
* Correspondence: jane_guojingjing@163.com (J.G.); sxszdh@126.com (P.Q.)

Abstract: The rapid increase in the number of sensors in the Internet of things (IoT) environment has resulted in the continuous generation of massive and rich data in Location-Based Social Networks (LBSN). In LBSN, the next point-of-interest (POI) recommendation has become an important task, which provides the best POI recommendation according to the user's recent check-in sequences. However, all existing methods for the next POI recommendation only focus on modeling the correlation between POIs based on users' check-in sequences but ignore the significant fact that the next POI recommendation is a time-subtle recommendation task. In view of the fact that the attention mechanism does not comprehensively consider the influence of the user's trajectory sequences, time information, social relations and geographic information of Point-of-Interest (POI) in the next POI recommendation field, a Context Geographical-Temporal-Social Awareness Hierarchical Attention Network (CGTS-HAN) model is proposed. The model extracts context information from the user's trajectory sequences and designs a Geographical-Temporal-Social attention network and a common attention network for learning dynamic user preferences. In particular, a bidirectional LSTM model is used to capture the temporal influence between POIs in a user's check-in trajectory. Moreover, In the context interaction layer, a feedforward neural network is introduced to capture the interaction between users and context information, which can connect multiple context factors with users. Then an embedded layer is added after the interaction layer, and three types of vectors are established for each POI to represent its sign-in trend so as to solve the heterogeneity problem between context factors. Finally reconstructs the objective function and learns model parameters through a negative sampling algorithm. The experimental results on Foursquare and Yelp real datasets show that the AUC, precision and recall of CGTS-HAN are better than the comparison models, which proves the effectiveness and superiority of CGTS-HAN.

Keywords: context awareness; attention network; dynamic user preferences; next POI recommendation; IoT

Citation: Liu, X.; Guo, J.; Qiao, P. A Context Awareness Hierarchical Attention Network for Next POI Recommendation in IoT Environment. *Electronics* **2022**, *11*, 3977. https://doi.org/10.3390/electronics11233977

Academic Editor: Juan M. Corchado

Received: 17 October 2022
Accepted: 26 November 2022
Published: 30 November 2022

1. Introduction

The rapid increase in the number of sensors in IoT environment has resulted in the continuous generation of massive and rich data in LBSN, which has greatly promoted the development of POIs recommendations. It is an important and challenging task to understand the mobility of users and recommend the next POI to users. The research on POIs recommendation considers user check-in data as a whole, which only considers the relationship between target users and points of interest but ignores the relationship between points of interest. In fact, there is a strong correlation between the user's current POI and the POI to be accessed next. For example, when users obtain off work at night, they may go to restaurants, bars and other places instead of traveling. Different from ordinary POI recommendations, the next POI recommendation is to recommend the POI to visit next based on the user's historical trajectory information and the current POI location. The next POI recommendation can enhance the user's travel experience but also help real economists push advertisements to the target user groups. Because the next

POI recommendation is of great significance to users and enterprises, the improvement of the next POI recommendation system is popular in the academic community, and much research focuses on enhancing the recommendation performance of the next POI.

Modeling according to the check-in sequence of users is the basic work for the next POI recommendation, and POIs tend to have certain correlations among themselves [1–4]. Some previous studies adopted Markov Chain [1,5–7] matrix decomposition [2,3], tensor decomposition [8,9] or transformation model [10,11] to solve the next POI recommendation problem. In recent years, the Recurrent Neural Network (RNN) model in Deep Neural Networks [12–15] has shown a good recommendation effect in processing sequence data. MST-RNN [16] exploits the duration time dimension and semantic tag dimension of POIs in each layer of the neural network. Attention Network [17–20], as a branch of RNN, has strong recommendation performance when applied to a recommendation. In real life, the check-in sequences of users on different dates will have different time characteristics in their historical trajectories. For example, users usually check in at POIs near the company on weekdays and go to shopping malls and tourist attractions on weekends. However, these studies did not explore the diversity of time sequence characteristics.

When the POI recommendation systems encounter new users, fewer user trajectories, or users in a new area, the cold-start problem may occur, and the recommendation system cannot effectively recommend the next POI for the user. Moreover, how to extract valuable contextual information [21] (e.g., geographical location, time, social relations) from the massive data generated by IoT sensors is particularly vital.

In order to enhance the accuracy of the next POI recommendation, this paper proposes the next POI recommendation model embedded with user context check-in information, which comprehensively considers the influence of geographical, temporal and social factors. This paper studies the next point of interest recommendation problem, and the main contributions are as follows:

(1) We propose the next POI recommendation model named CGTS-HAN. It uses the attention mechanism and the Bi-directional Long Short-Term Memory (BiLSTM) to establish a geo-temporal social attention network to learn user check-in sequences, which can simultaneously capture the user's social relationships, the temporal dependence of sequence patterns, and the geographical relationships between points of interest. Taking the influence of geography, time and social factors into account and embedding the user context check-in information can effectively reduce the problem of a cold start.

(2) We propose a recommendation algorithm suitable for a hierarchical model system. Incorporating the context interaction layer into CGTS-HAN, a feedforward neural network is used to learn feature intersections to describe the interaction effects of users and contexts.

(3) Combining the context interaction layer and CGTS-HAN model, a Co-attention Network is developed to learn the dynamic preferences of users so that CGTS-HAN can distinguish user preference degrees in its historical check-in trajectories.

(4) Experiments were conducted on two real-world datasets. The experimental results demonstrate that CGTS-HAN achieves better performance than other baseline comparison models in terms of AUC, precision and recall.

2. Related Work

The influence of sequential factors. Most next Points-of-Interest (POIs) recommendations rely on the sequence correlation in the user check-in. Wen et al. [5] calculated the global and individual transition probabilities between clusters according to the user's check-in sequence, then used multi-order Markov chains to discover and rank subsequent clusters, and finally combined with individual preferences to generate a ranking list. Although it can model time series well, the correlation between POIs is not high. Zhang et al. [1] used N-order Markov chains to implement POIs recommendations for higher-order sequences. The authors recommend coarse-grained areas to users, such as location recommendations based on city streets, rather than user personalization. The recommended points of interest are not targeted. Kong et al. [15] aimed to discover users' uncertain check-in points of interest; they extended Skip Gram to capture user preference transitions and then predicted the next POI for uncertain check-in users. Wang et al. [22] propose a location-aware POI recommendation system that user preferences by using user history trajectory and user review information. However, the collaborative filtering algorithm used in the article is likely to have a cold start problem.

The influence of geographical factors. Debnath et al. [7] mined sequential patterns from each user's check-in location, then used Markov chains to construct transition probability matrices and combined them with spatial influences to generate space-aware location recommendations. Liu et al. [2] improved the accuracy of POIs recommendations by using the conversion mode of the user's preference for the POIs category. Feng et al. [11] established a hierarchical binary tree according to the physical distance between POIs to reflect the influence of geographical location. However, previous efforts mainly consider the location information of check-in points as a whole and ignore their temporal relation. Using the information from users' location history, Yang et al. [23] proposed a location-aware POI recommendation system that uses information from users' location history and models user preferences based on their reviews. It aims to solve the user's POIs recommendation in new regions and cities without considering the impact of time context and other time dynamics.

The influence of time factors. Considering the spatiotemporal characteristics of LBSN, Cheng et al. [3] proposed FPMC with candidates Region Constraint (FPMC-LR) method and provided new POIs for users by combining individual Markov chains and local and regional constraints. Rendle et al. [10] used Factorized Personalized Markov Chains (FPMC) to predict the next check-in interest point by expressing the short-term and long-term preferences of users. Xiong et al. [8] proposed a Bayesian probability tensor decomposition model based on time context, which dynamically acquired the potential features of users, POIs and months and could learn the global evolution of potential features. However, it is too sparse to model the time factor by month, which made the results not ideal. However, in practical recommendation applications, the recommendation results obtained by these traditional recommendation methods lack the user's personalized requirements for POI. [24]. Liu et al. [14] used Skip Gram to train the temporal latent representation vector of POIs and proposed a time-aware POIs recommendation model. The spatio-temporal model TS-RNN proposed in this paper takes the spatio-temporal context elements into account in RNN mode to replace MF and FPMC, but the evaluation standard is still BPR.

With the continuous application of the Markov chain and factorization in the next POIs recommendation, both of them show their own limitations. The Markov chain is that it assumes strong independence among different factors, and the state of each POI in the first-order Markov chain is only related to the previous POI, which limits its performance. The limitation of tensor decomposition is that it is faced with a difficult problem which is called cold-start.

Some research work [5,9,12,13,17–19] shows that combining sequence, geography and time factors can obtain better recommendation results. Liu et al. [12] extended RNN and proposed a spatiotemporal recurrent neural network method. In this method, the time conversion matrix can be created with different time intervals to simulate the time context, and the distance conversion matrix can be created with different geographical distances to simulate the spatial context. Inspired by the Word2Vec framework, Zhao et al. [13] proposed the Geo-Teaser model, which embedded the time factor into the model to capture the time characteristics, and constructed the pairwise preference ranking at the geographical level. Then, POIs are ranked according to the preference score function, and the top-N POIs with the highest scores are recommended for users. In order to predict the access preference for the next POIs, Li et al. [17] introduced the time and multi-level context attention mechanism, which can dynamically select relevant check-in locations and distinguish context factors. The geographic-time awareness hierarchical attention network, which is developed by Liu et al. [18], can reveal the dependencies of the overall sequence and the relationship between POIs through the BiLSTM network while using the geographic factor. Huang et al. [19] proposed a context-based self-attention network for the next POIs recommendation, which used positional encoding matrices instead of time encodings to model dynamic contextual dependencies. Guo et al. [25] proposed DeepFM, which combined factorization machine and feature embedding and sharing strategy to recommend. Among them, feature embedding and sharing strategies can avoid the establishment of feature engineering. However, the invalid second-order combination features may bring noise and adversely affect the model performance.

However, each of the above models [13,17–19,25] does not deeply mine the distance and time relationship between POIs in the trajectory when obtaining the correlation between POIs and does not add user social information into the model or framework. Research work [26,27] shows that although the influence of social relations is far less than geographical factors and time factors, it can affect the user's check-in location selection that introduces the social factors into the next POIs recommendation.

The proposed CGTS-HAN model uses geographic factors to capture the features of POIs and their correlations in order to improve the recommendation performance of the next POIs recommendation.

3. Preliminaries

3.1. Problem Definition

This section presents five definitions [14] and specific problem statements related to the proposed next POIs recommendation problem.

Definition 1. *User set. The user sets represent a set of $|\mathcal{U}|$ users, denoted by $\mathcal{U} = \left\{u_1, u_2, \ldots, u_{|\mathcal{U}|}\right\}$.*

Definition 2. *POIs sets. The POIs sets represent a set of p points of interest, denoted by $\mathcal{P} = \left\{p_1, p_2, \ldots, p_{|\mathcal{P}|}\right\}$. Each object in set of POIs consists of a 2-tuple, denoted by $(\ell on_i, \ell at_i, t_i)$, where ℓon_i and ℓat_i represent the longitude and latitude of the point of interest, respectively, and t_i represents the check-in time.*

Definition 3. *Time state set. The time state set denoted by $\mathcal{T} = \left\{t_1, t_2, \ldots, t_{|\mathcal{T}|}\right\}$, which is used to indicate the time points of the user's check-in sorted by time in a day.*

Definition 4. *Check-in records. The check-in records denoted by $\mathcal{T}r_i$. A check-in record represents a record of a user's visit to a point of interest in one day.*

Definition 5. *Check-in history. The check-in history denoted by $\mathcal{T}r = \{\mathcal{T}r_1, \mathcal{T}r_2, \ldots, \mathcal{T}r_i, \ldots\}$. A user's check-in history contains all his check-in records. The check-in history represents a series of check-in POI records of the user arranged in chronological order.*

Problem statement: Given a user, denoted by u_i, and his check-in history, denoted by $\mathcal{T}r$, according to the user's check-in history and the current point of interest, recommend the p_n for the user to visit at the next moment from $\mathcal{P}$.

3.2. Symbols Definition

To illustrate, Table 1 summarizes the main symbols used in this paper as follows.

Table 1. Main Notation.

Symbols	Interpretation
ui, pi	Preference vectors of u_i, p_i
$\mathcal{T}r$	Sets of trajectory sequences for users
$\mathcal{T}r_i$	A set of check-ins for user u_i
c_i	Context of user u_i
ti	t_i latent semantic vector
Ui,So,Ti	Matrices of user preference, social relationship and t_i latent semantic
pr,su,pi	The geographical predecessor vector, the geographical successor vector, the preference vector of POI
Pr,Su,Pi	The geographical predecessor matrix, the geographical successor matrix, the preference matrix of POI
u_i^{dp}	Dynamic preferences of u_i
W_1,W_2,W_3,...,W_z^u	Weight matrices in the model
e_{u_i,c_i}	Feature vector of the interaction between user u_i and context c_i
p_{u_i}	Original feature vector of u_i
g_{c_i}	Context feature vector of c_i
$\mathbf{b}_1$, $\mathbf{b}_2$, $\mathbf{b}_3$,..., $\mathbf{b}_z^u$	Bias terms in the model

4. CGTS-HAN Model

The framework of the CGTS-HAN model proposed in this paper is shown in Figure 1.

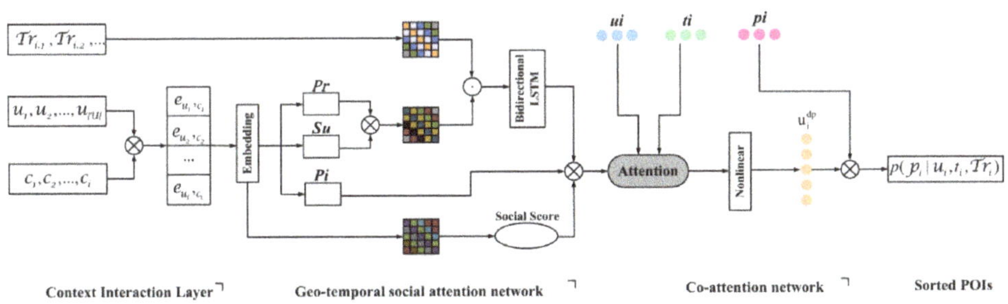

Figure 1. The Framework of CGTS-HAN Model.

The model mainly consists of a context interaction layer, a geo-temporal social attention network and a co-attention network. The context interaction layer models the interaction between each user and their context information in the context environment and obtains the influence of each context on the user. Embedding layers are used to address the heterogeneity that exists among recommendation factors. Afterward, the model introduces a geo-temporal-social attention network to model the geographic relationships, temporal dependencies, and users' social relationships among POIs of check-in sequences. The co-attention network is used to capture the dynamic preferences of users. Finally, we use a negative sampling algorithm to train the model. The next POI recommendation usually feeds back a sorted list of points of interest to the user, so this model first calculates the probability of the target user visiting the points of interest, then calculates the scores of candidate points of interest according to the Bayesian Equation, and finally sorts POIs to obtain an ordered list of top N POIs.

4.1. The Context Interaction Layer

In this paper, a feedforward neural network is introduced to simultaneously learn high-order features and low-order features to capture the interaction between the user and the context and obtain the influence of the context on the user. The eigenvectors are as follows.

$$e_{u_i,c_i} = f_1(p_{u_i}, g_{c_i}) \tag{1}$$

In the above formula, the $f_1(\cdot)$ is used to represent the feature interaction function; its input is the user and the context, which are represented by u_i and c_i, respectively. The e_{u_i,c_i} in the Equation represents the feature vector of the interaction between the user and context.

The input layer is responsible for receiving input and distributing it to the hidden layers (so called because they are invisible to the user). These hidden layers are responsible for the required calculations and output to the output layer, and the user can see the final output of the output layer. The modeling process of the context interaction layer feedforward neural network is shown in Figure 2.

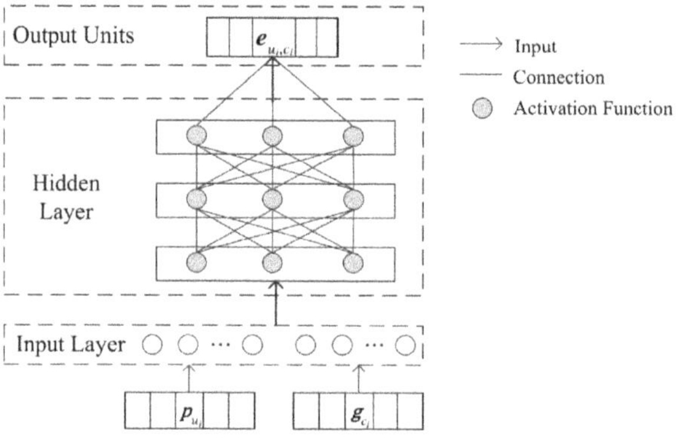

Figure 2. The Structure of Context Interaction Layer Module.

Since the user and the context belong to different feature types of input data, the model uses a nonlinear connection layer to map the user's original feature vector p_{u_i} and context feature vector g_{c_i} to the additional semantic space. The Equation is as follows:

$$y = RELU(W_{uz}p_{u_i} + W_{cz}g_{c_i} + \mathbf{b}_z) \tag{2}$$

In the above formula, $\mathbf{W}_{uz}$ and $\mathbf{W}_{cz}$ are the weight matrices of the nonlinear connection layer, $\mathbf{b}_z$ is the bias term; $RELU\ (\cdot)$ is the nonlinear activation function Linear Unit. After the input layer is multiplied by the weight, the result is often further processed; that is, the result is used as the input of the first layer of the hidden layer.

In order to enhance the interaction between the user and the context, the model builds three hidden layers on top of the nonlinear connection layer, which are specifically expressed as follows:

$$y_1 = RELU_1(W_1 y + \mathbf{b}_1) \tag{3a}$$

$$y_2 = RELU_2(W_2 y_1 + \mathbf{b}_2) \tag{3b}$$

$$y_3 = RELU_3(W_3 y_2 + \mathbf{b}_3) \tag{3c}$$

In the above formula, W_1, $\mathbf{b}_1$, and $RELU_1(\cdot)$ represent the weight matrix, bias term and RELU activation function of the first hidden layer, respectively. The pronouns in the second and third hidden layer Equations have meanings and so on.

y_1, y_2, y_3 represent the outgoing vectors of the first, second and third hidden layers, respectively.

The outgoing vector (y_3) of the third hidden layer in the model is passed to the output unit, and the output unit converts it into the feature vector of the context that acts on the user, which is expressed as follows:

$$e_{u_i,c_i} = (W_z^u y_3 + \mathbf{b}_z^u) \tag{4}$$

In the above formula, W_z^u and $\mathbf{b}_z^u$ represent the weight matrix and bias term of the output layer, respectively.

4.2. The Embedding Layer

The function of the embedding layer is to project the POIs into the latent semantic space, and use the matrix form to transform the geographic, temporal and social relationship factors of the points of interest in the context interaction layer, so as to relieve the heterogeneity that exists among the geographical, temporal and social factors.

Inspired by Transition [2], this paper firstly establishes a geographical predecessor vector, a geographical successor vector and a preference vector of points of interest for each POI, denoted as $pr \in R^{1\times d}$, $su \in R^{1\times d}$, $pi \in R^{1\times d}$, respectively, where d is the potential dimension. The predecessor vector is used to receive the check-in trend of other POIs, and the successor vector is used to reflect the check-in trend of this POI transferring to other POIs. After that, the embedding layer transforms the established POI vector into three matrices $Pr \in R^{|\mathcal{P}|\times d}$, $Su \in R^{|\mathcal{P}|\times d}$, $Pi \in R^{|\mathcal{P}|\times d}$, respectively. Similarly, we sequentially create a latent semantic matrix $Ti \in R^{|\mathcal{T}|\times d}$ for temporal states, a social relationship matrix $So \in R^{|\mathcal{U}|\times d}$ and a user preference matrix $Ui \in R^{|\mathcal{U}|\times d}$ for users.

4.3. Geographical–Temporal Social Attention Networks

4.3.1. Modeling for Geographic Factors

The geographical attention network adopts the Transformer [4] mechanism. The input vector is regarded as a key-value pair by the encoder. The output of the encoder is compressed into the query by the decoder. Finally, the output query is mapped to the set of keys and values. The transformer network structure is simple, based on a self-attention mechanism, and computation is executed in parallel, which makes the Transformer efficient and requires less training time. The attention function in Transformer is described as mapping a query and a set of key value pairs to the output. Queries, keys and values are vectors. The attention weight is calculated by computing the dot product attention for each word in the sentence. The final score is a weighted sum of these values. The transformer mechanism is divided into the following four steps.

Step 1: Take the dot product of the keys for each input vector in the query. The input data consists of a geographic predecessor matrix (Pr(query)), a geographic successor matrix (Su(key)) and a POI preference matrix (Pi(value)).

Step 2: Scale the dot product by dividing by the square root of the dimension of the key vector.

Step 3: Use softmax to normalize the scale values. After softmax is applied, all values are positive and add up to 1.

Step 4: Apply the dot product between the normalized fraction and the value vector $\frac{QK^{\mathrm{T}}}{\sqrt{\mathrm{n}}}$, and then calculate the sum.

A nonlinear transformation of shared parameters is used to map Pr and Su into the same semantic space and then compute their weight matrix as follows.

$$AG(Pr, Su) = \mathrm{softmax}(\frac{QK^{\mathrm{T}}}{\sqrt{\mathrm{n}}})Pi \tag{5}$$

$$Q = \mathrm{RELU}(PrW^Q + \mathbf{B}_q) \tag{6}$$

$$K = \mathrm{RELU}(SuW^K + \mathbf{B}_k) \tag{7}$$

Among them, Q, K are vectors of size $1 \times d$, $W^Q \in R^{d\times d}$, $W^K \in R^{d\times d}$, $\mathbf{B}_q \in R^{M\times d}$ and $\mathbf{B}_k \in R^{M\times d}$ are model parameters, $\sqrt{\text{n}}$ is used to scale the dot product value of QK^{T}, and the matrix output by Equation (5) represents the geographic relationship between M POIs.

Since the dot product cannot model the geographic distance between POIs, this paper uses a Power Law Function (PLF) [28] based on the geographic relationship to examine the effect of geographic distance between POIs. The PLF is defined as follows.

$$\text{P}(x) = \alpha x^{-\gamma} \tag{8}$$

In the above formula, x and $\text{P}(x)$ are positive random variables, and α and γ are constants greater than zero. In the model, the probability value of a user visiting a point of interest follows a PLF.

The model introduces the influence of geographical factors between adjacent POIs into the attention network, then embed Equation (8) into Equation (5), and the rewritten Equation (5) is as follows.

$$\widetilde{AG}(Pr, Su) = \text{softmax}(\frac{QK^{\text{T}}}{\sqrt{\text{n}}})Pi \times \text{P}(x) \tag{9}$$

To output the geographic impact to the next stage (i.e., the temporal impact), the matrix output by the geographical attention network from Equation (9) is defined as follows.

$$\widetilde{GAG}(Pr, Su, \text{P}(x)) = \widetilde{AG}(Pr, Su) \times \text{P}(x) \tag{10}$$

4.3.2. Modeling for Time Factors

This paper captures the user's long-term and recent interest features according to the bidirectional LSTM [29] based on the Attention mechanism. The output of the bidirectional LSTM $\widetilde{GAG}$ is used as the input for the time factor modeling, and the output is a vector set $H : [h_1, h_2, ..., h_i, ..., h_T]$ with dimension $M \times d$, where h_i represents the output at the time step i, T is the sentence length $i = 1, ..., T$. This part of the network includes two sub-networks, sequence and context, which transfer time information forward and backward, respectively. Then pass the set of output vectors H of the LSTM layer to the Attention layer to obtain a weight matrix for temporal factor modeling. Firstly, the model uses the tanh activation function to make the time factor available to the nonlinear model, as shown below.

$$\text{Z} = \tanh(H) \tag{11}$$

Then, the model process the values using the softmax function to transform Z into probabilities, as shown below.

$$\alpha_{\text{t}} = \text{softmax}(x^{\text{T}}\text{Z}) \tag{12}$$

Finally, the model obtains the weight matrix r based on the set H of output vectors and the probability α_{t}.

$$r = H\alpha_{\text{t}}^{\text{T}} \tag{13}$$

In the above formula, $H \in R^{M\times d}$, x^{T} is the transpose of the parameter vector, $\alpha_{\text{t}}^{\text{T}}$ is the transpose of the α_{t}. The representation r of the sentence is formed by a weighted sum of these output vectors. Finally, The resulting classification sequence pair is denoted as follows.

$$h^* = \tanh(r) \tag{14}$$

4.3.3. Modeling for Social Factors

The model uses a multi-layer sub-network to obtain the attention score, and the results are as follows.

$$\text{s}_1(u_i, u_m) = RELU(...RELU(RELU(W_1^u \cdot p_{u_i} + W_2^u \cdot e_m + \mathbf{b}_4) + \mathbf{b}_5)...) \tag{15}$$

In the above Equation, W_1^u and W_2^u are weight coefficients, $\mathbf{b}_4$ and $\mathbf{b}_5$ are bias terms, and e_m is the embedded data of the user's (u_i) neighbor user (u_m). After $s_1(u_i, u_m)$ is calculated by Equation (15), it is normalized by the softmax function used in Equation (16), and finally, the social influence score is obtained as follows.

$$\alpha_{u_i,u_m} = \text{softmax}(s_1(u_i, u_m)) = \frac{\exp(s_1(u_i, u_m))}{\sum\limits_{n \in So} \exp(s_1(u_i, u_n))} \tag{16}$$

4.4. Co-Attention Network

This paper creates a co-attention network to capture users' dynamic preferences. Specifically, the co-attention network uses a late fusion strategy to incorporate different weighted attention values (W^w) and nonlinear connection layers to learn the dynamic preferences of users. Given the context feature vector of u_i, the sequence pair of time factors and the social influence score, and then obtain the overall effect of the context on the target user by weighting and summing them, which is represented by the dynamic preference ($\mathrm{u}_\mathrm{i}^{\mathrm{dp}}$), which is calculated as follows.

$$\mathrm{u}_\mathrm{i}^{\mathrm{dp}} = \Psi(\alpha_i^{\mathrm{dp}} W^\alpha + \mathbf{b}_6) W^\mathrm{U} \tag{17}$$

$$\alpha_\mathrm{i}^{\mathrm{dp}} = \text{concat}([(ui, e_{u_i,c_i})W^\mathrm{W}, (ti, h^*)W^\mathrm{W}, (\alpha_{u_i,u_m})W^\mathrm{W}]) \tag{18}$$

In the above formula, $\mathbf{W}^\alpha \in R^{d \times d}$ and $\mathbf{W}^\mathrm{U} \in R^{d \times d}$ are model parameters, $\mathbf{b}_6$ is bias terms, $\Psi(\cdot)$ represents the nonlinear connection function and $\alpha_\mathrm{i}^{\mathrm{dp}}$ is the weighted joint score of the common attention network layers.

4.5. Learning and Optimization

After obtaining the user's dynamic preference, the model uses the softmax function to generate the conditional probability distribution of the next POI p_n, as follows.

$$\mathrm{P}(p_n | u_i, c_i, \mathcal{T}r) = \frac{\exp(\mathrm{u}_\mathrm{i}^{\mathrm{dp}}, pi^T)}{\sum\limits_{v=1}^{|\mathcal{P}|} \exp(\mathrm{u}_\mathrm{i}^{\mathrm{dp}}, pv^T)} \tag{19}$$

where $\mathrm{P}(p_n | u_i, c_i, \mathcal{T}r)$ is the probability distribution of user u_i's access to POI p_n, which is calculated according to the weighted average of user u_i's dynamic preference $\mathrm{u}_\mathrm{i}^{\mathrm{dp}}$'s attention weight. pi^T is the transpose matrix of preference vectors of POI p_i, and pv^T is the transpose matrix of preference vectors of random POI p_v.

Given a training dataset $\mathcal{X} = \{\mathcal{T}r, u_i, c_i, p_n\}$, its joint probability distribution is denoted as follows.

$$\mathrm{P}_\Theta(\mathcal{X}) = \prod_{x \in \mathcal{X}} \mathrm{P}(p_n | u_i, c_i, \mathcal{T}r) = \prod_{x \in \mathcal{X}} \frac{\exp(\mathrm{u}_\mathrm{i}^{\mathrm{dp}}, pi^T)}{\sum\limits_{v=1}^{|\mathcal{P}|} \exp(\mathrm{u}_\mathrm{i}^{\mathrm{dp}}, pv^T)} \tag{20}$$

In the above formula, $\mathcal{X}$ represents the training set, and the model parameter value is $\Theta = \{Pr, Su, Pi, So, Ui, Ti, W^\mathrm{Q}, W^\mathrm{K}, W^\mathrm{U}, W^\mathrm{W}\}$. By processing the regularization term of the above formula, the objective function is transformed into the following form.

$$\mathcal{L} = \sum_{x \in \mathcal{X}} \mathrm{P}(p_n | u_i, c_i, \mathcal{T}r) - \lambda\Theta \tag{21}$$

The computational cost of the above objective function will increase with the increase of POIs during optimization. Using the negative sampling method to optimize the objective function will reduce the training complexity, which can significantly improve computational efficiency. Therefore, the model rewrites Equation (18) using the Negative Sampling technique, and the results are shown below.

$$\mathcal{L} = \sum_{x\in\mathcal{X}} (\log\sigma(\mathrm{u}_{\mathrm{i}}^{\mathrm{dp}}, pi^{\mathrm{T}}) + \sum_{k=1,\overline{pi}\sim q}^{K} \log\sigma(-\mathrm{u}_{\mathrm{i}}^{\mathrm{dp}}, \overline{pi}^{\mathrm{T}}) - \lambda\Theta \tag{22}$$

In the above formula, $\sigma = \frac{1}{1+e^{-x}}$ is used to approximate the probability, and K is the number of negatively sampled POIs.

4.6. Generation of Recommendation List

In the case of known u_i and his check-in history (u_i), the model can calculate the ranking score of each candidate POI according to the Bayesian and then recommends top-ranked POIs to the user. The score is calculated as follows.

$$\hat{\mathrm{r}}^{\mathrm{P}}_{u_i} = \mathrm{P}(p_n|\mathcal{T}r) \propto \mathrm{P}(p_n)\mathrm{P}(\mathcal{T}r|p_n) = \mathrm{P}(p_n) \prod_{p_n\in\mathcal{T}r_i} \mathrm{P}(p_n{}'|p_n) \tag{23}$$

The ranking score of each POI in the final recommendation list of the model is formed according to the ranking score ($\hat{\mathrm{r}}^{\mathrm{P}}_{u_i}$) of the candidate POIs and their probability distribution.

$$\mathrm{S}\hat{\mathrm{r}}^{\mathrm{P}}_{u_i} = \hat{\mathrm{r}}^{\mathrm{P}}_{u_i} \times \mathrm{P}(p_n \,|\, u_i, c_i, Tr) \tag{24}$$

The following Algorithm 1 summarizes the learning algorithm flow of CGTS-HAN, which is mainly composed of three modules: data (lines 2–9), factor modeling (lines 10–15) and prediction (lines 16–19).

Algorithm 1: The Procedure of CGTS-HAN.

Input: $\mathcal{U}, \mathcal{P}, \mathcal{T}, \mathcal{X}, \Theta$.
Output: Final POIs recommendation list.
1. Initialization: $t_i = 0$.
// **Data Module**
2. for each $u_i \in \mathcal{U}$ do
3. Split check-in record $\mathcal{T}r_i$ by $\mathcal{T}r$;
4. Model preference vector u_i;
5. Model original feature vector p_{u_i};
6. Model context feature vector g_{c_i};
7. for each $u_i \in \mathcal{T}$ do
8. Split check-in time by day;
9. Model preference vector t_i;
// **Factor Module**
10. for each $u_i = 1$; $u_i \in \mathcal{U}$; u_i ++ do
11. Model α_{u_i,u_m};
12. for each $p_i = 1$; $p_i \in \mathcal{P}$; p_i ++ do
13. Model ***GAG***;
14. for each $t_i = 1$; $t_i \in \mathcal{T}$; t_i++ do
15. Model $\boldsymbol{h}^*$
// **Prediction Module**
16. for each $u_i \in \mathcal{U}$ do
17. Calculate the candidate POIs scores according to Equation (23);
18. Calculate the POIs scores according to Equation (24);
19. Rank POIs and select top-**N** individual POIs.

5. Experiment

5.1. Processing of Datasets

In this paper, we use the published Foursquare dataset and Yelp dataset for experiments. Among them, the Foursquare dataset selects the check-in data of New York users from 1 May to 30 June 2014. The Yelp dataset selects the activity data of New York users from 1 August 1 to 30 October 2017. Moreover, we remove inactive users with less than 10 check-in locations and POIs with less than 10 check-ins from the datasets.

Table 2 shows the dataset statistics after preprocessing. In order to make the model proposed in this paper more suitable for the check-in scenario of POIs, we take 80% of the check-in trajectories of each user in the two datasets as the training sets and 20% as the test sets.

Table 2. Statistics of Datasets.

Amount	Foursquare	Yelp
users	48,763	22,754
POIs	18,158	37,879
check-in records	1,287,429	248,5027
social ties	118,421	78,647

According to the research results in related works, the two important factors in the recommendation of the next POIs are the distance and time between POIs. Figure 3a and Figure 3b, respectively, represent the Cumulative Distribution Function (CDF) of the distance between two adjacent POIs checked in by each user in one day on the Foursquare datasets and Yelp datasets. The role of CDF is to help us understand the imbalance of distance distribution and find out which check-in distance accounts for the largest proportion of the total.

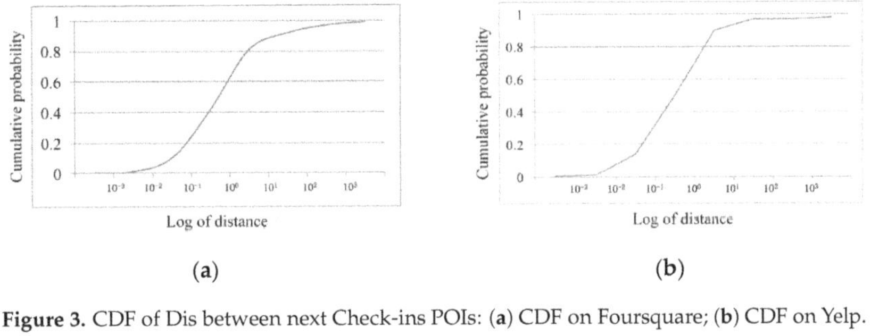

Figure 3. CDF of Dis between next Check-ins POIs: (**a**) CDF on Foursquare; (**b**) CDF on Yelp.

In Figure 4, the horizontal axis represents the check-in distance between POIs, and the vertical axis represents the distance distribution ratio. It can be seen that about 85% of the consecutive check-in distances in the Foursquare and Yelp datasets are within the range of 8.2 km and 7.8 km. It is well known that weekdays and weekends in a week have different effects on check-in locations, so we divide the check-in times in the datasets into two categories: weekdays and weekends. Similarly, in order to observe the influence of the time of day on the check-in location, we divided the time of day into six periods: Early Morning (04:01–08:00), Morning (08:01–12:00), noon (Noon, 12:01–16:00), afternoon (Afternoon, 16:01–20:00), evening (Night, 20:01–24:00) and late night (Wee, 00:01–04:00). The results in Figure 4. show that the distance between consecutive check-in points on weekends is slightly larger than that during weekdays, which means that people are more inclined to go to places with farther distances between POIs on weekends.

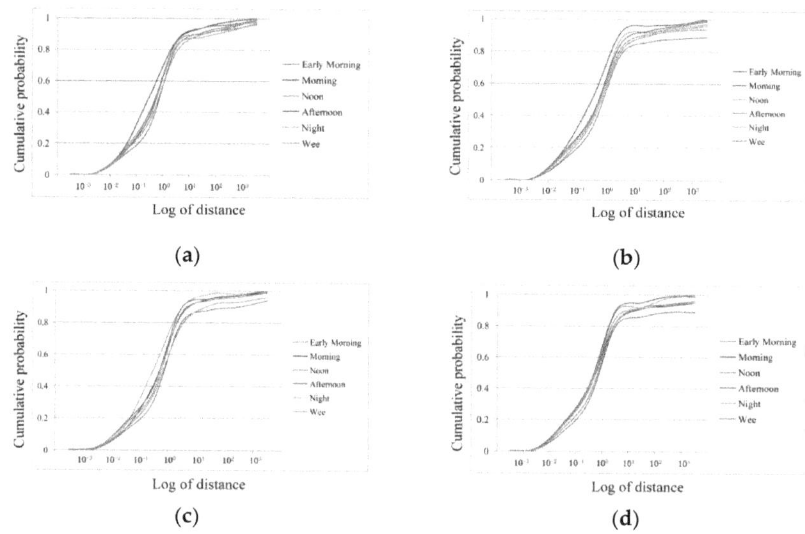

Figure 4. CDF of Dis between next Check-ins POIs on Weekday and Weekend: (**a**) Weekday's CDF on Foursquare; (**b**) Weekday's CDF on Yelp; (**c**) Weekend's CDF on Foursquare; (**d**) Weekend's CDF on Yelp.

5.2. Experimental Evaluation

To evaluate the performance of the recommendation algorithm, we use the Area Under the ROC curve (AUC), Precision and Recall as the evaluation indicators. AUC is a common indicator for evaluating the quality of ranking lists in machine learning, and it is calculated as follows.

$$\mathrm{AUC} = \frac{1}{\mathcal{U}} \sum_{u_i \in U} \frac{1}{|J|\,|J'|} \sum_{j \in |J|} \sum_{j \in |J'|} \delta(p_{u_i,j} > p_{u_i,j'}) \quad (25)$$

In the above formula, J represents the positive sample set, and J' represents the negative sample set, u_i represents the i'th user in the set $\mathcal{U}$. When the indicator function $\delta(p_{u_i,j} > p_{u_i,j'})$ returns true, it means that the predicted probability of u_i acting on positive samples is greater than the predicted probability of negative samples. The opposite is true if the instructed function $\delta(p_{u_i,j} > p_{u_i,j'})$ returns false. The higher the AUC, the stronger the ranking ability of the model.

Precision and Recall are used to compare the prediction results of all algorithms' bias sizes.

The calculation method of Precision@**N** is as follows.

$$\mathrm{Precision@}N = \frac{1}{|\mathcal{U}|} \sum_{u_i \in \mathcal{U}} \frac{\left|\mathcal{P}_{u_i}^{\mathrm{T}} \cap \mathcal{P}_{u_i}^{\mathrm{R}}\right|}{N} \quad (26)$$

The calculation method of Recall@**N** is as follows:

$$\mathrm{Recall@}N = \frac{1}{|\mathcal{U}|} \sum_{u_i=1}^{|\mathcal{U}|} \frac{\left|\mathcal{P}_{u_i}^{\mathrm{T}} \cap \mathcal{P}_{u_i}^{\mathrm{R}}\right|}{P_{u_i}^{\mathrm{T}}} \quad (27)$$

In the above formula, u_i represents the i'th user in the set $\mathcal{U}$, $\mathcal{P}_{u_i}^{\mathrm{R}}$ represents the set of POIs recommended to u_i in the training set, and $P_{u_i}^{\mathrm{T}}$ represents the set of POIs that u_i has checked in in the test set. **N** represents the number of test instances. Moreover, the higher the Precision@**N** and Recall@**N**, the more accurate and comprehensive the recommendation results are.

5.3. Compared Models

This paper compares the CGTS-HAN model proposed in this paper with the following four state-of-the-art models.

(1) Geo-Teaser [13]: Geo-Teaser introduces geographic factors based on the temporal POIs embedding model, which can capture the context of check-in sequences and various temporal features of different dates and establish a geographic-level preference ranking model.
(2) HST-LSTM [15]: This model combines the time factor with LSTM and adopts a hierarchical architecture to predict the next location by utilizing user historical check-in information.
(3) CSAN [19]: CSAN is a multi-modal content unified framework based on an attention mechanism, which projects users' heterogeneous behaviors into a common latent semantic space and then inputs the output results into a feature self-attention network to capture the polysemy of user behaviors.
(4) DeepFM [25]: DeepFM is a new neural network structure that combines the recommendation ability of factorization machines and the feature learning ability of deep learning.

Among these methods, CSAN and DeepFM do not introduce temporal effects, while Geo-Teaser and HST-LSTM do not introduce neural networks to learn user behavior features. The parameters of each method are set as follows:

(1) The weight of time and geographical influence in (2) is set to 0.5; (2), (3) and (4) the initial learning rate of the deep model is set to 0.01. In (2), the context window length and vector size of the hierarchical softmax algorithm are set to 8 and 200, respectively. The parameter Settings involved in the CGTS-HAN model proposed in this paper are the same as the above comparison methods, and CGTS-HAN uses three layers in the context iteration layer.

5.4. Experimental Results and Analysis

5.4.1. Analysis of the Effect

Table 3 shows the AUC experimental results of different models on the two datasets. It can be seen that the AUC of the CGTS-HAN model is consistently higher than all baseline models. The AUC of CGTS-HAN on both datasets is at least 21.4% and 14.5% higher than other models.

Table 3. Experimental Results of AUC.

Methods	Foursquare	Yelp
Geo-Teaser	0.422	0.473
HST-LSTM	0.498	0.520
CSAN	0.0584	0.0587
DeepFM	0.621	0.704
CGTS-HAN	0.835	0.849

5.4.2. Analysis of Effectiveness

This paper conduct experiments with different values of the recommendation list size and extracts the experimental results of N = {5,8,15,30} as samples. The experimental results of precision and recall are shown in Figures 5 and 6.

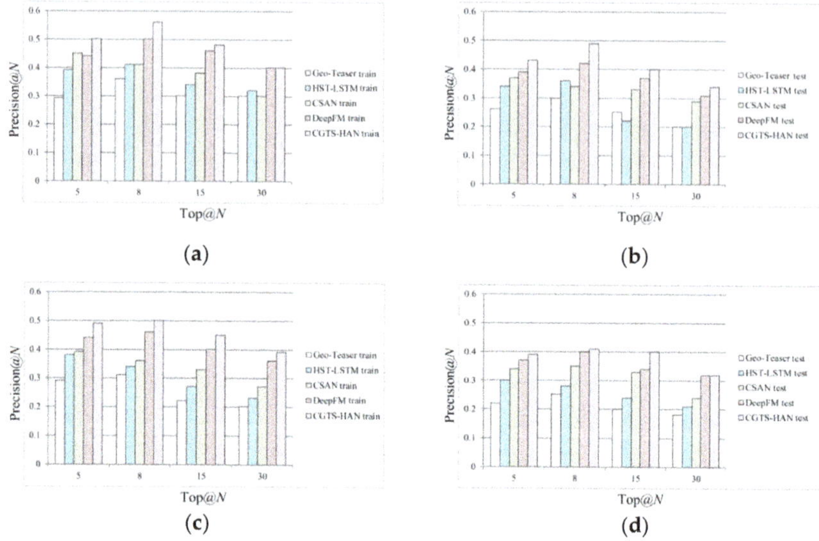

Figure 5. Comparison Results of Precision with N: (**a**) Precision of train on Foursquare; (**b**) Precision of test on Foursquare; (**c**) Precision of train on Yelp; (**d**) Precision of test on Yelp.

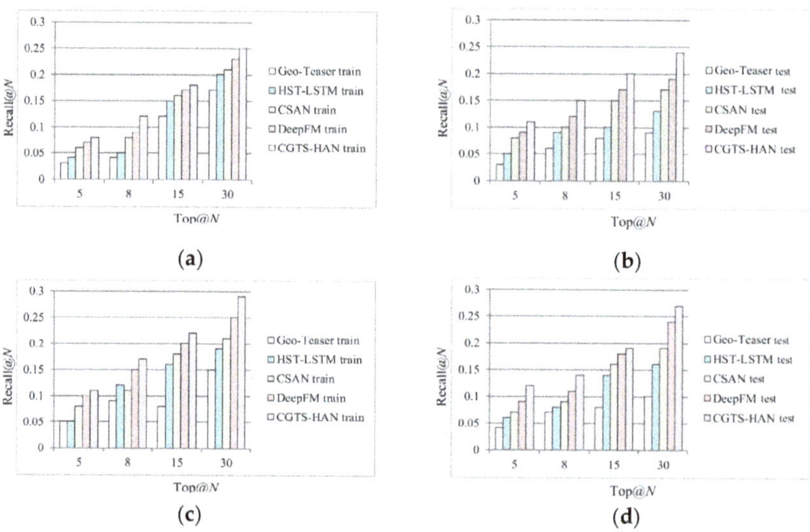

Figure 6. Comparison Results of Recall with N: (**a**) Recall of train on Foursquare; (**b**) Recall of test on Foursquare; (**c**) Recall of train on Yelp; (**d**) Recall of test on Yelp.

It can be seen from Figure 5 that, with the increase of N, the Precision of the rest of the models except the HST-LSTM model increases first and then gradually decreases. Moreover, it can see from Figure 6 that the Recall rate of all models increases with the increase of N. To balance the effects of precision and recall, we select the top 15 POIs for recommendation in the experiment. In this case, the precision of the proposed CGTS-HAN model on the Foursquare and Yelp datasets is 18% and 6% higher than that of DeepFM, and the Recall is 10% and 6% higher, respectively. This verifies the good effect of CGTS-HAN integrating sequence, geographic, time and social influencing factors. Experiments demonstrate that adding auxiliary information of social connections helps improve POIs recommendation performance and, at the same time, confirms the effectiveness of modeling with attention networks.

5.4.3. Parameter Analysis

In this experiment, the parameter sensitivities of time influence weight coefficient α, geographical influence weight coefficient β and learning rate η were quantitatively analyzed by the control variable method. This part of the parameter analysis uses the preprocessed data set and does not divide the training set and the test set.

From Figure 7, we can see how the parameters affect the precision and recall of CGTS-HAN. In order not to tune A, this paper integrates η into α and β in the tuning process (that is, $\alpha \times \eta \to \alpha$, $\beta \times \eta \to \beta$.) and then tunes the two parameters of α and β to weigh the influence of time and geographical factors. In order to ensure convergence, we make the values of α and β as small as possible in the experiment. At the beginning of the tuning, we assume that the values of α and η are equal and change the learning rate by tuning α. The experimental results show that when α increases, the precision and recall rate of the experimental model maintain an upward trend as a whole, and the growth rate slows down as the coefficient value increases. When α is equal to 0.05, the precision and recall performance of the model in this paper is balanced and reaches the best. Therefore, we assume that α is equal to 0.05 and remains unchanged by tuning the β value to observe the changes of $\frac{\beta}{\alpha}$ in the precision and recall rate of the model in the [0, 2] interval. Figure 7 shows that the CGTS-HAN can achieve the best recommendation performance when $\frac{\beta}{\alpha} \in [0.5, 1]$.

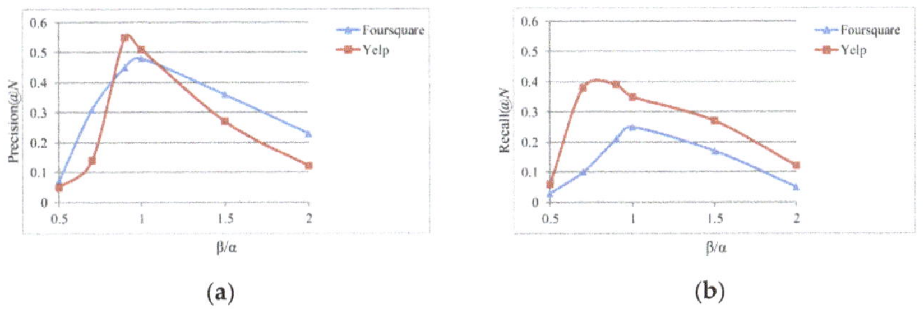

Figure 7. Parameter effect on α and β: (**a**) Precision on CGTS-HAN; (**b**) Recall of on CGTS-HAN.

5.4.4. Time Complexity Analysis

Complexity. For one check-in trajectory, learning the temporal embedding model costs $O(T{\cdot}K{\cdot}d)$, where T, K and d denote the context window size, the number of negative samples and the latent vector dimension, respectively. For the dynamic preference (u_i^{dp}) learning in Algorithm 1, we sample m unvisited POIs, which can generate maximum $\boldsymbol{O}(m^2)$. For each trajectory, the learning procedures cost $\boldsymbol{O}(\boldsymbol{m}^2{\cdot}\boldsymbol{d})$. Therefore, the complexity of CGTS-HAN is $\boldsymbol{O}(\boldsymbol{m}^2{\cdot}\boldsymbol{d}{\cdot}\mathcal{T}r)$, where $\mathcal{T}r$ denote the user trajectory length.

Scalability. Generally, POIs are sparse in LBSN, and the number of check-in POIs is greater than the number of unchecked POIs. Furthermore, in order to make the model more efficient, we use Negative Sampling for optimization. The calculation time of each iteration of Equation (21) is approximately $\boldsymbol{O}((\mathcal{T}r{\cdot}\boldsymbol{K})|\mathcal{X}|)$, where $\boldsymbol{K}$ is the number of negative samples, and $|\mathcal{X}|$ is the training set size. In fact, the values of $\mathcal{T}r$ and $\boldsymbol{K}$ both satisfy a relation ($\mathcal{T}r, \boldsymbol{K} << |\mathcal{X}|$). Therefore, as the check-in trajectories distribution of POIs in LBSNs follows PLF, the time complexity of CGTS-HAN proposed in this paper is linearly related to the number of check-in training sets $|\mathcal{X}|$, which also guarantees that CGTS-HAN to parallel the parameter updates and scalable on large datasets.

6. Conclusions and Future Directions

This work studies the use of contextual information and interest preferences of users' location-based social networks to recommend the next POI for users in the IoT environment. This paper proposed a new next POI recommendation model named CGTS-HAN, which can learn the contextual features of users' POI more accurately than other models. The model aims to recommend the next POI according to the user's historical trajectory, integrates the user's current geographical location and time and considers its social influence to model the user's dynamic preferences. In particular, the model can infer users' interest preferences based on their social information to provide high-quality recommendations in the IoT environment when users have little or no activity history. Experiments are conducted on two datasets based on geo-social information, and the results demonstrate the effectiveness and efficiency of the method.

At present, the CGTS-HAN model uses offline training data, and future work will consider developing effective online training data to improve the accuracy of recommendation results. At the same time, sophisticated distributed representation methods can be developed to improve the next POI recommendation task.

Author Contributions: Conceptualization, X.L. and J.G.; methodology, X.L.; software, X.L. and J.G.; validation, X.L., J.G. and P.Q.; formal analysis, X.L.; investigation, J.G. and P.Q.; resources, P.Q.; data curation, X.L.; writing—original draft preparation, X.L. and J.G., J.G. and P.Q.; writing—review and editing, X.L., J.G. and P.Q.; visualization, J.G. and P.Q.; supervision, P.Q.; project administration, X.L. All authors have read and agreed to the published version of the manuscript.

Funding: This research received no external funding.

Data Availability Statement: Data sharing not applicable. No new data were created or analyzed in this study. Data sharing is not applicable to this article.

Conflicts of Interest: The authors declare no conflict of interest.

References

1. Zhang, J.D.; Chow, C.Y.; Li, Y. Lore: Exploiting sequential influence for location recommendations. In Proceedings of the 22nd ACM SIGSPATIAL International Conference on Advances in Geographic Information Systems, Dallas, TX, USA, 4–7 November 2014; pp. 103–112.
2. Liu, X.; Liu, Y.; Aberer, K.; Miao, C. Personalized point-of-interest recommendation by mining users' preference transition. In Proceedings of the 22nd ACM international conference on Information & Knowledge Management, San Francisco, CA, USA, 27 October–1 November 2013; pp. 733–738.
3. Cheng, C.; Yang, H.; Lyu, M.R.; King, I. Where you like to go next: Successive point-of-interest recommendation. In Proceedings of the Twenty-Third international joint conference on Artificial Intelligence, Beijing, China, 3–9 August 2013; pp. 2605–2611.
4. Quba RC, A.; Hassas, S.; Fayyad, U.; Alshomary, M.; Gertosio, C. iSoNTRE: The Social Network Transformer into Recommendation Engine. In Proceedings of the 2014 IEEE/ACS 11th International Conference on Computer Systems and Applications (AICCSA), Doha, Qatar, 10–13 November 2014; pp. 169–175.
5. Wen, Y.; Zhang, J.; Zeng, Q.; Chen, X.; Zhang, F. *Loc2Vec-Based Cluster-Level Transition Behavior Mining for Successive POI Recommendation*; IEEE Access: Piscataway, NJ, USA, 2019; Volume 7, pp. 109311–109319.
6. Sarwat, M.; Mokbel, M.F. Differentially Private Location Recommendations in Geosocial Networks. In Proceedings of the IEEE 15th International Conference on Mobile Data Management, Brisbane, QLD, Australia, 14–18 July 2014; pp. 59–68.
7. Debnath, M.; Tripathi, P.K.; Elmasri, R. Preference-Aware Successive POI Recommendation with Spatial and Temporal Influence. In *International Conference on Social Informatics*; Springer: Berlin/Heidelberg, Germany, 2016; Volume 10046, pp. 347–360.
8. Xiong, L.; Chen, X.; Huang, T.K.; Schneider, J.; Carbonell, J.G. Temporal collaborative filtering with bayesian probabilistic tensor factorization. In Proceedings of the 2010 SIAM International Conference on Data Mining. Society for Industrial and Applied Mathematics, Columbus, OH, USA, 29 April–1 May 2010; pp. 211–222.
9. Feng, S.; Li, X.; Zeng, Y.; Cong, G.; Chee, Y.W.; Yuan, Q. Personalized ranking metric embedding for next new poi recommendation. In Proceedings of the Twenty-Fourth International Joint Conference on Artificial Intelligence, Buenos Aires, Argentina, 25–31 July 2015; pp. 2069–2075.
10. Rendle, S.; Freudenthaler, C.; Schmidt-Thieme, L. Factorizing personalized markov chains for next-basket recommendation. In Proceedings of the 19th International Conference on World Wide Web, Raleigh, NC, USA, 26-30 April 2010; pp. 811–820.
11. Feng, S.; Cong, G.; An, B.; Chee, Y.M. Poi2vec: Geographical latent representation for predicting future visitors. In Proceedings of the Thirty-First AAAI Conference on Artificial Intelligence, San Francisco, CA, USA, 4–9 February 2017; pp. 102–108.

12. Liu, X.; Liu, Y.; Li, X. Exploring the Context of Locations for Personalized Location Recommendations. In Proceedings of the Twenty-Fifth International Joint Conference on Artificial Intelligence, New York, NY, USA, 9–15 July 2016; pp. 1188–1194.
13. Zhao, S.; Zhao, T.; King, I.; Lyu, M.R. Geo-teaser: Geo-temporal sequential embedding rank for point-of-interest recommendation. In Proceedings of the 26th International Conference on World Wide Web Companion, Perth, Australia, 3–7 April 2017; pp. 153–162.
14. Liu, Q.; Wu, S.; Wang, L.; Tan, T. Predicting the next location: A recurrent model with spatial and temporal contexts. In Proceedings of the Thirtieth AAAI Conference on Artificial Intelligence, Phoenix, AZ, USA, 12–17 February 2016; pp. 194–200.
15. Kong, D.; Wu, F. HST-LSTM: A hierarchical spatial-temporal long-short term memory network for location prediction. In Proceedings of the Twenty-Seventh International Joint Conference on Artificial Intelligence, Stockholm, Sweden, 13–19 July 2018; Volume 18, pp. 2341–2347.
16. Li, C.; Li, D.; Zhang, Z.; Chu, D. MST-RNN: A Multi-Dimension Spatiotemporal Recurrent Neural Networks for Recommending the Next Point of Interest. *Mathematics* **2022**, *10*, 1838. [CrossRef]
17. Li, R.; Shen, Y.; Zhu, Y. Next point-of-interest recommendation with temporal and multi-level context attention. In Proceedings of the 2018 IEEE International Conference on Data Mining (ICDM), Singapore, 17–20 November 2018; pp. 1110–1115.
18. Liu, T.; Liao, J.; Wu, Z.; Wang, Y.; Wang, J. A geographical-temporal awareness hierarchical attention network for next point-of-interest recommendation. In Proceedings of the 2019 on International Conference on Multimedia Retrieval, Ottawa, ON, Canada, 10–13 June 2019; pp. 7–15.
19. Huang, X.; Qian, S.; Fang, Q. Csan: Contextual self-attention network for user sequential recommendation. In Proceedings of the 26th ACM International Conference on Multimedia, Seoul, Republic of Korea, 22–26 October 2018; ACM: New York, NY, USA, 2018; pp. 447–455.
20. Xie, Y.; Zhao, J.; Qiang, B.; Mi, L.; Tang, C.; Li, L. Attention mechanism-based CNN-LSTM model for wind turbine fault prediction using SSN ontology annotation. *Wirel. Commun. Mob. Comput.* **2021**, *2021*, 6627588. [CrossRef]
21. Ojagh, S.; Malek, M.R.; Saeedi, S.; Liang, S. An Internet of Things (IoT) Approach for Automatic Context Detection. In Proceedings of the 2018 IEEE 9th Annual Information Technology, Electronics and Mobile Communication Conference (IEMCON), Vancouver, BC, Canada, 1–3 November 2018; pp. 223–226.
22. Wang, X.; Liu, Y.; Zhou, X.; Wang, X.; Leng, Z. A Point-of-Interest Recommendation Method Exploiting Sequential, Category and Geographical Influence. *ISPRS Int. J. Geo-Inf.* **2022**, *11*, 80. [CrossRef]
23. Yang, X.; Zimba, B.; Qiao, T.; Gao, K.; Chen, X. Exploring IoT location information to perform point of interest recommendation engine: Traveling to a new geographical region. *Sensors* **2019**, *19*, 992. [CrossRef] [PubMed]
24. Wang, H.; Shen, H.; Ouyang, W.; Cheng, X. Exploiting POI-Specific Geographical Influence for Point-of-Interest Recommendation. In Proceedings of the Twenty-Seventh International Joint Conference on Artificial Intelligence, Stockholm, Sweden, 13–19 July 2018; pp. 3877–3883.
25. Guo, H.; Tang, R.; Ye, Y.; Li, Z.; He, X. DeepFM: A factorization-machine based neural network for CTR prediction. In Proceedings of the Twenty-Sixth International Joint Conference on Artificial Intelligence(IJCAI-17), Melbourne, Australia, 19–25 August 2017; pp. 1725–1731.
26. Li, J.; Sellis, T.; Culpepper, J.S.; He, Z.; Liu, C.; Wang, J. Geo-social influence spanning maximization. In *IEEE Transactions on Knowledge and Data Engineering*; IEEE: Piscataway, NJ, USA, 2017; Volume 29, pp. 1653–1666.
27. Haldar, N.; Li, J.; Reynolds, M.; Sellis, T.; Yu, J.X. Location prediction in large-scale social networks: An in-depth benchmarking study. *VLDB J.* **2019**, *5*, 623–648. [CrossRef]
28. Ye, M.; Yin, P.; Lee, W.C.; Lee, D.-L. Exploiting geographical influence for collaborative point-of-interest recommendation. In Proceedings of the 34th International ACM SIGIR Conference on Research and Development in Information Retrieval, Beijing, China, 24–28 July 2016; pp. 325–334.
29. Zhou, P.; Shi, W.; Tian, J.; Qi, Z.; Li, B.; Hao, H.; Xu, B. Attention-based bidirectional long short-term memory networks for relation classification. In Proceedings of the 54th Annual Meeting of the Association for Computational Linguistics, Berlin, Germany, 7–12 August 2016; Volume 2, pp. 207–212.

electronics

MDPI

Article

Cost-Sensitive Multigranulation Approximation in Decision-Making Applications

Jie Yang [1,2], Juncheng Kuang [2], Qun Liu [2] and Yanmin Liu [1,*]

[1] School of Physics and Electronic Science, Zunyi Normal University, Zunyi 563002, China
[2] Chongqing Key Laboratory of Computational Intelligence, Chongqing University of Posts and Telecommunications, Chongqing 400065, China
[*] Correspondence: yanmin7813@163.com; Tel.: +86-1508-606-1907

Abstract: A multigranulation rough set (MGRS) model is an expansion of the Pawlak rough set, in which the uncertain concept is characterized by optimistic and pessimistic upper/lower approximate boundaries, respectively. However, there is a lack of approximate descriptions of uncertain concepts by existing information granules in MGRS. The approximation sets of rough sets presented by Zhang provide a way to approximately describe knowledge by using existing information granules. Based on the approximation set theory, this paper proposes the cost-sensitive multigranulation approximation of rough sets, i.e., optimistic approximation and pessimistic approximation. Their related properties were further analyzed. Furthermore, a cost-sensitive selection algorithm to optimize the multigranulation approximation was performed. The experimental results show that when multigranulation approximation sets and upper/lower approximation sets are applied to decision-making environments, multigranulation approximation produces the least misclassification costs on each dataset. In particular, misclassification costs are reduced by more than 50% at each granularity on some datasets.

Keywords: multigranulation rough sets; optimistic approximation; pessimistic approximation; cost-sensitive; decision-making applications

Citation: Yang, J.; Kuang, J.; Liu, Q.; Liu, Y. Cost-Sensitive Multigranulation Approximation in Decision-Making Applications. *Electronics* **2022**, *11*, 3801. https://doi.org/10.3390/electronics11223801

Academic Editor: Domenico Ursino

Received: 25 October 2022
Accepted: 16 November 2022
Published: 18 November 2022

1. Introduction

As a human-inspired paradigm, granular computing (GrC) solves complex problems by utilizing multiple granular layers [1–4]. Zadeh [1] noted that information in granules refer to pieces, classes, and groups, into which complex information is divided in accordance with the characteristics and processes of understanding and decision-making. From the different views, GrC models mainly cover four types: fuzzy sets [5], rough sets [6], quotient spaces [7], and cloud models [8]. As representative models of GrC, rough sets describe uncertain concepts by upper and lower approximation boundaries, which have been applied to data mining [9,10], medical systems [11], attribute reductions [12,13], decision systems [14,15], and machine learning [16].

Regarding similarity, Zhang [17–20] presented the approximation set of rough sets, vague sets, rough fuzzy sets, rough vague sets, etc. These approximation models were developed by utilizing the existing equivalence classes to describe uncertain concepts. The approximation model has a higher similarity with the target concept than the upper/lower approximations. Furthermore, the approximation model has been applied in attribute reduction [21], image segmentation [22], the optimization algorithm [23], etc. Based on the approximation set theory, Yang [24,25] developed the approximation model of rough sets based on misclassification costs. In the process of cost-sensitive learning, the smaller misclassification costs will help to improve the decision-making qualities in real applications. Recently, from the perspective of three-way decisions [26–29], Yao [30] constructed a symbol–meaning–value (SMV) model for data analysis. In the three-way decision model, the equivalence classes in a boundary region will produce misclassification costs when they are used as approximation sets. Hence, the approximation model that is constructed from the perspective of similarity is no longer applicable to cost-sensitive scenarios. To minimize the misclassification costs of constructing the approximation set, we proposed the multigranulation approximation, i.e., the optimistic approximation model and pessimistic approximation model. Moreover, to search the optimal approximation layer for multigranulation rough sets [31] under the constraints, the algorithm of the cost-sensitive multigranulation approximation selection is further proposed to be applied to decision-making environments.

The following sections are arranged as follows: Section 2 presents the related works. Section 3 introduces the relevant definitions of the multigranulation rough set and approximation set. Section 4 introduces an approximate representation of the rough sets. Section 5 presents the cost-sensitive multigranulation approximations of rough sets and further introduces the optimal multigranulation approximation algorithm. To verify the availability of the proposed model, the related experiments and discussion are presented in Section 6. Ultimately, in Section 7, the conclusions are presented.

2. Literature Review

Rough sets are typically constructed based on a single binary relation. However, in many cases, they may be described in multiple granularity structures. In order to extend single granularity to multi-granularity in rough approximations, Qian [31] proposed the multigranulation rough set model (MGRS), where the upper/lower approximations were defined by multi-equivalence relations (multiple granulations) in the universe [32,33]. For the lower approximation of optimistic MGRS, at least one granular space was obtained, such that objects completely belonged to the target concept. For the lower approximation of pessimistic MGRS, objects completely belong to target concepts in each granular space. MGRS has two advantages: (1) In the process of decision-making applications, the decision of each decision maker may be independent of the same project (or an element) in the universe [34]. In this situation, the intersection operations between any two granularity structures will be redundant for decision-making [35]. (2) Extract decision rules from distributive information systems and groups of intelligent agents by using rough set approaches [34,36].

There are many works [33–35,37–42] on multigranulation rough sets. To extend the MGRS to the neighborhood information system, Hu [43,44] presented matrix-based incremental approaches to update knowledge about neighborhood information systems by changing the granular structures. From the perspective of uncertainty measure, Sun [39] proposed a feature selection based on fuzzy neighborhood multigranulation rough sets. Xu [38] proposed a dynamic approximation update mechanism of a multigranulation neighborhood rough set from a local viewpoint. Liu [35] introduced a parameter-free multi-granularity attribute reduction scheme, which is more effective for microarray data than other well-established attribute reductions. Based on the three-way decision theory, She [40] presented a five-valued logic approach for the multigranulation rough set model. From the above, however, the method of approximately describing the target concept with existing information granules is not given, which limits the application of the multigranulation rough set theory. Li [41] presented two kinds of local multigranulation rough set models in the ordered decision system by extending the single granulation environment to a multigranulation case. Zhang [42] constructed hesitant fuzzy multigranulation rough sets to handle the hesitant fuzzy information and group decision-making for person–job fit.

3. Preliminaries

In this section, some necessary definitions related to the multigranulation rough set and approximation set are reviewed to facilitate the framework of this paper. Let $S = (U, C \cup D, V, f)$ be a decision information table, where U is a non-empty finite set of objects, C is a non-empty finite set of condition attributes, D is a set of decision attributes, V is the set of all attribute values, and $f : U \times C$ is an information function.

Definition 1 ((Rough Sets) [6]). *Let $S = (U, C \cup D, V, f)$ be a decision information table, $A \subseteq C$ and $X \subseteq U$, the lower and upper approximation sets of X are given as follows:*

$$\underline{A}(X) = \{x \in U | [x]_A \subseteq X\},$$
$$\overline{A}(X) = \{x \in U | [x]_A \cap X \neq \phi\}.$$

where $[x]_A$ denotes the equivalence class induced by U/A, namely, $U/A = \{[x]_R\} = \{[x]_1, [x]_2, \cdots, [x]_l\}$.

Based on the lower and upper approximations, the universe U can be divided into three disjoint regions, which are expressed as follows:

$$POS_A(X) = \underline{A}(X),$$
$$BND_A(X) = \overline{A}(X) - \underline{A}(X),$$
$$NEG_A(X) = U - \overline{A}(X).$$

Definition 2 ((Optimistic multigranulation rough sets) [36]). *Let $S = (U, C \cup D, V, f)$ be a decision information table, $A_1, A_2, \ldots, A_m \subseteq C$ and $X \subseteq U$, then the lower and upper approximation sets of X related to $A_1, A_2, \ldots, A_m$ are given as follows:*

$$\underline{\sum_{i=1}^{m} A_i^O}(X) = \{x | [x]_{A_1} \subseteq X \vee [x]_{A_2} \subseteq X \vee \ldots \vee [x]_{A_m} \subseteq X, x \in U\}, \tag{1}$$

$$\overline{\sum_{i=1}^{m} A_i^O}(X) = \sim \underline{\sum_{i=1}^{m} A_i^O}(\sim X). \tag{2}$$

Then, $(\underline{\sum_{i=1}^{m} A_i^O}(X), \overline{\sum_{i=1}^{m} A_i^O}(X))$ is called optimistic multigranulation rough sets. The lower and upper approximation sets of X in optimistic multigranulation rough sets are presented by multiple independent approximation spaces. The boundary regions are defined as follows:

$$BND^O_{\sum_{i=1}^{m} A_i}(X) - \overline{\sum_{i=1}^{m} A_i^O}(X) - \underline{\sum_{i=1}^{m} A_i^O}(X). \tag{3}$$

Definition 3 ((Pessimistic multigranulation rough sets) [36]). *Let $S = (U, C \cup D, V, f)$ be a decision information table, $A_1, A_2, \ldots, A_m \subseteq C$, and $X \subseteq U$. The lower and upper approximation sets of X related to $A_1, A_2, \ldots, A_m$ are given as follows:*

$$\underline{\sum_{i=1}^{m} A_i^P}(X) = \{x | [x]_{A_1} \subseteq X \wedge [x]_{A_2} \subseteq X \wedge \ldots \wedge [x]_{A_m} \subseteq X, x \in U\}, \tag{4}$$

$$\overline{\sum_{i=1}^{m} A_i^P}(X) = \sim \underline{\sum_{i=1}^{m} A_i^P}(\sim X). \tag{5}$$

Then, $(\underline{\sum_{i=1}^{m} A_i^P}(X), \overline{\sum_{i=1}^{m} A_i^P}(X))$ is called pessimistic multigranulation rough sets. The lower and upper approximation sets of X in pessimistic multigranulation rough sets are presented by *multiple independent approximation spaces. However, the strategy is different from optimistic multigranulation rough sets. The boundary region is defined as follows:*

$$BND^{P}_{\sum_{i=1}^{m} A_i}(X) = \overline{\sum_{i=1}^{m} A_i^P(X)} - \underline{\sum_{i=1}^{m} A_i^P(X)}. \tag{6}$$

Definition 4 ((Approximation of rough sets) [17]). *Let $S = (U, C \cup D, V, f)$ be a decision information table, $A \subseteq C$ and $X \subseteq U$. $U/A = \{[x]_1, [x]_2, \cdots, [x]_l\}$ is a granularity layer on U, then the α-approximation of X on U/A is defined as follows:*

$$A_\alpha(X) = \cup\{[x]_i | \overline{\mu}([x]_i) \geq \alpha, [x]_i \subseteq U\}. \tag{7}$$

where $0 \leq \alpha \leq 1$, $\overline{\mu}([x]_i) = \frac{|[x]_i \cap X|}{|[x]_i|}$ denotes the membership degree of the equivalence class $[x]_i$ belongs to X.

Example 1. *Let Table 1 be a decision information table, $A_1, A_2, A_3 \subseteq C$ and $X \subseteq U$. For the element x_4, the equivalence classes $[x_4]_i (i = 1, 2, 3)$ belonging to X in the multigranulation approximation space are as follows:*

$$[x_4]_1 = \{x_1, x_2, x_3, x_4\};$$
$$[x_4]_2 = \{x_3, x_4, x_7, x_8\};$$
$$[x_4]_3 = \{x_2, x_4, x_6, x_8\}.$$

Accordingly, the membership degrees are computed:

$$\overline{\mu}([x_4]_1) = \frac{0+0+0+1}{4} = 0.25;$$
$$\overline{\mu}([x_4]_2) = \frac{0+1+1+1}{4} = 0.75;$$
$$\overline{\mu}([x_4]_3) = \frac{0+1+1+1}{4} = 0.75.$$

Table 1. A decision information table.

	A_1	A_2	A_3	X
x_1	0	0	0	0
x_2	0	0	1	0
x_3	0	1	0	0
x_4	0	1	1	1
x_5	1	0	0	0
x_6	1	0	1	1
x_7	1	1	0	1
x_8	1	1	1	1

If α is set to 0.5, considering the optimistic approximation, element x_4 will be classified into the optimistic lower approximation sets of X due to one of its membership degrees being greater than 0.5. However, if considering the pessimistic approximation, element x_4 will only be classified into the pessimistic upper approximation sets of X.

Based on the given conditions, we have:

$$X = \frac{0.25+0.25+0.25}{x_1} + \frac{0.25+0.25+0.75}{x_2} + \frac{0.25+0.75+0.05}{x_3} + \frac{0.25+0.75+0.75}{x_4} + \frac{0.75+0.25+0.25}{x_5} + \frac{0.75+0.25+0.75}{x_6} + \frac{0.75+0.75+0.25}{x_7} + \frac{0.75+0.75+0.75}{x_8}.$$

Then, the results of the optimistic approximations are shown as follows:

$$\underline{\sum_{i=1}^{m} A_i^O}(X) = \{x_2, x_3, x_4, x_5, x_6, x_7, x_8\},$$
$$\overline{\sum_{i=1}^{m} A_i^O}(X) = \{x_1, x_2, x_3, x_4, x_5, x_6, x_7, x_8\},$$
$$BND^O_{\sum_{i=1}^{m} A_i}(X) = \{x_1\}.$$

Moreover, the results of the pessimistic approximations, in this case, are changed as follows:

$$\underline{\sum_{i=1}^{m} A_i^P}(X) = \{x_8\},$$
$$\overline{\sum_{i=1}^{m} A_i^P}(X) = \{x_1, x_2, x_3, x_4, x_5, x_6, x_7, x_8\},$$
$$BND^P_{\sum_{i=1}^{m} A_i}(X) = \{x_1, x_2, x_3, x_4, x_5, x_6, x_7\}.$$

4. Cost-Sensitive Approximation Methods of the Rough Sets

Let $S = (U, C \cup D, V, f)$ be a decision information table, $A \subseteq C$ and $X \subseteq U$. $U/A = \{[x]_1, [x]_2, \cdots, [x]_l\}$ is a granularity layer on U. λ_{12} represents the cost generated by taking an element belonging to X as the approximation, λ_{21} represents the cost generated by taking an element that does not belong to X as the approximation. Furthermore, misclassification costs incurred by the equivalence classes in characterizing X on U/A are given in the following:

$$\lambda^Y = \lambda_{12}(1 - \overline{\mu}([x]_i))|[x]_i|. \tag{8}$$

Misclassification costs incurred by the equivalence classes when not characterizing X on U/A are given in the following:

$$\lambda^N = \lambda_{21}\overline{\mu}([x]_i)|[x]_i|. \tag{9}$$

Herein, $\overline{\mu}([x]_i)$ $(i = 1, 2, \ldots, l)$ denotes the membership degree of $[x]_i$ belonging to X. Based on the Bayesian decision procedure, the minimum cost decision rules are expressed as follows:

(P) If $\lambda^Y \leq \lambda^N$, $[x]_i \subseteq A(X)$;
(N) If $\lambda^Y > \lambda^N$, $[x]_i \not\subseteq A(X)$.

It is clear that the above rules are only relevant to the loss function $\overline{\mu}([x]_i)$. From Formulas (8) and (9), the decision rules are re-expressed in the following:

(P1) If $\overline{\mu}([x]_i) \geq \frac{\lambda_{12}}{\lambda_{12}+\lambda_{21}}$, $[x]_i \subseteq A(X)$;
(N1) If $\overline{\mu}([x]_i) < \frac{\lambda_{12}}{\lambda_{12}+\lambda_{21}}$, $[x]_i \not\subseteq A(X)$.

Supposing $\gamma = \frac{\lambda_{12}}{\lambda_{12}+\lambda_{21}}$, then we have the following decision rules:

(P2) If $\overline{\mu}([x]_i) \geq \gamma$, $[x]_i \subseteq A(X)$;
(N2) If $\overline{\mu}([x]_i) < \gamma$, $[x]_i \not\subseteq A(X)$.

Definition 5. *Let $S = (U, C \cup D, V, f)$ be a decision information table, $A \subseteq C$ and $X \subseteq U$. $U/A = \{[x]_1, [x]_2, \cdots, [x]_l\}$ is a granularity layer on U, then the cost-sensitive approximation model (CSA) of rough sets is defined as follows:*

$$A(X) = \cup\left\{[x]_i \middle| \overline{\mu}([x]_i) \geq \frac{\lambda_{12}}{\lambda_{12}+\lambda_{21}}, [x]_i \subseteq U\right\}. \tag{10}$$

Suppose $0 \le \lambda_{12}, \lambda_{21} \le 1$, boundary region I and boundary region II are denoted by $BN1(X) = \{[x]_i | \frac{\lambda_{12}}{\lambda_{12}+\lambda_{21}} \le \overline{\mu}([x]_i) < 1\}$and $BN2(X) = \{[x]_i | 0 < \overline{\mu}([x]_i) < \frac{\lambda_{12}}{\lambda_{12}+\lambda_{21}}\}$, respectively, then $BN(X) = BN1(X) \cup BN2(X)$, and $A(X) = BN1(X) \cup POS(X)$. Figure 1 shows the CSA of rough sets, where $BN1(X)$ is the dark blue region, which denotes the region in the boundary region used as the approximation. $BN2(X)$ is the light blue region, which denotes the region in the boundary region not used as the approximation. Therefore, the region surrounded by the green broken line in Figure 1 constructs the approximations of rough sets, and the misclassification costs come from two uncertain regions, defined as follows:

$$DC(A(X)) = \sum_{[x]_i \in BN1(X)} \lambda^Y + \sum_{[x]_i \in BN2(X)} \lambda^N. \tag{11}$$

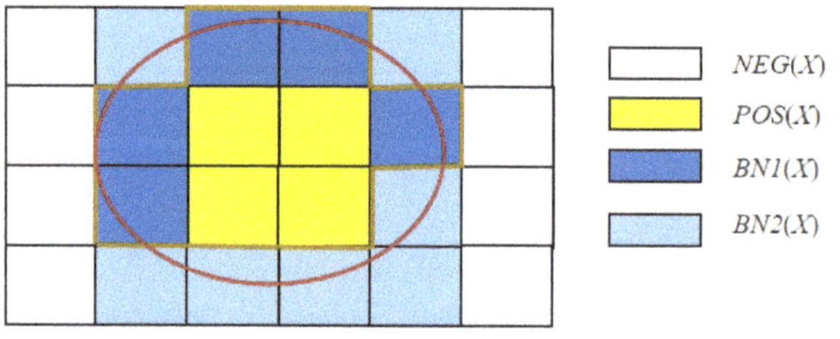

Figure 1. The approximation of rough sets (surrounded by the broken line).

Theorem 1. *Let $S = (U, A \cup D, V, f)$ be a decision information table, $A_1, A_2, \ldots, A_m \subseteq C$, then $DC(A_1(X)) \ge DC(A_2(X))$.*

Proof of Theorem 1. Let U be a non-empty finite domain, $U/A_1 = \{E_1, E_2, \cdots, E_l\}$ and $U/A_2 = \{F_1, F_2, \cdots, F_m\}$. Because $A_1 \subseteq A_2, U/A_2 \preccurlyeq U/A_1$. According to the condition, for simplicity, supposing only one granule E_1 can be subdivided into two finer sub-granules by $\Delta A = A_2 - A_1$ (the more complicated cases can be transformed into this case, so we will not repeat them here). Without loss of generality, let $E_1 = F_1 \cup F_2$, $E_2 = F_3$, $E_3 = F_4, \ldots, E_l = F_m$ $(m = l + 1)$, namely, $U/A_2 = \{F_1, F_2, E_2, E_3, \ldots, E_l\}$. There are two cases to prove the theorem as follows:

(1) Suppose $\bar{\mu}(E_1) \ge \gamma$, obviously, $E_1 \subseteq A_1(X)$.

Case 1. $\bar{\mu}(F_1) \ge \gamma$ and $\bar{\mu}(F_2) \ge \gamma$. Namely, $F_1 \subseteq A(X)$ and $F_2 \subseteq A(X)$. Case 1, in which the granules are subdivided in $BN1(X)$, can be shown in Figure 2a, then

$$\begin{aligned}\Delta DC_{A_1-A_2}(X) &= DC(A_1(X)) - DC(A_2(X)) \\ &= \lambda_{12}(1-\overline{\mu}(E_1))|E_1| - \lambda_{12}(1-\overline{\mu}(F_1))|F_1| - \lambda_{12}(1-\overline{\mu}(F_2))|F_2|. \\ &= (|E_1| - |F_1| - |F_2| + \sum_{x_i \in F_1} \mu(x_i) + \sum_{x_i \in F_2} \mu(x_i) - \sum_{x_i \in E_1} \mu(x_i))\lambda_{12}.\end{aligned}$$

Because $\sum_{x_i \in E_1} \mu(x_i) = \sum_{x_i \in F_1} \mu(x_i) + \sum_{x_i \in F_2} \mu(x_i)$ and $|E_1| = |F_1| + |F_2|$, then $\Delta DC_{A_1-A_2} = 0$. Therefore, $DC(A_1(X)) = DC(A_2(X))$.

Case 2. $\bar{\mu}(F_1) \ge \gamma$ and $\bar{\mu}(F_2) < \gamma$. Namely, $F_1 \subseteq A(X)$ and $F_2 \not\subset A(X)$. Case 2, in which the granules are subdivided in $BN1(X)$, can be shown in Figure 2b, then

$$\begin{aligned}\Delta DC_{A_1-A_2} &= DC(A_1(X)) - DC(A_2(X)) \\ &= \lambda_{12}(1-\overline{\mu}(E_1))|E_1| - \lambda_{12}(1-\overline{\mu}(F_1))|F_1| - \lambda_{21}\overline{\mu}(F_2)|F_2| \\ &= |F_2|(\lambda_{12} - \bar{\mu}(F_2)(\lambda_{21} + \lambda_{12}))\end{aligned}$$

Because $\bar{\mu}(F_2) < \gamma = \frac{\lambda_{12}}{\lambda_{12}+\lambda_{21}}$ and $|E_1| = |F_1| + |F_2|$, then $\Delta DC_{A_1-A_2}(X) \ge 0$. Therefore, $DC(A_1(X)) > DC(A_2(X))$.

(2) Suppose $\bar{\mu}(E_1) < \gamma$, obviously, $E_1 \not\subset A_1(X)$.

Case 1. $\bar{\mu}(F_1) \geq \gamma$ and $\bar{\mu}(F_2) < \gamma$. Namely, $F_1 \subseteq A(X)$ and $F_2 \not\subset A(X)$. Case 1, in which the granules are subdivided in $BN2(X)$, can be shown in Figure 2c, then

$$\begin{aligned}\Delta DC_{A_1-A_2} &= DC(A_1(X)) - DC(A_2(X)) \\ &= \bar{\mu}(E_1)|E_1|\lambda_{21} - \bar{\mu}(F_2)|F_2|\lambda_{21} - (1-\bar{\mu}(F_1))|F_1|\lambda_{12} \\ &= |F_1|\bar{\mu}(F_1)((\lambda_{21}+\lambda_{12}) - \lambda_{12}).\end{aligned}$$

Because $\bar{\mu}(F_1) \geq \gamma = \frac{\lambda_{12}}{\lambda_{12}+\lambda_{21}}$, then $\Delta DC_{A_1-A_2} \geq 0$. Therefore, $DC(A_1(X)) \geq DC(A_2(X))$.

Case 2. $\bar{\mu}(F_1) < \gamma$ and $\bar{\mu}(F_2) < \gamma$. Namely, $F_1 \not\subset R(X)$ and $F_2 \not\subset R(X)$. Case 2, in which the granules are subdivided in $BN2(X)$, can be shown in Figure 2d, then

$$\begin{aligned}\Delta DC_{A_1-A_2} &= DC(A_1(X)) - DC(A_2(X)) \\ &= \bar{\mu}(E_1)|E_1|\lambda_{21} - \bar{\mu}(F_1)|F_1|\lambda_{21} - \bar{\mu}(F_2)|F_2|\lambda_{21} \\ &= (\sum_{x_i \in E_1} \mu(x_i) - \sum_{x_i \in F_1} \mu(x_i) - \sum_{x_i \in F_2} \mu(x_i))\lambda_{21}.\end{aligned}$$

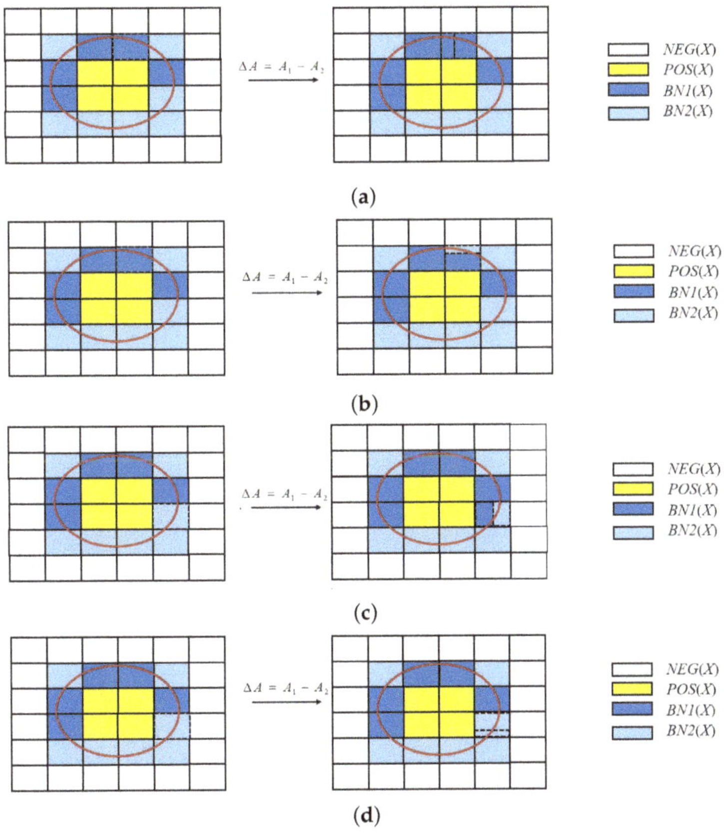

Figure 2. The granules subdivided in $BN1(X)$ and $BN2(X)$ of the cost-sensitive approximation model of rough sets. And all the red circles in the figure represent the set X. In addition, (**a**) shows the case 1 in which the granules are subdivided in $BN1(X)$; (**b**) shows the case 2 in which the granules are subdivided in $BN1(X)$; (**c**) shows the case 1 in which the granules are subdivided in $BN2(X)$; (**d**) shows the case 2 in which the granules are subdivided in $BN2(X)$.

Because $\sum_{x_i \in E_1} \mu(x_i) = \sum_{x_i \in F_1} \mu(x_i) + \sum_{x_i \in F_2} \mu(x_i)$, then $\Delta DC_{A_1 - A_2} = 0$. Therefore, $DC(A_1(X)) = DC(A_2(X))$. □

Theorem 1 shows that the misclassification costs in the approximation model will monotonously decrease with the changing approximation space, in accordance with human cognitive mechanisms.

5. Cost-Sensitive Multigranulation Approximations and Optimal Granularity Selection Method

The multigranulation rough set model (MGRS) [43] extends single granularity to multi-granularity in rough approximations to describe an uncertain concept. MGRS is an expansion of the classical rough set, and the target concept is characterized by optimistic and pessimistic upper/lower approximation boundaries in MGRS, respectively. However, there is a lack of an approximate description of an uncertain concept by utilizing the equivalence classes in MGRS. In this section, based on the model proposed in Section 3, we further construct the approximations of MGRS.

5.1. Cost-Sensitive Multigranulation Approximations of Rough Sets

Definition 6. *Let $S = (U, C \cup D, V, f)$ be a decision information table, $A_1, A_2, \ldots, A_m \subseteq C$ and $X \subseteq U$. The optimistic membership degree of $x \in U$ related to $A_1, A_2, \ldots, A_m$ is given as follows:*

$$\overline{\mu}^{O}_{\sum_{i=1}^{m} A_i}(x) = \max\{\overline{\mu}([x]_{A_i}) | i = 1, 2, \ldots, m\}. \tag{12}$$

Definition 7. *Let $S = (U, C \cup D, V, f)$ be a decision information table, $A_1, A_2, \ldots, A_m \subseteq C$ and $X \subseteq U$. The approximation model of the optimistic MGRS of X related to $A_1, A_2, \ldots, A_m$ is given as follows:*

$$\sum_{i=1}^{m} A_i^{O}(X) = \{x | \overline{\mu}([x]_{A_1}) \geq \gamma \vee \overline{\mu}([x]_{A_2}) \geq \gamma \vee \ldots \vee \overline{\mu}([x]_{A_m}) \geq \gamma, x \in U\}. \tag{13}$$

From the perspective of the optimistic membership degree, $\sum_{i=1}^{m} A_i^{O}(X)$ can be =expressed as follows:

$$\sum_{i=1}^{m} A_i^{O}(X) = \{x | \overline{\mu}^{O}_{\sum_{i=1}^{m} A_i}(x) \geq \gamma, x \in U\}. \tag{14}$$

The corresponding decision regions are defined as follows:

$$BN1^{O}_{\sum_{i=1}^{m} A_i}(X) = \{x | 1 > \overline{\mu}^{O}_{\sum_{i=1}^{m} A_i}(x) \geq \gamma, x \in U\},$$
$$BN2^{O}_{\sum_{i=1}^{m} A_i}(X) = \{x | 0 < \overline{\mu}^{O}_{\sum_{i=1}^{m} A_i}(x) < \gamma, x \in U\},$$
$$POS^{O}_{\sum_{i=1}^{m} A_i}(X) = \{x | \overline{\mu}^{O}_{\sum_{i=1}^{m} A_i}(x) = 1, x \in U\},$$
$$NEG^{O}_{\sum_{i=1}^{m} A_i}(X) = \{x | \overline{\mu}^{O}_{\sum_{i=1}^{m} A_i}(x) = 0, x \in U\}.$$

We have

$$\sum_{i=1}^{m} A_i^{O}(X) = BN1^{O}_{\sum_{i=1}^{m} A_i}(X) \cup POS^{O}_{\sum_{i=1}^{m} A_i}(X). \tag{15}$$

The misclassification costs of approximations of optimistic MGRS come from two uncertain regions $BN1^{O}_{\sum_{i=1}^{m} A_i}(X)$ and $BN2^{O}_{\sum_{i=1}^{m} A_i}(X)$, which are defined in the following:

$$DC(\sum_{i=1}^{m} A_i^{O}(X)) = \sum_{x \in BN1^{O}_{\sum_{i=1}^{m} A_i}(X)} \lambda^{Y} + \sum_{x \in BN2^{O}_{\sum_{i=1}^{m} A_i}(X)} \lambda^{N}. \tag{16}$$

Theorem 2. *Let $S = (U, C \cup D, V, f)$ be a decision information table, $A_1, A_2, \ldots, A_m \subseteq C$, $X \subseteq U$ and $X_1, X_2, \ldots, X_n \subseteq U$. The following properties hold:*

(1) $\sum_{i=1}^{m} A_i^{O}(X) = \bigcup_{i=1}^{m} A_i(X)$;
(2) $\sum_{i=1}^{m} A_i^{O}(\bigcap_{j=1}^{n} X_j) = \bigcup_{i=1}^{m} (\bigcap_{j=1}^{n} A_i^{O}(X_j))$;
(3) $\sum_{i=1}^{m} A_i^{O}(\bigcap_{j=1}^{n} X_j) \subseteq \bigcap_{j=1}^{n} (\sum_{i=1}^{m} A_i^{O}(X_j))$;
(4) $\sum_{i=1}^{m} A_i^{O}(\bigcup_{j=1}^{n} X_j) \supseteq \bigcup_{j=1}^{n} (\sum_{i=1}^{m} A_i^{O}(X_j))$;
(5) $\underline{\sum_{i=1}^{m} A_i^{O}}(X) \subseteq \sum_{i=1}^{m} A_i^{O}(X) \subseteq \overline{\sum_{i=1}^{m} A_i^{O}}(X)$.

Proof of Theorem 2.

(1) From Formula (14), $\sum_{i=1}^{m} A_i^{O}(X) = \bigcup_{i=1}^{m} A_i(X)$ obviously holds.
(2) $\sum_{i=1}^{m} A_i^{O}(\bigcap_{j=1}^{n} X_j) = \bigcup_{i=1}^{m} A_i^{O}(\bigcap_{j=1}^{n} X_j) = \bigcup_{i=1}^{m} (\bigcap_{j=1}^{n} A_i^{O}(X_j))$.
(3) $\sum_{i=1}^{m} A_i^{O}(\bigcap_{j=1}^{n} X_j) = \bigcup_{i=1}^{m} (\bigcap_{j=1}^{n} A_i(X_j))$
$= \bigcap_{j=1}^{n} (\bigcup_{i=1}^{m} A_i(X_j) \cap \cdots$
$= \bigcap_{j=1}^{n} (\sum_{i=1}^{m} A_i^{O}(X_j)) \cap \cdots$
$= \bigcap_{j=1}^{n} (\sum_{i=1}^{m} A_i^{O}(X_j))$.
(4) From $X_j \subseteq \bigcup_{j=1}^{n} X_j$, we have $\sum_{i=1}^{m} A_i^{O}(X_j) \subseteq \sum_{i=1}^{m} A_i^{O}(\bigcup_{j=1}^{n} X_j)$. Therefore, $\sum_{i=1}^{m} A_i^{O}(\bigcup_{j=1}^{n} X_j) \supseteq \bigcup_{j=1}^{n} (\sum_{i=1}^{m} A_i^{O}(X_j))$.
(5) It is easy to prove by Formulas (1), (2), and (14).
□

Definition 8. *Let $S = (U, C \cup D, V, f)$ be a decision information table, $A_1, A_2, \ldots, A_m \subseteq C$ and $X \subseteq U$. The pessimistic membership degree of $x \in U$ related to $A_1, A_2, \ldots, A_m$ is given as follows:*

$$\overline{\mu}^{P}_{\sum_{i=1}^{m} A_i}(x) = \min\{\overline{\mu}([x]_{A_i}) | i = 1, 2, \ldots, m\}. \tag{17}$$

Definition 9. *Let $S = (U, C \cup D, V, f)$ be a decision information table, $A_1, A_2, \ldots, A_m \subseteq C$ and $X \subseteq U$. The approximation model of pessimistic MGRS of X related to $A_1, A_2, \ldots, A_m$ is given as follows:*

$$\sum_{i=1}^{m} A_i^{P}(X) = \{x | \overline{\mu}([x]_{A_1}) \geq \gamma \wedge \overline{\mu}([x]_{A_2}) \geq \gamma \wedge \ldots \wedge \overline{\mu}([x]_{A_m}) \geq \gamma, x \in U\}. \tag{18}$$

From the perspective of the pessimistic membership degree, $\sum\limits_{i=1}^{m} A_i^P(X)$ *can be expressed as follows:*

$$\sum_{i=1}^{m} A_i^P(X) = \{x|\overline{\mu}^P_{\sum\limits_{i=1}^{m} A_i}(x) \geq \gamma, x \in U\}. \tag{19}$$

The corresponding decision regions are expressed in the following:

$$\begin{aligned}
BN1^P_{\sum\limits_{i=1}^{m} A_i}(X) &= \{x|1 > \overline{\mu}^P_{\sum\limits_{i=1}^{m} A_i}(x) \geq \gamma, x \in U\}, \\
BN2^P_{\sum\limits_{i=1}^{m} A_i}(X) &= \{x|0 < \overline{\mu}^P_{\sum\limits_{i=1}^{m} A_i}(x) < \gamma, x \in U\}, \\
POS^P_{\sum\limits_{i=1}^{m} A_i}(X) &= \{x|\overline{\mu}^P_{\sum\limits_{i=1}^{m} A_i}(x) = 1, x \in U\}, \\
NEG^P_{\sum\limits_{i=1}^{m} A_i}(X) &= \{x|\overline{\mu}^P_{\sum\limits_{i=1}^{m} A_i}(x) = 0, x \in U\}.
\end{aligned}$$

We have

$$\sum_{i=1}^{m} A_i^P(X) = BN1^P_{\sum\limits_{i=1}^{m} A_i}(X) \cup POS^P_{\sum\limits_{i=1}^{m} A_i}(X)\,. \tag{20}$$

The misclassification costs of approximations of optimistic MGRS come from two uncertain regions $BN1^P_{\sum\limits_{i=1}^{m} A_i}(X)$ *and* $BN2^P_{\sum\limits_{i=1}^{m} A_i}(X)$ *, which are defined in the following:*

$$DC(\sum_{i=1}^{m} A_i^P(X)) = \sum_{x \in BN1^P_{\sum\limits_{i=1}^{m} A_i}(X)} \lambda^Y + \sum_{x \in BN1^P_{\sum\limits_{i=1}^{m} A_i}(X)} \lambda^N. \tag{21}$$

Theorem 3. *Let* $S = (U, C \cup D, V, f)$ *be a decision information table,* $A_1, A_2, \ldots, A_m \subseteq C$, $X \subseteq U$ *and* $X_1, X_2, \ldots, X_n \subseteq U$. *The following properties hold:*

(1) $\sum\limits_{i=1}^{m} A_i^P(X) = \bigcap_{i=1}^{m}(A_i(X))$;

(2) $\sum\limits_{i=1}^{m} A_i^P(\bigcap_{j=1}^{n} X_j) = \bigcap_{j=1}^{n}(\sum\limits_{i=1}^{m} A_i^P(X_j))$;

(3) $\sum\limits_{i=1}^{m} A_i^P(\bigcup_{j=1}^{n} X_j) \supseteq \bigcup_{j=1}^{n}(\sum\limits_{i=1}^{m} A_i^P(X_j))$;

(4) $\underline{\sum\limits_{i=1}^{m} A_i^P(X)} \subseteq \sum\limits_{i=1}^{m} A_i^P(X) \subseteq \overline{\sum\limits_{i=1}^{m} A_i^P(X)}$.

Proof of Theorem 3.

(1) $\forall x \in \sum\limits_{i=1}^{m} A_i^P(X)$, according to Definition 9, $\overline{\mu}([x]_{A_i}) \geq \gamma$ holds, $i = 1, 2, \cdots, m$. According to Definition 5, $x \in A_i(X)$, $i = 1, 2, \cdots, m$ and $x \in \bigcap_{i=1}^{m}(A_i(X))$ holds, i.e., $\sum\limits_{i=1}^{m} A_i^P(X) \subseteq \bigcap_{i=1}^{m}(A_i(X))$. $\forall x \in \bigcap_{i=1}^{m}(A_i(X))$, $\overline{\mu}([x]_{A_i}) \geq \gamma$, $i = 1, 2, \cdots, m$. According to Definition 7, $x \in \sum\limits_{i=1}^{m} A_i^P(X)$ holds, i.e., $\bigcap_{i=1}^{m}(A_i(X)) \subseteq \sum\limits_{i=1}^{m} A_i^P(X)$. Therefore, we have $\sum\limits_{i=1}^{m} A_i^P(X) = \bigcap_{i=1}^{m}(A_i(X))$.

(2) From the proof of (1), $\sum_{i=1}^{m} A_i^P(\bigcap_{j=1}^{n} X_j) = \bigcap_{i=1}^{m} A_i(\bigcap_{j=1}^{n} X_j)$. According to Definition 7, $\bigcap_{i=1}^{m} A_i(\bigcap_{j=1}^{n} X_j) = \bigcap_{i=1}^{m}\bigcap_{j=1}^{n} A_i(X_j)$. Because $\bigcap_{i=1}^{m} A_i(X_j) = \sum_{i=1}^{m} A_i^P(X_j)$, $\sum_{i=1}^{m} A_i^P(\bigcap_{j=1}^{n} X_j) = \bigcap_{i=1}^{m} A_i(\bigcap_{j=1}^{n} X_j) = \bigcap_{j=1}^{n}\bigcap_{i=1}^{m} A_i(X_j) = \bigcap_{j=1}^{n}(\sum_{i=1}^{m} A_i^P(X_j))$.

(3) $\forall x \in \bigcup_{j=1}^{n}(\sum_{i=1}^{m} A_i^P(X_j)), \exists X_k(k \in \{1,2,\cdots,n\}), x \in \sum_{i=1}^{m} A_i^P(X_j)$. According to Definition 7, $\forall X_j, j = 1,2,\cdots,n, \overline{\mu}([x]_{A_i}) \geq \gamma, , i = 1,2,\cdots,m, x \in \sum_{i=1}^{m} A_i^P(\bigcup_{j=1}^{n}(X_j))$ holds. Therefore, $\sum_{i=1}^{m} A_i^P(\bigcup_{j=1}^{n} X_j) \supseteq \bigcup_{j=1}^{n}(\sum_{i=1}^{m} A_i^P(X_j))$.

(4) It is easy to prove by Formulas (4), (5), and (19).

□

Theorem 4. *Let $S = (U, C \cup D, V, f)$ be a decision information table, $A_1, A_2, \ldots, A_m \subseteq C$ and $X \subseteq U$, $A = A_1 \cup A_2 \cup \ldots \cup A_m$ and $X_1, X_2, \ldots, X_n \subseteq U$. The following properties hold:*

(1) $\sum_{i=1}^{m} A_i^P(X) \subseteq \sum_{i=1}^{m} A_i^O(X) \subseteq A(X)$;

(2) $\sum_{i=1}^{m-1} A_i^O(X) \subseteq \sum_{i=1}^{m} A_i^O(X)$ *and* $\sum_{i=1}^{m-1} A_i^P(X) \supseteq \sum_{i=1}^{m} A_i^P(X)$.

Proof of Theorem 4.

(1) According to Definition 6, we only need to prove $\sum_{i=1}^{m} A_i^P(X) \subseteq \sum_{i=1}^{m} A_i^O(X)$. $\forall x \in \sum_{i=1}^{m} A_i^P(X)$, according to Definition 7, $\overline{\mu}([x]_{A_i}) \geq \gamma$. From Definition 5, we have $x \in \sum_{i=1}^{m} A_i^O(X)$.

(2) It is easy to prove according to Definitions 5 and 7.

□

Lemma 1. *Let $S = (U, C \cup D, V, f)$ be a decision information table, $A \subseteq C$ and $X \subseteq U$, $U/A = \{E_1, E_2, \ldots, E_l\}$ is a granularity layer on U. The following properties hold:*

(1) $\sum_{E_i \in BN1(X)} \lambda_{E_i}^Y \leq \sum_{E_i \in BN1(X)} \lambda_{E_i}^N$;

(2) $\sum_{E_i \in BN2(X)} \lambda_{E_i}^N \leq \sum_{E_i \in BN2(X)} \lambda_{E_i}^Y$. *Here, $E_i \in U/A$ $(i = 1,2,\ldots,l)$.*

Proof of Lemma 1.

(1) $\lambda_{E_i}^Y - \lambda_{E_i}^N = \lambda_{12}(1 - \overline{\mu}(E_i))|E_i| - \lambda_{21}\overline{\mu}(E_i)|E_i| = |E_i|(\lambda_{12} - (\lambda_{12} + \lambda_{21}))\overline{\mu}(E_i)$, because $E_i \in BN1(X)$, we have $\frac{\lambda_{12}}{\lambda_{12}+\lambda_{21}} \leq \overline{\mu}(E_i) < 1$, then $\lambda_{E_i}^Y \leq \lambda_{E_i}^N$, therefore $\sum_{E_i \in BN1(X)} \lambda_{E_i}^Y \leq \sum_{E_i \in BN1(X)} \lambda_{E_i}^N$.

(2) $\lambda_{E_i}^Y - \lambda_{E_i}^N = \lambda_{12}(1 - \overline{\mu}(E_i))|E_i| - \lambda_{21}\overline{\mu}(E_i)|E_i| = |E_i|(\lambda_{12} - (\lambda_{12} + \lambda_{21}))\overline{\mu}(E_i)$. Because $E_i \in BN2(X)$, we have $0 < \overline{\mu}(E_i) < \frac{\lambda_{12}}{\lambda_{12}+\lambda_{21}}$, then $\lambda_{E_i}^Y \geq \lambda_{E_i}^N$. Therefore, $\sum_{E_i \in BN2(X)} \lambda_{E_i}^N \leq \sum_{E_i \in BN2(X)} \lambda_{E_i}^Y$.

□

Lemma 1 shows that the misclassification costs incurred by the equivalence classes in characterizing X are not more than the misclassification costs incurred by the equivalence classes when they do not characterize X in $BN1(X)$. Moreover, misclassification costs incurred by the equivalence classes when they do not characterize X are not more than the misclassification costs incurred by the equivalence classes in characterizing X in $BN2(X)$.

Theorem 5. *Let $S = (U, C \cup D, V, f)$ be a decision information table, $A \subseteq C$ and $X \subseteq U$, the following properties hold:*

(1) $\sum\limits_{x \in BN1^{O}_{\sum_{i=1}^{m} A_i}(X)} \lambda^{Y} \leq \sum\limits_{x \in BN1^{O}_{\sum_{i=1}^{m} A_i}(X)} \lambda^{N}$ *and* $\sum\limits_{x \in BN1^{P}_{\sum_{i=1}^{m} A_i}(X)} \lambda^{Y} \leq \sum\limits_{x \in BN1^{P}_{\sum_{i=1}^{m} A_i}(X)} \lambda^{N}$;

(2) $\sum\limits_{x \in BN2^{O}_{\sum_{i=1}^{m} A_i}(X)} \lambda^{N} \leq \sum\limits_{x \in BN2^{O}_{\sum_{i=1}^{m} A_i}(X)} \lambda^{Y}$ *and* $\sum\limits_{x \in BN1^{P}_{\sum_{i=1}^{m} A_i}(X)} \lambda^{N} \leq \sum\limits_{x \in BN1^{P}_{\sum_{i=1}^{m} A_i}(X)} \lambda^{Y}$.

Proof of Theorem 5. From Lemma 1, Theorem 5 holds. □

Theorem 5 shows that misclassification costs incurred by the equivalence classes in characterizing X are not more than misclassification costs incurred by the equivalence classes when they do not characterize X in $BN1^{O}_{\sum_{i=1}^{m} A_i}(X)$ and $BN1^{P}_{\sum_{i=1}^{m} R_i}(X)$. Moreover, misclassification costs incurred by the equivalence classes when they do not characterize X are not more than misclassification costs incurred by the equivalence classes in characterizing X in $BN2^{O}_{\sum_{i=1}^{m} A_i}(X)$ and $BN2^{P}_{\sum_{i=1}^{m} A_i}(X)$.

$DC(\sum_{i=1}^{m} A_i^{O}(X))$, $DC(\underline{\sum_{i=1}^{m} A_i^{O}}(X))$ and $DC(\overline{\sum_{i=1}^{m} A_i^{O}}(X))$ denote the misclassification costs generated when $\sum_{i=1}^{m} A_i^{O}(X)$, $\underline{\sum_{i=1}^{m} A_i^{O}}(X)$ and $\overline{\sum_{i=1}^{m} A_i^{O}}(X)$ are approximated to X, respectively. Then, the following theorem holds:

Theorem 6. *Let $S = (U, C \cup D, V, f)$ be a decision information table, $A \subseteq C$ and $X \subseteq U$, Then, $DC(\sum_{i=1}^{m} A_i^{O}(X)) \leq DC(\underline{\sum_{i=1}^{m} A_i^{O}}(X))$ and $DC(\sum_{i=1}^{m} A_i^{O}(X)) \leq DC(\overline{\sum_{i=1}^{m} A_i^{O}}(X))$.*

Proof of Theorem 6. When $\underline{\sum_{i=1}^{m} A_i^{O}}(X)$ is taken as the approximation of X, $DC(\underline{\sum_{i=1}^{m} A_i^{O}}(X)) = \sum\limits_{x \in BN(X)} \lambda^{N}$; when $\overline{\sum_{i=1}^{m} A_i^{O}}(X)$ is considered the approximation of X, $DC(\overline{\sum_{i=1}^{m} A_i^{O}}(X)) = \sum\limits_{x \in BN(X)} \lambda^{Y}$; when $DC(\sum_{i=1}^{m} A_i^{O}(X))$ is considered the approximation of X, $DC(\sum_{i=1}^{m} A_i^{O}(X)) = \sum\limits_{x \in BN1(X)} \lambda^{Y} + \sum\limits_{x \in BN2(X)} \lambda^{N}$.

Because $BN(X) = BN1(X) \cup BN2(X)$, we have:

$$DC(\underline{\sum_{i=1}^{m} A_i^{O}}(X)) = \sum_{x \in BN1(X)} \lambda^{N} + \sum_{x \in BN2(X)} \lambda^{N},$$

$$DC(\overline{\sum_{i=1}^{m} A_i^{O}}(X)) = \sum_{x \in BN1(X)} \lambda^{Y} + \sum_{x \in BN2(X)} \lambda^{Y}.$$

Therefore, according to Theorem 5,

$$DC(\sum_{i=1}^{m} A_i^{O}(X)) \leq DC(\underline{\sum_{i=1}^{m} A_i^{O}}(X)),$$

$$DC(\sum_{i=1}^{m} A_i^{O}(X)) \leq DC(\overline{\sum_{i=1}^{m} A_i^{O}}(X)).$$

Theorem 6 indicates that when $DC(\sum_{i=1}^{m} A_i^O(X))$, $DC(\underline{\sum_{i=1}^{m}} A_i^O(X))$ and $DC(\overline{\sum_{i=1}^{m} A_i^O(X)})$ are used as approximations of X, respectively, $DC(\sum_{i=1}^{m} A_i^O(X))$ generates the least misclassification costs. □

$DC(\sum_{i=1}^{m} A_i^P(X))$, $DC(\underline{\sum_{i=1}^{m}} A_i^P(X))$ and $DC(\overline{\sum_{i=1}^{m} A_i^P(X)})$ denote the misclassification costs generated when $\sum_{i=1}^{m} A_i^P(X)$, $\underline{\sum_{i=1}^{m}} A_i^P(X)$ and $\overline{\sum_{i=1}^{m} A_i^P(X)}$ are approximated to X, respectively.

Theorem 7. *Let $S = (U, C \cup D, V, f)$ be a decision information table, $A \subseteq C$ and $X \subseteq U$. Then, $DC(\sum_{i=1}^{m} A_i^P(X)) \leq DC(\underline{\sum_{i=1}^{m}} A_i^P(X))$ and $DC(\sum_{i=1}^{m} A_i^P(X)) \leq DC(\overline{\sum_{i=1}^{m} A_i^P(X)})$.*

Proof of Theorem 7. Similar to Theorem 6, Theorem 7 is easy to prove. □

From Theorem 7, when $DC(\sum_{i=1}^{m} A_i^P(X))$, $DC(\underline{\sum_{i=1}^{m}} A_i^P(X))$ and $DC(\overline{\sum_{i=1}^{m} A_i^P(X)})$ are used as approximations of X, respectively, $DC(\sum_{i=1}^{m} A_i^P(X))$ generates the least misclassification costs. Theorems 6 and 7 reflect the advantages of the multigranulation approximation sets that are used for approximating the target concept.

5.2. The Optimal Multigranulation Approximation Selection Method

The objects in the boundary region may be reclassified under different granularities. As a result, the equivalence classes used to represent the approximation set will be changed in the boundary region. In practical applications, the optimal approximation selection should consider both the misclassification and test costs. In MGRS, uncertain concepts characterized in a finer approximation layer result in lower misclassification costs, and test costs increase with the added attributes. Therefore, it is essential to find a balance between misclassification and test costs.

Lemma 2. *Let $S = (U, C \cup D, V, f)$ be a decision information table, $A_1, A_2, \ldots, A_m \subseteq C$, $A_1 \subseteq A_2 \subseteq, \ldots, \subseteq A_m$, and $X \subseteq U$, then $\forall x \in U$, $\overline{\mu}^O_{\sum_{i=1}^{m-1} A_i}(x) \leq \overline{\mu}^O_{\sum_{i=1}^{m} A_i}(x)$.*

Lemma 3. *Let $S = (U, C \cup D, V, f)$ be a decision information table, $A_1, A_2, \ldots, A_m \subseteq C$, $A_1 \subseteq A_2 \subseteq, \ldots, \subseteq A_m$, and $X \subseteq U$, then $\forall x \in U$, $\overline{\mu}^P_{\sum_{i=1}^{m-1} A_i}(x) \geq \overline{\mu}^P_{\sum_{i=1}^{m} A_i}(x)$.*

Theorem 8. *Let $S = (U, C \cup D, V, f)$ be a decision information table, $A_1, A_2, \ldots, A_m \subseteq C$, $A_1 \subseteq A_2 \subseteq, \ldots, \subseteq A_m$ and $X \subseteq U$, only when $\overline{\mu}(x) (x \in BN1^O_{\sum_{i=1}^{m-1} A_i}(X))$ changed with the attribute increased, then $DC(\sum_{i=1}^{m-1} A_i^O(X)) \geq DC(\sum_{i=1}^{m} A_i^O(X))$.*

Proof of Theorem 8.

$$\begin{aligned} DC(\sum_{i=1}^{m} A_i^O(X)) &= \sum_{x \in BN1^O_{\sum\limits_{i=1}^{m} A_i}(X)} \lambda^Y + \sum_{x \in BN2^O_{\sum\limits_{i=1}^{m} A_i}(X)} \lambda^N \\ &= \sum_{x \in BN1^O_{\sum\limits_{i=1}^{m} A_i}(X)} \lambda_{12}(1-\overline{\mu}(x)) + \sum_{x \in BN2^O_{\sum\limits_{i=1}^{m} A_i}(X)} \lambda_{21}\overline{\mu}(x). \end{aligned}$$

According to Lemma 2, we have

$$\overline{\mu}^O_{\sum\limits_{i=1}^{m-1} A_i}(x) \leq \overline{\mu}^O_{\sum\limits_{i=1}^{m} A_i}(x).$$

Obviously, $DC(\sum\limits_{i=1}^{m-1} A_i^O(X)) \geq DC(\sum\limits_{i=1}^{m} A_i^O(X))$. □

From Theorem 7, for optimistic MGRS, to reduce the misclassification costs of the approximation, we can add the attribute that only changes the membership of objects in $x \in BN1^O_{\sum\limits_{i=1}^{m-1} A_i}(X)$.

Theorem 9. *Let $S = (U, C \cup D, V, f)$ be a decision information table, $A_1, A_2, \ldots, A_m \subseteq C$ and $A_1 \subseteq A_2 \subseteq, \ldots, \subseteq A_m$, $X \subseteq U$, only when $\overline{\mu}(x)$ ($x \in BN2^P_{\sum\limits_{i=1}^{m-1} A_i}(X)$) changed with the attributes increased, then $DC(\sum\limits_{i=1}^{m-1} A_i^P(X)) \geq DC(\sum\limits_{i=1}^{m} A_i^P(X))$.*

Proof of Theorem 9.

$$\begin{aligned} DC(\sum_{i=1}^{m} A_i^P(X)) &= \sum_{x \in BN1^P_{\sum\limits_{i=1}^{m} A_i}(X)} \lambda^Y + \sum_{x \in BN2^P_{\sum\limits_{i=1}^{m} A_i}(X)} \lambda^N \\ &= \sum_{x \in BN1^P_{\sum\limits_{i=1}^{m} A_i}(X)} \lambda_{12}(1-\overline{\mu}(x)) + \sum_{x \in BN2^P_{\sum\limits_{i=1}^{m} A_i}(X)} \lambda_{21}\overline{\mu}(x). \end{aligned}$$

According to Lemma 3, we have

$$\overline{\mu}^P_{\sum\limits_{i=1}^{m-1} A_i}(x) \geq \overline{\mu}^P_{\sum\limits_{i=1}^{m} A_i}(x).$$

Obviously, $DC(\sum\limits_{i=1}^{m-1} A_i^P(X)) \geq DC(\sum\limits_{i=1}^{m} A_i^P(X))$. □

From Theorem 9, for pessimistic MGRS, to reduce the misclassification costs of the approximation, we can add the attribute that only changes the membership of objects in $x \in BN2^O_{\sum\limits_{i=1}^{m-1} R_i}(X)$.

In practical applications, on the one hand, the factors included in the test cost, such as money, time, environment, etc., are hard to evaluate objectively. On the other hand, these factors are hard to be integrated because of their different dimensions. In this section, we will evaluate test costs in an attribute-driven form, which are more objective.

Definition 10. *Let $S = (U, C \cup D, V, f)$ be a decision information table, $a \in C$ and $X \subseteq U$, the significance of a is defined as follows:*

$$Sig(a, C, D) = DC_{C-\{a\}} - DC_C. \tag{22}$$

Definition 11. *Let $S = (U, C \cup D, V, f)$ be a decision information table, $a \in C$, $A \subseteq C$ and $X \subseteq U$; the test cost to construct $A(X)$ is defined as follows:*

$$TC_A = \sum_{a \in A} Sig(a, C, D). \tag{23}$$

In this paper, for simplicity, to present the optimal granularity selection of the multi-granulation approximation, we only use the optimistic MGRS as an example.

Definition 12. *Let $S = (U, C \cup D, V, f)$ be a decision information table, $A_1, A_2, \ldots, A_m \subseteq C$ and $X \subseteq U$; the test cost to construct $\sum_{i=1}^{m} A_i^O(X)$ can be defined as follows:*

$$TC_{\sum_{i=1}^{m} A_i^O} = \sum_{i=1}^{m} TC_{A_i}. \tag{24}$$

In this paper, the misclassification and test costs for user requirements are represented as DC_u and TC_u, respectively. A multigranulation approximation $\sum_{i=1}^{k} A_i^O(X)$ is selected to meet the constraints $DC_{\sum_{i=1}^{k} A_i^O(X)} \leq DC_u$ and $TC_{\sum_{i=1}^{k} A_i^O(X)} \leq TC_u$, then the related decision are made on $\sum_{i=1}^{k} A_i^O(X)$. Figure 3 presents the optimal multigranulation approximation selection of optimistic MGRS. Herein, $\sum_{i=1}^{3} A_i^O(X)$ complies with the requirements of the misclassification costs and fails to comply with the requirements of the test costs; $A_1^O(X)$ complies with the requirements of the test cost and fails to comply with the requirements of misclassification costs; $\sum_{i=1}^{2} A_i^O(X)$ complies with both requirements of misclassification costs and test costs, enabling effective calculations according to granularity optimization. Similarly, the optimal approximation selection of pessimistic MGRS is the same. We formalize the computation as an optimization problem:

$$\arg\min_{k} Cost_{\sum_{i=1}^{k} A_i^O(X)}, \tag{25}$$

$$\begin{aligned} s.t.\quad & \\ & \xi DC_{\sum_{i=1}^{k} A_i^O(X)} \leq DC_u; \\ & TC_{\sum_{i=1}^{k} A_i^O(X)} \leq TC_u. \end{aligned}$$

where $Cost_{\sum_{i=1}^{k} A_i^O(X)} = \xi DC_{\sum_{i=1}^{k} A_i^O} + TC_{\sum_{i=1}^{k} A_i^O}$, and $Cost_{\sum_{i=1}^{k} A_i^O(X)}$ denotes the total cost for constructing $\sum_{i=1}^{k} A_i^O(X)$. $\xi = \frac{|U|}{\left|\sum_{i=1}^{m} A_i^O(X)\right|} * \frac{1}{DC(A_m(X))}$ reflects the contribution degree of the multigranulation approximation layer for the misclassification costs of $\sum_{i=1}^{k} A_i^O(X)$.

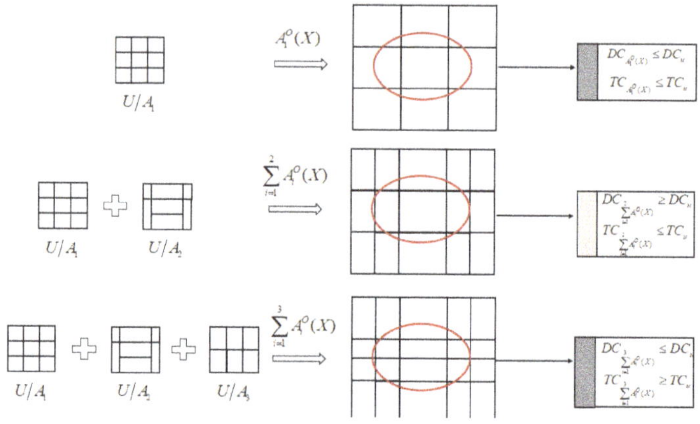

Figure 3. The optimal granularity selection of the optimistic multigranulation approximation. And the red circles in the figure represent the set X.

5.3. Case Study

Table 2 is an evaluation form of the company venture capital given by five experts. $U = \{x_1, x_2, \ldots, x_{900}\}$ is 900 investment plans, which are evaluated by 5 experts. The risk level is divided into three categories, including high and low. Suppose that $DC_u = 5$ and $TC_u = 2.5$.

(1) According to the above conditions, the attribute significance can be computed by Formula (25), which is shown in Table 3.

(2) Each multigranulation approximation is obtained by adding attributes in ascending order of attribute significance, $E_4 \to E_3 \to E_2 \to E_1 \to E_5$ in each stage, respectively. We represent the attributes as follows: $A_1 = E_4$, $A_2 = E_3$, $A_3 = E_2$, $A_4 = E_1$, and $A_5 = E_5$.

(3) For each multigranulation approximation layer, $DC_{\sum_{i=1}^{k} A_i^O}$, $TC_{\sum_{i=1}^{k} A_i^O}$, and $Cost_{\sum_{i=1}^{k} A_i^O(X)}$ are computed by Formulas (16), (24), and (25), respectively, where $k = 1, 2, \ldots, 5$ and the results are displayed in Table 4.

Table 2. Evaluation form of the company's venture capital.

Firm	E_1	E_2	E_3	E_4	E_5	D
x_1	3	3	3	3	1	High
x_2	2	1	2	3	2	High
x_3	2	1	2	1	2	High
.	.	.	.	.	.	.
.	.	.	.	.	.	.
.	.	.	.	.	.	.
x_{898}	3	3	2	2	3	Low
x_{899}	3	1	3	3	1	Low
x_{900}	1	1	3	3	1	Low

Table 3. Description of attribute significance.

Attribute	E_1	E_2	E_3	E_4	E_5
$Sig(a, C, D)$	0.74	0.56	0.54	0.36	0.77

Table 4. The description of cost in each multigranulation approximation.

	$A_1^O(X)$	$\sum_{i=1}^{2} A_i^O(X)$	$\sum_{i=1}^{3} A_i^O(X)$	$\sum_{i=1}^{4} A_i^O(X)$	$\sum_{i=1}^{5} A_i^O(X)$
ξDC	8.3	5.9	4.7	3.6	3.5
TC	0.36	0.9	1.46	2.2	2.97
$Cost$	8.66	6.8	6.16	5.8	6.47

$Cost_{\sum_{i=1}^{k} A_i^O(X)}$ changes with the increased attributes and only $Cost_{\sum_{i=1}^{3} A_i^O(X)}$ and $Cost_{\sum_{i=1}^{4} A_i^O(X)}$ satisfy $DC_{\sum_{i=1}^{k} A_i^O(X)} \leq DC_u$ and $TC_{\sum_{i=1}^{k} A_i^O(X)} \leq TC_u$ at the same time. According to the Formula (25), we choose the multigranulation approximation layer with the lowest total cost from the above layers; its corresponding approximation layer is $\sum_{i=1}^{3} A_i^O(X)$. Therefore, $\sum_{i=1}^{3} A_i^O(X)$ is the optimal multigranulation approximation used for deciding investment plans, because it possesses lower misclassification costs, i.e., from the perspective of optimistic MGRS, E_4, E_3, and E_2 are reasonable expert sets. The analysis of the case study shows that the proposed method can search for a reasonable approximation under the constraint conditions.

6. Simulation Experiment and Result Analysis

6.1. Simulation Experiment

In this section, the effectiveness and rationality of our model are demonstrated as shown by illustrative experiments. The computer used in the experiments was a WIN 10 operating system with 3.10-GHz CPU and 16.0 GB RAM, and the programming software was MATLAB R2022a. The capability of the proposed model was evaluated on six UCI datasets, which are shown in Table 5. In our experiments, we randomly took away some known attribute values from datasets 10–12 to create incomplete decision systems. The missing values are randomly distributed on all conditional attributes.

Table 5. The description of datasets.

ID	Dataset	Attribute Characteristics	Instances	Condition Attributes
1	Bank	Integer	39	12
2	Breast-Cancer	Integer	699	9
3	Car	Integer	1728	6
4	ENB2012data	Real	768	8
5	Mushroom	Integer	8124	22
6	Tic	Integer	958	9
7	Air Quality	Real	9358	12
8	Concrete	Real	1030	8
9	Hcv	Real	569	10
10	Wisconsin	Real	699	9
11	Zoo	Integer	101	16
12	Balance	Integer	625	4

From Figure 4, for classical rough sets, the misclassification costs of the approximation model monotonously decrease with the granularity being finer, which complies with human cognitive habits.

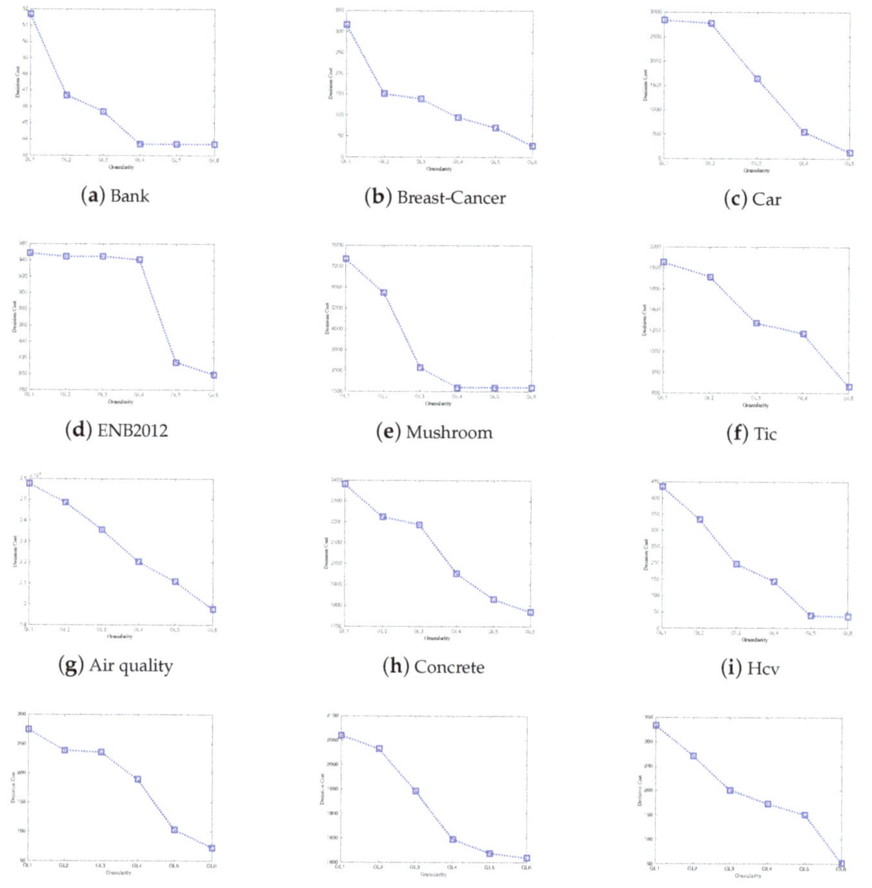

(a) Bank **(b)** Breast-Cancer **(c)** Car

(d) ENB2012 **(e)** Mushroom **(f)** Tic

(g) Air quality **(h)** Concrete **(i)** Hcv

(j) Wisconsin **(k)** Zoo **(l)** Balance

Figure 4. The misclassification cost with the changing granularity on each dataset.

In Figure 5, $O\,DC1$, $O\,DC2$, $O\,DC3$ and $O\,DC4$ represent $\sum_{x\in BN1^{O}_{\sum_{i=1}^{m}A_i}(X)} \lambda^{Y}$, $\sum_{x\in BN1^{O}_{\sum_{i=1}^{m}A_i}(X)} \lambda^{N}$, $\sum_{x\in BN2^{O}_{\sum_{i=1}^{m}A_i}(X)} \lambda^{N}$ and $\sum_{x\in BN2^{O}_{\sum_{i=1}^{m}A_i}(X)} \lambda^{Y}$, respectively. $P\,DC1$, $P\,DC2$, $P\,DC3$ and $P\,DC4$ represent $\sum_{x\in BN1^{P}_{\sum_{i=1}^{m}A_i}(X)} \lambda^{Y}$, $\sum_{x\in BN1^{P}_{\sum_{i=1}^{m}A_i}(X)} \lambda^{N}$, $\sum_{x\in BN2^{P}_{\sum_{i=1}^{m}A_i}(X)} \lambda^{N}$ and $\sum_{x\in BN1^{P}_{\sum_{i=1}^{m}A_i}(X)} \lambda^{Y}$, respectively. From Figure 5, under different granular layers, misclassification costs incurred by the equivalence classes in approximating X are always less than or equal to misclassification costs incurred by the equivalence classes when they do not characterize X in $BN1^{O}_{\sum_{i=1}^{m}A_i}(X)$ and $BN1^{P}_{\sum_{i=1}^{m}A_i}(X)$. Moreover, misclassification costs incurred by equivalence classes when they do not characterize X are less than or equal to the misclassification costs incurred by the equivalence classes in approximating X in $BN2^{O}_{\sum_{i=1}^{m}A_i}(X)$ and $BN2^{P}_{\sum_{i=1}^{m}A_i}(X)$. This is consistent with Theorem 4.

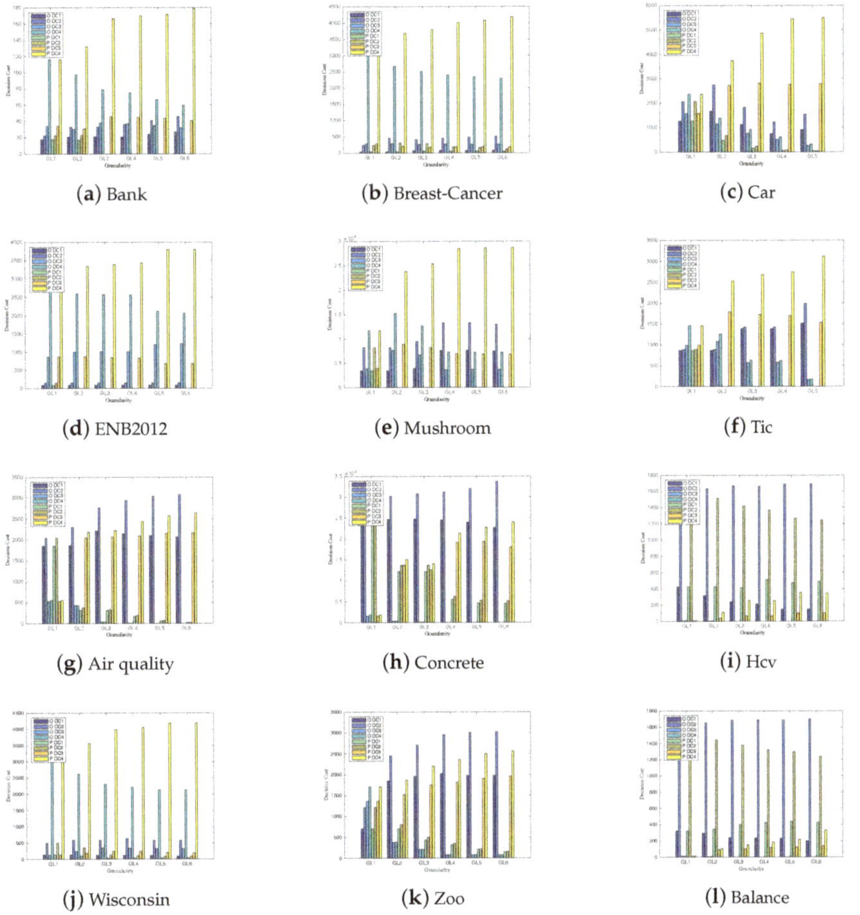

(a) Bank (b) Breast-Cancer (c) Car (d) ENB2012 (e) Mushroom (f) Tic (g) Air quality (h) Concrete (i) Hcv (j) Wisconsin (k) Zoo (l) Balance

Figure 5. The misclassification cost of two boundary regions under different granularities.

In Figure 6, the horizontal and vertical axes denote the granularity and misclassification costs, respectively. *O DC_lower*, *O DC_upper* and *O DC* represent $DC(\underline{\sum_{i=1}^{m} A_i^O(X)})$, $DC(\overline{\sum_{i=1}^{m} A_i^O(X)})$ and $DC(\sum_{i=1}^{m} A_i^O(X))$. *P DC_lower*, *P DC_upper* and *P DC* represent $DC(\underline{\sum_{i=1}^{m} A_i^P(X)})$,$DC(\overline{\sum_{i=1}^{m} A_i^P(X)})$ and $DC(\sum_{i=1}^{m} A_i^P(X))$, respectively, namely, the misclassification costs generated when $\underline{\sum_{i=1}^{m} A_i^O(X)}$, $\overline{\sum_{i=1}^{m} R_i^O(X)}$ and $\sum_{i=1}^{m} A_i^O(X)$ are approximated to X and the misclassification costs generated when $\underline{\sum_{i=1}^{m} A_i^P(X)}$, $\overline{\sum_{i=1}^{m} A_i^P(X)}$ and $\sum_{i=1}^{m} A_i^P(X)$ are approximated to X. Obviously, compared with $\underline{\sum_{i=1}^{m} A_i^O(X)}$ and $\overline{\sum_{i=1}^{m} A_i^O(X)}$, the misclassification costs of $\sum_{i=1}^{m} A_i^O(X)$ are the least on each granularity. Similarly, compared with $\underline{\sum_{i=1}^{m} A_i^P(X)}, \overline{\sum_{i=1}^{m} A_i^P(X)}$, the misclassification costs of $\sum_{i=1}^{m} A_i^P(X)$ are the least on each granularity. This is consistent with Theorems 5 and 6.

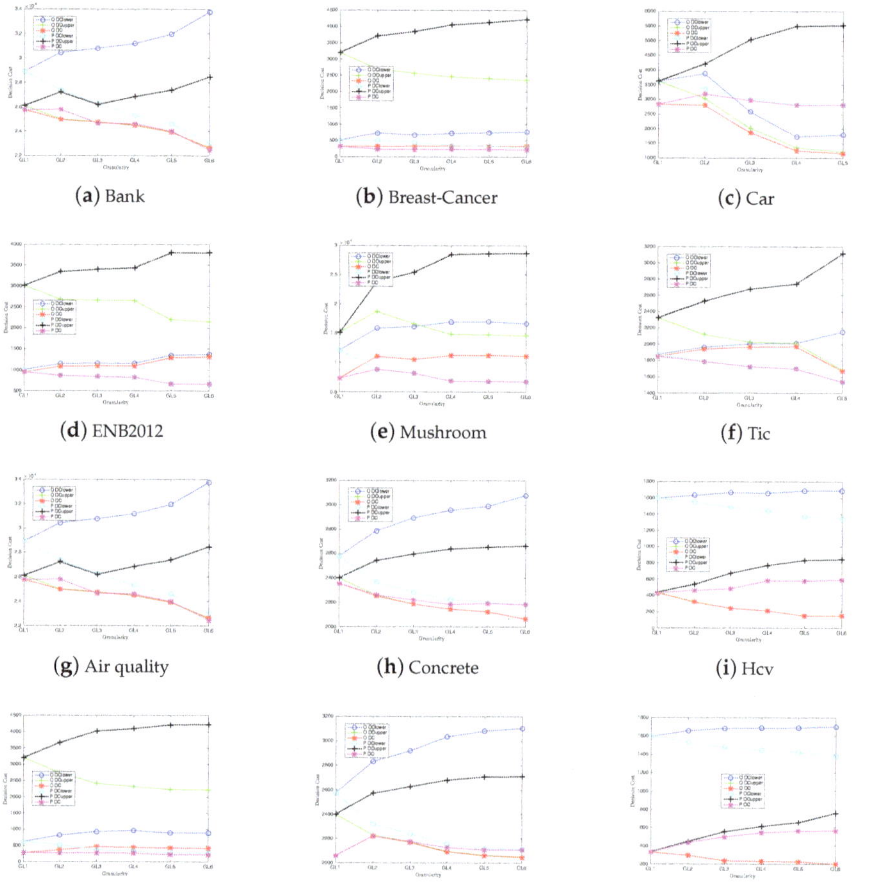

(**a**) Bank (**b**) Breast-Cancer (**c**) Car (**d**) ENB2012 (**e**) Mushroom (**f**) Tic (**g**) Air quality (**h**) Concrete (**i**) Hcv (**j**) Wisconsin (**k**) Zoo (**l**) Balance

Figure 6. The misclassification costs of $\underline{\sum_{i=1}^{m} A_i^O(X)}$, $\overline{\sum_{i=1}^{m} A_i^O(X)}$, $\sum_{i=1}^{m} A_i^O(X)$, $\underline{\sum_{i=1}^{m} A_i^P(X)}$, $\overline{\sum_{i=1}^{m} A_i^P(X)}$, $\sum_{i=1}^{m} A_i^P(X)$ with the changing granularity of each dataset.

6.2. *Results and Discussions*

According to the above experiments, compared with the upper/lower approximation sets, we can conclude that the multigranulation approximations have the following advantages when applied to decision-making environments:

(1) The misclassification costs of the approximation model monotonously decrease with the granularity being finer;

(2) In multigranulation approximations, under different granular layers, the misclassification costs incurred by the equivalence classes in approximating X are less than or equal to the misclassification costs incurred by the equivalence classes when they do not characterize X in boundary region I of optimistic and pessimistic rough sets. Moreover, the misclassification costs incurred by equivalence classes when they do not characterize X are less than or equal to the misclassification costs incurred by

the equivalence classes in approximating X in boundary region II of optimistic and pessimistic rough sets;

(3) Compared with the upper/lower approximation sets, the misclassification costs of the multigranulation approximations are the least on each granularity.

7. Conclusions

In MGRS, optimistic and pessimistic upper/lower approximation boundaries are utilized to characterize uncertain concepts. They still cannot take advantage of the known equivalence classes to establish the approximation of an uncertain concept. To handle the problem, cost-sensitive multigranulation approximations of rough sets were constructed. Furthermore, an optimization mechanism of the multigranulation approximations is proposed, which selects the optimal approximation to obtain the minimum misclassification costs under the conditions. The case study shows that the proposed algorithm is capable of searching for a rational approximation under restraints. Finally, the experiments demonstrate that the multigranulation approximations possess the least misclassification costs. In particular, our models apply to the decision-making environment where each decision-maker is independent. Moreover, our models are useful for extracting decision rules from distributive information systems and groups of intelligent agents through rough set approaches [34,36]. Figure 7 presents a diagram that summarizes the works conducted in this paper. Herein, we present the process of the cost-sensitive multigranulation approximations of rough sets; according to different granulation mechanisms, our approach can be extended to uncertainty models, i.e., vague sets, shadow sets, and neighborhood rough sets. These results will be important to contribute to the progress of the GrC theory.

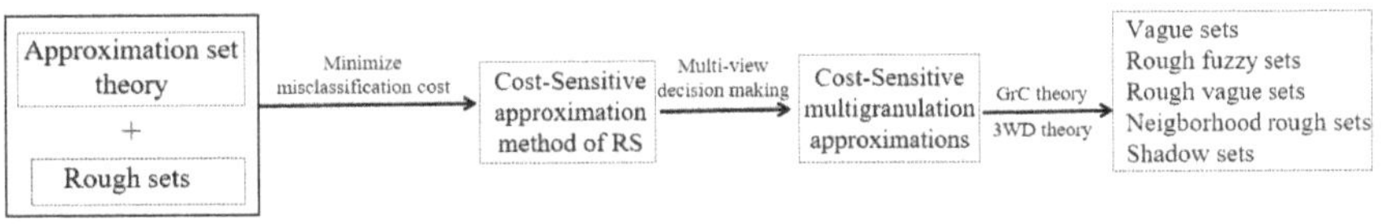

Figure 7. Diagram of works conducted in this paper.

Our future work will focus on the following two aspects: (1) We hope to build a more reasonable three-way decision model based on our model from optimistic and pessimistic perspectives; (2) we wish to combine the model with the cloud model theory to extend our model to construct a multigranulation approximation model with bidirectional cognitive computing. This will offer more cognitive advantages and benefits in application fields with uncertainty from multiple perspectives, i.e., image segmentation, clustering, and recommendation systems.

Author Contributions: Conceptualization, J.Y.; methodology, J.Y., J.K. and Q.L.; writing—original draft, J.Y. and J.K.; writing—review and editing, J.Y., J.K., Q.L. and Y.L.; data curation, J.Y., Q.L. and Y.L.; supervision, Y.L. All authors have read and agreed to the published version of the manuscript.

Funding: This work is supported by the National Science Foundation of China (no. 6206049), Excellent Young Scientific and Technological Talents Foundation of Guizhou Province (QKH-platform talent (2021) no. 5627), the Key Cooperation Project of Chongqing Municipal Education Commission (HZ2021008), Guizhou Provincial Science and Technology Project (QKH-ZK (2021) General 332), Science and Technology Top Talent Project of Guizhou Education Department (QJJ2022(088)), Key Laboratory of Evolutionary Artificial Intelligence in Guizhou (QJJ[2022] No. 059) and the Key Talens Program in digital economy of Guizhou Province, Electronic Manufacturing Industry University Research Base of Ordinary Colleges and Universities in Guizhou Province (QJH-KY Zi (2014) no. 230-2).

Institutional Review Board Statement: Not applicable.

Informed Consent Statement: Not applicable.

Data Availability Statement: Not applicable.

Acknowledgments: This study was mainly completed at the Chongqing Key Laboratory of Computational Intelligence, Chongqing University of Posts and Telecommunications, and the authors would like to thank the laboratory for its assistance.

Conflicts of Interest: The authors declare no conflict of interest.

References

1. Zadeh, L.A. Toward a theory of fuzzy information granulation and its centrality in human reasoning and fuzzy logic. *Fuzzy Sets Syst.* **1997**, *90*, 111–127. [CrossRef]
2. Bello, M.; Nápoles, G.; Vanhoof, K.; Bello, R. Data quality measures based on granular computing for multi-label classification. *Inf. Sci.* **2021**, *560*, 51–67. [CrossRef]
3. Pedrycz, W.; Chen, S. *Interpretable Artificial Intelligence: A Perspective of Granular Computing*; Springer Nature: Berlin/Heidelberg, Germany, 2021; Volume 937.
4. Li, J.; Mei, C.; Xu, W.; Qian, Y. Concept learning via granular computing: A cognitive viewpoint. *IEEE Trans. Fuzzy Syst.* **2015**, *298*, 447–467. [CrossRef] [PubMed]
5. Zadeh, L. Fuzzy sets. *Inf. Control.* **1965**, *8*, 338–353. [CrossRef]
6. Pawlak, Z. Rough sets. *Int. J. Comput. Inf. Sci.* **1982**, *11*, 341–356. [CrossRef]
7. Zhang, L.; Zhang, B. The quotient space theory of problem solving. In Proceedings of the International Workshop on Rough Sets, Fuzzy Sets, Data Mining, and Granular-Soft Computing, Chongqing, China, 26–29 May 2003; pp. 11–15.
8. Li, D.Y.; Meng, H.J.; Shi, X.M. Membership clouds and membership cloud generators. *J. Comput. Res. Dev.* **1995**, *32*, 15–20.
9. Colas-Marquez, R.; Mahfouf, M. Data Mining and Modelling of Charpy Impact Energy for Alloy Steels Using Fuzzy Rough Sets. *IFAC-Pap.* **2017**, *50*, 14970–14975. [CrossRef]
10. Hasegawa, K.; Koyama, M.; Arakawa, M.; Funatsu, K. Application of data mining to quantitative structure-activity relationship using rough set theory. *Chemom. Intell. Lab. Syst.* **2009**, *99*, 66–70. [CrossRef]
11. Santra, D.; Basu, S.K.; Mandal, J.K.; Goswami, S. Rough set based lattice structure for knowledge representation in medical expert systems: Low back pain management case study. *Expert Syst. Appl.* **2020**, *145*, 113084. [CrossRef]
12. Chebrolu, S.; Sanjeevi, S.G. Attribute Reduction in Decision-Theoretic Rough Set Model using Particle Swarm Optimization with the Threshold Parameters Determined using LMS Training Rule. *Procedia Comput. Sci.* **2015**, *57*, 527–536. [CrossRef]
13. Abdolrazzagh-Nezhad, M.; Radgohar, H.; Salimian, S.N. Enhanced cultural algorithm to solve multi-objective attribute reduction based on rough set theory. *Math. Comput. Simul.* **2020**, *170*, 332–350. [CrossRef]
14. Beaubier, S.; Defaix, C.; Albe-Slabi, S.; Aymes, A.; Galet, O.; Fournier, F.; Kapel, R. Multiobjective decision making strategy for selective albumin extraction from a rapeseed cold-pressed meal based on Rough Set approach. *Food Bioprod. Process.* **2022**, *133*, 34–44. [CrossRef]
15. Landowski, M.; Landowska, A. Usage of the rough set theory for generating decision rules of number of traffic vehicles. *Transp. Res. Procedia* **2019**, *39*, 260–269. [CrossRef]
16. Tawhid, M.; Ibrahim, A. Feature selection based on rough set approach, wrapper approach, and binary whale optimization algorithm. *Int. J. Mach. Learn. Cybern.* **2020**, *11*, 573–602. [CrossRef]
17. Zhang, Q.H.; Wang, G.Y.; Yu, X. Approximation sets of rough sets. *J. Softw.* **2012**, *23*, 1745–1759. [CrossRef]
18. Zhang, Q.H.; Wang, J.; Wang, G.Y. The approximate representation of rough-fuzzy sets. *Chin. J. Comput. Jisuanji Xuebao* **2015**, *38*, 1484–1496.
19. Zhang, Q.; Wang, J.; Wang, G.; Yu, H. The approximation set of a vague set in rough approximation space. *Inf. Sci.* **2015**, *300*, 1–19. [CrossRef]
20. Zhang, Q.H.; Zhang, P.; Wang, G.Y. Research on approximation set of rough set based on fuzzy similarity. *J. Intell. Fuzzy Syst.* **2017**, *32*, 2549–2562. [CrossRef]
21. Zhang, Q.H.; Yang, J.J.; Yao, L.Y. Attribute reduction based on rough approximation set in algebra and information views. *IEEE Access* **2016**, *4*, 5399–5407. [CrossRef]
22. Yao, L.Y.; Zhang, Q.H.; Hu, S.P.; Zhang, Q. Rough entropy for image segmentation based on approximation sets and particle swarm optimization. *J. Front. Comput. Sci. Technol.* **2016**, *10*, 699–708.
23. Zhang, Q.H.; Liu, K.X.; Gao, M. Approximation sets of rough sets and granularity optimization algorithm based on cost-sensitive. *J. Control. Decis.* **2020**, *35*, 2070–2080.
24. Yang, J.; Yuan, L.; Luo, T. Approximation set of rough fuzzy set based on misclassification cost. *J. Chongqing Univ. Posts Telecommun. (Nat. Sci. Ed.)* **2021**, *33*, 780–791.
25. Yang, J.; Luo, T.; Zeng, L.J.; Jin, X. The cost-sensitive approximation of neighborhood rough sets and granular layer selection. *J. Intell. Fuzzy Syst.* **2022**, *42*, 3993–4003. [CrossRef]
26. Siminski, K. 3WDNFS—Three-way decision neuro-fuzzy system for classification. *Fuzzy Sets Syst.* **2022**, *in press*. [CrossRef]
27. Subhashini, L.; Li, Y.; Zhang, J.; Atukorale, A.S. Assessing the effectiveness of a three-way decision-making framework with multiple features in simulating human judgement of opinion classification. *Inf. Process. Manag.* **2022**, *59*, 102823. [CrossRef]
28. Subhashini, L.; Li, Y.; Zhang, J.; Atukorale, A.S. Integration of semantic patterns and fuzzy concepts to reduce the boundary region in three-way decision-making. *Inf. Sci.* **2022**, *595*, 257–277. [CrossRef]

29. Mondal, A.; Roy, S.K.; Pamucar, D. Regret-based three-way decision making with possibility dominance and SPA theory in incomplete information system. *Expert Syst. Appl.* **2023**, *211*, 118688. [CrossRef]
30. Yao, Y.Y. Symbols-Meaning-Value (SMV) space as a basis for a conceptual model of data science. *Int. J. Approx. Reason.* **2022**, *144*, 113–128. [CrossRef]
31. Qian, Y.H.; Liang, J.Y.; Dang, C.Y. Incomplete multigranulation rough set. *IEEE Trans. Syst. Man-Cybern.-Part Syst. Humans* **2009**, *40*, 420–431. [CrossRef]
32. Huang, B.; Guo, C.X.; Zhuang, Y.L.; Li, H.X.; Zhou, X.Z. Intuitionistic fuzzy multigranulation rough sets. *Inf. Sci.* **2014**, *277*, 299–320. [CrossRef]
33. Li, F.J.; Qian, Y.H.; Wang, J.T.; Liang, J. Multigranulation information fusion: A Dempster-Shafer evidence theory-based clustering ensemble method. *Inf. Sci.* **2017**, *378*, 389–409. [CrossRef]
34. Liu, X.; Qian, Y.H.; Liang, J.Y. A rule-extraction framework under multigranulation rough sets. *Int. J. Mach. Learn. Cybern.* **2014**, *5*, 319–326. [CrossRef]
35. Liu, K.; Li, T.; Yang, X.; Ju, H.; Yang, X.; Liu, D. Hierarchical neighborhood entropy based multi-granularity attribute reduction with application to gene prioritization. *Int. J. Approx. Reason.* **2022**, *148*, 57–67. [CrossRef]
36. Qian, Y.H.; Liang, X.Y.; Lin, G.P.; Guo, Q.; Liang, J. Local multigranulation decision-theoretic rough sets. *Int. J. Approx. Reason.* **2017**, *82*, 119–137. [CrossRef]
37. Qian, Y.H.; Zhang, H.; Sang, Y.L.; Liang, J. Multigranulation decision-theoretic rough sets. *Int. J. Approx. Reason.* **2014**, *55*, 225–237. [CrossRef]
38. Xu, W.; Yuan, K.; Li, W. Dynamic updating approximations of local generalized multigranulation neighborhood rough set. *Appl. Intell.* **2022**, *52*, 9148–9173. [CrossRef]
39. Sun, L.; Wang, L.; Ding, W.; Qian, Y.; Xu, J. Feature selection using fuzzy neighborhood entropy-based uncertainty measures for fuzzy neighborhood multigranulation rough sets. *IEEE Trans. Fuzzy Syst.* **2020**, *29*, 19–33. [CrossRef]
40. She, Y.H.; He, X.L.; Shi, H.X.; Qian, Y. A multiple-valued logic approach for multigranulation rough set model. *Int. J. Approx. Reason.* **2017**, *82*, 270–284. [CrossRef]
41. Li, W.; Xu, W.; Zhang, X.Y.; Zhang, J. Updating approximations with dynamic objects based on local multigranulation rough sets in ordered information systems. *Artif. Intell. Rev.* **2021**, *55*, 1821–1855. [CrossRef]
42. Zhang, C.; Li, D.; Zhai, Y.; Yang, Y. Multigranulation rough set model in hesitant fuzzy information systems and its application in person-job fit. *Int. J. Mach. Learn. Cybern.* **2019**, *10*, 717–729. [CrossRef]
43. Hu, C.; Zhang, L.; Wang, B.; Zhang, Z.; Li, F. Incremental updating knowledge in neighborhood multigranulation rough sets under dynamic granular structures. *Knowl.-Based Syst.* **2019**, *163*, 811–829. [CrossRef]
44. Hu, C.; Zhang, L. Dynamic dominance-based multigranulation rough sets approaches with evolving ordered data. *Int. J. Mach. Learn. Cybern.* **2021**, *12*, 17–38. [CrossRef]

electronics

MDPI

Article

Relative Knowledge Distance Measure of Intuitionistic Fuzzy Concept

Jie Yang [1,2,*], Xiaodan Qin [1], Guoyin Wang [1], Xiaoxia Zhang [1] and Baoli Wang [3]

1 Chongqing Key Laboratory of Computational Intelligence, Chongqing University of Posts and Telecommunications, Chongqing 400065, China
2 School of Physics and Electronic Science, Zunyi Normal University, Zunyi 563002, China
3 College of Mathematics and Information Technology, Yuncheng University, Yuncheng 044000, China
* Correspondence: yj530966074@foxmail.com

Abstract: Knowledge distance is used to measure the difference between granular spaces, which is an uncertainty measure with strong distinguishing ability in a rough set. However, the current knowledge distance failed to take the relative difference between granular spaces into account under the given perspective of uncertain concepts. To solve this problem, this paper studies the relative knowledge distance of intuitionistic fuzzy concept (IFC). Firstly, a micro-knowledge distance (md) based on information entropy is proposed to measure the difference between intuitionistic fuzzy information granules. Then, based on md, a macro-knowledge distance (MD) with strong distinguishing ability is further constructed, and it is revealed the rule that MD is monotonic with the granularity being finer in multi-granularity spaces. Furthermore, the relative MD is further proposed to analyze the relative differences between different granular spaces from multiple perspectives. Finally, the effectiveness of relative MD is verified by relevant experiments. According to these experiments, the relative MD has successfully measured the differences in granular space from multiple perspectives. Compared with other attribute reduction algorithms, the number of subsets after reduction by our algorithm is in the middle, and the mean-square error value is appropriate.

Keywords: intuitionistic fuzzy concept; rough set; multi-granularity; relative macro-knowledge distance

Citation: Yang, J.; Qin, X.; Wang, G.; Zhang, X.; Wang, B. Relative Knowledge Distance Measure of Intuitionistic Fuzzy Concept. *Electronics* **2022**, *11*, 3373. https://doi.org/10.3390/electronics11203373

Academic Editor: Miin-shen Yang

Received: 20 September 2022
Accepted: 14 October 2022
Published: 19 October 2022

1. Introduction

Granular computing (GrC) [1–4] is a new type of computing used to solve problems by simulating the cognitive mechanism of humans. Information granule is the fundamental element in GrC for constructing granular spaces. A granular space consists of several information granules and their relationships, while a granular structure consists of many granular spaces and their relationships. By fusing the structure and optimization approach of granularity, Pedrycz [2] introduced the notion of justifiable granularity. Yao [5,6] examined the two fields of three-way decision and GrC, as well as their interplay. Wang [7,8] reviewed the GrC work from three aspects, including granularity optimization, granularity switching, and multi-granulation computing.

As the main GrC model, rough set [9] is a useful tool for handling uncertain knowledge by utilizing existing information granules. Uncertainty measure is a crucial tool for data analysis in a rough set. Wang [10] introduced a series of uncertainty measures for selecting the optimal features effectively. Li [11] offered the axiom definition of uncertainty measure for covering information systems by using its information structures. Sun [12] investigated the fuzzy neighborhood multigranulation rough set model to construct uncertainty measures. In generalized rough set models, Wang [13] described new uncertainty measures from the perspectives of the upper and lower approximations. Nevertheless, these uncertainty measures struggle to distinguish the differences between granular spaces when they possess the same uncertainty. To address this issue, Qian [14,15] first introduced the concept of knowledge distance, and there have been several works on knowledge distance in recent years. Li [16] proposed an interval-valued intuitionistic fuzzy set to describe fuzzy granular structure distance, and proved that knowledge distance is a special form of intuitionistic fuzzy granular structure distance. Yang [17,18] proposed a partition-based knowledge distance based on the Earth Mover's Distance and further established the fuzzy knowledge distance. Chen [19] presented a new measure formula of knowledge distance by using Jaccard distance to replace set similarity. To measure the uncertainty derived from the disparities between local upper and lower approximation sets, Xia [20] introduced the local knowledge distance.

In practical applications, the target concept may be vague or uncertain. As a classical soft computing tool, the intuitionistic fuzzy set [21] extends the membership from single value to interval value. For uncertain information, the intuitionistic fuzzy set has more powerful ability than the fuzzy set [22], and it is currently extensively applied in different fields, i.e., decision-making [23–25], pattern recognition [26,27], control and reasoning [28,29], and fuzzy reasoning [30,31]. In rough set, an intuitionistic fuzzy concept (IFC) can be characterized by a pair of lower and upper approximation fuzzy sets. There are many research works [32–37] on the combination between rough set and the intuitionistic fuzzy set. In particular, the uncertainty measure of IFC in granular spaces becomes a basic issue. A novel concept of an intuitionistic fuzzy rough set based on two universes was proposed by Zhang [32] along with a specification of the associated operators. On the basis of the rough set, Dubey [35] presented an intuitionistic fuzzy c-means clustering algorithm and applied it to the segmentation of the magnetic resonance brain images. Zheng [36] proposed an improved roughness method to measure the uncertainty of covering-based rough intuitionistic fuzzy sets. These works indicate that intuitionistic fuzzy set and rough set are suitable mathematical methods for studying vagueness and uncertainty. Current uncertainty measures failed to distinguish different rough granular spaces with the same uncertainty when they are used to describe an IFC; that is, it is difficult to reflect on the differences between them. However, in some situations, such as attribute reduction or granularity selection, the different rough granular spaces for describing an IFC are necessary to distinguish. To solve this problem, based on our previous works [17,18], two-layer knowledge distance measures—that is, micro-knowledge distance (md) and macro-knowledge distance (MD)—are constructed to reflect the difference between granular spaces for describing an IFC. Finally, in order to analyze the relative differences between rough granular spaces under certain prior granular spaces, the concept of relative MD applied to data analysis is also proposed.

The following are the main contributions of our paper: (1) Based on information entropy, md is designed to measure the difference among intuitionistic fuzzy information granules. (2) On the basis of md, MD with strong distinguishing ability is further constructed, which can calculate the difference between rough granular spaces for describing an IFC. (3) The relative MD is proposed to analyze the relative difference between two rough granular spaces from multiple perspectives. (4) An algorithm of attribute reduction based on MD or relative MD is presented, and its effectiveness is verified by relevant experiments.

The rest of this paper is arranged as follows. Section 2 introduces related preliminary concepts. In Section 3, the two types of information entropy-based distance measure (md and MD) are presented. Section 4 presents the concept of relative MD. The relevant experiments are reported in Section 5. Finally, in Section 6, conclusions are formed.

2. Preliminaries

This part will go through some of the core concepts. Let $S = (U, C \cup D, V, f)$ be an information system, where U, C, D and V represent the universe of discourse, condition attribute set, decision attribute set and attribute value set corresponding to each object, respectively, and $f : U \times C$ is an information function that specifies the property value of each object x in U.

Definition 1 (Intuitionistic fuzzy set). *Assume that U is the universe of discourse, the following is the definition of an intuitionistic fuzzy set I on U:*

$$I = \{< x, \gamma_I(x), v_I(x) > |x \in U\}$$

where $\gamma_I(x)$ and $v_I(x)$ denote two nonempty finite sets on the interval [0, 1], which refer to the set of degrees of membership and non-membership of x on I, respectively, and satisfy the conditions: $\forall x_i \in U,\ 0 \le \gamma_I(x_i) + v_I(x_i) \le 1$.

Note: For convenience, all I below are represented as intuitionistic fuzzy sets on U.

Definition 2 (Average step intuitionistic fuzzy set [38]). *Assume that in $S = (U, C \cup D)$, $R \subseteq C$ and $U/R = \{[x]_R\} = \{[x]_1, [x]_2, \cdots, [x]_l\}$, where $\forall x \in [x]_i$, $i = 1, 2, \cdots, l$, then,*

$$\overline{I}_R(x) = [\gamma_{\overline{I}_R}(x),\ 1 - v_{\overline{I}_R}(x)]$$

where, $\gamma_{\overline{I}_R}(x) = \frac{\sum\limits_{x\in[x]_i} \gamma_{I_R}(x)}{|[x]_i|}$, $v_{\overline{I}_R}(x) = \frac{\sum\limits_{x\in[x]_i} v_{I_R}(x)}{|[x]_i|}$. $\overline{I}_R(x)$ is therefore referred to as an average step intuitionistic fuzzy set on U/R.

As well known, the information entropy as an uncertainty measure is proposed in rough set theory,

$$E(x) = \sum_{x_i \in U} e(x_i)$$

where, $e(x_i) = -2\mu(x_i)\log_2\mu(x_i)$. Let U be a nonempty universe and I be an intuitionistic fuzzy set on U.

The information entropy of I can be expressed as follows:

(1) When U is continuous.

$$E_I(x) = \int_a^b e_I(x_i) \tag{1}$$

(2) When U is discrete.

$$E_I(x) = \sum_{x_i \in U} e_I(x_i)$$

where, $e_I(x_i) = -2\int_{\gamma_I(x_i)}^{1-v_I(x_i)} \mu\log_2\mu d\mu$, and μ denotes the membership degree of x_i belongs to the intuitionistic fuzzy set I.

To measure the information entropy of the rough granular space U/R of the IFC, this paper further proposed the definition of average information entropy as follows:

(1) When U is continuous, the average information entropy of the rough granular space of I can be denoted by:

$$E_{\overline{I}_R}(x) = \int_a^b e_{\overline{I}_R}(x)$$

(2) When U is discrete, the average information entropy of the rough granular space of I can be denoted by:

$$E_{\overline{I}_R}(x) = \sum_{x \in U} e_{\overline{I}_R}(x) \tag{2}$$

where, $e_{\overline{I}_R}(x) = -2\int_{\gamma_{\overline{I}_R}(x)}^{1-v_{\overline{I}_R}(x)} \mu\log_2\mu d\mu$, and μ denotes the membership degree of x belongs to the intuitionistic fuzzy set I, $\overline{I}_R(x)$ is the average step intuitionistic fuzzy set of U/R.

Example 1. *Assume that in* $S = (U, C \cup D)$, $R \subseteq C$, $I = \frac{[0.1,0.3]}{x_1} + \frac{[0.3,0.5]}{x_2} + \frac{[0.5,0.7]}{x_3} + \frac{[0.6,0.7]}{x_4} + \frac{[0.8,0.9]}{x_5}$ *and* $U/R = \{\{x_1, x_2, x_3\}, \{x_4, x_5\}\}$.

$$\bar{I}_R(x_1) = \bar{I}_R(x_2) = \bar{I}_R(x_3) = [\frac{0,1+0,3+0.5}{3}, \frac{0.3+0.5+0.7}{3}] = [0.3, 0.5]$$
$$\bar{I}_R(x_4) = \bar{I}_R(x_5) = [\frac{0,6+0,8}{2}, \frac{0.7+0.9}{2}] = [0.7, 0.8]$$
$$\bar{I}_R = \frac{[0.3,0.5]}{x_1} + \frac{[0.3,0.5]}{x_2} + \frac{[0.3,0.5]}{x_3} + \frac{[0.7,0.8]}{x_4} + \frac{[0.7,0.8]}{x_5}$$

Then

$$e_{\bar{I}_R}(x_1) = e_{\bar{I}_R}(x_2) = e_{\bar{I}_R}(x_3) = -2\int_{0.3}^{0.5} \mu log_2 \mu d\mu = 0.209$$
$$e_{\bar{I}_R}(x_4) = e_{\bar{I}_R}(x_5) = -2\int_{0.7}^{0.8} \mu log_2 \mu d\mu = 0.062$$

By Formula (2),

$$E_{\bar{I}_R}(x) = \sum_{x \in U} e_{\bar{I}_R}(x) = 0.209 \times 3 + 0.062 \times 2 = 0.751$$

Definition 3 (Distance measure [39]). *Assume that U is the universe of discourse; Y, P and Q are three finite sets on U. When $d(\cdot,\cdot)$ meets the following criteria, it is considered a distance measure,*
(1) Positive: $d(P, Q) \geq 0$;
(2) Symmetric: $d(P, Q) = D(Q, P)$;
(3) Triangle inequality: $d(Y, P) + d(P, Q) \geq d(Y, Q)$.

Definition 4 (Granularity measure [40]). *Assume that in* $S = (U, C \cup D)$, *G is a mapping from the power set of C to the real number set. For any* $R_1, R_2 \subseteq C$, *when G meets the following criteria, it is considered a granularity measure,*
(1) $G(R_1) \geq 0$;
(2) $U/R_1 \prec U/R_2 \Rightarrow G(R_1) < G(R_2)$;
(3) $U/R_1 = U/R_2 \Rightarrow G(R_1) = G(R_2)$.

Definition 5 (Information measure [41]). *Assume that in* $S = (U, C \cup D)$, *H is a mapping from the power set of C to the real number set. For any* $R_1, R_2 \subseteq C$, *when H meets the following criteria, it is considered an information measure,*
(1) $H(R_1) \geq 0$;
(2) $U/R_1 \prec U/R_2 \Rightarrow H(R_1) > H(R_2)$;
(3) $U/R_1 = U/R_2 \Rightarrow H(R_1) = H(R_2)$.

3. Information-Entropy-Based Two-Layer Knowledge Distance Measure

Although there are many research works [42–45] on distance measures between intuitionistic fuzzy sets from different perspectives, when an IFC is characterized by different rough granular spaces, respectively, the present fuzzy set distance measures failed to capture the differences between these granular spaces. In addition, as explained in Section 1, when an IFC is defined by two granular spaces, the measure result (fuzziness or information entropy) may be the same. Nevertheless, this does not mean that these two granular spaces are absolutely equal, and the difference between them for characterizing an IFC cannot be reflected. To tackle the difficulties listed above, this paper proposed micro-knowledge distance and macro-knowledge distance based on information entropy, which construct the two-layer knowledge distance measure in this section.

Example 2. *Assume that in* $S = (U, C \cup D)$, $R_1, R_2 \subseteq C$, $I = \frac{[0.1,0.3]}{x_1} + \frac{[0.3,0.5]}{x_2} + \frac{[0.5,0.7]}{x_3} + \frac{[0.6,0.7]}{x_4} + \frac{[0.8,0.9]}{x_5}$, $U/R_1 = \{\{x_2\}, \{x_1, x_3\}, \{x_4, x_5\}\}$ *and* $U/R_2 = \{\{x_1, x_2, x_3\}, \{x_4, x_5\}\}$.

By Formula (2),

$$E_{I_{R_1}}(x) = E_{I_{R_2}}(x) = 0.751$$

It shows that calculating the average information entropy does not necessarily distinguish and describe two different rough granular spaces. Although the average information entropy values of U/R_1 and U/R_2 are the same, U/R_2 is superior to U/R_1 in terms of granularity selection, since U/R_2 has a coarser granularity and has a stronger generalization ability for describing IFC.

Assume $S = (U, C \cup D)$, A is a finite set on U. Then, we call the intuitionistic fuzzy set generated by A as the intuitionistic fuzzy information granule (FIG), abbreviated as FIG_A.

Example 3 (Continued example 1). *Let* $P = \{x_1, x_2, x_3\}$ *and* $Q = \{x_4, x_5\}$, *then:*

$$IFG_P = \frac{[0.1, 0.3]}{x_1} + \frac{[0.3, 0.5]}{x_2} + \frac{[0.5, 0.7]}{x_3}$$
$$IFG_Q = \frac{[0.6, 0.7]}{x_4} + \frac{[0.8, 0.9]}{x_5}$$

Definition 6 (Micro-knowledge distance). *Assume in* $S = (U, C \cup D)$, IFG_P *and* IFG_Q *are two intuitionistic fuzzy information granules on* U, *hence, the following is the definition of the md formula:*

$$md(P, Q) = \frac{\sum_{x_i \in U} e_{IFG_P \cup IFG_Q}(x_i) - \sum_{x_i \in U} e_{IFG_P \cap IFG_Q}(x_i)}{E_I(x)}$$

Theorem 1. $md(\cdot, \cdot)$ *is a distance measure.*

Proof of Theorem 1. Let IFG_Y, IFG_P and IFG_Q be three intuitionistic fuzzy information granules. Let:

$$a = \sum_{x_i \in U} e_{IFG_Y \cup IFG_P}(x_i) - \sum_{x_i \in U} e_{IFG_Y \cap IFG_P}(x_i)$$
$$b = \sum_{x_i \in U} e_{IFG_P \cup IFG_Q}(x_i) - \sum_{x_i \in U} e_{IFG_P \cap IFG_Q}(x_i)$$
$$c = \sum_{x_i \in U} e_{IFG_Y \cup IFG_Q}(x_i) - \sum_{x_i \in U} e_{IFG_Y \cap IFG_Q}(x_i)$$

Because $(Y \cup P - Y \cap P) + (P \cup Q - P \cap Q) \geq Y \cup Q - Y \cap Q$, then $a + b \geq c$.

$$\frac{\sum_{x_i \in U} e_{FIG_Y \cup FIG_P}(x_i) - \sum_{x_i \in U} e_{FIG_Y \cap FIG_P}(x_i)}{E_I(x)} + \frac{\sum_{x_i \in U} e_{FIG_P \cup FIG_Q}(x_i) - \sum_{x_i \in U} e_{FIG_P \cap FIG_Q}(x_i)}{E_I(x)}$$
$$\geq \frac{\sum_{x_i \in U} e_{FIG_Y \cup FIG_Q}(x_i) - \sum_{x_i \in U} e_{FIG_Y \cap FIG_Q}(x_i)}{E_I(x)}$$

Then $md(Y, P) + md(P, Q) \geq md(Y, Q)$.

According to Definition 3, conditions (1) and (2) are obviously satisfied, Therefore, $md(\cdot, \cdot)$ is a distance measure. □

Example 4. *Assume that in* $S = (U, C \cup D)$, $R_1, R_2 \subseteq C$, $I = \frac{[0.3,0.6]}{x_1} + \frac{[0.2,0.5]}{x_2} + \frac{[0.5,0.7]}{x_3} + \frac{[0.7,0.9]}{x_4} + \frac{[0.6,0.8]}{x_5} + \frac{[0.1,0.2]}{x_6} + \frac{[0.2,0.4]}{x_7}$ *is an intuitionistic fuzzy set on* U, $A = \{x_1, x_3, x_4, x_6\}$ *and* $B = \{x_3, x_4, x_5, x_7\}$ *are two finite sets on* U.

$$FIG_A = \frac{[0.3, 0.6]}{x_1} + \frac{[0.5, 0.7]}{x_3} + \frac{[0.7, 0.9]}{x_4} + \frac{[0.1, 0.2]}{x_6}$$
$$FIG_B = \frac{[0.5, 0.7]}{x_3} + \frac{[0.7, 0.9]}{x_4} + \frac{[0.6, 0.8]}{x_5} + \frac{[0.2, 0.4]}{x_7}$$

Then,

$$e_I(x_i) = -2 \int_{\gamma_I(x_i)}^{1-v_I(x_i)} \mu log_2 \mu d\mu$$
$$\sum_{x_i \in U} e_{FIG_A \cup FIG_B}(x_i) = 1.009$$
$$\sum_{x_i \in U} e_{FIG_A \cap FIG_B}(x_i) = 0.277$$
$$E_I(x) = \sum_{x_i \in U} e_I(x_i) = 1.318$$

From Definition 6,

$$md(A, B) = \frac{\sum_{x_i \in U} e_{FIG_A \cup FIG_B}(x_i) - \sum_{x_i \in U} e_{FIG_A \cap FIG_B}(x_i)}{E_I(x)} = \frac{1.009 - 0.277}{1.318} = 0.555$$

Theorem 2. *Let* Y, P *and* Q *be three intuitionistic fuzzy sets on* U. *If* $Y \subseteq P \subseteq Q$, *then* $md(Y, P) \le md(Y, Q)$.

Proof of Theorem 2. According to condition, because $Y \subseteq P \subseteq Q$, obviously,

$$\sum_{x_i \in U} e_{FIG_Y \cup FIG_P}(x_i) = \sum_{x_i \in U} e_{FIG_P}(x_i) \le \sum_{x_i \in U} e_{FIG_Y \cup FIG_Q}(x_i) = \sum_{x_i \in U} e_{FIG_Q}(x_i)$$

$$\sum_{x_i \in U} e_{FIG_Y \cap FIG_P}(x_i) = \sum_{x_i \in U} e_{FIG_Y \cap FIG_Q}(x_i)$$

Then,

$$\frac{\sum_{x_i \in U} e_{FIG_Y \cup FIG_P}(x_i) - \sum_{x_i \in U} e_{FIG_Y \cap FIG_P}(x_i)}{E_I(x)} \le \frac{\sum_{x_i \in U} e_{FIG_Y \cup FIG_Q}(x_i) - \sum_{x_i \in U} e_{FIG_Y \cap FIG_Q}(x_i)}{E_I(x)}$$

Therefore, $md(Y, P) \le md(Y, Q)$ holds. Similarly, it is easy to get $md(P, Q) \le md(Y, Q)$. □

Theorem 3. *Let* Y, P *and* Q *be three intuitionistic fuzzy sets on* U. *If* $Y \subseteq P \subseteq Q$, *then* $md(Y, Q) = md(Y, P) + md(P, Q)$.

Proof of Theorem 3. Theorem 3 obviously holds.

$$\sum_{x_i \in U} e_{FIG_Y \cup FIG_P}(x_i) = \sum_{x_i \in U} e_{FIG_P}(x_i), \sum_{x_i \in U} e_{FIG_Y \cap FIG_P}(x_i) = \sum_{x_i \in U} e_{FIG_Y}(x_i),$$
$$\sum_{x_i \in U} e_{FIG_P \cup FIG_Q}(x_i) = \sum_{x_i \in U} e_{FIG_Q}(x_i), \sum_{x_i \in U} e_{FIG_P \cap FIG_Q}(x_i) = \sum_{x_i \in U} e_{FIG_P}(x_i),$$
$$\sum_{x_i \in U} e_{FIG_Y \cup FIG_Q}(x_i) = \sum_{x_i \in U} e_{FIG_Q}(x_i), \sum_{x_i \in U} e_{FIG_Y \cap FIG_Q}(x_i) = \sum_{x_i \in U} e_{FIG_Y}(x_i).$$

Then

$$
\begin{aligned}
& md(Y,P) + md(P,Q) \\
= & \frac{\sum\limits_{x_i \in U} e_{FIG_Y \cup FIG_P}(x_i) - \sum\limits_{x_i \in U} e_{FIG_Y \cap FIG_P}(x_i)}{E_I(x)} + \frac{\sum\limits_{x_i \in U} e_{FIG_P \cup FIG_Q}(x_i) - \sum\limits_{x_i \in U} e_{FIG_P \cap FIG_Q}(x_i)}{E_I(x)} \\
= & \frac{\sum\limits_{x_i \in U} e_{FIG_Q}(x_i) - \sum\limits_{x_i \in U} e_{FIG_Y}(x_i)}{E_I(x)} \\
= & md(Y,Q)
\end{aligned}
$$

Therefore, $md(Y,Q) = md(Y,P) + md(P,Q)$. □

Based on md, this research further created MD, which is formulated as follows, to express the difference between two rough granular spaces for characterizing an IFC.

Definition 7 (Macro-knowledge distance). *Assume that in* $S = (U, C \cup D)$, R_1, $R_2 \subseteq C$, $U/R_1 = \{g_1, g_2, ..., g_n\}$ *and* $U/R_2 = \{g_1', g_2', ..., g_m'\}$ *are two granular spaces induced by* R_1 *and* R_2, *respectively. Then, the following is the definition of MD between* U/R_1 *and* U/R_2.

$$
MD(U/R_1, U/R_2) = \frac{1}{|U|} \sum_{i=1}^{n} \sum_{j=1}^{m} md_{ij} f_{ij} \tag{3}
$$

where, $md_{ij} = md(g_i, g_j')$ *and* $f_{ij} = \left| g_i \cap g_j' \right|$. *Figure 1 shows the relationship between md and MD.*

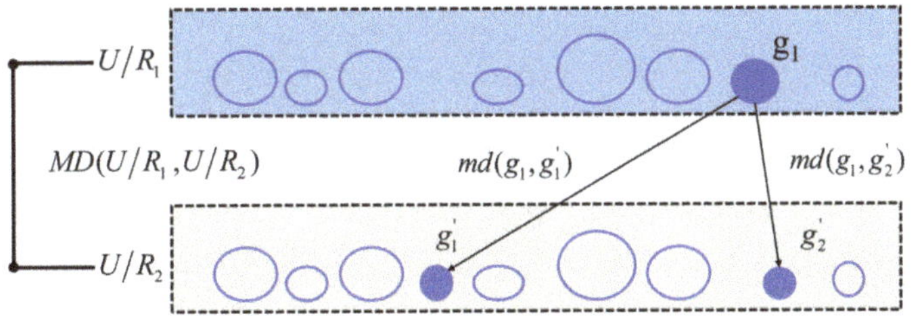

Figure 1. The relationship between md and MD.

Suppose that $U/R_1 = \{g_1, g_2, ..., g_n\}$ is a granular space on U induced by R_1, where $g_i = \{x_{i1}, x_{i2}, \cdots, x_{i|g_i|}\}$, Then, $g_i = s_{R_1}(x_{i1}) = s_{R_1}(x_{i2}) = \cdots = s_{R_1}(x_{i|g_i|})$.

For example,

$$
\begin{aligned}
& U/R_1 = \{\{x_1, x_2\}, \{x_3\}, \{x_4, x_5\}\} = \{g_1,\ g_2,\ g_3\} \\
& g_1 = s_{R_1}(x_1) = s_{R_1}(x_2) = \{x_1, x_2\} \\
& g_2 = s_{R_1}(x_3) = \{x_3\} \\
& g_3 = s_{R_1}(x_4) = s_{R_1}(x_5) = \{x_4, x_5\}
\end{aligned}
$$

Theorem 4. $MD(\cdot,\cdot)$ *is a distance measure.*

Proof of Theorem 4. Assume that in $S = (U, C \cup D), R_1, R_2, R_3 \subseteq C, U/R_1 = \{g_1, g_2, ..., g_n\}$, $U/R_2 = \{g_1^{'}, g_2^{'}, ..., g_m^{'}\}$ and $U/R_3 = \{g_1^{''}, g_2^{''}, ..., g_l^{''}\}$ are three granular spaces induced by R_1, R_2 andR_3, respectively. Obviously, $MD(\cdot, \cdot)$ is positive and symmetric.

$$MD(U/R_1, U/R_2) + MD(U/R_2, U/R_3) = \frac{1}{|U|}\sum_{i=1}^{n}\sum_{j=1}^{m} md_{ij}f_{ij} + \frac{1}{|U|}\sum_{j=1}^{m}\sum_{k=1}^{l} md_{ij}f_{ij}$$

$$= \frac{1}{|U|}\sum_{x_i \in U}(md(s_{R_1}(x_i), s_{R_2}(x_i)) + md(s_{R_2}(x_i), s_{R_3}(x_i)))$$

$$MD(U/R_1, U/R_3) = \frac{1}{|U|}\sum_{i=1}^{n}\sum_{j=1}^{l} md_{ij}f_{ij} = \frac{1}{|U|}\sum_{x_i \in U} md(s_{R_1}(x_i), s_{R_3}(x_i)) \quad (4)$$

From Theorem 1, $md(s_{R_1}(x_i), s_{R_2}(x_i)) + md(s_{R_2}(x_i), s_{R_3}(x_i)) \geq md(s_{R_1}(x_i), s_{R_3}(x_i))$ Then,

$$\frac{1}{|U|}\sum_{x_i \in U}(md(s_{R_1}(x_i), s_{R_2}(x_i)) + md(s_{R_2}(x_i), s_{R_3}(x_i))) \geq \frac{1}{|U|}\sum_{x_i \in U} md(s_{R_1}(x_i), s_{R_3}(x_i))$$

$$MD(U/R_1, U/R_2) + MD(U/R_2, U/R_3) \geq MD(U/R_1, U/R_3)$$

Therefore, from Definition 3, $MD(\cdot, \cdot)$ is a distance measure. □

In fact, md measures the difference between two sets, and MD measures the difference between two rough granular spaces, which integrates the md of all sets of the two granular spaces. According to Theorem 1, Theorem 4 and Formula (4), as long as md in MD is a distance measure, then MD is a distance measure.

Example 5 (Continued Example 2). *According to Formula (3),*

$$\begin{aligned} MD(U/R_1, U/R_2) &= \frac{1}{|U|}\sum_{i=1}^{3}\sum_{j=1}^{2} md_{ij}f_{ij} \\ &= \frac{md_{11} + md_{21} * 2 + md_{32} * 2}{5} \\ &= \frac{0.356 + 0.209 \times 2 + 0 \times 2}{5 \times 0.686} \\ &= 0.226 \end{aligned}$$

Theorem 5. *Assume that in $S = (U, C \cup D)$, $R_1, R_2, R_3 \subseteq C$. If $R_1 \subseteq R_2 \subseteq R_3$, then $MD(U/R_1, U/R_2) \leq MD(U/R_1, U/R_3)$.*

Proof of Theorem 5. As shown in Figure 2, suppose $U/R_1 = \{g_1, g_2, ..., g_n\}$, $U/R_2 = \{g_1^{'}, g_2^{'}, ..., g_m^{'}\}$ and $U/R_3 = \{g_1^{''}, g_2^{''}, ..., g_l^{''}\}$ are three granular spaces induced by R_1, R_2 and R_3, respectively. Because $R_1 \subseteq R_2 \subseteq R_3$, so $U/R_3 \preceq U/R_2 \preceq U/R_1$. For simplicity, supposing only one granule g_1 can be subdivided into two finer sub-granules $g_1^{'}$ and $g_2^{'}$ by $\Delta R = R_2 - R_1$ and only one granule $g_1^{'}$ can be subdivided into two finer sub-granules $g_1^{''}$ and $g_2^{''}$ by $\Delta R = R_3 - R_2$ (Because more sophisticated examples may be turned into this scenario, this essay will not go through them again.).

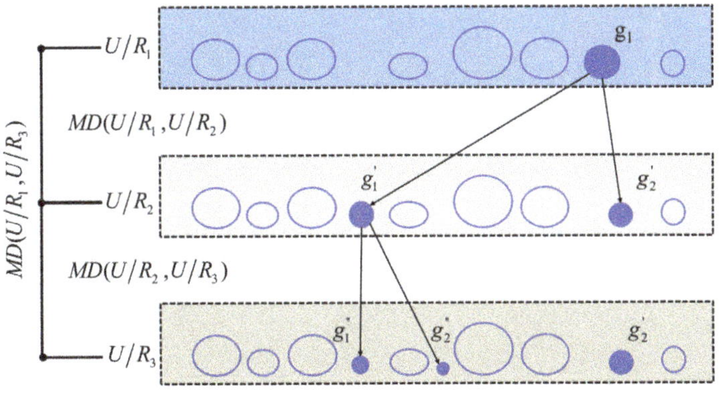

Figure 2. The relationship of MD among three granular spaces.

According to the above assumption, $g_1 = g_1^{'} \cup g_2^{'}, g_2 = g_3^{'}, g_3 = g_4^{'}, \cdots, g_n = g_m^{'}$ $(m = n+1)$, $g_1^{'} = g_1^{''} \cup g_2^{''}, g_2^{'} = g_3^{''}, g_3^{'} = g_4^{''}, \cdots, g_m^{'} = g_l^{''}$ $(l = m+1)$, namely, $U/R_2 = \{g_1^{'}, g_2^{'}, g_2, \cdots, g_n\}$ and $U/R_3 = \{g_1^{''}, g_2^{''}, g_2^{'}, \cdots, g_m^{'}\}$. Then, from Definition 7,

$$MD(U/R_1, U/R_2) = \frac{1}{|U|}\sum_{i=1}^{n}\sum_{j=1}^{m} md_{ij}f_{ij} = \frac{1}{|U|}(md(g_1, g_1^{'})\left|g_1^{'}\right| + md(g_1, g_2^{'})\left|g_2^{'}\right|)$$

$$\begin{aligned} MD(U/R_1, U/R_3) &= \frac{1}{|U|}\sum_{i=1}^{n}\sum_{j=1}^{l} md_{ij}f_{ij} \\ &= \frac{1}{|U|}(md(g_1, g_1^{''})\left|g_1^{''}\right| + md(g_1, g_2^{''})\left|g_2^{''}\right| + md(g_1, g_3^{''})\left|g_3^{''}\right|) \end{aligned}$$

Because $g_1^{'} = g_1^{''} \cup g_2^{''}, g_2^{'} = g_3^{''}$

$$\begin{aligned} &MD(U/R_1, U/R_3) - MD(U/R_1, U/R_2) \\ &= \frac{1}{|U|}((md(g_1, g_1^{''}) - md(g_1, g_1^{'}))\left|g_1^{''}\right| + (md(g_1, g_2^{''}) - md(g_1, g_1^{'}))\left|g_2^{''}\right|) \end{aligned}$$

According to Theorem 2, because $md(g_1, g_1^{'}) \le md(g_1, g_1^{''})$ and $md(g_1, g_1^{'}) \le md(g_1, g_2^{''})$, then $MD(U/R_1, U/R_3) - MD(U/R_1, U/R_2) \ge 0$.

Therefore, $MD(U/R_1, U/R_2) \le MD(U/R_1, U/R_3)$ holds. Similarly, it is easy to get $MD(U/R_2, U/R_3) \le MD(U/R_1, U/R_3)$. □

In this paper, the finest and coarsest granular spaces are represented by ω and σ, respectively. The following corollaries derive from Theorem 5:

Corollary 1. *Assume that in $S = (U, C \cup D)$, $R_1, R_2 \subseteq C$. If $R_1 \subseteq R_2 \subseteq C$, then $MD(U/R_1, \omega) \ge MD(U/R_2, \omega)$.*

Corollary 2. *Assume that in $S = (U, C \cup D)$, $R_1, R_2 \subseteq C$. If $R_1 \subseteq R_2 \subseteq C$, then $MD(U/R_1, \sigma) \le MD(U/R_2, \sigma)$.*

Theorem 6. *Assume that in $S = (U, C \cup D)$, $R \subseteq C$. According to Definition 4, $MD(U/R, \omega)$ is a granularity measure.*

Proof of Theorem 6. Suppose that $R_1, R_2 \subseteq C$,
(1) From Theorem 4, $MD(U/R,\omega) \geq 0$;
(2) When U/R_1=U/R_2, obviously, $MD(U/R_1,\omega) = MD(U/R_2,\omega)$;
(3) From Corollary 1, if $R_1 \subseteq R_2$, then $MD(U/R_1,\omega) \geq MD(U/R_2,\omega)$. □

Theorem 7. *Assume that in $S = (U, C \cup D)$, $R \subseteq C$. According to Definition 5, $MD(U/R,\sigma)$ is an information measure.*

Proof of Theorem 7. Suppose that $R_1, R_2 \subseteq C$,
(1) From Theorem 4, $MD(U/R,\sigma) \geq 0$;
(2) When U/R_1=U/R_2, obviously, $MD(U/R_1,\sigma) = MD(U/R_2,\sigma)$;
(3) From Corollary 2, if $R_1 \subseteq R_2$, then $MD(U/R_1,\sigma) \leq MD(U/R_2,\sigma)$. □

Theorem 8. *Assume that in $S = (U, C \cup D)$, $R_1 \subseteq R_2 \subseteq R_3 \subseteq C$, then $MD(U/R_1, U/R_3) = MD(U/R_1, U/R_2) + MD(U/R_2, U/R_3)$.*

Proof of Theorem 8. For simplicity, based on the proof of Theorem 5,

$$MD(U/R_1, U/R_2) = \frac{1}{|U|}(md(g_1, g_1^{'})\left|g_1^{'}\right| + md(g_1, g_2^{'})\left|g_2^{'}\right|)$$
$$MD(U/R_1, U/R_3) = \frac{1}{|U|}(md(g_1, g_1^{''})\left|g_1^{''}\right| + md(g_1, g_2^{''})\left|g_2^{''}\right| + md(g_1, g_3^{''})\left|g_3^{''}\right|)$$

Similarly,

$$MD(U/R_2, U/R_3) = \frac{1}{|U|}(md(g_1^{'}, g_1^{''})\left|g_1^{''}\right| + md(g_1^{'}, g_2^{''})\left|g_2^{''}\right|)$$

Because $g_1 = g_1^{'} \cup g_2^{'}, g_1^{'} = g_1^{''} \cup g_2^{''}$ and $g_2^{'} = g_3^{''}$.
According to Theorem 3,

$$\begin{aligned}
&MD(U/R_1, U/R_2) + MD(U/R_2, U/R_3) \\
&= \frac{1}{|U|}(md(g_1, g_1^{''})\left|g_1^{''}\right| + md(g_1, g_2^{''})\left|g_2^{''}\right| + md(g_1, g_2^{'})\left|g_2^{'}\right|) \\
&= \frac{1}{|U|}(md(g_1, g_1^{''})\left|g_1^{''}\right| + md(g_1, g_2^{''})\left|g_2^{''}\right| + md(g_1, g_3^{''})\left|g_3^{''}\right|) \\
&= MD(U/R_1, U/R_3)
\end{aligned}$$

□

According to Theorem 8, from the perspective of distance, the granular spaces in hierarchical granular structure are linearly additive, which can be explained by Figure 2 intuitively. Moreover, the following corollaries hold:

Corollary 3. *Assume that in $S = (U, C \cup D)$, $R_1, R_2 \subseteq C$. If $R_1 \subseteq R_2 \subseteq C$, then $MD(U/R_1, U/R_2) = MD(U/R_1,\omega) - MD(U/R_2,\omega)$.*

Corollary 4. *Assume that in $S = (U, C \cup D)$, $R_1, R_2 \subseteq C$. If $R_1 \subseteq R_2 \subseteq C$, then, $MD(U/R_1, U/R_2) = MD(U/R_2,\sigma) - MD(U/R_1,\sigma)$.*

Corollary 5. *Assume that in $S = (U, C \cup D)$, $R \subseteq C$. Then $MD(U/R,\omega) + MD(U/R,\sigma) = \frac{|U|-1}{|U|}$.*

Proof of Corollary 5.

$$MD(U/R,\omega) = \frac{1}{|U|}\sum_{i=1}^{n}\sum_{j=1}^{|U|} md_{ij}f_{ij} = \frac{1}{|U|}\sum_{x_i\in U} md(s_R(x_i),\{x_i\})$$
$$= \frac{1}{|U|}\sum_{x_i\in U}\frac{\sum e_{s_R(x_i)-\{x_i\}}(x)}{E_I(x)}$$
$$MD(U/R,\sigma) = \frac{1}{|U|}\sum_{i=1}^{n}\sum_{j=1}^{1} md_{ij}f_{ij} = \frac{1}{|U|}\sum_{x_i\in U} md(s_R(x_i),U)$$
$$= \frac{1}{|U|}\sum_{x_i\in U}\frac{\sum e_{C_U s_R(x_i)}(x)}{E_I(x)}$$
$$MD(U/R,\omega) + MD(U/R,\sigma)$$
$$= \frac{1}{|U|}\sum_{x_i\in U}\frac{\sum e_{s_R(x_i)-\{x_i\}}(x)}{E_I(x)} + \frac{1}{|U|}\sum_{x_i\in U}\frac{\sum e_{C_U s_R(x_i)}(x)}{E_I(x)}$$
$$= \frac{1}{|U|}\sum_{x_i\in U}\frac{\sum e_{s_R(x_i)-\{x_i\}}(x) + \sum e_{C_U s_R(x_i)}(x)}{E_I(x)}$$
$$= \frac{1}{|U|}\sum_{x_i\in U}\frac{\sum e_{U-\{x_i\}}(x)}{E_I(x)} = \frac{1}{|U|}\sum_{x_i\in U}\frac{E_I(x)-e(x_i)}{E_I(x)}$$
$$= \frac{1}{|U|}\times\frac{|U|\times E_I(x) - E_I(x)}{E_I(x)} = \frac{|U|-1}{|U|}$$

Therefore, $MD(U/R,\omega) + MD(U/R,\sigma) = \frac{|U|-1}{|U|}$ holds. □

From Corollary 3 and Theorem 6, for an IFC, the larger the granularity difference between granular spaces in hierarchical granular structure, the larger MD between them. From Corollary 4 and Theorem 7, for an IFC, the larger the information difference between granular spaces in hierarchical granular structure, the larger MD between them. From Corollary 5, the larger the information measure, the smaller the granularity measure, and one measure value can be deduced from another.

Note: By using a suitable md in Formula (3), the method of this paper is able to extend to quantify the difference between any types of granular spaces. These specifics are outside the scope of this paper's discussion.

4. Relative Macro-Knowledge Distance

Section 3 has constructed an MD based on md, which described the difference between two rough granular spaces of IFC. We regarded this knowledge distance as absolute. Because in data analysis, sometimes some conditions are known, it is necessary to analyze the differences between rough granular spaces at this time; that is, to analyze the differences between rough granular spaces under different prior granular spaces. Inspired by Wang [46], this section proposes the concept of relative MD and analyzes its properties.

Definition 8 (Relative macro-knowledge distance). *Assume that in $S = (U, C\cup D)$, $R_1, R_2 \subseteq C$, U/R is the prior granular space on U, $U/R_1 = \{g_1, g_2, \cdots, g_n\}$ and $U/R_2 = \{g_1', g_2', \cdots, g_m'\}$ are two granular spaces induced by R_1 and R_2, respectively. Then, the relative MD of U/R_1 and U/R_2 under U/R is defined as:*

$$RMD((U/R_1, U/R_2)/(U/R)) = \frac{1}{|U|}\sum_{x_i\in U} md(s_{R_1/R}(x_i), s_{R_2/R}(x_i)) \quad (5)$$

where, $s_{R_1/R}(x_i) = s_{R_1}(x_i)\cap s_R(x_i)$ and $s_{R_2/R}(x_i) = s_{R_2}(x_i)\cap s_R(x_i)$.

Based on the original MD, this definition adds prior granular space U/R, which reflects the relative differences between two rough granular spaces from different perspectives.

Theorem 9. *$RMD(\cdot,\cdot/\cdot)$ is a distance measure.*

Proof of Theorem 9. Assume that in $S=(U,C\cup D)$, U/R is the prior granular space on U. $R_1,R_2,R_3\subseteq C$, $U/R_1=\{g_1,g_2,\cdots,g_n\}$, $U/R_2=\{g_1^{'},g_2^{'},\cdots,g_m^{'}\}$ and $U/R_3=\{g_1^{''},g_2^{''},\cdots,g_l^{''}\}$ are three granular spaces induced by R_1, R_2 andR_3, respectively. Obviously, $RMD(\cdot,\cdot/\cdot)$ is positive and symmetric.

$$
\begin{aligned}
&RMD((U/R_1,U/R_2)/(U/R))+RMD((U/R_2,U/R_3)/(U/R))\\
&=\frac{1}{|U|}\sum_{x_i\in U}md(s_{R_1/R}(x_i),s_{R_2/R}(x_i))+\frac{1}{|U|}\sum_{x_i\in U}md(s_{R_2/R}(x_i),s_{R_3/R}(x_i))\\
&RMD((U/R_1,U/R_3)/(U/R))=\frac{1}{|U|}\sum_{x_i\in U}md(s_{R_1/R}(x_i),s_{R_3/R}(x_i))
\end{aligned}
$$

According to Theorem 1,

$$
md(s_{R_1/R}(x_i),s_{R_2/R}(x_i))+md(s_{R_2/R}(x_i),s_{R_3/R}(x_i))\geq md(s_{R_1/R}(x_i),s_{R_3/R}(x_i))
$$

Then,

$$
\begin{aligned}
&\frac{1}{|U|}\sum_{x_i\in U}md(s_{R_1/R}(x_i),s_{R_2/R}(x_i))+\frac{1}{|U|}\sum_{x_i\in U}md(s_{R_2/R}(x_i),s_{R_3/R}(x_i))\\
&\geq\frac{1}{|U|}\sum_{x_i\in U}md(s_{R_1/R}(x_i),s_{R_3/R}(x_i))\\
&RMD((U/R_1,U/R_2)/(U/R))+RMD((U/R_2,U/R_3)/(U/R))\\
&\geq RMD((U/R_1,U/R_3)/(U/R))
\end{aligned}
$$

Therefore, from Definition 3, $RMD(\cdot,\cdot/\cdot)$ is a distance measure. □

Example 6. *Assume that in $S=(U,C\cup D)$, $R_1,R_2\subseteq C$, $I=\frac{[0.1,0.3]}{x_1}+\frac{[0.3,0.5]}{x_2}+\frac{[0.5,0.7]}{x_3}+\frac{[0.6,0.7]}{x_4}+\frac{[0.8,0.9]}{x_5}$. $U/R_1=\{\{x_2\},\{x_1,x_3\},\{x_4,x_5\}\}$ and $U/R_2=\{\{x_1,x_2,x_3\},\{x_4,x_5\}\}$. Under the prior granular spaces $U/R_3=\{\{x_1,x_3,x_5\},\{x_2,x_4\}\}$ and $U/R_4=\{\{x_1,x_5\},\{x_2,x_3,x_4\}\}$, the relative MD of U/R_1 and U/R_2 are as follows:*

$$
\begin{aligned}
&RMD((U/R_1,U/R_2)/(U/R_3))=\frac{1}{|U|}\sum_{x_i\in U}md(s_{R_1/R_3}(x_i),s_{R_2/R_3}(x_i))=0\\
&RMD((U/R_1,U/R_2)/(U/R_4))=\frac{1}{|U|}\sum_{x_i\in U}md(s_{R_1/R_4}(x_i),s_{R_2/R_4}(x_i))\\
&=\frac{1}{5}\times(0+\frac{0.175}{0.686}+\frac{0.209}{0.686}+0+0)=0.112
\end{aligned}
$$

From Examples 5 and 6, after adding the prior granular space, the difference between the two rough granular spaces may change, and when the prior granular space is different, the obtained results may also be different.

Theorem 10. *Assume that in $S=(U,C\cup D)$, U/R is the prior granular space on U. If $R_1\subseteq R_2\subseteq R_3\subseteq C$, then $RMD((U/R_1,U/R_2)/(U/R))\leq RMD((U/R_1,U/R_3)/(U/R))$.*

Proof of Theorem 10. According to the conditions, $U/R_3 \preceq U/R_2 \preceq U/R_1$, then $s_{R_3}(x_i) \subseteq s_{R_2}(x_i) \subseteq s_{R_1}(x_i)$, $x_i \subseteq U$.

$$s_{R_1/R}(x_i) = s_{R_1}(x_i) \cap s_R(x_i)$$
$$s_{R_2/R}(x_i) = s_{R_2}(x_i) \cap s_R(x_i)$$
$$s_{R_3/R}(x_i) = s_{R_3}(x_i) \cap s_R(x_i)$$

So, $s_{R_3/R}(x_i) \subseteq s_{R_2/R}(x_i) \subseteq s_{R_1/R}(x_i)$.
According to Theorem 2,

$$md(s_{R_1/R}(x_i), s_{R_2/R}(x_i)) \le md(s_{R_1/R}(x_i), s_{R_3/R}(x_i))$$
$$\frac{1}{|U|}\sum_{x_i \in U} md(s_{R_1/R}(x_i), s_{R_2/R}(x_i)) \le \frac{1}{|U|}\sum_{x_i \in U} md(s_{R_1/R}(x_i), s_{R_3/R}(x_i))$$

Therefore, $RMD((U/R_1, U/R_2)/(U/R)) \le RMD((U/R_1, U/R_3)/(U/R))$ holds. Similarly, it is easy to get
$RMD((U/R_2, U/R_3)/(U/R)) \le RMD((U/R_1, U/R_3)/(U/R))$. □

Theorem 11. *Assume that in $S = (U, C \cup D)$, U/R is the prior granular space on U. If $R_1 \subseteq R_2 \subseteq R_3 \subseteq C$, then $RMD((U/R_1, U/R_3)/(U/R)) = RMD((U/R_1, U/R_2)/(U/R)) + RMD((U/R_2, U/R_3)/(U/R))$.*

Proof of Theorem 11. According to Theorem 10, it can be deduced $s_{R_3/R}(x_i) \subseteq s_{R_2/R}(x_i) \subseteq s_{R_1/R}(x_i)$ from condition $R_1 \subseteq R_2 \subseteq R_3 \subseteq C$. Additionally, according to Theorem 3, $md(s_{R_1/R}(x_i), s_{R_3/R}(x_i)) = md(s_{R_1/R}(x_i), s_{R_2/R}(x_i)) + md(s_{R_2/R}(x_i), s_{R_3/R}(x_i))$. Then,

$$\frac{1}{|U|}\sum_{x_i \in U} md(s_{R_1/R}(x_i), s_{R_3/R}(x_i))$$
$$= \frac{1}{|U|}\sum_{x_i \in U} md(s_{R_1/R}(x_i), s_{R_2/R}(x_i)) + \frac{1}{|U|}\sum_{x_i \in U} md(s_{R_2/R}(x_i), s_{R_3/R}(x_i))$$

Therefore, $RMD((U/R_1, U/R_3)/(U/R))$
$= RMD((U/R_1, U/R_2)/(U/R)) + RMD((U/R_2, U/R_3)/(U/R))$ holds. □

From Theorem 11, under the same prior granular space, the relative MD is linearly additive.

Theorem 12. *Assume that in $S = (U, C \cup D)$, U/R_3 and U/R_4 are two prior granular spaces on U, respectively, $R_1, R_2 \subseteq C$. If $U/R_3 \preceq U/R_4$, then $RMD((U/R_1, U/R_2)/(U/R_3)) \le RMD((U/R_1, U/R_2)/(U/R_4))$.*

Proof of Theorem 12.

$$RMD((U/R_1, U/R_2)/(U/R_3))$$
$$= \frac{1}{|U|}\sum_{x_i \in U} md(s_{R_1}(x_i) \cap s_{R_3}(x_i), s_{R_2}(x_i) \cap s_{R_3}(x_i))$$
$$= \frac{1}{|U|}\sum_{x_i \in U} \frac{\sum_{x \in U} e_{(s_{R_1}(x_i) \cap s_{R_3}(x_i)) \cup (s_{R_2}(x_i) \cap s_{R_3}(x_i))}(x) - \sum_{x \in U} e_{(s_{R_1}(x_i) \cap s_{R_3}(x_i)) \cap (s_{R_2}(x_i) \cap s_{R_3}(x_i))}(x)}{E_I(x)}$$
$$= \frac{1}{|U|}\frac{a}{E_I(x)}$$

$$
\begin{aligned}
a &= \sum_{x_i \in U} \Big(\sum_{x \in U} e_{(s_{R_1}(x_i) \cap s_{R_3}(x_i)) \cup (s_{R_2}(x_i) \cap s_{R_3}(x_i))}(x) - \sum_{x \in U} e_{(s_{R_1}(x_i) \cap s_{R_3}(x_i)) \cap (s_{R_2}(x_i) \cap s_{R_3}(x_i))}(x) \Big) \\
&= \sum_{x_i \in U} \Big(\sum_{x \in U} e_{((s_{R_1}(x_i) \cup s_{R_2}(x_i)) \cap s_{R_3}(x_i))}(x) - \sum_{x \in U} e_{(s_{R_1}(x_i) \cap s_{R_2}(x_i) \cap s_{R_3}(x_i))}(x) \Big) \\
&= \sum_{x_i \in U} \sum_{x \in U} e_{((s_{R_1}(x_i) \cup s_{R_2}(x_i) - s_{R_1}(x_i) \cap s_{R_2}(x_i)) \cap s_{R_3}(x_i))}(x)
\end{aligned}
$$

Similarly,
$RMD((U/R_1, U/R_2)/(U/R_4)) = \sum_{x_i \in U} \sum_{x \in U} e_{((s_{R_1}(x_i) \cup s_{R_2}(x_i) - s_{R_1}(x_i) \cap s_{R_2}(x_i)) \cap s_{R_4}(x_i))}(x)$

According to $U/R_3 \preceq U/R_4$, $s_{R_3}(x_i) \subseteq s_{R_4}(x_i)$, $x_i \subseteq U$. Obviously,

$$
\begin{aligned}
&((s_{R_1}(x_i) \cup s_{R_2}(x_i) - s_{R_1}(x_i) \cap s_{R_2}(x_i)) \cap s_{R_3}(x_i)) \\
\subseteq\ &((s_{R_1}(x_i) \cup s_{R_2}(x_i) - s_{R_1}(x_i) \cap s_{R_2}(x_i)) \cap s_{R_4}(x_i)) \\
&\sum_{x_i \in U} \sum_{x \in U} e_{((s_{R_1}(x_i) \cup s_{R_2}(x_i) - s_{R_1}(x_i) \cap s_{R_2}(x_i)) \cap s_{R_3}(x_i))}(x) \\
\leq &\sum_{x_i \in U} \sum_{x \in U} e_{((s_{R_1}(x_i) \cup s_{R_2}(x_i) - s_{R_1}(x_i) \cap s_{R_2}(x_i)) \cap s_{R_4}(x_i))}(x)
\end{aligned}
$$

Therefore, $RMD((U/R_1, U/R_2)/(U/R_3)) \leq RMD((U/R_1, U/R_2)/(U/R_4))$ holds. □

Corollary 6. *Assume that in $S = (U, C \cup D)$, $R_1, R_2 \subseteq C$, σ is the prior granular space on U, then $RMD((U/R_1, U/R_2)/\sigma) = MD(U/R_1, U/R_2)$.*

Proof of Corollary 6.

$$
\begin{aligned}
RMD((U/R_1, U/R_2)/\sigma) &= \frac{1}{|U|} \sum_{x_i \in U} md(s_{R_1}(x_i) \cap U, s_{R_2}(x_i) \cap U) \\
&= \frac{1}{|U|} \sum_{x_i \in U} md(s_{R_1}(x_i), s_{R_2}(x_i)) = MD(U/R_1, U/R_2)
\end{aligned}
$$

□

Corollary 7. *Assume that in $S = (U, C \cup D)$, $R_1, R_2 \subseteq C$, ω is the prior granular space on U, then $RMD((U/R_1, U/R_2)/\omega) = 0$.*

Proof of Corollary 7.

$$
\begin{aligned}
RMD((U/R_1, U/R_2)/\omega) &= \frac{1}{|U|} \sum_{x_i \in U} md(s_{R_1}(x_i) \cap \{x_i\}, s_{R_2}(x_i) \cap \{x_i\}) \\
&= \frac{1}{|U|} \sum_{x_i \in U} md(\{x_i\}, \{x_i\}) = 0
\end{aligned}
$$

□

Note: From Example 6, when the prior granular space is not the most refined, the relative MD may also be zero. Therefore, the prior granular space is the most refined granular space, which is only a sufficient condition for the relative MD to be zero, not a necessary condition.

According to Corollary 6, the absolute MD is the relative MD without any prior granular space; that is, the absolute MD can be viewed as a special case of the relative MD. By Corollary 7, when the prior granular space is fine enough, the relative MD between two different rough granular spaces has been infinitely reduced or even to zero. Combining

Theorem 12, it follows that $RMD((U/R_1, U/R_2)/\omega)) \le RMD((U/R_1, U/R_2)/(U/R)) \le RMD((U/R_1, U/R_2)/\sigma)$ is true when $\omega \preceq U/R \preceq \sigma$.

That is, $0 \le RMD((U/R_1, U/R_2)/(U/R)) \le MD(U/R_1, U/R_2)$.

Theorem 13. *Assume that in* $S = (U, C \cup D)$, $R_1, R_2 \subseteq C$. *Then* $RMD((U/R_1, U/R_2)/(U/R_1)) + RMD((U/R_1, U/R_2)/(U/R_2)) = MD(U/R_1, U/R_2)$.

Proof of Theorem 13.

$$
\begin{aligned}
& RMD((U/R_1, U/R_2)/(U/R_1)) + RMD((U/R_1, U/R_2)/(U/R_2)) \\
&= \frac{1}{|U|} \sum_{x_i \in U} (md(s_{R_1}(x_i) \cap s_{R_1}(x_i), s_{R_2}(x_i) \cap s_{R_1}(x_i))) \\
&+ \frac{1}{|U|} \sum_{x_i \in U} (md(s_{R_1}(x_i) \cap s_{R_2}(x_i), s_{R_2}(x_i) \cap s_{R_2}(x_i))) \\
&= \frac{1}{|U|} \sum_{x_i \in U} (md(s_{R_1}(x_i), s_{R_2}(x_i) \cap s_{R_1}(x_i))) + \frac{1}{|U|} \sum_{x_i \in U} (md(s_{R_1}(x_i) \cap s_{R_2}(x_i), s_{R_2}(x_i)))
\end{aligned}
$$

According to Theorem 3,

$$
\begin{aligned}
& \frac{1}{|U|} \sum_{x_i \in U} (md(s_{R_1}(x_i), s_{R_2}(x_i) \cap s_{R_1}(x_i))) + \frac{1}{|U|} \sum_{x_i \in U} (md(s_{R_1}(x_i) \cap s_{R_2}(x_i), s_{R_2}(x_i))) \\
&= \frac{1}{|U|} \sum_{x_i \in U} md(s_{R_1}(x_i),\ s_{R_2}(x_i)) = MD(U/R_1, U/R_2)
\end{aligned}
$$

Therefore, $RMD((U/R_1, U/R_2)/(U/R_1)) + RMD((U/R_1, U/R_2)/(U/R_2)) = MD(U/R_1, U/R_2)$ holds. □

From Theorem 13, an absolute MD was divided into the sum of two unidirectional relative MD in different directions. That is, the absolute MD of the two granular spaces is equal to the relative MD of the two granular spaces when the prior granular space is one of the two granular spaces, plus the relative MD of the two granular spaces when the prior granular space is the other granular space of the two granular spaces. This theoretically explains the dialectical unity of relative MD and absolute MD.

5. Experiment and Analysis

This section verifies that MD has a good advantage when describing IFC in multi-granularity space through relevant experiments. The experimental environments are Windows 10, Intel Core (TM) I5-10500 CPU (3.10 GHz) and 16GB RAM. The experimental platform is MATLAB 2022a. We filtered out nine datasets with decision attributes and a sufficient number of conditional attributes from UCI [47] and Dryad. Meanwhile, we removed attributes from some datasets that are completely independent of the decision attributes, such as serial number and date. The dataset's basic information is recorded in Table 1, and experiments will use the following formula [48] to convert numerical values to discrete values. For convenience, the ID numbers in Table 1 will be used to represent the datasets.

$$\alpha^1(x) = \lfloor (\alpha(x) - \min_\alpha)/\sigma_\alpha \rfloor$$

where $\alpha(x)$ represent the attribute value, $\min_\alpha$ represent the minimum value of $\alpha(x)$ and σ_α represent the standard deviation of the attribute.

Table 1. Experimental dataset.

ID	Dataset	Instances	Condition Attributes
1	Hungarian Chickenpox Cases Dataset	521	19
2	Data from: Relative importance of chemical attractiveness to parasites for susceptibility to trematode infection [49]	67	7
3	Waterlow score on admission in acutely admitted patients aged 65 and over [50]	839	11
4	Data from: Salivary gland ultrasonography as a predictor of clinical activity in Sjögren's syndrome [51]	70	10
5	Data from: Development and validation of a postoperative delirium prediction model for patients admitted to an intensive care unit in China: a prospective study [52]	300	13
6	Data from: Age of first infection across a range of parasite taxa in a wild mammalian population [53]	140	12
7	Air Quality	9538	10
8	Concrete	1030	8
9	ENB2012	768	8

5.1. Monotonicity Experiment

In this experiment, some attributes of the dataset in Table 1 were selected. Suppose $GL = (GL_1, GL_2, GL_3, GL_4, GL_5)$ is a hierarchical quotient space structure consisting of five granularity layers. $GL_i = (U_i,\ R_i \cup D,\ V,\ f)$, R_i represent attribute sets, and $R_1 \subset R_2 \subset R_3 \subset R_4 \subset R_5$. As shown in Figure 3, this figure shows that the behavior of each dataset is similar; that is, MD increases with the increase in the granularity difference between two granular spaces, and conversely, MD decreases with the decrease in the granularity difference between two granular spaces. Table 2 summarizes the changes in the two measures (granularity measure and information measure) based on MD in a hierarchical quotient space structure as the granularity layer becomes finer. The findings indicate that these two measures can provide additional information for assessing the uncertainty of fuzzy concepts. These findings support Theorems 6 and 7. The granularity measure decreases as the available information increases, while the information measure increases as the available information increases. According to Table 2, Corollary 5 can also be verified, the sum of granularity measure and information measure is fixed, and the result is $\frac{|U|-1}{|U|}$.

5.2. Attribute Reduction

The so-called attribute reduction is to delete the irrelevant or unimportant attributes under the condition that the classification ability of the knowledge base remains unchanged. In data analysis, deleting unnecessary attributes can greatly improve the efficiency of data analysis, and the subset derived from attribute reduction with prior granular space may be different from the subset derived from attribute reduction without prior granular space. Aiming at this problem, this section makes a comparative experiment of attribute reduction based on relative MD in different prior granular spaces and attribute reduction based on absolute MD; in this paper, the attributes that divide the prior granular space are called prior conditions.

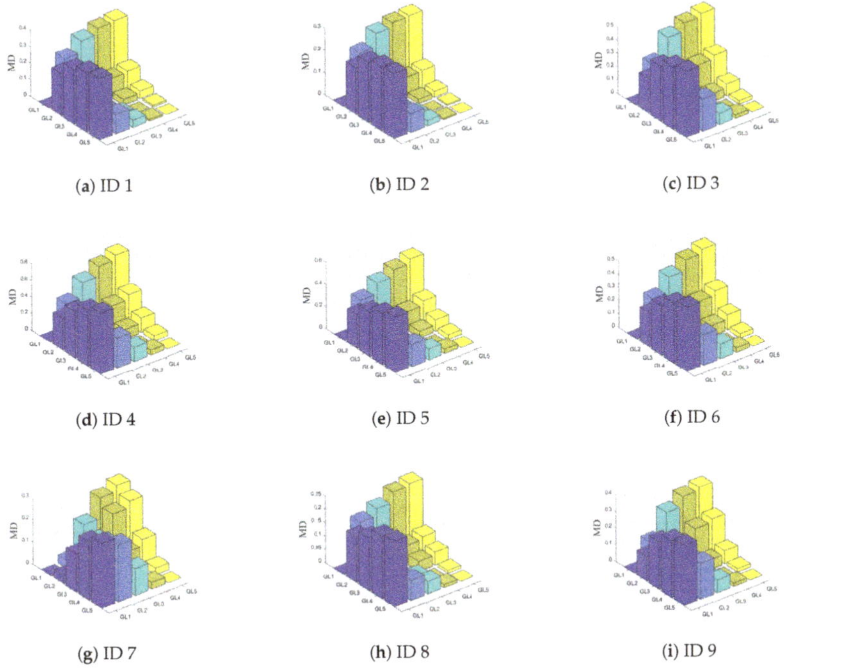

Figure 3. The change of MD between different granular spaces. Each dataset is represented by ID number.

Table 2. Granularity measure and Information measure.

ID (Dataset)	Measure	GL1	GL2	GL3	GL4	GL5
1	Granularity measure	0.4052	0.1523	0.0790	0.0453	0.0357
	Information measure	0.5928	0.8458	0.9190	0.9528	0.9624
2	Granularity measure	0.2952	0.0853	0.0253	0.0060	0.0012
	Information measure	0.6899	0.8998	0.9598	0.9791	0.9839
3	Granularity measure	0.5156	0.2604	0.1040	0.0451	0.0191
	Information measure	0.4832	0.7384	0.8948	0.9537	0.9797
4	Granularity measure	0.7516	0.3777	0.1916	0.0805	0.0199
	Information measure	0.2341	0.6080	0.7942	0.9052	0.9658
5	Granularity measure	0.5249	0.2229	0.1233	0.0653	0.0244
	Information measure	0.4717	0.7737	0.8734	0.9314	0.9723
6	Granularity measure	0.4975	0.2463	0.1267	0.0493	0.0155
	Information measure	0.4954	0.7466	0.8661	0.9436	0.9773
7	Granularity measure	0.3788	0.3438	0.2087	0.1170	0.0872
	Information measure	0.6211	0.6561	0.7911	0.8828	0.9127
8	Granularity measure	0.2557	0.0869	0.0588	0.0217	0.0112
	Information measure	0.7433	0.9121	0.9402	0.9773	0.9878
9	Granularity measure	0.3675	0.2237	0.0829	0.0271	0.0066
	information measure	0.6312	0.7750	0.9157	0.9716	0.9921

Some attributes of the dataset in Table 1 were selected in the experiment. Taking the calculation of attribute reduction based on relative MD as an example, Algorithm 1 is the algorithm used in the experiment. Attribute reduction based on absolute MD only needs to change the fourth step in Algorithm 1 to delete the first and last items in $conT$; that is, without any prior conditions. In this paper, Algorithm 2 is used to represent attribute reduction based on absolute MD. Suppose an information system $S = (U,\ C \cup D,\ V,\ f)$, then the calculation formula of attribute importance id is as follows:

$$id = MD(U/C, U/\{C - c_i\}) \tag{6}$$

As shown in Figure 4, the attribute importance represents MD between the granular space after removing attribute i in the dataset and the granular space without removing this attribute. The larger the distance, the higher the attribute importance degree, and c_i represents attribute i. Therefore, in this paper, the attributes with the largest and smallest attribute importance of the dataset are selected as prior conditions, and attribute reduction based on relative MD is carried out. Moreover, attribute reduction based on absolute MD is also performed without any prior conditions.

Algorithm 1 Attribute reduction based on relative MD

Input: An information system $S = (U,\ C \cup D,\ V,\ f)$
Output: Attribute subset R obtained after attribute reduction
1: Let $R = C$ and $conT = C$
2: Calculate the information entropy of each instance by Formula (1)
3: Calculate the attribute importance of all attributes in $conT$ by Formula (6), and sort this in ascending order, the result is recorded as $conT\_rank$
4: Take $conT\_rank(1)$ or $conT\_rank(length(conT))$ as a prior condition, that is, delete the last or first item in $conT\_rank$ and ensure that the first item or last item in $conT\_rank$ always exists
5: **while** $conT\_rank \neq \emptyset$ **do**
6: $\quad conT\_rank = conT\_rank - \{c\}$, where c is the first element in $conT\_rank$
7: $\quad$ According to Formula (3), calculate MD from the granular space obtained by the remaining conditions after $R - \{c\}$ to the granular space obtained by all conditions
8: $\quad$ **if** $MD < \xi$ **then**
9: $\quad\quad R = R - \{c\}$
10: $\quad$ **end if**
11: **end while**
12: Return R

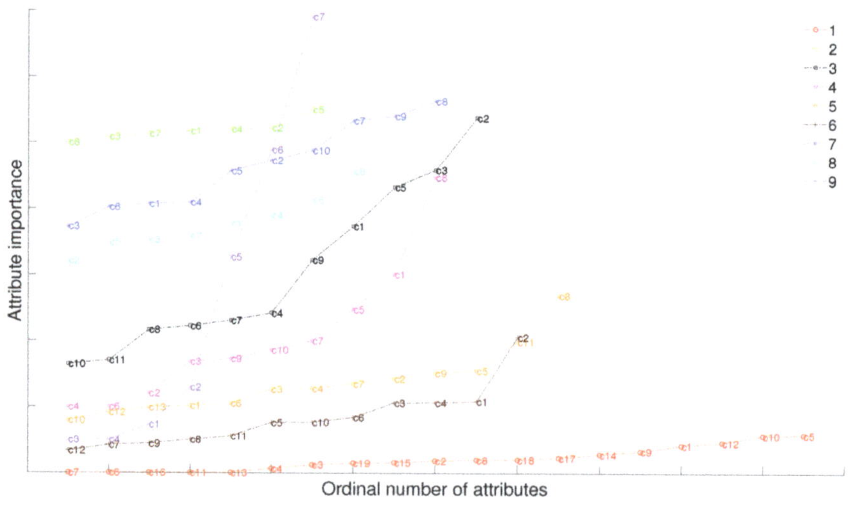

Figure 4. Attribute importance of different attributes.

(Note: Figure 4 is only used to analyze the importance of the conditional attribute of a single system, so there is no correlation between the height of the line graph of different systems).

As shown in Table 3, in the attribute reduction based on absolute MD, ξ is the maximum absolute MD between the granular space divided by attribute subsets after attribute reduction and the granular space divided by all attributes. In attribute reduction based on relative MD, ξ is the maximum relative MD between the granular space divided by attribute subsets after attribute reduction and the granular space divided by all attributes. This paper sets ξ to 0.003 and 0.006 for comparison. In the table, numbers are directly used to represent the serial numbers of the conditional attributes.

According to the analysis in Figure 4 and Table 3:

(1) When the prior conditions are more important attributes, the number of attributes is significantly reduced compared to the attribute reduction based on absolute MD, which shows that selecting more important properties increases the cognitive ability of the system, consistent with Theorem 12.

(2) When the prior condition is an unimportant attribute, compared with the prior condition is an important attribute, the number of subsets after attribute reduction is usually more, which also indicates that the more important the prior condition is, the more cognitive ability of the attribute to the system can be improved.

(3) When ξ is different—that is, the maximum MD between the granular space remains conditionally divided after attribute reduction and the granular space divided without reduction changes—the subsets after attribute reduction may be different, which illustrates the efficiency of this algorithm. The algorithm will obtain different attribute subsets as the requirements increase and decrease.

(4) The reduced attributes are all attributes with low attribute importance, which shows the effectiveness of this algorithm in calculating attribute importance.

Table 3. Attribute reduction on each dataset based on different situations.

ID (Dataset)	The Original Attributes (Number)	Attribute Reduction Based on MD (In Parentheses Is the Number of Attributes after Attribute Reduction)		
1	6,16,11,13,4,3,19,15,2, 8,18,17,14,9,1,12,10 (17)	Absolute MD	$\xi = 0.003$	15,2,8,18,17,14,9,1,12,10 (10)
			$\xi = 0.006$	8,18,17,14,9,1,12,10 (8)
		Relative MD with attribute 7 as a prior condition	$\xi = 0.003$	15,2,8,18,17,14,9,1,12,10 (10)
			$\xi = 0.006$	8,18,17,14,9,1,12,10 (8)
		Relative MD with attribute 5 as a prior condition	$\xi = 0.003$	2,8,18,17,14,9,1,12,10 (9)
			$\xi = 0.006$	8,18,17,14,9,1,12,10 (8)
2	3,7,1,4,2 (5)	Absolute MD	$\xi = 0.003$	3,7,1,4,2 (5)
			$\xi = 0.006$	7,1,4,2 (4)
		Relative MD with attribute 6 as a prior condition	$\xi = 0.003$	7,1,4,2 (4)
			$\xi = 0.006$	1,4,2 (3)
		Relative MD with attribute 5 as a prior condition	$\xi = 0.003$	7,1,4,2 (4)
			$\xi = 0.006$	7,1,2 (3)
3	11,8,6,7,4,9,1,5,3 (9)	Absolute MD	$\xi = 0.003$	8,6,7,4,9,1,5,3 (8)
			$\xi = 0.006$	6,7,4,9,1,5,3 (7)
		Relative MD with attribute 10 as a prior condition	$\xi = 0.003$	8,6,7,4,9,1,5,3 (8)
			$\xi = 0.006$	6,7,4,9,1,5,3 (7)
		Relative MD with attribute 2 as a prior condition	$\xi = 0.003$	6,7,4,9,1,5,3 (7)
			$\xi = 0.006$	7,4,9,1,5,3 (6)

Table 3. *Cont.*

ID (Dataset)	The Original Attributes (Number)	Attribute Reduction Based on MD (In Parentheses Is the Number of Attributes after Attribute Reduction)		
4	6,2,3,9,10,7,5,1 (8)	Absolute MD	$\xi = 0.003$ $\xi = 0.006$	6,2,3,9,10,7,5,1 (8) 6,3,9,10,7,5,1 (7)
		Relative MD with attribute 4 as a prior condition	$\xi = 0.003$ $\xi = 0.006$	6,3,9,10,7,5,1 (7) 6,3,10,7,5,1 (6)
		Relative MD with attribute 8 as a prior condition	$\xi = 0.003$ $\xi = 0.006$	3,9,10,7,5,1 (6) 3,10,7,5,1 (5)
5	12,13,1,6,3,4,7,2,9, 5,11 (11)	Absolute MD	$\xi = 0.003$ $\xi = 0.006$	1,6,3,4,7,2,9,5,11 (9) 6,3,4,7,2,9,5,11 (8)
		Relative MD with attribute 10 as a prior condition	$\xi = 0.003$ $\xi = 0.006$	1,6,3,4,7,2,9,5,11 (9) 3,4,7,2,9,5,11 (7)
		Relative MD with attribute 8 as a prior condition	$\xi = 0.003$ $\xi = 0.006$	3,4,7,2,9,5,11 (7) 3,4,2,9,5,11 (6)
6	7,9,8,11,5,10,6,3,4, 1 (10)	Absolute MD	$\xi = 0.003$ $\xi = 0.006$	9,8,5,10,6,3,4,1 (8) 8,5,10,6,3,4,1 (7)
		Relative MD with attribute 12 as a prior condition	$\xi = 0.003$ $\xi = 0.006$	8,11,5,10,6,3,4,1 (8) 8,5,10,6,3,4,1 (7)
		Relative MD with attribute 2 as a prior condition	$\xi = 0.003$ $\xi = 0.006$	11,5,10,6,3,4,1 (7) 5,10,6,3,4,1 (6)
7	6,1,4,5,2,10,7,9 (8)	Absolute MD	$\xi = 0.003$ $\xi = 0.006$	1,4,5,2,10,7,9 (7) 1,4,5,2,10,7,9 (7)
		Relative MD with attribute 3 as a prior condition	$\xi = 0.003$ $\xi = 0.006$	1,4,5,2,10,7,9 (7) 1,5,2,10,7,9 (6)
		Relative MD with attribute 8 as a prior condition	$\xi = 0.003$ $\xi = 0.006$	1,5,2,10,7,9 (6) 4,5,2,10,7,9 (6)
8	5,3,7,1,4,6 (6)	Absolute MD	$\xi = 0.003$ $\xi = 0.006$	3,7,1,4,6 (5) 3,7,1,4,6 (5)
		Relative MD with attribute 2 as a prior condition	$\xi = 0.003$ $\xi = 0.006$	3,7,1,4,6 (5) 7,1,4,6 (4)
		Relative MD with attribute 8 as a prior condition	$\xi = 0.003$ $\xi = 0.006$	3,7,1,4,6 (5) 7,1,4,6 (4)
9	4,1,2,5,6 (5)	Absolute MD	$\xi = 0.003$ $\xi = 0.006$	1,2,5,6 (4) 2,5,6 (3)
		Relative MD with attribute 3 as a prior condition	$\xi = 0.003$ $\xi = 0.006$	2,5,6 (3) 2,5,6 (3)
		Relative MD with attribute 7 as a prior condition	$\xi = 0.003$ $\xi = 0.006$	2,5,6 (3) 2,5,6 (3)

As shown in Figure 5, Algorithm 1 and Algorithm 2, respectively, represent the attribute reduction based on absolute MD and the attribute reduction based on relative MD with important condition as a prior condition proposed in this paper. Algorithm 3, Algorithm 4, and Algorithm 5 denote three attribute reduction algorithms based on Mi's fuzziness, entropy-based fuzziness, and secondary fuzziness, respectively, where the left figure shows the case of $\xi = 0.003$, and the right figure shows the case of $\xi = 0.006$. From Figure 5, the number of remaining attribute subsets in Algorithm 1 after attribute reduction is appropriate. In addition, after adding the prior conditions, the number of subsets obtained after attribute reduction is significantly reduced in Algorithm 2, indicating that our algorithm greatly improves the cognitive ability of the system. Figure 6 shows the average value of the mean-square error before and after attribute reduction in three different regression models (random forest regression, decision tree regression and GBDT regularization) after the normalization of the nine datasets. In the figure, the prior condition of relative MD1 is the least important attribute, and the prior condition of relative MD2 is the most important attribute. After attribute reduction, we discover that the mean-squared error does not significantly change, and sometimes even decreases. This shows the feasibility of our algorithm and also shows that the algorithm can be effectively used in data analysis.

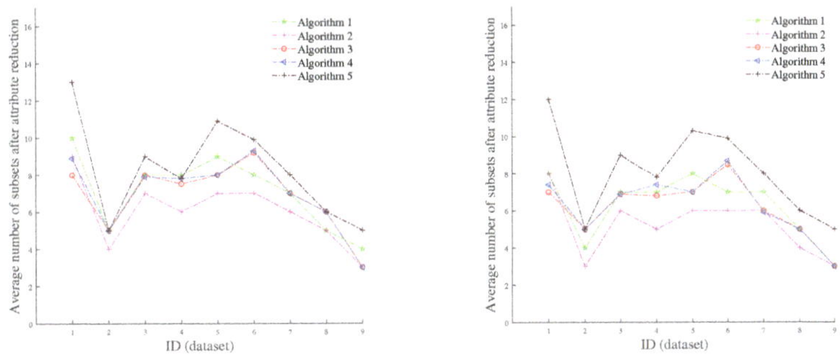

Figure 5. Average number of attribute subsets after five different attribute reductions.

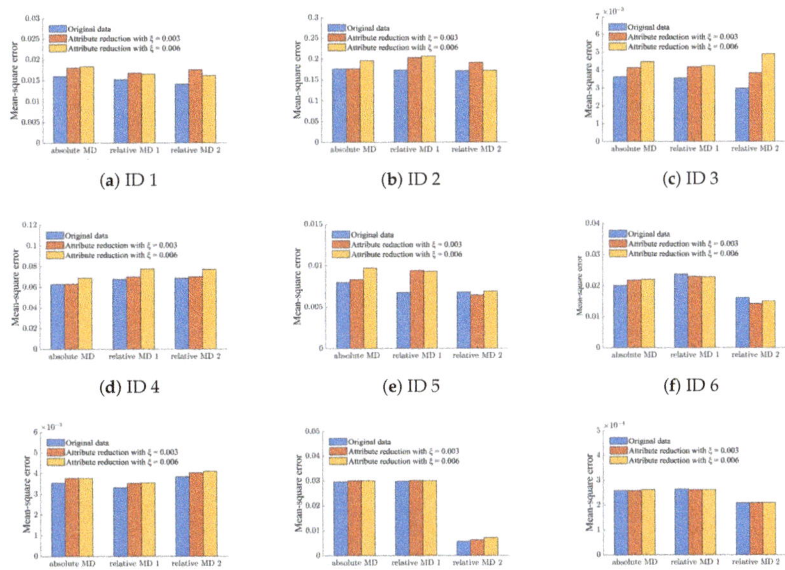

(**a**) ID 1 (**b**) ID 2 (**c**) ID 3

(**d**) ID 4 (**e**) ID 5 (**f**) ID 6

(**g**) ID 7 (**h**) ID 8 (**i**) ID 9

Figure 6. Average value of the mean-square error on each dataset. Each dataset is represented by ID number.

This paper also conducted a series of comparative experiments using the five algorithms mentioned above to compare the mean-square error values following the reduction of five different attributes. The experimental results are shown in Figure 7. In order to unify the standard, $\xi = 0.003$ is used. Except for the datasets with ID 8 and ID 9, the mean-square error values obtained by our algorithm are in the middle. From Figure 6, after attribute reduction of dataset ID 8 and dataset ID 9, the mean-square error does not change much. Therefore, the reason for this result is that the correlation between some attributes of the dataset itself and the decision attributes is too large or too small. There is still room for improvement in this regard.

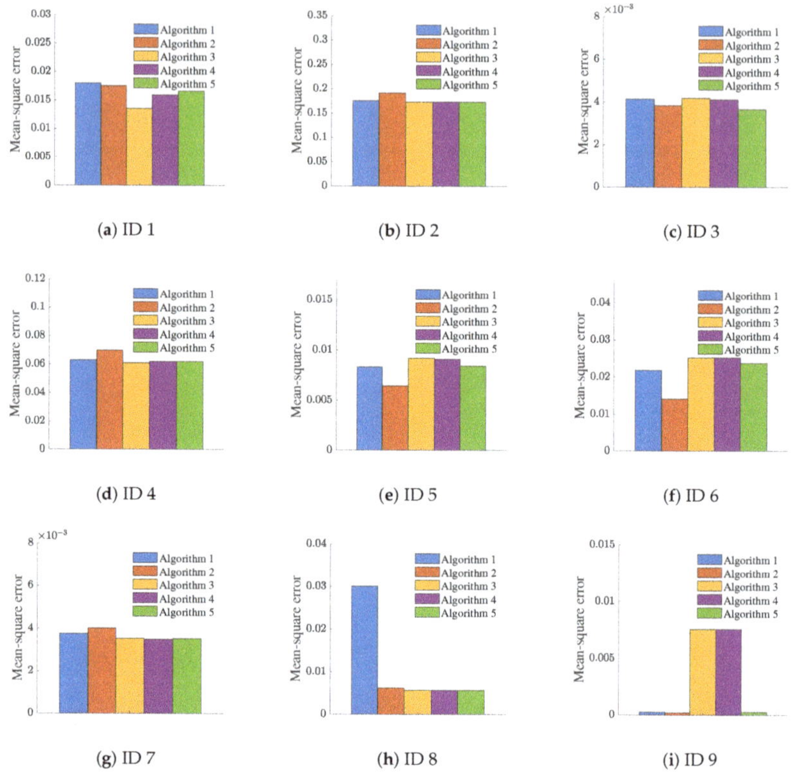

Figure 7. Average value of the mean-square error of five different attribute reductions. Each dataset is represented by ID number.

6. Conclusions and Discussion

In this paper, the macro-knowledge distance of intuitionistic fuzzy sets is proposed to measure the difference between granular spaces effectively. Under the given perspective of uncertain concepts, the current knowledge distance failed to account for the relative difference between granular spaces. As a result, we further propose the relative macro-knowledge distance and demonstrate its practicability through relative attribute reduction experiments. These results provide a new perspective to current knowledge distance research by measuring the relative differences between different granular spaces under prior granular spaces. The conclusions are as follows:

(1) Macro-knowledge distance increases with the increase in the granularity difference between two granular spaces, and vice versa. The sum of granularity measure and information measure is always $\frac{|U|-1}{|U|}$.

(2) After attribute reduction, the number of subsets obtained by our algorithm is appropriate, and in comparison to other algorithms, our mean square error is suitable. In the analysis of data, the more important the prior condition is, the more it can improve the cognitive ability of the attributes.

Under specific circumstances, the relative macro-knowledge distance is able to remove unnecessary attributes in practical applications, which can significantly increase the accuracy of attribute reduction and the effectiveness of data analysis. The characteristics of the data will be more thoroughly understood during the attribute reduction process.

Author Contributions: Conceptualization, J.Y. and X.Z.; methodology, J.Y., X.Q. and X.Z.; writing—original draft, J.Y. and X.Q; writing—review and editing, J.Y., X.Q., G.W. and B.W.; data curation, G.W., X.Z. and B.W.; supervision, X.Z. All authors have read and agreed to the published version of the manuscript.

Funding: This work is supported by the National Science Foundation of China (No. 62066049), Excellent Young Scientific and Technological Talents Foundation of Guizhou Province (QKH-platform talent[2021] No. 5627), National Science Foundation of Chongqing (cstc2021ycjh-bgzxm0013), Guizhou Provincial Science and Technology Project (QKH-ZK [2021]General 332), Science Foundation of Guizhou Provincial Education Department (QJJ2022[088]), The Applied Basic Research Program of Shanxi Province (No. 201901D211462), Electronic Manufacturing Industry-University-Research Base of Ordinary Colleges and Universities in Guizhou Province (QJH-KY-Zi [2014] No. 230-2).

Institutional Review Board Statement: Not applicable.

Informed Consent Statement: Not applicable.

Data Availability Statement: Not applicable.

Acknowledgments: This study was completed at the Chongqing Key Laboratory of Computational Intelligence, Chongqing University of Posts and Telecommunications, and the authors would like to thank the laboratory for its assistance.

Conflicts of Interest: The authors declare no conflict of interest.

References

1. Yao, Y.Y. The art of granular computing. In Proceedings of the International Conference on Rough Sets and Intelligent Systems Paradigms, Warsaw, Poland, 28–30 June 2007; pp. 101–112.
2. Bargiela, A.; Pedrycz, W. Toward a theory of granular computing for human-centered information processing. *IEEE Trans. Fuzzy Syst.* **2008**, *16*, 320–330. [CrossRef]
3. Yao, J.T.; Vasilakos, A.V.; Pedrycz, W. Granular computing: Perspectives and challenges. *IEEE Trans. Cybern.* **2013**, *43*, 1977–1989. [CrossRef] [PubMed]
4. Yao, Y.Y. Granular computing: Basic issues and possible solutions. In Proceedings of the 5th Joint Conference on Information Sciences, Atlantic City, NJ, USA, 27 February–3 March 2000; Volume 1, pp. 186–189.
5. Yao, Y.Y. Set-theoretic models of three-way decision. *Granul. Comput.* **2021**, *6*, 133–148. [CrossRef]
6. Yao, Y.Y. Tri-level thinking: Models of three-way decision. *Int. J. Mach. Learn. Cybern.* **2020**, *11*, 947–959. [CrossRef]
7. Wang, G.Y.; Yang, J.; Xu, J. Granular computing: From granularity optimization to multi-granularity joint problem solving. *Granul. Comput.* **2017**, *2*, 105–120. [CrossRef]
8. Wang, G.Y. DGCC: Data-driven granular cognitive computing. *Granul. Comput.* **2017**, *2*, 343–355. [CrossRef]
9. Pawlak, Z. Rough sets. *Int. J. Comput. Inf. Sci.* **1982**, *11*, 341–356. [CrossRef]
10. Wang, C.Z.; Huang, Y.; Shao, M.W.; Hu, Q.H.; Chen, D.G. Feature selection based on neighborhood self-information. *IEEE Trans. Cybern.* **2019**, *50*, 4031–4042. [CrossRef]
11. Li, Z.W.; Zhang, P.F.; Ge, X.; Xie, N.X.; Zhang, G.Q. Uncertainty measurement for a covering information system. *Soft Comput.* **2019**, *23*, 5307–5325. [CrossRef]
12. Sun, L.; Wang, L.Y.; Ding, W.P.; Qian, Y.H.; Xu, J.C. Feature selection using fuzzy neighborhood entropy-based uncertainty measures for fuzzy neighborhood multigranulation rough sets. *IEEE Trans. Fuzzy Syst.* **2020**, *29*, 19–33. [CrossRef]
13. Wang, Z.H.; Yue, H.F.; Deng, J.P. An uncertainty measure based on lower and upper approximations for generalized rough set models. *Fundam. Informaticae* **2019**, *166*, 273–296. [CrossRef]
14. Qian, Y.H.; Liang, J.Y.; Dang, C.Y. Knowledge structure, knowledge granulation and knowledge distance in a knowledge base. *Int. J. Approx. Reason.* **2009**, *50*, 174–188. [CrossRef]
15. Qian, Y.H.; Li, Y.B.; Liang, J.Y.; Lin, G.P.; Dang, C.Y. Fuzzy granular structure distance. *IEEE Trans. Fuzzy Syst.* **2015**, *23*, 2245–2259. [CrossRef]
16. Li, S.; Yang, J.; Wang, G.Y.; Xu, T.H. Multi-granularity distance measure for interval-valued intuitionistic fuzzy concepts. *Inf. Sci.* **2021**, *570*, 599–622. [CrossRef]
17. Yang, J.; Wang, G.Y.; Zhang, Q.H.; Wang, H.M. Knowledge distance measure for the multigranularity rough approximations of a fuzzy concept. *IEEE Trans. Fuzzy Syst.* **2020**, *28*, 706–717. [CrossRef]
18. Yang, J.; Wang, G.Y.; Zhang, Q.H. Knowledge distance measure in multigranulation spaces of fuzzy equivalence relations. *Inf. Sci.* **2018**, *448*, 18–35. [CrossRef]

19. Chen, Y.M.; Qin, N.; Li, W.; Xu, F.F. Granule structures, distances and measures in neighborhood systems. *Knowl.-Based Syst.* **2019**, *165*, 268–281. [CrossRef]
20. Xia, D.Y.; Wang, G.Y.; Yang, J.; Zhang, Q.H.; Li, S. Local Knowledge Distance for Rough Approximation Measure in Multi-granularity Spaces. *Inf. Sci.* **2022**, *605*, 413-432. [CrossRef]
21. Atanassov, K.T. Intuitionistic fuzzy sets. *Fuzzy Sets Syst.* **1986**, *20*, 87–96. [CrossRef]
22. Zadeh, L.A. Fuzzy sets. *Inf. Control* **1965**, *8*, 338–353. [CrossRef]
23. Yang, C.C.; Zhang, Q.H.; Zhao, F. Hierarchical three-way decisions with intuitionistic fuzzy numbers in multi-granularity spaces. *IEEE Access* **2019**, *7*, 24362–24375. [CrossRef]
24. Zhang, Q.H.; Yang, C.C.; Wang, G.Y. A sequential three-way decision model with intuitionistic fuzzy numbers. *IEEE Trans. Syst. Man, Cybern. Syst.* **2019**, *51*, 2640–2652. [CrossRef]
25. Boran, F.E.; Genç, S.; Kurt, M.; Akay, D. A multi-criteria intuitionistic fuzzy group decision making for supplier selection with TOPSIS method. *Expert Syst. Appl.* **2009**, *36*, 11363–11368. [CrossRef]
26. Garg, H.; Rani, D. Novel similarity measure based on the transformed right-angled triangles between intuitionistic fuzzy sets and its applications. *Cogn. Comput.* **2021**, *13*, 447–465. [CrossRef]
27. Liu, H.C.; You, J.X.; Duan, C.Y. An integrated approach for failure mode and effect analysis under interval-valued intuitionistic fuzzy environment. *Int. J. Prod. Econ.* **2019**, *207*, 163–172. [CrossRef]
28. Akram, M.; Shahzad, S.; Butt, A.; Khaliq, A. Intuitionistic fuzzy logic control for heater fans. *Math. Comput. Sci.* **2013**, *7*, 367–378. [CrossRef]
29. Atan, Ö.; Kutlu, F.; Castillo, O. Intuitionistic Fuzzy Sliding Controller for Uncertain Hyperchaotic Synchronization. *Int. J. Fuzzy Syst.* **2020**, *22*, 1430–1443. [CrossRef]
30. Debnath, P.; Mohiuddine, S. *Soft Computing Techniques in Engineering, Health, Mathematical and Social Sciences*; CRC Press: Boca Raton, FL, USA, 2021.
31. Mordeso, J.N.; Nair, P.S. *Fuzzy Mathematics: An Introduction for Engineers and Scientists*; Physica Verlag: Heidelberg, Germany, 2001.
32. Zhang, X.H.; Zhou, B.; Li, P. A general frame for intuitionistic fuzzy rough sets. *Inf. Sci.* **2012**, *216*, 34–49. [CrossRef]
33. Zhou, L.; Wu, W.Z. Characterization of rough set approximations in Atanassov intuitionistic fuzzy set theory. *Comput. Math. Appl.* **2011**, *62*, 282–296. [CrossRef]
34. Jiang, Y.C.; Tang, Y.; Wang, J.; Tang, S.Q. Reasoning within intuitionistic fuzzy rough description logics. *Inf. Sci.* **2009**, *179*, 2362–2378. [CrossRef]
35. Dubey, Y.K.; Mushrif, M.M.; Mitra, K. Segmentation of brain MR images using rough set based intuitionistic fuzzy clustering. *Biocybern. Biomed. Eng.* **2016**, *36*, 413–426. [CrossRef]
36. Zheng, T.T.; Zhang, M.Y.; Zheng, W.R.; Zhou, L.G. A new uncertainty measure of covering-based rough interval-valued intuitionistic fuzzy sets. *IEEE Access* **2019**, *7*, 53213–53224. [CrossRef]
37. Huang, B.; Guo, C.X.; Li, H.X.; Feng, G.F.; Zhou, X.Z. Hierarchical structures and uncertainty measures for intuitionistic fuzzy approximation space. *Inf. Sci.* **2016**, *336*, 92–114. [CrossRef]
38. Zhang, Q.H.; Wang, J.; Wang, G.Y.; Yu, H. The approximation set of a vague set in rough approximation space. *Inf. Sci.* **2015**, *300*, 1–19. [CrossRef]
39. Lawvere, F.W. Metric spaces, generalized logic, and closed categories. *Rend. Del Semin. Matématico E Fis. Di Milano* **1973**, *43*, 135–166. [CrossRef]
40. Liang, J.Y.; Chin, K.S.; Dang, C.Y.; Yam, R.C. A new method for measuring uncertainty and fuzziness in rough set theory. *Int. J. Gen. Syst.* **2002**, *31*, 331–342. [CrossRef]
41. Yao, Y.Y.; Zhao, L.Q. A measurement theory view on the granularity of partitions. *Inf. Sci.* **2012**, *213*, 1–13. [CrossRef]
42. Du, W.S.; Hu, B.Q. Aggregation distance measure and its induced similarity measure between intuitionistic fuzzy sets. *Pattern Recognit. Lett.* **2015**, *60*, 65–71. [CrossRef]
43. Du, W.S. Subtraction and division operations on intuitionistic fuzzy sets derived from the Hamming distance. *Inf. Sci.* **2021**, *571*, 206–224. [CrossRef]
44. Ju, F.; Yuan, Y.Z.; Yuan, Y.; Quan, W. A divergence-based distance measure for intuitionistic fuzzy sets and its application in the decision-making of innovation management. *IEEE Access* **2019**, *8*, 1105–1117. [CrossRef]
45. Jiang, Q.; Jin, X.; Lee, S.J.; Yao, S.W. A new similarity/distance measure between intuitionistic fuzzy sets based on the transformed isosceles triangles and its applications to pattern recognition. *Expert Syst. Appl.* **2019**, *116*, 439–453. [CrossRef]
46. Wang, T.; Wang, B.L.; Han, S.Q.; Lian, K.C.; Lin, G.P. Relative knowledge distance and its cognition characteristic description in information systems. *J. Bohai Univ. Sci. Ed.* **2022**, *43*, 151–160.
47. UCI Repository. 2007. Available online: Http://archive.ics.uci.edu/ml/(accessed on 10 June 2022).
48. Li, F.; Hu, B.Q.; Wang, J. Stepwise optimal scale selection for multi-scale decision tables via attribute significance. *Knowl.-Based Syst.* **2017**, *129*, 4–16. [CrossRef]
49. Langeloh, L.; Seppälä, O. Relative importance of chemical attractiveness to parasites for susceptibility to trematode infection. *Ecol. Evol.* **2018**, *8*, 8921–8929. [CrossRef]
50. Wang, J.W. Waterlow score on admission in acutely admitted patients aged 65 and over. *BMJ Open* **2019**, *9*, e032347. [CrossRef]
51. Fidelix, T.; Czapkowski, A.; Azjen, S.; Andriolo, A.; Trevisani, V.F. Salivary gland ultrasonography as a predictor of clinical activity in Sjögren's syndrome. *PLoS ONE* **2017**, *12*, e0182287. [CrossRef]

52. Xing, H.M.; Zhou, W.D.; Fan, Y.Y.; Wen, T.X.; Wang, X.H.; Chang, G.M. Development and validation of a postoperative delirium prediction model for patients admitted to an intensive care unit in China: A prospective study. *BMJ Open* **2019**, *9*, e030733. [CrossRef]
53. Combrink, L.; Glidden, C.K.; Beechler, B.R.; Charleston, B.; Koehler, A.V.; Sisson, D.; Gasser, R.B.; Jabbar, A.; Jolles, A.E. Age of first infection across a range of parasite taxa in a wild mammalian population. *Biol. Lett.* **2020**, *16*, 20190811. [CrossRef]

MDPI
St. Alban-Anlage 66
4052 Basel
Switzerland
www.mdpi.com

Electronics Editorial Office
E-mail: electronics@mdpi.com
www.mdpi.com/journal/electronics

www.ingramcontent.com/pod-product-compliance
Lightning Source LLC
Chambersburg PA
CBHW041933260926
51850CB00018B/118
* 9 7 8 3 0 3 9 2 8 6 1 5 7 *